SYMMETRIES IN SCIENCE XI

Symmetries in Science XI

Edited by

Bruno J. Gruber

College of Science,
Southern Illinois University at Carbondale,
Carbondale, Illinois, U.S.A.

Giuseppe Marmo

Università di Napoli "Federico II",
Instituto Nazionale di Fisica Nucleare,
Napoli, Italy

and

Naotaka Yoshinaga

Saitama University,
Saitama, Japan

KLUWER ACADEMIC PUBLISHERS
DORDRECHT / BOSTON / LONDON

A C.I.P. Catalogue record for this book is available from the Library of Congress.

ISBN 1-4020-2633-1 (HB)
ISBN 1-4020-2634-X (e-book)

Published by Kluwer Academic Publishers,
P.O. Box 17, 3300 AA Dordrecht, The Netherlands.

Sold and distributed in North, Central and South America
by Kluwer Academic Publishers,
101 Philip Drive, Norwell, MA 02061, U.S.A.

In all other countries, sold and distributed
by Kluwer Academic Publishers,
P.O. Box 322, 3300 AH Dordrecht, The Netherlands.

Printed on acid-free paper

Printed in the Netherlands.

This volume of the proceedings "Symmetries in Science XI" is dedicated to my colleagues and friends

Akito Arima,
Francesco Iachello,
Marcos Moshinsky,

and to my wife

Burghilde,

all of whom have significantly contributed, in various ways, to the series of symposia "Symmetries in Science".

Bruno J. Gruber, Chairman and Organizer

Contents

List of Figures

List of Tables

Preface

The symposium "Symmetries in Science XIII" was held at the Mehrerau, Bregenz, Austria, during the period July 20 - 24, 2003. On the occasion of the symposium three outstanding scientists were honored who have contributed significantly to the success of the series of symposia "Symmetries in Science". The honored scientists are

- **Professor Akito Arima**, The House of Councillors, Japan
- **Professor Francesco Iachello**, J.W. Gibbs Professor of Physics and Chemistry, Yale University
- **Professor Marcos Moshinsky**, Universidad Nacional Autonoma de Mexico

We all, but in particular one of us (B.J. G.), wish to thank the authorities of the Land Vorarlberg and the Landeshauptstadt Bregenz for their generous and continuous support of the symposia series. On part of the Landeshauptstadt Bregenz we wish to thank Mag. Michael Rauth and the mayor of the Landeshauptstadt, Dipl. Ing. Markus Linhart, as well as his predecessor as mayor, Dipl. Vw. Siegfried Gasser. On part of the Land Vorarlberg our thanks go to Mag. Gabriela Duer, Dr. Hubert Regner, and former Landesrat Dr. Guntram Lins.

While the support given by these officials to the symposia series is thankfully appreciated, it was the continuous, consistent, cooperation and support given to the symposia series by Dr. Hubert Regner during a period of some 20 years which was the key to the success of 13 symposia. Again, at this point, one of us (B.J. G.) wishes to express his sincere thanks to Dr. Hubert Regner.

The preparation of the submitted manuscripts for publication was done at Saitama University and the University of Naples, where the totality of articles was assembled. We wish to thank Professor Wolfgang Bentz of Tokai University for helping out with the manuscripts submitted to Saitama University. Our special thanks go to Guido Celentano of the University of Naples who had to deal with the totality of articles.

We also wish to thank Professor Michael Ramek of the Technical University Graz who, as scientific secretary to a number of symposia, has contributed to the smooth day to day operation of the symposia series.

BRUNO J. GRUBER, CHAIR OF THE ORGANIZING COMMITTEE

GIUSEPPE MARMO, MEMBER

NAOTAKA YOSHINAGA, MEMBER

WHY SYMMETRY?

Some personal reflections

P. Roman

Boston University, Emeritus

Abstract • What is symmetry? • Why is symmetry important in science? • Historical developments. • Mathematical characterization of Symmetry. • Basic areas where symmetry principles are used. • Some special topics.

Thanks to the enthusiasm and administrative skills of Bruno Gruber, and to the adroitness of his assistants, as well as the generosity of many sponsors, for thirty years now we have been regaled with a periodic sequence of inspiring, exciting, pleasurable encounters centered on the topic "Symmetries in Science". Since this is, unfortunately, the last occasion when this group of colleagues (nay friends) meet, perhaps it will be not amiss to distance ourselves, briefly, from details of our field, and spend some time on contemplating the deeper, perhaps we may say philosophical aspects of Symmetry.

What follows will certainly not be a scholarly, exhaustive, authoritative treatment of the topic. I can only transmit to you very individualistic, almost personal thoughts (or rather sentiments) about the topic in question. There have been, virtually, whole libraries written on symmetry, and I cannot add more wisdom. I shall not even be systematic in my exposition, and won't attempt to give credit to the workers in the field. If I quote opinion of authors, these will be haphazard and far from comprehensive. Further, I'll take the viewpoint of a physicist, neglecting, for instance, crystallography, chemistry, biology. Even within physics, my treatment will be prejudiced by the views of the quantum theory of "fundamental particles" and interactions.

Let us now start with the question: **What is symmetry**? To answer in the broadest sense, it will be well to go back in time as far as the early stage of mankind's awakening. Perplexed and troubled by the apparent diversity, complexity, and unpredictability of nature, man conceived and took solace in the notion of an all-embracing, ultimate harmony of the Universe. "Harmony", that is congruence of parts, balance and unification of elements, is but one of the

B. Gruber (ed.), Symmetries in Science XI, 1–12.

synonyms we use for "symmetry". In fact, only the belief in some underlying symmetry makes it possible for us to develop science. We shall come back to this point later; right now I'd like to quote Hermann Weyl, one of the four greatest mathematicians of the 20th century, to succinctly sum up these thoughts. "Symmetry is one of the ideas" - Weyl notes - "by which man throughout the ages has tried to comprehend and create order, beauty, and perfection." This applies to science, art, and human conduct in general.

Well then: **why is symmetry important for science**? Once again, we must delve deeper and ask first *what is science*? Contrary to what is taught in most junior high schools, science is not "the explanation of Nature". Nature, be it even objective reality, just *is*. It cannot be "explained", at least not as far as science is concerned. Existence is a primary category, including, by the way, ourselves, too. (Which would imply that the explainer himself must be explained.) And certainly science is much more than "the description of Nature". That alone would be ad hoc, incidental, utterly unsatisfying. We have gone far beyond such a casual phenomenology and even empiricism. We want to "understand", and we have in part succeeded. Indeed, as Anatole France has put it: "The wonder is not that the field of stars is so vast, - but that man has measured it". And Einstein went even further: "The most incomprehensible thing about our Universe is that it can be comprehended", says he. Comprehended? What do we mean by that? Perhaps surprisingly, several humanists of the past came near to a comprehensive characterization of science. Goethe says: "Herein consists the scientific method: that we show the concept of a single phenomenon in its connection with the rest of the world of ideas." And the twentieth century German writer Hermann Hesse tells us: "Every science is ... a kind of ordering, simplifying; an attempt to make digestible for the spirit that which is indigestible."

Indeed, we are safe to say: *Science is the attempt to correlate individual phenomena and events into a coherent framework* (or systems of such frameworks). The correlation of part-entities into a coherent framework must satisfy two criteria (at least): 1. It should be systematic, comprehensible, attractive, nay: beautiful; 2. it should have predictive power, that is, the framework should encompass special items that are extensions of the already encompassed ones.

A minute's reflection then tells us: the essential feature of a scientific theory is *structure*. And the framework for studying, analyzing, understanding, enjoying structure is *mathematics*. That is why Wigner spoke of the "unreasonable effectiveness of mathematics in the natural sciences" and came to the conclusion that "mathematics does play a sovereign role in physics", "it is, in a very real sense, the correct language" of science.

Finally, we are back at the concept of symmetry. Mathematics deals with structures in two basic ways: a) by topology, b) by algebra. Topological structures use mostly analysis; algebraic structures are much more varied and use mostly construction and composition. *The concept of symmetry is a central part of algebraic systems in revealing and classification of structures.* This is the answer to our question: why is symmetry so important for science.

Now that we have clarified the basic role of symmetry in scientific thinking, we may come back for a moment to reconsider our earlier casual observation, viz. that symmetry is a crucial element in the perception of beauty. The connection to science goes both ways. We recognize willingly and with ease a structure that, analyzed in terms of symmetry, is beautiful. Conversely, a structure scrutinized by symmetry-analysis becomes acceptable to us only if we find that it is beautiful. It will be well to remember Einstein's dictum: "A theory is acceptable to us only if it is beautiful". A similar statement was made by Dirac when someone, in public, questioned him as to why he chose precisely his experimentally then unsupported equation out of other possible ones. "Because it was beautiful" he answered. And, of course, beauty is created, assessed, enjoyed via symmetry. Thus the circle closes.

We shall now remark on the **historical development of the idea of symmetry**. Already to prehistoric man obvious, natural, concrete, geometric symmetry in Nature was manifest: he recognized the patterns on sea-shells or the multifarious forms of snowflake crystals. Later these geometric symmetries became consciously imitated and applied in art - to be soon abstracted to more sophisticated manifestations (such as balance of colors), and applied to music (both in the guise of harmony as well as rhythmic patterns). These roles of symmetry continued and were amplified up to the present, but a discussion would lead us too far.

Concerning rather the evolution of symmetry-ideas in science, we observe that the development was slower. If we disregard such fancies as the orders of Celestial Spheres and their music, it appears that the implanting of symmetry-ideas into physics begins only with the late renaissance. But Galileo, and later even Newton, relied on symmetry principles only unconsciously and implicitly. Nevertheless, Newton made a giant step forward. He realized that in the study of phenomena one must make a clear distinction between the underlying *dynamical law* on the one hand and the *initial conditions* on the other. The former are rigorously structured; the latter are entirely irrelevant and haphazard, in that they are not encompassed by the law. This separation made analytical science at all possible. The for us at present important thing here is that the set of possible initial conditions is obtained by applying onto the system certain symmetry transformations. For example, subjecting the system to a translation in space, we obtain a (shifted) initial coordinate. Subjecting the system to a Galilei transformation, we obtain an initial state with a changed velocity. Furthermore, if we know the relation between two initial specifications (given by a symmetry transformation), then the resulting dynamical development in the second case can be obtained from that pertaining to the first case by means of a certain code (given by a symmetry transformation) which does not depend on the particular nature of the relevant specifications. Finally, the mathematical form of the dynamical law cannot depend on the specific nature of the actual initial specifications: this means that the dynamical law is covariant (form invariant) under the pertinent symmetry transformations. We now see what fundamental role symmetry considerations play in the very foundations of "doing science" - but of course Newton did not use this language: he relied on his ingenious instinct.

In the century or so following Newton, symmetries of known dynamical laws were noted and described - but just as an interesting afterthought or observation. Also, various *conservation laws* were established (derived sometimes tortuously, from the equations of motion), but without understanding (or even noting) their relation to symmetries.

The turning point came at the beginning of the twentieth century. Two great breakthroughs occurred at that time. The first was the establishment of relativity theory. Einstein was faced with a problem. There was a discrepancy between the symmetries of mechanics and those of electrodynamics. The laws of mechanics possessed Galilean symmetry (as we now call it), those of electromagnetism had Lorentz symmetry. With his unerring insight and intuition, Einstein chose the latter as the *guiding principle of physics*. Thus was the theory of special relativity born. For the first time in history, *a symmetry consideration became a guiding principle*. From then on, it was not acceptable just to propose, by trial and error or otherwise, a law of nature (and then examine its properties, including symmetry), but rather it became obligatory to insist that any law of nature should be formulated so that it exhibits Lorentz covariance. (We now speak more generally of Poincaré covariance.) Thus postulates of symmetry became guiding principles of exploration. That this was a radical change in outlook may be illustrated by the fact that the Galilei symmetry group which Einstein replaced by the Poincaré symmetry, had not ever been consciously explored earlier; only in the 1960's were we treated to a systematic study of the Galilei group - even the name was not common knowledge. (Parenthetically it should be mentioned that the inhomogeneous Galilei and Poincaré groups are not closely related: the latter is purely kinematical, the former dynamical, because it contains the equations of motion, too.)

At about the same time that Einstein radically changed our views on symmetry transformations, Hilbert and his school in Göttingen made great progress in the mathematical handling and application of symmetry. Here the connection between symmetry and *conservation laws* was clarified, and by Noether's theorem the derivation of such laws was rendered almost automatic (for continuous symmetries, at least).

The second great breakthrough in the history of symmetries occurred in the 1920's when modern quantum theory was born. Symmetries play a much more fundamental role in quantum theory than in classical physics. The reason for this is the linear structure of quantum theory: that is, the superposition principle. More explicitly: the set of states of a system corresponds to a set of vectors in a suitable Hilbert space, and so symmetry operations connecting different states give rise to linear operations in Hilbert space. *These linear operations thus carry a representation of the symmetry in the Hilbert space* of the system. Later we shall say a bit more about symmetries in quantum theory, but right now we shall just recall two names: Eugene Wigner and Hermann Weyl. These giants recognized very early the never-before-thought-of power of symmetry principles in the quantum world.

Wigner's 1928 book, applying representations of symmetry groups to atomic and molecular physics, opened a new world and revolutionized spectroscopy as well as chemistry. And when in 1939 he classified the unitary representations of the Poincaré group in Hilbert space, he gave thereby a definition of elementary particles which, essentially, is still valid. In fact, as Heisenberg remarked in 1973: "We will have to give up ... the concept of fundamental elementary particles [and] should rather accept the concept of *fundamental symmetry*".

On the other hand, Weyl, among several achievements in quantum symmetry theory, discovered the basic ideas of gauge symmetry, which eventually grew into the present fundamental theory of fields and particles, the Standard Model.

Next I will turn briefly to the **mathematical characterization of symmetry**. When we talk of a symmetry, we mean thereby an *automorphism of a given system* onto an equally possible description of the system. The details (both of the specification of the system and of the particular features of the mapping) vary enormously from case to case, but the principle is the same. Systems that are related by a symmetry transformation form an equivalence-class. Thus, *a symmetry transformation leads to an equivalent alternative description of the system*, i.e. it is a canonical transformation in classical physics and a unitary transformation in quantum theory.

Most frequently we recognize not an isolated symmetry but a (finite or infinite) set of them which, as an algebraic system, satisfies the axioms of a group. Why such a conglomeration of closely related symmetries occur is not at all clear to me, except for space-time groups.

It is amusing to note that, for continuous space-time symmetries (and only for those) one may take not the active but instead the passive view of description. This consists of the following. Instead of saying that a state with transformed data is a possible state of the same system as seen by a different observer, we could also say that the system has been physically transformed into another one, and the two descriptions are both given by the same observer. It is not clear to me whether this is a triviality or whether it has some deeper meaning.

Often the terms "symmetry" and "invariance" are used interchangeably. This is a mistake. Any symmetry transformation can be performed on any system, and it gives an alternative description of the system. The question is: do some selected *features* of the system remain unchanged or not? In particular, for a given physical system, is the *dynamical behavior* of the transformed system the same as for the original? *If* yes, then we can say: the dynamical law is invariant under the symmetry transformation, we have an invariant law, we have an "invariance". This must be distinguished from covariance of an equation, which simply means that the *form* of *some* equation doesn't change. In addition, the term "an invariant" has also several different meanings. Mostly, a selected quantity whose numerical value does not change under a symmetry transformation is called "an invariant of the system". Also, Casimir operators (see later) of a symmetry group are "invariants". In addition, a state which belongs to the trivial one-dimensional (scalar) representation of some symmetry group, is also often called "an invariant state".

A piece of art with perfect symmetry may appear boring. Nature seems to know this: indeed, very often we encounter **broken symmetries**. There are several mechanisms operative here. First, if the system is not isolated, properties of the environment may break the symmetry (such as the vertical gravitational field on the surface of the Earth). We can safely disregard these as trivial cases. Important (and not fully understood) are what one may call dynamical (or explicit) symmetry breakings. For a set of circumstances a symmetry holds, but for other dynamical circumstances (other forces) only a subset of these are maintained. (Example: Systems governed fully by "strong interaction dynamics" have iso-$SU(2)$ symmetry, but systems governed by electrodynamical dynamics [or which simultaneously are subject to strong and electromagnetic forces] exhibit only $U(1)$ symmetry.) Often a clash of symmetries causes symmetry breaking. These phenomena lead sometimes to uneasy situations which we call "approximate symmetries". But apart from the explicit symmetry breaking, more fundamental and more interesting is the case of spontaneous symmetry breaking. We have here a situation where the equation of motion (the dynamics)is invariant under a certain symmetry, but there are solutions which do not conform with the symmetry. This is a typical quantum phenomenon. The cause of such behavior is that the ground state of the system ("the vacuum state") is not invariant under the symmetry. In fact, it is degenerate. A related spontaneous quantum symmetry breaking occurs in field theory when a so-called "triangular (or similar) anomaly" occurs. Here the associated renormalization procedure causes the effective breaking of symmetry. Spontaneous symmetry breaking has enormous importance both in condensed matter physics as well as in the quantum theory of fundamental particles. In the latter case, for example, it leads to gauge particles' achieving of mass, so that alone with this spontaneous symmetry breaking becomes the gauge theory of the electroweak interactions possible.

Broken symmetries provide again an occasion to point out the difference between symmetry and invariance. Even if a conservation law (invariance) does not hold, symmetry may still be a useful and even powerful computational concept. A good example is provided by current algebras. In the presence of weak interactions the iso-$SU(2)$ vector current is no longer conserved. Yet the $SU(2)$ algebra of the vector current holds, and leads to important physical conclusions. More than this: the $SU(2)$ axial-current is never conserved, but the axial $SU(2)$ symmetry algebra holds and leads to very deep physical results.

As a final remark to the topic of symmetry breaking we note that (in all types of it) the breaking is not haphazard and disorderly, but is subject again to some symmetry argument.

Above we have repeatedly pointed out the fundamental roles of symmetries in physics. As a way of summary and overview, we will now explicitly list the major **areas where symmetry principles are used** in the every day praxis of physics.

1. Symmetry principles provide a most valuable *heuristic guide* in the search for dynamical laws. We believe that all fundamental laws of nature share certain symmetry properties, and specific branches of physics or specific systems may exhibit additional symmetries. (We do not yet have, and may never have, a "theory of everything".) We intuit, from masses of observations, particular symmetry properties, and then formulate laws so as to satisfy these symmetries in a general and unified frame.

2. Once the appropriate fundamental equations have been found, symmetry properties furnish powerful *tools for their solution.* This topic has two major aspects:

2.a) The symmetries restrict the forms the solutions can take. Roughly speaking, all admissible solutions will be classified by their symmetry character. This is why, for example, the solutions of Lorentz covariant equations can be only tensors or spinors. Similarly, the possible state vectors of a quantized system must be and are labelled by appropriate characteristics (viz, eigenvalues of generators and values of Casimir operators) of the symmetry group allowed for by the dynamical equations.

2.b) Invariance of the dynamical law under some symmetry gives rise to *conservation laws*: constants of motion can be constructed. The existence of such constants then leads to *selection rules*: processes that would connect states with different values of the conserved quantity are forbidden.

The treatment is particularly striking in the framework of quantum theory. It is easy to show that the (self-adjoint generator of the) unitary operator realizing a symmetry is a constant of motion if and only if it commutes with the Hamiltonian (or S-matrix). This means that then its expectation value is time independent; and if a state is an eigenstate of it with some eigenvalue at a given time, then the same state at a later time will be still an eigenstate belonging to the same eigenvalue. It is important to note here that invariance of the Hamiltonian and dynamical invariance of the system are equivalent statements. In particular, we have a conservation law if and only if there exists a symmetry (which, specifically, leaves the Hamiltonian invariant). In praxis, we encounter mainly symmetry *groups*, and the physically interesting entities (conserved observables) are the self-adjoint generators corresponding to the infinitesimal unitary transformations. Besides these generators (all conserved, but of course not all simultaneously measurable) we have also certain polynomials of these generators, the so-called Casimir invariants in the enveloping algebra, which automatically commute not only with the Hamiltonian (i.e. are conserved) but also with the generators. This explains why eigenvalues of the Casimir operators plus those of a selected set of commuting generators can be used as a complete set of state labelling parameters.

We further note that invariance of the dynamics under a symmetry transformation manifests itself in quantum theory also in the *equality of transition amplitudes* for the original and the transformed pair of systems; a very useful facet, both for establishing a symmetry and for exploiting its consequences.

3. Established symmetry properties greatly facilitate the *computation of specific quantities* that are of interest. For example, a lengthy calculation of matrix elements can be shortened by invoking some symmetry property. Further, matrix elements pertaining to different processes become related by symmetries (cf. branching relations). In particular, if a symmetry holds, transition probabilities between pairs of different states (i.e. rates of different processes) can be expressed in terms of a small number of invariant amplitudes. (In fact, experimentally observed relations between cross sections may be utilized to spot the symmetry that underlies the processes.)

Quantum numbers labelling states of composite systems can be easily computed from those of the constituents, if a symmetry holds.

Perturbational calculations are also facilitated in the presence of a known symmetry. (For example, a symmetry imposes restrictions on admissible trial-functions.)

Finally, symmetry principles often give, without any detailed calculation, the general pattern of a perturbation's effect when applied to an unperturbed state or system. Actually, about 40 years ago attempts were made to reveal the entire level system of an object by using only symmetry arguments, assuming the operation of some very powerful symmetry group – this without explicitly solving the energy eigenvalue equation. These attempts in particle theory went under the name of "dynamical groups" and "spectrum generating algebras".

In the rest of this individualistic survey I would like to pinpoint **some isolated special topics** which, to me, appear interesting and not too outdated.

1. There is, first of all, the topic of superselection rules, discovered by Wigner, Wightman and Wick in the 1950's. Suppose there is a generator of some symmetry which is simultaneously measurable (commutes) not only with the Hamiltonian but also with *all* observables of a system. (A typical example is the operator of electric charge.) Then this observable is not only conserved but has a tremendous structural effect on the system's quantum theory. It forces the Hilbert space to decompose into incoherent subspaces: the superposition principle becomes restricted. Matrix elements of transition operators between these incoherent subspaces are automatically zero – hence the name "super*selection rule*". Even the concept of an observable becomes restricted if a superselection rule operates. Thus, a symmetry property can influence the entire structure of a quantal system. I feel that this topic has not yet been given sufficient consideration.

2. Particle theorists used to distinguish between *space-time symmetries and internal symmetries.* (The Poincaré group, its possible conformal extension, the de Sitter group etc. represent the first category; isospin-$SU(2)$, flavor-$SU(3)$ etc. the second. Of course, there are further symmetries too, e.g. permutation symmetry, which do not really fit in any category.) Somewhat misleadingly, Wigner used to call internal symmetries "dynamical symmetries" because they seem to be connected to specific forces. In the 1960's beautiful attempts were made to find some large symmetry group that nontrivially combines and contains both the space-time and the internal symmetries, and reveals their connection. Unfortunately, it soon turned out that such attempts, however successful they seemed to be in particular aspects, are doomed because of deep lying algebraic reasons pertinent to the Poincaré algebra (O'Raiffertaigh's theorem). So the issue had to be dropped. But then, less than ten years later, *gauge theories* triumphed. The gauge symmetry idea lingered in the minds for decades. The wonderful facet of gauge theories is that the form and even the relative strengths of interactions becomes determined by local gauge invariance. This is an entirely new aspect of invariance, the key being the local character of the transformations. When the vexing problem of how to obtain masses for gauge fields was solved by spontaneous symmetry breaking, and when the difficult formalism of renormalization in gauge theories was mastered, the unification of weak and electromagnetic forces was accomplished. Soon, strong interaction dynamics was also encompassed by an (unbroken) color-$SU(3)$ gauge invariance of the quark-gluon system. The necessary ingredients of confinement and asymptotic freedom are also connected to the specifics of the gauge group. More than that: an important ingredient in the renormalizability of the quantum field theory is precisely gauge invariance. Thus, finally, the relation of space-time to "internal" symmetries acquired a new and satisfying aspect.

In the fiber bundle representation of gauge theories, the fiber built at each base space-time point "contains" the appropriate "internal" gauge symmetry, relative to that point. This is implied by the locality principle underlying the gauge symmetry concept.

In rapid sequel to these breakthroughs, courageous efforts were made to unify the electroweak and strong interactions in some Grand Unified Theory (GUT). Several gauge groups have been considered, primarily an $SU(5)$ system. Notwithstanding the formal attractiveness and some numerical successes of GUT, several serious problems blemish the picture. First, because quarks and leptons appear in the same representation of the Big Gauge Group, baryon and lepton numbers are not conserved. In particular, baryon decays are predicted. The calculated lifetime of the proton, unfortunately, is about two orders of magnitude shorter than the experimentally allowed lower limit. Second, the calculated relations between the masses of the fundamental fermions do not appear to be correct. Then there is the obnoxious hierarchy problem: roughly speaking, why is there a stable enormous gap between the energy scales of symmetry breaking of the GUT group to the Standard Model symmetry, and the breaking of the letter to the $SU(3)$ [color] $\times$ $U(1)$[e.m.] world we live in? This difficulty is connected to uncertainties, ambiguities, and technical problems pertaining to the symmetry breaking Higgs scalars. It is possible that supersymmetry (see below) may alleviate these problems. On the other hand, personally, I also feel uneasy about the subtleties of the various renormalization schemes and regularization methods that must be used to get any numerical results at all from GUTs. In any case, the last word about GUT has not yet been spoken.

It is interesting to observe that the "greater" the symmetry, the more unification of particles and forces we obtain. The earlier the Universe, the hotter and more energetic, the higher is the symmetry. Thus, it may well be that the dreamland of an "ultimate unification" will be achieved by finding the "primeval" (pre-Planck state) symmetry of the Universe.

3. Yet another relatively novel symmetry idea in particle theory is that of *supersymmetry*. This implies quasi a symmetric role between fermions and bosons. According to such schemes, to every fermion there corresponds a boson and vice versa. Mathematically, supersymmetry may be realized by enlarging the Poincaré group Lie algebra to a graded algebra. This is done by adjoining to the tensorial generators of the Poincaré group one (or more) 4-spinor generators, with appropriate commutators and anticommutators. One may reformulate the theory by considering the fields as functions not only of the usual vectorial Minkowski coordinates but also of suitable spinorial "coordinates". If such an attractive symmetry of Nature really exists, it must be badly broken, because the "new" accompanying particles (s-lepton and s-quark bosons, gauginofermions,

etc.) ought to have an enormous mass compared to their customary partners. So far there is no evidence of s-particles, and personally I find it odd to have a basic symmetry which, to manifest itself, must be badly broken to start with. I say this despite the fact that supersymmetry, made into a local gauge theory in the spinor coordinates, may perhaps include quantum gravity. Even this "11-dimensional supergravity theory" seems by now to have been subsumed (together with all super-string theories) into what is fancifully called "M-theory". But the discussion would take us into even more unsafe waters.

4. There is one system that, by definition, is unique: this is our Universe. Yet, even in *cosmology* symmetries play a basic role. It is generally accepted (and used even as a guiding principle) that the Universe is endowed by the basic symmetries of spatial isotropy and spatial homogeneity. (The latter may be even crucial for us to do science, since it guarantees that laws of nature are everywhere valid.) These symmetries are often called the "cosmological principle". Any deviation from these symmetries, global or even local, have tremendous cosmological and astrophysical consequences. At one time (the 1950's) the idea was put forward that the Universe possesses also temporal homogeneity (in the sense of stationarity). This was referred to by Hoyle and Bondi as the "perfect cosmological principle". Its consequence was shocking: it lead to the beautifully attractive steady-state-theory of the cosmos. But observation of the thermal cosmic background radiation refuted this convenient model and gave way to the big bang picture. In the latter, a problem is encountered: the traditional dissociation of initial conditions and laws becomes obscure.

Concomitantly, the role and application of symmetries pertaining to the pre-Planck period ought to be thought over more carefully. There is also another area of cosmology where symmetry (more accurately symmetry breaking) plays a substantial role, to which we already hinted above. If some GUT theory of all interactions and matter prevailed at the "earliest time" (when time could be defined at all), then, to end up with the present features of the world, it must have been broken in successive steps. Each step can be described as a phase transition. These, again, are ruled by symmetry considerations. Together with these phase transitions of GUT, some "space-time" symmetries are also affected. In particular, PC is presumably broken - this permits the over-preponderance of matter versus anti-matter. However, the most amazing discrete space-time symmetry, TCP, remains unbroken. This symmetry has deep consequences: it ensures that to every particle there belongs an antiparticle with the same mass and lifetime. Here again we see the world-shaping role of symmetries.

Instead of summarizing, let us ask one more question: Why do we love symmetries? Because symmetries are such a basic feature of nature, including our mental apparatus, that they enable us to discover, explore, analyze - even, to some extent, understand - structure. And structure, recognized and properly contemplated, allows us to enjoy and adore the miracle of the Creation.

J-PAIRING INTERACTIONS OF FERMIONS IN A SINGLE-J SHELL

A. Arima
The House of Councilors,
2-1-1 Nagata-cho, Chiyoda-ku,
Tokyo 100-8962, Japan

Abstract In this talk I shall introduce our recent works on general pairing interactions and pair truncation approximations for fermions in a single-j shell, including the spin zero dominance, and features of eigenvalues of fermion systems in a single-j shell interacting by a J-pairing interaction.

Keywords: J-pairing interaction, sum-rules, spin zero dominance

1. Introduction

It is my great pleasure to talk to you here. I would like to thank the organizers, especially Dr. Bruno Gruber. I am extremely glad to see many of my friends again today in this beautiful city Bregenz.

My talk consists of four subjects:

- Spin 0^+ ground state dominance
- Pair approximations for fermions in a single-j shell
- Regularities of states in the presence of $J_{\max}$-pairing
- Solutions for cases of $n = 3$ and 4 with H_J

2. 0^+ ground state dominance

A preponderance of 0^+ ground states was discovered by Johnson, Bertsch and Dean in 1998 [1] using the two-body random ensemble (TBRE), and was related to a reminiscence of generalized seniority by Johnson, Bertsch, Dean and Talmi in 1999 [2]. These phenomena have been confirmed in different systems [3, 4].

Let us take a simple system consisting of four particles in a single-j shell. The Hamiltonian that we use is as follows:

B. Gruber (ed.), Symmetries in Science XI, 13–21.

Table 1. The angular momenta which give the lowest eigenvalues for 4 fermions in single-j shells when $G_J = -1$ and all other parameters are 0.

$2j$	G_0	G_2	G_4	G_6	G_8	G_{10}	G_{12}	G_{14}	G_{16}	G_{18}	G_{20}
7	0	4	2	8							
9	0	4	0	0	12						
11	0	4	0	4	8	16					
13	0	4	0	2	2	12	20				
15	0	4	0	2	0	0	16	24			
17	0	4	6	0	4	2	0	20	28		
19	0	4	8	0	2	8	2	16	24	32	
21	0	4	8	0	2	0	0	0	20	28	36

$$H = \sum_J G_J A^{J\dagger} \cdot A^J \equiv \sum_J G_J \sqrt{2J+1} \left[A^{J\dagger} \times A^J \right]^{(0)} ,$$

$$A_M^{J\dagger} = \frac{1}{\sqrt{2}} \left[a_j^\dagger \times a_j^\dagger \right]_M^{(J)} , \quad A_M^J = -(-1)^M \frac{1}{\sqrt{2}} \left[\tilde{a}_j \times \tilde{a}_j \right]_M^{(J)} , \quad (1)$$

where G_J is given by

$$G_J = \langle j^2 J | V | j^2 J \rangle .$$

Here V is a two-body interaction.

We have used a two-body random ensemble to confirm the interesting phenomenon of 0^+ ground state dominance, and discovered an empirical method to predict the probability of the ground state to have a spin I [5]. We keep only one G_J to be -1 and all others 0 :

$$G_J = -\delta_{JJ'} .$$

We then diagonalize the Hamiltonian to find the angular momenta which give the lowest eigenvalues. They are shown in Table I. We count how many different G_J's give the lowest eigenvalue to an angular momentum I. The number is denoted as $\mathcal{N}_I$. For example for $j = \frac{21}{2}$ and $n = 4$, $\mathcal{N}_0$=5, $\mathcal{N}_2$=$\mathcal{N}_4$=$\mathcal{N}_8$ = $\mathcal{N}_{20}$=$\mathcal{N}_{28}$=$\mathcal{N}_{36}$=1 and all others are equal to 0. The total number of different G_J's is $N = \frac{2j+1}{2}$. Then the I g.s. probability is approximately predicted as

$$P^{\text{pred}}(I) = \mathcal{N}_I / N. \quad (2)$$

Fig. 1 shows a comparison between $P^{\text{pred}}(0)$ and $P^{\text{TBRE}}(0)$, which is obtained by diagonalization of a TBRE Hamiltonian for four fermions in a single-j shell. Fig. 2 shows a comparison between $P^{\text{pred}}(I)$ and $P^{\text{TBRE}}(I)$ for examples of various systems.

One can see that the agreements between the $P^{\text{pred}}(I)$ and $P^{\text{TBRE}}(I)$ are very good. It is therefore important to diagonalize H with $G_J = -\delta_{JJ'}$. For this purpose we introduce the J-pair approximation for low-lying states.

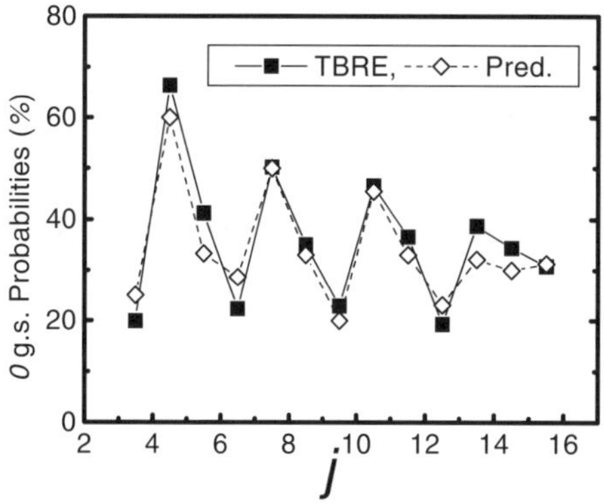

Figure 1. Comparison between $P^{\rm pred}(0)$ and $P^{\rm TBRE}(0)$ of four fermions in a single-j shell. The solid squares are obtained by 1000 runs of a TBRE Hamiltonian and the open squares are predicted by Eq. (2).

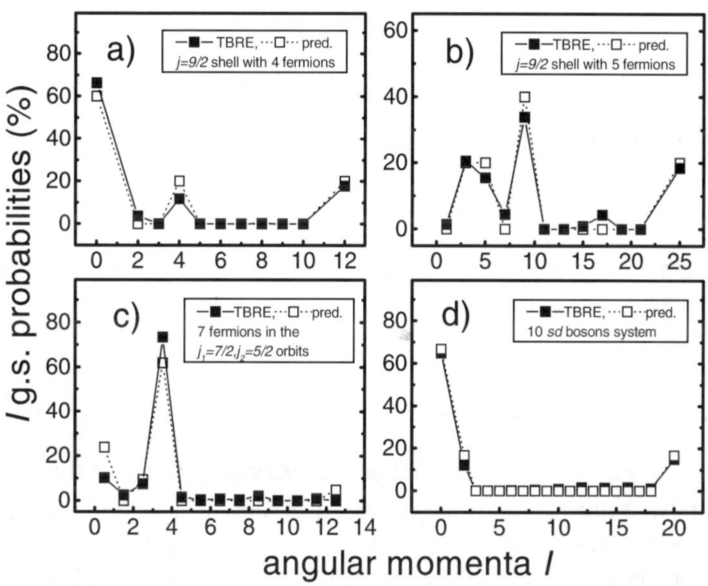

Figure 2. Comparison between $P^{\rm pred}(I)$ and $P^{\rm TBRE}(I)$ for more complicated systems. The solid squares are obtained by 1000 runs of a TBRE Hamiltonian and the open squares are predicted by Eq. (2).

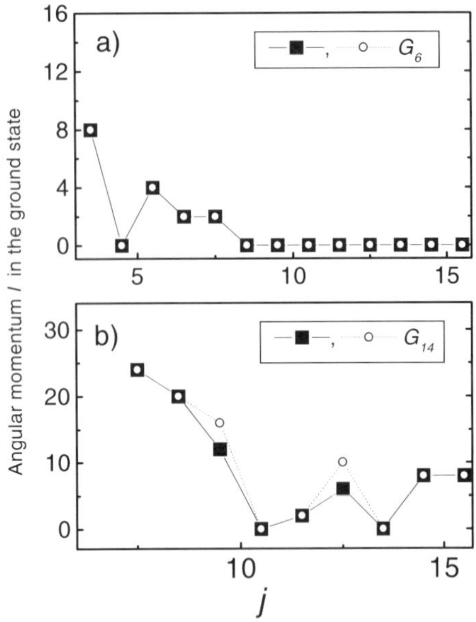

Figure 3. Ground state spin I for four fermions in a single-j shell for $J = 6$ in (a) and 14 in (b) as a function of j. The solid squares are obtained by diagonalizing H_J in the full shell-model space, and open circles are obtained by truncating the space to two pairs with spin J only.

3. Pair Approximation for Fermions in a single-j shell

Our Hamiltonian is defined as

$$H_J = -A^{J\dagger} \cdot A^J \,. \tag{3}$$

We first point out that the low-lying eigenvalues of H_J can be approximated by wavefunctions of pairs with spin J:

$$\Phi(I) = \frac{1}{\sqrt{N}} \left[A^{J\dagger} \times A^{J\dagger} \times \cdots \times A^{J\dagger} \right]^{(I)} |0\rangle \,, \tag{4}$$

where $\frac{1}{\sqrt{N}}$ is the normalization factor. It is very easy to prove that the J-pair truncation (with one pair and one particle) describes the low-lying states exactly in three-body systems.

Fig. 3(a) shows the spin of the ground state of j^4 configuration for $J = 6$. The ground states with spin 0 are obtained by exact shell-model calculations and by the J-pair approximation. Fig. 3 (b) shows the similar thing for $J = 14$. Fig. 4 shows energy levels obtained by the shell-model calculation and by the J-pair approximation when $j = 25/2$, $J = 14$ and $n = 4$. For the low-lying states, the pair approximation is very good. Giving the four low-lying states, two of

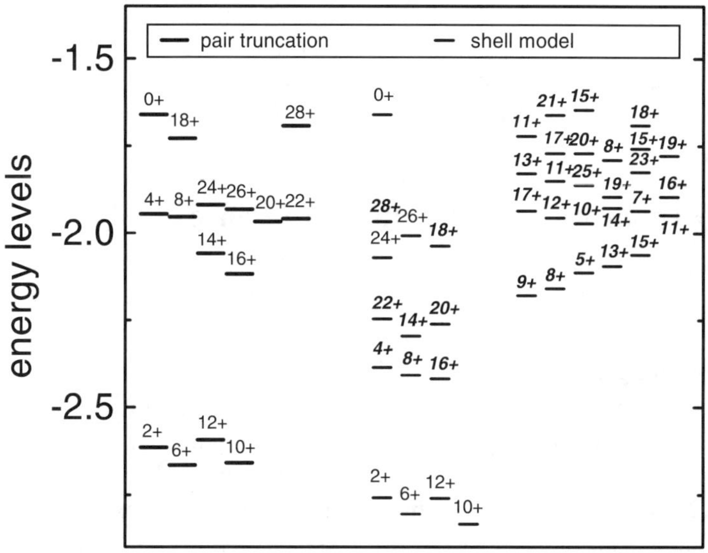

Figure 4. A comparison of low-lying spectra, obtained from two pairs with spin $J = 14$ (the column on the left hand side) and by a diagonalization of the full space (the columns in the middle and on the right hand side) for the case of four nucleons in a single-j ($j = 25/2$) shell. The middle column plots the shell-model states which are well reproduced by the two $J = 14$ pairs, and the right column plots the shell-model states which are not well reproduced by two $J = 14$ pairs. All the levels below 0_1^+ in the full shell-model space are included. One sees that the low-lying states with $I = 2_1^+, 6_1^+, 12_1^+$, and 10_1^+ are well reproduced.

them (6_1^+ and 10_1^+) compete to be the ground state. Their energies are almost the same in both the exact shell-model calculation and the pair approximation. This is why we failed to predict the ground state in this case, see Fig. 3 (b). For the n=5 and 6 cases that we have examined, the low-lying states are reasonably well approximated by the J-pair truncation.

So far J is general, between 0 and $2j-1$. Now let us take a very special value, $J_{\rm max} = 2j-1$. For $H = H_{\rm max} = H_{2j-1}$, the $I = I_{\rm max} = 4j-6$ corresponds to the lowest state, and $I = I_{\rm max} - 2$ state to the second lowest. These two states can be constructed by using pairs with angular momenta of either $J_{\rm max}$ or $J_{\rm max} - 2$. However, pairs with angular momentum $J_{\rm max} - 2$ do not present a good approximation of the other I states, while those with angular momentum $J_{\rm max}$ do. For example, for $n = 4$, $|J_{\rm max}^2, I = 0\rangle$ is exact but $|(J_{\rm max}-2)^2, I = 0\rangle$ is not exact, $|J_{\rm max}^2, I(\leq j)\rangle$ is almost exact ($\simeq -2$) but $|(J_{\rm max}-2)^2, I(\leq j)\rangle$ are not.

4. Regularities of states in the presence of $H_{J_{\rm max}}$

We first point out that eigenvalues of low I states ($n = 3, 4, 5$) are approximately integers. This can be proved in terms of six-j symbols for $n = 3$ [6]. For $n = 4$, one can prove this in terms of nine-j symbols [7].

Another regularity may be exemplified below by $j = 21/2$ and $n = 3$ and 4. Among many states of $n = 4$ with the same I, the lowest eigenvalue is expressed as $\mathcal{E}_I$ (obtained by a shell-model diagonalization). The $\mathcal{E}_I$ of four fermions in a single-j ($j = 21/2$) shell with I between 18 to 25 are shown in Table II. Note that there is no eigenvalue lower than -2 when I is smaller than 18. The eigenvalue of the $I_{\rm max}^{(3)} (= 3j-3)$ state with three fermions in the same j shell is $-\frac{59}{26}$=-2.26923076923077 (denoted as $E_{I_{\rm max}^{(3)}}$). From Table II, one sees that the $\mathcal{E}_I$'s of $n = 4$ with 18$\leq I \leq 25$ are very close to $E_{I_{\rm max}^{(3)}}$ and also very close to that of an I state constructed by $\Psi_I = \left[a_j^\dagger \times \left[a_j^\dagger \times a_j^\dagger \times a_j^\dagger \right]^{(I_{\rm max}^{(3)})} \right]^{(I)}$ (denoted as F_I). This indicates that the last added particle behaves just like a spectator and do not contribute to the total energy of the system.

We have calculated overlaps between the above states of $n = 4$ and the Ψ_I. They are almost 1 within a precision of 10^{-5}. This phenomenon has been confirmed for n up to 6 ($j \geq 11/2$).

Table 2. A comparison between eigen-energies obtained by diagonalizing $H_{J_{\max}}$ in the full shell-model space (the column "(SM)") and matrix elements $\langle\Psi_I|H|\Psi_I\rangle$ (column "F_I") for $n = 4$ and $j = 21/2$.

I	$\mathcal{E}_I$ (SM)	F_I (coupled)
18	-2.26923076925915	-2.26923076923498
19	-2.26923076930701	-2.26923076930702
20	-2.26923078555239	-2.26923077167687
21	-2.26923078386646	-2.26923078385375
22	-2.26923245245008	-2.26923102362432
23	-2.26923165420128	-2.26923165276669
24	-2.26930608933736	-2.26924197057701
25	-2.26925701778767	-2.26925692933680

5. Solutions for the case of $n = 3$

We take the following basis for three fermions

$$|j^3[jJ]I, M\rangle = \frac{1}{\sqrt{N^{(I)}_{jJ;jJ}}}\left[a_j^\dagger \times A^{J\dagger}\right]^{(I)}_M |0\rangle,$$

where $N^{(I)}_{jJ;jJ}$ is the diagonal matrix element of the normalization matrix

$$N^{(I)}_{jJ';jJ} = \langle 0| \left(\left[a_j^\dagger \times A^{J'\dagger}\right]^{(I)}_M\right)^\dagger \left[a_j^\dagger \times A^{J\dagger}\right]^{(I)}_M |0\rangle.$$

In general this basis is over-complete and the normalization matrix may have zero eigenvalues for a given I. Here J is not necessarily equal to $J_{\max}$.

The $N^{(I)}_{jJ';jJ}$ and $\langle j^3[jK']I, M|H_J|j^3[jK]I, M\rangle$ can be evaluated analytically:

$$N^{(I)}_{jJ';jJ} = \delta_{J',J} + 2\hat{J}\hat{J}'\left\{\begin{array}{ccc} J & j & I \\ J' & j & j. \end{array}\right\},$$

$$\langle j^3[jK']I, M|H_J|j^3[jK]I, M\rangle = -\frac{1}{\sqrt{N^{(I)}_{jK';jK'}N^{(I)}_{jK;jK}}} N^{(I)}_{jK';jJ}N^{(I)}_{jJ;jK},$$

where $\hat{L}$ is a short hand notation of $\sqrt{2L+1}$.

For a fixed J and any I, we construct one state $|j^3J : I\rangle = |j^3[jJ]I\rangle$ and all other states $|j^3K : I\rangle$, which are orthogonal to $|j^3J : I\rangle$, as follows:

$$|j^3K : I\rangle = |j^3[jK]I\rangle - \frac{N^{(I)}_{jK;jJ}}{\sqrt{N^{(I)}_{jJ;jJ}N^{(I)}_{jK;jK}}}|j^3[jJ]I\rangle, \ (K \neq J),$$

$$|j^3J : I\rangle = |j^3[jJ]I\rangle.$$

One easily confirms that all matrix elements of the Hamiltonian, $\langle j^3K' : I|H_J|j^3K : I\rangle$, are zero, except for $K' = K = J$:

$$\langle j^3[jJ]I|H_J|j^3[jJ]I\rangle = -N^{(I)}_{jJ;jJ} = -1 - 2(2J+1)\left\{ \begin{array}{ccc} J & j & I \\ J & j & j. \end{array} \right\} .$$

Thus, all the eigenvalues of H_J for $n = 3$ with any angular momentum I are zero except for the state with one pair of spin J, which has the eigenvalue $E^{J(j)}_I = -N^{(I)}_{jJ;jJ}$.

As by-products, we obtain a number of sum rules for six-j symbols. The procedure to derive these sum rules is straightforward. As is well known, the summation of all eigenvalues with a fixed I is equal to $\frac{n(n-1)}{2}$ times the number of I states, where n is the particle number. For $n = 3$, the number of states can be expressed by a compact formula [8].

If we use $E^{J(j)}_I$ to denote the non-zero eigenvalue of $H = H_J$ for any I , we have

$$\begin{aligned} \sum_J E^{J(j)}_I &= \sum_J \left[-1 - 2(2J+1)\left\{ \begin{array}{ccc} j & I & J \\ j & j & J \end{array} \right\}\right] \\ &= -\frac{n(n-1)}{2} D(j^3, I) , \end{aligned}$$

where $D(j^3, I)$ denotes the dimension of states with spin I for $n = 3$.

For $I \le j$ (j is a half integer), using the explicit expression for $D(j^3, I)$, we have

$$\sum_{J=\text{even}} 2(2J+1)\left\{ \begin{array}{ccc} j & I & J \\ j & j & J \end{array} \right\} = 3\left[\frac{2I+3}{6}\right] - I - \frac{1}{2} ,$$

where [ ] means to take the largest integer not exceeding the value inside.

Our new sum rules of six-j symbols will be given in [7] in detail.

6. Summary

In this talk, I discussed four interesting aspects concerning general pairing interactions and pair truncation approximations for fermions in a single-j shell. I first discussed an empirical rule to predict the spin I ground state probability. I then showed that pairs with spin J are reasonable building blocks for the low-lying states of a Hamiltonian with an attractive J-pairing interaction only. I also presented two interesting regularities of eigenvalues of a Hamiltonian with $J_{\max}$-pairing interaction: for low I states of n up to 5 we found that the eigenvalues are asymptotic integers; some of $n = 4$ states may be traced back to $n = 3$. Finally I proved for the case of $n = 3$ that the eigenvalues are written in terms of six-j symbols. This result leads to new sum rules for six-j symbols.

Acknowledgments

I would like to thank Drs. Y.M. Zhao, J.N. Ginocchio, and N. Yoshinaga for their collaborations in this work and also thank Dr. W. Bentz for reading the manuscript.

References

[1] C.W. Johnson, G.F. Bertsch, D.J. Dean, *Phys. Rev. Lett.* **80**, 2749 (1998).

[2] C.W. Johnson, G.F. Bertsch, D.J. Dean, and I. Talmi, *Phys. Rev. C* **61**, 014311 (1999).

[3] R. Bijker and A. Frank, *Phys. Rev. Lett.* **84**, 420(2000); *Phys. Rev. C* **62**, 14303(2000).

[4] D. Kusnezov, N.V. Zamfir, and R.F. Casten, *Phys. Rev. Lett.* **85**, 1396(2000).

[5] Y.M. Zhao, A. Arima, and N. Yoshinaga, *Phys. Rev.* **C66**, 034302 (2002).

[6] Y.M. Zhao, A. Arima, J.N. Ginocchio, and N. Yoshinaga, nucl-th/0305095, *Phys. Rev.* **C68**, in press (2003).

[7] Y.M. Zhao and A. Arima, (preprint) to be published.

[8] Y.M. Zhao and A. Arima, nucl-th/0304038, *Phys. Rev.* **C68**, in press (2003).

SUPERSYMMETRY IN NUCLEI

F. Iachello
Center for Theoretical Physics, Sloane Physics Laboratory,
Yale University, New Haven, CT 06520-8120

Abstract The concept of spectrum generating superalgebras and associated dynamic supersymmetries is introduced and placed within the general context of symmetry in physics. Applications of this concept to the study of spectra of atomic nuclei are presented.

1. Introduction

In the last 40 years the concept of spectrum generating algebras and dynamic symmetries has been extensively used to study physical systems. In the late 1970's this concept was enlarged to spectrum generating superalgebras and associated supersymmetries. In this article, dynamic symmetries are first placed within the context of symmetries in physics and applications to the structure of atomic nuclei are reviewed. Subsequently, the concept of dynamic supersymmetries is introduced and placed within the context of supersymmetry in physics. Applications to the study of spectra of nuclei are reviewed.

2. Symmetries

Symmetry is a wide-reaching concept used in a variety of ways.

2.1 Geometric symmetries

These symmetries are the first to be used in physics. They describe the arrangement of constituent particles into a structure. An example of symmetries of this type is the arrangement of atoms in a molecule. The mathematical framework needed to describe these symmetries is finite groups, sometimes called point groups. In Fig.1, the molecule C_{60} is shown as an example. The symmetry of this molecule is I_h. Geometric symmetries are used to reduce the complexity of the equations describing the system through the construction of a symmetry adapted basis. The Hamiltonian matrix in this basis is block diagonal.

B. Gruber (ed.), Symmetries in Science XI, 23–35.

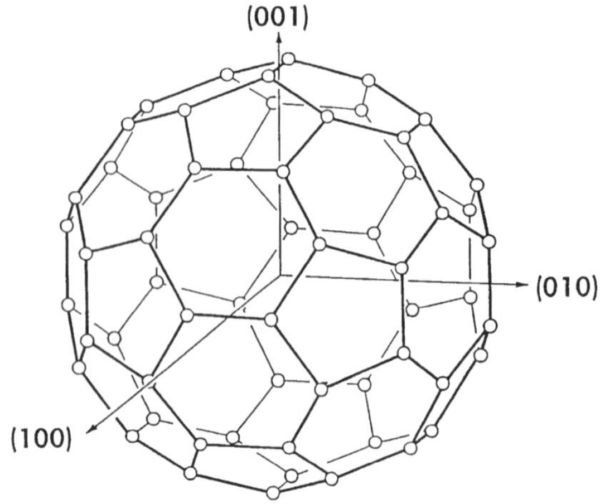

Figure 1. The fullerene molecule C_{60} is shown as an example of geometric symmetry, I_h.

2.2 Space-time symmetries

These symmetries fix the form of the equations governing the motion of the constituent particles. An example is provided by Lorentz invariance that fixes the form of the Dirac equation to be

$$(i\gamma^{\mu}\partial_{\mu} - m)\psi(x) = 0. \tag{1}$$

The mathematical framework needed to describe these symmetries is continuous groups, in particular Lie groups, here the Lorentz group $SO(3,1)$.

2.3 Gauge symmetries

These symmetries fix the form of the interaction between constituent particles and/or with external fields. An example is provided by the Dirac equation in the presence of an external electromagnetic field

$$[\gamma^{\mu}(i\partial_{\mu} - eA_{\mu}) - m]\psi(x) = 0. \tag{2}$$

Electrodynamics is invariant under the gauge transformation $A'_{\mu}(x) \rightarrow A_{\mu}(x) - \partial_{\mu}\Lambda(x)$. Also here the mathematical framework is Lie groups, in the case of electrodynamics being $U(1)$. In view of the fact that the strong and weak forces appear to be guided by gauge principles, gauge symmetries have become in recent years, one of the most important tools in physics.

2.4 Dynamic symmetries

These symmetries fix the interaction between constituent particles and/or external fields (hence the name dynamic). They determine the spectral proper-

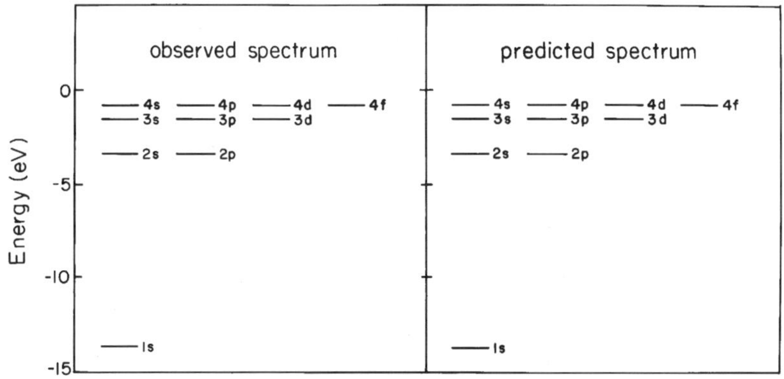

Figure 2. The spectrum of the non-relativistic hydrogen atom is shown as an example of dynamic symmetry of the Schrödinger equation, $SO(4)$.

ties of quantum systems (patterns of energy levels). They are described by Lie groups.

The earliest example of this type of symmetry is provided by the non-relativistic hydrogen atom. The Hamiltonian of this system can be written in terms of the quadratic Casimir operator C_2 of $SO(4)$ as [1]

$$\begin{aligned} H &= \frac{p^2}{2m} - \frac{e^2}{r} \\ &= -\frac{A}{C_2(SO(4)) + 1}, \end{aligned} \tag{3}$$

where A is a constant that depends on m and e. As a result, the energy eigenvalues can be written down explicitly in terms of quantum numbers

$$E(n, \ell, m) = -\frac{A}{n^2} \tag{4}$$

providing a straightforward description of the spectrum, Fig.2.

Another example is provided by hadrons. These can be classified in terms of a flavor $SU_f(3)$ symmetry [2]. The mass operator for hadrons can be written in terms of the Casimir operators of isospin, $SU(2)$, and hypercharge, $U(1)$, as

$$M = a + b\,[C_1(U(1)] + c\left[C_2(SU(2)) - \frac{1}{4}C_1^2(U(1))\right] \tag{5}$$

leading to the mass formula [4]

$$M(Y, I, I_3) = a + bY + c[I(I+1) - \frac{1}{4}Y^2]. \tag{6}$$

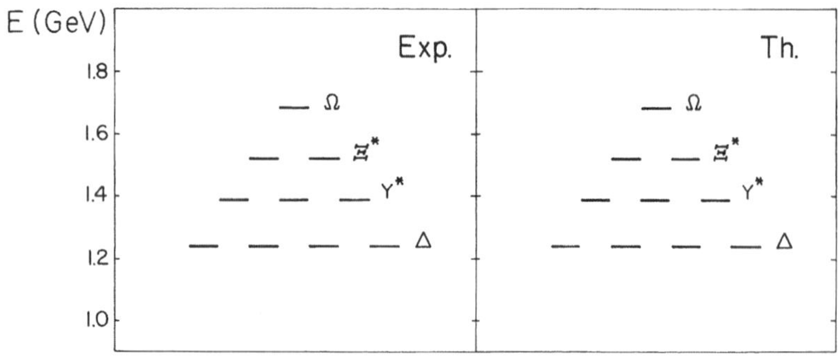

Figure 3. The spectrum of the baryon decuplet is shown as an example of dynamic symmetry of the mass operator, $SU_f(3)$.

This mass formula provides a very realistic description of hadron spectra, Fig.3.

The concept of dynamic symmetry was introduced implicitly by Pauli in the above mentioned paper [1], expanded by Fock [5], and, reintroduced in explicit form, by Dothan, Gell-Mann and Ne'emann [6] and Barut and Böhm [7]. It has been used extensively in the last 25 years and has produced many important discoveries. A mathematical definition is given in [8].

A dynamic symmetry is that situation in which:

(i) The Hamiltonian H is written in terms of elements, G_α, of an algebra G, called spectrum generating algebra (SGA), $G_\alpha \in G$.

(ii) H contains only invariant (Casimir) operators, C_i, of a chain of algebras $G \supset G' \supset G'' \supset \ldots$

$$H = f(C_i). \tag{7}$$

When a dynamic symmetry occurs, all observables can be written in explicit analytic form. For example, the energy levels are

$$E = \langle H \rangle = \alpha_1 \langle C_1 \rangle + \alpha_2 \langle C_2 \rangle + \ldots \tag{8}$$

One of the best studied cases is that of atomic nuclei, to be described in the following section.

3. Dynamic symmetries of the Interacting Boson Model

Atomic nuclei with an even number of nucleons can be described as a collection of correlated pairs with angular momentum $J = 0$ and $J = 2$. When the pairs are highly correlated they can be treated as bosons, called s and d. The corresponding model description is called Interacting Boson Model [9]. The spectrum generating algebra (SGA) of the Interacting Boson Model can be easily constructed by introducing six boson operators

$$b_\alpha(\alpha = 1, \ldots, 6) \equiv s, d_\mu(\mu = 0, \pm 1, \pm 2) \tag{9}$$

which span a six-dimensional space. The corresponding algebraic structure is that of $U(6)$. The elements of $U(6)$ can be written as bilinear products of creation and annihilation operators

$$G_{\alpha\beta} = b^\dagger_\alpha b_\beta \quad (\alpha, \beta = 1, \ldots, 6). \tag{10}$$

Since we are dealing with a system of bosons, the basis $\mathcal{B}$ is the totally symmetric irreducible representation of $U(6)$ with Young tableau

$$[N] \equiv \square\square \ldots \square \tag{11}$$

where N is the total number of bosons (pairs).

The model Hamiltonian and other operators are written in terms of creation and annihilation operators

$$H = E_0 + \sum_{\alpha\beta} \varepsilon_{\alpha\beta} b^\dagger_\alpha b_\beta + \sum_{\alpha\alpha'\beta\beta'} v_{\alpha\alpha'\beta\beta'} b^\dagger_\alpha b^\dagger_{\alpha'} b_\beta b_{\beta'}. \tag{12}$$

It can be rewritten in terms of elements of $U(6)$ as

$$H = E_0 + \sum_{\alpha\beta} \varepsilon'_{\alpha\beta} G_{\alpha\beta} + \sum_{\alpha\alpha'\beta\beta'} v_{\alpha\alpha'\beta\beta'} G_{\alpha\beta} G_{\alpha'\beta'}. \tag{13}$$

The fact that $U(6)$ is the SGA of this problem becomes then obvious.

The dynamic symmetries of the Interacting Boson Model can be constructed by considering all possible breakings of $U(6)$.

There are three and only three dynamic symmetries that include the angular momentum algebra $SO(3)$ as a subalgebra, corresponding to the breakings:

$$\begin{array}{lll} U(6) & \supset & U(5) \supset O(5) \supset O(3) \supset O(2) \;\; (I), \\ U(6) & \supset & SU(3) \supset O(3) \supset O(2) \qquad\quad (II), \\ U(6) & \supset & O(6) \supset O(5) \supset O(3) \supset O(2) \;\; (III). \end{array} \tag{14}$$

When a dynamic symmetry occurs, all properties can be calculated in explicit analytic form. In particular, the energies of the states are given in terms of quantum numbers by [10, 11, 12]

$$\begin{aligned} E^{(I)}(N, n_d, v, n_\Delta, L, M_L) &= E_0 + \varepsilon n_d + \alpha n_d(n_d + 4) \\ &\quad + \beta v(v + 3) + \gamma L(L + 1) \\ E^{(II)}(N, \lambda, \mu, K, L, M_L) &= E_0 + \kappa(\lambda^2 + \mu^2 + \lambda\mu + \\ &\quad + 3\lambda + 3\mu) + \kappa' L(L + 1) \\ E^{(III)}(N, \sigma, \tau, \nu_\Delta, L, M_L) &= E_0 + A\sigma(\sigma + 4) \\ &\quad + B\tau(\tau + 3) + CL(L + 1) \end{aligned} \tag{15}$$

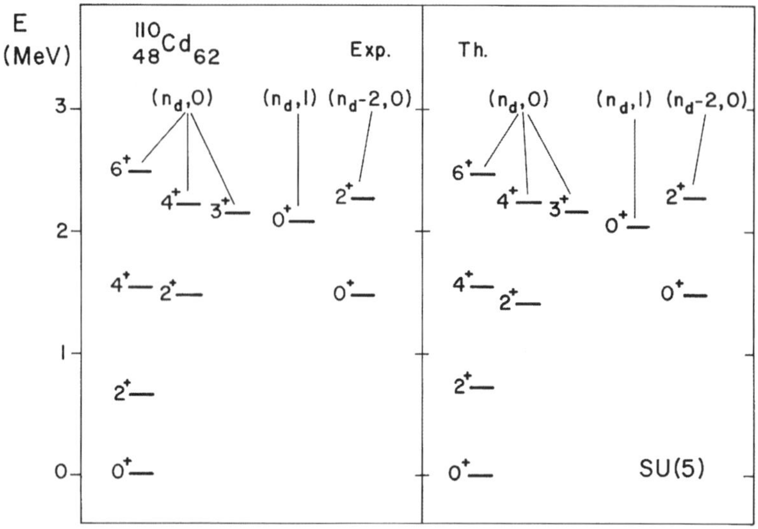

Figure 4. An example of $U(5)$ dynamic symmetry in nuclei: ^{110}Cd.

where the various terms are the eigenvalues of the Casimir operators in the appropriate irreducible representations.

Several examples of dynamic symmetries in nuclei have been found. Three of these examples are shown in Figs. 4,5,6. In the last 25 years, theInteracting Boson Model has provided a classification of spectra of even-even nuclei.

4. Supersymmetry

In the 1970's a new concept, supersymmetry, was introduced in physics, originally for applications to particle physics. This concept was implicitly introduced by Miyazawa [13]. Later Ramond [14] and Neveu and Schwartz [15] introduced it within the context of dual models. The concept became very popular after the work of Volkov and Akulov [16] and, especially, Wess and Zumino [17]. Supersymmetry, a symmetry that involve bosons and fermions, has become in the last 20 years one of the most important concepts in physics and has today wide reaching applications.

4.1 Geometric supersymmetries

Contrary to the case of geometric symmetries, this subject has not been much studied. An example is the introduction of a superlattice [18]. The

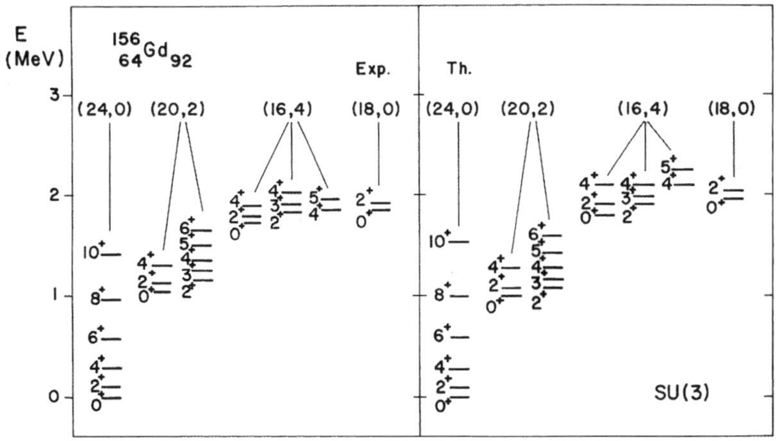

Figure 5. An example of $SU(3)$ dynamic symmetry in nuclei: ^{156}Gd.

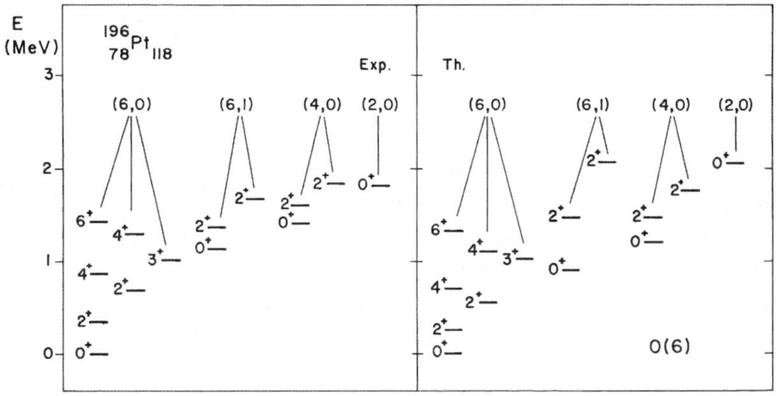

Figure 6. An example of $SO(6)$ dynamic symmetry in nuclei: ^{196}Pt.

mathematical framework to describe it is point supergroups, that is discrete subgroups of supergroups.

4.2 Space-time supersymmetries

These supersymmetries fix the form of the equation governing the motion of mixed systems of bosons and fermions. An example is the original Wess-

Zumino Lagrangian [17]

$$\begin{aligned}\mathcal{L} \;=\; & -\frac{1}{2}\left(\partial_\mu A(x)\right)^2 - \frac{1}{2}\left(\partial_\mu B(x)\right)^2 - \frac{1}{2} i\bar{\psi}(x)\gamma^\mu \partial_\mu \psi(x) \\ & -\frac{1}{2} m^2 [A^2(x) + B^2(x)] - \frac{1}{2} i m \bar{\psi}(x)\psi(x) \\ & -gmA(x)\left[A^2(x) + B^2(x)\right] - \frac{1}{2} g^2 \left[A^2(x) + B^2(x)\right] \\ & -ig\bar{\psi}(x)\left[A(x) - \gamma_5 B(x)\right]\psi(x).\end{aligned}$$

The mathematical framework here is continuous supergroups, as for example the SuperPoincaré group.

4.3 Gauge supersymmetries

These fix the form of interactions. For example in a supersymmetric gauge theory one has the occurrence of both bosonic and fermionic gauge fields with related properties.

4.4 Dynamic supersymmetries

These symmetries fix the interaction between constituent particles. They produce patterns of energy levels for mixed systems of bosons and fermions. They are a very ambitious unifying concept. A mathematical definition of dynamic supersymmetries is given in [19].

A dynamic supersymmetry is that situation in which:

(i) The Hamiltonian H is written in terms of elements G^*_α of a graded algebra G^*.

(ii) H contains only Casimir operators of a chain of algebras $G^* \supset G^{*\prime} \supset G^{*\prime\prime} \supset \ldots$ The subalgebras can be either graded or not.

One of the best studied cases is again that of atomic nuclei, where supersymmetries were introduced in 1980 [19], as described in the following section.

5. Dynamic Supersymmetries of the Interacting Boson-Fermion Model

In nuclei with an odd number of nucleons at least one is unpaired. Furthermore at higher excitation energies, some of the pairs may be broken. A comprehensive description of nuclei requires a simultaneous treatment of correlated pairs (bosons) and of fermions [20]. The corresponding model has been called Interacting Boson-Fermion Model [22]. The building blocks in this model are:

$$\begin{aligned} \text{Bosons} &: \quad s(J=0); d_\mu(J=2; \mu=0,\pm1,\pm2) \\ \text{Fermions} &: \quad a_{j,m}(m=\pm j, \pm(j-1), \ldots, \pm\frac{1}{2} \end{aligned} \tag{16}$$

The model Hamiltonian can be written as

$$H = H_B + H_F + V_{BF} \tag{17}$$

with

$$\begin{aligned} H_B &= E_0 + \sum_{\alpha\beta} \varepsilon_{\alpha\beta} b_\alpha^\dagger b_\beta + \sum_{\alpha\alpha'\beta\beta'} v_{\alpha\alpha'\beta\beta'} b_\alpha^\dagger b_{\alpha'}^\dagger b_\beta b_{\beta'} \\ H_F &= E_0' + \sum_{ik} \eta_{ik} a_i^\dagger a_k + \sum_{ii'kk'} u_{ii'kk'} a_i^\dagger a_{i'}^\dagger a_k a_{k'} \\ V_{BF} &= \sum_{\alpha\beta ik} w_{\alpha\beta ik} b_\alpha^\dagger b_\beta a_i^\dagger a_k. \end{aligned} \tag{18}$$

In order to study the possible symmetries of a mixed system of bosons and fermions, a new mathematical framework is needed, namely that of graded Lie algebras (also called superalgebras).

A set of elements

$$G_\alpha, F_i \tag{19}$$

are said to form a Lie superalgebra if they satisfy the following commutation relations

$$\begin{aligned} [G_\alpha, G_\beta] &= c_{\alpha\beta}^\gamma G_\gamma \\ [G_\alpha, F_i] &= d_{\alpha i}^j F_j \\ \{F_i, F_j\} &= g_{ij}^\alpha G_\alpha \end{aligned} \tag{20}$$

plus super Jacobi identities. [Graded semisimple Lie algebras with Z_2 grading have been classified by V. Kac [22]]. By inspection of Eq.(18) one can see that the combined boson-fermion Hamiltonian can be written in terms of elements of the graded superalgebra $G^* \equiv U(n/m)$

$$\begin{aligned} G_{\alpha\beta} &= b_\alpha^\dagger b_\beta \\ G_{ij} &= a_i^\dagger a_j \\ F_{\alpha i}^\dagger &= b_\alpha^\dagger a_i \\ F_{i\alpha} &= a_i^\dagger b_\alpha \end{aligned} \tag{21}$$

These elements can be arranged in matrix form

$$\begin{pmatrix} b_\alpha^\dagger b_\beta & b_\alpha^\dagger a_i \\ a_i^\dagger b_\alpha & a_i^\dagger a_j \end{pmatrix}. \tag{22}$$

The basis upon which the elements act is the totally supersymmetric irrep of $U(n/m)$ with Young supertableau

$$[\mathcal{N}\} \equiv \boxtimes\boxtimes\ldots\boxtimes \,. \tag{23}$$

For applications to Nuclear Physics, $\mathcal{N} = N_B + N_F$, $n = 6$ and $m = \sum_j (2j + 1) \equiv \Omega$, where Ω is the total degeneracy of the fermionic shell. A dynamic supersymmetry occurs whenever the Hamiltonian of Eq.(18) can be written in terms only of the Casimir operator of $U(n/m)$ and its subalgebras.

5.1 Supersymmetry in nuclei found

One of the consequences of supersymmetry is that if bosonic states are known, one can predict fermionic states. Both are given by the same energy formula. Indeed all properties of the fermionic system can be found from a knowledge of those of the bosonic system. Supersymmetry has thus a predictive power that can be tested by experiment. After its introduction in the 1980's, several examples of spectra with supersymmetric properties were found, relating spectra of even-even nuclei with those of odd-even nuclei (odd proton or odd neutron). In the first example, $j = 3/2$ fermions were coupled to s and d bosons. States were classified then in terms of the group $U(6/4)$ [23]. An example is shown in Fig.7, referring to the pair of nuclei Os-Ir. Other cases were subsequently found, for example $j = 1/2, 3/2, 5/2$ fermions with s and d bosons, described algebraically by $U(6/12)$ [24].

5.2 Supersymmetry in nuclei confirmed

Supersymmetry in nuclei has been recently confirmed in a series of experiments involving several laboratories. The confirmation relates to an improved description of nuclei in which proton and neutron degrees of freedom are explicitly introduced. The model with proton and neutron bosons is called Interacting Boson Model-2. The basic building blocks of this model are boson operators $b_{\alpha\pi}, b_{\alpha\nu}(\alpha = 1, \ldots, 6)$, where the index $\pi(\nu)$ refers to proton (neutron). The boson operators span a (six+six)-dimensional space with algebraic structure $U_\pi(6) \oplus U_\nu(6)$. Consequently, when going to nuclei with unpaired particles, one has a model with two types of bosons (proton and neutron) and two types of fermions (proton and neutron), called Interacting Boson-Fermion Model-2. If supersymmetry occurs for this very complex systems one expects now to have supersymmetric partners composed of a quartet of nuclei, even-even, even-odd, odd-even and odd-odd, for example

$$\begin{array}{cc} ^{194}Pt & ^{195}Pt \\ ^{195}Au & ^{196}Au \end{array}$$

Spectra of even-even and even-odd nuclei have been known for some time. However, spectra of odd-odd nuclei are very difficult to measure, since the

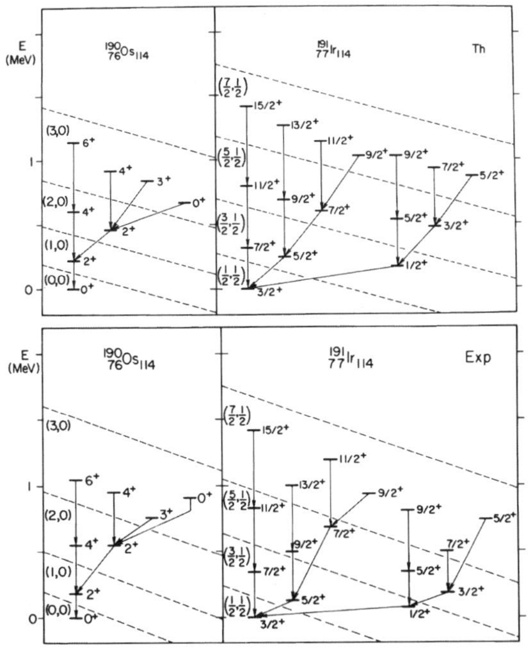

Figure 7. An example of supersymmetry in nuclei: the pair of nuclei $^{190}Os - ^{191}Ir, U(6/4)$.

density of states in these nuclei is very high and the energy resolution of most detectors is not sufficiently good. In a major effort that has involved several laboratories for several years it has now been possible to measure spectra of odd-odd nuclei. In particular, the magnetic spectrometer at the Ludwig-Maximilians Universität in Münich, Germany can separate levels only a few keV apart. It has thus been possible to measure the spectrum of ^{196}Au, the missing supersymmetric partner of ^{194}Pt, ^{195}Pt and ^{196}Au [25, 26, 27].

6. Implications of supersymmetry in nuclei

(a) Particle Physics

Supersymmetry has been sought in Particle Physics for decades. The confirmation of supersymmetry in nuclei indicates that this very complex type of symmetry can occur in Nature. It gives hope that, although badly broken, supersymmetry may occur in particle physics. However, supersymmetry in nuclear physics is a symmetry that relates composite objects (pairs) with fundamental objects (nucleons). Can it be the same in particle physics?

(b) Condensed matter physics

Some supersymmetric theories have been constructed in condensed matter physics [28]. Nambu has suggested that supersymmetry may occur in Type II superconductors [29].

Recently, it has been suggested that cuprate materials (high-T_c superconductors) may display supersymmetry. This case is being investigated at the present time [30].

7. Conclusions

A form of supersymmetry has been found and confirmed in Nuclei!

Acknowledgments

This work was supported in part under USDOE Contract No. DE-FG-02-91ER-40608.

References

[1] W. Pauli, *Z. Physik* **36**, 336 (1926).

[2] M. Gell-Mann, *Phys. Rev.* **125**, 1067 (1962).

[3] Y. Néman, *Nucl. Phys.* **26**, 222 (1962).

[4] S. Okubo, *Progr. Theor. Phys.* **27**, 949 (1962).

[5] V. Fock, *Z. Physik* **98**, 145 (1935).

[6] Y. Dothan, M. Gell-Mann, and Y. Ne'eman, *Phys. Lett.* **17**, 148 (1965).

[7] A.O. Barut and A. Böhm, *Phys. Rev. B* **139**, 1107 (1965).

[8] F. Iachello, in *Group Theoretical Methods in Physics*, Lecture Notes in Physics, **Vol. 94**, A. Böhm ed., Lange Springer, Berlin, p.420-429 (1079)

[9] For a review, see F. Iachello and A. Arima, *The Interacting Boson Model*, Cambridge University Press, Cambridge, England (1987).

[10] A. Arima and F. Iachello, *Ann. Phys. (N.Y.)* **99**, 253 (1976).

[11] A. Arima and F. Iachello, *Ann. Phys. (N.Y.)* **111**, 201 (1978).

[12] A. Arima and F. Iachello, *Ann. Phys. (N.Y.)* **123**, 468 (1979).

[13] H. Miyazawa, *Progr. Theor. Phys.* **36**, 1266 (1966).

[14] P. Ramond, *Phys. Rev. D* **3**, 2415 (1971).

[15] A. Neveu and J. Schwarz, *Nucl. Phys. B* **31**, 86 (1971).

[16] D.V. Volkov and V.P. Akulov, *Phys. Lett.* **46B**, 109 (1973).

[17] J. Wess and B. Zumino, *Nucl. Phys.* **70**, 39 (1974).

[18] V.A. Kostelecky and J.M. Rabin, in *Supersymmetry in Physics*, edited by V.A. Kostelecky and D.K. Campbell, *Physica D* **15**, 213 (1985).

[19] F. Iachello, *Phys. Rev. Lett.* **44**, 772 (1980).

[20] For a review see, F. Iachello and P. van Isacker, *The Interacting Boson-Fermion Model*, Cambridge University Press, Cambridge, England (1991).

[21] F. Iachello and O. Scholten, *Phys. Rev. Lett.* **43**, 679 (1979).

[22] V.C. Kac, *Functional Analysis Appl.* **9**, 91 (1975).

[23] A.B. Balantekin, I. Bars and F. Iachello, *Phys. Rev. Lett.* **47**, 19 (1981).

[24] A.B. Balantekin, I. Bars, R. Bijker and F. Iachello, *Phys. Rev.C* **27**, 2430 (1983).

[25] A. Metz, J. Jolie, G. Graw, R. Hertenberger, J. Groger, Ch. Gunther, N. Warr and Y. Eisermann, *Phys. Rev. Lett.* **83**, 1542 (1999).

[26] A. Metz, Y. Eisermann, A. Gollwitzer et al., *Phys. Rev.C* **61**, 064313 (2000).

[27] J. Groger, J. Jolie, R. Krucken, et al., *Phys. Rev.C* **62**, 064304 (2000).

[28] G. Parisi and N. Sourlas, *Phys. Rev. Lett.* **43**, 744 (1979).

[29] Y. Nambu, in *Supersymmetry in Physics*, edited by V.A. Kostelecky and D.K. Campbell, *Physica D* **15**, 147 (1985).

[30] F. Iachello, in preparation.

THE RELATIVISTIC MANY BODY PROBLEM IN QUANTUM MECHANICS

M. Moshinsky*
Instituto de Física, UNAM
Apartado Postal 20-364;
01000 México, D.F. MEXICO
moshi@fisica.unam.mx

Abstract It is well known that the analysis of a relativistic n-body problem invariant under the transformations of the Poincaré group and involving one time has only been done for $n = 1$. For $n > 1$ one uses the second quantization formalism of field theory. In this paper we state it in the ordinary space time coordinates associated with the n-bodies as Dirac did it in the one body case. We apply the formalism first to the two body problem if the interaction is of an harmonic oscillator form and then extend it to n particles.

1. Introduction

In non-relativistic quantum mechanics the passage from the single body to a many body problem, for non-interacting particles, is a trivial summation of the single body Hamiltonians.

In the relativistic case if we want to use a similar procedure and keep the problem invariant under Poincaré transformations, we have to deal with the times associated with each body.

In this paper we show first that for a system of n-particles, with the same mass and spin $(1/2)$ we can, with the help of appropriate matrices, formulate the many body problem in the center of mass reference frame, with the appearance of only one time.

While our procedure applies to arbitrary n-body problems, the algebra increases as a function of n. Thus the simplest problem we can attack is the two-body problem for which we give a procedure to determine its spectra start-

*Member of El Colegio Nacional and Sistema Nacional de Investigadores and Catalina Espinoza

B. Gruber (ed.), Symmetries in Science XI, 37–43.

ing from its non-relativistic formulation and then we indicate the generalization to n-particles.

2. A formulation of the relativistic many body problem

A many body problem usually starts with the case of non-interacting particles and, for simplicity, of the same mass.

In the non relativistic case the wave equation can be written as

$$\left(\frac{1}{2m}\sum_{s=1}^{n} p_s^2\right)\psi = i\frac{\partial\psi}{\partial t} \tag{1}$$

In the relativistic case, and taking particles of spin $\frac{1}{2}$, an obvious generalization from the Dirac equations would be

$$\sum_{s=1}^{n}(\gamma_s^\mu p_{\mu s} + 1)\psi = 0 \tag{2}$$

where repeated indices μ are summed with respect to the values $\mu = 0, 1, 2, 3$ and the index $\mu = 0$ refers to the time, and where we use the units $\hbar = m = c = 1$. The $p_{\mu s}$ is a covariant energy-momentum four vector of the s particle and γ_s^μ are the contravariant matrices related to

$$\gamma_s^0 = \beta_s, \quad \gamma_s^i = \beta_s\alpha_{is}, \quad i = 1, 2, 3; \quad s = 1, 2, \ldots n \tag{3}$$

with β and $\boldsymbol{\alpha}$ having the usual definitions[1] .

The Eq. (2) is certainly an invariant of the Poincaré group but is not satisfactory because it introduces n times through $p_{0s} = -i\partial/\partial x_s^0$.

How can we find a many body problem, still invariant under the Poincaré group but, in an appropriate system of reference, involving only one time?[2]; [3].

We start by denoting by u_μ a unit time like four vector which means that there is a reference frame in which it takes the form

$$(u_\mu) = (1, 0, 0, 0). \tag{4}$$

With the help of the four vector (3) we can define the Lorentz scalars[3]

$$\Gamma = \prod_{r=1}^{n}\left(\gamma_r^\mu u_\mu\right), \quad \Gamma_s = \left(\gamma_s^\mu u_\mu\right)^{-1}\Gamma, \tag{5}$$

where $(\gamma_s^\mu u_\mu)^{-1}$ eliminates the corresponding term in Γ and Γ_s is still in product form.

We now propose that instead of Eq. (2) we have the Lorentz invariant one [3]

$$\sum_{s=1}^{n} \Gamma_s(\gamma_s^\mu p_{\mu s} + 1)\psi = 0 \tag{6}$$

With the help of the total energy momentum four vector

$$P_\mu = \sum_{s=1}^{n} p_{\mu s}, \quad \mu = 0, 1, 2, 3, \tag{7}$$

we show that, in the frame of reference where $(u_\mu) = (1, 0, 0, 0)$, Eq. (6) takes the form

$$\left[\Gamma^0 \sum_{s=1}^{n} p_{0s} + \sum_{s=1}^{n} \Gamma_s^0(\boldsymbol{\gamma}_s \cdot \mathbf{p}_s + 1)\right]\psi = 0, \tag{8}$$

where boldface letters mean three dimensional vectors and

$$\Gamma^0 \equiv \prod_{r=1}^{n} \gamma_r^0, \quad \Gamma_s^0 \equiv (\gamma_s^0)^{-1}\Gamma^0, \tag{9}$$

Multiplying Eq. (8) by Γ^0 and making use of Eqs. (2,7,9) we obtain

$$\left[-P^0 + \sum_{s=1}^{n} (\boldsymbol{\alpha}_s \cdot \mathbf{p}_s + \beta_s)\right]\psi = 0 \tag{10}$$

where we used a metric in which $P_0 = -P^0$ and the latter is the zero component of P^μ *i.e.* the total energy of the system.

We would like that Eq. (6) represents the system of particles where the center of mass is at rest and this can be achieved if we define

$$u_\mu = P_\mu(-P_\tau P^\tau)^{-\frac{1}{2}} \tag{11}$$

as when $P_i = 0, i = 1, 2, 3$, we have $u_i = 0, u_0 = 1$.

For interactions depending on the relative coordinates

$$x_\mu^{st} \equiv x_{\mu s} - x_{\mu t}, \tag{12}$$

we can define

$$x_{\perp\mu}^{st} \equiv x_\mu^{st} - (x_\tau^{st} u^\tau)u_\mu \tag{13}$$

and thus suppressing the indices s, t we have that

$$r^2 \equiv (x_{\perp\mu} x_\perp^\mu) \tag{14}$$

is a Poincaré invariant and this is also true for any function of it.

3. The Hamiltonian of the n-body relativistic problem and its Foldy-Wouthuysen transformation

Continuing with the analysis of the n-particle relativistic non-interacting system in Eq. (10) we see that involves both positive and negative energies as $\boldsymbol{\alpha}_s$ merges the large and small components. We are only interested in the positive ones so we make a unitary transformation, well known for the one body case, *i.e.*the Foldy-Wouthuysen one[4]

$$(\boldsymbol{\alpha} \cdot \mathbf{p} + \beta) \rightarrow \beta\sqrt{1+p^2} \quad , \quad p^2 = \mathbf{p} \cdot \mathbf{p} \tag{15}$$

where β is a diagonal matrix with eigenvalues either 1 or -1. For the positive energy case β should be replaced by 1 and thus the Hamiltonian in Eq. (10) becomes

$$\sum_{s=1}^{n} \sqrt{1+p_s^2} \quad , \quad \mathbf{P} = \sum_{s=1}^{n} \mathbf{p}_s = 0 \tag{16}$$

We have now to add a potential interaction and the most convenient way is to do it in the frame of reference in which the non-interacting Hamiltonian has the form of Eq. (16), and assume that there it is of the oscillator type *i.e.*

$$V = \frac{\omega^2}{2n} \sum_{s>t=1}^{n} (\mathbf{r}_s - \mathbf{r}_t)^2 \tag{17}$$

where we note that we are using the frame of reference where $\hbar = m = c = 1$ so all the variables involved in Eqs. (16-17) are dimensionless.

It is now convenient to pass to the Hamilton-Jacobi variables, which will be denoted by a dot above them *i.e.*

$$\dot{\mathbf{r}}_s = [s(s+1)]^{-\frac{1}{2}} \left[\sum_{t=1}^{s} \mathbf{r}_t - s\mathbf{r}_{s+1} \right], \; 1 \leq s \leq n-1, \quad \dot{\mathbf{r}}_n = n^{-\frac{1}{2}} \sum_{t=1}^{n} \mathbf{r}_t \tag{18}$$

and similarly for $\dot{\mathbf{p}}_s$. The hamiltonian becomes

$$H = \sum_{s=1}^{n} \sqrt{1 + (\mathbf{p}_s - n^{-1}\mathbf{P})^2} + \frac{\omega^2}{2} \sum_{s=1}^{n-1} \dot{r}_s^2 \tag{19}$$

because from Eq. (16) $\mathbf{P} = 0$, while $(\mathbf{p}_s - n^{-1}\mathbf{P})$ is translation invariant and can be expressed only in terms of the first $n-1$ variables $\dot{\mathbf{p}}_s$

4. The particular case when $n = 2$

In Eq. (19) we proposed a simple general expression for the Hamiltonian of the n-body relativistic problem. To understand its properties better we shall discuss in this section the case of $n = 2$.

From (18) we see that the dotted momentum variables for $n = 2$ become

$$\dot{\mathbf{p}}_1 = \frac{1}{\sqrt{2}}(\mathbf{p}_1 - \mathbf{p}_2), \quad \dot{\mathbf{p}}_2 = \frac{1}{\sqrt{2}}(\mathbf{p}_1 + \mathbf{p}_2) \tag{20}$$

and in particular

$$[\mathbf{p}_1 - \tfrac{1}{2}(\mathbf{p}_1 + \mathbf{p}_2)] = -[\mathbf{p}_2 - \tfrac{1}{2}(\mathbf{p}_1 + \mathbf{p}_2)] = \frac{1}{\sqrt{2}}\dot{\mathbf{p}}_1 \tag{21}$$

so that we can write (19) as

$$H = 2\sqrt{1 + (\dot{p}_1^2/2)} + \frac{\omega^2}{2}\dot{r}_1^2 \tag{22}$$

Developing the square root by the binomial rule we get the wave equation corresponding to the Hamiltonian (22) in the form

$$\{2 + \tfrac{1}{2}[\dot{p}_1^2 + \omega^2\dot{r}_1^2] - \tfrac{1}{4}\dot{p}_1^4 + \ldots\}\psi = E\psi \tag{23}$$

Clearly then we can use harmonic oscillator states as our starting point in a variational numerical calculation

We shall not follow the procedure of Eq. (23) as we wish to developed a method that can be applied, in principle, to the n-body problem. We thus return to Eq. (22) and do the canonical transformation

$$\boldsymbol{\rho} = -\frac{1}{\sqrt{\omega}}\dot{\mathbf{p}}_1, \quad \boldsymbol{\pi} = \sqrt{\omega}\dot{\mathbf{r}}_1 \tag{24}$$

The Hamiltonian H of Eq. (22) then becomes

$$H = \frac{\omega}{2}(\pi^2 + \rho^2) + 2\sqrt{1 + \frac{\omega\rho^2}{2}} - \frac{\omega}{2}\rho^2 \tag{25}$$

where now a coordinate appears under the square root. We can then use the harmonic oscillator wave functions of frequency 1 to denote

$$|nlm >= R_{nl}(\rho)Y_{lm}(\theta, \varphi) \tag{26}$$

with

$$R_{nl}(\rho) = A_{nl}\rho^l L_n^{l+\frac{1}{2}}(\rho^2)e^{-\rho^2/2}, \quad A_{nl} = \left[\frac{2n!}{\Gamma(n + l + \frac{3}{2})}\right]^{\frac{1}{2}} \tag{27}$$

to obtain a numerical matrix whose elements are

$$< n'lm|H|nlm > \tag{28}$$

where we note that the matrix is diagonal in l, m as L^2, L_0 are integrals of motion. The matrix element (28) reduces then to a numerical term plus a radial integral *i.e.*

$$
\begin{aligned}
&< n'lm|H|nlm > \\
&= \frac{\omega}{2}\Big\{ - \Big[n(n+l+\tfrac{1}{2})\Big]^{\frac{1}{2}} \delta_{n'n-1} + (2n+l+\tfrac{3}{2})\delta_{n'n} \\
&- \Big[(n+1)(n+l+\tfrac{3}{2})\Big]^{\frac{1}{2}} \delta_{n'n+1}\Big\} \\
&+2\int_0^\infty R_{n'l}(\rho)\sqrt{1+\frac{\omega}{2}\rho^2} R_{nl}(\rho)\rho^2 d\rho
\end{aligned} \tag{29}
$$

The last integral in Eq. (29) can then be evaluated with the help of the generator functions of Laguerre polynomial and gives rise to Whittaker functions.

The integrations involve now harmonic oscillator functions of frequency 1 and thus are also of order 1, so the magnitude of the effect, as compared with the separation ω of the energy levels of the oscillators, has to do with the difference in behavior when $\omega \ll 1$ and $\omega \gg 1$.

5. Conclusion

We discussed the n-body relativistic Hamiltonian that starts with the formulation of the problem for non interacting particles with spin $\frac{1}{2}$ and the same mass, which is Poincaré invariant but involves only one time. This problem is analyzed with the help of the Foldy-Wouthuysen transformation (15) so that it is restricted only to the positive energy part, which is the one we are interested. We then, for simplicity, add a potential part of the harmonic oscillator type (17) to arrive finally at the Hamiltonian (19).

In Eq. (19) the relative moment appear under a square root while the coordinates are quadratic. Thus a canonical transformation of the type (24) exchanges the momenta and coordinates and thus we are confronted with a a Hamiltonian quadratic in the momenta and with square roots in the position coordinates. The last problem is a standard one for many bodies in which we can use as starting functions those of the n-body translationally invariant non-relativistic Hamiltonian. In this way we reduce our problem to the one discussed normally in the non-relativistic case[5].

Acknowledgments

We would like to thank to Consejo Nacional de Ciencia y Tecnología CONACYT (México), for the support through project 40527-F.

References

[1] P.A.M. Dirac, "The principles of quantum mechanics" 3rd. edn. (Oxford: Clarendon) pp. 252-274. ; L.I. Schiff "Quantum mechanics" (New York: McGraw Hill) pp. 311-314.

[2] A.D. Barut and G.L. Strobel, *Few body systems* **1**, 167 (1986).

[3] M. Moshinsky and G. Loyola, *Found. Phys* **23**, 197 (1993).

[4] L.L. Foldy and S.A. Wouthuysen, *Phys. Rev.* **78**, 29 (1950); B. J. Bjorken and S. Drell "Relativistic quantum mechanics" (New York: McGraw Hill) pp. 46-48.

[5] M. Moshinsky, Yu. F. Smirnov, *"The Harmonic Oscillator in Modern Physics"* (Hardwood Academic Publishers, The Netherlands). pp.102-109.

GAUGE INVARIANCE AND THE E1 SUM RULE IN NUCLEI

W. Bentz
Dept. of Physics, School of Science,
Tokai University,
1117 Kita-Kaname, Hiratsuka 259-1207, Japan
bentz@keyaki.cc.u-tokai.ac.jp

A. Arima
House of Councillors,
2-1-1 Nagata-cho,
Tokyo 100-8962, Japan

Abstract The connection between the enhancement factor $(1 + \kappa)$ of the photonuclear E1 sum rule and the orbital angular momentum g-factor (g_ℓ) of a bound nucleon is investigated in the framework of the Landau-Migdal theory for isospin asymmetric nuclear matter. Special emphasis is put on the role of gauge invariance to establish the $\kappa - g_\ell$ relation.

Keywords: Nuclear Magnetic Moments, Giant Resonances, Sum Rules

1. Introduction

The enhancement factor $(1 + \kappa)$ of the photonuclear E1 sum rule and the orbital angular momentum g-factor (g_ℓ) of a bound nucleon have attracted the attention of nuclear theorists as well as experimentalists for a long time, since these quantities reflect the presence of exchange forces and mesonic degrees of freedom in nuclei [1]. More than 30 years ago, Fujita and Hirata [2] used the isospin symmetric Fermi gas as a model for an N=Z nucleus to derive the simple relation $1 + \kappa = 2\, g_{\ell,\mathrm{IV}}$ between κ and the isovector (IV) part of g_ℓ in first order perturbation theory. Later it has been shown [3] that, because of the presence of correlations between the nucleons, only a part of the total κ is related to $g_{\ell,\mathrm{IV}}$. It has been argued [4] that this part of κ is related to the sum of the E1 strength in the region of the isovector giant dipole resonance (GDR). In more recent years [5], this modified $\kappa - g_\ell$ relation has been used to analyse

B. Gruber (ed.), Symmetries in Science XI, 45–54.

the results of photo-neutron experiments [6] and photon scattering experiments [7], in particular for nuclei in the lead region.

On the other hand, as early as 1965, Migdal and collaborators [8] used an approach based on a gas of quasiparticles to relate κ to the parameters characterizing the interaction between the quasiparticles (the Landau-Migdal parameters). Combining this relation with the more general one between $g_{\ell,\mathrm{IV}}$ and the Landau-Migdal parameters [9], their approach suggested that the relation $1+\kappa = 2\,g_{\ell,\mathrm{IV}}$ holds more generally without recourse to perturbation theory. The fact that their result involves the total κ instead of just a part of it reflects the quasiparticle gas approximation.

The main advantage of the Landau-Migdal theory [9], which is based on the Fermi liquid approach due to Landau [10], is that symmetries, like gauge invariance and Galilei invariance, are incorporated rigorously. The Fermi liquid approach to discuss sum rules in nuclear matter has therefore turned out to be very fruitful, and has been used in several papers on giant resonances [11]. However, to the best of our knowledge, a general discussion of the κ-g_ℓ relation is still lacking. In view of this fact, and also in view of the strong recent interest in nuclear giant resonances [12] and the recent attempts to extend the range of applicability of the Landau-Migdal theory [13], in this paper we will present a general discussion on the κ-g_ℓ relation in isospin asymmetric nuclear matter. The aims of our work are as follows: First, we will extend the relations obtained previously for the orbital g-factor [14] and the E1 enhancement factor [2] to the case of $N \neq Z$, putting special emphasis on the role of gauge invariance. Second, we will identify the physical processes which are taken into account in the κ-g_ℓ relation. For more detailed discussions and formal derivations of the results we refer to ref.[15].

2. The orbital g-factor

Consider a nucleon (quasiparticle) in nuclear matter at the Fermi surface. Its orbital angular momentum g-factor may be defined in terms of its electromagnetic current $\boldsymbol{j}$ by the relation[1]

$$\boldsymbol{j}(q=0) \equiv \frac{\boldsymbol{p}}{M}\, g_\ell\,, \tag{1}$$

where M is the free nucleon mass and $\boldsymbol{p}$ the momentum. In the Landau-Migdal theory, the current j is, up to a normalization factor, equivalent to the effective electromagnetic vertex Γ_{eff}, which is defined as that part of the full vertex Γ which is irreducible in the particle-hole (ph) channel. That is, the integral

[1] This identification, which holds in nuclear matter, follows directly from the definition of the magnetic moment, see ref.[14].

equation for the full vertex Γ can be written in the form (see Fig.1)

$$\Gamma = \Gamma_{\text{eff}} - i\, T_{\text{eff}}\, A\, \Gamma \tag{2}$$

where A is the pole part of the ph propagator, and the vertex Γ_{eff} as well as the interaction T_{eff} are by definition irreducible in the ph channel[2]. The effective

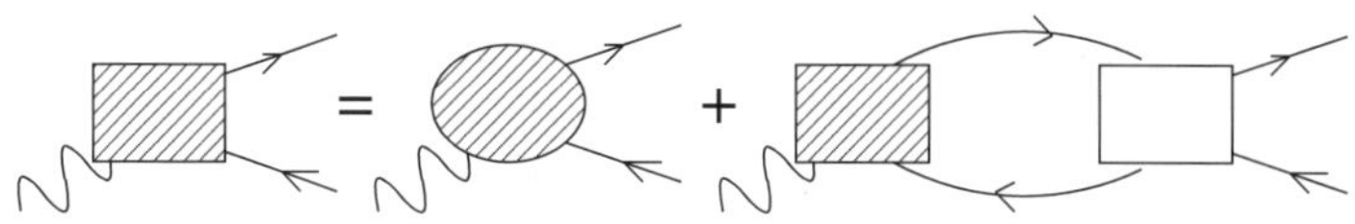

Figure 1. Integral equation for the vertex (shaded square) in terms of the ph-irreducible vertex (shaded circle) and the ph-irreducible interaction (white square).

vertex Γ_{eff} is renormalized by meson exchange currents, 2p-2h excitations, $N\overline{N}$ excitations, etc. Some examples are shown in Fig.2. Similarly, the effective interaction T_{eff} is also renormalized by processes like 2p-2h excitations, see Fig. 3 for examples.

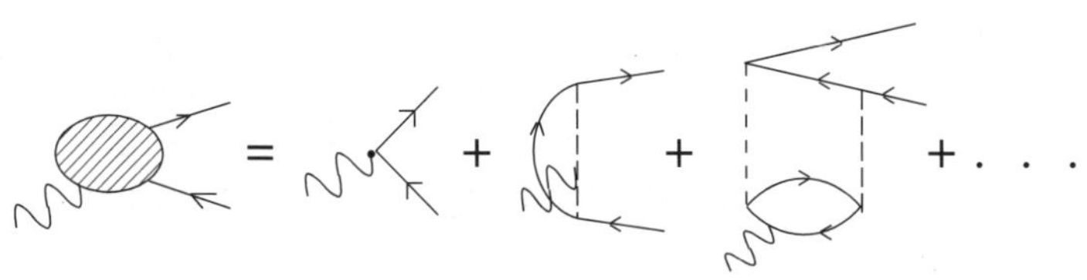

Figure 2. Examples for diagrams contained in the effective vertex Γ_{eff}. The first one represents the noninteracting part, the second a meson exchange current part, and the third one shows a 2p-2h excitation process.

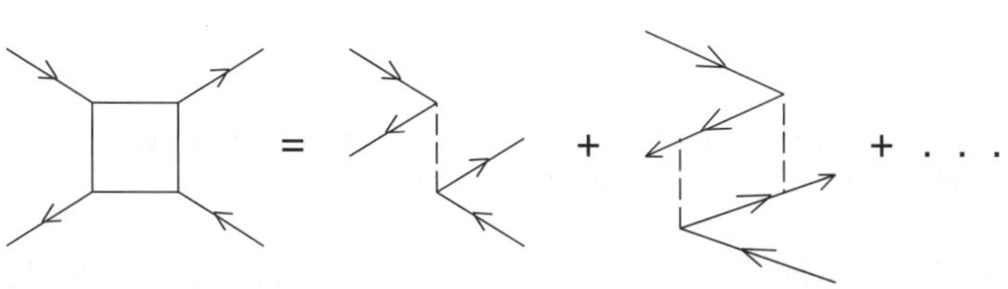

Figure 3. Examples for diagrams contained in the effective interaction T_{eff}. The first one represents the "bare" ph interaction, and the second one a 2p-2h excitation process. If the external particles are on the Fermi surface, T_{eff} becomes the Landau-Migdal interaction.

[2]More precisely, they are irreducible with respect to states which have ph cuts, see ref.[15] for details. In the Landau-Migdal theory [9] these quantities are usually denoted as $\Gamma^{(\omega)}$ and $T^{(\omega)}$.

The use of gauge invariance and Lorentz invariance (or Galilei invariance in the nonrelativistic case), combined with the integral equation (2), allows the determination of the quasiparticle current $\boldsymbol{j}(q = 0)$. The arguments, which are very general and hold also in relativistic field theory [14], are explained in detail in ref.[15]. In terms of the orbital g-factor of eq.(1), the results for the proton (p) and neutron (n) can be expressed as follows:

$$g_\ell(p) = \frac{M}{\mu_p} - \frac{Mv_F}{3p_F} F_1(pn) \frac{\mu_n}{\mu_p} \frac{N}{A} \beta \quad (3)$$

$$g_\ell(n) = \frac{Mv_F}{3p_F} F_1(pn) \frac{Z}{A} \beta . \quad (4)$$

Here the quantity $\beta = \left[1 - \left(\frac{N-Z}{A}\right)^2\right]^{-\frac{1}{3}}$ expresses the neutron excess, μ_p and μ_n are the chemical potentials (i.e., Fermi energies including the rest masses) of protons and neutrons, p_F and v_F are the Fermi momentum and Fermi velocity for the case N=Z, and $F_1(pn)$ is the dimensionless $\ell = 1$ Landau-Migdal parameter for the pn-interaction at the Fermi surface [9][3].

In spite of the simplicity of the results (3) and (4), they are very general and include all possible effects of meson exchange currents and configuration mixings. Fig.4 shows some examples for diagrams, drawn in the particle-particle channel, which contribute to the renormalization of the orbital g-factors. We can make several important observations from eqs.(3) and (4): First, since

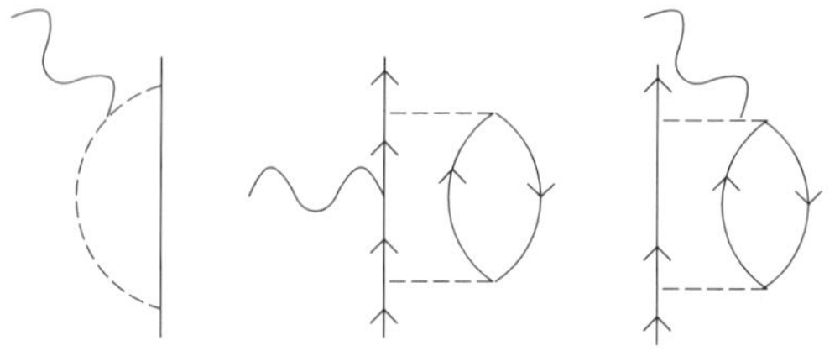

Figure 4. Examples for meson exchange current and configuration mixing processes which contribute to the renormalization of the orbital g-factors.

$F_1(pn) < 0$, the orbital g-factor of the proton is enhanced, while that of the neutron is negative. Second, one can show that $F_1(pn)$ depends only weakly on the neutron excess, and up to $\mathcal{O}\left(\frac{N-Z}{A}\right)$, the dependence on the neutron excess is as shown by the factors N/A and Z/A in (3) and (4). In symmetric

[3]The relation to the more familiar parameters F_1 and F_1' is $F_1(pn) = F_1 - F_1'$.

nuclear matter and in the nonrelativistic limit ($\mu_p = \mu_n = M$), the corrections to the g-factors are purely isovector, while in matter with neutron excess the proton g-factor is renormalized more strongly than the neutron one. Third, the g-factors satisfy the following relation:

$$\frac{Z}{A}\frac{\mu_p}{M}\, g_\ell(p) + \frac{N}{A}\frac{\mu_n}{M}\, g_\ell(n) = \frac{Z}{A}\,, \tag{5}$$

which is the extension to $N \neq Z$ of the well known fact [14, 16] that the isoscalar orbital g-factor is renormalized exclusively by relativistic effects.

3. The E1 sum rule and the $\kappa - g_\ell$ relation

The strength function (cross section) for absorption of unpolarized photons by a nucleus in its ground state $|0>$ is given by

$$\sigma(\omega) \;=\; 4\pi^2 \sum_n |\langle n|\hat{\jmath}(\boldsymbol{q})\frac{1}{\sqrt{\omega}}|0\rangle|^2\, \delta(\omega - \omega_{n0}) \tag{6}$$

$$\overset{\text{LWL}}{\longrightarrow}\; 4\pi^2 \sum_n |\langle n|\hat{\jmath}(\boldsymbol{q}=0)\frac{1}{\sqrt{\omega}}|0\rangle|^2\, \delta(\omega - \omega_{n0}), \tag{7}$$

where $\hat{\jmath}$ denotes the current operator[4], $\omega_{n0} = E_n - E_0$ is the excitation energy of the state $|n>$, and the long wave length limit (LWL) indicated in (7) holds if $|\boldsymbol{q}|R << 1$, where R is the nuclear radius. The (energy non-weighted) sum rule then becomes:

$$S \;\equiv\; \int_0^\infty \mathrm{d}\omega\, \sigma(\omega) = 4\pi^2 \sum_n \frac{1}{\omega_{n0}}\, |\langle n|\hat{\jmath}(\boldsymbol{q}=0)|0\rangle|^2 \tag{8}$$

$$= \; -2\pi^2 \langle 0|\, [[H, D]\,, D]\, |0\rangle, \tag{9}$$

where D is the dipole operator. This sum rule, however, is not directly observable, because the LWL contradicts the integration up to $\omega = \infty$. (Note that for a real photon we have $|\boldsymbol{q}| = \omega$.) It is therefore important to investigate whether it is possible to identify a part of the sum rule which is valid in the region of the GDR, where the LWL is justified. For this purpose, we again use the methods based on gauge invariance and the Landau-Migdal theory: The cross section in the LWL can be expressed in terms of the current-current correlation function $\Pi(\boldsymbol{q}, \omega)$ as

$$\sigma(\omega) = \frac{4\pi}{\omega}\, \mathrm{Im}\, \Pi(\boldsymbol{q} = 0, \omega) \qquad (\omega > 0). \tag{10}$$

---

[4] In order to eliminate the spurious effect of the center of mass motion, the current operator in this context is defined in terms of the effective charges $q_p = e_p\, N/A$ and $q_n = -e_p\, Z/A$, see ref. [15] for details. We also note that, for the case of unpolarized photons, the current operator $\hat{\jmath}$ as well as the dipole operator D in eq.(9) refer to any space component, e.g., the 3-component. With this choice, the current-current correlation functions Π of this section refer to the 33-component.

Gauge invariance can then be used to derive a Ward-Takahashi identity for the correlation function [15], which gives in the low energy limit

$$S = 2\pi^2\ \Pi(\boldsymbol{q} = 0, \omega = 0). \tag{11}$$

In nuclear matter, one must take the limit $\omega \to 0$ first before setting $\boldsymbol{q} = 0$. Because the correlation function can be expressed in terms of the full vertex Γ, one can use the integral equation (2) to split Π into 2 pieces:

$$\Pi = i\Gamma_{\text{eff}}\, A\, \Gamma + i\Gamma_0\, B\, \Gamma_{\text{eff}} \equiv \Pi_A + \Pi_B\,, \tag{12}$$

where the quantity A is the same as in eq.(2) and represents the product of the pole parts of particle and hole propagators, and B is the rest (product of particle-particle propagators, etc.) in the decomposition of the product of two propagators ($SS = A + B$). The part Π_A is shown graphically in Fig. 5, and an example of a graph which contributes exclusively to Π_B is shown in Fig. 6.

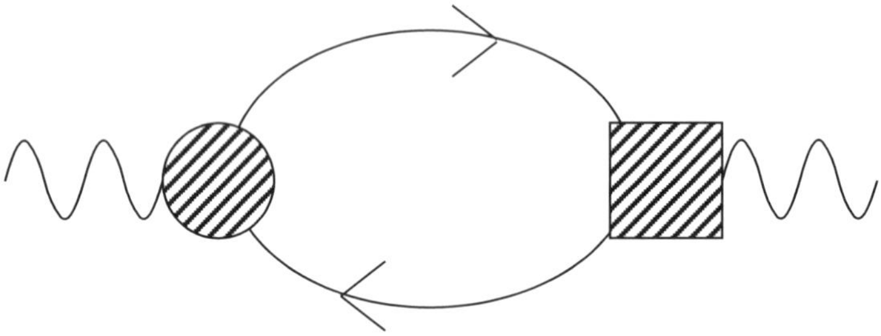

Figure 5. Graphical representation of the part Π_A in eq. (12). The symbols are the same as used in Fig.1.

The difference between Π_A and Π_B is that the former contains ph cuts while the latter has only cuts due to higher excited states like 2p-2h etc. If we use the RPA equation for the vertex (eq.(2)), we see that Π_A agrees with the familiar correlation function of the RPA theory, which is the starting point of almost all calculations of response functions, but processes like meson exchange currents, 2p-2h configuration mixings etc., are included in the definition of the effective vertex and the effective interaction.

In nuclear matter, one can now again use gauge invariance to completely specify the part Π_A and the associated part S_A of the sum rule (11). The result can be expressed by the following relation:

$$S_A = (\text{TRK})\,(1 + \kappa_A)\ , \tag{13}$$

where (TRK)=$2\pi^2 e_p^2 \dfrac{NZ}{AM}$ is the Thomas-Reiche-Kuhn sum rule value, and

$$1 + \kappa_A = g_\ell(p) - g_\ell(n), \tag{14}$$

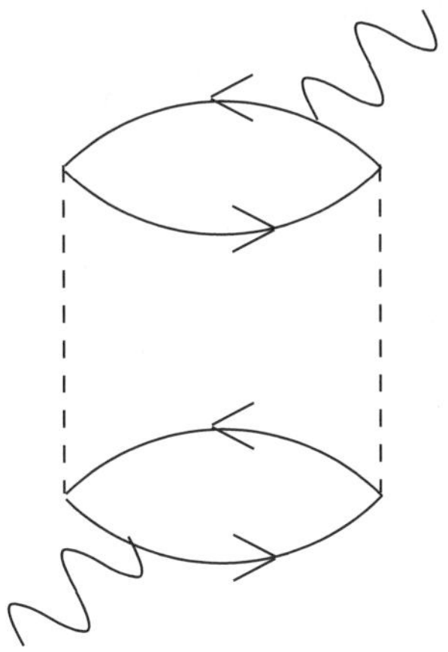

Figure 6. Example of a diagram which contributes exclusively to Π_B of eq.(12).

where the orbital g-factors are given in (3) and (4). An equivalent way to express the same result is as follows,

$$g_\ell(p) = 1 + \frac{N}{A}\kappa_A \tag{15}$$

$$g_\ell(n) = -\frac{Z}{A}\kappa_A \tag{16}$$

with

$$\kappa_A = -\frac{Mv_F}{3p_F}\, F_1(pn)\left[1 - \left(\frac{N-Z}{A}\right)^2\right]^{-\frac{1}{3}} . \tag{17}$$

The remaining part of the LWL sum rule, $S_B \equiv (\mathrm{TRK})\,\kappa_B$, which originates from Π_B, has no connection to the orbital g-factors and cannot be specified by using only symmetry principles.

One can now argue that the quantity κ_A can be approximately identified with the enhancement factor κ_{GDR}, which is extracted from the area under the Lorentzian of the GDR curve. The details are discussed in ref.[15], and the essential points of the argument are as follows:

First, only the part Π_A in eq.(12) involves the full vertex Γ, which is a solution to the RPA equation (2). Therefore, the associated strength function $\sigma_A(\omega)$ is the result of a renormalized RPA calculation, which produces a resonance structure due to the collective superposition of ph pairs. The part Π_B has no such resonance structure.

Second, in general the contribution $\Pi^{(i)}$ of a given time ordered diagram i to the correlation function can be split into its A and B-parts according to $\Pi^{(i)} = \Pi_A^{(i)} + \Pi_B^{(i)}$. Examples for diagrams which have both A and B-parts are shown in Fig. 7. The diagrams $\Pi^{(i)}$ in these examples involve

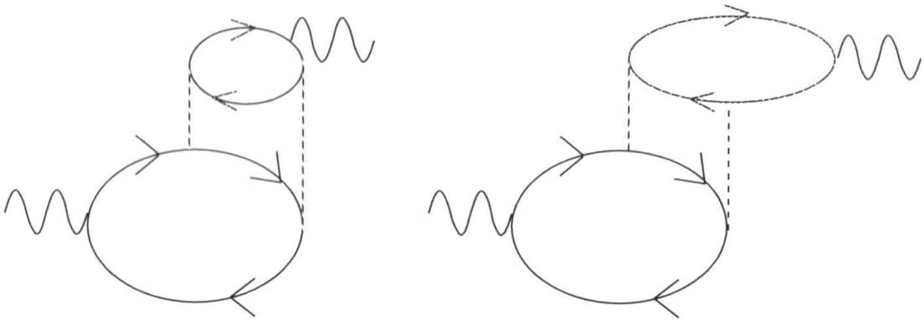

Figure 7. Example for diagrams with 2p-2h admixtures originating from the effective vertex (left) and from the effective interaction (right). These diagrams contribute to both Π_A and Π_B.

"high energy denominators" of the form $(\omega - \Delta E(2p-2h))^{-1}$. Their A-parts are obtained from $\Pi^{(i)}$ by replacing these high energy denominators by $(\Delta E_{\rm eff}(1p-1h) - \Delta E(2p-2h))^{-1}$, where $\Delta E_{\rm eff}(1p-1h)$ refers to the excitation energy of the ph state entering the vertex in the first diagram, and to the average of the ph states entering and leaving the interaction in the second diagram of Fig. 7. As long as, on the average, $\Delta E(2p-2h) >> \Delta E_{\rm eff}(1p-1h)$, we have the approximate relation

$$\Pi^{(i)}(\boldsymbol{q}=0, \omega=0) \simeq \Pi_A^{(i)}(\boldsymbol{q}=0, \omega=0)\,, \tag{18}$$

which indicates that the A-part gives the dominant contribution of the diagram i to the sum rule: $S^{(i)} \simeq S_A^{(i)}$. Under this condition, the B-part, which gives rise to the spreading width of ph-states in the low energy region, gives only a minor contribution to the sum rule. This in turn implies that the main contribution from Π_B to the LWL sum rule comes from those diagrams which involve no ph cuts at all, like the example shown in Fig. 6. It is known that, because of the short range nature of the tensor force, these diagrams involve large excitation energies, on the average several hundred MeV [17]. Therefore, these diagrams, for which the LWL is unjustified anyhow, will contribute to the sum rule only in the high energy region well beyond the GDR.

In conclusion of the above discussion, we can say that the approximate relation $\kappa_A \simeq \kappa_{\rm GDR}$ holds, although it is difficult to specify it more quantitatively without using model calculations.

For the sake of illustration of the $\kappa - g_\ell$ relation (14), let us discuss the nuclei in the ^{208}Pb region. The empirical values of $g_\ell(p)$ and $g_\ell(n)$ for nuclei in the lead region have been derived by Yamazaki (see ref.[1]) as $g_\ell(p) \simeq 1 + 0.13$, $g_\ell(n) \simeq -0.08$. These values are very close to the ones calculated in ref. [16]

from configuration mixing and meson exchange current effects, see table 7.12 of ref. [16].

From eq. (14) one then obtains the estimate $\kappa_A \simeq 0.21$. In the analysis of ref. [7], which uses the experimentally measured scattering cross section to extract the total photoabsorption cross section via dispersion relations, it was concluded that "any reasonable prescription gives (experimental) values of $\kappa_{\rm GDR}$ between 0.2 and 0.3", where $\kappa_{\rm GDR}$ was extracted from the area under a Lorentzian curve fitted to the GDR. This indicates at least a qualitative consistency between theory and experiment, since our κ_A can be identified with $\kappa_{\rm GDR}$ as discussed above.

4. Summary and conclusions

In this work we used the Landau-Migdal theory to discuss the orbital g-factor of a quasiparticle and the E1 sum rule for isospin asymmetric nuclear matter. The relations obtained for the orbital g-factors are in principle exact and hold also in relativistic field theory. For the E1 sum rule, we had to restrict ourselves to a nonrelativistic framework because of the problems arising from the center of mass motion.

We have split the strength function into two parts, where one comes from the p-h cuts including the effects of the higher excited states and meson exchange currents via their real parts, and the other comes from cuts at higher excitation energies. We have shown generally that the former part is related to the orbital g-factors, while the latter part has no relation to them. The former part has a close relation to the collective excitations of the system, i.e., the zero sound modes in infinite systems and the giant resonances in finite nuclei. We have discussed the importance of the $\kappa - g_\ell$ relation, which effectively separates the observable part of the LWL sum rule, which is related to the strength function in the low energy region, from the rest. Our discussions, which do not rely on perturbation theory, can serve to put many previous investigations on the $\kappa - g_\ell$ relation on a theoretically firm basis.

Concerning possible extensions, we would like to remark the following points: First, the methods used here to relate κ to g_ℓ refer to infinite nuclear matter, and it would be interesting to investigate to what extent they can be applied also to finite nuclei. Second, as we mentioned in the Introduction, very interesting attempts are now being made to extend the range of applicability of the Landau-Migdal theory to give a more general description of nuclear collective vibrations. The basic idea is to generalize the definition of the quantity A, which appears for example in the equation for the vertex (2), so as to include also more complicated configurations. It would be very interesting to see whether the results derived in this paper can be extended according to these lines.

References

[1] J.S. Levinger and H.A. Bethe, Phys. Rev. **78** (1950) 115;
H. Miyazawa, Prog. Theor. Phys. **6** (1951) 801;
J.-I. Fujita and M. Ichimura, Mesons in Nuclei, Vol. II (D.H. Wilkinson and M. Rho, eds.), North-Holland, Amsterdam 1979, p. 625;
T. Yamazaki, Mesons in Nuclei, Vol. II (D.H. Wilkinson and M. Rho, eds.), North-Holland, Amsterdam 1979, p.651.

[2] J.-I. Fujita and M. Hirata, Phys. Lett. **37 B** (1971) 237.

[3] A. Arima, G.E. Brown, H. Hyuga and M. Ichimura, Nucl. Phys. **A 205** (1973) 27.

[4] J.-I. Fujita, S. Ishida and M. Hirata, Progr. Theor. Phys. Suppl. **60** (1976) 73.

[5] M. Schumacher, A.I. Milstein, H. Falkenberg, K. Fuhrberg, T. Glebe, D. Häger and M. Hütt, Nucl. Phys. **A 576** (1994) 603.

[6] B.L. Berman and S.C. Fultz, Rev. Mod. Phys. **47** (1975) 713.

[7] D.S. Dale, A.M. Nathan, F.J. Federspiel, S.D. Hoblit, J. Hughes and D. Wells, Phys. Lett. **B 214** (1988) 329.

[8] A.B. Migdal, A.A. Lushnikov and D.F. Zaretsky, Nucl. Phys. **66** (1965) 193.

[9] A.B. Migdal, Theory of finite Fermi systems and applications to atomic nuclei, Wiley, New York, 1967.

[10] L.D. Landau, JETP **3** (1957) 920; **5** (1957) 101; **8** (1959) 70;
G. Baym and Ch. Pethick, The Physics of Liquid and Solid Helium, Part II (Wiley, 1978), p. 1;
P. Nozieres, Theory of Interacting Fermi Systems (W.A. Benjamin, 1964).

[11] E. Lipparini and S. Stringari, Phys. Rep. **175** (1989) 103;
J. Speth and J. Wambach, in *Electric and magnetic giant resonances* (ed. J. Speth), World Scientific (1991) 1.

[12] See the recent proceedings on conferences on Giant Resonances: Nucl. Phys. **A 569** (1994), **A 599** (1996), **A 649** (1999), **A 687** (2001);
M. Harakeh and A. can der Woude, Giant Resonances (Clarendon Press, Oxford, 2001).

[13] S. Kamerdzhiev, J. Speth and G. Tertychny, Extended Theory of Finite Fermi Systems: Collective Vibrations in Closed Shell Nuclei, preprint (2003).

[14] W. Bentz, A. Arima, H. Hyuga, K. Shimizu and K. Yazaki, Nucl. Phys. **A 436** (1985) 593.

[15] W. Bentz and A. Arima, The relation between the photonuclear E1 sum rule and the effective orbital g-factor, preprint, June 2003.

[16] A. Arima, K. Shimizu, W. Bentz and H. Hyuga, Adv. Nucl. Phys. **18** (1987) 1.

[17] M. Ichimura, H. Hyuga and G.E. Brown, Nucl. Phys. **A 196** (1972) 17.

THE HEISENBERG GROUP IN CLASSICAL AND QUANTUM INFORMATION TRANSMISSION

E. Binz and S. Pods
Lehrstuhl für Mathematik I,
Universität Mannheim,
68131 Mannheim, Germany.
binz@math.uni-mannheim.de
pods@math.uni-mannheim.de

Abstract Starting from an oriented three–dimensional Euclidean space as a subspace of the quaternions we introduce a simple transmission of information. The Euclidean space is split into a plane carrying information and a line of information transmission, thus having the structure of a Heisenberg algebra. The plane together with an additional periodic time scale is a Heisenberg group. This structure is given by any $2+1$–splitting of the three–dimensional Euclidean space. The centers of the Heisenberg groups form the special unitary group of $\mathbb{H}$. Schrödinger representations and spin $\frac{1}{2}$–representations as well as links between them are discussed, and the results are applied to the transmission of information.

Based on the tangent bundle on S^2 and the Hopf fibration, a natural principal and a complex line bundle are constructed for a smooth vector field X in three–space. These are bundles of normed states respectively bundles of states of qubits attached to the points in the domain of the vector fields. The fibres represent carriers of quantum information. A natural connection form yields the horizontal flow on both bundles. Quantum information is transmitted along this flow. This transmission yields a one–parameter group of unitary operators and a Schrödinger equation governing the evolution of the information. In addition this concept of quantum information contains classical information. Thus quantum information transmission is directly related to quantum information transmission. This method is generalized to transmit entangled states along two possibly different field lines of X.

The appendix deals with the Hopf fibration in detail.

Keywords: Heisenberg group, Schrödinger representation, spin representation, quantum information, information transmission, spin structures, Hopf fibration

B. Gruber (ed.), Symmetries in Science XI, 55–103.

1. Introduction

Our primary interest in these notes is a (mathematical) description of the transmission of information and quantum information. Both sorts of information rely on the inscription of information in a two–dimensional real plane, the information carrier. In order to link the geometry of the information carrier with the transmission of information, the latter is formulated in a geometric fashion. The ingredient unifying the inscription of information and its transmission along an information channel is a Heisenberg algebra $\mathcal{G}$. The channel together with the carrier of information form a three–space. A symplectic structure on the carrier plane turns the three–space into a Heisenberg algebra. The symplectic structure yields a natural scalar product on the three–space. Naturally associated with $\mathcal{G}$ is the (reduced) Heisenberg group G, a Lie group with $\mathcal{G}$ as its Lie algebra.

To treat both $\mathcal{G}$ and G on the same basis, a four–dimensional space is needed, namely the skew field of quaternions $\mathbb{H}$. Both $\mathcal{G}$ and $\mathbb{H}$ determine each other in a unique fashion. Since $\mathbb{H}$ carries a natural Minkowski metric (evolving from the geometry on S^3) with the center $\mathbb{R} \cdot e$ of $\mathbb{H}$ as time axis, time is implemented on both the center of $\mathcal{G}$, the information channel, and the center of G, a group $U^a(1)$ isomorphic to $U(1)$. The transmission of the information in the carrier plane of $\mathcal{G}$ and hence of G is practically performed by modulating each bit onto signals by means of the Schrödinger representation.

The common treatment of the Heisenberg algebra, the Heisenberg group, the quaternions and $SU(2)$ offers a link between these mathematical notions and signal theory.

The means to describe this link is the geometry of $\mathbb{H}$ reflected by the spin $\frac{1}{2}$–representation, which carries a natural two–by–two splitting of $\mathbb{H}$. This splitting together with the information channel causes $\mathcal{G}$ and G and, moreover, naturally yields the Schrödinger representation. The spin $\frac{1}{2}$–representation together with the symplectic structure resemble the geometry of a constant magnetic field in three–space with field lines parallel to the information channel. In fact, this sort of implementation of the geometry of $\mathbb{H}$ is used in magnetic resonance imaging (cf. [14]).

Since one of the simplest forms of information transmission follows the field lines of a vector field X in an open set O of a Euclidean three–space, we consider a series of natural bundles caused by this field assumed to be a singularity free gradient field. These bundles globally reflect the interplay between the quaternions and Heisenberg algebras. One of these bundles is a complex line bundle $\mathbb{F}^a$ based on the two–by–two splitting mentioned above. The points of the complex line are called *internal variables*. The points serve as carriers of bits of classical information. Restricting this bundle to the image of a field line yields via the geodesic flow on it a transmission of initially prescribed classical information along a field line. A more restrictive but simpler way is to use the horizontal flow of a prescribed connection form on the complex line bundle to transmit this information.

The complex line bundle $\mathbb{F}^a$ mentioned can be obtained by pulling back the tangent planes on S^2 by the so–called Gauss–map ε (assigning to each point $x \in O$ the unit vector of the value of the vector field at x).

This observation is the key to relate quantum information to the vector field (and, in turn, to classical information). Instead of pulling back the tangent planes of S^2 by ε, the fibres over S^2 of the Hopf fibration of S^3 are pulled back by ε. These fibres can naturally be regarded as normed states up to a global phase factor of the Hilbert space $\mathbb{H}$. These normed states up to a global phase factor are nothing else but qubits. Thus the points on S^2 are qubits and the elements in the fibres over them are normed states. Pulling back the Hopf fibration by ε to O naturally associates with each point on any field line a fibre of states. In this way the points of field lines correspond to qubits. The fibres are collected together to the quantum state bundle $\mathbb{K}$. The fibres of $\mathbb{K}$ over each point in O naturally yield complex lines and hence a complex quantum line bundle $\mathbb{L}$ containing the quantum state bundle $\mathbb{K}$.

There is a natural bundle projection mapping the *quantum* line bundle $\mathbb{L}$ fibrewisely onto $\mathbb{F}^a$, the complex line bundle encoding *classical* information. In this sense $\mathbb{F}^a$ can be complemented to $\mathbb{L}$. This complementation is directly linked to the two–by–two splitting of $\mathbb{H}$.

A connection form on $\mathbb{L}$ yields a horizontal flow on $\mathbb{L}$. Initial quantum information is prescribed in the fibre over an initial point on the field line. The horizontal flow provides a unitary evolution of quantum information along the field line. This evolution is called the transmission of quantum information along the field line.

Using the projection from $\mathbb{L}$ to $\mathbb{F}^a$ yields the transmission of classical information, as well.

The above outlined construction of a transmission of quantum information is extended to a transmission of entangled quantum information possibly following two different integral curves of the vector field. The Hilbert space in this case is $\mathbb{H} \otimes \mathbb{H}$. The 2–sphere, being the complex projection space $\mathbb{C}^a P(\mathbb{H})$ of $\mathbb{H}$ is replaced by the complex projection space $\mathbb{C}^a P(\mathbb{H} \otimes \mathbb{H})$ of $\mathbb{H} \otimes \mathbb{H}$. Instead of the Hopf fibration the tautological complex line bundle on $\mathbb{C}^a P(\mathbb{H} \otimes \mathbb{H})$ is used. The Gauss map is generalized to a map from O to $\mathbb{C}^a P(\mathbb{H} \otimes \mathbb{H})$ in order to construct the pull–back of the tautological bundle to O. A natural connection form yields a horizontal flow allowing to transmit entangled information along two possibly different field lines of X.

2. The Geometric Setting

2.1 The Quaternions

As mentioned in the introduction, we are interested in a geometric description of information transmission in a three–dimensional oriented Euclidean space. However, since time shall also be considered, we need to have four dimensions. Therefore, the *quaternions*form the basis of our investigation.

DEFINITION 1. *We define the* quaternions $\mathbb{H}$ *as*

$$\mathbb{H} := \left\{ l \in End\left(\mathbb{C}^2\right) \mid l + \tilde{l} = (tr\ l) \cdot id_{\mathbb{C}^2} \right\}. \tag{1}$$

A natural scalar product on $\mathbb{H}$ *is given by*

$$< l_1, l_2 >_{\mathbb{H}} := \frac{1}{2}\ tr\ (l_1 \circ \tilde{l}_2) \qquad \forall\ l_1, l_2 \in \mathbb{H} \tag{2}$$

and we fix an orientation $\mu_{\mathbb{H}}$ *on* $\mathbb{H}$.

There are several ways of linking the quaternions with the oriented Euclidean space $E, < , >_E, \mu_E$. On the one hand, if e denotes the neutral element in $\mathbb{H}$, we have an orthogonal splitting

$$\mathbb{H} = \mathbb{R} \cdot e \oplus E.$$

On the other hand, we may also start with the Euclidian space E and orthogonally extend it by $\mathbb{R} \cdot e$ to get $\mathbb{H}$.

The scalar product $< , >_{\mathbb{H}}$ on $\mathbb{H}$ becomes the scalar product on E when restricting it to elements in E, and, vice versa, the extension of the scalar product $< , >_E$ on E to $\mathbb{H}$ gives us $< , >_{\mathbb{H}}$ again.[1] Hence in the following we will denote both the scalar product on $\mathbb{H}$ and the one on E by $< , >$.

With this splitting the quaternions form a *skew field* under the multiplication

$$u \cdot v := (\lambda\mu - < u_1, v_1 >)e + \lambda v_1 + \mu u_1 + u_1 \times v_1 \tag{3}$$

for all $u = \lambda e + u_1, v = \mu e + v_1 \in \mathbb{H}$. Here $\times$ denotes the cross product in E given by the orientation μ_E.

From this multiplication we see that $\mathbb{R} \cdot e$ is the center of $\mathbb{H}$.

[1] The same is true for the orientations on E and $\mathbb{H}$.

The central part of the multiplication (3) can be used to define a natural Minkowski metric g_M on $\mathbb{H}$:

DEFINITION 2. *For any two vectors $u, v \in \mathbb{H}$ we set*

$$g_M(u, v) = g_M(\lambda e + u_1, \mu e + v_1) := -(\lambda\mu - < u_1, v_1 >).$$

Defined like this, g_M is a natural Minkowski metric on $\mathbb{H}$. Hence we can refer to the center $\mathbb{R} \cdot e$ of $\mathbb{H}$ as a *time axis*.

Considering the description of information transmission, we can use this splitting of $\mathbb{H}$ and take the spatial components to be elements of E, while time is described by the time axis $\mathbb{R} \cdot e$.

Another aspect we will make frequent use of in the sequel is the fact that the complex numbers can be embedded into the quaternions. Of course, this embedding is not unique. We choose $a \in S^2 \subset E$ as an imaginary unit and define

$$\mathbb{C}^a := \mathbb{R} \cdot e \oplus \mathbb{R} \cdot a.$$

Looking at the multiplication in $\mathbb{H}$, we see that a is indeed the imaginary unit in $\mathbb{C}^a$:

$$a^2 = - < a, a > \cdot e = -e.$$

With the identification $1 \mapsto e, i \mapsto a$ we thus show that $\mathbb{C}^a$ is isomorphic to the complex line $\mathbb{C}$. From this isomorphism we also see that there are as many embeddings of $\mathbb{C}$ into $\mathbb{H}$ as there are elements $a \in S^2$.

$\mathbb{C}^a$ contains the time axis $\mathbb{R} \cdot e$. The multiplication by a transfers the time axis $\mathbb{R} \cdot e$ into $\mathbb{R} \cdot a$.

Notation:. The unitary group of $\mathbb{C}^a$ is denoted by U^a.
It provides us with a periodic notion of time by means of the exponential map (cf. (7) in section 4).

From these embeddings of the complex numbers into the quaternions we can see that

$$\bigcup_{a \in S^2} U^a = S^3 \equiv SU(2).$$

This allows us to regard *any* quaternion as a complex number:
Given $k \in \mathbb{H}$, we have $\frac{k}{|k|} \in S^3$ and there is an element $a \in S^2$ and $t \in [0, 2\pi)$ so that

$$k = |k| \cdot \frac{k}{|k|} = |k| \cdot e^{ta}. \tag{4}$$

For the Lie algebra $su(2)$ of $SU(2)$ we find

$$su(2) \equiv E$$

as linear spaces. The cross product in E is the Lie bracket.

Finally, f^a denotes the orthogonal complement of $\mathbb{C}^a$ in $\mathbb{H}$ so that

$$\mathbb{H} =: \mathbb{C}^a \oplus f^a. \tag{5}$$

This gives us the following splittings of $\mathbb{H}$ and E:

$$\boxed{\begin{array}{lllllll} \mathbb{H} & = & \mathbb{R}\cdot e & \oplus & & E & \\ & = & \mathbb{R}\cdot e & \oplus & \mathbb{R}\cdot a & \oplus & f^a \\ & = & & \mathbb{C}^a & & \oplus & f^a. \end{array}}$$

Hence f^a is also the orthogonal complement of a in E.

2.2 The Symplectic Plane

We study the plane f^a in more detail: The complex line $\mathbb{C}^a$ acts on f^a from the right. Thus f^a is a $\mathbb{C}^a$–linear space and thus $\mathbb{C}$–linear and — because of its dimension — a complex line.

The set of all complex lines f^a, $a \in S^2$, gives us the tangent space of S^2, i.e.

$$TS^2 = \bigcup_{a\in S^2} \{a\} \times f^a.$$

We want to use the plane f^a as a two–dimensional carrier of (classical) information. In order to have a measure of information, we need a volume form on f^a (cf. [3]). Since the real dimension of f^a is two, the following definition of a *symplectic form* on f^a provides us with the required volume form:

DEFINITION 3. *By*

$$\begin{array}{rcl} \omega^a : f^a \times f^a & \longrightarrow & \mathbb{R} \\ \omega^a(v,w) & := & < v \times a, w > \end{array}$$

we define a symplectic structure *on* f^a.

Clearly

$$\omega^a(v,w) = \mu_E(v,a,w).$$

It can be shown that

LEMMA 4. *Starting from the symplectic plane* (f^a, ω^a) *we can reconstruct the quaternions* $\mathbb{H}$ *with their scalar product* $< , >_{\mathbb{H}}$.

For a proof see [2] or [14].

Now we have a *carrier of information* f^a with a volume form ω^a on it. The axis $\mathbb{R} \cdot a$ serves as the *channel of information transmission* while $\mathbb{R} \cdot e$ can be considered as a *time axis*, so our starting point is given by the following framework:

$$\mathbb{H} = \mathbb{R} \cdot e \oplus \mathbb{R} \cdot a \oplus f^a \ = \mathbb{R} \cdot e \oplus E$$

$\mathbb{R} \cdot e$	time axis
$\mathbb{R} \cdot a$	channel of information transmission
f^a	plane carrying information, symplectic plane

3. Information Transmission along Integral Curves of Vector Fields

In this section we consider the case that the transmission of information follows the trajectories of a vector field. The vector field looked at here is taken to be a singularity free gradient field

$$\begin{array}{rcl} X : O & \longrightarrow & O \times E \\ x & \longmapsto & (x, \mathbf{a}(x)) \end{array}$$

defined on an open set $O \subset E$, where $\mathbf{a}$ denotes the principal part.

The field vector $\mathbf{a}(x)$ now takes the role of $a \in S^2$ as we described it in section 2. For every $\mathbf{a}(x) \in E$ we have an orthogonal splitting

$$E = \mathbb{R} \cdot \mathbf{a}(x) \oplus \mathbb{F}_x^{\mathbf{a}}$$

where $\mathbb{F}_x^{\mathbf{a}}$ again is the orthogonal complement of $\mathbf{a}(x)$ in E.[2] If x varies in O, this gives rise to a *complex line bundle*

$$\mathbb{F}^{\mathbf{a}} := \bigcup_{x \in O} \{x\} \times \mathbb{F}_x^{\mathbf{a}} \quad \subset O \times E.$$

Each of the fibres carries a symplectic structure

$$\omega^{\mathbf{a}}(x; h, k) :=< h \times \mathbf{a}(x), k > \quad \forall h, k \in \mathbb{F}_x^{\mathbf{a}}.$$

We have:

LEMMA 5. *The complex line bundle $\mathbb{F}^{\mathbf{a}}$ completely determines the underlying vector field X, i.e. its principal part* $\mathbf{a}$.

[2]Hence $\mathbb{F}_x^a$ is the analogue of F^a introduced in section 2.

For a proof we refer to [5], [14] and [2].

In view of this lemma the elements of the fibres of the complex line bundle are also called *internal variables of the vector field.* They encode *bits of (classical) information.*

We now have a bundle of planes each functioning as a two–dimensional carrier of information. Via $\mathbb{R} \cdot \mathbf{a}(x)$ each plane is equipped with a channel of information transmission, but we do not yet have a transmission of information from one plane to another.

In order to have such a kind of information transmission we will look at the $U(1)$–principal bundle associated to $\mathbb{F}^{\mathbf{a}}$.

DEFINITION 6. *Let* $\mathbb{P}^{\mathbf{a}}_x$ *be the circle in* $\mathbb{F}^{\mathbf{a}}_x$ *centered around* 0 *with radius* $|\mathbf{a}(x)|^{-\frac{1}{2}}$. *Then*

$$\mathbb{P}^{\mathbf{a}} := \bigcup\nolimits_{x \in O} \{x\} \times \mathbb{P}^{\mathbf{a}}_x$$

is a submanifold of $\mathbb{F}^{\mathbf{a}}$ *and a* $U(1)$*–principal bundle.*

Here the right action of $U(1)$ on the fibres $\mathbb{P}^{\mathbf{a}}_x$ of $\mathbb{P}^{\mathbf{a}}$ is realized by the action of $U^{\mathbf{a}}_x(1) \subset \mathbb{C}^{\mathbf{a}}_x$, where $U^{\mathbf{a}}_x(1)$ is the unitary group of $\mathbb{C}^{\mathbf{a}}_x$ in analogy to $U^a \subset \mathbb{C}^a$ and the isomorphism between $\mathbb{C}$ and $\mathbb{C}^a$ (cf. subsection 2.1).

We will now consider a specific example, namely the *Coulomb field*:

DEFINITION 7. *The* Coulomb field *is defined as*

$$\begin{aligned} X : O &\rightarrow O \times E \\ x &\mapsto (x, \mathbf{a}(x)) \end{aligned}$$

with

$$\mathbf{a}(x) = \gamma m M \cdot \frac{x}{|x|^3}.$$

For reasons of simplicity we set the Coulomb constant γ *and the charges* m *and* M *equal to 1 and thus consider the vector field with principal part* $\mathbf{a}(x) = \frac{x}{|x|^3}$.

For this field, the fibres $\mathbb{P}^{\mathbf{a}}_x$ are circles with radius $|x|$. This means that along straight lines in O, the principal bundle $\mathbb{P}^{\mathbf{a}}$ has the form of a double cone without its vertex with a cone angle of $45°$.

An integral curve $\beta : I \rightarrow O$ of X with $\beta(1) = x_0$ for a given $x_0 \in O$ with $|x_0| = 1$ has the form

$$\beta(t) = x_0 \cdot |3t - 2|^{\frac{1}{3}}$$

(cf. [14]).

We want to describe an information transmission on $\mathbb{P}^{\mathbf{a}}|_\beta$. One way is to look at the horizontal lifts of $\mathbb{P}^{\mathbf{a}}|_\beta$. For a detailed discussion of this approach see [4] and section 13 of these notes. Another way is to look at the geodesics on $\mathbb{P}^{\mathbf{a}}|_\beta$, which we will do here. First of all we need a Clairaut chart of $\mathbb{P}^{\mathbf{a}}|_\beta$.

The restriction $\mathbb{P}^{\mathbf{a}}|_\beta$ of the principal bundle can be parametrized by the Clairaut chart $\mathbf{x} : \mathcal{U} \to E$ given by

$$\mathbf{x}(u,v) := (3\,v(s)-2)^{\frac{1}{3}} \cdot e^{u\frac{x}{|x|}} \cdot v_{x0} + (3\,v(s)-2)^{\frac{1}{3}} \cdot \frac{x}{|x|}$$

for a given initial vector $v_{x_0} \in \mathbb{P}^{\mathbf{a}}|_\beta$. Here $\mathcal{U} = \{(u,v) \in \mathbb{R}^2 |\, 0 < u < 2\pi + \mu\}$ with $\mu > 0$, $\nu < v < \infty$ for some $\frac{2}{3} \leq \nu < 1$.

For the geodesics α on $\mathbb{P}^{\mathbf{a}}|_\beta$ we find

$$\begin{aligned}\alpha(s) &= \mathbf{x}(u(s), v(s)) \\ &= (3\,v(s)-2)^{\frac{1}{3}} \cdot e^{u(s)\frac{x}{|x|}} \cdot v_{x0} + (3\,v(s)-2)^{\frac{1}{3}} \cdot \frac{x}{|x|} \end{aligned} \tag{6}$$

with

$$u(s) = \sqrt{2}\arctan\left(\frac{s}{\sqrt{2}c} + \frac{c_1}{c}\right) + c_2$$

and

$$v(s) = \pm\frac{1}{3}\left(\left(\frac{1}{\sqrt{2}}s + c_1\right)^2 + c^2\right)^{\frac{3}{2}} + \frac{2}{3}. \tag{7}$$

The constant c gives the slope of the geodesic in the Clairaut chart, the integration constant c_1 determines the "forward movement" on the cone to the point where α starts, and the second integration constant c_2 fixes the starting point of the rotation, i.e. the phase. The choice of the sign in (7) determines the orientation of the geodesic.

We now have:

LEMMA 8. *The geodesics described by (5) provide us with a means of transmitting information between the fibres of* $\mathbb{P}^{\mathbf{a}}|_\beta$.

As a conclusion of this section, we want to add a remark on how this example might be generalized (cf. [14]). We start with an arbitrary singularity free gradient field on $O \subset E$ with principal part b and an integral curve γ with

$$\dot\gamma(t) = b(\gamma(t)) \qquad \text{and} \qquad \gamma(1) = x_0$$

for some initial $x_0 \in O$. If there is a solution $\gamma(t)$ of form $\gamma(t) = x_0 \cdot \delta(t)$ with $\delta(1) = 1$, this solution curve can be lifted to a horizontal lift γ^{hor} of γ with $\gamma^{hor}(t) = v_{x_0} \cdot \delta(t)$. Here v_{x_0} denotes an initial vector in $\mathbb{P}^{\mathbf{a}}|_\beta$ at x_0 (see also section 13).

4. Vector Fields and Heisenberg Groups Linked to Information Transmission

4.1 Heisenberg Algebras and Heisenberg Groups

In the preceding section we discussed the transmission of information from a completely spatial point of view. Now we want to introduce time. As we already mentioned in section 2.1, one natural way is to take $\mathbb{R} \cdot e$ as a time axis. However, we may also regard $\mathbb{R} \cdot \mathbf{a}(x)$ generated by the field vector $\mathbf{a}(x)$ as a time axis in E. To see this, we look at the multiplication in the quaternions. There we have

$$(\mathbb{R} \cdot e \oplus E) \cdot \mathbf{a}(x) = (\mathbb{R} \cdot e \oplus \mathbb{R} \cdot \mathbf{a}(x) \oplus \mathbb{F}_x^{\mathbf{a}}) \cdot \mathbf{a}(x) = \mathbb{R} \cdot \mathbf{a}(x) \oplus \mathbb{R} \cdot e \oplus \mathbb{F}_x^{\mathbf{a}}$$

as well as

$$\mathbf{a}(x) \cdot (\mathbb{R} \cdot e \oplus E) = \mathbb{R} \cdot \mathbf{a}(x) \oplus \mathbb{R} \cdot e \oplus \mathbb{F}_x^{\mathbf{a}}.$$

Hence $\mathbb{R} \cdot \mathbf{a}(x)$ is a *linear* time axis. We may also define a *periodic* concept of time, i.e. a "watch". In the $2+2$–splitting $\mathbb{H} = \mathbb{C}_x^{\mathbf{a}} \oplus \mathbb{F}_x^{\mathbf{a}}$ we have the unitary group $U_x^{\mathbf{a}} \subset \mathbb{C}_x^{\mathbf{a}}$. This allows us to link $\mathbb{R} \cdot \mathbf{a}(x)$ and $U_x^{\mathbf{a}}$ via the exponential map

$$\begin{aligned} \exp_{\mathbb{C}_x^{\mathbf{a}}} : \mathbb{R} \cdot \mathbf{a}(x) \quad &\longrightarrow \quad U_x^{\mathbf{a}} \subset \mathbb{C}_x^{\mathbf{a}} \\ \exp_{\mathbb{C}_x^{\mathbf{a}}}(t \cdot \mathbf{a}(x)) \quad &:= \quad \cos t \cdot e + \sin t \cdot \mathbf{a}(x). \end{aligned} \tag{8}$$

Linking both the linear and the periodic time concept with the plane of information transmission $\mathbb{F}_x^{\mathbf{a}}$ gives rise to the concepts of Heisenberg algebras and Heisenberg groups:

DEFINITION 9. *We define the* Heisenberg algebra $\mathcal{G}_x^{\mathbf{a}}$ *as*

$$\mathcal{G}_x^{\mathbf{a}} := \mathbb{R} \cdot \mathbf{a}(x) \oplus \mathbb{F}_x^{\mathbf{a}}$$

together with the Lie bracket

$$[\lambda_1 \cdot \mathbf{a}(x) + h_1, \lambda_2 \cdot \mathbf{a}(x) + h_2] := \omega^{\mathbf{a}}(x; h_1, h_2) \cdot \mathbf{a}(x).$$

Since we have $\mathbb{F}_x^{\mathbf{a}}$ and the symplectic structure $\omega^{\mathbf{a}}(x; \ldots, \ldots)$, in analogy to lemma 4 we know that $\mathcal{G}_x^{\mathbf{a}}$ allows the reconstruction of the quaternions.

DEFINITION 10. *The* Heisenberg group

$$G_x^{\mathbf{a}} := |\mathbf{a}(x)|^{-\frac{1}{2}} \cdot U_x^{\mathbf{a}}(1) \oplus \mathbb{F}_x^{\mathbf{a}}$$

is defined with the rescaled multiplication

$$\begin{aligned} (k_1 + h_1) \cdot (k_2 + h_2) \quad &= \quad \left(|\mathbf{a}(x)|^{-\frac{1}{2}} e^{t_1 \frac{x}{|x|}} + h_1\right) \cdot \left(|\mathbf{a}(x)|^{-\frac{1}{2}} e^{t_2 \frac{x}{|x|}} + h_2\right) \\ &:= \quad |\mathbf{a}(x)|^{-\frac{1}{2}} e^{(t_1+t_2)\frac{x}{|x|}} e^{\frac{1}{2}\omega^{\mathbf{a}}(x;h_1,h_2)\frac{x}{|x|}} + h_1 + h_2. \end{aligned}$$

The multiplication first rescales any two elements so that they are in $U_x^{\mathbf{a}}(1) \oplus \mathbb{F}_x^{\mathbf{a}}$ *and then again rescales them after the multiplication. This method allows us to inherit the group structure of* $U_x^{\mathbf{a}}$ *to circles in* $\mathbb{C}_x^{\mathbf{a}}$ *with radius not equal to 1.*

The center of $G_x^{\mathbf{a}}$ is $U_x^{\mathbf{a}}$.

It can be shown that $\mathcal{G}_x^{\mathbf{a}}$ is the Heisenberg algebra belonging to the Heisenberg group $G_x^{\mathbf{a}}$, since

$$T_e G_x^{\mathbf{a}} = \mathbb{R} \cdot \mathbf{a}(x) \oplus \mathbb{F}_x^{\mathbf{a}} = \mathcal{G}_x^{\mathbf{a}}.$$

Moreover, we now see that there is a natural link between $\mathcal{G}_x^{\mathbf{a}}$ and $\mathbb{H}$. In constructing $\mathbb{H}$ out of $\mathcal{G}_x^{\mathbf{a}}$, the choice of e is by no means artificial, but rather natural (cf. lemma 4).

Having in mind the transmission of information, we look at the *Schrödinger representation* ρ_x^ν, which allows a modulation of information on signals $\psi \in \mathcal{S}(\mathbb{R}, \mathbb{C})$.

To do so, we first need to define a coordinate system on $\mathbb{F}_x^{\mathbf{a}}$. We choose a (unit) vector v in $\mathbb{F}_x^{\mathbf{a}}$ and define the coordinate system via $e_q := v$ and $e_p := e_q \cdot \frac{\mathbf{a}(x)}{|\mathbf{a}(x)|} = e_q \times \frac{\mathbf{a}(x)}{|\mathbf{a}(x)|}$, since the multiplication with $\frac{\mathbf{a}(x)}{|\mathbf{a}(x)|} = \frac{x}{|x|}$ amounts to a rotation about $\pi/2$. Then we can define:

DEFINITION 11. *The* Schrödinger representation with frequency ν *is given by*

$$\begin{aligned} \rho_x^\nu : G_x^{\mathbf{a}} &\longrightarrow U(L^2(\mathbb{R}, \mathbb{C})) \\ \rho_x^\nu(z+h)(\psi)(\xi) &:= e^{-\nu t i} \cdot e^{-\frac{1}{2}|\mathbf{a}(x)|\nu p_x q_x i} \cdot e^{|\mathbf{a}(x)|\nu p_x \xi i} \cdot \psi(\xi - q_x) \end{aligned}$$

with $z + h = e^{t\frac{x}{|x|}} + q_x e_q + p_x e_p \in G_x^{\mathbf{a}}, \xi \in \mathbb{R}$ *and* $\nu \in \mathbb{R}\backslash\{0\}$.

There are two characteristics determining the Schrödinger representation. The first one is the frequency ν that has to be chosen. The second one is the coordinate system e_q, e_p we had to fix in $\mathbb{F}_x^{\mathbf{a}}$, i.e. the vector $v \in \mathbb{F}_x^{\mathbf{a}}$ we chose.

The *theorem of Stone–von Neumann* gives us further information about the classification of the Schrödinger representations:

THEOREM 12. *The representation* ρ^ν *is irreducible for any* $\nu \in \mathbb{R}\backslash\{0\}$. *Each irreducible representation* $\tilde{\rho}$ *of a Heisenberg group* G^a *which is identical with* ρ^ν *on its center is equivalent to* ρ^ν.

This means that the equivalence classes of the Schrödinger representations are determined by ν. As has been shown by Kirrilov (cf. [15]), for a given ν we get *all* representations of the equivalence class by varying the choice of $v \in \mathbb{F}_x^{\mathbf{a}}$.

4.2 Vector Fields and Heisenberg Groups

Now consider a vector field $\mathbf{a}(x)$ for which we can describe the geodesics on $\mathbb{P}^{\mathbf{a}}|_\beta$ as in section 3. These geodesics have the form

$$\alpha(s) = \mathbf{x}(u(s), v(s)) = \varphi(v(s)) \cdot e^{u(s)\frac{\mathbf{a}(x)}{|\mathbf{a}(x)|}} \cdot v_{x0} + \psi(v(s)) \cdot \frac{\mathbf{a}(x)}{|\mathbf{a}(x)|}$$

where $u(s)$ fixes the rotation on the rotational surface $\mathbb{P}^{\mathbf{a}}|_\beta$ by means of the factor $e^{u(s)\frac{\mathbf{a}(x)}{|\mathbf{a}(x)|}}$ at the point given by $v(s)$. The parameter $v(s)$ determines the progress between the planes, i.e. the fibres given by $\mathbf{a}(x)$.

These geodesics may now be used to link the single Schrödinger representations. The field vector $\mathbf{a}(x)$ fixes a Heisenberg group $G_x^{\mathbf{a}}$. Let $z \in U_x^{\mathbf{a}}$, $\psi \in \mathcal{S}(\mathbb{R}, \mathbb{C})$ and $\xi \in \mathbb{R}$. Then $\varphi(v(s)) \cdot e^{u(s)\frac{\mathbf{a}(x)}{|\mathbf{a}(x)|}}$ fixes the coordinates q_x and p_x in $\mathbb{F}_x^{\mathbf{a}}$ and we have

$$\begin{aligned} &\rho_x^\nu\left(z + q_x e_q^x + p_x e_p^x\right)(\psi)(\xi) : \\ &= e^{-\nu t i} e^{-|\mathbf{a}(x)|\nu\frac{u(s)v(s)}{2}i} e^{|\mathbf{a}(x)|\nu u(s)\xi i}\psi(\xi - v(s)). \end{aligned}$$

The frequency ν determines the speed of progress.

Thus by now we have described two ways of information transmission:

PROPOSITION 13. *The transmission of information can be given*

(1) by the geodesics α on $\mathbb{P}^{\mathbf{a}}|_\beta$

and

(2) via the modulation of information on signals by means of the Schrödinger representations ρ_x^ν.

5. The Spin Group and Heisenberg Groups

In this section, we will leave the context of vector fields to discuss Heisenberg groups and Schrödinger representations in conjunction with the spin group $SU(2)$ and the spin representations, especially the spin $\frac{1}{2}$–representation.

To do so,the Heisenberg group under consideration is no longer determined by the vector $\mathbf{a}(x)$ of a given vector field with principal part $\mathbf{a}$, but we simply choose an element of the 2–sphere, i.e. we choose $a \in S^2$. This is equivalent to considering a vector field with $\mathbf{a}(x) \equiv a \in S^2$ for all $x \in O$.

The resulting Heisenberg groups and Heisenberg algebras are given by $G^a = U^a \oplus f^a$ and $\mathcal{G}^a = \mathbb{R} \cdot a \oplus f^a$, respectively, if we choose $a \in S^2$; the Schrödinger representation with frequency ν is given by

$$\rho^\nu(e^{ta} + qe_q + pe_p)(\psi)(\xi) = e^{-\nu t i} e^{-\nu\frac{pq}{2}i} e^{\nu p\xi i}\psi(\xi - q)$$

it follows that

$$-\omega^b(a_1, a_2) = < a_1 \times a_2, b > = < a_1 \times a_2, \frac{a_1 \times a_2}{|a_1 \times a_2|} > = |a_1 \times a_2| = \sin\varphi.$$

From

$$\cos\varphi = \pm\sqrt{1 - \sin^2\varphi}$$

we deduce that

$$< a_1, a_2 > = \pm\sqrt{1 + (\omega^b(a_1, a_2))^2},$$

so that $< a_1, a_2 >$ is also a datum given by the Heisenberg groups.

As a result we have:

PROPOSITION 16. *By means of the inner automorphisms τ_k, the Heisenberg groups can be transformed into each other. Hence the group structure of $SU(2)$ can be described by only* one *Heisenberg group together with the inner automorphisms τ_k.*

COROLLARY 17. *Starting from $SU(2) \equiv S^3$ we get any sphere S^3_R with radius R by a rescalation. Hence any sphere S^3_R in $\mathbb{H}$ can be reconstructed from a Heisenberg group.*

5.1.2 The spin group $SU(2)$ determines the Heisenberg groups. To describe how the spin group determines the Heisenberg groups, we again choose an element $a \in S^2$. In reconstructing G^a we split it into the parts that need to be given by $SU(2)$:

- The center U^a of G^a is given by the meridian in $SU(2)$ which runs through the chosen $a \in S^2$.

- The symplectic plane f^a is defined as the tangent space of S^2 (the equator of S^3) at the point a.

- The symplectic form ω^a is given by

$$\omega^a(h_1, h_2) := - < a, \frac{1}{2}[h_1, h_2] > \qquad \forall\, h_1, h_2 \in f^a. \tag{9}$$

  Here $< , >$ is the scalar product in $\mathbb{H}$ and

$$[h_1, h_2] := h_1h_2 - h_2h_1$$

  is the commutator in $\mathbb{H}$, which is also the Lie bracket in $su(2)$.

Proof of equation (9).

For any two $h_1, h_2 \in f^a$ we have

$$\begin{aligned} -< a, \frac{1}{2}[h_1, h_2] > &= -< a, \frac{1}{2}(h_1 h_2 - h_2 h_1) > \\ &= -< a, \frac{1}{2}h_1 \times h_2 > + < a, \frac{1}{2}h_2 \times h_1 > \\ &= -< h_1 \times h_2, a > \\ &= -\mu_E(h_1, h_2, a) \\ &= \mu_E(h_1, a, h_2) \\ &= < h_1 \times a, h_2 > \\ &= \omega^a(h_1, h_2), \end{aligned}$$

as was originally defined on f^a.

5.2 Spin $\frac{1}{2}$–Representations

Before we can extend this link between $SU(2)$ and the Heisenberg groups to their representations, we have to take a look at the spin $\frac{1}{2}$–representation and define it within our setting. As a representation space we want to take the quaternions so we start with a remark on the linear structure of $\mathbb{H}$.

The quaternions $\mathbb{H}$ form a real four–dimensional algebra. However, by considering the splitting

$$\mathbb{H} = \mathbb{C}^a \oplus f^a$$

we get a complex structure on $\mathbb{H}$, since both $\mathbb{C}^a$ and f^a are $\mathbb{C}^a$–linear spaces. Thus $\mathbb{H}$ is a complex two–dimensional space. We will use this complex structure on $\mathbb{H}$ to define a spin representation. Because of $\dim_{\mathbb{C}^a} \mathbb{H} = 2$, we will look at the *spin $\frac{1}{2}$–representation*:

DEFINITION 18. *The* spin $\frac{1}{2}$–representation r *on* $\mathbb{H} = \mathbb{C}^a \oplus \mathbb{F}^a$ *is defined as*

$$\begin{aligned} r : SU(2) &\longrightarrow U(\mathbb{C}^a \oplus f^a) \\ u &\longmapsto r(u) \end{aligned}$$

with

$$\begin{aligned} r(u) : \mathbb{C}^a \oplus f^a &\longrightarrow \mathbb{C}^a \oplus f^a \\ z' + h &\longmapsto u(z' + h) = uz' + uh. \end{aligned}$$

Because of the inner automorphisms $\tau_k, k \in S^3$, and the orthogonality of the splitting $\mathbb{C}^a \oplus \mathbb{F}^a$, this definition is independent of the choice of $a \in S^2$. In addition, there is a Hermitian product $< \mid >$ on $\mathbb{H}$ inherited from the one on $\mathbb{C}^a$ (i.e. the one on $\mathbb{C}$).

Clearly, the spin $\frac{1}{2}$–representation is unitary and irreducible.

There is another way of writing this representation.

Let $u \in SU(2)$. Then $u = e^{ta} \in U^a(1)$ for some $a \in S^2$. Thus we can write

$$r(u)(z' + h) = r(e^{ta})(z' + h) = e^{ta}z' + he^{-ta} \tag{10}$$

for all $z' + h \in \mathbb{C}^a \oplus f^a$. Here we used the fact that $z \cdot h = h \cdot \overline{z}$ for any $z \in \mathbb{C}^a, h \in f^a$, where $\overline{z}$ is the complex conjugate of z in analogy to the conjugation in $\mathbb{C}$ (cf. [2]).

Note that here a is the same vector both in the splitting $\mathbb{C}^a \oplus \mathbb{F}^a$ and in the meridian U^a. For another splitting $\mathbb{C}^b \oplus F^b$, the unitary group $U^a(1)$ does not operate on F^b as $U^b(1)$ does.

The restriction of r to the meridian U^a gives us a reducible representation $r|_{U^a}$. From (10) we can easily see that there are two invariant subspaces, namely $\mathbb{C}^a$ and f^a.

The representation $r|_U^a$ also allows us to easily read off the character χ_r of r at $e^{ta} \in U^a$:

$$\chi_r(e^{ta}) = e^{ta} + e^{-ta},$$

which is in accordance with the "usually known" character of the spin $\frac{1}{2}$–representation.

5.3 The Spin $\frac{1}{2}$–Representation r and the Schrödinger Representations ρ^ν

5.3.1 Construct r out of ρ^ν. Now we have all the prerequisites to show how the spin $\frac{1}{2}$–representation can be constructed out of the Schrödinger representations ρ^ν. One important aspect is that the center of any Heisenberg group G^a is contained in the spin group $SU(2)$.

We choose an element $z \in SU(2)$; we wish to construct $r(z)$.

The element z can be written in the form $z = e^{ta}$ for a suitable $a \in S^2$. This fixes the Heisenberg group the representation of which we want to use, namely G^a.

In a next step we choose the frequencies $\nu_1 = 1$ and $\nu_2 = -1$. Then we have two representations $\rho^1|_U^a$ and $\rho^{-1}|_U^a$ of the center U^a of G^a which read

$$\rho^{\pm 1}|_U^a(e^{ta})(\psi)(\xi) = e^{\mp ti} \cdot \psi(\xi).$$

We will establish the required link by means of the characters. The characters of ρ^1 and ρ^{-1}, respectively, are

$$\chi_{\rho^1}(e^{ta}) = e^{-ti} \qquad \text{and} \qquad \chi_{\rho^{-1}}(e^{ta}) = e^{ti}.$$

We set

$$\chi := \chi_{\rho^1} + \chi_{\rho^{-1}}, \qquad \chi(e^{ta}) = e^{-ti} + e^{ti}.$$

This

is the character of the spin $\frac{1}{2}$–representation $r|_U^a(e^{ta})$. An induction process provides us with $r|_U^a$ itself (up to isomorphy).

This procedure can be repeated for all meridians $U^b(1), b \in S^2$. Since $r|_U^a$ and $r|_{U^b(1)}$ are linked via τ_k we thus get the whole representation r. Hence we have shown:

THEOREM 19. *The spin $\frac{1}{2}$–representation r is determined by the family of Schrödinger representations $\rho^{\pm 1}$ of the Heisenberg groups $G^a, a \in S^2$.*

5.3.2 Construct $\rho^{\pm 1}$ out of r. We now start with the spin $\frac{1}{2}$–representation r. By choosing an element $a \in S^2$ we determine the Heisenberg group G^a whose representation shall be constructed.

Let $e^{ta} + h \in G^a$; we want to describe $\rho^{\pm 1}(e^{ta} + h)$.

The character χ_r of the spin $\frac{1}{2}$–representation r, given by

$$\chi_r(e^{ta}) = e^{-ti} + e^{ti},$$

defines two representations of U^a, the center of G^a, namely

$$\chi_r^1(e^{ta}) := e^{-ti} \qquad \text{and} \qquad \chi_r^2(e^{ta}) := e^{ti}.$$

By the theorem of Stone–von Neumann we know that the equivalence classes of ρ^1 and ρ^{-1} are fixed by their restrictions to the center.

We choose $v \in F^a$ with $|v| = 1$ and thus get a coordinate system $v, v \cdot a$ on f^a. In the same manner we choose a vector $-v \in F^{-a}$ and so have a coordinate system on F^{-a}.

Hence the two representations ρ^1 and ρ^{-1} are uniquely determined. Again, their explicit form is given by an induction process as it is described for example by Schempp ([15]) or in [14].

As a result we have:

THEOREM 20. *The Schrödinger representations ρ^1 and ρ^{-1} of the Heisenberg group G^a can be reconstructed starting from the spin $\frac{1}{2}$–representation $r|_U^a$.*

So far, we only got the two representations ρ^1 and ρ^{-1}. In the next section we will briefly discuss how other Schrödinger representations may be gained from spin representations.

5.3.3 Link between spin s–representations and ρ^ν for $\nu \in \mathbb{Z}$. What the spin $\frac{1}{2}$–representation cannot do is to give us Schrödinger representations with frequencies unequal to ± 1. Hence we make use of the spin s–representation with character

$$\chi_s(e^{ta}) = e^{-2sti} + e^{(-2s+2)ti} + \ldots + e^{2sti}$$

for any $e^{ta} \in U^a$. In the same way we described above, the character of the spin s–representation determines the Schrödinger representations given by

$$\rho^{-2s}|_U^a,\ \rho^{-2s+2}|_U^a,\ \ldots,\ \rho^{2s-2}|_U^a,\ \rho^{2s}|_U^a \tag{11}$$

Vice versa we can show that the sum of the characters of the Schrödinger representations in (11) gives us the character of the spin s–representation and hence provides us – up to isomorphy – with the spin s–representation itself.

5.4 The Spin $\frac{1}{2}$–Representation and Signals

We will look at the geometry of the 3–sphere more closely; in particular the great circles of S^3 are one focus of interest.

DEFINITION 21. *A great circle $K \subset S^3$ is a circle on S^3 with center 0. This means that both K and S^3 have the same center.*

The great circles of S^3 are the left (right) cosets of the operation of U^a on S^3.

We will establish a link between this geometry of the 3–sphere and the spin $\frac{1}{2}$–representation:[4]

LEMMA 22. *The left cosets of the operation of U^a on S^3 are given by*

$$U^a(1) \cdot k = r(U^a(1))(k) \qquad \forall k \in S^3.$$

PROOF 1. *For $k = z' + h \in S^3 \subset \mathbb{C}^a \oplus \mathbb{F}^a$ we have*

$$\begin{aligned} U^a(1) \cdot k &= U^a(1) \cdot (z' + h) \\ &= \{z(z' + h) \mid z \in U^a(1)\} \\ &= \{zz' + zh \mid z \in U^a(1)\} \\ &= \{r(z)(z' + h) \mid z \in U^a(1)\} \\ &= r(U^a(1))(z' + h) \\ &= r(U^a(1))(k). \end{aligned}$$

This means that the left cosets consist of

$$r(U^a(1))(k),\ k \in S^3.$$

[4] See also [2] and [14].

The analogous result for right cosets is shown accordingly.

Any great circle can be split up into a circle in $\mathbb{C}^a$ and a circle in f^a. The radii of these circles are unequal to 1, in general. Hence it seems that unless the coset $U^a \cdot k = U^a$, we *cannot* use the Schrödinger representation to modulate the information contained in $U^a \cdot k$ onto signals.

However, there is a related approach, which allows just this modulation of the information on the great circles onto signals. We again look at the spin $\frac{1}{2}$–representation $r(z)(k) = zz' + hz^{-1}$ of $z \in U^a$. Since $|k| = |z' + h| = 1$ and since the splitting $\mathbb{C}^a \oplus \mathbb{F}^a$ is orthogonal, by knowing $|h|$, we have also $|z'|$ given. This means that if we have h, we know on which circle in $\mathbb{C}^a$ the vector $z \cdot z'$ is. This information is enough as z runs through all of U^a. In particular, the circle

$$h \cdot U^a \subset f^a$$

is representative for $U^a \cdot k$ and thus also for $r(U^a)(k)$.

The spin $\frac{1}{2}$–representation can be generated by $\rho^1 + \rho^{-1}$, and since $(\rho^1 + \rho^{-1})(h \cdot U^a)$ is a subset of $\mathcal{S}(\mathbb{R}, \mathbb{C})$, we have:

PROPOSITION 23. *The information inscribed in the great circles of the 3–sphere S^3 can be modulated onto signals.*

This subject will be taken on in section 13 again.

6. A First Resummee

Let us hold for a moment and recollect what we got so far.

We started with splittings of the quaternions. One of them was the $1 + 3$–splitting, which provided us with a notion of a linear concept of time together with a three–dimensional Euclidean space based on special relativity.

We then continued by splitting up the Euclidean space E. This $1+2$–splitting provided us with a Heisenberg algebra setting.

Thus we have a description of an information transmission with both space– and time components in three and four dimensions.

In a next step we established a link between the spin group $SU(2)$ and Heisenberg groups G^a; a link we extended to comprise also the spin $\frac{1}{2}$– and the Schrödinger representations.

This means that we have a link between a representation coming from a *classical* setting, namely the Schrödinger representation, and a representation used in *quantum* mechanics, i.e. the spin $\frac{1}{2}$–representation. Thus we have entered a discussion of connecting a classical point of view with a quantum point of view. (Hence this link is also a link between a macroscopic and a microscopic approach.) This will be discussed in detail in sections 8 and 13.

7. Ingredients of Quantum Information — a Geometric Setting

We now want to focus on quantum information and a description of its transmission. In particular, we want to investigate the link between the classical description of information transmission as given in earlier sections and quantum information transmission as we will describe it here. We start with a short clarification of the terms we will use in the following.

The objects of interest in *quantum information* are quantum bitsqubits (*qubits*) and their states; in particular the *normed states* of qubits. We will begin with a traditional description of these notions.

The setting is provided by a two–dimensional *Hilbert space* over the complex numbers $\mathbb{C}$. For simplicity we take this Hilbert space to be $\mathbb{C}^2$ with the *natural Hermitian product* $< \mid >$. Then the collection of normed states is given by the 3–sphere $S^3 \subset \mathbb{C}^2$. A *state* $\psi \in \mathbb{C}^2$ can be represented for example as $\psi = z_1 \cdot (1, 0) + z_2 \cdot (0, 1)$ with $z_1, z_2 \in \mathbb{C}$ and $(1, 0), (0, 1)$ the canonical basis of $\mathbb{C}^2$ (in literature ψ is often referred to as a qubit).

For a *normed state* $\psi \in S^3$, the product $\psi \cdot \overline{\psi}$ is a probability density. This density is invariant under phase factors of ψ. Thus we consider ψ *up to a (global) phase factor*. This is what we mean by a *qubit*. Hence for a description of the qubit itself (whose normed state is given by ψ), the relevant entity is the coset $\psi \cdot U(1)$. Thus we refer to the *coset* $\psi \cdot U(1)$ as a qubit.

The quotient $S^3/U(1)$, i.e. the orbit space of the $U(1)$–action on S^3, leads us to the *complex projective space* $\mathbb{C}P_2$ of $\mathbb{C}^2$ (cf. appendix 3.4). $\mathbb{C}P_2$ is diffeomorphic to the 2–sphere $S^2 \subset S^3$. Thus a qubit is an element in the complex projective space $\mathbb{C}P_2$ and vice versa.

On the other hand $\mathbb{C}^2$ can be given the structure of a skew field, the quaternions (cf. section 5.2). Hence $S^3 = SU(2)$ and therefore

$$\mathbb{C}P_2 = SU(2)/U(1). \tag{12}$$

In order to link the Hilbert space with signal transmission as formulated in proposition 13 we realize it in terms of thequaternion *quaternions* $\mathbb{H}$ generated by a three-dimensional oriented Euclidean space E as done in section 2.1.

The fibration $\mathbb{C}P_2 = SU(2)/U(1)$ (Hopf fibration) is studied in more detail in a geometric fashion in the appendix 3.4.

As general references on quantum information and quantum computing see [7], [11] and [13] as well as the literature cited there.

8. The Quantum State Bundle of a Gradient Field

The vector field X under consideration here is the same as in section 3, i.e. a non-vanishing gradient field of a smooth $\mathbb{R}$–valued function f defined on an open set O with values in a Euclidean $\mathbb{R}$–linear oriented space E of dimension three. The principal part of X is denoted by $\mathbf{a}$. It assigns the gradient of f to any $x \in O$. The *flow* of grad f will be brought in connection with quantum information.

First of all we will link X with S^2 in order to have access to the space of qubits:

Assigning the unit vector

$$\varepsilon(x) := \frac{\mathbf{a}(x)}{|\mathbf{a}(x)|}$$

to each $x \in O$ yields a smooth map

$$\varepsilon : O \longrightarrow S^2,$$

the *Gauss map*, which is the direction field of X.

Given a level surface S_x through $x \in O$, the vector $\varepsilon(x)$ is the oriented unit normal to S_x in x.

Hence the vector field X is naturally linked to the space S^2 of *qubits*, the equator of the 3–sphere S^3 in $\mathbb{C}^2$, which is the space of normed states in the Hilbert space $\mathbb{C}^2$.

The aim of this section is to construct a $U(1)$–principal bundle $\mathbb{K}$ over O for which each fibre $\mathbb{K}_x$ is the normed state space of the qubit $\varepsilon(x) \in S^2$.

In order to link the geometry of E to the space S^2 of qubits and hence to X, we now fix an element $a \in S^2$ as a reference point and replace $\mathbb{C}^2$ by the skew field of quaternions $\mathbb{H}$ split according to (5) into $\mathbb{H} = \mathbb{C}^a \oplus \mathbb{F}^a$.

The skew field $\mathbb{H}$ is turned into a $\mathbb{C}^a$–linear space by setting

$$F^a = q_0 \cdot \mathbb{C}^a$$

for a fixed unit vector $q_0 \in F^a$. Clearly $q_0 \cdot z = \overline{z} \cdot q_0$ for all $z \in \mathbb{C}^a$. Thus

$$\mathbb{H} \;=\; \mathbb{C}^a \oplus \mathbb{C}^a \cdot q_0 \;=\; \mathbb{C}^a \oplus q_0 \cdot \mathbb{C}^a,$$

a left–linear (right–linear) space. The Hermitian form is here defined with respect to the left–linear structure as

$$< z_1 + z_2 \cdot q_0 \mid z_1' + z_2 \cdot q_0 > := \; \bar{z}_1 \cdot z_1' + \bar{z}_2 \cdot z_2'$$

for all $z_1, z_2, z_1', z_2' \in \mathbb{C}^a$ (cf. section 5.2).

The relation between qubits and their state spaces mentioned in section 7 is here given by the *Hopf fibration*

$$SU(2)/U^a(1) = S^2$$

with the projection

$$\mathrm{pr}_{\mathrm{Hopf}} : SU(2) \longrightarrow S^2$$

defined by

$$\mathrm{pr}_{\mathrm{Hopf}}(k) := \tau_k(a).$$

Here τ_k is an inner automorphism of $\mathbb{H}$ (cf. definition 14 and the appendix, which deals with the Hopf projection in detail (especially appendix 3.1)). In appendix 1 we show that τ_k is an isometry and hence respecting any orthogonal splitting of $\mathbb{H}$. If $\tau_k(a) = b$, the state space of a qubit $b \in S^2$ is hence

$$\tau_k^{-1}(b) = k \cdot U^a(1)$$

providing us with a relation between the state spaces of different qubits (cf. section 5.4). Furthermore we point out that the vector k in $\tau_k(a) = b$ can naturally be constructed out of a and b if $b \neq \pm a$ as shown in theorem 43 in appendix 3.2.

The *quantum state bundle* $\mathbb{K}$ will now be constructed as a subbundle of $O \times S^3$:

DEFINITION 24. *The fibre* $\mathbb{K}_x$ *of the* quantum state bundle $\mathbb{K}$ *is defined as*

$$\mathbb{K}_x := pr_{Hopf}^{-1}(\varepsilon(x)) \qquad \forall x \in O.$$

Thus the bundle $\mathbb{K}$ *consists of*

$$\mathbb{K} := \bigcup_{x \in O} \mathbb{K}_x.$$

Clearly, $\mathbb{K} \subset O \times S^3$ is a smooth submanifold if equipped with the subspace topology. It contains *normed states* only.

Describing the link between $\mathbb{K}$ and the Hopf fibration more closely, for all $k_x \in \mathbb{K}_x$ and for any $x \in O$ we embed the fibres of $\mathbb{K}$ into $SU(2)$ by

$$\begin{aligned} \widehat{\varepsilon} : \mathbb{K} &\longrightarrow SU(2) \\ \widehat{\varepsilon}(k_x) &:= k_{\varepsilon(x)}. \end{aligned}$$

Then $\mathbb{K}$ is the pull–back of the Hopf fibration over S^2 and hence a U^a–principal bundle over O for which the diagram

$$\begin{array}{ccc} \mathbb{K} & \xrightarrow{\widehat{\varepsilon}} & SU(2) \\ \mathrm{pr}_O \downarrow & & \downarrow \mathrm{pr}_{\mathrm{Hopf}} \\ O & \xrightarrow{\varepsilon} & S^2 \end{array} \qquad (13)$$

is commutative. pr_O is the projection to O assigning to any $k_x \in \mathbb{K}$ the point $x \in O$. This again reveals the smooth manifold structure of $\mathbb{K}$. All the maps in diagram (13) are smooth.

$SU(2)$ consists of all *normed states* of the collection S^2 of all qubits. Each state in $SU(2)$ can be considered to describe quantum information. Hence $\mathbb{K}$ is a bundle of states over the collection O of *points identified with qubits via* ε.

This provides us with the following result (cf. section 3):

PROPOSITION 25. $\mathbb{K}$ *can be viewed as a collection of internal variables which encode quantum information.*

We now shall illustrate the position of the fibres on S^3 with respect to the splitting (5):

Any coset

$$\mathbb{K}_{\tau_k(a)} := k \cdot U^a(1)$$

of normed states is a great circle on $SU(2)$ for any $k \in SU(2)$ (cf. lemma 22). The splitting $\mathbb{H} = \mathbb{C}^a \oplus f^a$ for a fixed $a \in S^2 \subset SU(2)$ yields the components

$$\mathbb{K}^{\mathbb{C}^a}_{\tau_k(a)} \subset \mathbb{C}^a \qquad \text{and} \qquad \mathbb{K}^{f^a}_{\tau_k(a)} \subset f^a$$

of $\mathbb{K}_{\tau_k(a)}$ obtained by the orthogonal projections $\mathrm{pr}_{\mathbb{C}^a}$ and pr_{f^a} from $\mathbb{H}$ to $\mathbb{C}^a$ and f^a, respectively. Thus $\mathbb{K}_{\tau_k(a)}$ is on the torus

$$\mathbb{T}_k := \mathbb{K}^{\mathbb{C}^a}_{\tau_k(a)} \oplus \mathbb{K}^{f^a}_{\tau_k(a)}.$$

There are two extreme cases, namely if there is either no f^a–component or no $\mathbb{C}^a$–component. Of course, the coset $U^a(1)$ is in the *degenerate torus*

$$\mathbb{T}_e := U^a(1) + \{0\}.$$

Let us investigate under which circumstances the other extreme case, namely

$$\mathbb{K}_{\tau_k(a)} \subset f^a \tag{14}$$

can hold true, i.e.

$$k \cdot U^a(1) \subset f^a.$$

Assuming that equation (14) holds is the same as

$$k \cdot z \in f^a \qquad \forall\, z \in U^a(1)$$

which implies

$$k \in f^a.$$

Since, moreover,

$$\tau_k(a) = k \cdot a \cdot k^{-1} = -a,$$

we obtain

LEMMA 26. *Given $a \in S^2$, the fibres $\mathbb{K}_a$ and $\mathbb{K}_{-a}$ are in $\mathbb{C}^a$. Moreover,*

$$\mathbb{K}_{\tau_k(a)} \subset f^a \qquad \Longleftrightarrow \qquad k \in f^a.$$

Thus $k \in f^a$ is the necessary and sufficient condition for $\mathbb{K}_{\tau_k(a)}$ to be on the degenerate torus

$$\mathbb{T}_a = \{0\} + q_0 \cdot U^a(1).$$

for the unit vector $q_0 \in f^a$. In all other cases with $k \neq e$ the tori are not degenerate.

9. A Natural Connection Form on $\mathbb{K}$

9.1 A Natural Connection Form on $SU(2)$

The principal bundle $(SU(2), \mathrm{pr}_{\mathrm{Hopf}}, S^2, U^a(1))$ (the Hopf fibration) admits a natural *connection form* α^{S^2}:

LEMMA 27. *The $\mathbb{R}$–valued one–form*

$$\alpha^{S^2}(k;\zeta) := < pr_{Hopf}(k) \cdot k, \zeta > \qquad \forall\, k \in SU(2) \text{ and } \forall\, \zeta \in T_k SU(2).$$

is U^a–equivariant and hence a connection form on $SU(2)$.

PROOF 2. *In fact, the form α^{S^2} is U^a–invariant, i.e.*

$$\alpha^{S^2}(kz;\zeta z) \;= < pr_{Hopf}(kz) \cdot kz, \zeta z > = < pr_{Hopf}(k) \cdot k, \zeta >$$

for all $z \in U^a$, any $k \in SU(2)$ and any $\zeta \in T_k SU(2)$.

Moreover,

$$\alpha^{S^2}(k;ka) \;= < kak^{-1} \cdot k, ka > = < ka, ka > = \; 1$$

since $ka \in S^3$. This shows that

$$\alpha^{S^2}(k;\zeta) = \zeta \qquad \forall\, \zeta \in T_k\mathbb{K}_x \text{ and } \forall\, k \in \mathbb{K}_x.$$

Let us have a closer look at α^{S^2}. Since for any $\zeta \in T_k\mathbb{K}_x$ we obviously have

$$\zeta = \zeta' \cdot k$$

for some $\zeta' \in su(2) = T_e SU(2)$, the connection form α^{S^2} can be rewritten as

$$\alpha^{S^2}(k;\zeta) = < \mathrm{pr}_{\mathrm{Hopf}}(k) \cdot k, \zeta' \cdot k >$$

and therefore, due to $|k| = 1$, as

$$\alpha^{S^2}(k;\zeta) =< \tau_k(a), \zeta' >=< \mathrm{pr}_{\mathrm{Hopf}}(k), \zeta' > \tag{15}$$

for all $k \in SU(2)$ and all $\zeta' \in su(2)$.

The connection form α^{S^2} allows us to describe the *horizontal space* of the Hopf fibration more explicitly: Splitting $su(2) = E$ into

$$su(2) = \mathbb{R} \cdot \mathrm{pr}_{\mathrm{Hopf}}(k) \oplus T_{\mathrm{pr}_{\mathrm{Hopf}}(k)} S^2,$$

which may be reformulated into

$$su(2) \cdot k = \mathbb{R} \cdot \tau_k(a) \cdot k \oplus T_{\tau_k(a)} S^2 \cdot k,$$

due to (15) we deduce

$$\ker \alpha^{S^2}(k; ...) = \left(T_{\tau_k(a)} S^2\right) \cdot k \qquad \forall\, k \in SU(2). \tag{16}$$

This is to say that the horizontal space

$$\mathrm{Hor}_k := \ker \alpha^{S^2}(k; ...)$$

has a geometric meaning, namely

$$\mathrm{Hor}_k = \left(T_{\tau_k(a)} S^2\right) \cdot k = T_k(S^2 \cdot k).$$

$S^2 \cdot k$ is a sphere in S^3 of constant latitude $\frac{\vartheta}{2}$ if $k = e^{\frac{\vartheta}{2} \cdot a}$ (cf. equation (A.14) in the appendix).

The horizontal distribution $\mathrm{Hor} := \bigcup_{k \in \mathbb{K}} \mathrm{Hor}_k$ is U^a–equivariant since α^{S^2} is U^a–invariant. Moreover, for any smooth family $k(t) \in \mathrm{pr}^{-1}_{\mathrm{Hopf}}(b)$ with $k(0) = k$ and $b \in S^2$,

$$(k(t) \cdot z)^{\cdot} = \dot{k}(0) \cdot z$$

holds true. Hence from

$$\mathrm{pr}_{\mathrm{Hopf}}(k(t) \cdot z) = \mathrm{pr}_{\mathrm{Hopf}}(k(t)) = b$$

we conclude that

$$T\mathrm{pr}_{\mathrm{Hopf}}(\dot{k}(0) \cdot z) = T\mathrm{pr}_{\mathrm{Hopf}}(\dot{k}(0)).$$

Since $\mathrm{Hor}_k \oplus T_k\mathbb{K}_x = T_k\mathbb{K}$ and $\ker T\mathrm{pr}_{\mathrm{Hopf}} = T\mathbb{K}_x$, it is obvious that

$$T\mathrm{pr}_{\mathrm{Hopf}}\mathrm{Hor}_k = T_{\tau_k(a)} S^2. \tag{17}$$

9.2 A Natural Connection Form on $\mathbb{K}$

Now our aim is to define a natural connection form on the bundle $\mathbb{K}$ of normed quantum states, which would then allow us to describe a transmission of information.

DEFINITION 28. *The connection form on $\mathbb{K}$ is given by the pull–back*

$$\alpha^{\mathbb{K}} := \hat{\varepsilon}^* \alpha^{S^2}$$

which is written in more detail as

$$\alpha^{\mathbb{K}}(k_x;\zeta) := \alpha^{S^2}(k_{\varepsilon(x)}; T\widehat{\varepsilon}(\zeta))$$

for any $x \in O$, any $k_x \in \mathbb{K}$ and any $\zeta \in T_{k_x}\mathbb{K}$.
Obviously, $\alpha^{\mathbb{K}}$ is U^a–invariant.

If we map the horizontal subspace

$$\mathrm{Hor}_{k_x} := \ker \alpha^{\mathbb{K}}(k_x; ...) \qquad \forall\, k_x \in \mathbb{K} \text{ and } \forall\, x \in O$$

by $T\mathrm{pr}_{\mathrm{Hopf}} \circ T\widehat{\varepsilon}$, it is identified by diagram (13) as

$$\begin{aligned} T\mathrm{pr}_{\mathrm{Hopf}} \circ T\widehat{\varepsilon}\,(\mathrm{Hor}_{k_x}) &= T\varepsilon \circ T\mathrm{pr}_O\,(\mathrm{Hor}_{k_x}) \\ &= T_{\varepsilon(x)} S^2. \end{aligned}$$

We therefore want to determine $\ker T_x\varepsilon$ at any $x \in O$. To do so let us consider a smooth curve

$$\sigma : (-\vartheta, \vartheta) \longrightarrow O$$

with fixed positive $\vartheta \in \mathbb{R}$ and $\sigma(0) = x$. Then

$$T\varepsilon(\dot{\sigma}) = \left(\frac{\mathbf{a}(\sigma)}{|\mathbf{a}(\sigma)|}\right)^{\cdot}$$

and hence

$$T\varepsilon(\dot{\sigma}) = \frac{d\mathbf{a}(\dot{\sigma})}{|\mathbf{a}(\sigma)|} - d\ln|\mathbf{a}(\sigma)|(\dot{\sigma}) \cdot \frac{\mathbf{a}(\sigma)}{|\mathbf{a}(\sigma)|}$$

or

$$T\varepsilon(\dot{\sigma}) = \frac{d\mathbf{a}(\dot{\sigma})}{|\mathbf{a}(\sigma)|} - d\ln|\mathbf{a}(\sigma)|(\dot{\sigma}) \cdot \varepsilon(\sigma).$$

Thus

$$T\varepsilon(\dot{\sigma}) = 0 \qquad \Longleftrightarrow \qquad d\mathbf{a}(\dot{\sigma}) = d\ln|\mathbf{a}(\sigma)|(\dot{\sigma}) \cdot \mathbf{a}(\sigma).$$

In general, a curve σ satisfying

$$T\varepsilon(\dot{\sigma}) = 0$$

is not identical with the integral curve β_x of $\mathbf{a}$ passing through x at 0. In fact

$$T\varepsilon(\dot{\beta}) = 0 \qquad \Longleftrightarrow \qquad \varepsilon(\beta_x) = \text{const.}$$

The condition $(\varepsilon(\beta_x))^{\cdot} = 0$ obviously is identical with the condition that the image of β_x is a straight line segment.

If S_x denotes the level surface of f (for which grad $f = \mathbf{a}$) through $x \in O$

$$T_x\varepsilon(h) = W_{S_x}(h) \qquad \forall\, h \in T_x S_x,$$

where W_{S_x} is the Weingarten map, since $\varepsilon(y)$ is the unit normal at S_x for any $y \in S_x$. Hence for any $x \in O$

$$T_x\varepsilon(h) = W_{S_x}(h) - d\ln|a(x)|(h) \cdot \varepsilon(x)$$

holds true. This yields

$$T_x\varepsilon(v) = 0 \qquad \Longleftrightarrow \qquad d\mathbf{a}(x)(v) = d\ln|\mathbf{a}(x)|(v) \cdot \mathbf{a}(x)$$

for all $v \in E$. Obviously,

$$T\mathrm{pr}_O : \mathrm{Hor}_k \longrightarrow T_{\mathrm{pr}_k} O \qquad \forall\, k \in \mathbb{K} \tag{18}$$

is an isomorphism as can be shown accordingly to (17).

The vector field X on O is lifted horizontally to $T\mathbb{K}$ by using the inverse of (18). The lifted vector field is denoted by X^{hor}.

For any integral curve β_x with $\beta(0) = x$ there is exactly one integral curve $\beta_{k_x}^{\mathrm{hor}}$ of X^{hor} such that for any given $x \in O$ and prescribed $k_x \in \mathbb{K}_x$

$$\dot{\beta}_{k_x}^{\mathrm{hor}}(t) \in \mathrm{Hor}_{\beta_{k_x}^{\mathrm{hor}}(t)} \qquad \forall\, t$$

and

$$\beta_{k_x}^{\mathrm{hor}}(0) = k_x.$$

$\beta_{k_x}^{\mathrm{hor}}$ is called the *horizontal lift* of β_x with initial condition k_x. Thus there is a flow $\Phi^{\mathbb{K}}$ on $\mathbb{K}$ which restricted to $\mathbb{K}|_{\mathrm{im}\beta_x}$ propagates initial data in $\mathbb{K}_x$ with the velocity given by β_x due to (18). Hence $\beta_{k_x}^{hor}$ and β_x have the same speed. The flow $\Phi^{\mathbb{K}}$ we call the *horizontal flow* over β_x. This will be our basic ingredient for quantum information transmission (cf. section 12).

10. The Quantum Line Bundle on O

We now want to extend the normed state bundle $\mathbb{K}$ to a complex line bundle, the *quantum line bundle* consisting of all possible states of the qubits $\varepsilon(x) \in S^2$ such that the quantum state bundle $\mathbb{K}$ is contained in this quantum line bundle. It is the quantum analogue of $\mathbb{F}^a$ introduced in section 3.

For any $x \in O$, the circle $\mathbb{K}_x \subset \mathbb{K}$ determines a complex line $\mathbb{L}_x$ in $\mathbb{H}$ given by

$$\mathbb{L}_x := k \cdot \mathbb{C}^a$$

for each $k \in \mathbb{K}_x$. Since $U^a(1)$ operates on $\mathbb{K}$ from the right, the field $\mathbb{C}^a$ operates on $\mathbb{L}_x$ from the right, as well. Thus

$$\mathbb{L} := \mathbb{K} \cdot \mathbb{C}^a \quad \subset O \times SU(2)$$

is naturally equipped with the structure of a complex line bundle, called the *quantum line bundle*. The canonical projection from $\mathbb{L}$ to O is also denoted by pr_O.

Obviously, $\mathbb{L}_x$ consists of all states of the qubit $\varepsilon(x) \in S^2$ and $\mathbb{K}_x \subset \mathbb{L}_x$ is the collection of all normed states for any $x \in O$.

For any $v \in S^2$ there is a unique complex line $\mathbb{L}_v^{S^2} \subset \mathbb{H}$ containing $\mathrm{pr}_{\mathrm{Hopf}}^{-1}(v)$ and

$$\mathbb{L}^{S^2} := \bigcup_{v \in S^2} \mathbb{L}_v^{S^2} \subset S^2 \times \mathbb{H}$$

is a complex line bundle, a submanifold of $S^2 \times \mathbb{H}$. Obviously

$$\mathbb{L} = \varepsilon^* \mathbb{L}^{S^2}.$$

However, $\mathbb{L} \neq \varepsilon^* TS^2 = \mathbb{F}^{\mathrm{a}}$.

10.1 A Horizontal Flow on the Quantum Line Bundle

The quantum line bundle $\mathbb{L}$ is *associated* to $\mathbb{K}$ with typical fibre $\mathbb{C}^a$. More precisely, $\mathbb{L}$ is the quotient of the action

$$\begin{aligned} (\mathbb{K} \times \mathbb{C}^a) \times U^a &\longrightarrow \mathbb{K} \times \mathbb{C}^a \\ ((k, z'), z) &\longmapsto (k_x \cdot z, z^{-1} \cdot z'). \end{aligned}$$

The reason for this is that the multiplication map

$$m : (\mathbb{K} \times \mathbb{C}^a) \times U^a \longrightarrow \mathbb{K} \cdot \mathbb{C}^a = \mathbb{L}$$

fibrewisely defined by

$$(k_x, z') \cdot z \longmapsto k_x z \cdot z^{-1} z' = k_x \cdot z'$$

factors over the quotient $\mathbb{K} \times \mathbb{C}^a / U^a$ and, therefore, yields a bundle isomorphism. Here k_x, z' and z vary over $\mathbb{K}, \mathbb{C}^a$ and U^a, respectively.

The horizontal distribution Hor in $T\mathbb{K}$ constructed in the previous section is mapped by Tm to $T\mathbb{L}$ and yields a U^a*–invariant distribution* in $T\mathbb{L}$, the *horizontal distribution* $\text{Hor}^{\mathbb{L}}$, which is a vector bundle. Moreover, for each $k_x \in \mathbb{K}_x \subset \mathbb{L}_x$ and any $x \in O$,

$$\text{Hor}^{\mathbb{L}}_{k_x} \cap T_{k_x}\mathbb{K} = \text{Hor}_{k_x}. \tag{19}$$

By a standard procedure $\text{Hor}^{\mathbb{L}}$ yields a connection on $\mathbb{L}$ (cf. [10] or [6]). The map $T\text{pr}_O$ restricted to $\text{Hor}^{\mathbb{L}}_{k_x}$ is an isomorphism onto T_xO for any $k_x \in \mathbb{L}$, as can be easily seen. Hence by the map

$$T\text{pr}_O^{-1} : TM \longrightarrow \text{Hor}^{\mathbb{L}}$$

the vector field X is uniquely lifted to $T\text{pr}_O^{-1} \circ X \circ T\text{pr}_O : \mathbb{L} \to \text{Hor}^{\mathbb{L}}$, which will also be called X^{hor}. It is called the *horizontal lift* of X. Its flow, denoted by Φ again, is called the *horizontal flow*, which, restricted to $\mathbb{K}$, yields the horizontal flow $\Phi^{\mathbb{K}}$ on $\mathbb{K}$ constructed in the previous section.

10.2 A Connection to the Magnetic Monopole

The bundle $\mathbb{L}$ is the associated bundle to the identity representation of $U^a(1)$ on $\mathbb{C}^a$. Of course, choosing the representation assigning with $e^{t \cdot a}$ the number $e^{m \cdot t \cdot a}$ for a fixed $m \in \mathbb{Z}$ for any t yields another complex line bundle $\mathbb{L}^m$, say. This line bundle plays a fundamental role in the geometric description of the magnetic monopole (cf. [9]).

11. Classical Versus Quantum Information Geometrically Formulated

The bundles $\mathbb{P}^{\mathbf{a}}, \mathbb{F}^{\mathbf{a}}, \mathbb{K}$ and $\mathbb{L}$ provide us with bundles of *internal variables* of the vector field. Internal variables are the carriers of information. Classical information is encoded in $\mathbb{P}^{\mathbf{a}}$ and $\mathbb{F}^{\mathbf{a}}$ (cf. section 3) while quantum information is encoded in $\mathbb{K}$ (in terms of normed states) and in $\mathbb{L}$.

Let us link these two sorts of descriptions of information. In fact, the link will be based on the 2+2–splitting (5) of the quaternions introduced in section 2. We will see that classical information enters quantum information as a component.

The complex lines in $\mathbb{F}^{\mathbf{a}}$ are precisely all $\mathbb{F}^{\mathbf{a}}_x$ where x varies in O. This is to say that

LEMMA 29.

$$\mathbb{F}^{\mathbf{a}} = \varepsilon^* TS^2.$$

We will next show how classical information can be extracted from quantum information in $\mathbb{L}$. For each $x \in O$,

$$\mathbb{C}_x^{\mathbf{a}} \oplus \mathbb{F}_x^{\mathbf{a}} = \mathbb{H},$$

where $\mathbb{C}_x^{\mathbf{a}} = \mathbb{C}^{\varepsilon(x)}$. This is to say that

$$\mathbb{C}^{\mathbf{a}} \oplus \mathbb{F}^{\mathbf{a}} = O \times \mathbb{H}$$

for $\mathbb{C}^{\mathbf{a}} = \bigcup_{x \in O} \{x\} \times \mathbb{C}_x^{\mathbf{a}}$. Quantum information is fibrewisely encoded in $\mathbb{L}$. Clearly

$$\mathbb{L} \subset \mathbb{C}^{\mathbf{a}} \oplus \mathbb{F}^{\mathbf{a}}.$$

The bundle projection

$$\Pr|_{\mathbb{L}} : \mathbb{L}|_O \longrightarrow \mathbb{F}^{\mathbf{a}}|_O \tag{20}$$

coincides fibrewisely with the bundle projection

$$\Pr|_{\mathbb{L}_x} : \mathbb{L}_x \longrightarrow \mathbb{F}_x^{\mathbf{a}} \qquad \forall x \in O$$

which is the restriction of the orthogonal bundle projection of

$$\mathbb{C}^{\mathbf{a}} \oplus \mathbb{F}^{\mathbf{a}}|_O \xrightarrow{\Pr} \mathbb{F}^{\mathbf{a}}|_O$$

to $\mathbb{L}_x$. The extraction of classical information out of quantum information is hence performed by (20).

We thus investigate under what circumstances $\Pr|_{\mathbb{L}_x}$ is not surjective. This happens precisely if $\mathbb{L}_x \cap \mathbb{C}^{\varepsilon(x)} \neq \{0\}$.

Given $k_x \cdot z \in \mathbb{L}_x \cap \mathbb{C}^{\varepsilon(x)}$ for some $z \in \mathbb{C}^a$, we deduce for $k_x \cdot z \in \mathbb{C}^{\varepsilon(x)}$ that

$$k_x \cdot z = \tau_k(z')$$

for some $z' \in \mathbb{C}^a$, since $\mathbb{C}^{\varepsilon(x)} = \tau_{k_x}(\mathbb{C}^a)$ (cf. section 5). Thus

$$k_x = z' \cdot z^{-1} \in \mathbb{C}^a,$$

and hence

$$k_x \cdot z' \in \mathbb{C}^{\varepsilon(x)} \qquad \Longleftrightarrow \qquad \varepsilon(x) = \frac{\mathbf{a}(x)}{|\mathbf{a}(x)|} = a.$$

Therefore we may state

LEMMA 30. $\Pr|_{\mathbb{L}_x}$ *is an isomorphism for all* $x \in O \backslash \{\varepsilon^{-1}(\pm a)\}$.

The following corollary is immediate:

COROLLARY 31. *Any $k_x \in \mathbb{L}_x$ is of the form*

$$k_x = \big(k_x - \operatorname{Pr}|_{\mathbb{L}_x}(k_x)\big) + \operatorname{Pr}|_{\mathbb{L}_x}(k_x).$$

Hence any $\operatorname{Pr}|_{\mathbb{L}_x}(k_x)$ may be regarded as a classical component of the bit of quantum information k_x.

Thus we may regard classical information as part of quantum information.

Concerning the quantum state bundle $\mathbb{K}$, corollary 31 implies:

PROPOSITION 32. *Quantum information encoded in $\mathbb{K}_x$ (and $\mathbb{L}_x$) is expressed in terms of information encoded in $U^a(1)$ (and $\mathbb{C}^a$) and classical information encoded in $\mathbb{P}_x^{\mathbf{a}}$ (and $\mathbb{F}_x^{\mathbf{a}}$) for any $x \in O$.*

This indeed shows geometrically that classical information can be complemented to quantum information (cf. proposition 23) and extracted from quantum information.

12. The Transmission of Quantum Information

Transmission of quantum information from one fibre in $\mathbb{L}$ to another one is a unitary transformation, i.e. the *multiplication by a phase factor*. Along an integral curve β we factor this transformation in a unitary one $\widetilde{U}^t$ on the initial fibre followed by another unitary transformation (another phase factor) caused by parallel transport ([3]). We here focus on the *parallel transport* only.

Let us consider the horizontal flow Φ on $\mathbb{L}|_{\mathrm{im}\beta}$. Quantum information is transmitted from one fibre to another one by means of Φ. It is also known as parallel transport. Its restriction to $\mathbb{K}$is the horizontal flow $\Phi^{\mathbb{K}}$ on $\mathbb{K}$. Both are $U^a(1)$–equivariant, since the horizontal distribution in $T\mathbb{L}$ is equivariant. This is to say

$$\Phi(t)(k_x \cdot z) = \Phi(t)(k_x) \cdot z \qquad \forall\, z \in U^a$$

for any initial value on $\mathbb{L}_{\beta(0)}$, say, and all t in the open domain D of β_x. Clearly $\Phi(t)$ is the multiplication by a *phase factor* on $\mathbb{L}_{\beta_x(t)}$ and thus $\Phi(t)$ is *unitary*. We set

$$\Phi(t)(k_x) := \beta_x^{\mathrm{hor}}(t) \quad \text{with } \beta_x^{\mathrm{hor}}(0) = k_x$$

for all $k_x \in \mathbb{K}$ and any $t \in D$. Here x is an initial point of β_x.

The integral curve β_x passing through x at $t = 0$ satisfies

$$\beta_x(t_1 + t_2) = \beta_{\beta_x(t_1)}(t_2) \tag{21}$$

for all t_1, t_2 for which $t_1 + t_2$ is in the domain D of definition of β_x. Clearly there is an open interval I centered at zero for which $I + I$ is in D.

Obviously Φ shares the property (21) with β_x.

To describe the transmission entirely on $\mathbb{H}$, we will extend $\Phi(t)$ to a map on all of $\mathbb{H}$. Due to (5) and $F^a = \mathbb{C}^a \cdot q_0$ for a fixed unit vector $q_0 \in F^a$,

$$k_x(t) \cdot (\mathbb{C}^a \oplus \mathbb{C}^a \cdot q_0) = \mathbb{H} \tag{22}$$

holds true for all t and hence allows to decompose any quaternion in terms of the left–hand side of (22) in a unique fashion. Thus we define

$$U^t(k_x(t))(z_1 + z_2 \cdot q_0) \quad := \quad \Phi(t)(k_x)(z_1 + z_2 \cdot q_0)$$

for all $z_1, z_2 \in \mathbb{C}^a$ and any t in the domain of β_x. Since $\Phi(t)$ acts on $k_x \in \mathbb{L}_x$ by multiplication of a phase factor, U^t is unitary on $\mathbb{H}$.

Therefore $\Phi(t)$ yields a *one–parameter group*

$$U^t : \mathbb{H} \longrightarrow \mathbb{H} \qquad \forall t \in D \tag{23}$$

generated by H_x, say, where H_x is a *Hermitian* $\mathbb{C}^a$*-linear* map on $\mathbb{H}$.

Thus the *transmission* of any state $k \in \mathbb{L}_x$ to a state in $\mathbb{L}_{\beta_x(t)}$ at $t \in D$ is given by

$$k(t) := U^t(k) \qquad \forall t \in D.$$

In that sense the one–parameter group U^t describes the *transmission of quantum information.*

Obviously

$$\frac{dU^t}{dt} = iH_x \circ U^t \qquad \forall t \in \mathbb{R}.$$

Clearly

$$H_x \in su(2) = E.$$

A normed state $k_x \in \mathbb{K}_x$ evolves along β_x by

$$k_{\beta_x(t)}(t) := e^{iH_x \cdot t}(k_x) \qquad \forall x \in O$$

for which obviously

$$\dot{k}_{\beta_x(t)}(t) = iH_x \cdot k_{\beta_x(t)}(t) \qquad \forall t \in \mathbb{R}$$

holds true. Specifying a level surface S of f we observe that

$$\begin{array}{ccc} S & \xrightarrow{H} & \text{Herm}\,\mathbb{H} \\ y & \longmapsto & H_y \end{array}$$

is smooth.

Clearly the operator $H_x(t) \in SU(2)$ maps $\mathbb{L}_x$ into $\mathbb{L}_{\beta_x(t)}$ for any $t \in D$ and any $x \in O$ by construction.

Due to (19) the one–parameter group (23) governs the transmission of quantum information of $\mathbb{K}$, as well.

Note that so far $\mathbb{L}_x$ is not regarded as a phase space of any sort.

Given $x \in O$ with $\varepsilon(x) \neq \pm a$, we choose a reference point $v \in \mathbb{L}_x$ (with $|v| = 1$, say,), so that $\mathbb{L}_x = \mathbb{R} \cdot v \oplus \mathbb{R} \cdot v \cdot a$ and $\mathbb{L}_x \neq \mathbb{C}^a$. Then $\mathbb{L}_x$ is a phase space and $\mathbb{R} \cdot a + \mathbb{L}_x$ can naturally be turned into a Heisenberg algebra with ω^a as symplectic structure.

By means of proposition 23 we have:

LEMMA 33. *For any $x \in O$ with $\varepsilon(x) \neq \pm a$, the transmission of quantum information by means of Φ can be modulated on signals, provided $\mathbb{L}_x$ is turned into a phase space.*

13. Transmission of Classical out of Quantum Information

The aim of this section is to link classical information transmission with quantum information transmission. Both transmissions follow the trajectories of a singularity free vector field with principal part $\mathbf{a}$. However they do so in different fashions, as we will see now. Again we merely concentrate on the parallel transport.

The complex line bundle $\mathbb{F}^{\mathbf{a}}$ shall now be equipped with a natural connection form α^{cl}. It is defined by

$$\alpha^{cl}(h_x;\zeta) := <\varepsilon(x)\cdot h_x, \zeta> \qquad \forall\, h_x \in \mathbb{F}^{\mathbf{a}}_x \text{ and } \forall\, \zeta \in T_{h_x} O$$

for any $x \in O$. [5]

The one–form $\mathrm{Pr}^*\,\alpha^{cl}$ is $U^a(1)$–equivariant and extracts from each $k_x \in \mathbb{L}_x$ the classical part in the sense of corollary 31.

Setting

$$\mathrm{Hor}^{cl}_{h_x} := \ker \alpha^{cl}(h_x, \ldots) \qquad \forall\, h_x \in \mathbb{F}^a \text{ and } \forall\, x \in O$$

implies

$$\mathrm{Pr}\,(\mathrm{Hor}_{k_x}) = \mathrm{Hor}^{cl}_{\mathrm{Pr}\,k_x} \qquad \forall\, k_x \in \mathbb{L}_x \text{ and } \forall\, x \in O\backslash\{\varepsilon(\pm a)\}.$$

The link between quantum and classical information expressed in corollary 31 can be extended to comprise also the transmission of classical information encoded in $\mathbb{P}^a$: Since

$$T\mathbb{K}_x \subset \lambda_x T U^{\varepsilon(x)}(1) \oplus \mu_x T\mathbb{P}^a_x \qquad \forall\, x \in O$$

[5] α^{cl} differs from the connection form determined by $\mathbb{P}^{\mathbf{a}}$ as presented in [5] and [2]. There the field strength explicitly enters. However, for sake of simplicity we here work with α^{cl}.

for suitable $\lambda_x, \mu_x \in \mathbb{R}$, it follows that

LEMMA 34. *For all $k_x \in \mathbb{K}_x$ and for all $x \in O\backslash\{\varepsilon^{-1}(\pm a)\}$,*

$$Hor_{k_x} = \lambda_x \cdot Hor_{k_x}^{U^{\varepsilon(x)}(1)} \oplus \mu_x Hor_{k_x}^{\mathbb{P}_x^a}$$

where $Hor_{k_x}^{U^{\varepsilon(x)}(1)}$ and $Hor_{k_x}^{\mathbb{P}_x^a}$ denote the orthogonal projections of Hor_{k_x} onto $TU^a(1)$ and $Hor_{k_x}^{\mathbb{P}_x^{\mathbf{a}}}$, respectively.

Clearly

$$\mu_x \cdot \mathrm{Hor}_{k_x}^{\mathbb{P}_x^a} \subset T\mathbb{P}_x^{\mathbf{a}}.$$

This finally shows:

PROPOSITION 35. *The transmission of classical information is completed by transmission of information in $\lambda_x \cdot U^{\varepsilon(x)}(1)$ to be turned into a transmission of quantum information encoded in state spaces of qubits.*

The transmission of classical information is performed by the horizontal flow Φ^{cl} of the horizontal lift of $\dot{\beta}_{x_0}$ of a field line with initial condition x_0. Indeed

$$\Pr \Phi(t) = \Phi^{cl} \qquad \forall t.$$

Let us return to the unitary transformation $\widetilde{U}^t$ introduced in the beginning of section 12. and choose some smooth L^2–map

$$\psi : \mathbb{L}_{x_0} \longrightarrow \mathbb{C}^a$$

for which $\int \psi\overline{\psi}\omega_{x_0} = 1$ with the symplectic structure ω_{x_0} caused by the multiplication by a in $\mathbb{C}^a$. We might regard

$$\psi \cdot \overline{\psi} : \mathbb{L}_{x_0} \longrightarrow \mathbb{R}$$

as a smooth density function for the distribution of quantum information. Since the transmission of quantum information along an integral curve β_{x_0} is unitary, the map ψ evolves by

$$\psi_t := \psi \circ U^{-t} \qquad \forall t. \tag{24}$$

We hence may write

$$\psi_t = \widetilde{U}^t(\psi_{x_0})$$

where $\widetilde{U}^t$ is the unitary operator on $L^2(\mathbb{L}_{x_0}, \mathbb{C}^a)$ determined by (24). $\widetilde{U}^t$ is a continuous one–parameter group. Using Stone's theorem (cf. [1]), $\widetilde{U}^t$ can be written as

$$\widetilde{U}^t = e^{itHa}$$

for some Hermitian operator H, and thus the evolution of ψ is governed by a Schrödinger type of equation

$$\frac{\partial}{\partial t}\psi_t = i \cdot H \cdot \psi_t.$$

The projection from $\mathbb{L}_{x_0}$ to $\mathbb{F}^{\mathbf{a}}$ is a $\mathbb{C}^a$–linear isomorphism for $x \in O \backslash \{\varepsilon(\pm a)\}$. We may thus transfer the classical density $\psi_t \overline{\psi}_t$ via this projection to a function $\psi_t \cdot \overline{\psi}_t$ on $\mathbb{F}^{\mathbf{a}}_{x_0}$. A continuity equation for $\psi_t \cdot \overline{\psi}_t$ and a symplectic transformation (replacing the unitary transformation in the quantum case) in the initial fibre together with the representation of the metaplectic group then yield the quantization of homogenous quadratic polynomials on $\mathbb{F}^{\mathbf{a}}_{x_0}$. The extension of this quantization to all inhomogeneous quadratic polynomials is achieved by the metaplectic representation and the Schrödinger representation (cf. [4]) of the semi–direct product of the metaplectic group and the Heisenberg group $G^{\mathbf{a}}_{x_0}$. The inhomogeneous quadratic polynomials are identical with the Hamiltonian on $\mathbb{F}^{a}_{x_0}$ to be quantized by this method (cf. [3]).

14. The Transmission of Entangled States

As an outlook, we briefly sketch the transmission of *two–fold entangled* states on a simple example.[6] We intend to transmit two entangled states along two possibly different trajectories of a vector field.

Following the idea of the transmission of states along vector fields as expressed in section 13, we need to replace $\mathbb{H}$ by $\mathbb{H} \otimes \mathbb{H}$, the 3–sphere S^3 by S^7 and S^2 by an appropriate manifold. Finally ε has to be generalized accordingly. At first we begin by treating two integral curves of the vector field X simultaneously.

The above setting in section 13 is generalized by considering the vector field $X \times X$ with principal part

$$\mathbf{a} \times \mathbf{a} : O \times O \longrightarrow E \times E,$$

a smooth map. For any initial condition $(x_1, x_2) \in O \times O$ the map

$$(\beta_1, \beta_2) : (-\lambda, \lambda) \longrightarrow O \times O$$

is an integral curve of $X \times X$ provided β_1 and β_2 are field lines of X satisfying

$$\beta_1(0) = x_1 \quad \text{and} \quad \beta_2(0) = x_2.$$

Indeed, for any $t \in (-\lambda, \lambda)$

$$(\beta_1, \beta_2)^{\cdot}(t) = \dot{\beta}_1(t) \times \dot{\beta}_2(t) = \mathbf{a}(\beta_1(t)) \times \mathbf{a}(\beta_2(t))$$

[6]We will expand on it in [2].

holds true.

Next let us consider the analogue of the Hopf fibration used for the vector field X.

Any normed state $\chi \in \mathbb{H} \otimes \mathbb{H}$ is up to a global phase factor on the great circle $\chi \cdot \mathbb{C}^a \cap S^7$. Hence the complex projective space $\mathbb{C}^a P(\mathbb{H} \otimes \mathbb{H})$ is the collection of the states modulo global phase factors.

To talk of a pair of two single qubits along a trajectory of the vector field $X \times X$, we need to form $S^2 \times S^2$. Each element $v \in S^2$ determines a complex line F_v, say. By appendix 3.4 both F_v and F_w are fibres of $\mathbb{L}$, namely $\mathbb{L}_{x_1}$ and $\mathbb{L}_{x_2}$, respectively, if $\varepsilon(x_1) = v$ and $\varepsilon(x_2) = w$. Thus for any pair $(v, w) \in S^2 \times S^2$ we may form

$$F_v \otimes F_w,$$

a $\mathbb{C}^a$–complex line again. Therefore we have the map

$$\mathcal{S} : S^2 \times S^2 \to \mathbb{C}^a P(\mathbb{H} \times \mathbb{H})$$

defined by

$$\mathcal{S}\big((v_1, v_2)\big) = F_{v_1} \otimes F_{v_2}$$

for any $v_1, v_2 \in E$.

This observation allows to generalize ε to a map $\varepsilon_\otimes$, say, defined on $O \times O$. In fact

$$\varepsilon_\otimes := \mathcal{S} \circ (\varepsilon \times \varepsilon)$$

maps the field vector $\mathbf{a}(x_1) \times \mathbf{a}(x_2)$ of $X \times X$ into the $\mathbb{C}^a$–complex line

$$\mathcal{S}(\mathbf{a}(x_1) \times \mathbf{a}(x_2)) \in \mathbb{C}^a P(\mathbb{H} \otimes \mathbb{H})$$

for any $x_1, x_2 \in O$. On $\mathbb{C}^a P(\mathbb{H} \otimes \mathbb{H})$ we have the tautological bundle (cf. appendix 3.4). This bundle can be pulled back by $\varepsilon_\otimes$ to $O \times O$ yielding $\mathbb{L}^\otimes$, say. The fibre of $\mathbb{L}^\otimes$ at $(x_1, x_2) \in O \times O$ is $\mathbb{L}_{x_1} \otimes \mathbb{L}_{x_2}$.

The connection form $\alpha^{\mathbb{L}}$ on the quantum line bundle $\mathbb{L}^\otimes$ generalizes to $\alpha^\otimes$ on the tautological bundle over $\mathbb{C}^a P(\mathbb{H} \otimes \mathbb{H})$ by setting

$$\alpha^\otimes := \alpha^{\mathbb{L}} \otimes \alpha^{\mathbb{L}}.$$

Obviously $\alpha^\otimes$ is $U^a(1)$–equivariant, since $k \otimes z \in S^7$ for all $z \in \mathbb{C}^a$ and all $k \in S^7$.

The horizontal subspace of $\alpha^\otimes$ and the horizontal transmission are constructed analogously as for $\alpha^{\mathbb{L}}$.

Mapping S^2 into the diagonal of $S^2 \times S^2$ yields the transmission of two entangled states along one field line of X.

Appendix

In this appendix we will give a detailed description of the *Hopf fibration* and the *Hopf projection*. They provide us with a powerful tool which we have used frequently in this paper.

1. Inner Automorphisms

The *Hopf fibration* of S^3 over S^2 can geometrically be described most easily if *inner automorphisms* are used. In definition 14, an inner automorphism

$$\tau_k : \mathbb{H} \longrightarrow \mathbb{H}$$

for any $k \in \dot{\mathbb{H}}$ was given by the conjugation by k, i.e.

$$\tau_k(h) = k \cdot h \cdot k^{-1} \qquad \forall\, h \in \mathbb{H}.$$

Hence

$$\tau_k(k) = k \qquad \forall\, k \in \mathbb{H}.$$

Any automorphism of the skew field $\mathbb{H}$ is an inner automorphism. Since $\tau_{t\cdot k} = \tau_k$ for any $t \in \mathbb{R}, t \neq 0$,

$$\begin{aligned} \tau : \dot{\mathbb{H}} &\longrightarrow \text{Aut}\,\mathbb{H} \\ k &\longmapsto \tau_k \end{aligned} \tag{A.1}$$

yields a homomorphism of $SU(2)$ to Aut $\mathbb{H}$ with $\{\pm e\}$ as kernel as well as a natural action of $\dot{\mathbb{H}}$ on $\mathbb{H}$.

Obviously τ_k is an $\mathbb{R}-$algebra isomorphism of $\mathbb{H}$ satisfying $\tau_k(\mathbb{R}\cdot e) = \mathbb{R}\cdot e$ and $\tau_k(E) = \tau_k(su(2)) = su(2)$. The latter is due to

$$< e, h > = \text{tr } h = 0 \qquad \forall\, h \in E.$$

As in every Lie group the inner automorphism τ_k with $k \in SU(2) = S^3$ determines an $\mathbb{R}-$linear automorphism on the Lie algebra $su(2) = E$. It is the tangent map of τ_k at the identity $e \in SU(2) \subset \mathbb{H}$, that is

$$T_e\tau_k(h) = \frac{d}{dt}\left(k \cdot h(t) \cdot k^{-1}\right)\Big|_{t=0} = k \cdot h \cdot k^{-1} \qquad \forall\, h \in E$$

for a smooth family $h(t) \in S^3$ with $h(0) = e$ and $\dot{h}(0) = h$. Thus

$$T_e\tau_k = \tau_k \qquad \forall\, k \in SU(2), \tag{A.2}$$

due to the linearity of τ_k. Instead of $T_e\tau_k$ the notation Ad_k is common. Ad_k is called the *adjoint representation*. From (A.2) we read off that

$$\text{Ad}_k = \tau_k \qquad \forall\, k \in SU(2).$$

In fact $\tau_k \in SO(E)$ for all $k \in SU(2) = S^3$, as we see from the definition of the scalar product $<\ ,\ >$ on $\mathbb{H}$ (cf. (2) in definition 1):

$$< \tau_k(h), \tau_k(h') > = \frac{1}{2} \cdot \text{ tr } \tau_k(h) \cdot \widetilde{\tau_k(h')} = < h, h' > . \tag{A.3}$$

Because of (A.2) and (14) the assignment

$$k \mapsto \mathrm{Ad}_k|_E = \tau_k|_E \qquad \forall\, k \in SU(2)$$

determines the representation

$$\mathrm{Ad}^{SU(2)} : SU(2) \longrightarrow SO(E).$$

LEMMA 36. *The map $Ad^{SU(2)}$ is a surjection with $\{\pm e\}$ as kernel. Thus $SO(E)$ is diffeomorphic to the real projective space $\mathbb{R}P(E)$ of E.*

PROOF 3. *We consider $\tau_k(h)$ for $k \in SU(2)$ and $h \in E$ and expand it into*

$$\tau_k(h) = \frac{1}{|k|^2} \cdot (\lambda^2 \cdot h + \lambda \cdot [u, h] - u \cdot h \cdot u). \tag{A.4}$$

where $k = \lambda \cdot e + u$ with $\lambda \in \mathbb{R}$ and $u \in E$ and $k^{-1} = \frac{\overline{k}}{|k|^2}$ with $\overline{k} = \lambda \cdot e - u$. Here $[\ ,\ ]$ is the commutator in $\mathbb{H}$ again. More explicitly, we can write (15) as

$$\tau_k(h) = \frac{1}{|k|^2} \cdot (\lambda^2 \cdot h + 2\lambda \cdot u \times h - u \times h \times u + < u, h > \cdot u).$$

Since

$$\tau_k(u) = \frac{1}{|k|^2} \cdot (\lambda^2 \cdot u - u^2 \cdot u) = u, \tag{A.5}$$

the rotation axis of τ_k is $\mathbb{R} \cdot u \subset E$. Thus the rotation τ_k can be described in the plane $u^\perp \subset E$, the orthogonal complement of u in E. The plane $u^\perp$ is called the rotation plane of τ_k. *For a unit vector $h \in u^\perp$ we have*

$$< \tau_k(h), h >= \frac{1}{|k|^2} \cdot (\lambda^2 - |u|^2).$$

Thus, if ϑ_k denotes the angle of rotation of τ_k,

$$\cos \vartheta_k = \frac{\lambda^2 - |u|^2}{\lambda^2 + |u|^2}. \tag{A.6}$$

This shows that $\tau : SU(2) \longrightarrow SO(E)$ mapping any $k \in SU(2)$ to $\tau_k \in SO(E)$ is a surjection map with $\{\pm e\}$ as kernel, as we claimed.

Now let $k = \lambda \cdot e + u$ with $\lambda \in \mathbb{R}$ and a *non-vanishing* $u \in E$. The unit vector $\frac{u}{|u|}$ yields the splitting $E = \mathbb{R} \cdot \frac{u}{|u|} \oplus F^{\frac{u}{|u|}}$ where $F^{\frac{u}{|u|}}$ is the orthogonal complement of $\frac{u}{|u|}$ in E. Hence any $h \in E$ is split into $h = \zeta \cdot \frac{u}{|u|} + h'$ with $h' \in F^{\frac{u}{|u|}}$ and $\zeta =< h, \frac{u}{|u|} >$. Hence

$$\tau_k(h) = \zeta \cdot \frac{u}{|u|} + \tau_k(h')$$

since $\tau_k(u) = u$ according to (A.5). This means that $\tau_k(h)$ precesses about the *rotation axis* $\mathbb{R} \cdot u$ by the angle ϑ_k.

Hence we started to present the rotation $\tau_k \in SO(E)$ in detail. So far we determined the rotation axis and the rotation angle in the plane perpendicular to this axis. Without loss of generality, we can assume that k is a unit vector. In the next section we will relate these geometric entities with the vector k on the unit–sphere $S^3 \subset \mathbb{H}$.

2. Rotation Angle and Latitude of an Element in $SU(2)$

In order to compute the rotation angle in terms of the latitude of an element $k \in SU(2)$, we now want to rewrite the action of τ_k in terms of $\frac{k^2}{|k|^2}$. As in the previous section, we split any $h \in E$ into $h = \zeta \cdot \frac{u}{|u|} + h'$ with $h' \in F^{\frac{u}{|u|}}$ and $\zeta =< h, \frac{u}{|u|} >$. The image $\tau_k(h')$ of $h' \in F^{\frac{u}{|u|}}$ is computed as

$$\tau_k(h') = k \cdot h' \cdot k^{-1} = k \cdot h' \cdot \frac{\overline{k}}{|k|^2}$$

and therefore is

$$\tau_k(h') = \frac{k^2}{|k|^2} \cdot h' \tag{A.7}$$

since $h' \cdot \overline{k} = h' \cdot (\lambda \cdot e - u) = (\lambda \cdot e + u) \cdot h' = k \cdot h'$. Thus

$$\tau_k(h) = \zeta \cdot \frac{u}{|u|} + \frac{k^2}{|k|^2} \cdot h'$$

for $h = \zeta \cdot \frac{u}{|u|} + h'$.

Clearly $\frac{k^2}{|k|^2} \in S^3$ and $\left| \frac{k^2}{|k|^2} \cdot h' \right| = |h'|$. Any unit quaternion can obviously be represented in terms of some $k \in \dot{\mathbb{H}}$ as

$$\frac{k^2}{|k|^2} = \frac{k}{\overline{k}}$$

due to $|k|^2 = k \cdot \overline{k} = \overline{k} \cdot k$ for all $k \in \mathbb{H}$ and, therefore,

$$\tau_k(h) = \zeta \cdot \frac{u}{|u|} + \frac{\lambda \cdot e + u}{\lambda \cdot e - u} \cdot h' = \zeta \cdot \frac{u}{|u|} + \frac{k}{\overline{k}} \cdot h'.$$

The oriented rotation angle ϑ_k (called *rotation angle* in the sequel) of τ_k is determined by the equation

$$< \tau_k(h'), h' > = < \frac{k^2}{|k|^2} \cdot h', h' > = |h'|^2 \cdot \cos \vartheta_k$$

and hence

$$< \tau_k(h), h >= \zeta^2 + |h'|^2 \cdot \cos \vartheta_k.$$

ϑ_k shall now be evaluated in terms the components of $k \in \mathbb{H}$.

To this end and in view of (A.7) we observe that for any $k \in \mathbb{H}$ its square k^2 can be expressed as

$$k^2 = (\lambda \cdot e + u)^2 = (\lambda^2 - |u|^2) \cdot e + 2 \cdot \lambda \cdot u, \tag{A.8}$$

yielding

$$\frac{k^2}{|k|^2} = \frac{\lambda^2 - |u|^2}{\lambda^2 + |u|^2} \cdot e + 2 \cdot \frac{\lambda \cdot |u|}{|k|^2} \cdot \frac{u}{|u|} \tag{A.9}$$

and hence

$$\left(\frac{k}{|k|}\right)^2 = e \cdot \cos\vartheta_k + \frac{u}{|u|} \cdot \sin\vartheta_k = e^{\vartheta_k \cdot \frac{u}{|u|}} \tag{A.10}$$

with $\cos\vartheta_k = \frac{\lambda^2 - |u|^2}{\lambda^2 + |u|^2}$. (according to (A.6)) and thus $\sin\vartheta_k = 2\lambda \cdot \frac{u}{|k|^2} \cdot \frac{k^2}{|k|^2} = e^{\vartheta_k \cdot \frac{u}{|u|}}$ and, In turn,

$$\frac{k}{|k|} = e^{\frac{\vartheta_k}{2} \cdot \frac{u}{|u|}}. \tag{A.11}$$

In terms of the *Minkowski metric* g_M characterized by

$$-g_M(k,k) = \lambda^2 - |u|^2$$

for all $k \in S^3$ (cf. definition 2), equation (A.9) reads

$$\frac{k^2}{|k|^2} = \frac{-g_M(k,k)}{< k, k >} \cdot e + 2 \cdot \frac{\lambda \cdot |u|}{|k|^2} \cdot \frac{u}{|u|}. \tag{A.12}$$

This formula allows to express $|k|^2$ in terms of $g_M(k,k)$ as

$$g_M(k,k) = -|k|^2 \cdot \cos\vartheta_k,$$

which expresses the fact that $\tau : \dot{\mathbb{H}} \to \operatorname{Aut}\mathbb{H}$ is associated with a natural Minkowski metric (cf. (A.7)).

In terms of the oriented rotation angle ϑ_k of τ_k, the vector $\tau_k(h) \in E$ is revealed in terms of the geometry of $\mathbb{R} \cdot \frac{u}{|u|} + F^{\frac{u}{|u|}}$ as

$$\tau_k(h) = \zeta \cdot \frac{u}{|u|} + \frac{\lambda^2 - |u|^2}{\lambda^2 + |u|^2} \cdot h' + \frac{2 \cdot \lambda \cdot |u|}{|k|^2} \cdot \frac{u}{|u|} \times h'$$

(due to (A.7)). Equation (A.6) yields

$$\tau_k(h) = \zeta \cdot \frac{u}{|u|} + h' \cdot \cos\vartheta_k + \frac{u}{|u|} \times h' \cdot \sin\vartheta_k$$

or

$$\tau_k(h) = \mathrm{pr}_{\mathbb{R} \cdot \frac{u}{|u|}}(h) \cdot \frac{u}{|u|} + e^{\vartheta_k \cdot \frac{u}{|u|}} \cdot \mathrm{pr}_{F^{\frac{u}{|u|}}}(h) \qquad \forall\, h \in E \tag{A.13}$$

where $\mathrm{pr}_{\mathbb{R} \cdot \frac{u}{|u|}}$ and $\mathrm{pr}_{F^{\frac{u}{|u|}}}$ are the projections from E onto $\mathbb{R} \cdot \frac{u}{|u|}$ and $F^{\frac{u}{|u|}}$, respectively. Hence

$$\cos\vartheta_k = 0 \quad \Leftrightarrow \quad |\lambda| = |u|.$$

To identify the geometric meaning of the rotation angle ϑ_k in terms of the position of k in S^3 let us consider the plane spanned by $\lambda \cdot e$ and u. This plane is nothing else but $\mathbb{C}^{\frac{u}{|u|}}$. In fact k, $\lambda \cdot e$ and u form a right triangle. The oriented angle between k and $\lambda \cdot e$ shall be denoted by β_k. Then with (A.10) and (A.11) we have

$$\vartheta_k = 2\beta_k. \tag{A.14}$$

Thus we have shown:

THEOREM 37. *Any $k \in S^3$ given by $k = \lambda e + u$ with $u \in E$ is of the form $k = e^{\beta_k \cdot \frac{u}{|u|}}$, where β_k is the latitude of k on the 3–sphere. Then the axis of rotation of τ_k is $\mathbb{R} \cdot u$ and the rotation angle ϑ_k is twice the latitude β_k.*

Assuming the rotation angle ϑ_k is fixed, then the unit vector $\frac{k^2}{|k|^2}$ varies on all of the two-dimensional sphere $S^2_{\vartheta_k} \subset S^3$ given by

$$S^2_{\vartheta_k} := e \cdot \cos \vartheta_k + S^2 \cdot \sin \vartheta_k, \tag{A.15}$$

provided $\frac{u}{|u|}$ varies in S^2. The 2–sphere $S^2_{\vartheta_k}$ is the sphere of constant latitude $\frac{\vartheta_k}{2}$. This is to say that the rotation angle ϑ_k of τ_k fixes the latitude of the unit vector $\frac{k^2}{|k|^2} \in S^3$ within S^3 by (A.14). Two inner automorphisms with the same rotation angles may have different rotations axes and thus may rotate in different tangent planes of S^2.

Finally we want to compute the *Euler angles* of τ_k with $k \in S^3$. Splitting k into $k = \lambda \cdot e + u$ and expressing u in terms of a basis $e_1, e_2, e_3 \in S^2$ as $u = \sum_{i=1}^3 \xi^i e_i$ yields due to (A.11)

$$\tau_k = e^{\frac{\vartheta_k}{2} \cdot \sum_{i=1}^3 \xi^i e_i}$$

and

$$\tau_k = \prod_{i=1}^3 e^{\Phi_i e_i}$$

with the *Euler angles*

$$\Phi_i = \frac{\vartheta_k}{2} \cdot \xi^i \qquad i = 1, 2, 3.$$

3. The Hopf Fibration

The natural action

$$\underline{\psi} : SU(2) \times S^2 \longrightarrow S^2$$

of $SU(2)$ on the unit sphere $S^2 \subset E$ is defined by

$$\underline{\psi}(k, u) := \tau_k(u) \qquad \forall\, k \in SU(2) \text{ and } \forall\, u \in S^2.$$

Fixing some $c \in S^2$ allows us to write any $k \in SU(2)$ in the form $k = e^{tc}$ with $t \in \mathbb{R}$ (cf. (4)).

To compute the stable group

$$S_u := \{g \in SO(E) \mid g(u) = u\}$$

at $u \in S^2$ we investigate whether $e^{tc} \cdot u = u$ for all $t \in \mathbb{R}$. This is equivalent to

$$c \times u = 0,$$

which holds iff $c = \pm u$, showing that the stable group at u is the commutative group $U^u(1)$. Clearly $U(1)$ acts on $U^u(1)$ via the field isomorphism

$$i_u : \mathbb{C} \to \mathbb{C}^u$$

sending 1 and i to e and u, respectively. Thus we have shown:

LEMMA 38. *The stable group S_b of any $b \in S^2$ is isomorphic to $SO(F^b)$ where $F^b = b^{\perp_E}$. More exactly, the isomorphism maps any rotation $g \in S_b$ into $g|_{F^b}$.*

3.1 The Hopf Projection

We now fix $a \in S^2$ for the remainder of these notes. For each $b \in S^2$ there is some $g_b \in SO(E)$ such that

$$b = g_b(a).$$

Therefore, we have a projection

$$\begin{array}{ccc} SO(E) & \xrightarrow{\mathrm{pr}_a} & S^2 \\ g & \longmapsto & g(a). \end{array} \tag{A.16}$$

Let us look at the stable group more closely. Obviously

$$S_b = \left\{ \exp_{SO(E)} t \cdot b \mid t \in \mathbb{R} \right\}.$$

Moreover,

$$\mathrm{pr}_a^{-1}(b) = g_b \cdot \mathrm{pr}_a^{-1}(a) = g_b \cdot S_a \quad \text{if} \quad b = g_b(a).$$

Thus

$$S^2 = SO(E)/S_a.$$

Combining τ of (A.0) in appendix 1 with the projection pr_a yields the surjective map

$$\mathrm{pr}_a \circ \tau : SU(2) \longrightarrow S^2,$$

which will be called the *Hopf projection* in the sequel and will be denoted by $\mathrm{pr}_{\mathrm{Hopf}}$. Due to equation (A.15),

$$\mathrm{pr}_{\mathrm{Hopf}}(k) = \tau_k(a)$$

for any $k \in SU(2)$ and thus $\mathrm{pr}_{\mathrm{Hopf}}$ is smooth. Clearly

$$\tau^{-1}(S_a) = U^a(1).$$

The construction made so far is subsumed as follows:

THEOREM 39. *The following commutative diagram of smooth maps visualizes the above construction:*

$$\begin{array}{ccc} SU(2) & \xrightarrow{\tau} & SO(E) \\ pr_{Hopf} \searrow & & \swarrow pr_a \\ & S^2 & \end{array}$$

Moreover,

$$\left(SU(2), pr_{Hopf}, S^2, U^a(1) \right)$$

is a principal bundle, the Hopf fibration.

3.2 A Reconstruction Formula

In this subsection we start with $a, b \in S^2$ and will determine $k \in SU(2)$ for which $b = \tau_k(a)$. To this end we will split k into

$$k = \lambda e + \mu u \tag{A.17}$$

with $\lambda, \mu \in \mathbb{R}$ and $u \in S^2$. Obviously

$$\lambda^2 + \mu^2 = 1$$

According to (A.5), $\mathbb{R} \cdot u$ is the axis of rotation of τ_k. To see how a is rotated by τ_k about $\mathbb{R} \cdot u$ we decompose a into

$$a_u :=< a, u > u \quad \text{and} \quad a_\perp := a- < a, u > u.$$

Clearly

$$a_\perp \in F^u$$

is the component of a that is rotated while a_u is left invariant. We set

$$\cos\beta :=< a, u > \tag{A.18}$$

for some $\beta \in \mathbb{R}$ and observe that

$$\tau_k(a_\perp) = e^{\vartheta \cdot u} \cdot a_\perp$$

since F^u is the plane of rotation. Again, ϑ denotes the angle of rotation of τ_k (cf. (A.11)). Moreover

$$< \tau_k(a), u >=< a, \tau_k^{-1}(u) > .$$

Therefore we have

PROPOSITION 40.

$$< b, u >=< a, u >= \cos\beta. \tag{A.19}$$

Hence

$$b - a \in F^u.$$

Equation (A.19) expresses that a and b are on the same *cone* with vertex in $0 \in E$. The cone has β as cone angle and $\mathbb{R} \cdot u$ as axis of rotation.

Applying τ_k to $a_\perp$ yields

$$e^{\vartheta \cdot u} \cdot a_\perp + < a, u > \cdot u = b.$$

We define b_u and $b_\perp$ in analogy to a_u and $a_\perp$ and apply τ_k to a. This yields for $k^2 = e^{\vartheta \cdot u}$

$$\tau_k(a_u) = b_u$$

and, therefore,

$$\tau_k(a_\perp) = b_\perp.$$

From (A.11) we obtain

$$e^{\vartheta u} = b_\perp \cdot a_\perp^{-1}$$

or

$$e^{\vartheta u} = -\frac{1}{|a_\perp|^2}(b_\perp \times a_\perp - < b_\perp, a_\perp > \cdot e)$$

due to

$$a_\perp^{-1} = \frac{1}{|a_\perp|} \cdot a_\perp.$$

Since, moreover, $|a_\perp| = |b_\perp|$, we easily verify

$$e^{\vartheta \cdot u} =< \frac{a_\perp}{|a_\perp|}, \frac{b_\perp}{|a_\perp|} > \cdot e + \frac{a_\perp}{|a_\perp|} \times \frac{b_\perp}{|b_\perp|}.$$

Thus k^2 is entirely determined by the unit vectors $\frac{a_\perp}{|a_\perp|}$ and $\frac{b_\perp}{|b_\perp|}$ in F^u.

PROPOSITION 41. *The components of k^2 formed with respect to the splitting*

$$\mathbb{H} = \mathbb{R} \cdot e + E$$

are

$$(\lambda^2 - \mu^2) \cdot e =< \frac{b_\perp}{|b_\perp|}, \frac{a_\perp}{|a_\perp|} > \cdot e$$

and

$$2\lambda \cdot \mu \cdot u = \frac{a_\perp}{|a_\perp|} \times \frac{b_\perp}{|b_\perp|}$$

which are directly related to the rotation angle ϑ_k of τ_k as seen from (A.9).

Since $k \cdot U^a = \mathbb{C}^a \cap S^3$, the coset $K := k \cdot U^a(1)$ is a great circle inclined from $\mathbb{R} \cdot e$ if $k \neq e$. Let us remark the following:

LEMMA 42. *Given a right coset $k \cdot U^a(1)$ for some $k \in SU(2)$ and a fixed vector $a \in S^2$ there is some $k' \in k \cdot U^a(1)$ such that the axis of rotation $\mathbb{R} \cdot u'$ of $\tau_{k'}$ is perpendicular to a. Here $u' \in S^2$. If, moreover,*

$$b := \tau_{k'}(a) = pr_{Hopf}(k),$$

the axis of rotation of $\tau_{k'}$ is $\mathbb{R} \cdot \frac{a \times b}{|a \times b|}$.

PROOF 4. *To see this we represent any element in K with respect to a fixed $k \in K$ as*

$$k(t) := k \cdot e^{t \cdot a}$$

and split k according to (A.17). Thus

$$k(t) = \lambda \cdot \cos t + \lambda \cdot a \cdot \sin t + \mu u \cdot \cos t + \mu \cdot u \cdot a \sin t.$$

The rotation axis of $\tau_{k(t)}$ is

$$\mathbb{R} \cdot (\lambda a \cdot \sin t + \mu u \cdot \cos t + \mu u \cdot a \cdot \sin t)$$

according to (A.5). If this axis shall be perpendicular to a for some $t_0 \in \mathbb{R}$,

$$\lambda \cdot \sin t_0 + \mu \cdot < u, a > \cdot \cos t_0 = 0$$

has to hold. If $\lambda \neq 0$, the solution t_0 is hence determined by

$$\tan t_0 = -\frac{\mu}{\lambda} \cdot < a, u >$$

(Clearly there is some $k \in K$ for which $\lambda \neq 0$.) Since $k = e^{\frac{\vartheta}{2} \cdot u}$, the parameters λ and μ are identified as $\lambda = \cos \frac{\vartheta}{2}$ and $\mu = \sin \frac{\vartheta}{2}$ and we finally find

$$\tan t_0 = -\tan \frac{\vartheta}{2} \cdot < a, u >$$

or

$$\tan t_0 = -\tan \frac{\vartheta}{2} \cos \beta,$$

due to (A.18).

The preceding lemma allows us to solve the problem mentioned in the beginning of 3.2, namely to determine the vector k in $\tau_k(a) = b$ more closely for a given pair $a, b \in S^2$.

The solution is rather easy. Without loss of generality we may assume that the rotation axis of τ_k (where k still has to be constructed), is perpendicular to a due to lemma 42. Thus k has to be of the form

$$k = e^{\frac{\vartheta}{2} \frac{a \times b}{|a \times b|}}. \tag{A.20}$$

The rotation angle ϑ of τ_k is obtained up to 2π from $|a \times b|$ as

$$\sin \vartheta = |a \times b|$$

due to the formula

$$|a \times b| = |a| \cdot |b| \cdot \sin \vartheta = \sin \vartheta.$$

Therefore, as a summary of the preceding lemmas, we have the following theorem:

THEOREM 43. *For each fixed $a \in S^2$ the projection*

$$pr_{Hopf} : SU(2) \longrightarrow S^2$$

defined by

$$pr_{Hopf}(k) = \tau_k(a) \qquad \forall k \in SU(2)$$

is a (smooth) surjection with

$$pr^{-1}_{Hopf}(\tau_k(a)) = k \cdot U^a(1).$$

Any coset $k \cdot U^a(1)$ with $k \in SU(2)$ contains some k' such that the axis of rotation of $\tau_{k'} \in SO(E)$ is perpendicular to $\mathbb{R} \cdot a$. Hence given $b \in S^2$ the coset $pr^{-1}_{Hopf}(b) \subset SU(2)$ is of the form

$$pr^{-1}_{Hopf}(b) = e^{\frac{\vartheta}{2} \cdot \frac{a \times b}{|a \times b|}} \cdot U^a(1)$$

where ϑ satisfies

$$|a \times b| = \sin \vartheta.$$

3.3 The Hopf fibration and the geometry of the tangent bundle of the 2–sphere

In order to relate the Hopf fibration with the geometry of the tangent bundle of the 2–sphere we consider

$$\mathbb{S} := \bigcup_{x \in S_2} S'_x \subset TS^2,$$

the collection of unit circles S'_x formed with respect to the Riemannian metric the sphere S^2 inherits from the inclusion $S^2 \subset E$.

On the sphere S^2 we choose some point a, say. Moreover we fix some v_0 in the S'_x. This allows us to define an action

$$\Phi_{\mathbb{S}} : SO(E) \longrightarrow \mathbb{S}$$

as

$$\Phi_{\mathbb{S}}(g) = g(v_0) \qquad \forall\, g \in SO(E).$$

Clearly

$$\mathrm{pr}_a : SO(E) \quad \longrightarrow \quad S^2$$

mapping a into $g(a)$ for any $g \in SO(E)$ is compatible with $\Phi_{\mathbb{S}}$, in the sense that

$$\mathrm{pr}_{S^2} \circ \Phi_{\mathbb{S}}(g) = \mathrm{pr}_a(g) \qquad \forall\, g \in SO(E)$$

where $\mathrm{pr}_{S^2} : TS^2 \longrightarrow S^2$ is the canonical projection (stable groups of pr_a operate via $\Phi_{\mathbb{S}}$ on the tangent planes of S^2).

Since

$$E = \mathbb{R} \cdot a + T_a S^2,$$

the map $\Phi_{\mathbb{S}}$ is a bijection. In fact, it is a diffeomorphism. Moreover, $SO(T_a S^2)$ maps S'_a into itself. Thus the diagram

$$\begin{array}{ccc} SO(E) & \xrightarrow{\Phi_{\mathbb{S}}} & \mathbb{S} \\ \mathrm{pr}_a \searrow & & \swarrow \mathrm{pr}_{S^2 a} \\ & S^2 & \end{array}$$

commutes (cf. (A.15)). Hence we have:

THEOREM 44. *The Hopf fibration is related to* $\mathbb{S}$ *by the commuting diagram*

$$\begin{array}{ccccc} SU(2) & \xrightarrow{\tau} & SO(E) & \xrightarrow{\Phi_{\mathbb{S}}} & \mathbb{S} \\ & pr_{Hopf} \searrow & \downarrow pr_a & \swarrow pr_{S^2} & \\ & & S^2 & & \end{array}$$

This shows that the Hopf fibration is a spin structure *of* S^2.

3.4 Tautological Bundles

The Hopf fibration as presented in theorem 43 admits an interpretation in terms of a tautological bundle over the complex projective space $\mathbb{C}^a P(\mathbb{H})$, the manifold of all $\mathbb{C}^a$–lines in $\mathbb{H}$. To verify this we write any $\mathbb{C}^a$–line in $\mathbb{H}$ in the form $k \cdot \mathbb{C}^a$ with $k \in \mathbb{H}$. The projective space $\mathbb{C}^a P(\mathbb{H})$ will be mapped diffeomorphically onto S^2. At first we observe that

$$k \cdot \mathbb{C}^a \cap S^3 = k \cdot U^a \qquad \forall\, k \in S^3.$$

Thus the orbit space S^3/U^a of the $\mathbb{C}^a$–action on S^3 corresponds bijectively to $\mathbb{C}^a P(\mathbb{H})$. We may hence identify $\mathbb{C}^a P(\mathbb{H})$ with the orbit space S^3/U^a. Using the identification of $SU(2)$ with S^3, the $2:1$–map

$$\tau : S^3 \longrightarrow SO(E)$$

yields the desired correspondence between orbit spaces

$$\tau^a : S^3/U^a \longrightarrow SO(E)/SO(F^a) = S^2,$$

say, given by

$$\tau^a(k \cdot U^a) := \tau_k \cdot \tau(U^a) \qquad \forall\, k \in S^3.$$

It is a smooth one–to–one map of the orbit space S^3/U^a onto $SO(E)/SO(F^a)$. However, the latter is identical to S^2. With this identification,

$$k \cdot U^a = \tau_k(a)$$

holds true by (A.15).

The fibre at $b \in S^2$, say, of the principal bundle $\left(SU(2), \mathrm{pr}_{\mathrm{Hopf}}, S^2, U^a\right)$ is hence $k \cdot U^a$, where $\tau_k(a) = b$. Accordingly, the fibre over the coset $k \cdot U^a \in S^3/U^a$ is $k \cdot U^a$ itself. This is the reason why the Hopf fibration over S^2

$$\mathbb{C}^a P(\mathbb{H}) = S^3/U^a$$

is called the *tautological principal bundle* over $\mathbb{C}^a P(\mathbb{H})$.

Accordingly, the complex line bundle

$$\mathbb{L}^{\mathbb{C}^a P(\mathbb{H})} \longrightarrow \mathbb{C}^a P(\mathbb{H})$$

with the complex line $k \cdot U^a$ as fibre over $k \cdot \mathbb{C}^a \in \mathbb{C}^a P(\mathbb{H})$ over any $k \in \mathbb{H}$ is called the *tautological complex line bundle* over $\mathbb{C}^a P(\mathbb{H})$.

Acknowledgments

We want to heartily thank Bruno Gruber for his kind invitation to the symposium "Symmetries in Sciences XIII" held at Bregenz in July 2003 and the opportunity to present this paper in two lectures.

References

[1] Abraham, R., J.E. Marsden, and T. Ratiu: 1983, *Manifolds, Tensor, Analysis, and Applications*, No.2. In Global Analysis Pure and Applied B. London, Amsterdam, Don Mills (Ontario), Sydney, Tokyo: Addison-Wesley.

[2] Binz, E. and S. Pods, 'The Geometry of Heisenberg Groups in Information Theory, Quantization, Quantum Information and Holography' (in preparation).

[3] Binz, E. and S. Pods: (to appear), 'Classical and Quantum Information Transmission'. In: A. Khrennikov (ed.): *Proceedings of Quantum Theory: Reconsideration of Foundations - 2. Växjö (Sweden)* June 1-6, 2003.

[4] Binz, E., S. Pods, and W. Schempp: 2003, 'Heisenberg Groups - the Fundamental Ingredient to describe Information, its Transmission and Quantization'. *Journal of Physics A: Mathematical and General* **36**, 1-21.

[5] Binz, E. and W. Schempp: 2000, 'Vector Fields in Three Space, Natural Internal Degrees of Freedom, Signal Transmission and Quantization'. *Results in Mathematics* **37**, 226-245.

[6] Binz, E., Sniatycki, and H. Fischer. 1988, *The Geometry of Classical Fields*, Mathematical Studies / Notas de Matemàtica **154**. Amsterdam: North Holland.

[7] Bouwmeester, D., A. Ekert, and A. Zelinger (eds.): 2000, *Physics of Quantum Information: Quantum Cryptography, Quantum Teleportation, Quantum Computation*. Berlin, Heidelberg, New York: Springer.

[8] Cover, T.M. and J.A. Thomas: 1991, *Elements of Information Theory*. New York: John Wiley & Sons.

[9] Greub, W. and H. Petry: 1975, 'Minimal Coupling and Complex Line Bundles', *J. of Math. Phys.* **16**(6).

[10] Kobayashi, S. and K. Nomizu: 1996, *Foundations of Differential Geometry*, Vol. I. New York: John Wiley & Sons, classics library edition.

[11] Lomonaco, Jr., S.J. (ed.): 2001, 'Quantum Computation. A Grand Mathematical Challenge for the Twenthy-First Century and the Millennium', Proceedings of the Symposia of Applied Mathematics. Providence, Rhode Island: American Mathematical Society.

[12] Mosseri, R. and R. Dandaloff: 2001, 'Geometry of Entangled States, Bloch Spheres and Hopf Fibrations'. arXiv:quant-ph/010837 v1.

[13] Nielsen, M.A. and I.L. Chuang: 2000, *Quantum Computation and Quantum Information*. Cambridge University Press.

[14] Pods, S.: 2003, 'Die Theorie der Heisenberggruppen zur Beschriebung einer Informationsübertragung insbesondere bei der Bildgebung in der Magnetresonanztomographie'. Ph.D. thesis, Universität Mannheim, Logos-Verlag, Berlin.

[15] Schempp, W.J.: 1986, *Harmonic Analysis in the Heisenberg Nilpotent Lie Group with Applications to Signal Theory*, Pitman Research Notes in Mathematics Series **147**. Longman Scientific & Technical.

[16] Urbantke, H.: 1991, 'Two-Level Quantum Systems: States, Phases, and Holonomy', *American Journal of Physics* **59**(6), 503-509.

QUANTUM FIELD THEORY OF PARTICLE MIXING AND OSCILLATIONS

M. Blasone
The Blackett Laboratory, Imperial College London,
London SW7 2AZ, U.K. and
Institute für Theoretische Physik,
Freie Universität Berlin,
Arnimallee 14, D-14195 Berlin, Germany

G. Vitiello
Dipartimento di Fisica and INFN,
Università di Salerno, 84100 Salerno, Italy

Abstract We report on recent results on the Quantum Field Theory of mixed particles. The quantization procedure is discussed in detail, both for fermions and for bosons and the unitary inequivalence of the flavor and mass representations is proved. Oscillation formulas exhibiting corrections with respect to the usual quantum mechanical ones are then derived.

1. Introduction

The chapter of particle mixing and oscillations [1] is one of the most important and fascinating in the book of modern Particle Physics. This is especially true after the recent experimental results [2] which finally confirm, after a long search, the reality of neutrino oscillations [3, 4]: this represents indeed the first clear evidence for physics beyond the Standard Model.

Many unanswered questions about the physics of particle mixing are however still there, in particular from a theoretical point of view. Apart from the problem of the origin of mixing and of the small neutrino masses, difficulties arise already in the attempt to find a proper mathematical setting for the description of mixed particles in Quantum Field Theory (QFT).

B. Gruber (ed.), Symmetries in Science XI, 105–128.

It is indeed well known [5] that mixing of states with different masses is not allowed[1] in non-relativistic Quantum Mechanics (QM). In spite of this fact, the quantum mechanical treatment is the one usually adopted for its simplicity and elegance. A review of the problems connected with the QM of mixing and oscillations can be found in Ref.[7]. Difficulties in the construction of the Hilbert space for mixed neutrinos were pointed out in Ref.[8].

Only recently [9]-[22] a consistent treatment of mixing and oscillations in QFT has been achieved and we report here on these results.

The main point of our analysis [9] consists in the observation that a problem of representation (i.e. choice of the Hilbert space) may arise when we start to mix fields with different masses. This has to do with the peculiar mathematical structure of QFT, where unitarily inequivalent representations of the algebra of fields do exist [14, 24]: a classical example is the one of theories with spontaneous breakdown of symmetry. This situation is in contrast to the one of QM, which deals with systems with a finite number of degrees of freedom and where only one Hilbert space is admitted (von Neumann theorem).

On this basis, a careful analysis of the usual mixing transformations in QFT reveals a rich non-perturbative structure associated to the vacuum for mixed particles, which appears to be a condensate of particle-antiparticle pairs, both for fermions and bosons. The vacuum for the mixed fields is a generalized coherent state à la Perelomov [25].

The structure of flavor vacuum reflects into observable quantities: exact oscillation formulas [11, 18] are derived in QFT exhibiting corrections with respect to the usual QM ones. We also show that a geometric phase is associated to flavor oscillations [26].

The material here presented is organized in the following way:

In Section 2, the mixing transformations are studied in QFT, both for fermions and bosons, in the case of two flavors. The currents and charges for mixed fields are also introduced and then used in Section 3 to derive exact oscillation formulas for charged fields (bosons and fermions). The case of neutral fields is treated in Section 3.3.

The geometric phase for oscillating particles is studied in Section 4. In Section 5 the case of three flavor mixing is considered and the deformation of the associated algebra due to CP violation is discussed. Finally, in Section 6, a space dependent oscillation formula for neutrinos is derived using the relativistic flavor current.

[1] See however also Ref.[6].

2. Mixing transformations in Quantum Field Theory

In this Section we study the quantization of mixed fields both for Dirac fermions and for charged bosons [9, 10, 18]. For simplicity, we limit ourselves to the case of two generations (flavors) although the main results presented below have general validity [14]. Three flavor fermion mixing [19] is discussed in §5.

2.1 Fermion mixing

Let us consider[2] two flavor fields ν_e, ν_μ. The mixing relations are [3]

$$\begin{aligned} \nu_e(x) &= \cos\theta\, \nu_1(x) + \sin\theta\, \nu_2(x) \\ \nu_\mu(x) &= -\sin\theta\, \nu_1(x) + \cos\theta\, \nu_2(x)\,, \end{aligned} \tag{1}$$

Here ν_e, ν_μ are the (Dirac) neutrino fields with definite flavors. ν_1, ν_2 are the (free) neutrino fields with definite masses m_1, m_2, respectively. θ is the mixing angle. The fields ν_1 and ν_2 are expanded as

$$\nu_i(x) = \frac{1}{\sqrt{V}} \sum_{\mathbf{k},r} \left[u^r_{\mathbf{k},i}(t) \alpha^r_{\mathbf{k},i} + v^r_{-\mathbf{k},i}(t) \beta^{r\dagger}_{-\mathbf{k},i} \right] e^{i\mathbf{k}\cdot\mathbf{x}}, \qquad i = 1,2\,. \tag{2}$$

where $u^r_{\mathbf{k},i}(t) = e^{-i\omega_{k,i}t} u^r_{\mathbf{k},i}$ and $v^r_{\mathbf{k},i}(t) = e^{i\omega_{k,i}t} v^r_{\mathbf{k},i}$, with $\omega_{k,i} = \sqrt{\mathbf{k}^2 + m_i^2}$. The $\alpha^r_{\mathbf{k},i}$ and the $\beta^r_{\mathbf{k},i}$ $(r = 1,2)$, are the annihilation operators for the vacuum state $|0\rangle_{1,2} \equiv |0\rangle_1 \otimes |0\rangle_2$: $\alpha^r_{\mathbf{k},i}|0\rangle_{1,2} = \beta^r_{\mathbf{k},i}|0\rangle_{1,2} = 0$. The anticommutation relations are:

$$\{\nu_i^\alpha(x), \nu_j^{\beta\dagger}(y)\}_{t=t'} = \delta^3(\mathbf{x}-\mathbf{y})\delta_{\alpha\beta}\delta_{ij}\,, \qquad \alpha,\beta = 1,..,4\,, \tag{3}$$

$$\{\alpha^r_{\mathbf{k},i}, \alpha^{s\dagger}_{\mathbf{q},j}\} = \delta_{\mathbf{kq}}\delta_{rs}\delta_{ij}; \qquad \{\beta^r_{\mathbf{k},i}, \beta^{s\dagger}_{\mathbf{q},j}\} = \delta_{\mathbf{kq}}\delta_{rs}\delta_{ij}, \quad i,j = 1,2\,. \tag{4}$$

All other anticommutators are zero. The orthonormality and completeness relations are:

$$u^{r\dagger}_{\mathbf{k},i} u^s_{\mathbf{k},i} = v^{r\dagger}_{\mathbf{k},i} v^s_{\mathbf{k},i} = \delta_{rs}\,, \quad u^{r\dagger}_{\mathbf{k},i} v^s_{-\mathbf{k},i} = v^{r\dagger}_{-\mathbf{k},i} u^s_{\mathbf{k},i} = 0\,, \tag{5}$$

$$\sum_r (u^r_{\mathbf{k},i} u^{r\dagger}_{\mathbf{k},i} + v^r_{-\mathbf{k},i} v^{r\dagger}_{-\mathbf{k},i}) = 1\!\!1\,. \tag{6}$$

In QFT the basic dynamics, i.e. the Lagrangian and the resulting field equations, is given in terms of Heisenberg (or interacting) fields. The physical observables are expressed in terms of asymptotic in- (or out-) fields, also called

[2]We refer to neutrinos, but the discussion is clearly valid for any Dirac fields.

physical or free fields. In the LSZ formalism of QFT [14, 24], the free fields, say for definitiveness the in-fields, are obtained by the weak limit of the Heisenberg fields for time $t \to -\infty$. The meaning of the weak limit is that the realization of the basic dynamics in terms of the in-fields is not unique so that the limit for $t \to -\infty$ (or $t \to +\infty$ for the out-fields) is representation dependent.

Typical examples are the ones of spontaneously broken symmetry theories, where the same set of Heisenberg field equations describes the normal (symmetric) phase as well as the symmetry broken phase. Since observables are described in terms of asymptotic fields, unitarily inequivalent representations describe different, i.e. physically inequivalent, phases. It is therefore of crucial importance, in order to get physically meaningful results, to investigate with much care the mapping among Heisenberg or interacting fields and free fields, i.e. the dynamical map.

With this warnings, mixing relations such as the relations (103) deserve a careful analysis, since they actually represent a dynamical mapping. It is now our purpose to investigate the structure of the Fock spaces $\mathcal{H}_{1,2}$ and $\mathcal{H}_{e,\mu}$ relative to ν_1, ν_2 and ν_e, ν_μ, respectively. In particular we want to study the relation among these spaces in the infinite volume limit. As usual, we will perform all computations at finite volume V and only at the end we will put $V \to \infty$.

Our first step is the study of the generator of Eqs.(103) and of the underlying group theoretical structure. Eqs.(103) can be recast as [9]:

$$\nu_e^\alpha(x) = G_\theta^{-1}(t)\, \nu_1^\alpha(x)\, G_\theta(t) \tag{7}$$

$$\nu_\mu^\alpha(x) = G_\theta^{-1}(t)\, \nu_2^\alpha(x)\, G_\theta(t)\,, \tag{8}$$

where $G_\theta(t)$ is given by

$$G_\theta(t) = \exp\left[\theta \int d^3\mathbf{x} \left(\nu_1^\dagger(x)\nu_2(x) - \nu_2^\dagger(x)\nu_1(x)\right)\right]\,, \tag{9}$$

and is (at finite volume) an unitary operator: $G_\theta^{-1}(t) = G_{-\theta}(t) = G_\theta^\dagger(t)$, preserving the canonical anticommutation relations (104). Eq.(9) follows from $\frac{d^2}{d\theta^2}\nu_e^\alpha = -\nu_e^\alpha$, $\frac{d^2}{d\theta^2}\nu_\mu^\alpha = -\nu_\mu^\alpha$ with the initial conditions $\nu_e^\alpha|_{\theta=0} = \nu_1^\alpha$, $\frac{d}{d\theta}\nu_e^\alpha|_{\theta=0} = \nu_2^\alpha$ and $\nu_\mu^\alpha|_{\theta=0} = \nu_2^\alpha$, $\frac{d}{d\theta}\nu_\mu^\alpha|_{\theta=0} = -\nu_1^\alpha$.

Note that G_θ is an element of $SU(2)$ since it can be written as

$$G_\theta(t) = \exp[\theta(S_+(t) - S_-(t))]\,, \tag{10}$$

$$S_+(t) = S_-^\dagger(t) \equiv \int d^3\mathbf{x}\, \nu_1^\dagger(x)\nu_2(x)\,. \tag{11}$$

By introducing then

$$S_3 \equiv \frac{1}{2}\int d^3\mathbf{x} \left(\nu_1^\dagger(x)\nu_1(x) - \nu_2^\dagger(x)\nu_2(x)\right)\,, \tag{12}$$

the $su(2)$ algebra is closed (for t fixed):

$$[S_+(t), S_-(t)] = 2S_3 \quad , \quad [S_3, S_\pm(t)] = \pm S_\pm(t) \, . \tag{13}$$

The action of the mixing generator on the vacuum $|0\rangle_{1,2}$ is non-trivial and we have (at finite volume V):

$$|0(t)\rangle_{e,\mu} \equiv G_\theta^{-1}(t)\;|0\rangle_{1,2} \, . \tag{14}$$

$|0(t)\rangle_{e,\mu}$ is the *flavor vacuum*, i.e. the vacuum for the flavor fields. Note that $G_\theta^{-1}(t)$ is just the generator for generalized coherent states of $SU(2)$ [25]: the flavor vacuum is therefore an $SU(2)$ (time dependent) coherent state. Let us now investigate the infinite volume limit of Eq.(14). Using the Gaussian decomposition, G_θ^{-1} is written as [25]

$$\exp[\theta(S_- - S_+)] = \exp(-\tan\theta \; S_+)\;\exp(-2\ln\,\cos\theta\; S_3)\;\exp(\tan\theta\; S_-)$$

where $0 \leq \theta < \frac{\pi}{2}$. We then compute ${}_{1,2}\langle 0|0(t)\rangle_{e,\mu}$ and obtain

$${}_{1,2}\langle 0|0(t)\rangle_{e,\mu} = \prod_{\mathbf{k}} \left(1 - \sin^2\theta\;|V_{\mathbf{k}}|^2\right)^2 \equiv \prod_{\mathbf{k}} \Gamma(k) = e^{\sum_{\mathbf{k}} ln\,\Gamma(k)} . \tag{15}$$

where the function $|V_{\mathbf{k}}|^2$ is defined in Eq.(25) and plotted in Fig.1. Note that $|V_{\mathbf{k}}|^2$ depends on $|\mathbf{k}|$, it is always in the interval $[0, \frac{1}{2}[$ and goes to zero for $|\mathbf{k}| \rightarrow \infty$. By using the customary continuous limit relation $\sum_{\mathbf{k}} \rightarrow \frac{V}{(2\pi)^3}\int d^3\mathbf{k}$, in the infinite volume limit we obtain (for any t)

$$\lim_{V\to\infty}\;{}_{1,2}\langle 0|0(t)\rangle_{e,\mu} = \lim_{V\to\infty}\; e^{\frac{V}{(2\pi)^3}\int d^3\mathbf{k}\; ln\,\Gamma(k)} = 0 \tag{16}$$

since $\Gamma(\mathbf{k}) < 1$ for any value of $\mathbf{k}$ and of m_1 and m_2 (with $m_2 \neq m_1$).

Notice that (16) shows that the orthogonality between $|0(t)\rangle_{e,\mu}$ and $|0\rangle_{1,2}$ is due to the infrared contributions which are taken in care by the infinite volume limit and therefore high momentum contributions do not influence the result (for this reason here we do not need to consider the regularization problem of the UV divergence of the integral of $\ln\;\Gamma(\mathbf{k})$). Of course, this orthogonality disappears when $\theta = 0$ and/or when $m_1 = m_2$ (in this case $V_{\mathbf{k}} = 0$ for any $\mathbf{k}$).

Eq.(16) expresses the unitary inequivalence in the infinite volume limit of the flavor and the mass representations and shows the non-trivial nature of the mixing transformations (103), resulting in the condensate structure of the flavor vacuum. In Section 3 we will see how such a vacuum structure leads to phenomenological consequences in the neutrino oscillations, which may be possibly experimentally tested.

By use of $G_\theta(t)$, the flavor fields can be expanded as:

$$\nu_\sigma(x) = \sum_{r=1,2}\int \frac{d^3\mathbf{k}}{(2\pi)^{\frac{3}{2}}}\left[u^r_{\mathbf{k},i}(t)\alpha^r_{\mathbf{k},\sigma}(t) + v^r_{-\mathbf{k},i}(t)\beta^{r\dagger}_{-\mathbf{k},\sigma}(t)\right] e^{i\mathbf{k}\cdot\mathbf{x}} , \tag{17}$$

with $(\sigma, i) = (e,1), (\mu,2)$. The flavor annihilation operators are defined as $\alpha^r_{\mathbf{k},\sigma}(t) \equiv G^{-1}_\theta(t)\alpha^r_{\mathbf{k},i}G_\theta(t)$ and $\beta^{r\dagger}_{-\mathbf{k},\sigma}(t) \equiv G^{-1}_\theta(t)\beta^{r\dagger}_{-\mathbf{k},i}G_\theta(t)$. In the reference frame such that $\mathbf{k} = (0,0,|\mathbf{k}|)$, we have the simple expressions:

$$\alpha^r_{\mathbf{k},e}(t) = \cos\theta\; \alpha^r_{\mathbf{k},1} \;+\; \sin\theta\; \left(U^*_{\mathbf{k}}(t)\; \alpha^r_{\mathbf{k},2} \;+\; \epsilon^r\; V_{\mathbf{k}}(t)\; \beta^{r\dagger}_{-\mathbf{k},2}\right) \tag{18}$$

$$\alpha^r_{\mathbf{k},\mu}(t) = \cos\theta\; \alpha^r_{\mathbf{k},2} \;-\; \sin\theta\; \left(U_{\mathbf{k}}(t)\; \alpha^r_{\mathbf{k},1} \;-\; \epsilon^r\; V_{\mathbf{k}}(t)\; \beta^{r\dagger}_{-\mathbf{k},1}\right) \tag{19}$$

$$\beta^r_{-\mathbf{k},e}(t) = \cos\theta\; \beta^r_{-\mathbf{k},1} \;+\; \sin\theta\; \left(U^*_{\mathbf{k}}(t)\; \beta^r_{-\mathbf{k},2} \;-\; \epsilon^r\; V_{\mathbf{k}}(t)\; \alpha^{r\dagger}_{\mathbf{k},2}\right) \tag{20}$$

$$\beta^r_{-\mathbf{k},\mu}(t) = \cos\theta\; \beta^r_{-\mathbf{k},2} \;-\; \sin\theta\; \left(U_{\mathbf{k}}(t)\; \beta^r_{-\mathbf{k},1} \;+\; \epsilon^r\; V_{\mathbf{k}}(t)\; \alpha^{r\dagger}_{\mathbf{k},1}\right) \tag{21}$$

where $\epsilon^r = (-1)^r$ and

$$U_{\mathbf{k}}(t) \equiv u^{r\dagger}_{\mathbf{k},2}(t)u^r_{\mathbf{k},1}(t) = v^{r\dagger}_{-\mathbf{k},1}(t)v^r_{-\mathbf{k},2}(t) = |U_{\mathbf{k}}|\, e^{i(\omega_{k,2}-\omega_{k,1})t} \tag{22}$$

$$V_{\mathbf{k}}(t) \equiv \epsilon^r\; u^{r\dagger}_{\mathbf{k},1}(t)v^r_{-\mathbf{k},2}(t) = -\epsilon^r\; u^{r\dagger}_{\mathbf{k},2}(t)v^r_{-\mathbf{k},1}(t) = |V_{\mathbf{k}}|\; e^{i(\omega_{k,2}+\omega_{k,1})t} \tag{23}$$

$$|U_{\mathbf{k}}| = \frac{|\mathbf{k}|^2 + (\omega_{k,1}+m_1)(\omega_{k,2}+m_2)}{2\sqrt{\omega_{k,1}\omega_{k,2}(\omega_{k,1}+m_1)(\omega_{k,2}+m_2)}} \tag{24}$$

$$|V_{\mathbf{k}}| = \frac{(\omega_{k,1}+m_1) - (\omega_{k,2}+m_2)}{2\sqrt{\omega_{k,1}\omega_{k,2}(\omega_{k,1}+m_1)(\omega_{k,2}+m_2)}}\,|\mathbf{k}| \tag{25}$$

$$|U_{\mathbf{k}}|^2 + |V_{\mathbf{k}}|^2 = 1. \tag{26}$$

The condensation density of the flavor vacuum is given by

$$_{e,\mu}\langle 0(t)|\alpha^{r\dagger}_{\mathbf{k},i}\alpha^r_{\mathbf{k},i}|0(t)\rangle_{e,\mu} = \sin^2\theta\; |V_{\mathbf{k}}|^2, \qquad i = 1,2\,, \tag{27}$$

with the same result for antiparticles[3]. Note that the $|V_{\mathbf{k}}|^2$ has a maximum at $\sqrt{m_1 m_2}$ and $|V_{\mathbf{k}}|^2 \simeq \frac{(m_2-m_1)^2}{4|\mathbf{k}|^2}$ for $|\mathbf{k}| \gg \sqrt{m_1 m_2}$.

2.2 Boson mixing

Let us now consider boson mixing [10, 18] in the case of charged fields. We define the mixing relations as:

$$\begin{aligned} \phi_A(x) &= \cos\theta\; \phi_1(x) + \sin\theta\; \phi_2(x) \\ \phi_B(x) &= -\sin\theta\; \phi_1(x) + \cos\theta\; \phi_2(x) \end{aligned} \tag{28}$$

[3]In the case of three flavors [9, 19], the condensation densities are different for different i and for antiparticles (when CP violation is present)

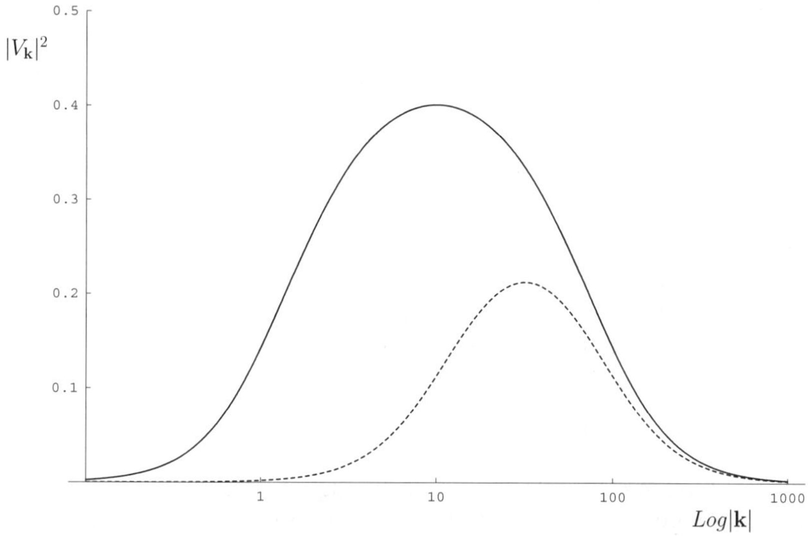

Figure 1. The fermion condensation density $|V_{\mathbf{k}}|^2$ as a function of $|\mathbf{k}|$ for $m_1 = 1, m_2 = 100$ (solid line) and $m_1 = 10$, $m_2 = 100$ (dashed line).

where generically we denote the mixed fields with suffixes A and B. Let the fields $\phi_i(x)$, $i = 1, 2$, be free complex fields with definite masses. Their conjugate momenta are $\pi_i(x) = \partial_0 \phi_i^\dagger(x)$ and the commutation relations are the usual ones:

$$[\phi_i(x), \pi_j(y)]_{t=t'} = \left[\phi_i^\dagger(x), \pi_j^\dagger(y)\right]_{t=t'} = i\delta^3(\mathbf{x}-\mathbf{y})\,\delta_{ij} \tag{29}$$

with $i, j = 1, 2$ and the other equal–time commutators vanishing. The Fourier expansions of fields and momenta are:

$$\phi_i(x) = \int \frac{d^3\mathbf{k}}{(2\pi)^{\frac{3}{2}}} \frac{1}{\sqrt{2\omega_{k,i}}} \left(a_{\mathbf{k},i}\, e^{-i\omega_{k,i}t} + b^\dagger_{-\mathbf{k},i}\, e^{i\omega_{k,i}t}\right) e^{i\mathbf{k}\cdot\mathbf{x}} \tag{30}$$

$$\pi_i(x) = i \int \frac{d^3\mathbf{k}}{(2\pi)^{\frac{3}{2}}} \sqrt{\frac{\omega_{k,i}}{2}} \left(a^\dagger_{\mathbf{k},i}\, e^{i\omega_{k,i}t} - b_{-\mathbf{k},i}\, e^{-i\omega_{k,i}t}\right) e^{i\mathbf{k}\cdot\mathbf{x}}\,, \tag{31}$$

where $\omega_{k,i} = \sqrt{\mathbf{k}^2 + m_i^2}$ and $\left[a_{\mathbf{k},i}, a^\dagger_{\mathbf{p},j}\right] = \left[b_{\mathbf{k},i}, b^\dagger_{\mathbf{p},j}\right] = \delta^3(\mathbf{k}-\mathbf{p})\delta_{ij}$, with $i, j = 1, 2$ and the other commutators vanishing.

We proceed in a similar way as for fermions and write Eqs.(28) as

$$\phi_\sigma(x) \;=\; G_\theta^{-1}(t)\, \phi_i(x)\, G_\theta(t) \tag{32}$$

with $(\sigma, i) = (A, 1), (B, 2)$, and similar expressions for π_A, π_B. We have

$$G_\theta(t) = \exp[\theta(S_+(t) - S_-(t))]\,. \tag{33}$$

The operators

$$S_+(t) = S_-^\dagger(t) \equiv -i \int d^3\mathbf{x}\, \left(\pi_1(x)\phi_2(x) - \phi_1^\dagger(x)\pi_2^\dagger(x)\right), \tag{34}$$

$$S_3 \equiv \frac{-i}{2} \int d^3\mathbf{x} \left(\pi_1(x)\phi_1(x) - \phi_1^\dagger(x)\pi_1^\dagger(x) - \pi_2(x)\phi_2(x) + \phi_2^\dagger(x)\pi_2^\dagger(x)\right) \tag{35}$$

close the $su(2)$ algebra (at a given t).

As for fermions, the action of the generator of the mixing transformations on the vacuum $|0\rangle_{1,2}$ for the fields $\phi_{1,2}$ is non-trivial and induces on it a $SU(2)$ coherent state structure [25]:

$$|0(t)\rangle_{A,B} \equiv G_\theta^{-1}(t)\; |0\rangle_{1,2}\,. \tag{36}$$

We will refer to the state $|0(t)\rangle_{A,B}$ as to the flavor vacuum for bosons. The orthogonality between $|0(t)\rangle_{A,B}$ and $|0\rangle_{1,2}$ can be proved [18]. The Fourier expansion for the flavor fields is:

$$\phi_\sigma(x) = \int \frac{d^3\mathbf{k}}{(2\pi)^{\frac{3}{2}}} \frac{1}{\sqrt{2\omega_{k,i}}} \left(a_{\mathbf{k},\sigma}(t)\, e^{-i\omega_{k,i}t} + b^\dagger_{-\mathbf{k},\sigma}(t)\, e^{i\omega_{k,i}t}\right) e^{i\mathbf{k}\cdot\mathbf{x}} \tag{37}$$

with $(\sigma, i) = (A, 1), (B, 2)$, and similar expressions for π_A, π_B.
The annihilation operators for the vacuum $|0(t)\rangle_{A,B}$ are defined $a_{\mathbf{k},A}(t) \equiv G_\theta^{-1}(t)\; a_{\mathbf{k},1}\; G_\theta(t)$, etc. We have:

$$a_{\mathbf{k},A}(t) = \cos\theta\; a_{\mathbf{k},1} + \sin\theta\; \left(U_{\mathbf{k}}^*(t)\; a_{\mathbf{k},2} + V_{\mathbf{k}}(t)\; b^\dagger_{-\mathbf{k},2}\right), \tag{38}$$

$$a_{\mathbf{k},B}(t) = \cos\theta\; a_{\mathbf{k},2} - \sin\theta\; \left(U_{\mathbf{k}}(t)\; a_{\mathbf{k},1} - V_{\mathbf{k}}(t)\; b^\dagger_{-\mathbf{k},1}\right), \tag{39}$$

$$b_{-\mathbf{k},A}(t) = \cos\theta\; b_{-\mathbf{k},1} + \sin\theta\; \left(U_{\mathbf{k}}^*(t)\; b_{-\mathbf{k},2} + V_{\mathbf{k}}(t)\; a^\dagger_{\mathbf{k},2}\right), \tag{40}$$

$$b_{-\mathbf{k},B}(t) = \cos\theta\; b_{-\mathbf{k},2} - \sin\theta\; \left(U_{\mathbf{k}}(t)\; b_{-\mathbf{k},1} - V_{\mathbf{k}}(t)\; a^\dagger_{\mathbf{k},1}\right). \tag{41}$$

These operators satisfy the canonical commutation relations (at equal times). As for the case of the fermion mixing, the structure of the flavor ladder operators Eqs.(38)-(41) is recognized to be the one of a rotation combined with a Bogoliubov transformation. Indeed, in the above equations appear the Bogoliubov coefficients:

$$U_{\mathbf{k}}(t) \equiv |U_{\mathbf{k}}|\; e^{i(\omega_{k,2}-\omega_{k,1})t} \quad , \quad V_{\mathbf{k}}(t) \equiv |V_{\mathbf{k}}|\; e^{i(\omega_{k,1}+\omega_{k,2})t} \tag{42}$$

$$|U_{\mathbf{k}}| \equiv \frac{1}{2}\left(\sqrt{\frac{\omega_{k,1}}{\omega_{k,2}}} + \sqrt{\frac{\omega_{k,2}}{\omega_{k,1}}}\right), \quad |V_{\mathbf{k}}| \equiv \frac{1}{2}\left(\sqrt{\frac{\omega_{k,1}}{\omega_{k,2}}} - \sqrt{\frac{\omega_{k,2}}{\omega_{k,1}}}\right) \tag{43}$$

$$|U_{\mathbf{k}}|^2 - |V_{\mathbf{k}}|^2 = 1\,, \tag{44}$$

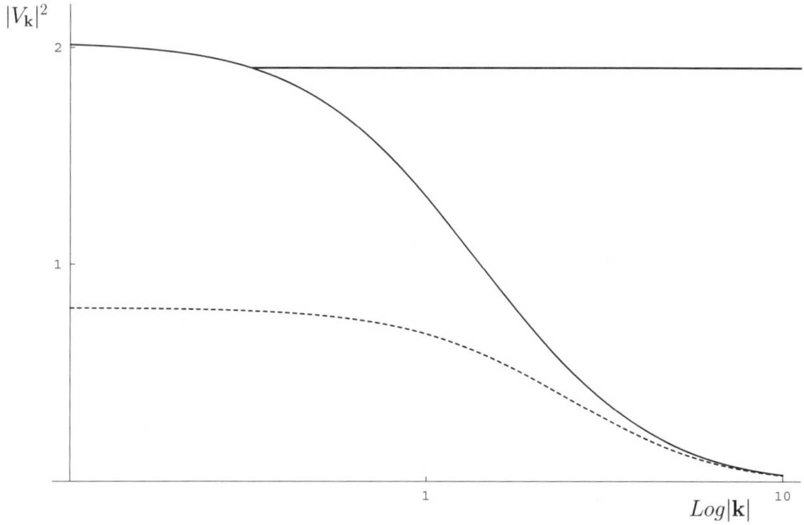

Figure 2. The boson condensation density $|V_{\mathbf{k}}|^2$ as a function of $|\mathbf{k}|$ for $m_1 = 1$, $m_2 = 10$ (solid line) and $m_1 = 2$, $m_2 = 10$ (dashed line).

Note the difference with respect to the fermionic case Eq.(26).

The condensation density of the flavor vacuum is given for any t by

$$ {}_{A,B}\langle 0(t)|a^{\dagger}_{\mathbf{k},i}a_{\mathbf{k},i}|0(t)\rangle_{A,B} = \sin^2\theta \, |V_{\mathbf{k}}|^2, \qquad i = 1,2\,, \tag{45} $$

with same result for antiparticles. The function $|V_{\mathbf{k}}|^2$ is maximal at $|\mathbf{k}| = 0$ ($|V_{max}|^2 = \frac{(m_1-m_2)^2}{4m_1m_2}$) and $|V_{\mathbf{k}}|^2 \simeq \left(\frac{\Delta m^2}{4|\mathbf{k}|^2}\right)^2$ for $|\mathbf{k}|^2 \gg \frac{m_1^2+m_2^2}{2}$. A plot is given in Fig.2 for sample values of the masses.

2.3 Currents and charges for mixed fields

We now study the transformations acting on a doublet of free fields with different masses. The results of this Section clarify the meaning of the $su(2)$ algebraic structure found before and will be useful in the discussion of neutrino oscillations.

2.3.1 Fermions. Let us consider the Lagrangian for two free Dirac fields, with masses m_1 and m_2:

$$ \mathcal{L}(x) = \bar{\Psi}_m(x)\,(i\,\partial\!\!\!/ - M_d)\,\Psi_m(x) \tag{46} $$

where $\Psi_m^T = (\nu_1, \nu_2)$ and $M_d = diag(m_1, m_2)$. We introduce a subscript m to denote quantities which are in terms of fields with definite masses.

$\mathcal{L}$ is invariant under global $U(1)$ phase transformations of the type $\Psi'_m = e^{i\alpha}\,\Psi_m$: as a result, we have the conservation of the Noether charge $Q =$

$\int d^3\mathbf{x}\, I^0(x)$ (with $I^\mu(x) = \bar{\Psi}_m(x)\,\gamma^\mu\,\Psi_m(x)$) which is indeed the total charge of the system (i.e. the total lepton number). Consider then the global $SU(2)$ transformation [16]:

$$\Psi'_m(x) \;=\; e^{i\alpha_j \tau_j}\,\Psi_m(x), \qquad\qquad j \;=\; 1,2,3. \tag{47}$$

with $\tau_j = \sigma_j/2$ and σ_j being the Pauli matrices. For $m_1 \neq m_2$, the Lagrangian is not generally invariant under the above transformations. We have indeed:

$$\delta\mathcal{L}(x) \;=\; i\alpha_j\,\bar{\Psi}_m(x)\,[\tau_j, M_d]\,\Psi_m(x) \;=\; -\alpha_j\,\partial_\mu J^\mu_{m,j}(x) \tag{48}$$

$$J^\mu_{m,j}(x) \;=\; \bar{\Psi}_m(x)\,\gamma^\mu\,\tau_j\,\Psi_m(x), \qquad\qquad j \;=\; 1,2,3. \tag{49}$$

Explicitly:

$$J^\mu_{m,1}(x) \;=\; \frac{1}{2}\left[\bar{\nu}_1(x)\,\gamma^\mu\,\nu_2(x) \;+\; \bar{\nu}_2(x)\,\gamma^\mu\,\nu_1(x)\right] \tag{50}$$

$$J^\mu_{m,2}(x) \;=\; \frac{i}{2}\left[\bar{\nu}_1(x)\,\gamma^\mu\,\nu_2(x) \;-\; \bar{\nu}_2(x)\,\gamma^\mu\,\nu_1(x)\right] \tag{51}$$

$$J^\mu_{m,3}(x) \;=\; \frac{1}{2}\left[\bar{\nu}_1(x)\,\gamma^\mu\,\nu_1(x) \;-\; \bar{\nu}_2(x)\,\gamma^\mu\,\nu_2(x)\right] \tag{52}$$

The charges $Q_{m,j}(t) \equiv \int d^3\mathbf{x}\, J^0_{m,j}(x)$ satisfy the $su(2)$ algebra (at equal times): $[Q_{m,j}(t), Q_{m,k}(t)] = i\,\epsilon_{jkl}\,Q_{m,l}(t)$. Note that $2Q_{m,2}(t)$ is indeed the generator of mixing transformations introduced in §2.1. Also note that Casimir operator is proportional to the total (conserved) charge: $C_m = \frac{1}{2}Q$ and that, since $Q_{m,3}$ is conserved in time, we have

$$Q_1 \equiv \frac{1}{2}Q + Q_{m,3} \quad , \quad Q_2 \equiv \frac{1}{2}Q - Q_{m,3} \tag{53}$$

$$Q_i = \sum_r \int d^3\mathbf{k}\left(\alpha^{r\dagger}_{\mathbf{k},i}\alpha^r_{\mathbf{k},i} - \beta^{r\dagger}_{-\mathbf{k},i}\beta^r_{-\mathbf{k},i}\right) \quad , \quad i = 1,2. \tag{54}$$

These are nothing but the Noether charges associated with the non-interacting fields ν_1 and ν_2: in the absence of mixing, they are the flavor charges, separately conserved for each generation.

2.3.2 Bosons. The above analysis can be easily extended to the boson case. We consider the Lagrangian

$$\mathcal{L}(x) \;=\; \partial_\mu\Phi^\dagger_m(x)\partial^\mu\Phi_m(x) \;-\; \Phi^\dagger_m(x)M_d\Phi_m(x) \tag{55}$$

with $\Phi^T_m = (\phi_1, \phi_2)$ being charged scalar fields and $M_d = diag(m_1^2, m_2^2)$.

We have now [16]

$$\Phi'_m(x) = e^{i\alpha_j \tau_j} \Phi_m(x) \tag{56}$$

$$\delta\mathcal{L}(x) = i\,\alpha_j\,\Phi_m^\dagger(x)\,[\tau_j\,,\,M_d]\,\Phi_m(x) = -\alpha_j\,\partial_\mu\,J^\mu_{m,j}(x)\,, \tag{57}$$

$$J^\mu_{m,j}(x) = i\,\Phi_m^\dagger(x)\,\tau_j\,\overleftrightarrow{\partial^\mu}\,\Phi_m(x)\,, \qquad j = 1,2,3. \tag{58}$$

Again, the corresponding charges $Q_{m,j}(t)$ satisfy the $su(2)$ algebra and the mixing generator for bosons is proportional to $Q_{m,2}(t)$.

2.4 Generalization of mixing transformations

We have seen in §2.1 how the fields ν_e and ν_μ can be expanded in the same bases as ν_1 and ν_2, see Eq.(17). As observed in Ref.[13], however, such a choice is actually a special one, and a more general possibility exists. Indeed, in the expansion Eq.(17) one could use eigenfunctions with arbitrary masses μ_σ and write the flavor fields as [13]:

$$\nu_\sigma(x) = \sum_{r=1,2} \int \frac{d^3\mathbf{k}}{(2\pi)^{\frac{3}{2}}} \left[u^r_{\mathbf{k},\sigma} \widetilde{\alpha}^r_{\mathbf{k},\sigma}(t) + v^r_{-\mathbf{k},\sigma} \widetilde{\beta}^{r\dagger}_{-\mathbf{k},\sigma}(t) \right] e^{i\mathbf{k}\cdot\mathbf{x}}, \tag{59}$$

where u_σ and v_σ are the eigenfunctions with mass μ_σ ($\sigma = e, \mu$). We denote by a tilde the generalized flavor operators introduced in Ref.[13]. The expansion Eq.(59) is more general than the one in Eq.(17) since the latter corresponds to the particular choice $\mu_e \equiv m_1$, $\mu_\mu \equiv m_2$. The relation between the general flavor operators of Eq.(59) and those of Eq.(17) is

$$\begin{pmatrix} \widetilde{\alpha}^r_{\mathbf{k},\sigma}(t) \\ \widetilde{\beta}^{r\dagger}_{-\mathbf{k},\sigma}(t) \end{pmatrix} = J^{-1}(t) \begin{pmatrix} \alpha^r_{\mathbf{k},\sigma}(t) \\ \beta^{r\dagger}_{-\mathbf{k},\sigma}(t) \end{pmatrix} J(t)\ , \tag{60}$$

$$J(t) = \prod_{\mathbf{k},r} \exp\left\{ i \sum_{(\sigma,j)} \xi^{\mathbf{k}}_{\sigma,j} \left[\alpha^{r\dagger}_{\mathbf{k},\sigma}(t) \beta^{r\dagger}_{-\mathbf{k},\sigma}(t) + \beta^r_{-\mathbf{k},\sigma}(t) \alpha^r_{\mathbf{k},\sigma}(t) \right] \right\}.$$

where $\xi^{\mathbf{k}}_{\sigma,j} \equiv (\chi_\sigma - \chi_j)/2$ with $\cot\chi_\sigma = |\mathbf{k}|/\mu_\sigma$ and $\cot\chi_j = |\mathbf{k}|/m_j$.

Thus the Hilbert space for the flavor fields is not unique: an infinite number of vacua can be generated by introducing the arbitrary mass parameters μ_σ. It is obvious that physical quantities must not depend on these parameters. Similar results are valid for bosons, see Ref.[18].

3. Flavor oscillations in QFT

As an application of the theoretical scheme above developed, we study flavor oscillations, both for fermions and for bosons. The QFT treatment leads to exact oscillation formulas exhibiting corrections with respect to the usual QM ones.

3.1 Neutrino oscillations

Let us now return to the Lagrangian Eq.(46) and write it in the flavor basis (subscript f denotes here flavor)

$$\mathcal{L}(x) = \bar{\Psi}_f(x)\,(i\,\partial\!\!\!/ - M)\,\Psi_f(x) \tag{61}$$

where $\Psi_f^T = (\nu_e, \nu_\mu)$ and $M = \begin{pmatrix} m_e & m_{e\mu} \\ m_{e\mu} & m_\mu \end{pmatrix}$. Obviously, $\mathcal{L}$ is still invariant under $U(1)$. We then consider the $SU(2)$ transformation [16]:

$$\Psi'_f(x) = e^{i\alpha_j \tau_j}\,\Psi_f(x), \tag{62}$$

$$\delta\mathcal{L}(x) = i\alpha_j\,\bar{\Psi}_f(x)\,[\tau_j, M]\,\Psi_f(x) = -\alpha_j\,\partial_\mu J^\mu_{f,j}(x)\,, \tag{63}$$

$$J^\mu_{f,j}(x) = \bar{\Psi}_f(x)\,\gamma^\mu\,\tau_j\,\Psi_f(x), \qquad j = 1,2,3. \tag{64}$$

The charges $Q_{f,j}(t) \equiv \int d^3\mathbf{x}\, J^0_{f,j}(x)$ satisfy the $su(2)$ algebra. Note that, because of the off–diagonal (mixing) terms in the mass matrix M, $Q_{f,3}$ is not anymore conserved. This implies an exchange of charge between ν_e and ν_μ, resulting in the phenomenon of flavor oscillations.

Let us indeed define the *flavor charges* for mixed fields as

$$Q_e(t) \equiv \int d^3\mathbf{x}\,\nu_e^\dagger(x)\nu_e(x) = \frac{1}{2}Q + Q_{f,3}(t) \tag{65}$$

$$Q_\mu(t) \equiv \int d^3\mathbf{x}\,\nu_\mu^\dagger(x)\nu_\mu(x) = \frac{1}{2}Q - Q_{f,3}(t) \tag{66}$$

where $Q_e(t) + Q_\mu(t) = Q$. They are related to the Noether charges as

$$Q_\sigma(t) = G_\theta^{-1}(t)\,Q_i\,G_\theta(t) \tag{67}$$

with $(\sigma, i) = (e,1), (\mu,2)$. From Eq.(67), it follows that the flavor charges are diagonal in the flavor ladder operators:

$$Q_\sigma(t) = \sum_r \int d^3\mathbf{k}\left(\alpha^{r\dagger}_{\mathbf{k},\sigma}(t)\alpha^r_{\mathbf{k},\sigma}(t) - \beta^{r\dagger}_{-\mathbf{k},\sigma}(t)\beta^r_{-\mathbf{k},\sigma}(t)\right), \tag{68}$$

with $\sigma = e, \mu$. We work in the Heisenberg picture and define the state for a particle with definite (electron) flavor, spin and momentum as[4]:

$$|\alpha^r_{\mathbf{k},e}\rangle \equiv \alpha^{r\dagger}_{\mathbf{k},e}(0)|0\rangle_{e,\mu} = G_\theta^{-1}(0)\alpha^{r\dagger}_{\mathbf{k},1}|0\rangle_{1,2}\,, \tag{69}$$

[4]Similar results are obtained for a muon neutrino state: $|\alpha^r_{\mathbf{k},\mu}\rangle \equiv \alpha^{r\dagger}_{\mathbf{k},\mu}(0)|0\rangle_{e,\mu}$.

where $|0\rangle_{e,\mu} \equiv |0(0)\rangle_{e,\mu}$. Note that the $|\alpha^r_{\mathbf{k},e}\rangle$ is an eigenstate of $Q_e(t)$, at $t=0$: $Q_e(0)|\alpha^r_{\mathbf{k},e}\rangle = |\alpha^r_{\mathbf{k},e}\rangle$. We thus have ${}_{e,\mu}\langle 0|Q_\sigma(t)|0\rangle_{e,\mu} = 0$ and

$$\begin{aligned}\mathcal{Q}_{\mathbf{k},\sigma}(t) &\equiv \langle \alpha^r_{\mathbf{k},e}|Q_\sigma(t)|\alpha^r_{\mathbf{k},e}\rangle \\ &= \left|\left\{\alpha^r_{\mathbf{k},\sigma}(t), \alpha^{r\dagger}_{\mathbf{k},\rho}(0)\right\}\right|^2 + \left|\left\{\beta^{r\dagger}_{-\mathbf{k},\sigma}(t), \alpha^{r\dagger}_{\mathbf{k},\rho}(0)\right\}\right|^2 \qquad (70)\end{aligned}$$

Charge conservation is ensured at any time: $\mathcal{Q}_{\mathbf{k},e}(t) + \mathcal{Q}_{\mathbf{k},\mu}(t) = 1$. The oscillation formulas for the flavor charges are then [11]

$$\begin{aligned}\mathcal{Q}_{\mathbf{k},e}(t) &= 1 - \sin^2(2\theta)\,|U_{\mathbf{k}}|^2\, \sin^2\left(\frac{\omega_{k,2}-\omega_{k,1}}{2}t\right) \\ &\quad + \sin^2(2\theta)\,|V_{\mathbf{k}}|^2\, \sin^2\left(\frac{\omega_{k,2}+\omega_{k,1}}{2}t\right), \qquad (71)\end{aligned}$$

$$\begin{aligned}\mathcal{Q}_{\mathbf{k},\mu}(t) &= \sin^2(2\theta)\,|U_{\mathbf{k}}|^2\, \sin^2\left(\frac{\omega_{k,2}-\omega_{k,1}}{2}t\right) \\ &\quad + \sin^2(2\theta)\,|V_{\mathbf{k}}|^2\, \sin^2\left(\frac{\omega_{k,2}+\omega_{k,1}}{2}t\right). \qquad (72)\end{aligned}$$

This result is exact. There are two differences with respect to the usual formula for neutrino oscillations: the amplitudes are energy dependent, and there is an additional oscillating term.

In the relativistic limit ($|\mathbf{k}| \gg \sqrt{m_1 m_2}$) we obtain ($\theta = \pi/4$):

$$\begin{aligned}\mathcal{Q}_{\mathbf{k},\mu}(t) \simeq \left(1 - \frac{(\Delta m)^2}{4|\mathbf{k}|^2}\right)\sin^2\left[\frac{\Delta m^2}{4|\mathbf{k}|}\,t\right] \\ + \frac{(\Delta m)^2}{4k^2}\sin^2\left[\left(|\mathbf{k}| + \frac{m_1^2+m_2^2}{4|\mathbf{k}|}\right)t\right]. \qquad (73)\end{aligned}$$

The usual QM formulas [3], are thus approximately recovered. Observe that for small times we have:

$$\mathcal{Q}_{\mathbf{k},\mu}(t) \simeq \frac{(m_2-m_1)^2}{4}\left(1 + \frac{m_1^2+m_2^2}{2|\mathbf{k}|^2} + \frac{(m_1+m_2)^2}{4|\mathbf{k}|^2}\right)t^2. \qquad (74)$$

Thus, even for the case of relativistic neutrinos, QFT corrections are in principle observable (for sufficiently small time arguments).

We also note that the above quantities are not interpreted as probabilities, rather they have a sense as *statistical averages*, i.e. as mean values. This is because the structure of the theory for mixed field is that of a many–body theory, where does not make sense to talk about single–particle states. This situation has a formal analogy with QFT at finite temperature, where only statistical averages are well defined.

We now show [15] that the above results are consistent with the generalization introduced in §2.4, i.e. that the exact oscillation probabilities are independent of the arbitrary mass parameters.

It can be indeed explicitly checked that

$$\langle \widetilde{\alpha}^{r}_{\mathbf{k},e} | \widetilde{Q}_{\sigma}(t) | \widetilde{\alpha}^{r}_{\mathbf{k},e} \rangle \;=\; \langle \alpha^{r}_{\mathbf{k},e} | Q_{\sigma}(t) | \alpha^{r}_{\mathbf{k},e} \rangle \tag{75}$$

which ensure the cancellation of the arbitrary mass parameters.

Note that the flavor charge operators $Q_\sigma(t)$ are *invariant* under the action of the Bogoliubov generator Eq.(60); however this is not sufficient to guarantee the result Eq.(75) which is non-trivial and provide a criterion for the selection of the observables for mixed fields [22]. As a matter of fact, the number operators for mixed fields are not good observables since their expectation values do depend on the arbitrary mass parameters. In §3.3 we will consider another observable, the momentum operator.

3.2 Meson oscillations

The bosonic counterpart of the above oscillation formulas can be derived in a similar way by use of the flavor charges for boson fields [18]. By defining the mixed bosonic state as:

$$|a_{\mathbf{k},A}\rangle \;\equiv\; a^{\dagger}_{\mathbf{k},A}(0)\,|0\rangle_{A,B} \tag{76}$$

and the flavor charges ($\sigma = A, B$):

$$Q_\sigma(t) = \int d^3\mathbf{k}\, \Big(a^{\dagger}_{\mathbf{k},\sigma}(t) a_{\mathbf{k},\sigma}(t) \;-\; b^{\dagger}_{-\mathbf{k},\sigma}(t) b_{-\mathbf{k},\sigma}(t) \Big)\,, \tag{77}$$

we obtain ${}_{A,B}\langle 0|\, Q_\sigma(t)\, |0\rangle_{A,B} \;=\; 0$ and

$$\begin{aligned} \mathcal{Q}_{\mathbf{k},\sigma}(t) \;&\equiv\; \langle a_{\mathbf{k},A}|\, Q_\sigma(t)\, |a_{\mathbf{k},A}\rangle \\ &=\; \Big|\Big[a_{\mathbf{k},\sigma}(t), a^{\dagger}_{\mathbf{k},A}(0)\Big]\Big|^2 \;-\; \Big|\Big[b^{\dagger}_{-\mathbf{k},\sigma}(t), a^{\dagger}_{\mathbf{k},A}(0)\Big]\Big|^2 . \end{aligned} \tag{78}$$

The conservation of the total charge gives $\sum_\sigma \mathcal{Q}_{\mathbf{k},\sigma}(t) = 1$ and the oscillation formulas are:

$$\begin{aligned}\mathcal{Q}_{\mathbf{k},A}(t) &= 1 - \sin^2(2\theta)\,|U_{\mathbf{k}}|^2 \sin^2\left(\frac{\omega_{k,2}-\omega_{k,1}}{2}t\right) \\ &\quad + \sin^2(2\theta)\,|V_{\mathbf{k}}|^2 \sin^2\left(\frac{\omega_{k,2}+\omega_{k,1}}{2}t\right), \qquad (79)\end{aligned}$$

$$\begin{aligned}\mathcal{Q}_{\mathbf{k},B}(t) &= \sin^2(2\theta)\,|U_{\mathbf{k}}|^2 \sin^2\left(\frac{\omega_{k,2}-\omega_{k,1}}{2}t\right) \\ &\quad - \sin^2(2\theta)\,|V_{\mathbf{k}}|^2 \sin^2\left(\frac{\omega_{k,2}+\omega_{k,1}}{2}t\right). \qquad (80)\end{aligned}$$

Thus also for bosons, the non-trivial structure of the flavor vacuum induces corrections to the usual QM expressions for flavor oscillations. The most obvious difference with respect to fermionic case is in the negative sign which makes it possible a negative value for the bosonic flavor charges. This only reinforces the statistical interpretation given above, i.e. we are not dealing anymore with probabilities for single particle evolution. As already noted for neutrinos, in the relativistic limit the usual QM formulas are (approximately) recovered.

3.3 Mixing and oscillations of neutral particles

The above scheme is only valid for charged fields, since in the case of neutral fermions (Majorana) and bosons, the (flavor) charges vanish identically. It is however possible to identify also in this case the relevant observables for the description of flavor oscillations.

As an example, let us consider the case of a neutral boson field, analogous treatment can be done for the Majorana field [20]: the notation is the same as in §2.2, the mixing relations being given by Eq.(28). The expansion for the neutral field is (with $x_0 \equiv t$):

$$\phi_i(x) = \int \frac{d^3\mathbf{k}}{(2\pi)^{\frac{3}{2}}} \frac{1}{\sqrt{2\omega_{k,i}}} \left(a_{\mathbf{k},i}\, e^{-i\omega_{k,i}t} + a^\dagger_{-\mathbf{k},i}\, e^{i\omega_{k,i}t}\right) e^{i\mathbf{k}\cdot\mathbf{x}}, \qquad (81)$$

with $i = 1, 2$ and a similar expansion for the conjugate momenta $\pi_i(x)$. The generator of the mixing transformations can be written as by $G_\theta(t) = \exp[\theta(S_+(t) - S_-(t))]$ with

$$S_+(t) \equiv -i \int d^3\mathbf{x}\, \pi_1(x)\phi_2(x), \quad S_-(t) \equiv -i \int d^3\mathbf{x}\, \pi_2(x)\phi_1(x) \qquad (82)$$

$$S_3 \equiv \frac{-i}{2} \int d^3\mathbf{x}\, (\pi_1(x)\phi_1(x) - \pi_2(x)\phi_2(x)) \qquad (83)$$

The $SU(2)$ structure is thus still present, although being not related to any flavor charges.

The flavor annihilation operators take now the following form [20]:

$$a_{\mathbf{k},A}(t) = \cos\theta \, a_{\mathbf{k},1} + \sin\theta \left(U^*_{\mathbf{k}}(t) \, a_{\mathbf{k},2} + V_{\mathbf{k}}(t) \, a^{\dagger}_{-\mathbf{k},2} \right), \tag{84}$$

$$a_{\mathbf{k},B}(t) = \cos\theta \, a_{\mathbf{k},2} - \sin\theta \left(U_{\mathbf{k}}(t) \, a_{\mathbf{k},1} - V_{\mathbf{k}}(t) \, a^{\dagger}_{-\mathbf{k},1} \right). \tag{85}$$

where the Bogoliubov coefficients coincide with those above defined for charged bosons.

We then consider the momentum operator, defined as [14]: $P^j \equiv \int d^3\mathbf{x} \Theta^{0j}(x)$, with $\Theta^{\mu\nu} \equiv \partial^\mu \phi \partial^\nu \phi - g^{\mu\nu} \left[\frac{1}{2}(\partial\phi)^2 - \frac{1}{2} m^2 \phi^2\right]$. For the free fields ϕ_i we have:

$$\mathbf{P}_i = \int d^3\mathbf{x} \; \pi_i(x) \nabla \phi_i(x) = \int d^3\mathbf{k} \, \frac{\mathbf{k}}{2} \left(a^{\dagger}_{\mathbf{k},i} a_{\mathbf{k},i} - a^{\dagger}_{-\mathbf{k},i} a_{-\mathbf{k},i} \right), \tag{86}$$

with $i = 1, 2$. The momentum operator for mixed fields is:

$$\mathbf{P}_\sigma(t) \equiv G^{-1}_\theta(t) \, \mathbf{P}_i \, G_\theta(t) = \int d^3\mathbf{k} \, \frac{\mathbf{k}}{2} \left(a^{\dagger}_{\mathbf{k},\sigma}(t) a_{\mathbf{k},\sigma}(t) - a^{\dagger}_{-\mathbf{k},\sigma}(t) a_{-\mathbf{k},\sigma}(t) \right), \tag{87}$$

with $\sigma = A, B$. Note that the total momentum is conserved in time: $\mathbf{P}_A(t) + \mathbf{P}_B(t) = \mathbf{P}_1 + \mathbf{P}_2 \equiv \mathbf{P}$. Let us now consider the expectation values of the momentum operator for flavor fields on the flavor state $|a_{\mathbf{k},A}\rangle_{A,B}$, defined as in Eq.(76). Obviously, this is an eigenstate of $\mathbf{P}_A(t)$ at time $t = 0$:

$$\mathbf{P}_A(0) \, |a_{\mathbf{k},A}\rangle = \mathbf{k} \, |a_{\mathbf{k},A}\rangle, \tag{88}$$

which follows from $\mathbf{P}_1 \, |a_{\mathbf{k},1}\rangle = \mathbf{k} \, |a_{\mathbf{k},1}\rangle$ by application of $G^{-1}_\theta(0)$.

At time $t \neq 0$, the expectation value of the momentum (normalized to the initial value) gives ${}_{A,B}\langle 0 | \mathbf{P}_\sigma(t) | 0 \rangle_{A,B} = 0$ and:

$$\begin{aligned} \mathcal{P}^{\mathbf{k}}_\sigma(t) &\equiv \frac{\langle a_{\mathbf{k},A} | \mathbf{P}_\sigma(t) | a_{\mathbf{k},A} \rangle}{\langle a_{\mathbf{k},A} | \mathbf{P}_\sigma(0) | a_{\mathbf{k},A} \rangle} \\ &= \left| \left[a_{\mathbf{k},\sigma}(t), a^{\dagger}_{\mathbf{k},A}(0) \right] \right|^2 - \left| \left[a^{\dagger}_{-\mathbf{k},\sigma}(t), a^{\dagger}_{\mathbf{k},A}(0) \right] \right|^2, \end{aligned} \tag{89}$$

with $\sigma = A, B$, which is of the same form as the expression one obtains for the charged field. The oscillation formulas coincide with those in Eqs.(79),(80). Similar results are valid for Majorana neutrinos [20].

4. Geometric phase for oscillating particles

Let us now see how the notion of geometric phase [27] enters the physics of mixing by considering the example of neutrino oscillations.

We consider here two flavor mixing in the Pontecorvo approximation [26], for an extension to three flavors see Ref.[28]. The flavor states are:

$$|\nu_e\rangle = \cos\theta\,|\nu_1\rangle + \sin\theta\,|\nu_2\rangle \tag{90}$$

$$|\nu_\mu\rangle = -\sin\theta\,|\nu_1\rangle + \cos\theta\,|\nu_2\rangle\,. \tag{91}$$

The electron neutrino state at time t is [3]

$$\begin{aligned} |\nu_e(t)\rangle &\equiv e^{-iHt}|\nu_e(0)\rangle \\ &= e^{-i\omega_1 t}\Big(\cos\theta\,|\nu_1\rangle + e^{-i(\omega_2-\omega_1)t}\,\sin\theta\,|\nu_2\rangle\Big), \end{aligned} \tag{92}$$

where $H|\nu_i\rangle = \omega_i|\nu_i\rangle$, $i = 1, 2$. The state $|\nu_e(t)\rangle$, apart from a phase factor, reproduces the initial state $|\nu_e(0)\rangle$ after a period $T = \frac{2\pi}{\omega_2-\omega_1}$:

$$|\nu_e(T)\rangle = e^{i\phi}|\nu_e(0)\rangle \qquad , \qquad \phi = -\frac{2\pi\omega_1}{\omega_2-\omega_1}\,. \tag{93}$$

We now show how such a time evolution does contain a purely geometric part. It is straightforward to separate the geometric and dynamical phases following the standard procedure [27]:

$$\begin{aligned} \beta_e &= \phi + \int_0^T \langle\nu_e(t)|\, i\partial_t\, |\nu_e(t)\rangle\, dt \\ &= -\frac{2\pi\omega_1}{\omega_2-\omega_1} + \frac{2\pi}{\omega_2-\omega_1}(\omega_1\,\cos^2\theta + \omega_2\,\sin^2\theta) = 2\pi\sin^2\theta\,. \end{aligned} \tag{94}$$

We thus see that there is indeed a non-zero geometrical phase β_e, related to the mixing angle θ, and that it is independent from the neutrino energies ω_i and masses m_i. In a similar fashion, we obtain the Berry phase for the muon neutrino state:

$$\beta_\mu = \phi + \int_0^T \langle\nu_\mu(t)|\, i\partial_t\, |\nu_\mu(t)\rangle\, dt = 2\pi\cos^2\theta\,. \tag{95}$$

Note that $\beta_e + \beta_\mu = 2\pi$.

Generalization to $n-$cycles is also interesting. Eq.(94) can be rewritten for the $n-$cycle case as

$$\beta_e^{(n)} = \int_0^{nT} \langle\nu_e(t)|\,(i\partial_t - \omega_1)\,|\nu_e(t)\rangle\, dt = 2\pi\, n\,\sin^2\theta\,, \tag{96}$$

Eq.(96) shows that the Berry phase acts as a "counter" of neutrino oscillations, adding up $2\pi \sin^2\theta$ to the phase of the (electron) neutrino state after each complete oscillation.

In Ref.[26], a gauge structure and a covariant derivative were introduced in connection with the above geometric structures.

The case of three flavor mixing has been analyzed in Ref.[28]. The above result also applies to other (similar) cases of particle oscillations, for example to Kaon oscillations. Finally, we note that a measurement of the above geometric phase would give a direct measurement of the mixing angle independently from the values of the masses.

5. Three flavor fermion mixing

We now consider some aspects of fermion mixing in the case of three flavors [9, 19]. This is particularly relevant because of the possibility of CP violation associated with it. Among the various possible parameterizations of the mixing matrix for three fields, we choose to work with the standard representation of the CKM matrix [1]:

$$\Psi_f(x) = \mathcal{U}\,\Psi_m(x) \tag{97}$$
$$\mathcal{U} = \begin{pmatrix} c_{12}c_{13} & s_{12}c_{13} & s_{13}e^{-i\delta} \\ -s_{12}c_{23} - c_{12}s_{23}s_{13}e^{i\delta} & c_{12}c_{23} - s_{12}s_{23}s_{13}e^{i\delta} & s_{23}c_{13} \\ s_{12}s_{23} - c_{12}c_{23}s_{13}e^{i\delta} & -c_{12}s_{23} - s_{12}c_{23}s_{13}e^{i\delta} & c_{23}c_{13} \end{pmatrix},$$

with $c_{ij} = \cos\theta_{ij}$ and $s_{ij} = \sin\theta_{ij}$, being θ_{ij} the mixing angle between ν_i, ν_j and $\Psi_m^T = (\nu_1, \nu_2, \nu_3)$, $\Psi_f^T = (\nu_e, \nu_\mu, \nu_\tau)$.

As shown in Ref.[9], the generator of the transformation (97) is:

$$\nu_\sigma^\alpha(x) \equiv G_\theta^{-1}(t)\,\nu_i^\alpha(x)\,G_\theta(t), \tag{98}$$

with $(\sigma, i) = (e,1), (\mu,2), (\tau,3)$, and

$$G_\theta(t) = G_{23}(t)G_{13}(t)G_{12}(t)\,, \tag{99}$$

where $G_{ij}(t) \equiv \exp\left[\theta_{ij}L_{ij}(t)\right]$ and

$$L_{12}(t) = \int d^3\mathbf{x}\left[\nu_1^\dagger(x)\nu_2(x) - \nu_2^\dagger(x)\nu_1(x)\right], \tag{100}$$

$$L_{23}(t) = \int d^3\mathbf{x}\left[\nu_2^\dagger(x)\nu_3(x) - \nu_3^\dagger(x)\nu_2(x)\right], \tag{101}$$

$$L_{13}(\delta, t) = \int d^3\mathbf{x}\left[\nu_1^\dagger(x)\nu_3(x)e^{-i\delta} - \nu_3^\dagger(x)\nu_1(x)e^{i\delta}\right]. \tag{102}$$

It is evident from the above form of the generators, that the phase δ is unavoidable for three field mixing, while it can be incorporated in the definition of the fields in the two flavor case.

In Ref.[19], the flavor vacuum and the flavor annihilation operators were studied for the above mixing relations. Oscillation formulas were derived exhibiting CP violation. Here we do not report on these results, rather we comment on the algebraic structure associated with the generator Eq.(99). Indeed, the generators Eqs.(100)-(102) can be obtained by acting on the triplet $\Psi_m^T = (\nu_1, \nu_2, \nu_3)$ with global phase transformations, in analogy with what has been done in §2.3.1. One then obtains the following set of charges [19]:

$$\tilde{Q}_{m,j}(t) = \int d^3\mathbf{x}\, \Psi_m^\dagger(x)\, \tilde{F}_j\, \Psi_m(x) \,, \qquad j = 1, 2, ..., 8. \qquad (103)$$

where $\tilde{F}_j \equiv \frac{1}{2}\tilde{\lambda}_j$ and the $\tilde{\lambda}_j$ are a generalization of the usual Gell-Mann matrices λ_j:

$$\tilde{\lambda}_1 = \begin{pmatrix} 0 & e^{i\delta_2} & 0 \\ e^{-i\delta_2} & 0 & 0 \\ 0 & 0 & 0 \end{pmatrix}, \qquad \tilde{\lambda}_2 = \begin{pmatrix} 0 & -ie^{i\delta_2} & 0 \\ ie^{-i\delta_2} & 0 & 0 \\ 0 & 0 & 0 \end{pmatrix}$$

$$\tilde{\lambda}_4 = \begin{pmatrix} 0 & 0 & e^{-i\delta_5} \\ 0 & 0 & 0 \\ e^{i\delta_5} & 0 & 0 \end{pmatrix}, \qquad \tilde{\lambda}_5 = \begin{pmatrix} 0 & 0 & -ie^{-i\delta_5} \\ 0 & 0 & 0 \\ ie^{i\delta_5} & 0 & 0 \end{pmatrix}$$

$$\tilde{\lambda}_6 = \begin{pmatrix} 0 & 0 & 0 \\ 0 & 0 & e^{i\delta_7} \\ 0 & e^{-i\delta_7} & 0 \end{pmatrix}, \qquad \tilde{\lambda}_7 = \begin{pmatrix} 0 & 0 & 0 \\ 0 & 0 & -ie^{i\delta_7} \\ 0 & ie^{-i\delta_7} & 0 \end{pmatrix},$$

$$\tilde{\lambda}_3 = \begin{pmatrix} 1 & 0 & 0 \\ 0 & -1 & 0 \\ 0 & 0 & 0 \end{pmatrix}, \qquad \tilde{\lambda}_8 = \frac{1}{\sqrt{3}}\begin{pmatrix} 1 & 0 & 0 \\ 0 & 1 & 0 \\ 0 & 0 & -2 \end{pmatrix}. \qquad (104)$$

These are normalized as $tr(\tilde{\lambda}_j\tilde{\lambda}_k) = 2\delta_{jk}$. Thus the matrix Eq.(97) is generated by $\tilde{Q}_{m,2}(t)$, $\tilde{Q}_{m,5}(t)$ and $\tilde{Q}_{m,7}(t)$, with $\{\delta_2, \delta_5, \delta_7\} \to \{0, \delta, 0\}$.

The interesting point is that the algebra generated by the matrices Eq.(104) *is not* $su(3)$ unless the condition $\Delta \equiv \delta_2 + \delta_5 + \delta_7 = 0$ is imposed: such a condition is however incompatible with the presence of a CP violating phase. When CP violation is allowed, then $\Delta \neq 0$ and the $su(3)$ algebra is deformed. To see this, let us introduce the raising and lowering operators, defined as [1]:

$$\tilde{T}_\pm \equiv \tilde{F}_1 \pm i\tilde{F}_2 \quad , \quad \tilde{U}_\pm \equiv \tilde{F}_6 \pm i\tilde{F}_7 \quad , \quad \tilde{V}_\pm \equiv \tilde{F}_4 \pm i\tilde{F}_5 \qquad (105)$$

We also define:

$$\tilde{T}_3 \equiv \tilde{F}_3 \quad , \quad \tilde{U}_3 \equiv \frac{1}{2}\left(\sqrt{3}\tilde{F}_8 - \tilde{F}_3\right) \quad , \quad \tilde{V}_3 \equiv \frac{1}{2}\left(\sqrt{3}\tilde{F}_8 + \tilde{F}_3\right) \qquad (106)$$

Then the deformed commutators are the following ones:

$$[\tilde{T}_+, \tilde{V}_-] = -\tilde{U}_- \, e^{2i\Delta \tilde{U}_3} \quad , \quad [\tilde{T}_+, \tilde{U}_+] = \tilde{V}_+ \, e^{-2i\Delta \tilde{V}_3} \, ,$$
$$[\tilde{U}_+, \tilde{V}_-] = \tilde{T}_- \, e^{2i\Delta \tilde{T}_3} \, , \tag{107}$$

all the others being identical to the ordinary $su(3)$ ones [1].

6. Neutrino oscillations from relativistic flavor current

A realistic description of neutrino oscillations requires to take into account the finite size of source and detector and the fact that in current experiments what is measured is the distance source-detector rather than the time of flight of (oscillating) neutrinos. Thus various approaches were developed, based on wave-packets and leading to a space-dependent oscillation formula [29]-[35].

Here we report about recent results, showing how an exact expression for QFT space-dependent oscillation formula can be found by using the above defined flavor states and relativistic flavor currents [21]. Such an approach was first proposed in Ref.[36] in the context of non-relativistic QM (see also Ref. [7]). We thus consider the flux of (electron) neutrinos through a detector surface

$$\Phi_{\nu_e \to \nu_e}(L) = \int_{t_0}^{T} dt \int_{\Omega} \langle \nu_e | J^i_e(\mathbf{x}, t) | \nu_e \rangle \; d\mathbf{S}^i \tag{108}$$

The neutrino state is described by a wave packet:

$$|\nu_e(\mathbf{x}_0, t_0)\rangle = A \int d^3\mathbf{k} \, e^{-i(\omega_{k,1} t_0 - \mathbf{k}\cdot\mathbf{x}_0)} f(\mathbf{k}) \, \alpha^{r\dagger}_{\mathbf{k},e}(t_0) \, |0(t_0)\rangle_{e,\mu} \tag{109}$$

The flavor current is: $J^\mu_e(x) = \bar{\nu}_e(x)\gamma^\mu \nu_e(x)$. In Ref.[21] it is shown that ${}_{e,\mu}\langle 0|J^\mu(\mathbf{x},t)|0\rangle_{e,\mu} = 0$ and

$$\langle \nu_e | J^\mu_e(\mathbf{x}, t) | \nu_e \rangle = \bar{\Psi}(\mathbf{x}, t) \, \Gamma^\mu \begin{pmatrix} 1 & 1 \\ 1 & 1 \end{pmatrix} \Psi(\mathbf{x}, t) \tag{110}$$

with

$$\Psi(\mathbf{x}, t) \equiv A \int \frac{d^3\mathbf{k}}{(2\pi)^{\frac{3}{2}}} \, e^{i\mathbf{k}\cdot\mathbf{x}} \, f(\mathbf{k}) \begin{pmatrix} u^r_{\mathbf{k},1} \, X_{\mathbf{k},e}(t) \\ \sum_s v^s_{-\mathbf{k},1} \, (\vec{\sigma}\cdot\mathbf{k})^{sr} \, Y_{\mathbf{k},e}(t) \end{pmatrix} \tag{111}$$

$$X_{\mathbf{k},e}(t) = \cos^2\theta e^{-i\omega_{k,1}t} + \sin^2\theta \left[e^{-i\omega_{k,2}t} |U_{\mathbf{k}}|^2 + e^{i\omega_{k,2}t} |V_{\mathbf{k}}|^2 \right]$$

$$Y_{\mathbf{k},e}(t) = \sin^2\theta |U_{\mathbf{k}}| \chi_1 \chi_2 \left[\frac{1}{\omega_{k,2} + m_2} - \frac{1}{\omega_{k,1} + m_1} \right] \left[e^{-i\omega_{k,2}t} - e^{i\omega_{k,2}t} \right]$$

where $\vec{\sigma}\cdot\mathbf{k} = \begin{pmatrix} k_3 & k_- \\ k_+ & -k_3 \end{pmatrix}$ and $\chi_i \equiv \left(\frac{\omega_{k,i} + m_i}{4\omega_{k,i}} \right)^{\frac{1}{2}}$.

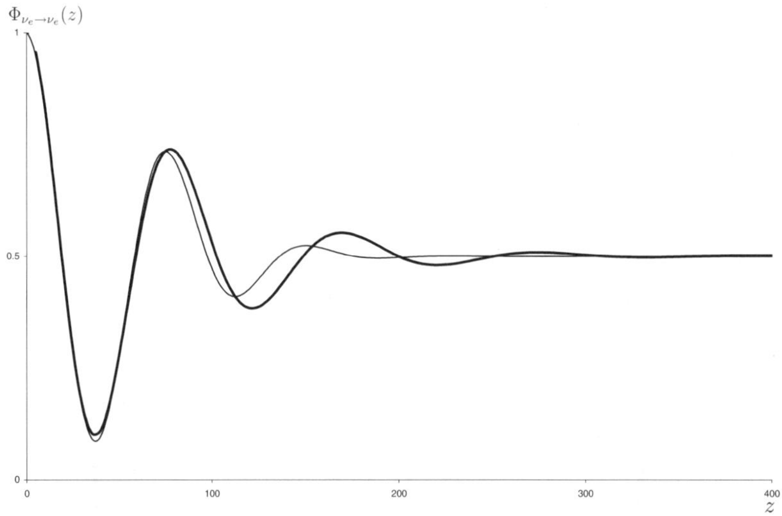

Figure 3. QFT flux (thick line) vs. standard formula (thin line) for $\theta = \pi/4$, $\sigma_k = 10$, $m_1 = 1, m_2 = 3, Q = 50$.

The expression in Eq.(110) contains the most general information about neutrino oscillations and can be explicitly evaluated once the form of the wave-packet is specified. A similar expression can be easily obtained for the other quantity of interest, namely $\langle \nu_e | J^{\mu}_{\mu}(\mathbf{x}, t) | \nu_e \rangle$.

An oscillation formula in space is then obtained in Ref.[21] in the case of spherical symmetry and by assuming a gaussian wave packet for the flavor state:

$$f_k = \frac{1}{(2\pi\sigma_k^2)^{\frac{1}{4}}} \exp\left[-\frac{(k-Q)^2}{4\sigma_k^2}\right] \tag{112}$$

Such an expression can be evaluated numerically (see Fig.(3)) and it reduces [21] to the standard formula [31, 30] in the relativistic limit:

$$\begin{aligned}\Phi_{\nu_e\to\nu_e}(z) &\simeq 1 - \frac{1}{2}\sin^2(2\theta) \\ &+ \frac{1}{2}\sin^2(2\theta)\cos\left(2\pi\frac{z}{L^{osc}}\right)\exp\left[-\left(\frac{z}{L^{coh}}\right)^2 - 2\pi^2\left(\frac{\sigma_x}{L^{osc}}\right)^2\right]\end{aligned} \tag{113}$$

with $L_{osc} = \frac{4\pi Q}{\Delta m^2}$ and $L_{coh} = \frac{L_{osc} Q}{\sqrt{2}\pi\sigma_k}$ being the usual oscillation length and coherence length [31, 30].

7. Summary

In this report we have discussed recent results in the area of field mixing and oscillations. We have shown that a consistent field theoretical treatment is possible, both for fermions and for bosons, once we realize the unitary inequivalence of the mass and flavor representations. The flavor Hilbert space is thus constructed and the flavor vacuum is shown to have the structure of a $SU(n)$ generalized coherent state, for the case of mixing among n generations[5]. We have then discussed the algebraic structure of the currents and charges associated with field mixing.

On the basis of these results, exact oscillation formulas have been calculated, exhibiting non-perturbative corrections with respect to the usual QM ones. The usual formulas are shown to be approximately valid in the relativistic region. Exact oscillation formulas in space can also be derived by use of the relativistic flavor currents.

We have also shown that a geometric phase is associated to flavor oscillations and discussed the role of the CP violating phase in connection with the algebra of currents associated to three flavor mixing.

For lack of space, we have omitted other interesting development, in particular we would like to mention the analysis, in the above framework, of the neutrino oscillations in matter (MSW effect) [37]. An interesting new line of research is the investigation of the issue of Lorentz invariance for the flavor states [38]: deformed dispersion relations for neutrino flavor states may be indeed incorporated into frameworks encoding the breakdown of Lorentz invariance [39].

Acknowledgments

We acknowledge the ESF Program COSLAB, EPSRC, INFN and INFM for partial financial support.

References

[1] T. Cheng and L. Li, *Gauge Theory of Elementary Particle Physics*, Clarendon Press, Oxford, 1989.

[2] J. Davis, D.S. Harmer and K.C. Hoffmann, *Phys. Rev. Lett.* **20** (1968) 1205. M. Koshiba, in "Erice 1998, From the Planck length to the Hubble radius", 170; S. Fukuda et al. [Super-Kamiokande collaboration], *Phys. Rev. Lett.* **86** (2001) 5656. Q. R. Ahmad et al. [SNO collaboration] *Phys. Rev. Lett.* **87** (2001) 071301; *Phys. Rev. Lett.* **89** (2002) 011301. K. Eguchi *et al.* [KamLAND Collaboration], *Phys. Rev. Lett.* **90** (2003) 021802 M.H. Ahn *et al.* [K2K Collaboration], *Phys. Rev. Lett.* **90** (2003) 041801

[3] B. Pontecorvo, *Zh. Eksp. Theor. Fiz.* **33** (1958) 549; *JEPT* **6** (1958) 429; Z. Maki, M. Nakagawa and S. Sakata, *Prog. Theor. Phys.* **28** (1962) 870; V. Gribov and B. Pontecorvo, *Phys. Lett. B* **28** (1969) 493; S.M. Bilenky and B. Pontecorvo, *Phys. Rep.* **41** (1978) 225.

[5] When no CP violating phases are present - see §5.

[4] R. Mohapatra and P. Pal, *Massive Neutrinos in Physics and Astrophysics*, (World Scientific, Singapore, 1991); J.N. Bahcall, *Neutrino Astrophysics*, (Cambridge Univ. Press, Cambridge, 1989).

[5] V. Bargmann, *Annals Math.* **59** (1954) 1; A. Galindo and P. Pascual, *Quantum Mechanics*, (Springer Verlag, 1990).

[6] D.M. Greenberger, *Phys. Rev. Lett.* **87** (2001) 100405.

[7] M. Zralek, *Acta Phys. Polon. B* **29** (1998) 3925.

[8] C. Giunti, C.W. Kim and U.W. Lee, *Phys. Rev. D* **45** (1992) 2414; C.W. Kim and A. Pevsner, *Neutrinos in Physics and Astrophysics*, Harwood Academic Press, Chur, Switzerland,1993.

[9] M. Blasone and G. Vitiello, *Annals Phys.* **244** (1995) 283 [Erratum-ibid. **249** (1995) 363].

[10] M. Blasone, P.A. Henning and G. Vitiello, in M. Greco Ed."La Thuile 1996, Results and perspectives in particle physics", INFN Frascati 1996, p.139-152 [hep-ph/9605335].

[11] M. Blasone, P.A. Henning and G. Vitiello, *Phys. Lett. B* **451** (1999) 140; M. Blasone, in A.Zichichi Ed. "Erice 1998, From the Planck length to the Hubble radius" (World Scientific) p.584, [hep-ph/9810329].

[12] M. Binger and C.R. Ji, Phys. Rev. D **60** (1999) 056005. C.R. Ji and Y. Mishchenko, *Phys. Rev. D* **64** (2001) 076004; *Phys. Rev. D* **65** (2002) 096015.

[13] K. Fujii, C. Habe and T. Yabuki, *Phys. Rev. D* **59** (1999) 113003 [Erratum-ibid. D **60** (1999) 099903]; *Phys. Rev. D* **64** (2001) 013011.

[14] K.C. Hannabuss and D.C. Latimer, *J. Phys. A* **36** (2003) L69; *J. Phys. A* **33** (2000) 1369.

[15] M. Blasone and G. Vitiello, *Phys. Rev. D* **60** (1999) 111302.

[16] M. Blasone, P. Jizba and G. Vitiello, *Phys. Lett. B* **517** (2001) 471.

[17] M. Blasone, A. Capolupo and G. Vitiello, in Yue-Liang Wu, editor, *Flavor Physics*, 425-433. World Scientific, Singapore 2002. [hep-th/0107125];

[18] M. Blasone, A. Capolupo, O. Romei and G. Vitiello, *Phys. Rev. D* **63** (2001) 125015.

[19] M. Blasone, A. Capolupo and G. Vitiello, *Phys. Rev. D* **66** (2002) 025033;

[20] M. Blasone and J.S. Palmer, [hep-ph/0305257]

[21] M. Blasone, P.P. Pacheco and H.W. Tseung, *Phys. Rev. D* **67** (2003) 073011.

[22] M. Blasone, P. Jizba and G. Vitiello, [hep-ph/0308009].

[23] C. Itzykson and J.B. Zuber, *Quantum Field Theory*, (McGraw-Hill, New York, 1980);

[24] H. Umezawa, *Advanced Field Theory: Micro, Macro and Thermal Physics* (American Institute of Physics, 1993)

[25] A. Perelomov, *Generalized Coherent States and Their Applications*, (Springer–Verlag, Berlin, 1986).

[26] M. Blasone, P. A. Henning and G. Vitiello, *Phys. Lett. B* **466** (1999) 262;

[27] Y. Aharonov and J. Anandan *Phys. Rev. Lett.* **58** (1987) 1593; *Phys. Rev. Lett.* **65** (1990) 1697.

[28] X.B. Wang, L.C. Kwek, Y. Liu and C.H. Oh, *Phys. Rev. D* **63** (2001) 053003.

[29] B. Kayser, *Phys. Rev. D* **24** (1981) 110; B. Kayser, F. Gibrat-Debut and D. Perrier, *The Physics of massive neutrinos,* World Scientific, 1989.

[30] M. Beuthe, *Phys. Rev. D* **66** (2002) 013003; *Phys. Rep.* **375** (2003) 105.

[31] C. Giunti, *JHEP* **0211** (2002) 017. C. Giunti and C.W. Kim, *Phys. Rev. D* **58** (1998) 017301.

[32] W. Grimus and P. Stockinger, *Phys. Rev. D* **54** (1996) 3414; W. Grimus, P. Stockinger and S. Mohanty, Phys. Rev. D **59** (1999) 013011.

[33] C.Y. Cardall, *Phys. Rev. D* **61** (2000) 073006; C.Y. Cardall and D.J.H. Chung, *Phys. Rev. D* **60** (1999) 073012.

[34] K. Kiers and N. Weiss, *Phys. Rev. D* **57** (1998) 3091.

[35] T. Yabuki and K. Ishikawa, *Prog. Theor. Phys.* **108** (2002), 347.

[36] B. Ancochea, A. Bramon, R. Munoz-Tapia and M. Nowakowski, *Phys. Lett. B* **389** (1996) 149.

[37] K. Fujii, C. Habe and M. Blasone, [hep-ph/0212076].

[38] M. Blasone, J. Magueijo and P. Pires-Pacheco, [hep-ph/0307205].

[39] J. Magueijo and L. Smolin, *Phys. Rev. D* **67** (2003) 044017; *Phys. Rev. Lett.* **88** (2002) 190403.

TWO-PHOTON INTERACTIONS IN CAVITY QED

S.K. Bose
Department of Physics
University of Central Florida
Orlando, FL 32816, USA

M. Alexanian
Department of Physics and Physical Oceanography
University of North Carolina at Wilmington
Wilmington, NC 28403, USA

Abstract Non-linear Jaynes-Cummings models, in particular two-photon interactions between one or two modes of the electromagnetic field and few-level atoms or molecules, are of great interest to the understanding of quantum electrodynamics, with many applications in spectroscopy and problems of biology. Experimentally, theories of quantum electrodynamics can be tested by allowing single atoms of a beam to be present in a high-Q cavity, and allowing interaction with radiation in the cavity. We present here a formalism by deriving a two-photon interaction Hamiltonian exactly without approximations and apply the formalism to study several two-photon processes in high-Q cavities.

1. Introduction

Cavity quantum electrodynamics (QED) is the study of the interaction of single atoms with one or two modes of the radiation field in a high Q cavity. In a typical experiment velocity selected Rydberg atoms are made to cross the cavity such that only one atom is in the cavity at a time. The atom is capable of interacting with the field before exiting the cavity. Experiments in high Q cavities allow test of fundamental atom-field theoretical models. One such model is the Jaynes-Cummings model (JCM) [1], dealing with the interaction of two-level atoms with a single-mode quantized electromagnetic field. This model has served as the fundamental tool to theoretically study single atom-photon interactions. One of the most interesting features of the JCM is the "collapse" and "revival" phenomena of the atomic population inversion [2]. In-

B. Gruber (ed.), Symmetries in Science XI, 129–144.

teresting developments are the single-photon and two-photon micromasers [3]. A unique feature of the micromaser is the generation of nonclassical radiation in the steady state.

There have been many efforts to consider nonlinear versions of the JCM, for instance a two-photon model [4], useful to study two-photon process of single atoms. Two-photon interactions between atoms and radiation give rise to interesting nonlinear effects such as the two-photon absorption and the two-photon micromaser. It has recently been shown that spatially confined photobleaching can be achieved by using two-photon excitation, the confinement being much sharper than that produced by one-photon excitation [5]. Two-photon excitation is also a valuable tool having great potential in imaging biological systems using laser-scanning microscopy [6].

In this paper, we present a brief review of two-photon cavity QED work done by us recently. The paper is organized as follows: in section 2 we present a discussion of various approaches to obtain two-photon interaction Hamiltonians of the JC type and in particular discuss a method of obtaining the Hamiltonian using a unitary transformation. In section 3 we discuss two-photon absorption and obtain a master equations for the photon probability. We also discuss coherence that follows from two-photon absorption. In section 4 we discuss trapping states in a two-photon micromaser and discuss the possibility of obtaining a photon number state. In section 5, we discuss possible macroscopic quantum superpositions and in section 6 we present some results using a Raman interaction model and discuss cloning of coherent states.

2. Two-Photon Hamiltonians

There are several methods to obtain two-photon Hamiltonians:

a adiabatic elimination of a level in a three-level atom [7],

b effective phenomenological Hamiltonians [8], and

c the solution of the quantum mechanical three-level atom problem [9].

All of these methods rely on the assumption of large detuning of the intermediate level of a three-level system in order to describe interactions involving two photons simultaneously. We proposed an unitary transformation [10], later generalized by Wu [11] to obtain exact two-photon interaction Hamiltonians, valid for all detunings including zero as well as high detunings. The procedure also allows the inclusion of dynamical Stark shifts in the formalism.

The interaction of a three-level atom and one mode of the radiation field is

$$H = \hbar\omega\, a^\dagger a + \sum_{i=1}^{3} E_i\sigma_{ii} + \hbar g_1\left(\sigma_{21}a + \sigma_{12}a^\dagger\right) + \hbar g_2\left(\sigma_{32}a + \sigma_{23}a^\dagger\right), \quad (1)$$

where $\sigma_{ij} = | i\rangle\langle j |$ and i, j refer to the atomic levels; g_1 and g_2 being the atom-photon coupling constants. $a^\dagger$ and a are the photon creation and annihilation operators. The atomic levels 1 and 3 have the

same parity, while level 2 has opposite parity. The unitary transformation, $U = \exp(S)$, where,

$$S = \alpha\,(\sigma_{21}a - \sigma_{12}a^\dagger) + \beta\,(\sigma_{32}a - \sigma_{23}a^\dagger) \tag{2}$$

with $\beta = -\frac{g_2}{g_1}\,\alpha$, leads to an effective Hamiltonian

$$H_{eff} = \exp(S)\,H\,\exp(-S)\,,$$

where,

$$H_{eff} = \hbar\omega\,N + E_0 + \hbar\mu\,\sigma_{33} + \hbar\eta\,\sigma_{11} + \hbar\lambda\,(\sigma_{31}\,a^2 + \sigma_{13}\,a^{\dagger 2})\,. \tag{3}$$

We note that level 2 is effectively eliminated and the transformed Hamiltonian is explicitly two-photon. Expressions for $N, E_0, \mu, \eta, \lambda$ can be found in reference [12]. The effective Hamiltonian can be diagonalized. The eigenvalues $E_n^\pm$ are given by

$$E_n^+ = \hbar\omega\,(n+1) + \tfrac{E_1+E_3}{2} - \tfrac{\Delta}{2} + \tfrac{1}{2}\sqrt{\Delta^2 + 4\hbar^2\left[g_1^2(n+2) + g_2^2(n+1)\right]}$$

$$E_n^- = \hbar\omega\,(n+1) + \tfrac{E_1+E_3}{2}\,. \tag{4}$$

The corresponding eigenfunctions are given by,

$$\begin{aligned} | \Psi_n^+\rangle &= \sin\theta_n\,|\,3,n\rangle + \cos\theta_n\,|\,1,n+2\rangle\,, \\ | \Psi_n^-\rangle &= \cos\theta_n\,|\,3,n\rangle - \sin\theta_n\,|\,1,n+2\rangle\,, \end{aligned} \tag{5}$$

where

$$\begin{aligned} \cos\theta_n &= \tfrac{r(n+2)^{1/2}}{[n(r^2+1)+2r^2+1]^{1/2}}\,, \\ \sin\theta_n &= \tfrac{(n+1)^{1/2}}{[n(r^2+1))+2r^2+1]^{1/2}}\,. \end{aligned} \tag{6}$$

Here $r = \frac{g_1}{g_2}$ and n is the photon number. $|\,\Psi_n^\pm\rangle$ represent dressed states of the atom-field system, expressed in terms of atom-photon states $|\,j,n\rangle = |\,j\rangle\otimes|\,n\rangle$. The dressed-state basis is convenient for our purpose.

3. Two-photon absorption

As an application of the Hamiltonian of eq. (3), consider the absorption of two photons of single frequency by an atom. A possible experimental arrangement is one in which a beam of monoenergetic atoms, obtained by a velocity selector, traverses a high-Q cavity containing a single-mode electromagnetic field, such that a single atom crosses the cavity at a time. Each atom interacts with the field and exits the cavity. Cavity damping and spontaneous decay processes are assumed to be negligible.

The absorption process is described by an equation for the reduced density operator of the photon field, obtained after tracing out the atomic variables. Analytic solutions are obtained from the equation describing the diagonal elements [13,14]. An analysis using the two-photon effective Hamiltonian was carried out and the resulting equation was solved in the high as well as zero detuning limits [15].

Suppose that initially atoms are prepared in the ground state and the field in the cavity is a superposition of photon number states. If the atom exits the cavity in the excited state, two photons would be absorbed. Following standard procedure [16], the following master equation is derived:

$$\begin{aligned}\frac{d\rho}{dt} &= -\frac{i}{\hbar}\left[\cos^2\theta_{n-2}E^+_{n-2} + \sin^2\theta_{n-2}E_{n-2}, \rho(t)\right] \\ &\quad -\frac{R\tau^2}{2\hbar^2}\Big[\sin^2\theta_{n-2}\cos^2\theta_{n-2}\left(E^+_{n-2} - E^-_{n-2}\right)^2\rho(t) \\ &\quad +\rho(t)\sin^2\theta_{n-2}\cos^2\theta_{n-2}\left(E^+_{n-2} - E^-_{n-2}\right)^2 \\ &\quad -\frac{2\cos\theta_n\sin\theta_n}{\sqrt{(n+1)(n+2)}}\left(E^+_n - E^-_n\right)a^2\rho(t)a^{\dagger 2} \\ &\quad \frac{\cos\theta_n\sin\theta_n}{\sqrt{(n+1)(n+2)}}\left(E^+_n - E^-_n\right)\Big]\,. \end{aligned} \tag{7}$$

In the above equation R is the average rate of injection of atoms in the cavity, $n = a^\dagger a$ and τ is the interaction time of each atom. Let us consider high and low detuning limits of the above equation.

3.1 High detuning limit master equation

In the large detuning limit the master equation takes the form,

$$\begin{aligned}\frac{d\rho}{dt} &= -\frac{i}{\hbar}\left[a^\dagger a(\hbar\omega + \frac{\hbar^2 g_1^2}{\Delta}), \rho(t)\right] \\ &\quad -K_L\left\{a^{\dagger 2}a^2\rho(t) + \rho(t)a^{\dagger 2}a^2 - 2a^2\rho(t)a^{\dagger 2}\right\}\end{aligned} \tag{8}$$

where $K_L = \left(R\tau^2 g_1^2 g_2^2 \hbar^2\right)/2\Delta^2$. The high-detuning limit master equation, except for a Stark shift term $\hbar^2\frac{g_1^2}{\Delta}a^\dagger a$ term, is the same as that obtained by Guerra et al. [16]. It may be noted that the two-photon coupling constant K_L is given in terms of known parameters and is not arbitrary.

3.2 Zero-Detuning limit master equation

Consider now the case of zero detuning. Obviously, theories based upon the high detuning limit is not applicable for this case. In our formalism genuine two-photon absorption is possible even for zero detuning. In this limit, the master equation for the photon probability, $P_n = \langle n \mid \rho \mid n \rangle$, becomes

$$\frac{dP_n}{dt} = -\frac{R\tau^2 g_1^2 n(n-1)}{n(r^2+1)-1} P_n + \frac{R\tau^2 g_1^2 (n+1)(n+2)}{n(r^2+1)+2r^2+1} P_{n+2} \,. \tag{9}$$

Defining $x = R\tau^2 g_1^2 t$ we have,

$$\frac{d\langle n \rangle}{dx} = -2\langle n(n+1)\rangle - \frac{(r^2+1)}{2r^2}\frac{d\langle n^2 \rangle}{dx} \tag{10}$$

One interesting feature of the rate equation is that absorption of photons is highly dependent on the statistics of the field, as can be seen from the result $\langle n(n-1)\rangle = \langle n \rangle^2 g^{(2)}(t)$, where $g^{(2)}(t)$ is the second-order correlation function. A particularly simple solution of the master equation can be obtained in the mean field approximation, i.e., $\langle n^2 \rangle = \langle n \rangle^2$, $\frac{|\langle n(t)\rangle - 1|^{1+2\gamma}}{\langle n(t) \rangle} = K \exp(-2x)$. Here, K is the initial value $\frac{|\langle n(0)\rangle - 1|^{1+2\gamma}}{\langle n(0) \rangle}$, and $\gamma = (1+r^2)/2r^2$. One feature of the master equation given by eq. (10) is the second term which has the effect of reducing absorption, since $\frac{d}{dx}\langle n^2 \rangle < 0$.

3.3 Coherence in two-photon absorption

Two photon absorption is capable of producing non-classical light, while one-photon absorption is known to destroy quantum features. As the two-photon absorbers cross the cavity, the residual radiation in the cavity will be found in either a vacuum state $\mid 0\rangle$ or in a one-photon state $\mid 1\rangle$. It has been known for sometime that the final state of the radiation is a coherent superposition [17], i.e.,

$$\mid \Psi \rangle = \alpha \mid 0 \rangle + \beta \mid 1 \rangle \,. \tag{11}$$

Coherence is demonstrated by non-zero off-diagonal matrix elements of the photon density operator, $\rho = \mid \Psi \rangle \langle \Psi \mid$. It is convenient to transform the density operator to the form

$$\rho_I(t) = e^{+iM} \rho(t)\, e^{-iM} \,,$$

where

$$M = \frac{t}{\hbar}\left(\cos^2\theta_{n-2}\, E_{n-2}^{+} + \sin^2\theta_{n-2}\, E_{n-2}^{-}\right) \,.$$

Taking off-diagonal matrix elements of eq. (7), we obtain,

$$\frac{1}{R}\frac{d}{dt}\langle n \mid \rho_I(t) \mid n+\mu \rangle = A_n \langle n \mid \rho_I(t) \mid n+\mu \rangle + B_n \langle n+2 \mid \rho_I(t) \mid n+\mu+2 \rangle \,, \tag{12}$$

where the coefficients are given by

$$\begin{aligned} A_n &= -\frac{\tau^2}{2\hbar^2}\left\{Y_n^2 + Y_{n+\mu}^2\right\} , \\ B_n &= \frac{\tau^2}{\hbar^2} Y_{n+2} Y_{n+\mu+2} , \\ Y_n &= \sin\theta_{n-2}\cos\theta_{n-2}(E_{n-2}^+ - E_{n-2}^-) . \end{aligned}$$

From eq. (12) it follows that $\sum_{n=0}^{\infty}\langle 2n \mid \rho_I(t) \mid 2n\rangle$ and, $\sum_{n=0}^{\infty}\langle 2n+1 \mid \rho_I(t) \mid 2n+1\rangle$ are constants of the motion.We now demonstrate that a linear combination $\sum_{n=0}^{\infty} D_{2n}\langle 2n \mid \rho_I(t) \mid 2n+1\rangle$ is also a constant of the motion, where the constants D_{2n} must satisfy the condition

$$D_{2n+2}/D_{2n} = -B_{2n}/A_{2n+2} . \tag{13}$$

To demonstrate this, multiply eq. (12) by D_{2n}, sum over both sides over n, and choose $\mu = 1$. We obtain,

$$\frac{1}{R}\sum_{n=0}^{\infty} D_{2n}\frac{d}{dt} < 2n \mid \rho(t) \mid 2n+1 >$$
$$= \sum_{n=0}^{\infty} D_{2n}\left[A_{2n}\langle 2n \mid \rho(t) \mid 2n+1\rangle + B_{2n}\langle 2n+2 \mid \rho \mid 2n+3\rangle\right] \tag{14}$$

In view of the fact that $A_0 = 0$, the right side of eq. (14) can be written as

$$\sum_{n=0}^{\infty}(D_{2n+2}\, A_{2n+2} + D_{2n}\, B_{2n})\,\langle 2n+2 \mid \rho(t) \mid 2n+3\rangle .$$

If the coefficients satisfy condition (13), $\sum_{n=0}^{\infty} D_{2n}\,\langle 2n \mid \rho(t) \mid 2n+1\rangle$ is a constant of the motion, i.e.,

$$\sum_{n=0}^{\infty} D_{2n}\,\langle 2n \mid \rho(t) \mid 2n+1\rangle = \sum_{n=0}^{\infty} D_{2n} < \langle 2n \mid \rho(0) \mid 2n+1\rangle . \tag{15}$$

It can be seen that in the steady state $\langle 2n|\rho_I(t)|2n+1\rangle \to 0$ as $t \to \infty$, for $n = 1, 2, 3, \cdots$. Hence,

$$\langle 0 \mid \rho(\infty) \mid 1\rangle = \sum_{n=0}^{\infty}\frac{D_{2n}}{D_0}\langle 2n \mid \rho(0) \mid 2n+1\rangle \tag{16}$$

Eq. (16) gives the residual coherence in terms of the initial field density operator. D_{2n}/D_0 can be calculated from eq. (13). The result is

$$\frac{D_{2n}}{D_0} = \frac{\Gamma(n+\frac{1}{2})\Gamma(\beta_+)\Gamma(\beta_-)}{\Gamma(n+\beta_+)\Gamma n+\beta_-)}\sqrt{\frac{(2n+1)\Gamma(2n+2\alpha)}{\pi\Gamma(2\alpha)2^{2n}}} , \tag{17}$$

where

$$\alpha = \frac{2r^2+1}{2(r^2+1)} ,$$

and,

$$\beta_{\pm} = 1 + \frac{(r^2-1)}{8(r^2+1)} \pm \frac{1}{8}\sqrt{8 + \frac{(r^2-1)^2}{(r^2+1)^2}} .$$

The residual coherence obtained as output of two-photon absorption, is quite different from that predicted in ref. [17]. A numerical evaluation using a initial coherent field, and large average photon number, gives a higher value of the asymptotic coherence than that obtained in ref. [17]

Recently, Ezaki et al. [18] proposed that a phase state of the form $\mid \psi_{output}\rangle = \frac{1}{\sqrt{2}}(\mid 0\rangle + e^{i\varphi} \mid 1\rangle$ can be obtained, from an initial coherent state $\mid \alpha\rangle, \alpha = \mid \alpha \mid e^{i\varphi}$, by a pure two-photon absorption process. Such a state could be important for quantum information processing. It has been shown, however, that although there is a residual coherence, the output state in strictly not a phase state [19].T he steady-state density operator used in ref.[18] does not satisfy the condition of a pure state $\text{tr}\rho^2 = \text{tr}\rho = 1$. Numerically, for $\mid \alpha \mid \gg 1$, $\text{tr}\rho^2(\infty) \simeq 0.818$. A calculation using the method proposed by the present authors yields [19], for $r = 1$ and $\mid \alpha \mid \gg 1$, the result $\text{tr}\rho^2(\infty) \simeq 0.940$. While coherence is enhanced, it still does not lead to a pure state.

4. Two photon micromaser

A micromaser provides experimental realizations of cavity QED, and allows tests of fundamental concepts of atom-field interactions. There is considerable interest in determining the statistical properties of the micromaser radiation. The radiation , under suitable conditions, demonstrates trapping, a state in which the atom cannot add any more photons, i.e., the radiation attains a maximum number of photons. Another interesting question is whether a photon number state can be realized within the cavity.

Suppose that atoms, prepared in the excited state, enter the cavity; the atom-cavity radiation interactions being the effective two-photon Hamiltonian given by eq. (3). A recursion relation for the steady-state photon probability can be derived [12] following the general method proposed by Filipowicz et al. [20]:

$$\begin{aligned} P_n = & \left[A + B\frac{(n+1)}{(2n+1)^2} \sin^2 D(\sqrt{C^2+2n+1} - C) \right] P_{n-1} \\ & + B\frac{(n-1)}{(2n-1)^2} \sin^2 D(\sqrt{C^2+2n-1} - C)\, P_{n-2} \,. \end{aligned} \tag{18}$$

The parameters are given by:

$$A = \frac{n_b}{(n_b+1)}\,, \quad B = \frac{4R}{\gamma(n_b+1)}\,, \quad C = \frac{\Delta}{2\hbar g}\,, \quad D = \frac{\tau g}{2}\,.$$

Here n_b is the average number of thermal photons and we assumed $g_1 = g_2 = g$. The cavity damping rate $\gamma = \frac{\omega}{2\pi Q}$, where Q is the quality factor of the cavity. Trapping states occur when, for a suitable interaction time τ, the photon probability P_n is truncated at some value of n. The excited atoms cannot add any more photons under the circumstance. From eq. (18) it follows that $n = 0$ is a trapping state, the vacuum trapping state, if

$$\left(\sqrt{C+3} - C\right) D = j\pi\,, \tag{19}$$

where j is an integer. Then $P_n = 0$ for $n \geq 1$. For zero detuning, the trapping condition becomes $\tau g = \frac{2\pi j}{\sqrt{3}}$. Trapping states can also be realized in the non-vacuum ($n \neq 0$) states provided the following two conditions are satisfied:

$$\left(\sqrt{C^2 + 2k - 1} - C\right) D = j\pi\,,$$

and

$$\left(\sqrt{C^2 + 2k + 1} - C\right) D = l\pi\,.$$

These can be written as:

$$C = \frac{l^2(2k-1) - j^2(2k+1)}{\sqrt{4lj(l-j)[j(2k+1) - l(2k-1)]}}\,, \quad D = \pi\sqrt{\frac{jl(l-j)}{j(2k+1) - l(2k-1)}}\,. \tag{20}$$

Since $D > 0$ we must satisfy $j(2k+1) > \ell(2k-1)$ and $\ell - 1 \geq j \geq k \geq 3$. For arbitrary C and D, it is difficult to find integer ℓ, j, k that would satisfy the above conditions. Consider a special case, $j = k$, $\ell = k+1$. In this case we have

$$C = \frac{(2k^2-1)}{2\sqrt{k(k+1)}}$$

and

$$D = \pi\sqrt{k(k+1)}\,.$$

With these expressions for C and D, and using the recursion relation (18), it is possible to numerically evaluate $\langle n \rangle$ as a function of k; trapped states will be indicated by dips in $\langle n \rangle$ as function of k. Corresponding values of the variance show both super-Poissonian and sub-Poissonian nature of the radiation field [12].

4.1 Photon Number States

A number state (Fock state), a state of a definite number of photons, is a highly desirable object since such a state would be ideal to carry information in quantum information processing, apart from being an interesting example of non-classical radiation. Several proposals have been made for the generation of photon number states utilizing atom-photon interactions in high Q microcavities. Among various schemes are state reduction [21,22], quantum non-demolition [23], and the combination of non-selective and conditional measurement schemes [24]. Generation of number states by utilizing two-photon interactions have also been suggested [25,26].

We now investigate the conditions under which a number state can be obtained in our model. Non-vacuum trapping can lead to the generation of a number state in the limit $B \to \infty$, which corresponds to $\gamma \to 0$ for given R. Consider the case of trapping with $n = k$ and $n_b = 0$. If trapping is to happen, we must have $P_k = P_{k+1} = P_{k+2} = \cdots = 0$, with non-zero $P_{k-1}, P_{k-2}, \cdots$. Using the recursion relation (18) it is straightforward to express P_s, $(s = 0, 1, 2, \cdots, k-2)$ in terms of P_0. Furthermore, in the $B \to \infty$ limit we have,

$$P_s = P_0 \prod_{n=1}^{s} B \frac{(n+1)}{(2n+1)^2} \sin^2 D \left(\sqrt{C^2 + 2n + 1} - C \right) \tag{21}$$

Using the normalization, $P_0 + P_1 + P_2 + \cdots + P_{k-2} = 1$, and the result P_s is $\mathcal{O}(B^{2-k+s})$, $s = 0, 1, 2, \cdots, k-3$, we have, in the limit of large B, only two nonzero probabilities, P_{k-1} and P_{k-2}, given by $P_{k-1} \to \frac{k-2}{2k-3}$, $P_{k-2} \to \frac{k-1}{2k-3}$.

Thus $B \to \infty$ limit reduces the number of available probabilities to two, a result that can be shown numerically [12]. In terms of these probabilities, the average number of photon $\langle n \rangle$ and the dispersion σ^2 are given by $\langle n \rangle = \frac{2(k-1)(k-2)}{2k-3}$, $\sigma^2 = \frac{1}{2(2k-3)}$.

Further, if k is also large, $\langle n \rangle \sim k$ and $\sigma \sim \frac{1}{2k^{1/2}}$ and the probabilities, P_{k-1} and P_{k-2} become the same, indicating the approach to a photon number state.

5. Macroscopic Quantum Superpositions

One of the most interesting results of cavity QED is the generation of macroscopic quantum superpositions, Schrödinger cat states. For example, the so-called "even" and "odd" states have been shown to arise in two-photon micromasers, in the high detuning limit, when trapping conditions are fulfilled [27]. Atom-field interactions also produce entanglement. The important question is whether under certain circumstances disentanglement can occur leading to was disentangled [28]. On the contrary, it has been shown that an initially disentangled state, with interactions described by the JCM , lead to entanglement followed by disentanglement, the latter occurring at precisely half the "revival" time, in the collapse-revival regime of the micromaser [29]. Our objective here is to reexamine the conditions under which "pure states" of the field can be generated in the interacting atom-field system.

We consider two-photon cavity QED in which a beam of atoms, each in a linear superposition of of its ground state $| 1 >$, and its excited state $| 3 >$, enters a high-Q cavity having a initial radiation field, a superposition of photon number states $| n >$, and interact through the Hamiltonian given by eq. (3). Each atom spends a fixed time, τ, inside the cavity. Since the cavity is assumed to be lossless, the time development of the cavity field will only take place when the atoms are in the cavity. The time evolution of the atom-field system is described by the evolution operator $U(t)$, given by $U(t) = \exp[i(H - H_0)t/\hbar]$, where $H_0 = \hbar\omega(a^\dagger a + \sigma_{33} - \sigma_{11}) + (E_1 + E_3)/2$.

The set of eigenstates $| \Psi_n^\pm\rangle$, with $n = 0, 1, 2 \cdots$, together with the states $| 1, 0\rangle = - | \Psi_{-2}^-\rangle$ and $| 1, 1\rangle =| \Psi_{-1}^+\rangle$ form a complete basis. Let at $t = 0$, as the first atom enters the cavity, the atom-photon state being given by

$$| \Psi(0)\rangle = \sum_{n=0}^{\infty} S_n | n\rangle \otimes \{\alpha | 1\rangle + \beta | 3\rangle\} \ . \tag{22}$$

Expressed in terms of the dressed states,

$$| \Psi\rangle = A | 1, 0\rangle + B | 1, 1\rangle + \sum_{n=0}^{\infty}\{A_n | \Psi_n^+\rangle + B_n | \Psi_n^-\rangle\} \ . \tag{23}$$

The coefficients in (23) are given by

$$\begin{aligned} A &= \alpha S_0 \ , \\ B &= \alpha S_1 \ , \\ A_n &= \beta \sin\theta_n S_n + \alpha \cos\theta_n S_{n+2} \ , \\ B_n &= \beta \cos\theta_n S_n - \alpha \sin\theta_n S_{n+2} \ . \end{aligned} \tag{24}$$

The time development of the state is governed by the $U(t)$. Hence, the state of the system at time t is,

$$\begin{aligned} | \Psi(t)\rangle \quad = \quad & A | 1, 0\rangle + B \exp(-i\omega_{-1}t) \ | 1, 1\rangle \\ & \sum_{n=0}^{\infty} A_n \exp(-\omega_n t) \ | \Psi_n^+\rangle + \sum_{n=0}^{\infty} B_n \ | \Psi_n^-\rangle \ . \end{aligned} \tag{25}$$

At $t = \tau$, as the atom exits the cavity and the atomic state is not measured, it is reasonable to assume that the state of the atom and that of the field in the cavity becomes disentangled, i.e.,

$$| \Psi(\tau)\rangle = \sum_{n=0}^{\infty} S'_n \mid n\rangle[\alpha' \mid 1 > +\beta' \mid 3\rangle] \,. \tag{26}$$

S'_n depend on τ. Using the dressed states in (26) and comparing with eq. (25), we have,

$$\begin{aligned} A &= \alpha' S'_0 \,, \\ B \exp(-i\omega_{-1}\tau) &= \alpha' S'_1 \,, \\ A_n \exp(-\omega_n \tau) &= \alpha' S'_{n+2} \cos\theta_n + \beta' S'_n \sin\theta_n \,, \\ B_n &= \beta' S'_n \cos\theta_n - \alpha' \sin\theta_n S'_{n+2} \,. \end{aligned} \tag{27}$$

Using eq. (24), we have,

$$\begin{aligned} \alpha S_0 &= \alpha' S'_o, \\ \alpha S_1 \exp(-i\omega_{-1}\tau) &= \alpha' S'_1, \\ [\alpha \cos\theta_n S_{n+2} + \beta \sin\theta_n S_n] \exp(-i\omega_n \tau) &= \alpha' S'_{n+2} \cos\theta_n + \beta' S'_n \sin\theta_n, \\ \beta \cos\theta_n S_n - \alpha \sin\theta_n S_{n+2} &= \beta' S'_n \cos\theta_n - \alpha' S'_{n+2} \sin\theta_n. \end{aligned} \tag{28}$$

From (28), we have,

$$S'_n = \frac{\beta}{\beta'}[\sin^2\theta_n \, e^{-i\omega_n \tau} + \cos^2\theta_n] \, S_n + \frac{\alpha}{\beta'} \sin\theta_n \cos\theta_n \, [\, e^{-i\omega_n \tau} - 1] \, S_{n+2}. \tag{29}$$

A similar relation will exist for the state at any time t_i and that at time $t_i + \tau$.

We now investigate how a steady-state of the radiation in the cavity can be achieved [27]. At $t = \tau$, the first atom just exits the cavity.The field inside the cavity at that instant is given by,

$$\rho_1^F(\tau) \; = \mathrm{tr}_A[U(\tau) \, \rho^F(0) \, \rho^A \, U^\dagger(\tau)] \,. \tag{30}$$

Here, $\rho^A = | \Psi_A\rangle\langle\Psi_A |$, where $| \Psi_A\rangle = \alpha \mid 1\rangle + \beta \mid 3\rangle$ and the trace over the atomic states indicates that the state of the exiting atom is not measured [28]. As more and more atoms cross the cavity, the field in the cavity is determined by iteration. After $k-1$ atoms cross, the field in the cavity is given by,

$$\rho_k^F = \mathrm{tr}_A[U(\tau) \, \rho_{k-1}^F \, \rho^A \, U^\dagger(\tau)] \,. \tag{31}$$

If the iterations tend to a limit, the field in the cavity will have a steady state, corresponding to a fixed point of the map. In the steady- state, the field will not change even as more atoms cross the cavity. If the steady-state is to be a "pure state", the field density operator will be of the form $\rho_F^{ss} = \sum_{n,m} S_n^{ss} S_m^{ss*} \mid n\rangle\langle m \mid$. This implies that the atom-field state must be disentangled and be of the form

$$| \Psi_{ss}\rangle = \sum_n S_n^{ss} \mid n\rangle \otimes (\alpha \mid 1\rangle + \beta \mid 3\rangle) \,. \tag{32}$$

In the steady-state, $\rho^F(t_i+\tau)=\rho^F(t_i)$. The photon state in the steady-state then can be obtained from eq. (29), by putting $\alpha^/ = \alpha$, $\beta^/ = \beta$. We thus have the recursion relation

$$S_{n+2} = -\frac{\beta}{\alpha}\,\tan\theta_n\, S_n = -\frac{\beta}{\alpha r}\sqrt{\frac{n+1}{n+2}}\, S_n\,. \tag{33}$$

The disentangled steady-state must satisfy the condition $\exp\{-i\omega_{-1}\tau\}=1$, i.e., $\omega_{-1}\tau = 2\pi j$, where $j=1,2,3,\cdots$. Thus the interaction time τ satisfies the condition

$$\left[-\frac{\Delta}{2}+\frac{1}{2}\sqrt{\Delta^2+4\hbar^2 g_1^2}\right]\frac{\tau}{\hbar} = 2\pi j\,. \tag{34}$$

Thus, the atom-field system undergoes cycles of entanglement followed by disentanglement. It is to be noted that eq. (34) is not a trapping condition, but rather a condition of disentanglement.

From the recursion relation (33) it follows that two distinct normalized states, $\mid\Psi_{even}\rangle$ and $\mid\Psi_{odd}\rangle$ are generated. These are,

$$\mid\Psi_{even}\rangle = \sum_{n=0}^{\infty} S_{2n}\mid 2n\rangle\,,\qquad \mid\Psi_{odd}\rangle = \sum_{n=0}^{\infty} S_{2n+1}\mid 2n+1\rangle\,. \tag{35}$$

The steady-state photonic state, therefore, in general, is a superposition of $\mid\Psi_{even}\rangle$ and $\mid\Psi_{odd}\rangle$. From (33), the even and odd expansion coefficients are given by:

$$S_{2n} = \left(-\frac{\beta}{\alpha r}\right)^n \frac{\sqrt{(2n)!}}{2^n n!}\, S_0 \tag{36}$$

$$S_{2n+1} = \left(-\frac{\beta}{\alpha r}\right)^n \frac{2^n n!}{\sqrt{(2n+1)!}}\, S_1\,, \tag{37}$$

where, $n=1,2,3,....$ We thus have,

$$\mid\Psi_{even}\rangle = \sum_{n=0}^{\infty}\left(-\frac{\beta}{\alpha r}\right)^n \frac{\sqrt{(2n)!}}{2^n n!}\, S_0\mid 2n\rangle \equiv S(\zeta)\mid 0\rangle\,, \tag{38}$$

where $S(\zeta)$ is the squeeze operator, and

$$\zeta = \sigma\, e^{i\varphi}\,,\quad S_0 = \frac{1}{\sqrt{\cosh\sigma}}\,,\quad \frac{\beta}{\alpha r} = e^{i\varphi}\tan\sigma\,.$$

The $\mid\Psi_{even}\rangle$ is thus a squeezed vacuum state. The odd-photon state $\mid\Psi_{odd}\rangle$ is also given by

$$\mid\Psi_{odd}\rangle = \sum_{n=0}^{\infty}\left(-\frac{\beta}{\alpha r}\right)^n \frac{2^n n!}{\sqrt{(2n+1)!}}\, S_1\mid 2n+1\rangle\,. \tag{39}$$

The properties of the above state has been studied in ref. [30].

6. Raman interactions, Quantum information, and cloning

The fundamental concept in quantum computation is the qubit, the quantum bit, usually a two-state system. Unlike the classical bit, a qubit is a linear superposition of the states. Recently, quantum information processing using continuous spectrum systems have been suggested [31]. Continuous systems could be easier to manipulate experimentally. For instance, a qubit could be formed by the superposition of $\mid \Psi_{even}\rangle$ and $\mid \Psi_{odd}\rangle$ discussed in the previous section. Such a possibility has been conjectured recently [30].

Another interesting system for quantum information processing is a cavity QED experiment involving Raman interactions of a three-level atom (in the Λ configuration), in a two-mode cavity. The Hamiltonian of such a system is a simple generalization of that of the two-photon single-mode case discussed earlier. The Hamiltonian can be written as $H = H_0 + H_{int}$, with

$$H_0 = \hbar\omega_1 N_1 + \hbar\omega_2(N_2 - 1) + E_1 \,, \tag{40}$$

and

$$H_{int} = E_0 - E_1 + \hbar\omega_2 + \frac{\hbar\lambda}{2}\left(a_1^\dagger a_2\sigma_{12} + a_2^\dagger a_1\sigma_{21}\right) + \frac{\hbar\omega}{2}\left(\sigma_{22} - \sigma_{11}\right) \,, \tag{41}$$

where, a_i and $a_i^\dagger$, $i = 1, 2$, are photon mode operators of frequencies ω_i. The operators $N_1 = a_1^\dagger a_1 + 1 - \sigma_{11}$ and $N_2 = a_2^\dagger a_2 + 1 - \sigma_{22}$ are constants of the motion. All parameters including ω can be found in ref. [10]. The eigenvalues and of H are:

$$\begin{aligned} E^+_{n_1,n_2} &= E_1 + \hbar\omega_1 n_1 + \hbar\omega_2 n_2 \,, \\ E^-_{n_1,n_2} &= E^+_{n_1,n_2} - \hbar\left[\sqrt{(\tfrac{\Delta}{2})^2 + g_1^2 n_1 + g_2^2(n_2+1)} \;-\; \tfrac{\Delta}{2}\right] . \end{aligned} \tag{42}$$

E_1 and E_2 are the energies of the lower and upper states respectively and Δ is the detuning. The eigenfunctions, denoted $\mid \Psi^\pm_{n_1 n_2}\rangle$, are given by

$$\begin{aligned} \mid \Psi^+_{n_1,n_2}\rangle &= -\sin\theta_{n_1,n_2} \mid 1; n_1, n_2\rangle + \cos\theta_{n_1,n_2} \mid 2; n_1 - 1, n_2 + 1\rangle \,, \\ \mid \Psi^-_{n_1,n_2}\rangle &= \cos\theta_{n_1,n_2} \mid 1; n_1, n_2\rangle + \sin\theta_{n_1 n_2} \mid 2; n_1 - 1, n_2 + 1\rangle \,, \end{aligned} \tag{43}$$

where, $\mid 1\rangle$ and $\mid 2\rangle$ are the lower and upper atomic states, respectively, and $\mid n_i\rangle$ is the photon number state. Furthermore,

$$\sin\theta_{n_1,n_2} = \frac{r\sqrt{n_2+1}}{\sqrt{[n_1 + r^2(n_2+1)]}} \,, \qquad \cos\theta_{n_1,n_2} = \frac{\sqrt{n_1}}{\sqrt{[n_1 + r^2(n_2+1)]}} \,.$$

We now discuss entanglement and disentanglement phenomena associated with the interaction of the atoms and the modes of radiation in the cavity. The time evolution of an arbitrary atom-field state is given by the unitary operator $U(t) = \exp\left[-i(H - H_0)\frac{t}{\hbar}\right]$.

Consider the disentangled atom-field state at $t = 0$,

$$\mid \Psi\rangle = \sum_{n_1,n_2=0}^{\infty} C_{n_1,n_2} \mid n_1, n_2 > \{\alpha \mid 1\rangle + \beta \mid 2\rangle\} . \tag{44}$$

The time development of $\mid \Psi\rangle$ is given by $\mid \Psi(t) >= U(t) \mid \Psi\rangle$. Using the iteration procedure discussed in Section 5, one obtains the following two-term recursion relation for the steady state [32],

$$C_{n_1,n_2} = -\left(\frac{\beta r}{\alpha}\right)\sqrt{\frac{n_2+1}{n_1}} C_{n_1-1,n_2+1}, \quad (n_1 = 1, 2, \cdots; n_2 = 0, 1, 2, \cdots) . \tag{45}$$

From this we obtain,

$$C_{n_1,n_2} = \left(-\frac{\alpha}{\beta r}\right)^{n_2} \sqrt{\frac{(n_1+n_2)!}{n_1! n_2!}} C_{n_1+n_2,0} . \tag{46}$$

The dynamics governed by the Raman Hamiltonian (40,41) conserves the total number $\hat{N} = a_1^\dagger a_1 + a_2^\dagger a_2$ of photons in the two modes. The operator $\delta_{\hat{N},N}$, which is equal to unity when the operator produces N, zero otherwise, is a constant of the motion. Accordingly, if the cavity fields, prior to the entrance of the first atom in the cavity, had probabilities $P_N = \langle \Psi(t) \mid \delta_{\hat{N},N} \mid \Psi(t)\rangle$ of $\hat{N}$ having the value N, then, $< \Psi(t) \mid \delta_{\hat{N},N} \mid \Psi(t)\rangle = \langle \Psi \mid \delta_{\hat{N},N} \mid \Psi\rangle$. Therefore, the coefficients $C_{n_1+n_2,0} = C_{N,0}$ are determined by the probabilities P_N of the initial state of radiation. Thus [32]

$$P_N = \sum_{n_1,n_2=0}^{\infty} \mid C_{n_1,n_2} \mid^2 \delta_{n_1+n_2,N} = \mid C_{N,0} \mid^2 \left[1 + \mid \frac{\alpha}{\beta r} \mid^2\right]^N . \tag{47}$$

As a particular example, consider the initial photonic state to be such that mode 1 is a coherent state $\mid \gamma\rangle$ and mode 2 is the vacuum state $\mid 0\rangle$. Then

$$P_N = \frac{\mid \gamma \mid^{2N}}{N!} e^{-|\gamma|^2} ,$$

and therefore, using Eeqs. (46) and (47), we obtain,

$$C_{n_1,n_2} = e^{-\frac{1}{2}|\gamma|^2} \frac{1}{\sqrt{n_1!}} \left(\frac{\gamma}{\sqrt{1 + \mid \frac{\alpha}{\beta r} \mid^2}}\right)^{n_1} \frac{1}{\sqrt{n_2!}} \left(\frac{-\frac{\alpha}{\beta r}\gamma}{\sqrt{1 + \mid \frac{\alpha}{\beta r} \mid^2}}\right)^{n_2} . \tag{48}$$

Accordingly, the steady state of the electromagnetic field inside the cavity is a two-mode coherent state and so the effect of the atom-field interactions is to give rise to the following relationship between the initial input state and the output state

$$| \gamma\rangle_1 \otimes | 0\rangle_2 \rightarrow \left| \frac{\gamma}{\sqrt{1+ | \frac{\alpha}{\beta r} |^2}}\right\rangle_1 \otimes \left| \frac{-\frac{\alpha\gamma}{\beta r}}{\sqrt{1+ | \frac{\alpha}{\beta r} |^2}}\right\rangle_2 , \tag{49}$$

where the subscripts indicate the modes of the radiation field. It is interesting to note that cloning of the coherent states can be achieved. In this sense, the cavity atom-field system is a cloning machine. Note that for the case $\frac{\alpha}{\beta r} = -1$, the two output states become identical signifying perfect cloning.The cloning is different from those considered previously in that the output cloned states are not identical to the input coherent state. However, one obtains a unique output for a given initial state. If one wants to clone the coherent state $| \gamma\rangle$, then one needs to have a coherent state $| \sqrt{2}\gamma\rangle$ initially inside the cavity. The single-atom cavity QED behaves actually like a quantum beam splitter. Details are discussed in [32].

References

[1] E.T. Jaynes, and F.W. Cummings (1963), *Proc. IEEE*, **51**, 89.

[2] J.H. Eberly, N.B. Narozhny, and J.J. Sánchez-Mondragon (1980), *Phys. Rev. Lett.*, **44**, 1323; H.I. Yoo, and J.H. Eberly (1985), *Phys. Rep.*, **118**, 239.

[3] G. Rempe, H. Walther, and N. Klein (1987), *Phys. Rev. Lett.*, **58**, 353;

D. Meschede, H. Walther, and G. Müller (1985), *Phys. Rev. Lett.*, **54**, 551; M. Brune, J.M. Raimond, P. Goy, L. Davidovich, and S. Haroche (1987), *Phys. Rev. Lett.*, **59**, 1899.

[4] S. Singh (1982), *Phys. Rev. A*, **25**, 3206; C.V. Sukumar and B. Buck (1981), *Phys. Lett. A*, **83**, 211.

[5] Y. Shen , J. Swiatkiewicz, P.N. Prasad, and R.A. Vaia (2001), *Opt. Communications*, **200**, 9; Y. Shen , J. Swiatkiewicz, D. Jakubczyk, F. Xu, P.N. Prasad, R.A. Vaia, and B.A. Reinhardt (2001), *Applied Opt.*, **40**, 938.

[6] W. Denk, D.W. Piston, and W.W. Webb in "Handbook of Biological Conformal Microscopy", (1995), second edition, ed. James B. Pawley, Plenum Press, N.Y.; T. Parasassi, E. Gratton, W. M. Yu, P. Wilson, and M. Levi (1997), *Biophys. Journal*, **72**, 2413.

[7] C.C. Gerry and J.H. Eberly (1990), *Phys. Rev. A*, **42**, 6805.

[8] S.J.D. Phoenix and P. L. Knight (1990), *J. Opt. Soc. Am. B*, **7**, 116; L. Gilles, B.M. Garraway, and P.L. Knight (1994), *Phys. Rev. A*, **49**, 2785.

[9] M. Brune, J. M. Raimond, and S. Haroche (1987), *Phys. Rev. A*, **35**, 154; L. Davidovich, J.M. Raimond, M. Brune, and S. Haroche (1987), *Phys. Rev. A*, **36**, 3771.

[10] M. Alexanian and S.K. Bose (1995), *Phys. Rev. A*, **52**, 2218.

[11] Y. Wu (1996), *Phys. Rev. A*, **54**, 1586.

[12] M. Alexanian , S.K. Bose, and L. Chow (1998), *J. Mod. Optics*, **45**, 2519; M. Alexanian, S.K. Bose, and L. Chow (1998), *J. Luminescence*, **76-77**, 677.

[13] K.J. McNeil and D.F. Walls (1974), *J. Phys. A*, **7**, 617.

[14] L. Gilles and P.L. Knight (1993), *Phys. Rev. A*, **48**, 1582.

[15] M. Alexanian and S.K. Bose (1999), *J. Luminescence*, **83-84**, 167.

[16] E.S. Guerra, B.M. Garraway, and P.L. Knight (1997), *Phys. Rev. A*, **55**, 3842.

[17] H.D. Simaan and R. Loudon (1978), *J. Phys. A*, **11**, 435.

[18] H. Ezaki, E. Hanamura, and Y. Yamamoto (1999), *Phys. Rev. Lett.*, **83**, 3558.

[19] M. Alexanian and S.K. Bose (2000), *Phys. Rev. Lett.*, **85**, 1136.

[20] P. Filipowicz, J. Javanainen, and P. Meystre (1986), *Phys. Rev. A*, **34**, 3077.

[21] J. Krause, M.O. Scully, and H. Walther (1987), *Phys. Rev. A*, **36**, 4547.

[22] J. Krause, M.O. Scully, T. Walther, and H. Walther (1989), *Phys. Rev. A*, **39**, 1915.

[23] M. Brune, S. Haroche, V. Lefevre, J.M. Raimond, and N. Zagury (1990), *Phys. Rev. Lett.*, **65**, 976.

[24] G. Harel and G. Kurizki (1996), *Phys. Rev. A*, **54**, 5410.

[25] B.M. Garraway, B. Sherman, H. Moya-Cessa, P.L. Knight, and G. Kurizki (1994), *Phys. Rev. A*, **49**, 535.

[26] A. Napoli and A. Messina (1997), *J. Mod. Optics*, **44**, 2093.

[27] M. Orzag, R. Ramirez, J.C. Retamal, and L. Roa (1992), *Phys. Rev. A*, **45**, 6717.

[28] S. Bose, I. Fuentes-Guridi, P.L. Knight, and V. Vedral (2001), *Phys. Rev. Lett.*, **87**, 050401.

[29] V. Buzek, H. Moya-Cesa, P.L. Knight, and S.J.D. Phoenix (1992), *Phys. Rev. A*, **45**, 8190.

[30] M. Alexanian, and S.K. Bose (2002), *Phys. Rev. A*, **65**, 033819.

[31] S.L. Braunstein (1998), *Phys. Rev. Lett.*, **80**, 4084; P. van Loock and S.L. Braunstein (2001), *Phys. Rev. Lett.*, **84**, 3482.

[32] M. Alexanian (2003), *Phys. Rev. A*, **67**, 033809.

LOW-DIMENSIONAL SPIN SYSTEMS: HIDDEN SYMMETRIES, CONFORMAL FIELD THEORIES AND NUMERICAL CHECKS

C. Degli Esposti Boschi
Unità di ricerca INFM
viale Berti-Pichat, 6/2, I-40127, Bologna, Italia
desposti@bo.infm.it

E. Ercolessi
Dipartimento di Fisica,
Università di Bologna and INFN,
viale Berti-Pichat, 6/2, I-40127, Bologna, Italia
ercolessi@bo.infn.it

G. Morandi
Dipartimento di Fisica,
Università di Bologna and INFN,
viale Berti-Pichat, 6/2, I-40127, Bologna, Italia
morandi@bo.infn.it

Abstract We review here some general properties of antiferromagnetic Heisenberg spin chains, emphasizing and discussing the rôle of hidden symmetries in the classification of the various phases of the models. We present also some recent results that have been obtained with a combined use of Conformal Field Theory and of numerical Density Matrix Renormalization Group techniques.

1. Introduction and Summary.

For quite some time low-dimensional magnetic systems (i.e. (quantum) spins on $1D$ and/or $2D$ lattices) have been considered essentially only as interesting models in Statistical Mechanics with no realistic counterpart. It is only in

B. Gruber (ed.), Symmetries in Science XI, 145–173.

recent times that systems that can be considered to a high degree of accuracy as assemblies of isolated or almost isolated spin chains and/or of spin ladders (a few chains coupled together) have began to be produced and have hence become experimentally accessible, thus renewing the interest in their study, which is by now one of the most active fields of experimental and theoretical research in Condensed Matter Physics.

In this paper we will discuss only some relevant properties of isolated spin chains, referring to the literature [15] for a general review of the properties of spin ladders.

More than one decade ago it was pointed out [20, 5] that *integer* spin chains (more specifically, spin-1 chains, but extensions to different values of the spin have also been devised in the literature [43]) possess unexpected and highly non-trivial hidden symmetries, whose spontaneous breaking manifests itself through the appearance of unusual and highly nonlocal "string" order parameters. The string order parameters, together with the more conventional magnetic order parameters, can be used to classify the various phases that the phase diagram of one-dimensional magnets can display.

In the present paper, which is a slightly enlarged version of the talk presented by one of us (G.M.) at the XIII Conference on "*Symmetries in Physics*" we will concentrate, without pretensions to full generality, on the discussion of a few models of antiferromagnetic Heisenberg chains, of their phase diagrams and on the rôle of hidden symmetries in their explanation. The paper is organized as follows. In Sect.2 we review some general facts concerning Heisenberg spin chains and discuss how in the continuum limit one can map a "standard" (see below for the terminology) Heisenberg chain onto an effective field theory described by a nonlinear sigma-model, and how the presence in the latter of a topological term can account for the radically different behaviors of integer versus half-odd-integer spin chains. In Sect.3, concentrating on spin-1 chains, we consider the effects of the addition to the "standard" model of biquadratic exchange terms and/or of Ising-like as well as of single-ion anisotropies, and how the addition of such terms can drive the model away from what is commonly called the "Haldane phase" (again, see below for an explanation) towards other phases. In this context we will introduce in a more explicit manner the notion of hidden symmetries and discuss their rôle. Sects.4 and 5 will be devoted to the discussion of more recent results that have been obtained by some of us [17] with a careful and combined use of analytical (effective actions and Conformal Field Theory) and numerical (Density Matrix Renormalization Group) techniques. The final Sect.6 is devoted to the conclusions and to some general comments.

2. General Features of Spin Chains.

Let us begin by discussing here what can be considered as the "standard" model of an isotropic antiferromagnetic (AFM) Heisenberg chain with nearest-neighbor (nn) interactions, which is described by the Hamiltonian:

$$\mathcal{H} = J\sum_{i=1}^{N} \overrightarrow{S}_i \cdot \overrightarrow{S}_{i+1} \equiv \mathcal{H}_\perp + \mathcal{H}_z \tag{1}$$

$$\mathcal{H}_\perp = \frac{1}{2}J_\perp \sum_{i=1}^{N} \left\{S_i^+ S_{i+1}^- + S_i^- S_{i+1}^+\right\};\ \mathcal{H}_z = J_z \sum_{i=1}^{N} S_i^z S_{i+1}^z;\ J_\perp = J_z = J \tag{2}$$

where, for each $i = 1, ..., N$, $\overrightarrow{S}_i$ is a spin operator[1]:

$$\left[S_i^\alpha, S_j^\beta\right] = i\hbar\delta_{ij}\varepsilon^{\alpha\beta\gamma}S_i^\gamma;\ \alpha,\beta,\gamma = x,y,z;\quad \overrightarrow{S}_i^2 = \hbar^2 S(S+1) \tag{3}$$

(S integer or half-odd integer) located at the $i - th$ site of a one-dimensional lattice of N sites, interacting with its neighbors with an AFM ($J > 0$) nn interaction of strength J. Later on we will consider more general models in which $J_\perp \neq J_z$ will be allowed[2].

It may be useful to define a vector $\overrightarrow{n}_i$ as: $\overrightarrow{n}_i =: \overrightarrow{S}_i/\hbar S$, whereby:

$$\left[n_i^\alpha, n_j^\beta\right] = \frac{i}{S}\varepsilon^{\alpha\beta\gamma}n_i^\gamma;\ \overrightarrow{n}^2 = 1 + \frac{1}{S} \tag{4}$$

Although one is ultimately interested in the thermodynamic ($N \to \infty$) limit, for finite N one can adopt either periodic boundary conditions (PBC's), by imposing:

$$\overrightarrow{S}_{i+N} = \overrightarrow{S}_i\ \forall i \tag{5}$$

by which the system is actually considered to "live" on a circle, or open boundary conditions (OBC's), where $\overrightarrow{S}_1$ is coupled only to $\overrightarrow{S}_2$ and $\overrightarrow{S}_N$ only to $\overrightarrow{S}_{N-1}$[3]. The Hamiltonian of Eq.(1) has an obvious (global) $O(3)$ symmetry and, for PBC's, it is also invariant under the (discrete) translation group of the lattice.

In the classical limit ($\hbar \to 0$ and $S \to \infty$ with: $\hbar S = const.$) the spins (the $\overrightarrow{n}_i$'s) become (see Eq.(4)) classical vectors (and $\overrightarrow{n}_i \in \mathbb{S}^2$, the unit

---

[1] and: $S_i^\pm = S_i^x \pm iS_i^y$.

[2] $J_\perp = 0$, in particular, corresponds to the one-dimensional Ising model, a trivially soluble *classical* model. Notice however that an Ising model in a *transverse* magnetic field becomes a genuinely quantum and nontrivial model.

[3] In which case the Hamiltonian should be actually rewritten as: $\mathcal{H} = J\sum_{i=1}^{N-1} \overrightarrow{S}_i \cdot \overrightarrow{S}_{i+1}$.

sphere in $\mathbb{R}^3$). The minimum-energy configuration of the spins corresponds to: $\overrightarrow{n}_i \cdot \overrightarrow{n}_{i+1} = const. = -1$. Neighboring spins are then aligned antiparallel to each other and, in the absence of any external magnetic field, can point in a common but otherwise arbitrary direction on the sphere. This is the *Néel state*. Let us remark that, at variance with the *ferro*magnetic ($J < 0$) case, in which neighboring spins are all aligned parallel, at the quantum level the Néel state is *not* an eigenstate of the Hamiltonian (1). This points to the fact that *quantum* fluctuations will play a much more relevant rôle in the (quantum) antiferromagnetic case than in the ferromagnetic one.

The classical energy of the Néel state is of course: $E_N = -JN(\hbar S)^2$. In this state the $O(3)$ symmetry is spontaneously broken down to $O(2)$[4], and the state exhibits *long-range order* (LRO).

Elementary excitations over the Néel state are well-known to be in the form of *spin waves* [38]: coherent deviations of the spins with a dispersion: $\omega(\overrightarrow{k}) \propto k$ in the long-wavelength limit ($ka \ll 1$, with a the lattice spacing). Hence, the (classical) spectrum of the Hamiltonian (1) is *gapless.* We would like to stress that nothing of what has been said hitherto depends on the value of the spin. *At the classical level, the spin S*[5] *can be simply reabsorbed into a redefinition of the coupling constant* ($J \to J(\hbar S)^2$) and will contribute only an essentially irrelevant and additional multiplicative overall scale factor.

All this is elementary and well known. Let us turn now to the quantum case[6]. In the early 30's Bethe [8] and Hulthén [30], employing what has been known since as the "Bethe-Ansatz", were able to show that the quantum $S = 1/2$ Heisenberg chain is actually an *integrable* model. We will not discuss here the Bethe-Ansatz in any detail [38], but will only summarize the main features of the solution of the $S = 1/2$ model. The (exact) ground state is nondegenerate, it exhibits only *short-range* AFM correlations, but *no* LRO. Parenthetically, this is in agreement with a general, and later, theorem [14]. The (staggered) static spin-spin correlation functions:

$$\mathcal{G}^{\alpha}(i-j) = (-1)^{|i-j|} \left\langle S_i^{\alpha} S_j^{\alpha} \right\rangle ; \ \alpha = x, y, z \tag{6}$$

where $\langle ... \rangle$ stands for the expectation value in the ground state, are all equal and decay *algebraically* to zero at large distances. We recall here that genuine LRO would imply (we omit here the index α):

$$\lim_{|i-j| \to \infty} \mathcal{G}(i-j) = \mathcal{O}_N \neq 0 \tag{7}$$

---

[4]Translational symmetry, if present is also broken, as the Néel state is not invariant under translations of a lattice spacing as the original Hamiltonian but only of *twice* the lattice spacing. This has important consequences on the location of the Goldstone mode [4, 40, 55] in momentum space, that we will not discuss here, however.

[5]Or better $\hbar S$.

[6]From now on we will set for simplicity $\hbar = 1$.

this defining the *Néel order parameter* $\mathcal{O}_N$ (actually the square of the equilibrium staggered (i.e. sublattice) magnetization). On the other extreme, an *exponential* decay of the correlations of the form, say:

$$\mathcal{G}(i-j) \underset{|i-j|\to\infty}{\approx} \exp\{-|i-j|/\xi\}P(|i-j|) \tag{8}$$

with $P(.)$ some inverse power of $|i-j|$ would imply a finite *correlation length* ξ and a mass gap (or, better, a spin gap) Δ in the excitation spectrum roughly given by: $\Delta \propto c\xi^{-1}$,with c a typical spin-wave velocity. Algebraic decay of correlations (formally corresponding to $\xi \to \infty$) implies then that the system is *gapless*. Summarizing, the main features of the $S = 1/2$ Heisenberg AFM chain are that it has a (quantum) disordered ground state, with only short-range AFM correlations, and that it is gapless. It is therefore a (actually the first) prototype of a (quantum disordered and) *quantum critical* system [48]. It can be said then that, as compared with the classical limit, the system remains gapless but *quantum fluctuations destroy* LRO.

About thirty years later Lieb, Schutz and Mattis [36] (LSM) proved an important theorem stating that an $S = 1/2$ chain has either a degenerate ground state or is gapless. No surprise that the Bethe solution obeys the Lieb-Schutz-Mattis theorem, which is however of much wider reach, as it can cover models that are more general than the "standard" nn chain, such as, e.g., the Majumdar-Ghosh [37] model, another integrable model that we will nor discuss here, though. The results of LSM were extended later on by other authors [3] beyond $S = 1/2$ to cover all the half-odd-integer values of the spin. One can then take as rigorously proven that (at $T = 0$) *isotropic half-odd-integer Heisenberg chains* (with constant nn interactions) *are all quantum disordered and quantum critical* (i.e. gapless). This result was thought for quite some time to be "generic", i.e. valid for chains of arbitrary spin until, in the early 80's, Haldane [28] put forward what has become known since as "*Haldane's conjecture*", according to which half-odd-integer spin chains should be quantum disordered and gapless but *integer* spin chains should instead exhibit a spin gap and an exponential decay of correlations. This implied that, contrary to what happens in the classical limit, the physical behavior of spin chains should be a *highly discontinuous* function of the value of the spin.

Completely rigorous proofs of (the second part of) Haldane's conjecture are still lacking. However, strong support to it comes from the analysis of the continuum limit of the Heisenberg chain, which we will briefly describe now, referring to the existing literature [1, 6, 23] for more details.

The canonical partition function for the Hamiltonian of Eq.(1) at temperature $T = (k_B\beta)^{-1}$ (with k_B the Boltzmann constant):

$$\mathfrak{Z} = Tr\left\{\exp\left[-\beta\mathcal{H}\right]\right\} \tag{9}$$

can be written as a spin coherent-state path-integral [32], whereby the spin variables get replaced, inside the path-integral, by classical variables according to:

$$\overrightarrow{S}_i \to S\widehat{\Omega}_i \tag{10}$$

with $\widehat{\Omega}_i$ a classical unit vector: $\left|\widehat{\Omega}_i\right| = 1$. The next (and perhaps the most important) step in Haldane's analysis is the parametrization of the $\widehat{\Omega}_i$'s as[7]:

$$\widehat{\Omega}_i = (-1)^i \widehat{n}_i \sqrt{1 - \left(\frac{\overrightarrow{l}_i}{S}\right)^2} + \frac{\overrightarrow{l}_i}{S} \tag{11}$$

with: $|\widehat{n}_i| = 1$ and: $\widehat{n}_i \cdot \overrightarrow{l}_i = 0$. The $\widehat{n}_i$'s are assumed to be slowly-varying (on the scale of the lattice spacing). In this way, capitalizing, so-to-speak, on the information gained from the Bethe-Ansatz solution of the $S = 1/2$ model, they incorporate the information that the system still retains some short-range AFM ordering, which would be global only for $\widehat{n}_i = const.$ (and $\overrightarrow{l}_i = 0$). The $\overrightarrow{l}_i$'s can be shown [1] to be the (local) generators of angular momentum. In the semiclassical (large S) limit, an expansion of the action in the path-integral up to lowest (second) order in the $\overrightarrow{l}_i$'s is justified. Taking then the continuum limit together with a gradient expansion, and integrating out the $\overrightarrow{l}_i$'s, one ends up with the following expression for the partition function:

$$\mathfrak{Z} = \int [\mathcal{D}\widehat{n}]\, \delta\left(\widehat{n}^2 - 1\right) \exp\{-S_E - iS_B\} \tag{12}$$

where $[\mathcal{D}\widehat{n}]$ stands for the functional measure and the δ inside the integral is a functional δ. The first term in the action is given by:

$$S_E = \int_0^L dx \int_0^\beta d\tau \left\{ \frac{1}{2g} \left[\frac{1}{c} |\partial_\tau \widehat{n}|^2 + c \left|\partial_x \widehat{n}\right|^2 \right] \right\};\ \widehat{n} = \widehat{n}(x, \tau) \tag{13}$$

where $L(= N \times$ lattice spacing) is the length of the chain, $g = 2/S$ is the coupling constant and: $c = 2JS$ is the spin-wave velocity. This is simply the Euclidean action of an $O(3)$ nonlinear sigma model [6, 23, 59] ($NL\sigma M$). The second term is the integral of a Berry phase [50], and is given by:

$$S_B = \frac{\theta}{4\pi} \int_0^L dx \int_0^\beta d\tau \widehat{n} \cdot (\partial_\tau \widehat{n} \times \partial_x \widehat{n}) \tag{14}$$

[7]This is what is known in the literature as "*Haldane's mapping*".

with: $\theta = 2\pi S$. The coefficient of θ is easily recognized to be the Pontrjagin index [11, 41, 44], or winding number, of the map:

$$\widehat{n} : \mathbb{R}^2_{comp} \mapsto \mathbb{S}^2 \tag{15}$$

from spacetime compactified to a sphere and the two-sphere where $\widehat{n}$ takes values, and it is an integer: S_B is therefore a topological term, and: $S_B = 2\pi n S$, $n \in \mathbb{Z}$. Therefore, $\exp\{-iS_B\} \equiv 1$ for integer S ($\theta = 0 \mod (2\pi)$), but: $\exp\{-iS_B\} = (-1)^n$ ($\theta = \pi \mod (2\pi)$) if the spin is half-odd integer. This will generate interference between the different topological sectors, and it is the at the heart of the different behaviors of the two types of chains.

The pure ($\theta = 0$ in our case) $(1+1)$ $O(3)$ $NL\sigma M$ is a completely integrable model [60]. It has a unique ground state, and the excitation spectrum is exhausted by a degenerate triplet of *massive* excitations that are separated from the ground state by a finite gap. On the contrary, the $\theta = \pi$ model was shown [45] to be *gapless*. Therefore, Haldane's conjecture is fully confirmed by the analysis of the continuum limit of the Heisenberg model.

We would like only to mention in passing that quite a similar behavior occurs in spin *ladders* [15, 19, 46], namely even-legged ladders are gapped, while odd-legged ladders are gapped for integer spin and gapless for half-odd-integer spin. This "*even-odd*" effect has been shown [19, 51] to have the same topological origin an in single chains.

How do these results compare with the gaplessness (irrespective of the value of S) of the $S \to \infty$ classical limit? The answer resides in the dependence of the spin gap on S. Already at the mean-field level, but more accurately from large-N expansions and/or renormalization group analyses [42], it turns out that the spin gap Δ behaves as: $\Delta \propto \exp\{-\pi S\}$ for large S [8]. Hence, integer-spin chains become exponentially gapless for large S, and the classical limit is recovered correctly.

3. More general Models. Hidden Symmetries and String Order Parameters.

In view of what has been said up to now, the second part of Haldane's conjecture is by far the most intriguing part of it. Therefore *integer* spin chains are the most interesting ones, and we will concentrate from now on on $S = 1$ chains.

What has been called in the previous Section the "standard" AFM Heisenberg model is actually a member of at least two larger families of models that we will illustrate briefly here. The first class of models, that we will call "*θ−models*", includes a biquadratic term in the spins, and is described, setting

[8] $\Delta \propto \exp\{-\pi n_l S\}$ for spin ladders [19], where n_l is the number of legs of the ladder.

$J = 1$, by the Hamiltonian:

$$\mathcal{H} = \sum_{i=1}^{N} \left\{ \cos\theta \left(\overrightarrow{S}_i \cdot \overrightarrow{S}_{i+1} \right) + \sin\theta \left(\overrightarrow{S}_i \cdot \overrightarrow{S}_{i+1} \right)^2 \right\} \tag{16}$$

with $\theta = 0$ corresponding of course to the "standard" model. Most of the phase diagram has been obtained [5] numerically, except for the points at $\theta = \pm\pi/4$, that correspond to integrable models. The point $\theta = \pi/4$ is the Sutherland model [53], while $\theta = -\pi/4$ is the integrable model [7, 54] of Babujian and Takhatajan[9]. Both models are gapless, while the entire region $-\pi/4 < \theta < \pi/4$ is known (numerically, again) to be *gapfull*. This whole region has been called the "*Haldane phase*". It includes a particularly interesting point that has been studied extensively by Affleck, Kennedy, Lieb and Tasaki [2] ($AKLT$), namely $\theta = \theta^*$, with: $\tan\theta^* = 1/3$. The corresponding Hamiltonian (omitting an irrelevant overall numerical factor) is given by:

$$\mathcal{H}_{AKLT} = \sum_{i=1}^{N} \left\{ \overrightarrow{S}_i \cdot \overrightarrow{S}_{i+1} + \frac{1}{3} \left(\overrightarrow{S}_i \cdot \overrightarrow{S}_{i+1} \right)^2 \right\} \tag{17}$$

This model is not completely integrable, but the ground state is known, it is unique in the thermodynamic limit and can be exhibited explicitly. The ultimate reason for this is that, apart from numerical constants, the $i - th$ term in curly brackets is just:

$$\overrightarrow{S}_i \cdot \overrightarrow{S}_{i+1} + \frac{1}{3} \left(\overrightarrow{S}_i \cdot \overrightarrow{S}_{i+1} \right)^2 = 2 \left[P_2(i, i+1) - \frac{1}{3} \right] \tag{18}$$

where $P_2(i, i+1)$ is the projector [39] onto the state of total spin $S_{tot} = 2$ of the pair of $S = 1$ spins located at sites i and $i + 1$. Therefore, the ground state of $\mathcal{H}_{AKLT}$ must lie in the sector of the Hilbert space that is annihilated by all the projectors. It was shown by $AKLT$ that the exact ground state (also called the "Valence-Bond-Solid" (VBS) state) can be constructed as a linear superposition of states Φ_σ that have the following characteristics. Let: $\sigma = \{\sigma_1, ...\sigma_N\}$ be a given spin configuration, with: $\sigma_i = 0, \pm 1$, $i = 1, ..., N$. Then, Φ_σ is such that:

$i)$ $S_i^z \Phi_\sigma = \sigma_i \Phi_\sigma$ and moreover: $ii)$ If a given spin is, say, $+1$, then the next *nozero* spin must be -1, and viceversa. Typical such states correspond therefore to spin configurations of the form:

$$\sigma = \{+00 - 0 + -000 + ...\} \tag{19}$$

[9]This point is also known familiarly as the "Armenian point".

In other words, "up" and "down" spins do alternate in Φ_σ, but their spatial distribution is completely *random*, as an arbitrary number of zeroes can be inserted between any two nonzero spins. So, if a given spin in nonzero, we can predict *what* the value of the next nonzero spin will be, but not *where* it will be located. There is therefore no long-range (Néel) order in any conventional sense in the VBS ground state, but a sort of "*Liquid Néel Order*" (LNO). Conventional Néel order would be characterized by a nonvanishing of (at least one of) the *Néel order parameters*:

$$\mathcal{O}_N^\alpha = \lim_{|i-j|\to\infty} (-1)^{|i-j|} \left\langle S_i^\alpha S_j^\alpha \right\rangle ; \quad \alpha = x, y, z \tag{20}$$

In the VBS state and (numerically) in the whole of the Haldane phase one finds instead [2, 5]: $\mathcal{O}_N^\alpha = 0, \alpha = x, y, z$, and this is consistent with the absence of a "rigid" Néel order.

There remains however what we have called the "liquid" Néel order, and it has been argued convincingly in the literature [20, 5] that this is connected with the nonvanishing of a novel class of order parameters that we will discuss now briefly. Let us begin by defining the *string correlation functions* as:

$$\mathcal{G}_S^\alpha(n) =: -\left\langle S_0^\alpha \exp\left[i\pi \sum_{l=1}^{n-1} S_l^\alpha \right] S_n^\alpha \right\rangle ; \quad \alpha = x, y, z; \; n > 0 \tag{21}$$

These are similar to the standard two-point correlation functions:

$$\mathcal{G}^\alpha(n) =: (-1)^n \left\langle S_0^\alpha S_n^\alpha \right\rangle \tag{22}$$

whose asymptotic ($n \to \infty$) limit yields the Néel order parameter(s), except that a *string* of exponentials of intermediate spins has been inserted between the leftmost and the rightmost spins.

The *string order parameters* (SOP's) $\mathcal{O}_S^\alpha$ are then defined as:

$$\mathcal{O}_S^\alpha = \lim_{n\to\infty} \mathcal{G}_S^\alpha(n) \tag{23}$$

It turns out [2, 25] that the string correlation functions are strictly *constant* in the $AKLT$ ground state, namely:

$$\mathcal{G}_S^\alpha(n) \equiv const. = \mathcal{O}_S^\alpha = \frac{4}{9} \tag{24}$$

The ground-state spin-spin correlation functions have also been evaluated exactly for the VBS state [2], and they turn out to be given by:

$$\mathcal{G}^\alpha(n) = \frac{4}{3}\left(\frac{1}{3}\right)^n \tag{25}$$

In other words: $\mathcal{G}^{\alpha}(n) \propto \exp\{-n/\xi_{AKLT}\}$, where the *correlation length* ξ_{AKLT} is given, in units of the lattice spacing, by:

$$\xi_{AKLT} = \frac{1}{\log 3} \simeq 0.91 \tag{26}$$

less than unity in units of the lattice spacing, implying a rather large spin gap.

So far for the ground state of the $AKLT$ model. String and ordinary correlation functions as well as Néel and string order parameters have also been evaluated (numerically away from the $AKLT$ point) for other points of the Haldane phase [25]. For example, at the Heisenberg point, exact diagonalization methods[10] have shown that the string correlation functions are not strictly constant, but still decay exponentially to a value of the string order parameter that is somewhat smaller ($\mathcal{O}_S^{\alpha} \simeq 0.36...$) than the $AKLT$ value ($\mathcal{O}_S^{\alpha} = 4/9 \simeq 0.44...$) but still nonzero. The spin correlation length was also found [25] to be slightly larger than the $AKLT$ value, but still finite. So, there is convincing evidence that the entire Haldane phase is characterized by *vanishing Néel order parameters* but by *nonzero SOP's*. There is also convincing numerical evidence [25] that the string order parameters vanish at the integrable boundaries of the Haldane phase (i.e. for $\theta = \pm\pi/4$).

That the nonvanishing of the SOP's is connected to the breaking of a symmetry, and hence to the onset of an ordering that is not apparent in the original Hamiltonian was clarified in a seminal paper by Kennedy and Tasaki [5] (KT). With reference to a given configuration $\{\sigma\}$, and defining $N(\sigma)$ as the number of odd sites at which the spins are zero, one defines a new configuration $\{\overline{\sigma}\}$ via:

$$\overline{\sigma}_i = \exp\left[i\pi\sum_{j=1}^{i-1}\sigma_j\right]\sigma_i \tag{27}$$

and then a unitary operator U via:

$$U\Phi_{\sigma} = (-1)^{N(\sigma)}\,\Phi_{\overline{\sigma}} \tag{28}$$

In a nutshell, the action of U amounts to leaving the first nonzero spin unchanged and to flipping every other nonzero spin proceeding to the right of the chain. For example:

$$\begin{array}{c}\{++0-+00+0-0++\} \mapsto \{+-0--00+0+0+-\} \\ \{0+-00+00-+00-\} \mapsto \{0++00+00++00+\}\end{array} \tag{29}$$

and so on. It is obvious that U is a unitary[11]. What is less obvious is that the unitary transformation is a *nonlocal* one, in the sense that U cannot be

[10] With the Lanczos method and for chains with up to no more than 14 sites.

[11] Notice also that: $N(\sigma) = N(\overline{\sigma})$, as zero spins are mapped into zero spins.

written as a product of unitary operators acting at each single site. This has the important consequence that symmetries that are local (in the above sense) for the Hamiltonian $\mathcal{H}$ will of course remain symmetries of the transformed Hamiltonian $\widetilde{\mathcal{H}}$ (as U is unitary) but need not survive as *local* symmetries of $\widetilde{\mathcal{H}}$. Specifically, the symmetry group of $\mathcal{H}$ is $SU(2)$, that includes a discrete $Z_2 \times Z_2$ subgroup of rotations of π around the coordinate axes. Explicitly, the transformed Hamiltonian has the form [5]:

$$\widetilde{\mathcal{H}} = \sum_i \left\{ \cos\theta h_i + \sin\theta \left(h_i^2 \right) \right\} \tag{30}$$

where:

$$h_i = -S_i^x S_{i+1}^x + S_i^y \exp\left\{ i\pi \left(S_i^z + S_{i+i}^z \right) \right\} S_{i+1}^y - S_i^z S_{i+1}^z \tag{31}$$

and it evident that $Z_2 \times Z_2$ is the only *local* surviving symmetry group of the transformed Hamiltonian $\widetilde{\mathcal{H}}$. Even more important is how the string order parameters transform. The result is [5]:

$$\mathcal{O}_S^\alpha (\mathcal{H}) = \mathcal{O}_{ferro}^\alpha \left(\widetilde{\mathcal{H}} \right) \tag{32}$$

where:

$$\mathcal{O}_{ferro}^\alpha \left(\widetilde{\mathcal{H}} \right) = \lim_{|i-j| \to \infty} \langle S_i^\alpha S_j^\alpha \rangle |_{\widetilde{\mathcal{H}}} \tag{33}$$

and the r.h.s stands here for an average taken w.r.t. the ground state of the transformed Hamiltonian. The transformed order parameter is now a *ferromagnetic* order parameter. Therefore: $\mathcal{O}_S^\alpha (\mathcal{H}) \neq 0 \Longrightarrow \mathcal{O}_{ferro}^\alpha \left(\widetilde{\mathcal{H}} \right) \neq 0$, and this implies the onset of a spontaneous ferromagnetic polarization in the $\alpha - th$ direction in the ground state of $\widetilde{\mathcal{H}}$. This in turns entails a partial (if $\mathcal{O}_{ferro}^\alpha \left(\widetilde{\mathcal{H}} \right) \neq 0$ for just one value of α) or total (if this happens in more than one direction) spontaneous breaking of the discrete $Z_2 \times Z_2$ symmetry. It is known [4, 55] that spontaneous breaking of a continuous symmetry is accompanied by massless excitations (the Goldstone modes), while breaking of a discrete symmetry usually implies the opening of a gap (the most conspicuous and familiar example being the $2D$ Ising model). Therefore, KT were led to consider the spontaneous breaking of the $Z_2 \times Z_2$ symmetry as the origin of the Haldane gap.

One has however to be a bit careful on this point. It appears to be true that spontaneous (partial or total) breaking of the $Z_2 \times Z_2$ symmetry implies the generation of a spin gap. But:

i) The converse need not be true. We will see that there are spin models that exhibit *gapped* phases[12] while retaining the full $Z_2 \times Z_2$ symmetry, and:

[12] The so-called "large-D" phases of the "$\lambda - D$" model to be discussed immediately below.

ii) The mere nonvanishing of (one or more) string order parameters is not enough to fully determine in which (gapped) phase the system is. It is the *full set* of order parameters, both string and Néel, that allows for a full characterization of the various phases. In particular, the Haldane phase is fully characterized by the vanishing of all the Néel order parameters and by all the three string parameters being nonzero.

We turn now to a different class of models, the so-called "$\lambda - D$" family of models[13]. They are described by the family of Hamiltonians (parametrized by two real parameters, λ and D):

$$\mathcal{H} = \sum_{i=1}^{N} \left\{ S_i^x S_{i+1}^x + S_i^y S_{i+1}^y + \lambda S_i^z S_{i+1}^z + D\,(S_i^z)^2 \right\} \tag{34}$$

The "standard" (isotropic) AFM Heisenberg model corresponds of course to $\lambda = 1$ and $D = 0$. $\lambda = -1$ (and $D = 0$) can be easily shown[14] to correspond to a (isotropic) ferromagnetic Heisenberg model. $|\lambda| \neq 1$ introduces an "Ising-like" anisotropy, while $D \neq 0$ introduces what is called "single-ion" anisotropy.

The model can be solved exactly for $S = 1/2$ [33], but no exact solutions are available for integer spin. There are obvious asymptotic limits when either λ (resp. D) is large and D (resp. λ) not too large, so that the "λ-term" (resp. "D-term") can be considered as a zeroth-order Hamiltonian and the rest as a perturbation:

i) $|\lambda| \gg 1$. The reference ground state is either a Néel AFM state ($\lambda > 0$) or a ferromagnetic ($\lambda < 0$) state.

ii) $|D| \gg 1$. For $D > 0$ (the so-called "large-D" phase) the reference state becomes a planar state with $S_i^z = 0$ for all i's, while for $D < 0$ the reference state is a state where $S_i^z = 0$ is excluded, hence a state where the $S = 1$ spins become effectively two-level systems, and a detailed map of the model into an effective spin-$1/2$ model [17, 47] can be successfully performed. For $|\lambda| \neq 1$ and $D \neq 0$ the symmetry group of the Hamiltonian is $O(2) \times Z_2$ (the Z_2 factor corresponding to a reflection in the $x - y$ plane: $S_i^z \to -S_i^z$).

Apart from these limiting cases, the model has been studied analytically [49] as well as numerically [10, 13, 26, 52] , and the corresponding phase diagram is displayed in Fig.16.

The various sectors of the phase diagram can be characterized as follows [13, 17, 22]:

i) the *Haldane phase.* The ground state is unique with total component $S_{tot}^z = 0$ of the spin. The order parameters are: $\mathcal{O}_N^\alpha = 0$, $\mathcal{O}_S^\alpha \neq 0, \alpha = x, y, z$. The (Haldane) gaps are in different spin channels according to the sign of D,

[13]This is the class of models on which the Bologna group is currently working.

[14]By performing a rotation of π around the z-axis on one of the two sublattices (i.e. on every other site).

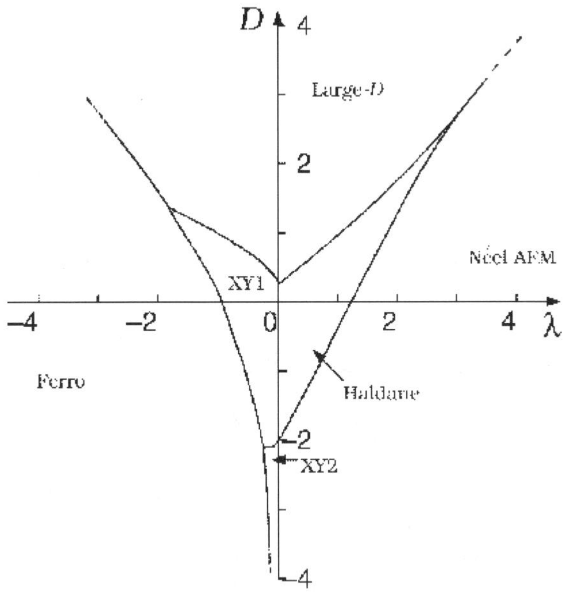

Figure 1. Phase diagram of the $\lambda - D$ spin-1 Hamiltonian of Eq. (34). The indicated regions are explained in the text.

but nonzero in any case. The isotropic, Heisenberg point $\lambda = 1, D = 0$ is in this phase, and lies on a line separating the two subphases, that are denoted as $H1$ and $H2$ in the literature [10].

$ii)$ The *Néel phase.* The ground state is doubly degenerate, and the order parameters are: $\mathcal{O}_N^\alpha = \mathcal{O}_S^\alpha = 0$ for $\alpha = x, y$, but: $\mathcal{O}_N^z, \mathcal{O}_S^z \neq 0$.

$iii)$ The *large-D phase*.The ground state is unique, it is *gapped*, but here:$\mathcal{O}_S^\alpha = \mathcal{O}_N^\alpha = 0\ \forall\alpha$.

$iv)$ The two *XY phases.* These are both *gapless* phases. They are distinguished by the nature of the low-lying spin excitations (spin-1 in the $XY1$ phase, spin-2 in the $XY2$ phase).

$v)$ The *ferromagnetic* phase. The ground state is doubly degenerate, with maximal magnetization: $S_{tot}^z = \pm N$, and the phase is gapped. In this case it is the *ferro*magnetic order parameter that is nonvanishing, and actually [6]: $\mathcal{O}_{ferro}^z = 1$ (the other two being zero). Also: $\mathcal{O}_S^z(j,k) = (-)^{j-k-1}$, while: $\mathcal{O}_S^{x,y} = 0$.

Anticipating some of the numerical results of Sect.5, we give below, in Figs.2 and 3, some examples [6] of the behavior of the various correlators and order parameters as functions of the parameters of the model.

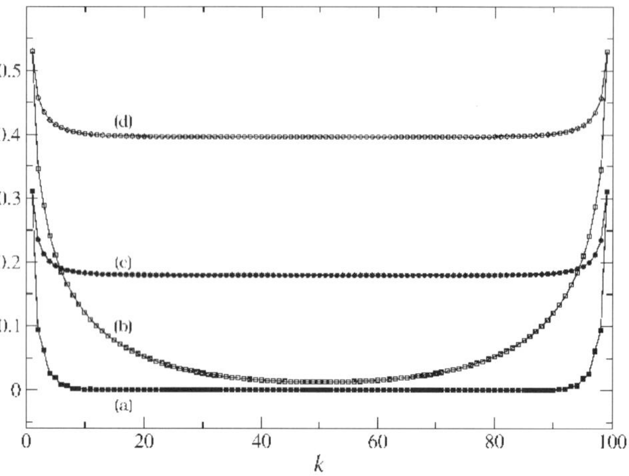

Figure 2. Ordinary and string correlation functions in the Haldane phase: (a) $\mathcal{G}^z(k)$, (b) $\frac{1}{2}(-)^k\langle S_0^+ S_k^-\rangle$, (c) $\mathcal{G}_S^z(k)$ and (d) $\mathcal{G}_S^x(k)$. Selected values of the parameters are $(D = 0.5, \lambda = 1)$. Note that with this choice the transverse correlation length is appreciably larger than the longitudinal one. The data have been obtained with finite-size $DMRG$ on a chain of $L = 100$ spins $(S = 1)$ with PBC and $M = 216$ states (Sect.5 for details).

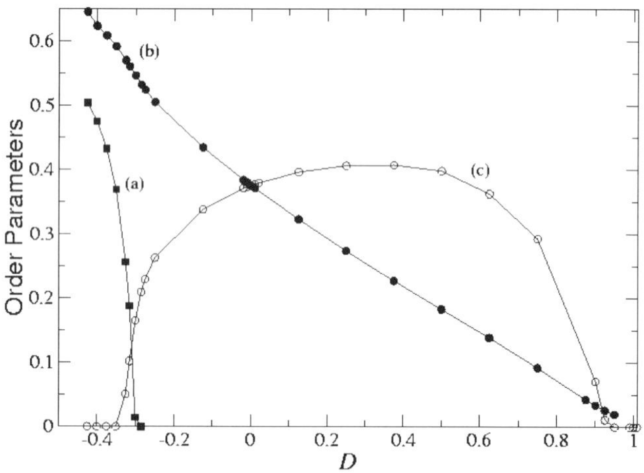

Figure 3. Order parameters relevant to the Néel-Haldane-large D transitions plotted *versus* the anisotropy coefficient D of Eq. (34) fixing $\lambda = 1$: (a) $\mathcal{O}_N^z$ defined in Eq. (20); (b) $\mathcal{O}_S^z$ and (c) $\mathcal{O}_S^x$ defined in Eq. (23). The asymptotic values are extrapolated using an algebraic best-fit function $\mathcal{O}_\infty + C/|i-j|^\gamma$ on the $DMRG$ data (same choices as in Fig.1). Near $D \simeq -0.3$ $\mathcal{O}_N^z$ and $\mathcal{O}_S^x$ do not vanish in the same point due to finite-size effects and to a moderate number of $DMRG$ states.

Concerning the nature of the transitions between the various phases [10, 13, 17], both the Haldane-large-D and the Haldane-Néel transition lines are critical (gapless) lines. The two critical lines merge at a tricritical point (at $D \simeq \lambda \simeq 3$), above which the Haldane phase disappears and the transition (a large-D-Néel transition,now) is first-order. The XY-ferromagnetic transition is instead a first-order one, as well as the large-D-ferromagnetic transition. Finally, The Haldane-XY transition is considered [13] to be of the Berezinskii-Kosterlitz-Thouless [7, 34] (BKT) type, as well as the XY-large-D transition.

The "$\lambda - D$" model has also been studied by KT. Applying the same nonlocal unitary transformation that was discussed previously, they showed that the transformed Hamiltonian, whose explicit form we will not give here, is still given in terms of the operators h_i (see Eq.(31)), and retains therefore $Z_2 \times Z_2$ as the only local symmetry,just as in the case of the Hamiltonian of Eq. (16). Therefore, the same conclusions as before apply concerning the connection of the nonvanishing of the string order parameters with the spontaneous breaking of the $Z_2 \times Z_2$ symmetry.

In the present paper we will address mainly to the detailed nature of the Haldane-large-D and Haldane-Néel critical transition lines. It is known that the (large distance) critical behavior of one-dimensional quantum systems is well described by Conformal Field Theory [12, 21, 24, 27, 33] (CFT). In the next Section we will report on a proposal of an effective CFT for the "$\lambda - D$" model on the Haldane-large-D critical line. This allows for the prediction of the operator content of the theory, and hence also for the prediction of the structure of the conformal tower of excited states above the ground state. To confirm the predictions, we will report also on extended numerical analyses, whose details will be reported elsewhere [18], that fully confirm the theoretical predictions.

4. Conformal Field Theory and Effective Actions.

Let us begin by recalling some basic results and examples of CFT that will be used in the forthcoming analysis of the critical properties of the spin-1 $\lambda - D$ chain.

It is well known [21, 27] that critical properties of two-dimensional systems are completely classified by CFT's: since in $2D$ the conformal group is infinite dimensional, the Hilbert space of a conformally invariant theory can be completely understood in terms of the irreducible representations of its algebra, the Virasoro algebra. We recall that the latter has an infinite number of generators, denoted with $L_n, \bar{L}_n$ ($n \in \mathbb{Z}$) for its holomorphic and antiholomorphic part respectively, satisfying the commutation relations:

$$[L_n, L_m] = (m - n)L_{m+n} + \frac{c}{12}(m^3 - m)\delta_{m+n,0} \tag{35}$$

and similarly for the $\bar{L}_n$. The constant c is called the central charge of the algebra or the conformal anomaly. Since we are interesting in a comparison between theoretical predictions and numerical data, which are performed on a finite lattice, we will consider a CFT defined on a cylinder with spatial dimension of finite length L. In this case [12, 21], the energy and the momentum operator are represented respectively by:

$$H = \frac{2\pi}{L}\left(L_0 + \bar{L}_0 - \frac{c}{12}\right) \tag{36}$$

$$P = \frac{2\pi}{L}\left(L_0 - \bar{L}_0\right) \tag{37}$$

In order for H to be bounded from below, we must restrict our attention to highest weight representations of the Virasoro algebra, for which there exists a highest weight (or primary) state $|\Delta, \bar{\Delta}\rangle$ satisfying:

$$L_0|\Delta, \bar{\Delta}\rangle = \Delta|\Delta, \bar{\Delta}\rangle \ , \ L_n|\Delta, \bar{\Delta}\rangle = 0 \quad \text{for } n > 0 \tag{38}$$

and analogous relations with respect to the $\bar{L}_n$ generators.

Each of these representations is thus identified by the values of the central charge c and of the couple $(\Delta, \bar{\Delta})$ (the conformal dimensions). They fix both the energy and the momentum of the primary state $|\Delta, \bar{\Delta}\rangle$, according to:

$$E^0_{\Delta,\bar{\Delta}} = \frac{2\pi}{L}(\Delta + \bar{\Delta} - \frac{c}{12}) \tag{39}$$

$$P^0_{\Delta,\bar{\Delta}} = \frac{2\pi}{L}(\Delta - \bar{\Delta}) \tag{40}$$

Notice that, in a finite geometry (with PBC), the vacuum state, corresponding to $\Delta = \bar{\Delta} = 0$, has a non zero energy (Casimir effect):

$$E^0_{vac} = -\frac{\pi c}{6L} \tag{41}$$

Also, the two-point correlation function of the operator creating a given primary state out of the vacuum ($|\Delta, \bar{\Delta}\rangle = \mathcal{O}_{\Delta,\bar{\Delta}}|0\rangle$) has an algebraic decay whose critical exponents are determined by the values of the conformal dimensions $(\Delta, \bar{\Delta})$: one has [21, 27]

$$\langle \mathcal{O}_{\Delta,\bar{\Delta}}(z,\bar{z})\mathcal{O}_{\Delta,\bar{\Delta}}(0,0)\rangle \propto \frac{1}{z^{2\Delta}\bar{z}^{2\bar{\Delta}}} \tag{42}$$

Finally, from the primary state $|\Delta, \bar{\Delta}\rangle$ one can obtain all excited (or secondary) states by applying strings of powers of $L_n, \bar{L}_n$ with $n < 0$. It is easy to see that, if $m, n < 0$, the commutation relations (35) imply: $L_0(L_m)^j|\Delta, \bar{\Delta}\rangle =$

$(\Delta + mj)|\Delta, \bar{\Delta}\rangle$, $\bar{L}_0(\bar{L}_n)^k|\Delta, \bar{\Delta}\rangle = (\bar{\Delta} + nk)|\Delta, \bar{\Delta}\rangle$, so that the secondary states have energies and momenta:

$$E^{(r,\bar{r})}_{\Delta,\bar{\Delta}} - E^0_{vac} = \frac{2\pi}{L}(\Delta + \bar{\Delta} + r + \bar{r}) , \tag{43}$$

$$P^{(r,\bar{r})}_{\Delta,\bar{\Delta}} = \frac{2\pi}{L}(\Delta - \bar{\Delta} + r - \bar{r}) \tag{44}$$

with $r, \bar{r} \in \mathbb{N}$ and a degeneracy that can be explicitly calculated for each representation. It may happen that some of these states have null norms. In this case the true (non-degenerate) Hilbert space of states is obtained after projecting out these null vectors, which therefore do not contribute to the operator content of the corresponding CFT. The quantity in brackets in the right hand side of Eq. (43) yields the coefficients with which the energy of the corresponding state scales to zero in the thermodynamic limit. It is therefore called "scaling dimension" and will be denoted by $d^{(r,\bar{r})}_{\Delta,\bar{\Delta}}$ in the sequel.

Let us examine some examples. We will consider only unitary theories, corresponding [21] to the following two sets of values of the central charge c:

$$c = 1 - \frac{6}{p(p+1)} , \; p = 3, 4, \ldots ; \tag{45}$$

$$c \geq 1 . \tag{46}$$

The first set of values corresponds to the so called minimal models [21, 27], whose primary states are of finite number. Their conformal dimensions are given by the formula:

$$\Delta_{rs}, \bar{\Delta}_{rs} = \frac{[(p+1)r - ps]^2 - 1}{4p(p+1)} , \; 1 \leq s \leq r \leq p-1 , \; r, s \in \mathbb{Z} \tag{47}$$

Theories with $c \geq 1$ have instead an infinite number of primary states.

The simplest case of a CFT corresponds to $c = 1/2$ ($p = 3$ in Eq. (45)) and describes the universality class of the two-dimensional Ising model. According to (47), there are only three primary operators: the identity $\mathbb{I}$ corresponding to the vacuum, $(\Delta, \bar{\Delta})_{\mathbb{I}} = (0, 0)$, the Ising spin σ with $(\Delta, \bar{\Delta})_\sigma = (1/16, 1/16)$ and the energy density ε with $(\Delta, \bar{\Delta})_\varepsilon = (1/2, 1/2)$. Notice that the spin-spin correlator $\langle\sigma(x)\sigma(0)\rangle$ decays with a critical exponent $\eta^z = 4\Delta_\sigma = 0.25$. In Table 1 we list the lowest conformal (primary and secondary) states, together with their scaling dimensions and momenta. As explained in the next section, a comparison with the numerical data given in the last column will allow us to conclude that the Haldane-Néel critical transition line is indeed of the Ising type.

Table 1. Columns 1-4 show the conformal dimensions $(\Delta, \bar{\Delta})$,$(r, \bar{r})$, the scaling dimensions $d^{(r,\bar{r})}_{\Delta,\bar{\Delta}}$ and the momenta $P^{(r,\bar{r})}_{\Delta,\bar{\Delta}}$ of the lowest conformal states in the $c = 1/2$ minimal model. The numerical results in the last column are explained in Sect.5.
Notice that the states with $\Delta = \bar{\Delta} = 0$, $(r, \bar{r}) = (1, 0), (0, 1)$ do not appear since they correspond to null vectors.

$(\Delta, \bar{\Delta})$, $(r, \bar{r})$	$d^{(r,\bar{r})}_{\Delta,\bar{\Delta}}$	$P^{(r,\bar{r})}_{\Delta,\bar{\Delta}}$	$d_{(num)}$
$(0, 0)\,;\,(0, 0)$	0	0	
$(1/16, 1/16)\,;\,(0, 0)$	$1/8$	0	0.1250 ± 0.0004
$(1/2, 1/2)\,;\,(0, 0)$	1	0	0.962 ± 0.001
$(1/16, 1/16)\,;\,(1, 0), (0, 1)$	$9/8$	$\pm 2\pi/L$	1.0959 ± 0.0008
			1.100 ± 0.003
$(1/2, 1/2)\,;\,(1, 0), (0, 1)$	2	$\pm 2\pi/L$	1.87 ± 0.02
			1.87 ± 0.02
$(0, 0)\,;\,(2, 0), (0, 2)$	2	$\pm 4\pi/L$	1.904 ± 0.004
			1.86 ± 0.01

Gaussian model with Lagrangian:

$$\mathcal{L} = \frac{1}{2}\left[\frac{1}{v}(\partial_\tau \Theta)^2 + v(\partial_x \Theta)^2\right] \tag{48}$$

where Θ represents an angular variable spanning a circle of a given radius R and the constant v, which has the dimension of a velocity, is called spin velocity. If we assume for Θ, and hence for its dual field Φ[15], periodic boundary conditions, the Hilbert space of the theory splits into a direct sum of distinct topological sectors labeled by the winding numbers $n, m \in \mathbb{Z}$ of the fields Θ and Φ respectively. The primary fields are then vertex operators of the form [21, 27]

$$V_{mn} = \exp\left(i\sqrt{4\pi K} n\Phi + i\sqrt{\pi/K} m\Theta\right) \tag{49}$$

whose scaling dimensions are given by

$$d_{mn} = \left(\frac{m^2}{4K} + n^2 K\right), \qquad K = \frac{\pi}{R^2} \tag{50}$$

Notice that the latter depend explicitly on the radius of compactification. Thus we obtain a different $c = 1$ theory for each value of R, i.e. of K. For example, $K = 1$ corresponds (via fermionization [21, 27]) to a $1D$ model of free Dirac (FD) fermions. The $K = 1/2$ point is said to be self-dual (SD) since it is invariant under the duality transformation $\Theta \Leftrightarrow \Phi$, $m \Leftrightarrow n$, while the point $K = 2$ corresponds to the BKT critical theory.

[15] If we decompose the field Θ in its holomorphic and antiholomorphic part, $\Theta(z, \bar{z}) = \Theta_h(z) + \Theta_{ah}(\bar{z})$, the dual field is defined as $\Phi = \Theta_h(z) - \Theta_{ah}(\bar{z})$.

We remark also that the energy operator $(\partial\Theta)^2$ has conformal dimension 2 for any value of R and hence it is always marginal. The effect of adding it to the Lagrangian (48) results only in a change of the coupling constant in front, which, in turn, can be absorbed into a rescaling of the radius of compactification of Θ. Thus we generate a continuous line of inequivalent critical $c = 1$ theories, corresponding to different values of K.

It is well known [27] that the Gaussian model (48) describes the continuum limit of the spin 1/2 XXZ chain with anisotropy parameter Δ, as long as $-1 \leq \Delta \leq 1$. From the exact Bethe-ansätz results, one can show [27] that the interesting cases $\Delta = -1, 0, 1$ corrrespond to the SD, FD and BKT points of the bosonic theory, respectively. We would like to show now, that the Gaussian model (48) describes also the critical properties of the spin-1 $\lambda - D$ Hamiltonian (34) on the Haldane-large-D transition line. In doing so, we will also establish a relationship between the coupling constants D, λ of the discrete model and those of the continuum theory, namely the spin-wave velocity v and the compactification radius. This will allow us to make quantitative theoretical predictions to be compared, in next section, to the numerical results.

In the spirit of Haldane'a mapping, we start from a classical solution, which for $D > \lambda - 1$, is a planar state where the unit vectors $\widehat{\Omega}_j(\tau)$ that represent our spins ($\overrightarrow{S}_j \rightarrow S\widehat{\Omega}_j(\tau)$, see Sect.2) are Néel ordered in the xy-plane: $\widehat{\Omega}_j(\tau) = (\cos(\theta_0 + j\pi), \sin(\theta_0 + j\pi), 0)$. Hence we make the Haldane-like ansätz:

$$\widehat{\Omega}_j(\tau) = (-1)^j \hat{n}_j(\tau)\sqrt{1 - \frac{l_j^2(\tau)}{S^2}} + \hat{z}\,\frac{l_j(\tau)}{S} \tag{51}$$

where $\hat{n}_j(\tau) = e^{i\theta_j(\tau)} \in O(2)_{xy}$, $\hat{z}$ is the unitary vector $(0, 0, 1)$, and the fluctuation field l_j is supposed to be small. Thus, as for the isotropic case, it is possible to obtain an effective Lagrangian that describes the low-energy physics of the Hamiltonian (34) in the continuum limit. Carrying out this calculation as explained in Sect.2, one obtains in this case a Gaussian model (48), where now $\Theta = \theta/\sqrt{g}$ and

$$g = \frac{1}{s}\sqrt{2\,(1 + D + \lambda)}; \qquad v = s\sqrt{2\,(1 + D + \lambda)} \tag{52}$$

In other words, we have a free theory for a bosonic field Θ, which is compactified along a circle of radius $1/\sqrt{g}$. Thus, the operator content of the theory can be read from Eq. (49): the list of primary operators is exhausted by the vertex oprators V_{mn} whose scaling dimensions are given by Eq. (50), with $K = \pi/g$.

In addition, the scaling dimensions (50) fix also the (non universal) critical exponents of the correlation functions. For instance it is easy to see that the transverse spin-spin correlator should decay according to:

$$\langle S^+(0)S^-(x)\rangle \approx \langle e^{i\theta(0)}e^{-i\theta(x)}\rangle \propto |x|^{-\eta} \text{ with } \eta = 2d_{10} = g/2\pi\ . \tag{53}$$

5. The Density Matrix Renormalization Group and Spin Chains.

The code we have used for density matrix renormalization group ($DMRG$) calculations follows rather closely the algorithms reported in White's seminal papers [56, 57], with the following points to be mentioned:

• The superblock geometry was chosen to be $[\mathrm{B}^s \bullet |\mathrm{B}^{s'}_{\mathrm{ref}} \bullet]$ with PBC, where $\mathrm{B}^{s'}_{\mathrm{ref}}$ is the (left $\leftrightarrow$ right) reflected of block $\mathrm{B}^{s'}$ with s' sites. The rationale for adopting this configuration is that, being effectively on a ring, the two blocks are always separated by a single site, for which the operators are small matrices that are treated exactly (no truncation) [57]. In this way we expect a better precision in the correlation functions calculated fixing one of the two point on these sites and moving the other one along the block. Moreover, whenever the system has an underlying antiferromagnetic structure (typically when a staggered field is switched on), this geometry seems to be the one that preserve it at best, both for even and odd values of s.

• We used the *finite-system algorithm* with three iterations. This prescription should ensure the virtual elimination of the so-called environment error [35], which is expected to dominate in the very first iterations for $L < L^*(m)$ (see below). Normally the correlations are computed at the end of the third iteration, once that the best approximation of the ground state is available. This has the advantage of using less memory during the finite-size iterations but requires the storage of all the matrices needed to represent, on the reduced basis of the last step, the operators entering the correlation functions of interest. At the moment, disk storage is the ultimate factor that limits the size of the systems that we are able to treat.

• We always exploit the conservation of S^z_{tot}. With the exception of the ferromagnetic phase, that we do not address now, the ground state(s) is (are) at $S^z_{tot} = 0$ [10]. In order to maximize their accuracy, the correlations are calculated targeting only the lowest-energy state within this sector. However, in order to analyze the energy spectrum, we had to target also the lowest-energy states in the other sectors $|S^z_{tot}| = 1, 2, \ldots$ and/or a few excited states within the $S^z_{tot} = 0$ sector, depending on the phase of interest. On the one hand, this requires a modification of the basic Lanczos method to go beyond the lowest eigenvalue of the superblock Hamiltonian. On the other hand, once the N_{t} eigenvalues of interest are found, one can build the block density matrix as the average (mixture) of the matrices associated with the corresponding N_{t} eigenvectors. At present we are not aware of any specific "recipe" other than that of equal weights.

Going back to the modified Lanczos routine, our DMRG code implements the so-called Thick Restart algorithm of Wu and Simon [58]. Once S^z_{tot} is fixed, in a given run we want to determine simultaneously the first N_{t} levels

$|S^z_{tot}; \mathtt{b}\rangle$ with b=0,1,2,..., $N_{\mathtt{t}} - 1$ (the ground state being identified by ($S^z_{tot} = 0$,b$= 0$)). Then, as in the conventional Lanczos scheme, we have to push the iteration until the norms of the residual vectors and/or the differences of the energies in consecutive steps are smaller than prescribed tolerances ($10^{-9} - 10^{-12}$ in our calculations). The delicate point to keep under control is that, once the lowest state $|S^z_{tot}; 0\rangle$ is found, if we keep iterating searching for higher levels the orthogonality of the basis may be lost, just because the eigenvectors corresponding to these levels tend to overlap again with the vector $|S^z_{tot}; 0\rangle$. As a result, the procedure is computationally more demanding to the extent that one has to re-orthogonalize the basis from time to time. Typically, we have seen that this part takes a 10-20% of the total time spent in each call to the Lanczos routine. We have also observed that if this re-orthogonalization is not performed, one of the undesired effects is that the excited doublets (generally due to momentum degeneracy) are not correctly computed. More specifically, it seems that while the two energy values are nearly the same in the asymmetric stages of the iterations, when the superblock geometry becomes symmetric ($s = s'$ in the notations of the preceding point) the double degeneracy is suddenly lost and only one of the two states appears in the numerical spectrum.

So far for the specific algorithm. Now, the crucial point to consider in accurate $DMRG$ calculations is the choice of M, that is, the number of optimized states. White argued [57] that the convergence of the ground state energy is almost exponential in M with a step-like behaviour, probably related to the successive inclusion of more and more complete spin sectors. Unfortunately, the effective accuracy gets poorer when we deal with energy differences and correlation functions, for which little is known about convergence. It must be told, however, that despite its name the $DMRG$ performs somehow better for systems with a definite gap rather than for gapless (critical) ones. We refer to the papers by Andersson, Boman and Östlund [5] and by Legeza and Fáth [35] where, for different systems and in terms of different observables, the following common feature emerges: Even if the quantum system is rigorously critical in the limit $L \to \infty$, the $DMRG$ truncation introduces a spurious length, $L^*(m)$, which, as expected, diverges as M is increased. (Our analysis of the accuracy of the energy levels in some selected points of the $\lambda - D$ chain near criticality leads to a similar conclusion [18]). Hence, even if we are technically able to deal with sizes $L > L^*(m)$ (at a given M), as far as criticality is concerned we cannot rely completely on the $DMRG$ data because the system experiences an effective length which should be absent in the critical regime.

Therefore, our strategy can be summarized as follows: We fix a rather high value of M such that the trustable values of L are sufficiently large to see the scaling limit of CFT, but not too large as compared to $L^*(m)$. In other words, even in the study of (supposed) critical systems, we prefer to exploit the computing resources to include as many $DMRG$ states as possible, and to refine the calculations with finite-size iterations, rather than trying to take naïvely the limit $L \to \infty$. In addition, to judge whether M is sufficiently large or not we checked the properties of translational and reflectional invariance that the correlation functions should have [16]. To be specific, if $\mathcal{G}(0,k)$ is a certain correlation function computed starting at $j=0$, we have always increased M (at the expenses of L) until the bound $|\mathcal{G}(\ell,\ell \pm k) - \mathcal{G}(0,k)|/\mathcal{G}(0,k) \lesssim 0.05$ was met for k varying from 0 to $\ell = L/2$, possibly with the exception of the ranges where $\mathcal{G}(0,k)$ is below numerical uncertainties (10^{-6}, say).

The quality of the numerical analysis of the critical properties depends heavily on the location of the critical points of interest. As far as the transitions from the Haldane phase are concerned, it is convenient to fix some representative values of λ and let D vary across the phase boundaries. This preliminary task of finding $D_{\rm c}(\lambda)$ turns out to be crucial for subsequent calculations and is divided in two steps. First, one has to get an approximate idea of the transition points using a direct extrapolation in $1/L$ of the numerical values of the gaps, computed at increasing L with a moderate number of $DMRG$ states. Clearly, one may want to explore a rather large interval of values and so the increments in D will not be particularly small (0.1, say). Then, the analysis must be refined around the minima of the curves ΔE-vs-D with smaller increments in D and a larger value of M. In our problem, the approach that seems to give better results is standard finite-size scaling (FSS) theory [26, 29] (for instance as compared to the phenomenological renormalization group).

Once the critical point is located, we take full advantage of the conformal structure by looking at the finite-size spectrum (see Eqs. (41) and (44)) of relevant and marginal operators. In practice, we select a number of states that tend to become degenerate with the ground state and look for straight lines in the ΔE-vs-L^{-1} plot. Then, from a best fit we expect to have a very small offset (ideally a zero gap in the thermodynamic limit) and a slope given by the scaling dimension d multiplied by the velocity prefactor, v, which is absent in the field-theoretical formulation but has to be determined (in terms of the microscopic parameters) in a lattice system. In the latter case, Eq. (41) should contain also a term $e_\infty L$, e_∞ being the energy density of the problem at hand. Actually, due to the prefactor v, we have to imagine a self-consistent procedure: Depending on the type of the transition we have in mind (that is, depending on the central charge c), we stick on one or more levels in the spectrum that have exactly $d=1$. Then the slope of these is nothing but v. Once the velocity is estimated, one uses Eq. (41) to best fit the product cv and see whether the value of c and the hypothesis on the universality class are self-consistent or not.

[16]In [18] it is shown that, in general, $\mathcal{G}_S^{x,y}(j,k)$ behaves nontrivially under $j \leftrightarrow k$, due to the fact that the ground state is not necessarily in the $S^z_{tot}=0$ sector.

To clarify the matter, let us start with the simpler case of the Haldane-Néel transition, that is thought to be in the $2D$ Ising universality class. Fixing $\lambda = 0.5$ we find $D_c(0.5) = -1.2$, and the β-function method [29] yields $\nu(0.5) = 1.023 \pm 0.009$, as far as the gap exponent, $\Delta E \propto (D - D_c)^\nu$ is concerned. Moreover, we observe the following nontrivial feature of the spectrum: The massless modes described by the CFT seem to be all and only the levels within $S^z_{tot} = 0$, while those with $S^z_{tot} \neq 0$ mantain a finite energy gap in the limit of large L. Hence, the reference state for the calculation of v will be the second excited state in $S^z_{tot} = 0$, corresponding to the primary field of conformal dimensions (1/2,1/2). Using quadratic extrapolations in $1/L$ we get $v = 2.44$, and consequently $e_\infty = -2.0011961 \pm 0.0000006$ and $c = 0.5008 \pm 0.0008$, thereby confirming the Ising universality class. The scaling dimensions can be estimated from the slopes of the straight lines in a plot like that of Fig.3. In Table 1 the theoretical values anticipated in Sect.4 are compared with these numerical estimates. The overall agreement is good (7 % in the worst case). Note that all the marginal operators have nonzero momentum and so they cannot represent a valid perturbation to the continuum Hamiltonian in as much as they would break translational invariance. The absence of marginal operators suggests that each point of the Haldane-Néel transition corresponds to the same $c = 1/2$ theory and the line in the phase diagram is "generated" by the mapping from the discrete spin model to the continuum CFT. Repeating the same passages at $\lambda = 1$ we get $D_c(1) = -0.315$, $\nu(1) = 1.003 \pm 0.006$ together with $v = 2.65$, $e_\infty = -1.62651$, $c = 0.498 \pm 0.002$, that is, again a $c = 1/2$ continuum theory.

We now pass to an example of numerical investigation of a $c = 1$ line of critical points, namely the transition from the Haldane to the large-D phase. In the past [20], a similarity with the critical fan of the Ashkin-Teller model has been suggested. The operator content of this model arises from Ginsparg's orbifold construction [24] and consists of a number of K-independent scaling dimensions plus the contributions coming from the pure Gaussian part (free boson) discussed in the previous Section. The fact that we do not observe K-independent dimensions (apart from trivial secondaries of the identity) indicates that the continuum description of our spin-1 Hamiltonian with PBC at the Haldane-large-D transition should be purely Gaussian rather than "orbifold-like".

In order to support this claim, we try again to match the whole spectrum of the relevant and marginal operators ($d \leq 2$). The difference with the $c = 1/2$ case is that here we have to fix not one but *two* nonuniversal parameters, v and K (see Eq. (50)). As regards the former, the velocity stems from the first and second excited states in $S^z_{tot} = 0$. Note that in choosing these levels we are assuming, self-consistently, that $K > 1$ so that the two secondaries of the identity ($d = 1$) come first than the primaries with $(m = 0, n = \pm 1)$, having $d_{0,\pm 1} = K$. As

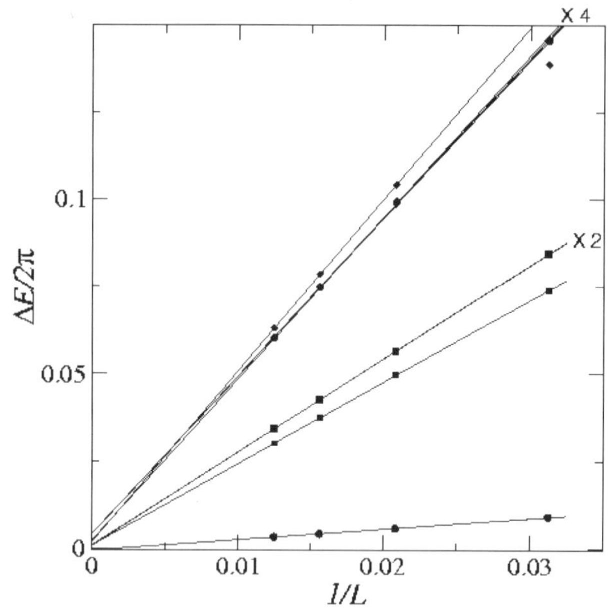

Figure 4. Energy differences, divided by 2π, plotted vs $1/L$ at the Ising transition ($\lambda = 0.5, D = -1.2$). Points represent the numerical values obtained with multi-target DMRG runs ($M = 216$) collecting nine excited states within $S^z_{tot} = 0$. Continuous lines are best-fit whose slopes are given in Table 1, together with the theoretical predictions of the scaling dimensions (the labels on the right indicate the multiplicities, all correctly met).

far as the Luttinger parameter K is concerned, we have to inspect the spectrum in other sectors of S^z_{tot} too. In particular, the first excited state lies in $|S^z_{tot}| = 1$, that corresponds to $m = \pm 1, n = 0$ in Eq. (50). The value of K is obtained from the slope $d_{\pm 1,0} = 1/4K$ in a plot similar to that of Fig.3. More generally, in order to check the self-consistency of the hypothesis $c = 1$, we have computed the finite-size spectrum of relevant and marginal operators in different sectors of S^z_{tot} for a couple of critical points on the Haldane-large-D line (first two rows of Table 2). Once that v and K are numerically determined, the structure of the Gaussian spectrum is correctly reproduced (including the multiplicities) and the overall comparison is satisfactory since, in worst cases, the relative difference does not exceed 3% (see plots and tables of Ref. [17]). The agreement with the theoretical predictions of the mapping in the planar regime is also remarkable. If we plug the coordinates of the critical points in the formulae of g and v for the Gaussian model derived above (Eq. (52)), we obtain $v = g = 2.07$,

Table 2. Velocity, central charge and ground state energy density for some critical points on the Haldane-large-D transition line. The numbers are the outcome of $DMRG$ calculations with $L = 16, 20, 24, 32, 48, 64$ and $M = 405$, for cases with $K > 1$, or $L = 16, 20, 24, 28, 32, 36, 40$ and $M = 400$, for cases with $K < 1$. The last two columns contain the numerical estimate of the nonuniversal parameter K, (according to the procedures described in the text) and the gap exponent obtained from the CFT formula $\nu = 1/(2 - K)$. The error on e_∞ is typically of one unit in the last digit or better.

$[\lambda, D_c(\lambda)]$	v	c	e_∞	K	ν
$(0.50, 0.65)$	2.197 ± 0.004	1.008 ± 0.003	-0.908765	1.580 ± 0.004	2.38
$(1.00, 0.99)$	2.588 ± 0.006	0.997 ± 0.003	-0.859152	1.328 ± 0.004	1.49
$(2.59, 2.30)$	3.70 ± 0.04	0.99 ± 0.01	-0.675099	0.85 ± 0.01	0.870
$(3.20, 2.90)$	4.445 ± 0.005	1.133 ± 0.006	-0.59132	0.526 ± 0.007	0.678

Enforced by these quantitative predictions, we try to approach the multicritical point where the $c = 1$ line meets the $c = 1/2$ one. Supposedly, the central charge at this point is $c = 3/2$ and it has been proposed [49] that the corresponding CFT is a $SU(2)_2$ Wess–Zumino–Witten–Novikov model. If this was true, the two lines should join at the point where the effective Gaussian theory has $K = 1$ [27] (FD point). Using $\lambda \simeq D$ in the expression of g we find that $K = \pi/g(\lambda) = 1$ is satisfied for $\lambda \simeq 2$, while it is believed [13] that the multicritical point lies at $\lambda \gtrsim 3$. We guess that the two lines join at $K < 1$, and in order to test this conjecture we study two more points: $(\lambda = 2.59, D = 2.30)$, again on the $c = 1$ line, and $(\lambda = 3.20, D = 2.90)$ proposed in [13] as the multicritical point itself. Altough the steps are conceptually the same as above, here we encounter two additional complications. First, due to the closeness (or almost coincidence in the multicritical case) of the Ising transition, we observe the merging of the two (quasi)critical spectra. Hence, we have to target more states and separate the ones belonging to $c = 1$ from the ones belonging instead to $c = 1/2$. Second, we observe sizeable finite-size corrections from irrelevant operators. In fact, our analysis shows that we are moving at values of K smaller than one towards $K = 1/2$ where certain irrelevant operators become marginal. As explained in [17], the last two rows of Table 2 are obtained by extracting K not from the first excited state, but rather from half the sum of the pair of levels with $m = 0, n = \pm 1$ in $S^z_{tot} = 0$, to get rid of finite-size corrections. As anticipated, moving to the right of the Haldane-large-D line the value of K keeps on decreasing towards 1/2 (SD point) where we argue that this line meets the Haldane-Néel one and a first order transition starts.

We close the section with a few comments on the hidden topological order measured by string order parameters (Eq. 23). It is expected that, leaving the Haldane phase, the $Z_2 \times Z_2$ symmetry is partially or totally restored. More precisely, when the $c = 1$ line is crossed, both $\mathcal{O}^z_S$ and $\mathcal{O}^{x,y}_S$ vanish. As customary, we can introduce two off-critical exponents that control the closure of these order parameters. For instance, fixing λ and varying D about $D_c(\lambda)$:

$$\mathcal{O}_S^z \propto (D_c - D)^{2\beta_S^z} \ , \ \ \mathcal{O}_S^x \propto (D_c - D)^{2\beta_S} \tag{54}$$

Now, according to FSS arguments (sec. 5.1 of [24]), β_S and β_S^z are related, via the gap exponent ν, to their counterparts at criticality, that is, the scaling dimensions of the operators entering the associated string correlation functions. These dimensions, in turn, can be extracted from the slopes, η_S and η_S^z, in the log-log plots of $\mathcal{O}_S^{x,z}(D = D_c)$ evaluated at half of the chain. Using the relation $2\beta_S = \nu\eta_S$ (and analogously for the z channel) we find the values reported in Table 3 for a couple of critical points already discussed above. We should observe that the scaling dimensions $\eta_S/2$ and $\eta_S^z/2$ are *not* contained in the $c = 1$ spectra cited above. However, we notice also that the numerical estimates of η_S^z are rather close to the values $\frac{2\, d_{0,\pm 1}}{4} = K/2$ and that these levels actually exist in the effective continuum theory provided that *half-integer* values of n are allowed in Eq. (50). In the XXZ spin-1/2 formulation this is known to correspond to twisted boundary conditions on the chain. Thus, considering that the calculations presented here for the spin-1 case are with PBC, it's not surprising that the scaling dimensions associated with $\mathcal{O}_S^{x,z}$ are absent in the numerical spectra. Nonetheless, we believe that the closeness to $K/2$ is not accidental and in Ref. [17] we speculated about the possibility that the longitudinal string correlation functions (Eq. (21) with $\alpha = z$) acquires, in the continuum limit, the asymptotic form

$$\mathcal{G}_S^z(r) \sim \langle \exp\left[\mp i\sqrt{\pi K}\Phi(0)\right] \exp\left[\pm i\sqrt{\pi K}\Phi(r)\right]\rangle \tag{55}$$

so that the lattice string $S_0^z \exp\left[i\pi \sum_{l=1}^{r-1} S_l^z\right]$ is somehow related to the continuum twist operator $\exp\left[\pm i\sqrt{\pi K}\Phi(r)\right]$.

$[\lambda, D_c(\lambda)]$	$2\beta_S$	$2\beta_S^z$	η_S	η_S^z
(0.50,0.65)	0.597 ± 0.009	1.91 ± 0.02	0.251 ± 0.002	0.804 ± 0.003
(1.00,0.99)	0.407 ± 0.002	1.10 ± 0.01	0.2733 ± 0.0006	0.741 ± 0.002

Table 3. Exponents associated with the vanishing string order parameters at the Gaussian transitions taking place at the points indicated in the first column (see text for definitions). All the numbers are obtained with FSS on the data at $L = 32, 48, 64, 80, 100$ and $M = 300$.

6. Conclusions.

In the present paper we have reviewed, to the best of our knowledge, part of the status-of the-art concerning Heisenberg spin chains, including biquadratic interaction terms and various kinds of anisotropies, concentrating on the rôle of hidden symmetries in the various families of spin models. We have discussed how the inclusion of anisotropy terms can drive the "standard" Heisenberg chain away from the Haldane phase and how hidden symmetries (and their spontaneous breaking) are of great help in classifying the "massive" (gapfull) phases of the model. The location of the critical lines of the model has been accurately obtained numerically, confirming and extending earlier predictions [13].

The combined use proposed here of analytical and numerical [(CFT) and $(DMRG)$, respectively] techniques to investigate the critical properties of the models has proved to be a rather successful strategy to clarify the nature and structure of the critical phases of the models. Numerical simulation techniques (Monte Carlo and $DMRG$, to quote only the most known ones) are of more and more frequent and extended use in almost all branches of Theoretical Physics. A blind use of them can however be more dangerous than helpful in understanding the physical properties of the systems for whose study they are employed. We believe instead that an "educated" use of numerical techniques in support of analytical approaches, as described here, can result in a powerful synergy that can be of great help in understanding the physics of many problems in Theoretical Physics.

Acknowledgments

The authors would like to thank F. Ortolani and M. Roncaglia for useful discussions and for a critical reading of the manuscript. One of us (G.M.) would like to thank the organizer of the XIII Conference on "*Symmetries in Physics*", Prof. Bruno Gruber, for inviting him to take part in the Bregenz Conference, where the main content of the present paper was presented.

References

[1] Affleck, I. in: *"Fields, Strings and Critical Phenomena"*. Brézin, E., Zinn-Justin, J. (Eds.). North-Holland, 1989

[2] Affleck, I., Kennedy, T., Lieb, E.H., Tasaki, H., *Comm. Math. Phys.* **115**,477,1988

[3] Affleck, I., Lieb, E.H., *Lett. Math. Phys.* **12**, 57, 1986

[4] Amit, D.J.: *"Field Theory, the Renormalization Group and Critical Phenomena"*. McGraw-Hill, 1978

[5] Andersson, M., Boman, M., Östlund, S., *Phys. Rev.* **B59**, 10493, 1999

[6] Auerbach, A.: *"Interacting Electrons and Quantum Magnetism"*. Springer, 1994

[7] Berezinskii, V.L., *Sov. Phys. JETP* **59**, 907, 1970 and **61**, 1144, 1971

[8] Bethe, H.Z., *Phys.* **71**,205,1931

[9] Babujian, H.M., *Phys. Lett.* **90A**, 479, 1982, and Nucl. Phys. **B215**, 317, 1983

[10] Botet, R., Julien, R., Kolb, M., *Phys. Rev.* **B28**,3914,1983

[11] Bott, R., Tu, L.: *"Differential Forms in Algebraic Topology"*. Springer,1982

[12] Cardy, J.L., in: see Ref. [1]

[13] Chen, W., Hida, K., Sanctuary, B.C., cond-mat/0209403 and: Phys. Rev. **B67**, 104401, 2003

[14] Coleman, S., *Comm. Math. Phys.* **31**, 259, 1973

[15] Dagotto, E., Rice, T.M., *Science* **271**, 618, 1996

[16] Degli Esposti Boschi, C. Unpublished

[17] Degli Esposti Boschi, C., Ercolessi, E., Ortolani, F., Roncaglia, M., cond-mat/0307396, to be published in *Eur. Phys. J. B*

[18] Degli Esposti Boschi, C., Ortolani, F. In preparation

[19] Dell'Aringa, S., Ercolessi, E., Morandi, G., Pieri, P., Roncaglia, M., *Phys. Rev. Lett.* **78**, 2457, 1997

[20] Den Nijs, M., Rommelse, K., *Phys. Rev.* **B40**, 4709, 1989

[21] Di Francesco, P., Mathieu, P., Séneéchal, D.: *"Conformal Field Theory"*. Springer,1997

[22] Ercolessi, E., in: Proceedings of the Conference "Space-time and fundamental interactions: quantum aspects", Vietri sul Mare (2003). Lizzi, F., Marmo, G., Sparano, G., Vilasi, G. (Eds.). To be published in *Mod. Phys. Lett. A*

[23] Fradkin, E.: *"Field Theories of Condensed-Matter Systems"*. Addison-Wesley, 1991

[24] Ginsparg, P., in: see Ref. [1]

[25] Girvin, S.M., Arovas, D.P., *Physica Scripta* **T27**, 156 (1989)

[26] Glaus, U., Schneider, T., *Phys. Rev.* **B30**, 215, 1984

[27] Gogolin, A.O., Nersesyan, A.A., Tsvelik, A.M.: *"Bosonization and Strongly Correlated Systems"*. C.U.P.,1998

[28] Haldane, F.D.M., *Phys. Rev. Lett.* **50**, 1153 (1983)

[29] Hamer, C.J., Barber, M.N., J. Phys. A: Math. Gen. **14**, 241, 1981

[30] Hulthén, L., *Ark. Mat. Astronom. Fysik* **26A**, Na.11, 1938

[31] Kennedy, T., Tasaki, H., *Comm. Math. Phys.* **147**, 431 (1992)

[32] Klauder, J.R., Skagerstam, B.S.: *"Coherent States"*. World Scientific,1984

[33] Korepin, V.E., Bogoliubov, N.M., Izergin, A.G.: *"Quantum Inverse Scattering Method and Correlation Functions"*. C.U.P., 1993

[34] Kosterlitz, J.M., Thouless,D.J., *J. Phys.* **C5**, L124, 1972 and **C6**, 1181, 1973

[35] Legeza, Ö., Fáth, G., *Phys. Rev.* **B53**, 14349, 1996

[36] Lieb, E., Schultz, T., Mattis, D.C., *Ann. Phys.* (NY) **16**, 407, 1961

[37] Majumdar, C.K., Ghosh, D.K., *J. Phys.* **C3**, 911, 1970

[38] Mattis, D.C.: *"The Theory of Magnetism"*. Springer, 1981

[39] Messiah, A.: *"Quantum Mechanics"*. North-Holland, 1961

[40] Morandi, G., *Nuovo Cim.* **66B**, 77, 1970

[41] Morandi, G.: *"The rôle of Topology in Classical and Quantum Physics"*. Springer, 1992

[42] Polyakov, A.M., *Phys. Lett.* **131B**, 121, 1975 and: *"Gauge Fields and Strings"*. Harwood,1993

[43] Qin, S., Lou, J., Sun, L., Chen, C., *Phys. Rev. Lett.* **90**, 067202, 2003

[44] Rajaraman, R.: *"Solitons and Instantons"*. North-Holland, 1982.

[45] Read, N., Shankar, S., *Nucl. Phys.* **B316**, 609, 1989

[46] Rice, T.M., Gopalan, S., Sigrist, M., *Europhys. Lett.* **23**, 445, 1993

[47] Roncaglia, M. Unpublished

[48] Sachdev, S.: *"Quantum Phase Transitions"*. C.U.P., 1999

[49] Schulz, H.J., *Phys. Rev.* **B34**, 6372, 1986

[50] Shapere., A., Wilczek, F. (Eds.): *"Geometric Phases in Physics"*. World Scientific, 1989

[51] Sierra, G. J., *Phys.* **A29**, 3299, 1966

[52] Sólyom, J., *Phys. Rev.* **B36**, 8642, 1987

[53] Sutherland, B., *Phys. Rev.* **B12**, 3795, 1975

[54] Takhatajan, L., *Phys. Lett.* **87A**, 479, 1982

[55] Wagner, H. Z., *Physik* **195**, 273, 1966

[56] White, S.R., *Phys. Rev. Lett.* **69**, 2863, 1992

[57] White, S.R., *Phys. Rev.* **B48**, 10345, 1993

[58] Wu, K., Simon, H., *SIAM J. Matrix Anal. Appl.* **22**, 602, 2000

[59] Zakrzewki, W.J.: *"Low Dimensional Sigma Models"*. Adam Hilger, 1989

[60] Zamolodchikov, A.B., Zamolodchikov, Al.B., *Nucl. Phys.* **B379**, 602, 1992

QUANTUM TOMOGRAPHY, WAVE PACKETS AND SOLITONS*

S. De Nicola
Istituto di Cibernetica "Eduardo Caianiello" - CNR
Comprensorio "A. Olivetti" Fabbr. 70,
Via Campi Flegrei, 34, I-80078 Pozzuoli (NA), Italy.

R. Fedele
Dipartimento di Scienze Fisiche,
Università "Federico II" di Napoli
and
Istituto Nazionale di Fisica Nucleare, Sezione di Napoli,
Complesso Universitario di Monte Sant'Angelo,
Via Cintia, I-80126 Napoli, Italy.

M.A. Man'ko and V.I. Man'ko
P.N. Lebedev Physical Institute,
Leninskii Prospect, 53,
Moscow 119991 Russia
mmanko@sci.lebedev.ru

Abstract The wave packets both linear and nonlinear such as solitons (signals) described by a complex time-dependent function are mapped onto positive probability distributions (tomograms). Quasidistributions, wavelets and tomograms are shown to have an intrinsic connection. Analysis is extended to signals obeying to the von Neumann-like equation. For solitons (nonlinear signals) obeying to the nonlinear Schrödinger equation, the tomographic probability representation is introduced. It is shown that in the probability representation the soliton satisfies to a nonlinear generalization of the Fokker–Planck equation. Solutions to the Gross–Pitaevskii equation corresponding to solitons in Bose–Einstein condensate are considered.

*lecture presented by Margarita Man'ko.

B. Gruber (ed.), Symmetries in Science XI, 175–208.

Keywords: Tomographic map, phase space, Wigner function, solitons, nonlinear Schrödinger equation, Bose–Einstein condensate.

1. Introduction

Quantumlike systems [1] were considered recently [2] because their completely classical behavior can be studied in detail using formalism and methods of quantum mechanics. Among these methods, there is quantum tomography [3–8] which uses as an important ingredient the Radon transform [9, 10] of a function of two variables. The physical systems which demonstrate the quantumlike behavior are, for example, the beams of charge particles [11], photon-beams propagating along the optical fibers [1], electric or acoustic signals depending on time and space coordinates [1, 2]. The electromagnetic waves and wave packets propagating paraxially through optical waveguides obey to the Schrödinger-like equation found by Fock and Leontovich [12]. The Fock–Leontovitch approximation was intensively used in fiber-optics problems [13–15]. The propagation of acoustic waves, in particular, in ocean is also described in the paraxial approximation by Schrödinger-like equation [16].

On the other hand, the thermal wave model of charged particle beams in accelerators was suggested [17] and developed [18]. Within the framework of this model, the charged particle beams are also described by the Schrödinger-like equation. Classical transmission line for electric signals was shown recently to be described by Schrödinger-like equation [19]. Another example of quantumlike approach to purely classical problems is the analysis of signals of different nature (electromagnetic, acoustic, seismic, biological, etc.). The signal is described by a function of time and this function is associated with the so-called analytic signal function [20–23]. The complex analytic signal $f(t)$ has exactly the same mathematical properties as the wave function of a quantum state $\psi(x)$ has. In view of this, in signal processing all known methods of quantum mechanics can be applied [24, 25]. In particular, some notion of quantum mechanics like entanglement can be applied to describe space–time correlations of electromagnetic signals in media [26].

The Schrödinger-like equation can be either linear or nonlinear. The linear equation corresponds to the standard Schrödinger equation [27] of quantum mechanics. Electromagnetic fields of high intensity, for example, in optical fibers have nonlinear properties. The nonlinear properties are described, at the lowest order, by the nonlinear Schrödinger equation with cubic nonlinearity. There exist other types of nonlinearities [28, 29]. The interesting solutions to many types of nonlinear equations correspond to solitary waves or solitons. There is huge mathematical literature on solitons (see, for example, [30]). The nonlinear Schrödinger equation governs the nonlinear dynamics of some other quantumlike systems, such as large-amplitude wave packet propagation in plasmas, large-amplitude dynamics of surface gravity waves, the transverse and longitudinal collective dynamics of intense high-energy charged-particle beams in both conventional and plasma-based accelerators.

Another important example of the nonlinear equation with soliton solutions is the Gross–Pitaevskii equation [31]. This equation describes the states of Bose–Einstein condensate (BEC). Different (bright, dark, grey) solitons in BEC have been experimentally observed [32–34].

All the systems mentioned are united by the feature that they are described by linear or nonlinear equations of the Schrödinger-like type. Also interesting solutions to these equations are either packets of different sorts in the linear case, e.g., coherent states [35–39], squeezed states [40, 41], correlated states [42] corresponding to minimization of Robertson–Schrödinger [35, 43] uncertainty relation (see also [44]), or to the solitary waves in nonlinear cases.

The tomographic map [7, 8, 45] provides the tool to map analytic signal or a solution to linear or nonlinear Schrödinger equation onto the standard probability distribution function. In view of the linear or nonlinear equations describing different variants of quantum or quantumlike systems, the tomographic probability representation, which we call *the probability representation*, can be written down. The advantage of this probability representation is that all different types of the objects like analytic signal, quantum wave packets, states of quantumlike photon beams and charged particle beams, solitary waves can be treated as standard positive probability distributions. The evolution equations for the probability distributions are similar in some sense to the Fokker–Planck equation of classical stochastical processes and we will call these equations *Fokker-Planck-type* equations. It is worthy noting that the tomographic map was shown [46] to be in one-to-one correspondence to a specific version of the star-product quantization scheme [47]. Also it was clarified recently [48] that the tomographic probability is related to known object of quantum information theory called *quantum probability measure* [49].

The aim of this article is to review the results on quantumlike systems related to linear charged-particle-beam transport [50] and nonlinear systems described by nonlinear Schrödinger equation [51] and by the Gross–Pitaevskii equation in the BEC case [52] using the tomographic probability representation. To do this, we also consider Wigner–Moyal [53, 54] phase space representation of the equations under discussion.

The paper is organized as follows.

In Section 2 we review the Weyl–Wigner map and in Section 3 we construct tomographic map, with the tomographic map for chirped soliton being elaborated in Section 4. In Section 5 we study the particle beam propagation and in Section 6 we consider the propagator of particle beam in the phase space. In Section 7 we show tomographic representation of particle beams and in Section 8 we develop perturbation approach to the tomographic evolution equation, with the example of quadrupole being given in Section 9. In Sections 10–13 nonlinear Schrödinger equations are studied. In Section 14 Bose–Einstein condensate is considered and in Section 15 the Gross–Pitaevskii equation is reviewed, while in Section 16 solitons of BEC are studied. A general relation among different transforms of analytic signal is given in Section 17. Conclusions and remarks are presented in Section 18.

2. Phase-space representation

In this section, we review the approach called symplectic tomography of the quantum state. The real meaning of this scheme is a map of the complex wave function $\psi(x)$ of a real variable x $(-\infty < x < \infty)$ onto a family of probability distributions $w(X, \mu, \nu)$ of a random real variable X $(-\infty < X < \infty)$ labeled by two real parameters μ $(-\infty < \mu < \infty)$ and ν $(-\infty < \nu < \infty)$.

We use dimensionless variables. The map can be realized by the following steps.

First, one constructs the density matrix [55] which is a complex function of two variables x and x' (although the quantities depend on the timelike variable, for simplicity, hereafter we adopt notations that do not indicate explicitly such a dependence)

$$\rho_\psi(x, x') = \psi(x)\psi^*(x'). \tag{1}$$

Then one uses the Weyl–Wigner map of the density matrix onto the real Wigner function on the phase space $W(q, p)$ [53] of two real variables p and q

$$W_\psi(q, p) = \frac{1}{2\pi} \int \rho_\psi\left(q + \frac{u}{2}, q - \frac{u}{2}\right) e^{ipu}\, du. \tag{2}$$

The Wigner function takes real values. If the wave function is normalized

$$\int |\psi(x)|^2\, dx = 1,$$

the Wigner function is also normalized

$$\int W_\psi(q, p)\, dq\, dp = 1.$$

Ville used this map in the analytic signal theory [56].

The inverse of the Fourier transform (37) defining the Wigner function in terms of the density matrix reads

$$\psi(x)\psi^*(x') = \int W_\psi\left(\frac{x + x'}{2}, p\right) e^{-ip(x-x')}\, dp. \tag{3}$$

One can see that the Wigner function determines the complex wave function up to a constant factor:

$$\psi^*(0)\psi(x) = \int W_\psi\left(\frac{x}{2}, p\right) e^{-ipx}\, dp. \tag{4}$$

Modulus of the constant factor $|\psi(0)|$ is determined by the relationship

$$|\psi(0)|^2 = \int W_\psi\,(0,p)\;dp. \tag{5}$$

We suppose that $\psi(0)$ is not equal to zero.

Thus, given the Wigner function one can reconstruct the complex wave function up to the constant phase factor. This means that the Wigner function contains the same information that the density matrix does. Also this means that the Wigner function contains the same information that the wave function $\psi(x)$ does (up to the constant phase factor). The Wigner function can be identified with the so-called Weyl symbol of the density operator describing the quantum state. There exist different kinds of symbols of operators. As it was shown recently [46] tomograms can be also identified with a specific symbol of a density operator.

3. State tomogram

Let us now construct the tomographic map.

To do this, we use the integral Radon transform of the Wigner function

$$w(X,\mu,\nu) = \int W_\psi(q,p)\delta(X-\mu q-\nu p)\,dq\,dp\,. \tag{6}$$

In this formula, the Dirac delta-function term $\delta(X-\mu q-\nu p)$ collects values of the Wigner function $W_\psi(q,p)$ from the line in the phase space which is described by the expression obtained by equating the argument of Dirac delta-function to zero. One can prove [57] that, for normalized Wigner function, the function $w(X,\mu,\nu)$ is a normalized probability distribution function of a random variable X (called tomogram of the Wigner function), i.e., one has

$$\int w(X,\mu,\nu)\,dX = 1, \tag{7}$$

and this distribution function can be used for the new formulation of quantum mechanics with the probability instead of the wave function. This tomogram was called symplectic tomogram [8], since it is related to linear symplectic transform in the phase space.

There is another form of the map (44) given in terms of the Fourier transform [we used the Fourier transform of delta-term in (44)]

$$w(X,\mu,\nu) = \int W_\psi(q,p)\exp\Big[ik(X-\mu q-\nu p)\Big]\,\frac{dk\,dq\,dp}{2\pi}. \tag{8}$$

The Fourier integral form of the tomogram (46) gives the possibility to get easily the inverse transform [7]

$$W_\psi(q,p) = \int w(X,\mu,\nu)\exp\Big[i(X-\mu q-\nu p)\Big]\,\frac{dX\,d\mu\,d\nu}{(2\pi)^2}\,. \tag{9}$$

Thus, given the symplectic tomogram $w(X, \mu, \nu)$ one can find the Wigner function $W(q,p)$. Using known relationships (36), (37) and (41) of the Wigner function $W(q,p)$ and the wave function $\psi(x)$, one can obtain the expression for tomograms in terms of the wave function

$$w(X, \mu, \nu) = \frac{1}{2\pi|\nu|} \left| \int \psi(y) \exp\left(\frac{i\mu}{2\nu} y^2 - \frac{iX}{\nu} y\right) dy \right|^2. \tag{10}$$

This formula shows that for a complex function $\psi(x)$ one can find the tomogram. Formula (10) was found for the quantum wave function and noncommutative tomography of analytic signal in [22]. But it can be used for other arbitrary aims as well. The goal which we are going to reach is to use this formula for the description of wave packets and solitons.

4. Fourier transform of a chirped packet

There are some properties of tomograms to be used. The homogeneity property [45] follows from relations (44) and (10)

$$w(\lambda X, \lambda\mu, \lambda\nu) = \frac{1}{|\lambda|} w(X, \mu, \nu). \tag{11}$$

This means that, in reality, the tomogram is the function of two real variables. For example, one can take

$$\mu = \cos\theta, \qquad \nu = \sin\theta. \tag{12}$$

In this case, the symplectic tomogram is given by the formula

$$w(X, \theta) = \int W(q,p)\delta(X - q\cos\theta - p\sin\theta) dq\, dp. \tag{13}$$

This relation of the Wigner function $W(q,p)$ and the tomographic probability $w(X,\theta)$ (called tomogram of the optical tomography scheme) was used to point out the possibility to reconstruct the phase-space distribution in terms of the optical tomogram in [3, 4]. The symplectic tomogram is simpler than the optical tomogram because the reconstruction formula for the Wigner function (9) does not contain the singular terms of the Radon transform (see [3, 4]). Nevertheless, the useful property of the optical tomogram is the fact that it depends on two variables only. The relation of the optical tomogram $w(X,\theta)$ to the wave function reads

$$w(X, \theta) = \frac{1}{2\pi|\sin\theta|} \left| \int \psi(y) \exp\left(\frac{i\cot\theta}{2} y^2 - \frac{iX}{\sin\theta} y\right) dy \right|^2. \tag{14}$$

The integrand in the above expression (14) is similar to the Green function of the quantum harmonic oscillator and the tomogram coincides with modulus squared of the fractional Fourier transform of $\psi(y)$ [23]. The tomogram $w(X,\theta)$ is used in quantum optics in the scheme of measuring quantum photon states by means of the so-called optical homodyne tomography [5, 58]. It was also discussed in the context of quantum-optics measurements and signal detection, respectively, in [6] and [59]. Thus the tomogram $w(X,\mu,\nu)$ determines completely the Wigner function and the complex function $\psi(x)$ (up to a phase factor), if it is known for real parameters satisfying the constraint

$$\mu^2+\nu^2=1. \tag{15}$$

But one can use other constraints too. Thus, the homogeneity property (11) implies that the particular values of symplectic tomogram, e.g., $w(1,\mu,\nu)$, $w(X,1,\nu)$ and $w(X,\mu,1)$ determine the whole tomogram and, consequently, the complex function $\psi(x)$ (up to the phase factor) and the Wigner function $W(q,p)$ completely.

Another property can be obtained by change of variables in (10)

$$\frac{y}{\nu}=z, \tag{16}$$

which gives the following expression for the tomogram

$$w(X,\mu,\nu)=\frac{|\nu|}{2\pi}\left|\int\widetilde{\psi}(z,\mu,\nu)e^{-iXz}dz\right|^2, \tag{17}$$

where the function $\widetilde{\psi}(z,\mu,\nu)$ is the function describing "a chirped soliton" [we mean that $\psi(y)$ in (10) is considered as a soliton solution of a nonlinear equation]

$$\widetilde{\psi}(z,\mu,\nu)=\psi(z,\nu)\exp\left(\frac{i}{2}\,\mu\nu z^2\right). \tag{18}$$

Expression (17) is convenient for numerical calculations because it gives the tomogram in terms of the standard Fourier transform of the chirped soliton. In terms of optical tomogram, the expression can be rewritten as

$$w(X,\theta)=\frac{|\sin\theta|}{2\pi}\left|\int\widetilde{\psi}(z,\theta)e^{-iXz}dz\right|^2, \tag{19}$$

where

$$\widetilde{\psi}(z,\theta)=\psi(z,\sin\theta)\exp\left(\frac{iz^2}{4}\,\sin 2\theta\right). \tag{20}$$

The formula obtained was used to make plots of soliton tomograms in [51].

5. Integrals of motion and propagator

Within the framework of the thermal wave model, a charged-particle beam as a whole is described by *the beam wave function* $\Psi(x,z)$ where (in the one-dimensional case) x and z are the transversal and longitudinal coordinates, respectively. The complex beam wave function satisfies a Schrödinger-like equation [17]

$$i\epsilon\,\frac{\partial\Psi}{\partial z}=-\frac{\epsilon^2}{2}\frac{\partial^2}{\partial x^2}\Psi+U(x,z)\Psi, \tag{21}$$

where ϵ is the beam emittance. The solution to equation (21) can be presented in terms of the propagator $G(x,x',z)$ which is the matrix element of the "evolution operator" $\widehat{U}(z)$, namely,

$$G(x,x',z)=\langle x\mid\widehat{U}(z)\mid x'\rangle, \tag{22}$$

and one has

$$\Psi(x,z)=\widehat{U}(z)\Psi(x,0), \tag{23}$$

or

$$\Psi(x,z)=\int G(x,x',z)\Psi(x',0)\,dx'. \tag{24}$$

Thus, the propagator satisfies the equation

$$\left[i\epsilon\,\frac{\partial}{\partial z}+\frac{\epsilon^2}{2}\frac{\partial^2}{\partial x^2}-U(x,z)\right]G(x,x',z)=i\epsilon\,\delta(z)\,\delta(x-x'). \tag{25}$$

In analogy with the proper quantum mechanics, a quantumlike integral of motion $\widehat{I}(z)$, which satisfies the following condition

$$\frac{\partial\widehat{I}(z)}{\partial z}+\frac{i}{\epsilon}\left[\widehat{H},\widehat{I}\right]=0, \tag{26}$$

where $\widehat{H}$ is the Hamiltonian of the thermal wave model, and in the coordinate representation reads

$$\widehat{H}=-\frac{\epsilon^2}{2}\frac{\partial^2}{\partial x^2}+U(x,z)\,, \tag{27}$$

with $U(x,z)$ being the potential-energy term, can be naturally introduced [18]. Consequently, according to the quantum-mechanical case [60, 61], the operators of the form

$$\widehat{I}(z)=\widehat{U}(z)\,\widehat{I}(0)\,\widehat{U}^{\dagger}(z) \tag{28}$$

are integrals of motion. The physical meaning of the integrals of motion (28) is the following. The integral of motion "remembers" the initial value of the observable $\widehat{I}(z)$ at the value of the longitudinal coordinate $z = 0$. There are two specific integrals of motion — initial momentum

$$\hat{p}_0(z) = \widehat{U}(z)\,\hat{p}\,\widehat{U}^\dagger(z) \tag{29}$$

and initial position

$$\hat{x}_0(z) = \widehat{U}(z)\,\hat{x}\,\widehat{U}^\dagger(z), \tag{30}$$

in which the position and momentum operators in the coordinate representation read

$$\hat{p} = -i\epsilon\frac{\partial}{\partial x}, \qquad \hat{x} = x. \tag{31}$$

One can check that the quantumlike propagator $G(x, x', z)$ satisfies the following equations, which are analogs of the equations for the Green function discussed in quantum mechanics [60, 61]

$$\hat{p}_0(z)G(x, x', z) = i\epsilon\,\frac{\partial}{\partial x'}\,G(x, x', z) \tag{32}$$

and

$$\hat{x}_0(z)G(x, x', z) = x'\,G(x, x', z), \tag{33}$$

where the operators $\hat{x}_0(z)$ and $\hat{p}_o(z)$ act on the coordinate x of the propagator.

6. The particle-beam propagator for Wigner function

In this section, we develop the concept of particle-beam propagator in the phase space, making use of the Wigner quasidistribution function.

According to [18] one can introduce the density operator, which in the phase-space representation is described by the Wigner function (we introduce the parameter ϵ and the longitudinal coordinate z)

$$\rho_w(x, p, z; \epsilon) = \frac{1}{2\pi\epsilon}\int_{-\infty}^{+\infty} \Psi^*\left(x + \frac{y}{2}, z\right)\Psi\left(x - \frac{y}{2}, z\right)\,\exp\left(i\,\frac{py}{\epsilon}\right)\,dy, \tag{34}$$

which is normalized

$$\int \rho_w(x, p, z; \epsilon)\,dx\,dp = 1. \tag{35}$$

Note that, in addition, for the beam state described by the beam wave function, the beam's Wigner function satisfies "the purity" condition

$$2\pi\epsilon\int \rho_w^2(x, p, z; \epsilon)\,dx\,dp = 1. \tag{36}$$

The Wigner function satisfies the Moyal-like evolution equation

$$\frac{\partial \rho_w}{\partial z} + p\,\frac{\partial \rho_w}{\partial x} + \frac{i}{\epsilon}\left[U\left(x + \frac{i\epsilon}{2}\frac{\partial}{\partial p}\right) - U\left(x - \frac{i\epsilon}{2}\frac{\partial}{\partial p}\right)\right]\rho_w = 0, \qquad (37)$$

which is the von Neumann equation for the density operator in the Wigner–Weyl representation.

One can introduce the propagator for the Moyal-like evolution equation by the formula

$$\rho_w(x, p, z; \epsilon) = \int \pi(x, p, x', p', z; \epsilon)\, \rho_w(x', p', 0; \epsilon)\, dx'\, dp'. \qquad (38)$$

The propagator $\pi(x, p, x', p', z; \epsilon)$ satisfies the equation

$$\begin{aligned}&\frac{\partial \pi}{\partial z} + p\,\frac{\partial \pi}{\partial x} + \frac{i}{\epsilon}\left[U\left(x + \frac{i\epsilon}{2}\frac{\partial}{\partial p}\right) - U\left(x - \frac{i\epsilon}{2}\frac{\partial}{\partial p}\right)\right]\pi \\ &= \delta(z)\delta(x - x')\delta(p - p'). \end{aligned} \qquad (39)$$

If one introduces the classical-like Liouville operator

$$\widehat{\mathcal{L}} \equiv \frac{\partial}{\partial z} + p\,\frac{\partial}{\partial x} - \left(\frac{\partial U}{\partial x}\right)\frac{\partial}{\partial p}, \qquad (40)$$

the Moyal-like equation can be rewritten as

$$\widehat{\mathcal{L}}\rho_w = \sum_{k=1}^{\infty} \frac{(-1)^k}{(2k+1)!}\left(\frac{\epsilon}{2}\right)^{2k} \frac{\partial^{2k+1} U}{\partial x^{2k+1}}\,\frac{\partial^{2k+1}\rho_w}{\partial p^{2k+1}}, \qquad (41)$$

and, according to the quantum formalism, it is easy to see that the propagator of the Wigner function is expressed in terms of the propagator of the beam wave function as follows:

$$\begin{aligned}\pi(q, p, q', p', z; \epsilon) \;=\; &\frac{1}{2\pi}\int G\left(q + \frac{u}{2}, q' + \frac{s}{2}, z\right) G^*\left(q - \frac{u}{2}, q' - \frac{s}{2}, z\right) \\ &\times \exp\left(-ipu + ip's\right)\, du\, ds. \end{aligned} \qquad (42)$$

7. Propagator in probability representation

Using the approach of the previous section, one can define the following tomographic-probability-distribution function for the particle beam [62],[63]

$$w(X, \mu, \nu, z; \epsilon) = \int \exp\left[-ik(X - \mu x - \nu p)\right] \rho_w(x, p, z; \epsilon)\, \frac{dk\, dx\, dp}{(2\pi)^2}, \qquad (43)$$

which contains all information on the Wigner quasidistribution ρ_w. The function $w(X,\mu,\nu,z;\epsilon)$, also called the *marginal distribution*, has all the features of a classical distribution function of the random variable X.

In the tomographic representation, one can obtain the Fokker–Planck-like equation for the marginal distribution $w(X,\mu,\nu,z;\epsilon)$ of the form

$$\frac{\partial w}{\partial z} - \mu\frac{\partial}{\partial \nu}w + \frac{i}{\epsilon}\left[U\left(-\frac{1}{\partial/\partial X}\frac{\partial}{\partial \mu} + i\frac{\nu\epsilon}{2}\frac{\partial}{\partial X}\right)\right.$$
$$\left. -U\left(-\frac{1}{\partial/\partial X}\frac{\partial}{\partial \mu} - i\frac{\nu\epsilon}{2}\frac{\partial}{\partial X}\right)\right] w = 0. \tag{44}$$

Following [64] one can represent equation (43) in the form

$$\widehat{L}w = \frac{2}{\epsilon}\sum_{n=1}^{\infty}\frac{(-1)^n}{(2n+1)!}U^{(2n+1)}(\hat{q})\left(\frac{\nu\epsilon}{2}\right)^{2n+1}\frac{\partial^{2n+1}w}{\partial X^{2n+1}}, \tag{45}$$

where

$$\widehat{L} \equiv \frac{\partial}{\partial z} - \mu\frac{\partial}{\partial \nu} - \nu U^{(1)}(\hat{q})\frac{\partial}{\partial X}, \tag{46}$$

with

$$U^{(2n+1)}(\hat{q}) \equiv \frac{\partial^{2n+1}}{\partial x^{2n+1}}U(x=\hat{q}) \quad \text{and} \quad \hat{q} = -\frac{1}{\partial/\partial X}\frac{\partial}{\partial \mu}.$$

One can introduce the propagator for the tomographic probability (below we take $\epsilon = 1$ and omit it)

$$w(X,\mu,\nu,z) = \int \Pi(X,\mu,\nu,X',\mu',\nu',z)\, w(X',\mu',\nu',0)\, dX'\, d\mu'\, d\nu' \tag{47}$$

and the propagator for the beam density matrix

$$\rho(X,X',z) = \Psi(X,z)\Psi^*(X',z), \tag{48}$$

such that

$$\rho(X,X',z) = \int K(X,X',Y,Y',z)\, \rho(Y,Y',0)\, dY\, dY',$$

which is related to the Green function (propagator) of the beam wave function as

$$K(X,X',Y,Y',z) = G(X,Y,z)\, G^*(X',Y',z). \tag{49}$$

The propagator for the density matrix is related to the propagator for the tomographic probability by the relationship [64]

$$\begin{aligned} K(X,X',Z,Z',z) &= \frac{1}{(2\pi)^2}\int \frac{1}{|\nu'|}\exp\left\{i\left(Y-\mu\,\frac{X+X'}{2}\right)\right. \\ &\left. -i\,\frac{Z-Z'}{\nu'}\,Y'+i\,\frac{Z^2-Z'^2}{2\nu'}\,\mu'\right\} \\ &\times\Pi(Y,\mu,X-X',Y',\mu',\nu',t)\,d\mu\,d\mu'\,dY\,dY'\,d\nu'\,. \end{aligned} \tag{50}$$

The tomographic propagator is expressed in terms of the propagator of the beam wave function as follows:

$$\begin{aligned} &\Pi(X,\mu,\nu,X',\mu',\nu',z) \\ &= \frac{1}{(4\,\pi)^2}\int k^2 G\left(a+\frac{k\nu}{2}\,,y\,,z\right)G^*\left(a-\frac{k\nu}{2}\,,\tilde{z}\,,z\right)\delta\left(y-\tilde{z}-k\nu'\right) \\ &\times\exp\left[ik\left(X'-X+\mu a-\mu'\,\frac{y+\tilde{z}}{2}\right)\right]\,dk\,dy\,d\tilde{z}\,da. \end{aligned} \tag{51}$$

8. First order Born approximation

Since in quantum mechanics the perturbation theory for solving the Schrö - din - ger equation is well developed, one can apply the results of this theory to problems of electron optics given in the quantumlike domain. Thus, the beam wave function $\Psi(x,z)$, being the solution to a Schrödinger-like equation with the Hamiltonian

$$H = H_0 + V(x,z)\,,$$

where $V(x,z)$ is a small perturbative potential (aberration potential) and H_0 is unperturbed Hamiltonian, can be written in the form (for simplicity, we take emittance $\epsilon = 1$)

$$\Psi(x,z) \approx \Psi_0(x,z) + \frac{1}{i}\int_0^z dz'\int dx'\,G_0(x,x',z,z')V(x',z')\Psi_0(x',z')\,dx'. \tag{52}$$

Here, $\Psi_0(x,z)$ is the solution of the Schrödinger-like equation with the Hamiltonian H_0 and the propagator $G_0(x,x',z,z')$ corresponds to evolution with the unperturbed Hamiltonian H_0, namely,

$$\Psi_0(x,z) = \int G_0(x,x',z,z')\,\Psi_0(x',z')\,dx'. \tag{53}$$

Formula (52) is a first Born approximation for the beam wave function in the case of small perturbative potential $V(x,z)$.

One can use relation (52) to express perturbed tomographic probability in terms of the beam wave function and the Green function of unperturbed Schrödinger-like equation. Analogously, for the perturbative solution of the Schödinger-like equation, one can obtain directly the perturbative solution of the tomographic evolution equation in the form

$$
\begin{aligned}
& w(X,\mu,\nu,z) = w_0(X,\mu,\nu,z) \\
& + \int_0^z dz' \int dX'\, d\mu'\, d\nu'\, \Pi_0(X,\mu,\nu,X',\mu',\nu',z,z') \\
& \times i \left[V\left(-\frac{1}{\partial/\partial X'}\frac{\partial}{\partial \mu'} - \frac{i\nu'}{2}\frac{\partial}{\partial X'}, z' \right) \right. \\
& \left. - V\left(-\frac{1}{\partial/\partial X'}\frac{\partial}{\partial \mu'} + \frac{i\nu'}{2}\frac{\partial}{\partial X'}, z' \right) \right] w_0(X',\mu',\nu',z'), \qquad (54)
\end{aligned}
$$

where the propagator Π_0 corresponds to the evolution of unperturbed tomographic probability.

The function $w_0(X,\mu,\nu,z)$ is the solution to the Fokker–Planck-like equation with unperturbed Hamiltonian H_0. The formula obtained corresponds to a first Born approximation of the perturbation series. The normalization factor $\left[\int w(X,\mu,\nu,z)\, dX\right]^{-1}$ should be taken into account in (53).

The Wigner quasidistribution in a first Born approximation has the form

$$
\begin{aligned}
\rho_\omega(x,p,z) &= \rho_\omega^{(0)}(x,p,z) - i\int_0^z dz' \int \pi_0(x,p,x',p',z,z') \\
&\times \left[V\left(x' + \frac{i}{2}\frac{\partial}{\partial p'}, z'\right) - V\left(x' - \frac{i}{2}\frac{\partial}{\partial p'}, z'\right) \right] \\
&\times \rho_\omega^{(0)}(x',p',z')\, dx'\, dp', \qquad (55)
\end{aligned}
$$

where $\rho_\omega^{(0)}$ is the solution of the Moyal-like equation for the Hamiltonian H_0 and the propagator π_0 corresponds to the unperturbed Hamiltonian.

9. Parametric oscillator model

The quadrupole case corresponds to the Hamiltonian (here we use dimensionless variables and put emittance $\epsilon = 1$)

$$
H = -\frac{1}{2}\frac{\partial^2}{\partial x^2} + \frac{x^2}{2}. \qquad (56)
$$

For this case, the propagator for the beam wave function is

$$
G(x,x',z) = \frac{1}{\sqrt{2\pi i \sin z}} \exp\left\{ \frac{i}{2}\left[\cot z \left(x^2 + x'^2\right) - \frac{2xx'}{\sin z} \right] \right\}. \qquad (57)
$$

Due to the relation of the propagator (57) to the kernel of fractional Fourier transform [23] (the reader interested to know how fractional Fourier transform is used in optics, we address, for example, to [65]), one can conclude that in electron optics the charged-particle-beam propagation realizes the fractional Fourier transform of the beam wave function.

The linear in x and $\partial/\partial x$ integrals of motion are

$$A(z) = \frac{e^{iz}}{\sqrt{2}}\left(x + \frac{\partial}{\partial x}\right), \qquad A^{\dagger}(z) = \frac{e^{-iz}}{\sqrt{2}}\left(x - \frac{\partial}{\partial x}\right). \tag{58}$$

The integrals of motion (29) and (30) read

$$\hat{p}_0(z) = \frac{A(z) - A^{\dagger}(z)}{i\sqrt{2}} \tag{59}$$

and

$$\hat{x}_0(z) = \frac{A(z) + A^{\dagger}(z)}{\sqrt{2}}\,. \tag{60}$$

In the case of z-dependent quadrupole potential

$$U(x,z) = \frac{k_1(z)}{2}\,x^2, \tag{61}$$

there exist linear integrals of motion for the electronic beam which have the explicit form

$$A(z) = \frac{i}{\sqrt{2}}\left[\xi(z)\left(-i\frac{\partial}{\partial x}\right) - \dot{\xi}(z)x\right], \tag{62}$$

$$A^{\dagger}(z) = -\frac{i}{\sqrt{2}}\left[\xi^*(z)\left(-i\frac{\partial}{\partial x}\right) - \dot{\xi}^*(z)x\right], \tag{63}$$

where we take units with $k_1(0) = 1$. Also the complex function $\xi(z)$ satisfies the equation

$$\ddot{\xi}(z) + k_1(z)\xi(z) = 0, \tag{64}$$

with the initial conditions

$$\xi(0) = 1, \qquad \dot{\xi}(0) = i. \tag{65}$$

The integrals of motion (29) and (30) for z-dependent quadrupole are given by (59) and (60) with $A(z)$ and $A^{\dagger}(z)$ given by (62) and (63).

The product of two linear integrals of motion (62) and (63)

$$J(z) = A^{\dagger}(z)A(z)$$

is the quadratic integral of motion which in the classical case is the Ermakov invariant [66] discussed in the quantum case in [67].

The propagator for the tomographic probability in the case of z-independent quadrupole potential $k_1 = 1$ reads

$$\begin{aligned}\Pi(X,\mu,\nu,X',\mu',\nu',z) &= \delta(X-X')\,\delta(\nu'-\nu\cos z+\mu\sin z)\\ &\quad\times\delta(\mu'-\nu\sin z-\mu\cos z).\end{aligned} \tag{66}$$

For the case of (61), one has the propagator

$$\begin{aligned}&\Pi_1(X,\mu,\nu,X',\mu',\nu',z)=\delta(X-X')\\ &\times\delta\left(\nu'-\frac{1}{2i}\,[\dot{\xi}-\dot{\xi}^*]\nu-\frac{1}{2i}\,[\xi-\xi^*]\mu\right)\\ &\times\delta\left(\mu'-\frac{1}{2}\,[\dot{\xi}+\dot{\xi}^*]\nu-\frac{1}{2}\,[\xi+\xi^*]\mu\right).\end{aligned} \tag{67}$$

Now we evaluate the influence of the aberration on the z-independent beam due to the quartic term

$$V(x)=\lambda x^4, \tag{68}$$

where λ is a small parameter. The discussion of polynomial aberrations in optics is presented in [68]. In our case, the influence of the aberration on the tomographic probability $w_0(X,\mu,\nu,z)$ provides perturbed tomographic probability of the form

$$\begin{aligned}&w(X,\mu,\nu,z)=w_0(X,\mu,\nu,z)+\lambda\int_0^z dz'\int dX'\,d\mu'\,d\nu'\;\delta(X-X')\\ &\times\delta\Big(\nu'-\nu\cos[z-z']+\mu\sin[z-z']\Big)\\ &\times\delta\Big(\mu'-\nu\sin[z-z']-\mu\cos[z-z']\Big)\\ &\times\left[\nu'\frac{\partial}{\partial\mu'}\left(\nu'^2\frac{\partial^2}{\partial X'^2}-\frac{1}{(\partial/\partial X')^2}\frac{\partial^2}{\partial\mu'^2}\right)\right]w_0(X',\mu',\nu',z').\end{aligned} \tag{69}$$

In order to calculate the influence of the aberration potential (68) in a first Born approximation, we used the following formula:

$$(a+b)^4-(a-b)^4=8\,ab\,(a^2+b^2),$$

where

$$a=-\frac{1}{\partial/\partial X'}\frac{\partial}{\partial\mu'}\,,\qquad b=-\frac{i\nu'}{2}\frac{\partial}{\partial X'}\,.$$

Analogously, one can find the influence of the aberration on the Wigner quasidistribution.

10. Phase-space form of nonlinear equations

Let us consider the following generalized nonlinear Schrödinger equation (NLSE):

$$i\frac{\partial\psi}{\partial s} = -\frac{1}{2}\frac{\partial^2\psi}{\partial x^2} + U\left[|\psi|^2\right]\psi, \tag{70}$$

where s and x are the time-like and space-like variables and $\psi = \psi(x,s)$ is a complex wave function describing the system's evolution in the configuration space; $U = U\left[|\psi|^2\right]$ is an arbitrary real functional of $|\psi|^2$.

In this section, we derive the evolution equations for solitons in the phase-space Weyl–Wigner–Moyal representation. For the case of cubic NLSE

$$i\,\frac{\partial\psi}{\partial s} = -\frac{1}{2}\,\frac{\partial^2\psi}{\partial x^2} + q_0|\psi|^2\psi, \tag{71}$$

the density matrix (36) satisfies the following evolution equation:

$$\begin{aligned} i\,\frac{\partial\rho(x,x',s)}{\partial s} &= -\frac{1}{2}\left(\frac{\partial^2}{\partial x^2} - \frac{\partial^2}{\partial x'^2}\right)\rho(x,x',s) \\ &\quad + \left\{U\left[\int\delta(x-y)\rho(x,y,s)\,dy\right]\right. \\ &\quad \left. -U\left[\int\delta(x'-y)\rho(x',y,s)\,dy\right]\right\}\rho(x,x',s), \end{aligned} \tag{72}$$

where

$$U\left[\int\delta(x-y)\rho(x,y,s)\,dy\right] = q_0\int\delta(x-y)\rho(x,y,s)\,dy \tag{73}$$

is the potential-energy functional of the density matrix for nonlinear Schrödinger equation (71) with cubic nonlinearity.

The transition to the evolution equation for the Wigner function can be done using the standard algebra, which provides the following recepie. One has to make the replacement $\rho \to W$ in (71) along with the following replacement [61]:

$$\begin{aligned} \frac{\partial}{\partial x}\,\rho\,(x,\,x') &\longrightarrow \left(\frac{1}{2}\,\frac{\partial}{\partial q} + i\,p\right)W\,(q,\,p), \\ \frac{\partial}{\partial x'}\,\rho\,(x,\,x') &\longrightarrow \left(\frac{1}{2}\,\frac{\partial}{\partial q} - i\,p\right)W\,(q,\,p), \\ x\,\rho\,(x,\,x') &\longrightarrow \left(q + \frac{i}{2}\,\frac{\partial}{\partial p}\right)W\,(q,\,p), \\ x'\,\rho\,(x,\,x') &\longrightarrow \left(q - \frac{i}{2}\,\frac{\partial}{\partial p}\right)W\,(q,\,p). \end{aligned} \tag{74}$$

This replacement can be easily obtained, in view of Fourier transform properties. For Fourier transform, the action on a function by different operators can be given in the form of action on the Fourier component of the function by the corresponding operators. It provides the following Moyal-like form of nonlinear equation (71) for the Wigner function

$$\frac{\partial W(q,p,s)}{\partial s} = -p\frac{\partial W(q,p,s)}{\partial q} + \frac{1}{i}\Big\{U\left[\rho\left(q+\frac{i}{2}\frac{\partial}{\partial p}, q+\frac{i}{2}\frac{\partial}{\partial p}, s\right)\right] - \text{c.c.}\Big\}W(q,p,s), \tag{75}$$

where the arguments of the potential energy are replaced by the operators which act onto the Wigner function. For the cubic NLSE

$$U(z) = q_0 z, \tag{76}$$

one has a simple form of the equation

$$\frac{\partial W(q,p,s)}{\partial s} + p\frac{\partial}{\partial q}W(q,p,s) - 2q_0 \operatorname{Im}\rho\left(q+\frac{i}{2}\frac{\partial}{\partial p}, q+\frac{i}{2}\frac{\partial}{\partial p}, s\right)W(q,p,s) = 0. \tag{77}$$

Using the relation

$$\rho(x,x) = \int W(x,p)\,\frac{dp}{2\pi}, \tag{78}$$

one has

$$\frac{\partial W(q,p,s)}{\partial s} + p\frac{\partial W(q,p,s)}{\partial q} - 2q_0 \operatorname{Im}\int W\left(q+\frac{i}{2}\frac{\partial}{\partial p}, P, s\right)\frac{dP}{2\pi}\,W(q,p,s) = 0. \tag{79}$$

In (74), (76), and (150) the arguments of the density matrix and Wigner function are replaced by operators and the operators act on the Wigner function itself. Equations (74) and (76) can be presented in the form of Moyal-like series [69]. It is easy to see that, for an arbitrary nonlinear potential $U(z)$, equation (74) can be written in terms of the following nonlinear-functional partial differential equation for the Wigner function only

$$\frac{\partial W(q,p,s)}{\partial s} + p\frac{\partial W(q,p,s)}{\partial q} - 2\operatorname{Im}\left\{U\left[\int W\left(q+\frac{i}{2}\frac{\partial}{\partial p}, P, s\right)\frac{dP}{2\pi}\right]\right\}W(q,p,s) = 0. \tag{80}$$

One can see that the above nonlinear equation in the Moyal representation has a specific form of the equation with nonlocal quadratic interaction of the Wigner function with itself.

11. Probability representation of nonlinear equations

Now we consider the relation of the density matrix, Wigner function and tomogram. For arbitrary (pure and mixed) states,

$$W(q,p) = \frac{1}{2\pi}\int w(X,\mu,\nu)\exp\left[-i(\mu q + \nu p - X)\right] d\mu\, d\nu\, dX. \tag{81}$$

One can also calculate the density matrix in the coordinate representation

$$\rho(X,X') = \frac{1}{2\pi}\int w(Y,\mu,X-X')\exp\left[i\left(Y - \mu\frac{X+X'}{2}\right)\right] d\mu\, dY. \tag{82}$$

In view of this relation, by using the following substitution rule

$$\begin{aligned}
\rho(x,x',s) &\longrightarrow w(X,\mu,\nu,s),\\
x\rho &\longrightarrow \left[-\left(\frac{\partial}{\partial X}\right)^{-1}\frac{\partial}{\partial\mu} + \frac{i}{2}\nu\frac{\partial}{\partial X}\right] w,\\
\frac{\partial\rho}{\partial x} &\longrightarrow \left[\frac{\mu}{2}\frac{\partial}{\partial X} - i\left(\frac{\partial}{\partial X}\right)^{-1}\frac{\partial}{\partial\nu}\right] w,\\
x'\rho &\longrightarrow \left[-\left(\frac{\partial}{\partial X}\right)^{-1}\frac{\partial}{\partial\mu} - \frac{i}{2}\nu\frac{\partial}{\partial X}\right] w,\\
\frac{\partial\rho}{\partial x'} &\longrightarrow \left[\frac{\mu}{2}\frac{\partial}{\partial X} + i\left(\frac{\partial}{\partial X}\right)^{-1}\frac{\partial}{\partial\nu}\right] w
\end{aligned} \tag{83}$$

in the evolution equation (71), one obtains the tomographic form of the nonlinear equation under consideration. The substitution rule can be easily obtained (see [70, 64]) using the same rules that we applied to derive nonlinear equation in the Moyal form. Since the symplectic tomography map is similar to Fourier transform, calculating the action of differential operators on the function and corresponding tomogram is straightforward

$$\begin{aligned}
&\frac{\partial w(X,\mu,\nu,s)}{\partial s} + \mu\frac{\partial w(X,\mu,\nu,s)}{\partial\nu} - 2\,\mathrm{Im}\,U\left\{\int w(y,\mu',0,s)\right.\\
&\times\exp\left[i\left(y + \mu'\left[\left(\frac{\partial}{\partial X}\right)^{-1}\frac{\partial}{\partial\mu} - \frac{i}{2}\nu\frac{\partial}{\partial X}\right]\right)\right]\left.\frac{dy\, d\mu'}{2\pi}\right\}\\
&\times w(X,\mu,\nu,s) = 0.
\end{aligned} \tag{84}$$

In the above equation (83), one has the integro-differential operator in the exponent, which acts on the tomographic probability function. The integral operator

$(\partial/\partial X)^{-1}$ is defined by the following action on the Fourier component of the function $f(X)$:

$$\left(\frac{\partial}{\partial X}\right)^{-1} f(X) = \left(\frac{\partial}{\partial X}\right)^{-1} \int \widetilde{f}(k) e^{ikX} dk = \int \frac{\widetilde{f}(k)}{ik}\, e^{ikX} dk.$$

For the case of cubic NLSE, one has

$$\frac{\partial w(X,\mu,\nu,s)}{\partial s} + \mu\,\frac{\partial w(X,\mu,\nu,s)}{\partial \nu} - 2q_0\,\mathrm{Im}\left\{\int w(y,\mu',0,s)\right.$$
$$\left.\times \exp\left[i\left(y + \mu'\left[\left(\frac{\partial}{\partial X}\right)^{-1}\frac{\partial}{\partial \mu} - \frac{i}{2}\nu\,\frac{\partial}{\partial X}\right]\right)\right]\frac{dy\,d\mu'}{2\pi}\right\}$$
$$\times w(X,\mu,\nu,s) = 0. \tag{85}$$

It should be pointed out that equation (84) has the solutions which in course of the evolution process preserve the positivity and normalization. Thus the soliton solutions of the nonlinear equations can be mapped onto probability distribution functions.

The meaning of the probability distributions is the following. If x is a coordinate and p is the momentum, the value $X = \mu x + \nu p$ is the position in a reference frame in the phase space (x, p), the reference frame being scaled and rotated. The parameters of the scaling λ and rotation θ are determined by the real parameters μ and ν, namely, $\mu = e^{\lambda}\cos\theta$ and $\nu = e^{-\lambda}\sin\theta$.

The probability distributions $w(X,\mu,\nu)$ determine soliton solutions $\psi(x)$ in the corresponding representation. Consequently, in the tomographic representation soliton solutions of nonlinear dynamic systems are solutions of generalized Fokker–Planck-type equations for the standard probability distributions. Such representation can be useful from mathematical point of view since the analysis of probabilities and their asymptotics can be additionally incorporated (using existing theorems on the behaviour of the probability distribution functions).

12. Solitons of cubic nonlinear Schrödinger equation

In this section, we study bright soliton in both tomographic and Weyl–Wigner representations. The envelope bright soliton of cubic NLSE (71) for $q_0 < 0$ is

$$\Psi_{\mathrm{b}}(x,s) = \left(\frac{2\,|E|}{|q_0|}\right)^{1/2} \mathrm{sech}\left[\sqrt{2\,|E|}\;\xi\right] \exp\left\{i\left[V_0 x - \left(E + \frac{V_0^2}{2}\right)s\right]\right\}, \tag{86}$$

where E is a negative real constant, V_0 is an arbitrary real constant and $\xi = x - V_0 s$ (see, for example, [71]). Thus, the corresponding optical tomogram can be obtained, in view of (19) and (20), and it is given by the formula

$$w_{\rm b}(X,\theta,s)=\frac{|E\sin\theta|}{|q_0|\pi}\left|\int\operatorname{sech}\left[\sqrt{2|E|}(y\sin\theta-V_0s)\right]\right.$$
$$\left.\times\exp\left\{i\left[V_0y\sin\theta-\left(E+\frac{V_0^2}{2}\right)s\right]+\frac{i\sin 2\theta}{4}y^2-iXy\right\}dy\right|^2.$$

The Wigner function of bright soliton for $V_0=0$ can be obtained using (36) and (37) and it is given by the formula

$$\begin{aligned}W_{\rm b}(x,p) &= \frac{|E|}{|q_0|}\int\operatorname{sech}\left[\sqrt{2|E|}\left(x+\frac{u}{2}\right)\right]\\ &\quad\times\operatorname{sech}\left[\sqrt{2|E|}\left(x-\frac{u}{2}\right)\right]e^{ipu}du.\end{aligned}\tag{87}$$

It has been recently shown that a modified NLSE (70) with

$$U[|\psi|^2]=q_0|\psi|^{2\beta},$$

i.e.,

$$i\frac{\partial\Psi}{\partial s}=-\frac{1}{2}\frac{\partial^2\Psi}{\partial x^2}+q_0|\Psi|^{2\beta}\Psi,\tag{88}$$

for $q_0<0$ and any real positive value of β, has the following envelope soliton-like solutions [72]:

$$\begin{aligned}\Psi(x,s) &= \left[\frac{|E|\,(1+\beta)}{|q_0|}\right]^{1/2\beta}\operatorname{sech}^{1/\beta}\left[\beta\sqrt{2\,|E|}\,\xi\right]\\ &\quad\times\exp\left\{i\left[V_0x-\left(E+\frac{V_0^2}{2}\right)s\right]\right\},\end{aligned}\tag{89}$$

where the real numbers V_0 and E are arbitrary and negative, respectively, and still $\xi=x-V_0s$ (V_0 is the soliton velocity). It should be noted that the case $\beta=1$ (ordinary envelope bright soliton of the cubic NLSE [73] can be easily recovered [72]).

Correspondingly, the tomogram and Wigner function of the soliton solution of generalized nonlinear Schrödinger equation are given by the formulas ($V_0=0$) which are obtained, in view of (19), (20) and (36), (37), respectively, and they read

$$\begin{aligned}w(X,\mu,\nu) &= \frac{1}{2\pi|\nu|}\left|\left[\frac{|E|(1+\beta)}{|q_o|}\right]^{1/2\beta}\int\operatorname{sech}^{1/\beta}\left[\beta\sqrt{2|E|}\,y\right]\right.\\ &\quad\left.\times\exp\left(\frac{i\mu}{2\nu}y^2-\frac{iXy}{\nu}\right)dy\right|^2\end{aligned}\tag{90}$$

and

$$W(x,p) = \left[\frac{|E|(1+\beta)}{|q_0|}\right]^{1/\beta} \int \operatorname{sech}^{1/\beta}\left[\beta\sqrt{2|E|}\left(x+\frac{u}{2}\right)\right] \times \operatorname{sech}^{1/\beta}\left[\beta\sqrt{2|E|}\left(x-\frac{u}{2}\right)\right] e^{ipu}du. \tag{91}$$

13. Measuring of space and amplitudes of electromagnetic field

In the previous sections, we have discussed the tomography of solitons considering the tomograms as additional characteristics of soliton solutions of nonlinear dynamic equations. But there exists another experimental aspect of tomograms because the tomograms can be directly measured in different situations. Thus, the problem of tomography of some phenomena which is described by a complex function $\psi(x)$ is equivalent to the problem of measuring the amplitude $|\psi(x)|$ and the phase $\varphi(x)$ of the complex function $\psi(x) = |\psi(x)| \exp i\varphi(x)$. The function $\psi(x)$ can describe a soliton but it can describe some signal connected with different processes, as well, e.g., in optical fibers, in plasma, etc. In all processes where one needs to measure the amplitude and phase by measuring experimentally only intensities, the tomography can be used as an instrument for achieving this aim. We consider two different possibilities which are based on symplectic tomogram (10). The density matrix can be reconstructed either by measuring the tomogram of the optical tomography scheme $w(X,\theta)$ or tomogram $w(X,1,\nu)$ (we call it Fresnel tomogram [74]).

Now we show that both tomograms can be obtained in two different and realizable processes. The optical tomogram can be rewritten in terms of the fractional Fourier transform (which is reduced to the Green function of the harmonic oscillator) [23]. In fact, one has

$$w(X,\Theta) = \left|\frac{1}{\sqrt{2\pi i \sin\Theta}} \int \psi(y) \exp\left[\frac{i\cot\Theta}{2}\left(y^2+X^2\right) - \frac{iXy}{\sin\Theta}\right] dy\right|^2. \tag{92}$$

In this formula, we take $\nu = \sin\Theta$ and $\mu = \cos\Theta$.
The phase factor $\exp\left[\left(iX^2/2\right)\cot\Theta\right]$ does not change the value of the tomogram.

The tomogram presented in such form coincides with the value of the wave function at the point x at the time moment t if the initial value of the wave function at the time moment $t = 0$ is equal to $\psi(y)$. This means that to reconstruct the initial value of the wave function $\psi(x)$, including both the amplitude $|\psi(x)|$ and phase $\varphi(x)$, $\psi(x) = |\psi(x)| \exp i\varphi(x)$, one can measure the tomogram, i.e., amplitude squared of the wave function which evolves in the quadratic potential well. This situation can be perfectly realized in optical fibers with a parabolic profile of the refractive index, so-called "selfoc" (linear propagation).

In fact, the light beams in optical fibers obey to the Schrödinger-like equation which follows from the Helmholtz equation in the Fock–Leontovich approximation [12]. But the time t in the Schrödinger equation is replaced by the longitudinal coordinate z and the Planck's constant is replaced by the wavelength. Thus, to measure the input field amplitude and phase, it is sufficient to measure the tomogram which is the field intensity in each cross-section of the fiber given by longitudinal coordinate $0 < z \leq 2\pi$.

Another possibility is related to the formula

$$w(X,1,\nu) = \left| \frac{1}{\sqrt{2\pi i \nu}} \int \exp \frac{i(X-y)^2}{2\nu} \psi(y)\, dy \right|^2 . \tag{93}$$

This formula is equivalent to formula (10) in which we put $\mu = 1$ and added nonessential phase factor $\exp\left(iX^2/2\nu\right)$. One can see that the tomogram is expressed in terms of Fresnel integral. Another type of Fresnel tomography was discussed recently in [75]. Thus the tomogram for "time moment" ν is equivalent to the intensity of free propagating signal. In fact, the kernel in (93) is the Green function of a free particle. Since due to homogeneity the Fresnel tomogram $w(X,1,\nu)$ is equivalent to the tomogram $w(X,\mu,\nu)$, while measuring the intensity of free propagating signal, one measures both the phase and amplitude of the input signal $\psi(y)$. If one measures the field in optical fiber, the structure of the output field can be evaluated by measuring free propagation of the light beam. There is a peculiarity in using formula (93). For complete reconstructing the amplitude, one needs to know the intensity for arbitrary large values of time (or longitudinal coordinate z). Practically the length or duration can be chosen to fit appropriate accuracy of the measurement (window).

Thus, the approach suggested has a potential to be applied to measuring the phase of the fields propagating nonlinearly, e.g., in optical fibers.

14. Solitons in Bose–Einstein condensate

The states of Bose–Einstein condensates (BEC) are described by solutions of nonlinear Gross–Pitaevskii equation [31, 76, 77]. In comparison with NLSE, this equation contains additional linear term which depends on the potential-energy term (e.g. a harmonic-oscillator potential energy of a trap). The tomographic probability distribution map was used in [57] to write down linear von Neumann equation for density matrix [55] in the form of classical-like equation for the standard positive probability density.

Our aim is to obtain a combination of both the tomographic approach to NLSE developed in [51] and the approach to von Neumann equation given in [57] and to apply a generalization of this approach to nonlinear Gross–Pitaevskii equation [52]. The solitons in BEC were observed experimentally in [32–34]. Within the framework of tomographic approach, the bright [78–80] and dark [81–85] solitons of Bose–Einstein condensate can be associated with the probability distribution functions, which describe completely the solitons in BEC. In the tomographic probability representation, the solutions to Gross–Pitaevskii equation have the form of positive probability distribution functions. This means that one can associate with solitons of BEC such characteristics as entropy, which is determined by the probability distribution and use all the mathematical tools of the probability theory.

15. Gross–Pitaevskii equation

The nonstationary Gross–Pitaevskii equation describes the Bo - se–Ein - stein condensate. It has the form (see, for example, [86])

$$i\frac{\partial\psi}{\partial s} = -\frac{1}{2}\frac{\partial^2\psi}{\partial x^2} + U(x)\psi + g|\psi|^2\psi. \tag{94}$$

One can see that this equation contains sum of linear and nonlinear terms. Using the change of variables given by the tomographic map, one can obtain the Gross–Pitaevskii equation in the tomographic form

$$\begin{aligned}
&\frac{\partial w(X,\mu,\nu,s)}{\partial s} + \mu\,\frac{\partial w(X,\mu,\nu,s)}{\partial \nu} - 2q_0\,\mathrm{Im}\left\{\int w(y,\mu',0,s)\right.\\
&\times\exp\left[i\left(y+\mu'\left[\left(\frac{\partial}{\partial X}\right)^{-1}\frac{\partial}{\partial\mu} - \frac{i}{2}\,\nu\,\frac{\partial}{\partial X}\right]\right)\right]\left.\frac{dy\,d\mu'}{2\pi}\right\} w(X,\mu,\nu,s)\\
&-2\,\mathrm{Im}\left\{U\left[\left(\frac{\partial}{\partial x}\right)^{-1}\frac{\partial}{\partial\mu} - \frac{i}{2}\nu\frac{\partial}{\partial X}\right]\right\} w(X,\mu,\nu,s) = 0.
\end{aligned} \tag{95}$$

One can see that the equation obtained contains two contributions. One contribution is nonlinear potential of BEC, the other one is related to an external potential. The equation obtained is one of the main results of this paper. The potential energy can be chosen as the harmonic oscillator potential

$$U = \frac{\omega^2 x^2}{2}. \tag{96}$$

One can also get the Moyal form of the Gross–Pitaevskii equation by the same method, it reads

$$\begin{aligned}
&\frac{\partial W(q,p,s)}{\partial s} + p\,\frac{\partial W(q,p,s)}{\partial q} - 2q_0\,\mathrm{Im}\int W\left(q+\frac{i}{2}\frac{\partial}{\partial p},P,s\right)\frac{dP}{2\pi}\\
&\times W(q,p,s) - 2\,\mathrm{Im}\left[U\left(q+\frac{i}{2}\frac{\partial}{\partial p}\right)\right] W(q,p,s) = 0.
\end{aligned} \tag{97}$$

The Moyal form of the obtained equation also has two contributions. One is related to nonlinear potential (cubic one). Another one contains an external potential (oscillator potential for trapped BEC).

16. Tomograms of solitons in Bose–Einstein condensate

In this section, we consider the 3D Gross–Pitaevskii equation describing mean field of the BEC (see, f.i., [86]):

$$\left\{-\frac{\hbar^2}{2m}\nabla^2 + gN|\psi(\vec{r},t)|^2 + \frac{1}{2}m\left[\omega_\perp^2(x^2+y^2)+\omega_z^2 z^2\right]\right\}\psi(\vec{r},t)$$
$$= i\hbar\frac{\partial}{\partial t}\psi(\vec{r},t), \tag{98}$$

where $g = 4\pi^2 a/m$, a is the s-wave scattering length, m is the atomic mass, N is the number of atoms in the condensate, and $\omega_\perp$ and ω_z are axial and longitudinal oscillation frequencies of the atoms in the trapping potential, respectively.

As it was shown in [80] this equation can be analyzed by using a variational procedure leading to the variational ansazts of a guessed stationary solution in the form of bright soliton

$$\psi_a(\vec{r}) = \frac{1}{\sqrt{2\pi\sigma_\perp^2 \ell_z}}\exp\left(-\frac{x^2+y^2}{2\sigma_\perp^2}\right)\text{sech}\left(\frac{z}{\ell_z}\right). \tag{99}$$

This solution is important for explosive potential for which $\omega_z^2 < 0$. Here $\sigma_\perp$ and ℓ_z are the variational parameters, which describe the transverse and axial widths of the wave function. Notice that solution (99) is factorized into two terms depending on the transverse (x, y) and longitudinal (z) coordinates. This allows us to use tomographic approach developed in the previous sections to analyze the longitudinally dependent part of the solution (99). Indeed, the quasi-1D limit of the 3D Gross–Pitaevskii equation, for $\sigma_\perp \ll \ell_z$ gives the following equation:

$$\left[-\frac{\hbar^2}{2m}\frac{\partial^2}{\partial z^2} + g_{1D}N|\phi|^2 + \hbar\omega_\perp + \frac{1}{2}m\omega_z^2 z^2\right]\phi = i\hbar\frac{\partial\phi}{\partial t}, \tag{100}$$

where $g_{1D} = 2a\hbar\omega_\perp$ is the renormalized quasi-1D coupling constant. The normalized solution $\phi(z,t)$ for the 1D Gross–Pitaevskii equation can be written in the form of bright-soliton-like wave, namely,

$$\phi(z,t) = \frac{1}{\sqrt{2\ell_z}}\text{sech}\left(\frac{z}{\ell_z}\right)\exp\left(-\frac{i\mu t}{\hbar}\right), \tag{101}$$

where μ is the chemical potential. The z-dependent part is essentially the longitudinal factor of the guessed variational ansatz for the stationary solution (99). To determine the tomographic probability distribution of the longitudinal motion let us use our equation (10), where now the complex function $\psi(y)$ is given in the above notation by equation (101) with the substitution $z \to yL$, where

$L = \sqrt{\hbar/m\omega_\perp}$ is the normalization length for the 1D Gross–Pitaevskii equation (100). It should be pointed out that the soliton solution under consideration is approximate solution to the initial 3D Gross–Pitaevskii equation.

The tomographic probability distribution is given by

$$w(X, \mu, \nu) = \frac{1}{2\pi|\nu|} \left| \int \sqrt{\frac{\gamma}{2}} \operatorname{sech}(\gamma y) \exp\left(\frac{i\mu}{2\nu} y^2 - \frac{iX}{\nu} y\right) dy \right|^2, \quad (102)$$

where the dependence on the parameter $\gamma = L/\ell_z$ governing the width of the longitudinal soliton distribution is shown.

To illustrate the behaviour of the tomogram, we take $\mu = \cos\theta$, $\nu = \sin\theta$.

The 3D plot of the tomogram of the bright-soliton-like solution (101) is displayed in the X–θ plane in Figure 1, while the corresponding density plot is shown in Figure 2 [52]. The spread of the tomographic map is basically governed by the dimensionless parameter $\gamma = L/\ell_z$. The value of γ in Figures 1 and 2 is adopted from experimental condition reported in [87], where $L = 1.4$ μm and the axial width distribution $\ell_z = 1.7$ μm.

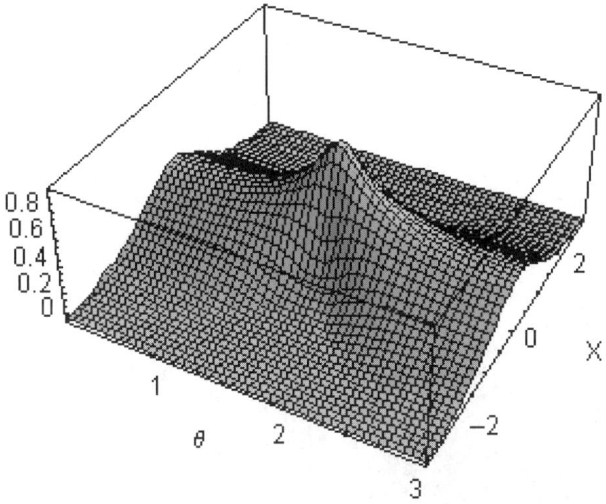

Figure 1. Plot of the tomogram of the bright soliton for $\gamma = L/l_z = 0.82$ ($L = 1.4$ μm and $\ell_z = 1.7$ μm) corresponding to the experimental conditions of BEC reported in [87].

3D plot of the tomogram of the bright soliton for various values of the parameter ℓ_z are shown in Figure 3, $\ell_z = 2\ \mu m$, in Figure 4, $\ell_z = 1.4\ \mu m$ and in Figure 5, $\ell_z = 1\ \mu m$.

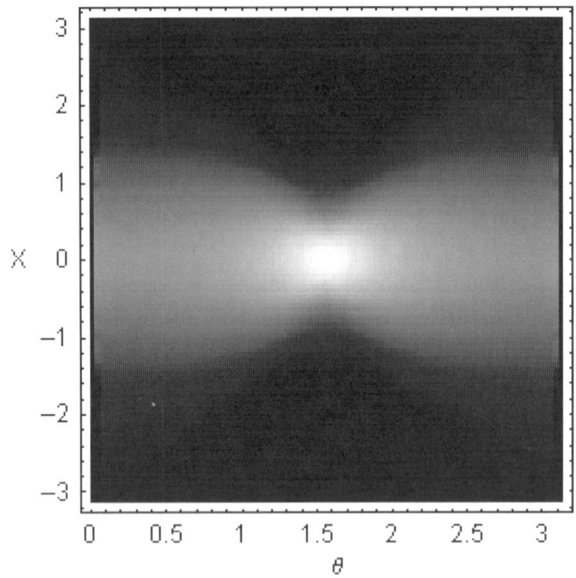

Figure 2. Density plot corresponding to the tomogram displayed in Figure 1.

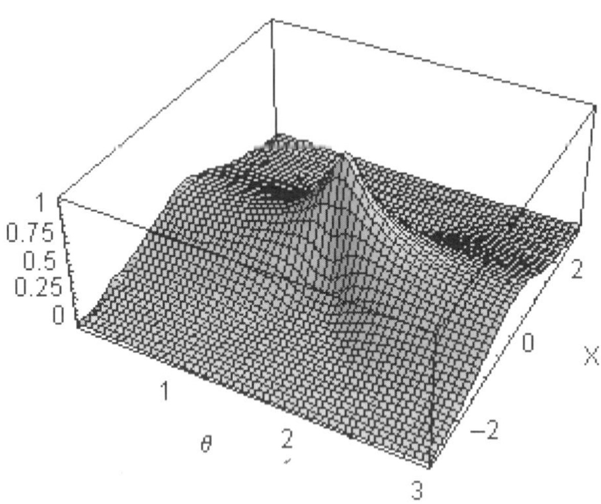

Figure 3. Plot of the tomogram of the bright soliton for $\gamma = L/l_z = 0.7$ ($L = 1.4$ μm and $\ell_z = 2$ μm).

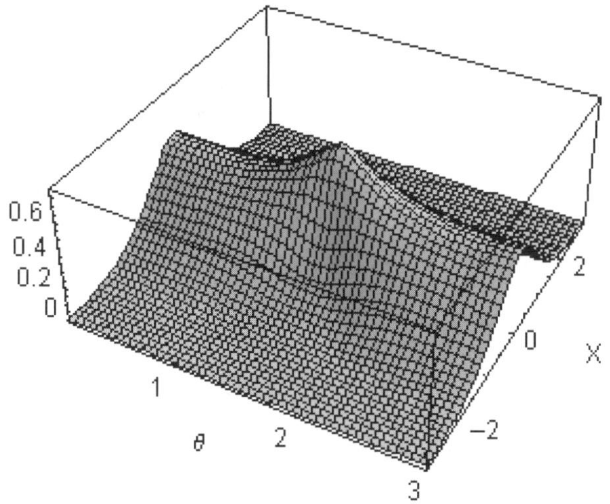

Figure 4. Plot of the tomogram of the bright soliton for $\gamma = L/l_z = 1$ ($L = 1.4$ μm and $\ell_z = 1.4$ μm).

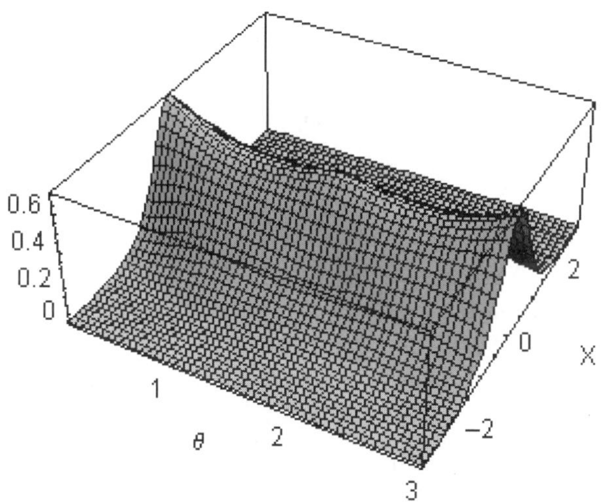

Figure 5. Plot of the tomogram of the bright soliton for $\gamma = L/l_z = 1.4$ ($L = 1.4$ μm and $\ell_z = 1$ μm).

17. Wavelet-like transforms, quasidistributions and tomograms

In this section, following [25] we present a unified general construction of three types of transforms. The first class consists of wavelet-type transforms, the second of quasidistributions, and in the third class the tomographic transforms are. Quasidistributions are transforms like the Wigner–Ville one. Husimi–Kano positive quasidistributions will be also discussed.

In quantum mechanics, quasidistributions describe a quantum state in terms of phase-space quasiprobability densities. In signal analysis, quasidistributions describe the structure of analytic signals in the time–frequency plane. There also exist quasidistributions characterizing the signal structure in the time–scale plane. We refer to quasiprobability densities because the corresponding functions are not conventional probabilities, being either complex or nonpositive. The corresponding observables do not commute and the uncertainty relation prevents the existence of a joint probability distribution function for noncommuting observables.

The general setting for our construction is as follows [25].

Signals $f(t)$ are considered to be vectors $\mid f\rangle$. With α being a set of parameters, $\{U(\alpha)\}$ is a family of operators.

In this setting, three types of transforms are defined.

Consider a reference vector $\mid h\rangle$ chosen in such a way that out of the set $\{U(\alpha)\mid h\rangle = \mid h\rangle\}$ a complete set of vectors can be chosen to serve as a basis. Completeness relation for the vectors $\mid h,\alpha\rangle$ means existence of measure $d\mu(\alpha)$ in the unity operator decomposition

$$\int \mid h,\alpha\rangle\langle h,\alpha\mid\, d\mu(\alpha) = \hat{\mathbf{1}}.$$

Two of the transforms considered are given by scalar products

$$W_f^{(h)}(\alpha) = \langle U(\alpha)\, h \mid f\rangle, \qquad Q_f(\alpha) = \langle U(\alpha)\, f \mid f\rangle. \tag{103}$$

If $U(\alpha)$ is operator of unitary irreducible representation of some Lie group, transform (103) is matrix element of the irreducible representation.

We will denote the transforms of the $W_f^{(h)}$-type as *wavelet-type* transforms and those of the Q_f-type as *quasidistribution* transforms.

In general, if $U(\alpha)$ are unitary operators, there are self-adjoint operators $B(\alpha)$ such that

$$W_f^{(h)}(\alpha) = \langle h \mid e^{iB(\alpha)} \mid f\rangle, \qquad Q_f^{(B)}(\alpha) = \langle f \mid e^{iB(\alpha)} \mid f\rangle. \tag{104}$$

In this case, because $B(\alpha)$ has a real valued spectrum, another transform may be defined by means of Dirac (or Kronecker) delta function

$$M_f^{(B)}(X) = \langle f \mid \delta\Big(B(\alpha) - X\Big) \mid f\rangle. \tag{105}$$

Equation (105) defines what we call the tomographic transform of analytic signal or *tomogram*. In contrast to the quasiprobabilities, the transform $M_f^{(B)}(X)$ is positive and it can be correctly interpreted as a probability distribution.

For a normalized vector $\mid f\rangle$, $\langle f \mid f\rangle = 1$, the tomogram is a normalized function

$$\int M_f^{(B)}(X)\, dX = 1$$

and, therefore, it may be interpreted as a probability distribution for the random variable X corresponding to the observable defined by the operator $B(\alpha)$.

The three classes of transforms are mutually related

$$\begin{aligned} M_f^{(B)}(X) &= \frac{1}{2\pi}\int Q_f^{(kB)}(\alpha)\, e^{-ikX}\, dk, \\ Q_f^{(B)}(\alpha) &= \int M_f^{(B/p)}(X)\, e^{ipX}\, dX. \end{aligned} \tag{106}$$

Wavelet-type transforms, quasidistributions, and tomograms are related by the formulae

$$Q_f^{(B)}(\alpha) = W_f^{(f)}(\alpha), \tag{107}$$

$$\begin{aligned} W_f^{(h)}(\alpha) &= \frac{1}{4}\int e^{iX}\Big[M_{f_1}^{(B)}(X) - iM_{f_2}^{(B)}(X) \\ &\quad -M_{f_3}^{(B)}(X) + iM_{f_4}^{(B)}(X)\Big]\, dX, \end{aligned} \tag{108}$$

where

$$\begin{aligned} \mid f_1\rangle &= \mid h\rangle + \mid f\rangle, & \mid f_3\rangle &= \mid h\rangle - \mid f\rangle, \\ \mid f_2\rangle &= \mid h\rangle + i \mid f\rangle, & \mid f_4\rangle &= \mid h\rangle - i \mid f\rangle. \end{aligned}$$

Another important case concerns operators $U(\alpha)$, which can be represented in the form

$$U(\alpha) = e^{ib(\alpha)} P_h e^{-ib(\alpha)},$$

with P_h being a projector on a reference vector $\mid h\rangle$. This creates a quasidistribution of the Husimi–Kano type

$$H_f^{(b)}(\alpha) = \langle f \mid U(\alpha) \mid f\rangle.$$

Tomograms and Wigner functions of solitons considered in previous sections can be described in view of the constructions presented above. Also wavelet transforms can be applied to the soliton solutions of nonlinear equations.

18. Conclusions

In this paper, we have reviewed the tomographic probability distribution associated to wave packets of linear equation for particle beams as well as to the soliton solutions of nonlinear equations.

Nonlinear dynamical equations like nonlinear Schrödinger equation or Gross–Pitaevskii equation have been presented in the form of equation (a nonlinear generalization of the Fokker–Planck equation) for the standard probability distribution function.

Specific cases of solitons for a family of modified nonlinear Schrödinger equation have been studied in the tomographic representation explicitly.

The possibility to use both tomograms (of the symplectic and Fresnel types) to reconstruct the phase of linear or nonlinear signals by measuring the signals' intensities has been discussed.

One has to point out that measuring the tomograms in experiments for determining the field states in optical fibers has to be elaborated. The mathematical aspects of solving the nonlinear equations in the tomographic representation have also to be better studied because till now the nonlocal interaction of unknown probability distributions in kinetic equations in the form found in this paper has not been discussed in the literature. On the other hand, since the soliton solutions are known and can be easily obtained in the standard representation, the solution of nonlinear equations in the form of kinetic equation for tomogram can be also obtained.

Acknowledgments

This study was supported by Universitá "Federico II" di Napoli and the Russian Foundation for Basic Research under Projects Nos. 01-02-17745 and 03-02-16408 and will appear in [88].

M.A.M. thanks the Organizers of XIII International Symposium "Symmetries in Science" for kind hospitality and the Russian Foundation for Basic Research for Travel Grant No. 03-02-26840.

References

[1] R. Fedele and P.K. Shukla (eds.) 1995 "Quantum-like Models and Coherent Effects" (Singapore: World Scientific)

[2] S. De Martino, S. De Nicola, S. De Siena, R. Fedele and G. Miele (eds.) 1997 "Quantumlike Description and Macroscopic Coherent Phenomenon" (Singapore: World Scientific)

[3] J. Bertrand and P. Bertrand 1987 *Found. Phys.* **17** 397

[4] K. Vogel and H. Risken 1989 *Phys. Rev.* A **40** 2847

[5] M.G. Raymer 1997 *Contemporary Phys.* **38** 343

[6] D.T. Smithey, M. Beck, M.G. Raymer and A. Faridani 1993 *Phys. Rev. Lett.* **70** 1244

[7] S. Mancini, V.I. Man'ko and P. Tombesi 1995 *Quantum Semiclass. Opt.* **7** 615

[8] G.M. D'Ariano, S. Mancini, V.I. Man'ko and P. Tombesi 1996 *Quantum Semiclass. Opt.* **8** 1017

[9] J. Radon 1917 "Über die bestimmung von funktionen durch ihre integralwerte längs gewisser mannigfaltigkeiten" *Berichte Sachsische Akademie der Wissenschaften* (Leipzig: Mathematishe-Physicalische Klasse) Vol. 69, S. 262–267

[10] A. Wünsche 1997 *J. Mod. Opt.* **44** 2293

[11] R. Fedele and V.I. Man'ko 1998 *Phys. Rev.* E **58** 992

[12] M.A. Leontovich 1944 *Izv. Akad. Nauk SSSR* **8** 16
M.A. Leontovich and V.A. Fock 1946 *Zh. Éksp. Teor. Fiz.* **16** 557

[13] D. Gloge and D. Marcuse 1969 *J. Opt. Soc.* A **59** 1629

[14] D. Marcuse 1972 "Light Transmission Optics" (New York: Van Nostrand)

[15] J.A. Arnaud 1976 "Beam and Fiber Optics" (New York: Academic)

[16] L.M. Brekhovskikh 1973 "Waves in Slab Media" (Moscow: Nauka) [in Russian]

[17] R. Fedele and G. Miele 1991 *Nuovo Cim.* D **13** 1527

[18] R. Fedele and G. Miele 1992 *Phys. Rev.* A **46** 6634
R. Fedele, F. Galluccio and G. Miele 1994 *Phys. Lett.* A **209** 263
R. Fedele, F. Galluccio, V.I. Man'ko and G. Miele 1995 *Phys. Lett.* A **209** 263

[19] R. Fedele, M.A. Man'ko, V.I. Man'ko and V. Vaccaro 2003 *Phys. Scr.* **68** 277

[20] D. Gabor 1946 "Theory of Communication" *IEE J.* **93** 429

[21] L. Mandel and E. Wolf 1995 "Optical Coherence and Quantum Optics" (Cambridge University Press)

[22] V. I. Man'ko and R. V. Mendes 1999 *Phys. Lett.* A **263** 53

[23] M. A. Man'ko 1999 *J. Russ. Laser Res.* **20** 226; 1999 "Optical tomography approach in signal analysis" in: D. Han, Y.S. Kim and S. Solimeno "VI International Conference on Squeezed States and Uncertainty Relations (Naples, Italy, May 1999)" *NASA Conference Proceedings in CD* 2000-209899 (Goddard Space Flight Center, Greenbelt, Maryland), SISSA-LANL E-print quant-ph/99060102000; 2000 "Fractional Fourier analysis and quantum propagator" in: H.-D. Doebner, V.K. Dobrev, J.-D. Henig and W. Lucke (eds.) "Proceedings of the International Symposium on Quantum Theory and Symmetries (Goslar, Germany, July 1999)" (Singapore: World Scientific) p. 226; 2000 *J. Russ. Laser Res.* **21** 411; 2000 "Quantum tomography method in information processing and fractional Fourier transform" in: A.N. Sissakyan, G.S. Pogosyan and L.G. Mardoyan (eds.) "XXIII International Colloquium on Group Theoretical Methods in Physics (Dubna, Russia, July–August 2000)" (Dubna: Joint Institute for Nuclear Research Press) p. 640

[24] L. Cohen 1996 *IEEE Trans. Signal Process.* **44** 1080

[25] M.A. Man'ko, V.I. Man'ko and R.V. Mendes 2001 *J. Phys. A: Math. Gen.* **34** 8321
M.A. Man'ko 2001 *J. Russ. Laser Res.* **22** 505; 2002 "Wavelets and transforms in information processing" in: E. Kapuścik and A. Horzela (eds.) "Proceedings of the Second International Symposium on Quantum Theory and Symmetries (Krakow, Poland, July 2001)" (Singapore: World Scientific) p. 473; 2002 "Wigner function and tomograms in signal processing and quantum computing" Talk at the Wigner Centennial Conference (Pecs, Hungary, July 2002); 2002 "Unified view of tomographic and other transforms in signal analysis" in: J.-P. Gaseau, R. Kerner, J.-P. Antoine, S. Métens and J.-Y. Thibon (eds.) "Proceedings of the XXIV International Colloquium on Group Theoretical Methods in Physics (Paris, July 2003)" (Bristol: Institute of Physics Publishers, in press)

[26] M.A. Man'ko 2001 *J. Russ. Laser Res.* **22** 168; 2002 "Quantum-tomography method in information processing" in: P. Tombesi and O. Hirota (eds.) "Quantum Communication, Computing, and Measurements 3. Proceedings of the Fifth International Conference on Quantum Communication, Measurements and Computing (Capri, Italy, July 2000)" (New York: Kluwer Academic/Plenum Publishers) p. 147; 2002 *J. Russ. Laser Res.* **23** 433

[27] E. Schrödinger 1926 *Ann. d. Phys.* (Leipzig) **79** 489

[28] A.N. Leznov and M.V. Savel'ev 1985 "Group Methods of Integration of Nonlinear Dynamic Systems" (Moscow: Nauka) [in Russian]

[29] H.A. Haus and S.W. Wong 1996 *Rev. Mod. Phys.* **68** 423

[30] D.J. Kaup 1990 *Phys. Rev.* A **42** 5689
C. Sulem and P.-L. Sulem 1999 "The Nonlinear Schrödinger Equation: Self-focusing and Wave Collapse" (New York: Springer)

[31] E.P. Gross 1961 *Nuovo Cim.* **20** 454
L.P. Pitaevskii 1961 *Sov. Phys. JETP* 1961 *13* 451

[32] M.H. Anderson, J.R. Ensher, M.R. Matthews, G.E. Wieman and E.A. Cornell 1995 *Science* **269** 198

[33] K.B. Davis, M.-O. Mewes, M.R. Andrews, N.J. van Druten, D.S. Durfee, D.M. Kurn and W. Ketterle 1995 *Phys. Rev. Lett.* **95** 3969

[34] C.C. Bradley, C.A. Sackett, J.J. Tollett and R.G. Hulet 1995 *Phys. Rev. Lett.* **75** 1687
C.C. Bradley, C.A. Sackett and R.G. Hulet 1997 *Phys. Rev. Lett.* **78** 985
C.A. Sackett, C.C. Bradley, M. Welling and R.G. Hulet 1997 *Appl Phys. B: Lasers Opt.* **65** 433
J.M. Gerton, D. Strekalov, I. Prodan and R.G. Hulet 2000 *Nature* (London) **408** 692

[35] E. Schrödinger 1930 *Ber. Kgl. Akad. Wiss.* **24** 296

[36] E. H. Kennard 1927 "Zur Quantenmechanik einfacher Bewegungstypen" *Z. Phys.* **44** 326

[37] R. Glauber 1963 *Phys. Rev. Lett.* **10** 84

[38] E.C.G. Sudarshan 1963 *Phys. Rev. Lett.* **10** 277

[39] J.R. Klauder 1964 *J. Math. Phys.* **5** 177

[40] J. Plebański 1954 "Classical properties of oscillator wave packets" *Bulletin de l' Académie Polonaise des Sciences* **2** 213
L. Infeld and J. Plebański 1955 "On a certain class of unitary transformations" *Acta Phys. Polon.* **14** 41
J. Plebański 1955 "On certain wave packets" *Acta Phys. Polon.* **14** 275; 1956 "Wave functions of a harmonic oscillator" *Phys. Rev.* **101** 1825

[41] H.P. Yuen 1976 *Phys. Rev.* A **13** 2226

[42] V.V. Dodonov, E.V. Kurmyshev and V.I. Man'ko 1980 *Phys. Lett.* A **79** 150

[43] H.P. Robertson 1930 *Phys. Rev.* **35** 122

[44] E.C.G. Sudarshan, C. B. Chiu and G. Bramathi 1995 *Phys. Rev.* A **52** 43

[45] V.I. Man'ko, L. Rosa and P. Vitale 1998 *Phys. Rev.* A **58** 3291

[46] O.V. Man'ko, V.I. Man'ko and G. Marmo 2001 *J. Phys. A: Math. Gen.* **35** 699

[47] F. Bayen, M. Flato, C. Fronsdal, A. Lichnerovicz and O. Sternheimer 1975 *Lett. Math. Phys.* **1** 521

[48] G.G. Amosov and V.I. Man'ko 2003 "Quantum probability measure and tomographic probability densities" Eprint quant/ph-0304182 v1; 2003 *Phys. Lett.* A **318** 287

[49] A.S. Holevo 1998 *Russ. Math. Surveys* **53(6)** 1295

[50] R. Fedele, M.A. Man'ko and V.I. Man'ko 2000 *J. Opt. Soc.* A **17** 2506

[51] S. De Nicola, R. Fedele, M.A. Man'ko and V.I. Man'ko 2003 *J. Opt. B: Quantum Semiclass. Opt.* **5** 95

[52] S. De Nicola, R. Fedele, M.A. Man'ko and V.I. Man'ko 2003 *European J. Phys. B* (in press)

[53] E.P. Wigner 1932 *Phys. Rev.* **40** 749
W. Yourgrau and A van der Merwe (eds.) 1979 "Perspectives in Quantum Theory" (New York: Dover)

[54] J.E. Moyal 1949 *Proc. Cambridge Philos. Soc.* **45** 99

[55] J. Von Neumann 1932 "Mathematische Grundlagen der Quantummechanik" (Berlin: Springer)

[56] J. Ville 1948 *Cables et Transmission* **2** 61

[57] S. Mancini, V.I. Man'ko and P. Tombesi 1996 *Phys. Lett.* A **213** 1; 1997 *Found. Phys.* **27** 801

[58] S. Schiller, G. Breitenbach, S.F. Pereira, T. Mikker and J. Mlynek 1996 *Phys. Rev. Lett.* **77** 2933

[59] M.G. Raymer, M. Beck and D.F. McAlister 1994 *Phys. Rev. Lett.* **72** 1137

[60] I.A. Malkin and V.I. Man'ko 1979 "Dynamical Symmetries and Coherent States of Quantum Systems" (Moscow: Nauka) [in Russian]

[61] V.V. Dodonov and V.I. Man'ko 1989 "Invariants and Evolution of Nonstationary Quantum Systems" *Proceedings of the P. N. Lebedev Physical Institute* (New York: Nova Science) Vol 183

[62] R. Fedele and V.I. Man'ko 1998 *Phys. Scr.* T **75** 283

[63] R. Fedele and V.I. Man'ko 1998 "Quantumlike corrections and tomography in beam physics," in: S. Meyers, L. Liljeby, Ch. Petit-Jean-Genaz, J. Poole and K.-G. Rensfelt (eds.) "Proceedings of the 6th European Particle Accelerator Conference (Stockholm, June 1998)" (Bristol & Philadelphia: Institute of Physics Publishers) p. 1268

[64] O. Man'ko and V.I. Man'ko 1997 *J. Russ. Laser Res.* **18** 407

[65] H.M. Ozaktas and D. Mendlovich 1995 *J. Opt. Soc. Am.* A **12** 743

[66] P. Ermakov 1880 "Differential equations of the second order and the conditions of integrability in finite form," Izv. Sant Vladimir University (Kiev), Vol. XX, No. 9, Pt. II, Sec. III (September 1880) pp. 1–25

[67] H.R. Lewis and W.B. Riesenfeld 1969 *J. Math. Phys.* **10** 1458

[68] A.L. Rivera, N.M. Atakishiyev, S.M. Chumakov and K.-B. Wolf 1977 *Phys. Rev.* A **55** 876

[69] R. Fedele and D. Anderson 2000 *J. Opt. B: Quantum Semiclass. Opt.* **2** 207
B. Hall, M. Lisak, D. Anderson, R. Fedele and V.E. Semenov 2002 *Phys. Rev.* E **65** 035602(R)
L. Helczynski, D. Anderson, R. Fedele, B. Hall and M. Lisak 2002 *IEEE J. Selected Topics in Quantum Electronics* **8** 408
M. Onorato, A. Osborne, R. Fedele and M. Serio 2003 *Phys. Rev.* E **67** 046305
R. Fedele, P.K. Shukla, M. Onorato, D. Anderson and M. Lisak 2002 *Phys. Lett.* A **303** 61

[70] V.I. Man'ko and R.V. Mendes 2000 *Physica* D **145** 330

[71] R. Fedele and H. Schamel 2002 *European J. Phys.* B **27** 313

[72] R. Fedele 2002 *Phys. Scr.* **65** 502

[73] V.I. Karpman 1975 "Nonlinear Waves in Dispersive Media" (Oxford: Pergamon)

[74] S. De Nicola, R. Fedele, M.A. Man'ko and V.I. Man'ko 2003 "Wigner picture and tomographic representation of envelope solitons" in: M. J. Ablowitz, M. Boiti and F. Pempinelli (eds.) "Proceedings of the International Workshop 'Nonlinear Physics: Theory and Experiment. II' (Gallipoli, Lecce, Italy, 27 June – 6 July 2002)" (Singapore: Worl Scientific) p. 372; 2002 "Tomographic analysis of envelope solitons: Concepts and applications" in "Abstracts of the International Workshop 'Optics in Computing'. International Optical Congress 'Optics XXI Centuary' (St. Petersburg, Russia, 14–18 November 2002)"

[75] P. Lougovski, E. Solano, Z.M. Zhang, H. Walter, H. Mack and W.P. Schleich 2002 "Fresnel transform: an operational definition of the Wigner function" Eprint quant-ph/0206083

[76] F. Dalfovo, S. Giorgini, L.P. Pitaevskii and S. Stringary 1999 *Rev Mod. Phys.* **71** 463

[77] Ph. Noziéres and D. Pines 1990 "The Theory of Quantum Liquids" (Reading MA: Addison-Wesley) Vol II

[78] W. Hai, Ch. Lee and G. Chong 2003 "Propagation, breathing and transition of mattar-wave packet trains" Eprint quant-ph/0301112 v1

[79] V.Y.F. Leung, A.G. Truscott and K.G.H. Baldwin 2002 *Phys. Rev.* A **66** 061602

[80] L.D. Carr and Y. Castin 2002 *Phys. Rev.* A **66** 063602

[81] D.L. Feder, M.S. Pindzola, L.A. Collins, B.I. Schneider and C.W. Clark 2002 *Phys. Rev.* A **62** 053606

[82] A. Muryshev, G.V. Shlyapnikov, W. Ertmer, K. Sengstock and M. Lewenstein 2002 *Phys. Rev. Lett.* **89** 110401

[83] J. Dziarmaga, Z.P. Karkuszewski and K. Sacha 2003 "Images of the dark soliton in a depleted condensate" Eprint cond-mat/0212492 v2

[84] V.A. Brazhnyi, A.M. Kamchatnov and V.V. Konotop 2003 "On creation and evolution of dark solitons in Bose–Einstein condensate" Eprint quant-ph/0301319 v1

[85] S. Burger, K. Bongs, S. Dettmer, W. Ertmer and K. Sengstock 1999 *Phys. Rev. Lett.* **83** 5198

[86] Z.P. Karkuszewski, K. Sacha and J. Zakrzewski J 2001 *Phys. Rev.* A **63** 061601(R)

[87] L. Khaykovich, F. Schreck, G. Ferrari et al 2003 *Science* **290** 1290

[88] S. De Nicola, R. Fedele, M.A. Man'ko and V.I. Man'ko 2004 *J. Russ. Laser Res.* **25** 1

BOREL QUANTIZATION AND NONLINEAR QUANTUM MECHANICS

A Review of Developments in the Series "Symmetries in Science" I – XIII

H.-D. Doebner
Department of Physics and Material Sciences
D–538670 Clausthal-Zellerfeld
asi@pt.tu-clausthal.de

J. Tolar
Faculty of Nuclear Sciences and Physical Engineering
Czech Technical University
CZ–115 19 Prague, Břehová 7
jiri.tolar@fjfi.cvut.cz

1. Introduction

In the first Symmetries in Science meeting in 1979 at Carbondale we presented preliminary results for a quantization of the classical kinematic for non–relativistic systems which are localized and moving on a smooth manifold M. Our paper [7] 'On Global Properties of Quantum Systems' was published in the Proceedings of Symmetries in Science series. We developed subsequently (with Bernd Angermann) [8, 9] a quantization method on smooth manifolds — the 'Quantum Borel Kinematics' (QBK); for a recent review see [10]. In 1992 a suitable time dependence was proposed (with Jerry Goldin) (see the review [11] and a more general 'Borel Quantization' (BQ) which emerged from geometrical and topological considerations; it indicated a nonlinear extension of quantum mechanics. We participated in some of the later editions of Symmetries in Science series, often together with members of the 'Clausthal group', e.g. Vlado Dobrev, Jerry Goldin, Wieland Groth, Jörg Hennig, Wolfgang Lücke, Hans-Jürgen Mann, Peter Nattermann, Wolfgang Scherer, Christoph Schulte, Pavel Šťovíček and Reidun Twarock. The results, different aspects and applications of Borel Quantization can be found the volumes of 'Symmetries in Science'.

B. Gruber (ed.), Symmetries in Science XI, 209–225.

Our interest in quantum mechanics on manifolds was connected with the following situation: During 1970–1980 some of our colleagues in quantum theory and in particle physics thought that Lie groups and their representations are a major key to model and to understand particle physics. In this context we worked e.g. on spectrum generating algebras and on embeddings of physical Lie algebras. Based on Mackey's theory of induced representations we wrote a paper [12] on a quantization of particles moving on homogeneous G–spaces. We realized that the geometry of the G–space does not contain 'enough' information for a time evolution on G. Furthermore, we failed to generalize Mackey's method to physical important non–homogeneous spaces. Group theory was obviously a very successful model, but it was too 'rigid': If one chooses the group and its representation, the complete mathematical framework is already given; there does not appear the flexibility which one wants for a description of physical systems. Hence those mathematical formalisms which are 'close' to group theory and which are in addition more 'flexible' became interesting. Among such formalisms are: nonlinear and non–integrable representations of Lie algebras and their deformations in the sense of Gerstenhaber. A further promising field for a geometric modelling are differential geometrical and algebraic notions on M. Here one views physical laws e.g. as relation between geometrical or algebraic objects living on M. Following the pioneering papers of George Mackey [13] and Irving Segal [14], we found a path to understand quantizations of a system on a topologically nontrivial configuration space and how the quantized system 'feels' the topology. This leads to Quantum Borel Kinematics characterized by topological quantum numbers and one additional quantum number D. This D is connected with the structure of the infinite–dimensional Lie algebra spanned by quantized kinematical operators.

In our kinematical design a physical interpretation of D was obscure. It became more transparent in 1988 when Jerry Goldin and one of the authors (HDD) realized that two approaches are equivalent: the quantized Borel kinematics on Euclidean configuration spaces and the representations of non–relativistic current algebras on multiparticle configuration spaces for indistinguishable objects found by Goldin and co-workers [15]. The quantum number D appears in front of an additional term in the generalized momentum operator as well as in the momentum current. Jerry Goldin and HDD introduced, based on this observation, a generic time dependence for pure states and derived a family of nonlinear Schrödinger equations [16] — DG equations — with nonlinear term proportional to D. Special generalizations to mixed states (von Neumann equations) are known [17]. A direct connection of the DG family to certain nonlinear gauge transformations [18, 19] and an interpretation of D through nonlinear transformation [20] was elaborated.

Some of these developments are reviewed and commented in this contribution.

2. Borel Kinematic

2.1 Classical Case

We start with a classical model for the kinematic of a system (particle) localized and moving on a smooth manifold M. The kinematic is characterized by the following set of observables:

Generalized positions

$$f \in C^{\infty}(M, R)$$

realized by real functions on M, and
generalized momenta

$$X \in Vect(M)$$

realized by smooth vectorfields on M.

We define the tuple

$$S(M) = (C^{\infty}(M, R), Vect(M))$$

as the generic KINEMATIC on M (or covariance algebra of M). $S(M)$ has the following properties:

a $Vect(M)$ is a (∞–dimensional) Lie algebra of a subgroup of the diffeomorphism group DIFF(M) of M;

b $C^{\infty}(M, R)$ can be viewed as an (∞–dimensional) Abelian Lie algebra;

c f and X defined on M form a semidirect sum

$$S(M) = C^{\infty}(M, R) \oplus_s Vect(M).$$

For physical reasons (see section 2.3) we restrict $Vect(M)$ to the subset of complete vectorfields $Vect_0(M)$; this subset spans a partial Lie algebra) (the Lie bracket of two complete vector fields may not be complete) and the corresponding kinematic $S_0\,(M)$ has partial Lie algebra structure.

2.2 Quantization of $S_0(M)$

To quantize the classical object $S_0(M)$ we construct a map from $S_0(M)$ into the set of essentially self-adjoint operators on a common dense domain in a separable Hilbert space H

$$\mathbb{Q} = (\mathsf{Q}, \mathsf{P}): \quad S_0(M) \longrightarrow SA(H)$$

sending

$$f \longrightarrow \mathsf{Q}(f), \quad X \longrightarrow \mathsf{P}(X)$$

We realize H as $L^2(M, \mathbb{C}, d\nu)$, i.e. the space of square integrable complex functions over M; $d\nu$ is a standard measure on M. We assume furthermore that there is no internal degree of freedom like spin. The map can be viewed as a representation of an infinite dimensional (partial) Lie algebra.

The following properties I.-III. are assumed for $\mathbb{Q}$ to be a *quantization map*:

I. In $L^2(M, \mathbb{C}, d\nu)$ operators $\mathsf{Q}(f)$ act as multiplication operators by f, i.e.

$$\mathsf{Q}(f)\Psi = f\Psi.$$

II. The Lie algebra structure of $S_0(M)$ survives, i.e. $\mathbb{Q}$ is a partial Lie algebra homomorphism.

III. $\mathsf{P}(X)$ is a local operator, i.e. $supp\ (\mathsf{P}(X)\Psi) \subset supp\ \Psi$.

These assumptions have the following background:

Ad **I.** Consider a set of localization regions $B \subset M$; choose for this set the Borel field $\mathfrak{B}(M)$ over M and define the quantization map:

$$\mathbb{Q} : B \in \mathfrak{B}(M) \longrightarrow E(B) \in SA(H).$$

The states of the system are given by normed positive trace–class operators W. The expectation value of a measurement of $E(B)$ in a state W is

$$Tr(WE(B)) = \mu_W(B) \tag{1}$$

$(Tr(.)$ denotes trace) and contains information on the probability of localization of the system in state W in the region B. Using properties of position measurements we assume that the r.h.s. of (36) is (elementary) spectral measure on $\mathfrak{B}(M)$. With our realization of the spectral theorem we find property **I.**

Ad **II.** The algebraic structure of $S_0(M)$ reflects that the classical system is localized and moving on M. Also the quantum system lives on M. Hence this algebraic structure should 'survive' under the quantization map. In this sense quantizations are based "on an algebra" [21].

Our later analysis shows that there are different possibilities for such maps which are related to unitarily inequivalent quantizations. Therefore the information encoded in the classical system is not sufficient to characterize its quantized form; one needs additional information — so called 'quantum information'.

Ad **III.** For $M = R^n$ we know that the momentum operator acts in H as a self-adjoint differential operator. It is plausible to expect also that $\mathsf{P}(X)$ acts locally on complex functions over M as differential operator. Hence in our design we have to define differential operators (of finite order) on functions in $L^2(M, \mathbb{C}, d\nu)$. For this we sketch two notions (**A, B**) and their relation (**C**):

A. To define derivatives of complex functions over M one needs a differentiable structure DS on the point set $M \times \mathbb{C}$. The restrictions of DS give the differentiable structure $DS(M)$ of M (smooth manifold)

$$DS(M \times \mathbb{C})|_M = DS(M)$$

and the restriction to $\mathbb{C}$ yields the standard differentiable structure of $\mathbb{C}$,

$$DS(M \times \mathbb{C})|_{\mathbb{C}} = DS(\mathbb{C})$$

Geometrical objects with these properties are complex line bundles on M with hermitian connection, $\mathfrak{L}(M \times \mathbb{C}, pr, \mathbb{C}, \langle . \rangle, \nabla)$. Some sections of $\mathfrak{L}$ are square integrable. The Hilbert space $L^2(M, \mathbb{C}, d\nu)$ can be viewed as the space of square–integrable sections of $\mathfrak{L}$.

The structure of the set $\{\mathfrak{L}\}$ of such bundles is known. Denoting the curvature of the connection ∇ in $\mathfrak{L}$ by R, one can construct such $\mathfrak{L}$ if and only if

$$\frac{1}{2\pi i} \int_S R \in \mathbb{Z}$$

for all closed 2-surfaces S in M. In terms of cohomology the de Rham class of $\frac{1}{2\pi i} R$ has to be integral,

$$\left[\frac{1}{2\pi i} R\right] \in H^2(M, \mathbb{Z}), \tag{2}$$

i.e. there is a strong bundle isomorphism between two complex line bundles, if and only if their Chern classes in $H^2(M, \mathbb{Z})$ coincide.

For each of these inequivalent classes there is a set of inequivalent connections labelled by

$$H^1(M, U(1)) = \pi_1^*(M),$$

i.e. by elements of the character group of the fundamental group of M.

These algebraic invariants classify the line bundles together with their differentiable structures and covariant derivatives ∇, we are looking for. We have no result whether the introduction of differentiable structures to our model via line bundles is unique. For internal degrees of freedom complex vector bundles can be used [22, 23].

B. For physical reasons we want to avoid nonlocal effects, i.e. we quantize the kinematic by local operators $\mathsf{Q}(f)$ and $\mathsf{P}(X)$. The position operators are local by construction; the locality of $\mathsf{P}(X)$ is equivalent to condition **III.** Note that we assume locality only for $\mathbb{Q}S_0(M)$; other operators representing observables could be nonlocal.

C. The locality condition is linked with differential operators defined via differentiable structures [24]: if there is a differentiable structure $DS(M \times \mathbb{C})$, then the locality of $\mathsf{P}(X)$ implies that $\mathsf{P}(X)$ is a differential operator of finite order with respect to $DS(M \times \mathbb{C})$.

The arguments in section 2.2 are related to our generalization of Mackey's imprimitivity theorem on homogeneous spaces. Our review [10] and references therein have utilized such a generalization.

2.3 A Classification Theorem for Quantization Maps

With the assumptions **I. – III.** we derived a classification theorem for the quantization maps

Theorem

The set $\{\mathbb{Q}\}$ of unitarily inequivalent quantization maps

$$\mathbb{Q}^{(.)}: \quad S_0(M) \longrightarrow SA(H)$$

on a Hilbert space H which is realized with square–integrable sections of a hermitian line bundle, $\mathbb{Q}(S_0(M))$, is labelled by the triple

$$(J, \alpha, D) \in H^2(M, \mathbb{Z}) \times \pi_1^*(M) \times R.$$

Here $H^2(M, \mathbb{Z})$ labels the set of closed two–forms J which satisfy the integrality condition (2), $\pi_1^*(M)$ denotes the character group of the fundamental group of M. For a fixed J we have classification by (α, D).

Explicit form of the map $\mathbb{Q}^{(\alpha,D)} = (\mathsf{Q}^{(\alpha,D)}), \mathsf{P}^{(\alpha,D)})$ can be found in [9, 10, 21]. Details for the case $M = R^3$ are given in section 2.4.

The operators $\mathsf{Q}^{(J,\alpha,D)}(f)$ and $\mathsf{P}^{(J,\alpha,D)}(X)$ are essentially self-adjoint (we used complete vector fields) on a common invariant domain; the representation $\mathbb{Q}^{(\cdot)}(S_0(M))$ is irreducible; α is a topological quantum number; D is independent of the topology and is related to the algebraic structure of $S_0(M)$. Hence there are inequivalent quantizations for systems on topologically trivial manifolds.

REMARKS

a Quantum Borel kinematics are based on a classical configuration space in contrast to geometric (pre–)quantization [25] which works on a symplectic space or more specially on the phase space. In both methods the topological quantum numbers play (with different motivations) an essential role. However, the quantum number D appears only in quantum Borel kinematics. This gap was closed recently: Jörg Hennig and Peter Nattermann showed in [26, 21] that geometric quantization of the kinematic corresponds to our approach, if one uses $(\frac{1}{2} - i\gamma)$–density instead of a $\frac{1}{2}$–density; the imaginary part of the density is proportional to the quantum number D.

b We sketch in paragraphs A, B, C formulations of quantum Borel kinematic for more general situations:

A. If the system has internal degrees of freedom like spin, the Hilbert space is spanned by vector–valued functions. There are two types of quantum maps: type 0 in which different vector components are not mixed; this type is described in [22]. In type 1 a mixing is allowed; Michael Drees gave some preliminary results [23]. A discussion of the quantum Borel kinematic for a spinning particle can be found in [22, 23].

B. If there exists an external field or potential on M, i.e. a closed 2-form B, an additional term in the commutator of the momentum observables appears. Hence a quantization map based on M exists only if B fulfils the integrality (admissibility) condition (2). We refer to [10, 21]. However, such external potentials were already included in an indirect way in our earlier discussion, since the quantum number J is responsible for the existence of both the line bundles and the admissible closed external one–forms.

C. The physical background of D cannot be completely clarified in a kinematical framework. One can calculate how expectation values of momenta depend on D. But this dependence can be analyzed only if it is explicitly known. We discuss this in section 3.

2.4 Applications of the Classification Theorem

The theorem shows that for topologically trivial as well as non–trivial manifolds 'different quantizations' exist. This means that the probabilities of certain observables measured in certain states depend on (J, α, D). One and the same classical system yields — after the quantization maps — a set of different quantum systems. Additional information — the already mentioned 'quantum information' — is necessary to choose or to determine (J, α, D). The source of this information can be, e.g., first principles or experimental results.

As the first example of inequivalent quantizations we consider a topologically trivial manifold $M = R^3$ with vector fields $X(\vec{g}) = \vec{g}(\vec{x}).\vec{\nabla}$ and with quantization map $\mathbb{Q}^{(D)}(S_0(R^3))$,

$$\mathsf{Q}(f) = f$$

$$\mathsf{P}^{(D)}(X) = -i\hbar\vec{g}.\vec{\nabla} + (-i\frac{\hbar}{2} + D)\ div\ \vec{g} \qquad (3)$$

Different D yield unitarily inequivalent representations and hence different quantum systems.

The representations of $S_0(R^n)$ for multiparticle configuration spaces for N indistinguishable objects can be viewed also as representations of non–relativistic inhomogeneous current algebras. Jerry Goldin and co-workers [15] constructed such representations; they derived the above result (38) and found independently the quantum number D.

Now let us mention some systems on topologically non–trivial smooth manifolds and their different quantizations (see also Table 1):

The physics of N indistinguishable particles — anyons — and distinguishable particles on a 2–dimensional Euclidean space (in the framework of cur-

Table 1. Examples of elementary quantum Borel kinematics [10].

$QS^{*)}$	M	$\pi_1(M)$	$H_1(M,Z)$	$H^2(M,Z)$	Topological quantum numbers
a	R^3	$\{e\}$	0	0	—
b	$R^3\backslash R$	Z	Z	0	$\vartheta \in [0,1)$
c	$R^3\backslash O = R_+ \times S^2$	$\{e\}$	0	Z	$n \in Z$
d	$R^3 \times R_+ \times S^2$	$\{e\}$	0	Z	$n \in Z$
e	$R^3 \times R_+ \times RP^2$	S_2	Z_2	Z_2	$m \in Z_2$
f	$R^3 \times SO(3)$	Z_2	Z_2	Z_2	$m \in Z_2$
g	S_2	$\{e\}$	0	Z	$n \in Z$
h	S^1	Z	Z	0	$\vartheta \in [0,1)$
i	K_p	$\pi_1(K_p)$	Z^{2p}	Z	$n \in Z$, $\vartheta_1 \ldots \vartheta_{2p} \in [0,1)$

*) QS — Quantum system:
[a] Spinless particle in R^3
[b] Aharonov-Bohm configuration
[c] Dirac's monopole
[d] 2 distinguishable particles in R^3
[e] 2 indistinguishable particles in R^3
[f] Rigid body
[g] Symmetric top
[h] Rotator with fixed axis
[i] Particle on orientable surface of genus p

rent algebra) was discussed by Jerry Goldin and co-workers. A review (in Borel quantization) of indistinguishable and distinguishable particles on 2–dimensional manifolds can be found in [24]. Parastatistics appears for N indistinguishable particles on manifolds with dimension > 3 [24]. Aharonov–Bohm situations were discussed as topological effects in [25]. The quantum maps for systems on non–orientable 2–dimensional manifolds (Möbius strip and Klein bottle) were treated in [26]. For quantizations on the trefoil knot see [27]. A recent review of many aspects in quantum Borel kinematics can be found in [10] together with some further examples of configuration spaces with nontrivial topology.

3. Borel Dynamics

3.1 Difficulties with $\mathbb{Q}^{(\cdot)}(S_0(M))$

Quantum Borel Kinematic holds for any fixed time t, it considers a 'frozen' system and it carries no direct information on a t–dependence. Hence a principle is needed to construct related dynamical equations. As explained before one hopes that this could be a key for a physical interpretation of the quantum number D.

As a plausible model for an evolution of pure states we choose a dynamical group $\{\mathfrak{D}_{t,g}; t \in R\}$ with a linear operator $\mathbb{D}_g$ as generator acting on H; g denotes a Riemannian structure on M. In the Heisenberg picture we relate the quantized momentum $\mathsf{P}^{(\cdot)}(X)$ to $\mathbb{D}_g$, i.e. we assume the existence of a map

$$C^\infty(M, R) \to Vect_0(M)$$

such that

$$[\mathbb{D}_g, \mathsf{Q}(f)] = -i\mathsf{P}^{(\cdot)}(X_f) \quad \text{for all} \quad f \in C^\infty(M, R).$$

In analogy to Hamiltonian mechanics in phase space we specialise this map with

$$X_f = grad_g\, f.$$

To analyze this ansatz consider the ∇–lift of the Laplace Beltrami operator Δ_g on M (with metric g) is a candidate for $\mathbb{D}$. Hence we write (with some operator $\mathbb{K} = \mathsf{Q}(V)$)

$$\mathbb{D}_g = -\frac{1}{2}\Delta_g^\nabla + \mathbb{K}.$$

The commutator between $\mathbb{D}_g$ and f has with (38) for all $f \in C^\infty(M, R)$ the form

$$[\mathbb{D}_g, f] = i\mathsf{P}(grad_g\, f) - iD\Delta_g f.$$

A comparison with the previous result yields $D = 0$. Thus our ansatz for a dynamical group (in the Heisenberg picture) fails; it leads to the trivial result $D = 0$ [28]. This failure is partly connected with the implicit assumption that evolutions of wave functions are linear.

3.2 Nonlinear evolutions from $\mathbb{Q}^{(D)}(S_0(R^3))$

An alternative method is to start with the assumption that the positional probability is conserved. Consider again a system in $M = R^3$. We assumed [5]

$$\frac{\partial}{\partial t}\int_{R^3} \varrho(x,t)d^3x = 0, \qquad \varrho(x,t) = \overline{\Psi}(x,t)\Psi(x,t).$$

This implies indirectly that a pure state remains a pure state. With a suitable behaviour of Ψ at infinity we are allowed to apply the Gauss theorem and find

$$\frac{\partial}{\partial t}\varrho(x,t) = -\nabla\vec{j}(x,t) \tag{4}$$

with a vector field density $\vec{j}$ depending on the wave function Ψ.

How to construct $\vec{j}$ in our model? Consider the above equation as an operator equation in the 1–particle sector $\mathfrak{F}_1$ of the Fock space generated from a cyclic vacuum $|0\rangle$. The generic operators $\mathsf{Q}(f)$, $\mathsf{P}^{(D)}(X)$ correspond to operator–valued densities ϱ, $\vec{j}^D$ in $\mathfrak{F}_1$:

$$\mathsf{Q}(f) = \int f(x)\varrho(x,t)d^3x$$
$$\mathsf{P}^{(D)}(X) = \int \vec{g}(x)\vec{j}^D(x,t)d^3x$$

We have already $\varrho(x,t) = \overline{\Psi}\Psi$. We get for $\vec{j}^D(x,t)$ from (38)

$$\vec{j}^D(x,t) = \vec{j}^0(x,t) - D\nabla\varrho(x,t), \quad \vec{j}^0(x,t) = \frac{\hbar}{2mi}(\overline{\Psi}\cdot\nabla\Psi - \overline{\nabla\Psi}\cdot\Psi)$$

and with (4)

$$\frac{\partial}{\partial t}\varrho(x,t) = -\nabla\vec{j}^0(x,t) + D\nabla\varrho(x,t).$$

This is a Fokker–Planck type equation. For $D = 0$ we have the quantum mechanical continuity equation with the usual quantum mechanical current. The term proportional to the quantum number D is a quantum mechanical diffusion current which is characteristic for the model.

Any ansatz for a time dependence of $\Psi(x,t)$ has to respect this Fokker–Planck equation. We use this fact to construct an evolution of first order in ∂_t, with the usual linear terms and with an additional term F depending e.g. on the wave function Ψ:

$$i\hbar\frac{\partial}{\partial t}\Psi(x,t) = (-\frac{\hbar^2}{2m}\Delta + V(x) + F\,[\Psi]\,\Psi$$

Inserting this ansatz into the Fokker–Planck equation, a non linear Schrödinger equation with a complex nonlinear term is obtained,

$$i\hbar\frac{\partial}{\partial t}\Psi(x,t) = (-\frac{\hbar^2}{2m}\Delta + V(x) + iIm\;F\,[\Psi] + Re\;F\,[\Psi])\Psi,$$

where

$$Im\;F\,[\Psi] = \hbar\frac{D}{2}\frac{\Delta\varrho}{\varrho}$$

is enforced through $\mathbb{Q}^{(D)}$, and $Re\ F[\Psi]$ is independent of D (arbitrary).

We see that the imaginary part of F is fixed by the quantization method, but there is no information on the real part. We assume for Re F a function of the wave function which is of the same type as ImF, i.e.

- complex homogeneous of order zero
- rational with derivatives not higher than second order in the numerator
- Euclidean invariant.

With these assumptions for $Re\ F$ Doebner and Goldin obtained a family of singular nonlinear Schrödinger equations (DG equations) for a particle with mass m, potential V parametrized by $\hbar$, D, c_1,..., c_5:

$$i\hbar\frac{d}{dt}\Psi = \left[-\frac{\hbar^2}{2m}\Delta + V(\vec{x}) + \frac{1}{2}\hbar D\frac{\Delta\varrho}{\varrho} + \hbar D'\sum_{i=1}^{5} c_i R_i[\Psi]\right]\Psi$$

$$R_1[\Psi] = \frac{m}{\hbar}\frac{\nabla\vec{j}^0}{\varrho}, \quad R_2[\Psi] = \frac{\Delta\varrho}{\varrho}, \quad R_3[\Psi] = \frac{m^2}{\hbar^2}\frac{(\nabla\vec{j}^0)^2}{\varrho^2}$$

$$R_4[\Psi] = \frac{m^2}{\hbar^2}\frac{(\nabla\vec{j}^0)^2}{\varrho^2}, \quad R_5[\Psi] = \frac{(\nabla\varrho^2)}{\varrho^2}.$$

The choice of the real nonlinearity corresponds to the 'gauge generalization' used in another derivation of the DG equations (see sections 4.1, 4.2).

Independently of the known fact (see e.g. the review [11]) that the usual framework of quantum mechanics does not allow fundamental nonlinear evolutions for pure states, one may argue that a small nonlinearity (small D) can be treated approximately with the usual methods. With this precaution some results for atomic spectra were derived; for the hydrogen atom present precession experiments show no difference to the linear theory. This leads to an upper bound for D [29]

$$D > 10^{-7}\frac{\hbar}{m}$$

There are discussions on DG type nonlinearities in quantum optics ('nonlinear photons') [29].

For an exact calculation of observable effects a new framework of quantum mechanics is necessary. There are indications how to formulate general requirements (e.g. [30]), but there is by no means a complete and mathematically acceptable theory which incorporates fundamental nonlinearities.

The mathematical structures and properties of DG equations and the mentioned precaution for their physical applications are partly known; we quote (the following list is incomplete):

Cauchy problem [31]; Lie symmetries [32]; solutions for stationary states [16], time dependent solutions for certain coefficients [33]; generalizations for: arbitrary smooth M [21], mixed states [17]; other methods for a derivation via: nonlinear gauge transformations [19], generalized Ehrenfest relations [17], stochastic processes [34]. Further applications are known for anti–particles [35] and the dynamics for D–branes.

4. Borel Kinematic and Nonlinear Structures

Our quantum number D yields a family of nonlinear evolution equations for pure states with a nonlinearity proportional to D. This indicates a hidden nonlinear structure in $\mathbb{Q}^{(0)}(S_0(R^3))$. We assume in the following $M = R^3$.

4.1 Nonlinear Gauge Transformations

In connection with the properties of stationary solutions of the DG family a group $\mathcal{G}$ of nonlinear gauge transformations was introduced [18]. They are invertible transformations $\mathbb{N}$

$$\mathbb{N} : \Psi \in H \longrightarrow \mathbb{N}\Psi = \mathbb{N}\left[\Psi\right] \in H \tag{5}$$

with $\mathbb{N}$ depending on $\Psi(x,t)$, x, t. They are restricted by the assumption that the positional probability density is invariant:

$$\overline{\mathbb{N}\left[\Psi\right]}\mathbb{N}\left[\Psi\right] = \overline{\Psi}\Psi.$$

The reason for this restriction is the following: $\mathbb{N}$ should transform a given system, i.e. a given Ψ, to a 'physically equivalent' one: "equivalent" means that the results of measurements on both systems are the same.

Behind the notion of 'physical equivalence' is the 'principle' that the positional density $\varrho(x,t)$ for all x and t determines the outcomes of the measurements of all observables [36]. Such $\mathbb{N}$ build a nonlinear gauge group $\mathcal{G}$.

Now applying $\mathbb{N}$ to the usual (linear) Schrödinger equation, i.e. to a system with linear evolution operator

$$\mathbb{D}_S = i\hbar\partial_t + \frac{\hbar^2}{2m}\Delta - V(x), \quad \mathbb{D}_S\Psi = 0,$$

a (family of) nonlinear Schrödinger equations is obtained,

$$\mathbb{D}_S \circ \mathbb{N}\left[\Psi\right] = 0.$$

The result is a subfamily in the DG family.

By construction it describes the physics of a system which is equivalent to the linear system. A generic procedure (gauge generalization) to construct 'new' systems is the gauge generalization. This is a generic procedure for families of partial differential equations depending on coefficients which are related to each other; the breaking of this relation is our model for a "gauge generalization" [18, 19]. This gauge generalization leads to the DG family; some of its members are inequivalent, they represent systems with new physical properties.

Sections 3.2 and 4.1 show that the DG equations can be derived from two different structures:

- From a geometric structure via a representation of an inhomogeneous diffeomorphism group acting on the configuration space and with a t–dependence from a conservation of positional probability density.
- From a nonlinear structure via nonlinear transformations of positional probability densities between physically equivalent systems applied to a t–dependence of the linear system and gauge generalization.

4.2 Nonlinear Tangent Map

Another method to describe a hidden nonlinear structure of $\mathbb{Q}^{(0)}\ (S_0(R^3))$ more directly was recently presented [20].

Let $\mathbb{N} \in \mathcal{N}$ as in (5) be a group of nonlinear transformations in H. To introduce convenient transformation properties for operators we consider physically interesting ones which are (often) essentially selfadjoint. They can be viewed as generators $i\mathbb{A}$ of a one parameter group U_ε of unitary transformations

$$U_t = \exp it\mathbb{A}.$$

Take a path $\{U_t\Psi,\ t \in [-\varepsilon, \varepsilon]\}$ in H. Then

$$\frac{d}{dt}(U_t\Psi)|_{t=0} = \frac{d}{dt}(\Psi + it\mathbb{A}\Psi)|_{t=0} = i\mathbb{A}\Psi.$$

Hence $i\mathbb{A}\Psi$ appears as a tangent map T of U_ε. Take now a transformed path $\mathbb{N}(U_\varepsilon)$ and define the transformed generator $i\mathbb{A}_\mathbb{N}$ by the tangent map $T(\mathbb{N})$ of $\mathbb{N}(U_\varepsilon)$:

$$\frac{d}{dt}\mathbb{N}(U_\varepsilon\Psi)|_{t=0} = i\mathbb{A}_\mathbb{N}\Psi.$$

Hence we have the $\mathbb{N}$-tangent map

$$T(\mathbb{N}) : \quad \mathbb{A} \to \mathbb{A}_\mathbb{N}.$$

For linear $\mathbb{N}$ we get the usual result. For nonlinear $\mathbb{N}$ the resulting operator $\mathbb{A}_\mathbb{N}$ is in general nonlinear. The $\mathbb{N}$-tangent map is a Lie algebra isomorphism. One can extend the method formally to non essentially selfadjoint operators.

We apply now the $\mathbb{N}$-tangent map to quantize kinematical observables. Similarly as in section 4.1 we restrict $\mathbb{N}$ such that the $\mathbb{N}$-tangent mapped elements in $\mathbb{Q}^{(D)}(S_0(R^3))$ are again linear and of order 0 or 1, i.e.

1. $\mathsf{Q}(f)$ is a linear multiplication operator, i.e. f
2. $\mathsf{P}^{(D)}(X)$ is a linear differential operator of order 1.

Condition 1. is fulfilled by construction; condition 2. is equivalent to the relation

$$\mathsf{P}^{(0)}_\mathbb{N}(\vec{g} \cdot \nabla) = \vec{g}_1 \cdot \nabla + g_0$$

with $\vec{g}_1(x)$, $g_0(x)$ depending on $\vec{g}(x)$. The last condition implies [14] a set $\mathbb{N}$ of non linear transformations. We write this (formally) in polar decomposition

$$\Psi = R \exp iS, \qquad \mathbb{N}[\Psi] = \mathbb{N}[R,S] = R_{\mathbb{N}}(R,S) \exp iS_{\mathbb{N}}(R,S)$$

The form of $\mathbb{N}$ is

$$\begin{aligned} R_{\mathbb{N}}(R,S) &= R^{\kappa+1} r(S) \\ S_{\mathbb{N}}(R,S) &= \gamma \ln R + t(S) + 1 \end{aligned}$$

with $\kappa, \gamma \in R$, and real functions $r(S)$ and $t(S)$. The functions $\vec{g}_1$, g_0 are

$$\begin{aligned} \vec{g}_1 &= +\frac{\hbar}{i}\vec{g} \\ g_0 &= (\frac{\hbar}{2i} + \frac{1}{4}\gamma) div\, \vec{g} \end{aligned}$$

The transformations $\mathbb{N}(\kappa, \gamma, r(S), t(S))$ build a group $\mathfrak{N}$.

4.3 Applications of the Nonlinear Tangent Map

We consider the behaviour of $\mathbb{Q}^{(D)}(S_0(R^3))$ under $\mathfrak{N}$. For $D = 0$, i.e. for $\mathbb{Q}^{(0)}(S_0(R^3))$, we have (see (38))

$$\begin{aligned} \mathsf{Q}(f) &= f \\ \mathsf{P}^{(0)}(X) &= \frac{\hbar}{i}\vec{g} \cdot \nabla - i\frac{\hbar}{2} div\, \vec{g}. \end{aligned}$$

Hence with $\gamma = 4D$

$$\mathbb{Q}^{(D)}(S_0(R^3)) = \mathbb{Q}_{\mathbb{N}}^{(0)}(S_0(R^3))$$

holds. Representations of $\mathbb{Q}^{(D)}(S_0(R^3))$ with different D which are inequivalent under linear unitary transformations are 'equivalent' with respect to certain non–unitary ones. It is interesting to apply the $\mathbb{N}$-tangent map to the linear Schrödinger equation. For $\mathbb{N}$ we find that

$$\mathbb{D}_S^0 \circ \mathbb{N}\,[\Psi] = 0, \qquad \mathbb{N} \in \mathfrak{N}$$

leads to a subfamily of generalized NLSE which contain (after gauge generalization) the DG family. After gauge generalization a "general" DG family is constructed.

If one applies $\mathbb{N}$ to an ordered polynomial generated from $\mathsf{P}^{(D)}(X)$, $\mathsf{Q}(f)$ one gets a nonlinear quantization of all observables of polynomial type. The partial Lie algebra structure of these nonlinear operators is known.

5. Summary and Outlook

We started with a geometrical framework to quantize the kinematic of a system living on a topologically nontrivial manifold (Quantum Borel Kinematic). We showed how the quantization depends on the topology through topological quantum numbers. Since inhomogeneous diffeomorphisms are used as models for the kinematic, a new quantum number of non–topological origin appears. If a time dependence is introduced to the quantized kinematic through conservation of probability, this quantum number is the root of nonlinear Schrödinger equations (DG equations) for pure states. Properties of stationary solutions of the DG equations lead to the introduction of nonlinear gauge transformations which transform a system in a physically equivalent one. Applied to the linear Schrödinger equation, nonlinear gauge transformations in $\mathcal{G}$ lead after gauge generalization to the DG equations. A more direct indication of an intrinsic structure of the quantum Borel kinematic utilizes $\mathbb{N}$–tangent maps, $\mathbb{N} \in \mathcal{N}$, which transform linear quantized kinematical operators into nonlinear ones. These $\mathbb{N}$–tangent maps can describe a nonlinear quantization of polynomial observables; for the Hamiltonian a generalized family of DG equations appears.

Topological effects in quantum mechanics and the topological quantum numbers are a well established field with few applications to real systems, e.g. indistinguishable particles in R, anyons, Bohm–Aharonov situations. Nonlinear quantum mechanics for pure states is an interesting but controversial field. On one hand it seems to be plausible that quantum theory is a linearization of a more involved theory and that nonlinear evolutions derived from first principles are a key stone for a formulation of such a framework. On the other hand we know that linear structures are deeply rooted in the mathematical and physical formulation of quantum theory. The present formalism does not allow nonlinear operators; only for approximations see [37]. A 'new' formalism is not yet developed. There is no experimental indication for a fundamental nonlinearity. However, deviations from usual quantum mechanics are discussed in connection with quantum mechanical precision experiments and with new possibilities for an experimental design. In this connection topological viewpoints as well as nonlinearities in the evolution are of interest.

Acknowledgments

One of the authors (HDD) acknowledges discussions with Alois Kopp.

References

[1] H.-D. Doebner and J. Tolar, in *Symmetries in Science,* eds. B. Gruber and R.S. Millman (Plenum, New York, 1980, 475); in *Symmetries in Science II,* eds. B. Gruber and R. Lenczewski (Plenum, New York 1986, 115)

[2] B. Angermann and H.D. Doebner, Physica 114A, 433 (1982)

[3] B. Angermann, H.D. Doebner and J. Tolar, Lecture Notes in Mathematics, Vol. 137 (Springer, Berlin, 1983, 171)

[4] H.D. Doebner, P. Šťovíček and J. Tolar, Rev. Math. Phys. 13, 799 (2001)

[5] H.D. Doebner and G.A. Goldin, 'Remarks and Recent Results on Nonlinear Extensions of Quantum Mechanics', J.Phys. A (in print) (2003)

[6] H.D. Doebner and J. Tolar, J. Math. Phys. 16, 975 (1975)

[7] G.W. Mackey, 'Mathematical Foundations of Quantum Mechanics', (W.A. Benjamin, New York, 1963)

[8] I.E. Segal, J.Math. Phys. 1, 468 (1960)

[9] G.A. Goldin, R. Menikoff and D.H. Sharp, J. Math. Phys. 21, 650 (1980), J. Phys. A: Math.Gen. 16, 1827 (1983)

[10] H.D. Doebner and G.A. Goldin, J. Phys. A: Math. Gen. 27, 1771 (1994)

[11] H.D. Doebner and J.D. Hennig, in *Symmetries in Science VIII,* ed. B. Gruber (Plenum Press, New York, 1995, 85)

[12] H.D. Doebner and G.A. Goldin, Phys. Rev. A 54, 3764 (1996)

[13] H.D. Doebner, P. Nattermann and G.A. Goldin, J. Math. Phys. 40, 49 (1999)

[14] H.D. Doebner and J.D. Hennig, 'Quantum Borel Kinematics and Nonlinear Quantisations', in preparation (2004)

[15] H.D. Doebner and J. Tolar, in *Symmetries in Science V,* eds. B. Gruber, L.C. Biedenharn and H.D. Doebner (Plenum Press, New York, 1990, 137)

[16] H.D. Doebner and U.A. Müller, J. Phys. A: Math. Gen, 26, 719 (1993)

[17] M. Drees, 'Zur Kinematik lokalisierbarer quantenmechanischer Systeme unter Berücksichtigung innerer Freiheitsgrade und äußerer Felder', Ph.D. Dissertation, TU Clausthal (1992)

[18] J. Peetre, Math. Scand. 7, 211 (1959) and 8, 116 (1960)

[19] N.M.J. Woodhouse, 'Geometric Quantisation', 2nd edition (Clarendon Press, Oxford, 1992)

[20] J.D. Hennig, in 'Nonlinear, Deformed and Irreversible Quantum Systems', eds. H.D. Doebner, V.K. Dobrev and P. Nattermann (World Scientific, Singapore, 1995, 155)

[21] P. Nattermann, 'Dynamics in Borel Quantisation: Nonlinear Schrödinger Equations vs. Master Equations', Ph.D. Dissertation, TU Clausthal (1997)

[22] H.D. Doebner, R. Zhdanov and A. Kopp, 'Nonlinear Dirac Equations and Nonlinear Gauge Transformations', in 'Symmetry in Nonlinear Mathematical Physics', Proceedings of the Institute of Mathematics of NAS of Ukraine (2004)

[23] H.D. Doebner and A. Kopp, 'Nonlinear Pauli Equations', in preparation (2004)

[24] H.D. Doebner, W. Groth and J.D. Hennig, J. Geom. Phys. 31, 35 (1998); H.D. Doebner, P. Šťovíček and J. Tolar, Czech. J. Phys. B32, 1240 (1982)

[25] C. Schulte, in *Symmetries in Science X,* eds. B. Gruber and M. Ramek (Plenum Press, New York, 1998, 357)

[26] C. Schulte, in *Symmetries in Science IX,* eds. B. Gruber and M. Ramek (Plenum Press, New York, 1997, 313)

[27] H.D. Doebner and W. Groth, J. Phys. A 30, L503 (1997)

[28] B. Angermann, 'Über Quantisierungen lokalisierter Systeme – Physikalisch interpretierbare mathematische Modelle', Ph.D. Dissertation, TU Clausthal (1983)

[29] H.D. Doebner, V.I. Manko and W. Scherer, Phys. Lett. A 268, 17 (2000)

[30] B. Mielnik, Commun. Math. Phys 15, 1 (1969)

[31] H. Teismann, in 'Group21', Vol. 1, eds. H.D. Doebner, P. Nattermann and W. Scherer (World Scientific, Singapore 1997, 433)

[32] P. Nattermann, Rep. Math. Phys. 36, 387 (1995)

[33] A.G. Ushveridze, Phys. Lett. A185, 123 and 128 (1994); V.V. Dodonov and S.S. Mizrahi, Ann. Phys. 37, 226 (1995); Physica A 214, 619 (1995)

[34] H.J. Mann, Rep. Math. Phys. 44, 143 (1999)

[35] H.D. Doebner and G.A. Goldin, 'Extensions of Quantum Mechanics — Nonlinear Schrödinger Equations for Particles and Antiparticles', in Proceedings of the 3rd Symposium 'Quantum Theory and Symmetries' (World Scientific, Singapore 2004)

[36] R.P. Feynman and A.R. Hibbs, 'Quantum Mechanics and Path Integral' (McGraw Hill, New York 1965)

[37] W. Lücke and P. Nattermann, in *Symmetries in Science X,* eds. B. Gruber and M. Ramek (Plenum Press, New York, 1997, 197);
N. Gisin, in 'Nonlinear, Deformed and Irreversible Quantum Systems', eds. H.D. Doebner, V.K. Dobrev and P. Nattermann (World Scientific, Singapore, 1995, 109)

SEEING SCIENCE THROUGH SYMMETRY

An Interdisciplinary Multimedia Course

L.I. Gould
Physics Department
University of Hartford
West Hartford, CT 06117 U.S.A.

Abstract Seeing Through Symmetry is a course that introduces non-science majors to the pervasive influence of symmetry in science. The concept of symmetry is used both as a link between subjects (such as physics, biology, mathematics, music, poetry, and art) and as a method within a subject. This is done through the development and use of interactive multimedia learning environments to stimulate learning. Computer-based labs enable the student to further explore the concept by being gently led from the arts to science. This talk is an update that includes some of the latest changes to the course. Explanations are given on methodology and how a variety of interactive multimedia tools contribute to both the lecture and lab portion of the course (created in 1991 and taught almost every semester since then, including one in Sweden).

1. Introduction

Symmetry is something that we are all probably aware of, for better or for worse, in our everyday lives: A desirable situation can occur in the supermarket, when a shopper attempts to find another tomato in the pile that looks just like the nice one already selected. In the faculty dining room, on the other hand, a colleague sometimes guesses which of two apparently identical metal dispensers contains the hot water for tea; and, failing the determination, releases coffee onto a tea bag!

The subject of symmetry has been written about extensively. There are numerous works on symmetry in science, art, mathematics, philosophy, music, poetry, and information processing. From the World Wide Web, two journals ([1], [3]), articles from conference proceedings [5], and collections of essays [2] can be found many examples covering all of those subjects... and more.

B. Gruber (ed.), Symmetries in Science XI, 227–238.

The term "symmetry" used here has no meaning apart from some *operation*. With this in mind one can put forth the following fairly standard definition: *An object is symmetric under a particular operation if it appears unchanged after that operation has been performed.* A very simple example: the square has rotational symmetry because after rotating it about its center through 90°, in its plane, the square appears as it did prior to the rotation. A more complicated example: In the figure below there are four letters "T" and one letter "O". The whole figure has reflectional symmetry about two perpendicular lines passing through the center of the "O", 2-fold rotational symmetry about the same center, and infinite-fold rotational symmetry of just the "O" (if it was a perfect circle) with respect to that center.

Trial restriction

Furthermore, if you begin at the center and go counterclockwise through the "T" at the upper right (UR), the "T" at the upper left (UL), and back to the "O", the word "OTTO" is spelled. The identical word is spelled again if you go clockwise (from the "O" to UL to UR and back to the "O"). Hence, one has what can be dubbed a *palindromic symmetry*, usually called a *palindrome* (the same palindrome occurs by going clockwise or counterclockwise using the bottom half of the figure). Finally, starting at the "O", going the counterclockwise route through the top half of the figure, followed by a clockwise route through the bottom half of the figure, will again bring you back to the "O"; thus tracing out a figure eight which, moreover, repeats under those combined operations. The Italian word for "eight" is "otto"!

If one defines an object's "image" as the result of having performed a particular operation on the object, then a more concise definition of symmetry is: *An object is symmetric under a particular operation if it is identical to its image.* Even briefer is: *Symmetry means invariance under change* (cf. [9],19).

2. Foundations for the course

Students taking the course work up to a definition of symmetry: starting in a qualitative way, by observing pleasing patterns in the arts and in the sciences, they go on to characterize symmetry using elementary mathematical notions. That gives them the raw material from which to abstract and formulate a general definition of symmetry as the course's key concept.

The concept of symmetry can be fundamentally regarded as coming from the area of *philosophy* — most basic of all the disciplines. It is therefore not so surprising that the concept of symmetry applies to such a diverse set of disciplines (see, e.g., Van Fraassen [10]). I would even argue that it applies to *all* disciplines. (Indeed, thinking is itself an activity that appears to require the identity of ideas under change of mental state.)

So a small but important part of the course is devoted to understanding the philosophical aspects of symmetry. One such aspect could be called the "epistemology of symmetry": How valid is our concept of symmetry? Where are its limitations? For example: Do we mean that an object is symmetric if it *appears* so? (Is the right side of your face the true mirror image of the left side?) Are there symmetries that lie below appearances? (Are all the laws of physics symmetric? If you drop an object at one place will its motion be the same as when you drop the object after having moved two feet to the right or after having waited for one minute?)

Educators at other universities[1] have also thought about the value of teaching symmetry. A particularly good description of symmetry's value for education in general and for science education in particular is given by P. Klein [8]:

> It aims at [an] *interdisciplinary* approach since it deals first with formal conditions of understanding applicable to all possible objects of experience;
>
> It relates objects of learning to each other, thus rendering possible *shaped, understanding* learning;
>
> All formal laws raise from and remain closely related to *sensual experience*;
>
> A deep feeling of comfort is raised by having symmetrical orders open to our senses; this affects our sense of *beauty*;

[1]Only a small number of schools have (or had) courses explicitly based on symmetry. Some examples are by G. Darvas at The Institute for Advanced Symmetry Studies in Budapest, I. Halpern at the University of Washington, J. Kapraff [6] at the New Jersey Institute of Technology, E. Merzbacher at Williams College, D. Nagy at Arizona State University, A. Rosenberg at Swarthmore College, M. Senechal at Smith College, and S. Wait at Rensselaer Polytechnic Institute. But those courses have not taken the same approach as the one discussed here with regard to the method used, kinds of disciplines discussed, and the laboratory-enriching "hands-on" experience.

> Learning with symmetries may be based on *action*, and action will continue to give a basis of understanding for complicated problems: the promoting unity of action, of sensual and intellectual activity in understanding will be experienced;
>
> These activities may be abstracted and *formalized* towards mathematics, simple enough, yet basic, thus *evolving* the mathematical interpretation of the world.

Symmetry, as utilized in the Seeing Through Symmetry course, is what I would call a "hub concept". It is as if it stood for the axis of a cylinder consisting of a multidisciplinary world with ties to disciplines constituting the surface, and with the disciplines, as a consequence, tied to each other. Through this hub many aspects of the scientific and artistic worlds can be better understood and appreciated. For example, a concept of symmetry common to mathematics, physics, and music can enable students to interrelate those disciplines — not only by acquiring some new understanding of each through the concept, but also by seeing a commonality of each through that concept; and, by using that commonality, it is possible for advances in one discipline to lead to advances in another. Explaining further: the concept of "translational symmetry", when given its precise mathematical formulation, can be used to explain the physics of sine waves, which in turn can be applied to understanding why a musical tone sounds the way it does. The physics of sound and the sound of music can then feed into each other, each area contributing to the other. (An interrelationship of such disciplines is perhaps strange for many academic specialists. But in the Middle Ages it was a matter of course: For example, *music* was part of the *mathematical sciences* called the "quadrivium", a division of the seven liberal arts. There music was personified by Pythagoras because he was able to relate pleasing sounds made by a taut string that was plucked, after having one part of it held down so as to divide the string's length into the ratio of lengths corresponding to the ratio of positive integers.)

Some details about the course

Seeing Through Symmetry was created in 1991 and taught almost every year since then; including during the summer of 2000, as a result of an invitation to teach the lecture portion at the University of Skövde, Sweden. Currently, a book and instructor's manual is being written to augment the course notes (handouts) and the laboratory manual.

Seeing Through Symmetry enables students to develop their quantitative abilities and analogical thinking by using symmetry both as a method within a discipline and as a bridge between disciplines. Starting with the topics of symmetry in art, in poetry, and in music, we then go on to display and interrelate those topics to symmetry in areas such as mathematics, physics, chemistry, biology, and cosmology. Thus students "see through" (i.e., understand) the concept of symmetry as well as "see the world" *through* (i.e., by means of employing) the concept of symmetry. They develop inter-relational and scientific abilities, in part, through the medium of a highly graphical laboratory experience; this utilizes computers in order to explore the many facets of symmetry including the generation of their own patterns through the use of software packages and elementary programming.

Integrative skills in this course develop in a variety of ways. A primary one is the use of communication skills for a term project, through speaking, writing, and class presentation. The term project incorporates course concepts of science/math and requires students to display their creative and analytical abilities. Throughout this project, as one of the course's integrative benefits, students not only see interrelations between subjects within the realm of science but also between that realm, the humanities in general, and their own discipline in particular. (Examples are given in the Term Project section.)

Lectures

These are multimedia interactions. Several films are shown. One of them is an easily understood general introduction to a variety of symmetries. Another stresses the glories of learning as seen through the mind of the prototype "Renaissance man", Leonardo da Vinci[2] An audio tape is employed to demonstrate the use of the Golden Ratio in Bartok's "Divertimento for Strings". A laser and "grating" exhibit the wave interference phenomenon called "diffraction"; and glow-discharge tubes of hydrogen and helium illustrate how the spectrum of each, seen through the grating, can lead us to an understanding of what stars are made of. A special feature of the course is a computer-animation-and-sound show.[3]

It illustrates symmetry in art, geometry, geophysics, and both cellular and molecular biology. It also dynamically illustrates "broken" symmetry through a portion of a motion picture showing the "sickling" (change of shape from circular to crescent moon) of a red blood cell — a manifestation of the disease called "sickle cell anemia".

The course has been changed more and more towards encouraging students to refer to their *own* experiences throughout the development of any topic. They are also able to try things out on the computer which illustrate ideas discussed in class. The point here is that science *begins* from an individual's observations

---

[2]Some Films: (a) Bobker, Lee R. 1992. "Leonardo: A Journey of the Mind". ©Vision Associates, Inc. for IBM. VHS. 40 minutes; (b) Bregman, Judith, Davisson, Richard, and Holden, Alan. 1967; "Symmetry" ©Polytechnic Institute of Brooklyn. New York: Contemporary Films/McGraw-Hill. 16 mm. 10.5 minutes. *) (c) Robinson, Peter. 1970. "Aspects of Symmetry". ©Polytechnic Institute of Brooklyn. 16 mm. 15 minutes.*)

*) I am grateful to A. Rosenberg, Emeritus professor from Swarthmore College, for bringing this film to my attention.

[3]Created by D.P. Buckley (Quinnipiac University) and L.I. Gould.

of the world. It is a highly imaginative probing into the workings of nature (not just a rigid compilation of facts and formulas).

Over the years the subject matter of the course has been broadened. Some of the more recent topics which Seeing Through Symmetry includes are: the nature of fractals, the structure and function of DNA, the nature of the chemical bond through explanations of electric forces, an introduction to the mathematical theory of groups with applications to design and to computer algorithms, the application of waves to some aspects of the strange world of quantum physics, and the very important issue of scientific methodology (as exhibited in part through the use of Venn diagrams to illustrate class inclusions and methods of concept formation).

During the early years of the course, a national grant[4] enabled a colleague enabled us to outfit a multimedia classroom. The room contained 12 nodes, each with a Mac computer (plus desk lamp) in a space that easily accommodated two or three students, an instructor node with another Mac tied to a projection system at the front center of the classroom, and a laser printer; all connected via an Ethernet network.[5] Because each node's computer screen faces the front of the room, the instructor could monitor its activity. There was also a VCR player connected to an overhead projector. The instructor node permitted access to each of the student nodes (e.g. for loading software or transferring files). The equipment was completed by a variety of hardware and software packages used both in teaching and in the laboratory portion of the course.[6]

Labs

Once-a-week labs give the student a deeper understanding of the course. Discussion is continually encouraged among the participants. Students normally work in groups of three or less. (A typed report is required for each lab and may come from one or more students in the group.) Earlier labs feed into later ones as lower-level abstractions feed into higher-level ones. Several of the labs are briefly described below in relation to the question: What is the activity and how

[4]National Science Foundation Instrument and Laboratory Improvement grant awarded in 1993 (No. DUE-9352670).

[5]The course can also be adapted to run on Windows machines, if the need arises.

[6]Some of the Hardware used in the Labs and Lecture (and supported by a range of software packages): A typical Apple computer used is an iMac. There is also the Universal Laboratory Interface (ULI) box plus Test Leads, Ultrasonic Motion Detector, Student Force Sensor, Microphone/Amplifier, Heart Rate Monitor, and Light Sensor (all obtained from Vernier Software, Oregon); headphones, tuning forks, masses, clamps, aluminum rods, 1.5 V batteries; Hi/Low Intensity Lamp, and polarizers. In addition there are diffraction gratings through which students can view the spectrum from different sources of light (the sun, incandescent bulbs, fluorescent bulbs, and from hydrogen and mercury discharge tubes). Other materials required for the course are meager but necessary for giving students the sense that the scientific enterprise is heavily dependent on quantitative measurements and computations (e.g., a calculator whose display should show at least 5 places to the right of the decimal point but need only perform the operations of addition, subtraction, multiplication, division, and taking the square root).

does it connect to other areas of the course so as to convey the interdisciplinary experience? (Subjects referred to are covered earlier during lecture.)

INTRICACIES OF COLORATION:
COORDINATES, TRANSLATIONS, GROUPS AND TESSELLATIONS — Students continue learning (from an earlier lab) how to use a simple computer algorithm (based on the Logo language) to see: (a) how coordinates of points can be represented by the computer; (b) how different colors are determining factors for the nature of translationally symmetric designs; (c) an instance from the mathematical theory of groups; and (d) how algorithms can be created to tessellate the screen. Hence students learn (or re-learn) the analytical-geometry basics of locating objects in space; see how this is related to design, to poetry, and to music through the creation of symmetries and "broken" symmetries; understand how art can be a manifestation of group theory in mathematics and how the latter can be used to create art; and glimpse "infinity" through symmetries that can go on and on through time and space.

PATTERNS IN MUSIC: SOUND AND SIGHT — At our School of Music each group of students gains a literal "hands-on" introduction to elementary ideas in music and music symmetry through exploration of the visual and aural aspects of the keyboard. Students: (a) learn about symmetrical patterns that can be associated with the keyboard, both visual and (what I would call) "aural temporal"; (b) relate the psychophysical concepts of pitch and frequency to each other and to the Fibonacci sequence; and (c) perform the broken symmetry of musical "rounds". As a result of this lab students can interrelate music, mathematics, art, and physics, with even a little psychoacoustics.

EXPERIENCING MOTION IN SPACE AND TIME — In this lab and the ones that follow, students see the value of technology for investigating aspects of certain natural phenomena including those which exist beyond the range of human vision and hearing. The computer with auxiliary devices attached to it "extends" our visual and auditory senses. This in turn makes it possible for us to understand the manner in which symmetric aspects of nature contribute to our sense of the world of motion, sound, and light.

Students build on their experience of graphical representations (as introduced in an earlier lab) through the visualization of scientific data obtained via the simultaneous monitoring and display of different phenomena by computer equipment. This enables them to: (a) see the value of technology's omnipresent concept of *voltage* through the display of battery outputs as a function of time; (b) understand aspects of *motion* through experiencing the movement of their hand, using an ultrasonic motion detector; (c) conceptualize some details of the periodic phenomenon, "simple harmonic motion", that has time-translational invariance, using a mass-on-a-spring connected to a force probe. Students can see the obvious tie to mathematics, as well as to epistemology (i.e., how we acquire knowledge about the world; in this case, scientific knowledge). They can also see connections to art, music, and poetry. For example, the stressed-and-unstressed pattern in poetry's sonorous iambic pentameter may be represented through the use of visual voltage steps exhibiting a similar periodicity in time.

BUILDING SYMMETRY FROM SYMMETRY: THE FOURIER SPECTRUM — In this introduction to the addition of symmetrical wave patterns, the student learns about: (a) "Fourier synthesis", as manifested in complex periodic sounds, with their corresponding shapes, occurring when two or more simple periodic sounds with their corresponding shapes, resembling "sine wave" shapes, are combined, or "synthesized" (although true sine-wave shapes are referred to as "Fourier components", our simple approximate shapes will also be so referred to); (b) "Fourier decomposition", a method showing how complex periodic shapes can be broken down (or "decomposed") into their Fourier components, enabling students to look at the "shape" of their voice and of their heartbeat; and (c) how to synthesize their voice and heartbeat from the Fourier components.

There are a variety of interconnections gleaned from this lab. Ties to mathematics are obvious. But in addition, connections can be made to the technology of "electronic" music and voice production, the visual arts, biology (e.g., how does the structure and function of the vocal chords and heart relate to the nature of the waves they produce?), and, most remarkably, to the limitations of a certain type of knowledge through the Heisenberg Uncertainty Relation in quantum physics (e.g., why is it that to be able to precisely locate a particle's position is to be unable to precisely locate its velocity?).

A list of Lectures and Labs appears below.

2.1 Recent syllabus for seeing through symmetry

Lectures

a Symmetry in Nature and in Art: An Overview

b Building Blocks for Symmetry: Points, Polygons, and Philosophy

c Tessellating Space: Constructions in the Plane, Diversions in Space

d Patterns in Flatland: Translations, Rotations, & Reflections ... to Escher and Fractals!

e Patterns of Motions in the Physical World: Sliding, Turning, Flipping, and the Theory of Groups

f Rhyme and Reason: Patterns in Poetry and Music

g Regularities in Waves: From Water through Sound . . . to Light and Chance!

h Symmetry through Space & Time: Projectiles and Planets

i Symmetry in Spacetime: Black Holes and the Cosmos

j Coordination in Chemical Structures, Crystals, and Life

k The Ambidextrous World of Light

l Symmetry Broken: Chaos . . . and the World of Elementary Particles

<u>Labs</u>

a Computer Drawing: Reflections, Rotations, and Designs

b Learning Algorithms through the Language of Logo

c Drawing & Hearing Patterns: Polygonal Symmetry and Fibonacci Tones

d Intricacies of Coloration: Coordinates, Translations, Groups, and Tessellations

e Finding an Iterated Function System (IFS) for Fractal Images

f Patterns in Music: Sound and Sight

g Experiencing Motion In Space & Time

h Symmetry of Oscillations: Sine Waves and Sound

i Building Symmetry from Symmetry: The Fourier Spectrum

j Waviness: In Water and Light

k On Balls and Bombs: The Geometry of Projectile Motion*

l Bounced, Flipped, Rotated & Decomposed Light: From Mirrors to Spectra *

m "Slipping" Symmetry: Crystals and Chaos*

* (in preparation)

2.2 Prerequisites for the course

The labs are explicit enough so that there is almost no necessity for instructor supervision, an important aspect of a single-instructor overseeing 24 students in a lab that must normally be completed within 2 hours.

There is no algebra needed beyond solving one equation in one unknown. As to geometry — none is expected beyond a small subset of concepts from high school (such as the idea of"a point", "a line", and "an angle"). Furthermore, there is always a brief review of the mathematical ideas when they are needed. Concepts from science, no matter how elementary, are also carefully presented so as to attempt to connect ideas to the student's normal experience. As a consequence, even some of the most profound concepts of mathematics (such as "group" and "limit") appear to have been understood. It seems — from class discussions, exams, and term projects — that students are adequately prepared for the course.

Museum Report

As a result of visiting several science museums in this country, in Canada, and overseas, I have experienced the joys of seeing creative exhibits which frequently employ a hands-on approach to science. Consequently, I ask my students to partake in such an experience by submitting a rather open-ended report of their experience. They are only asked to describe their observations, explaining what they saw, and to make remarks critiquing (pro or con) exhibits they found notable.

The report (to my surprise and joy, given its lack of structured requirements) has been one of the most successful parts of the course. It is not focused on symmetry *per se*, but on the scientific method (spoken about in class) of making observations and drawing conclusions. Students, often going with their friends in order to compare their experiences, are genuinely delighted to partake of the exhibits and critically examine them.

Term Project

This is the capstone experience of the course. The student is advised to start researching possible topics early in the course and invited to consult with the instructor throughout the course. In order to maximize the value of this experience, there are several stages spread out over half a semester. These consist, in order, of a report on their Preliminary Idea, a Progress Report, a Class Presentation, and the Final Report. The project must draw on a library search of the literature and incorporate quantitative aspects of the course.

Over the years there have been some outstanding projects. One was by a philosophy major on the application of symmetry principles to metaphysics. Another, by a business major, found concepts of symmetry in the stock market. One student did an in-depth search for her video-supported report on "The Symmetries in Synchronized Swimming", a sport in which she has also participated. Another student grew crystals and researched "Crystal Structure and Symmetry". A music student wrote an original composition to explain symmetry in tonal elements using set theory and vectors; he titled his project "The 4 Arms of Chenrezi" because of the manner in which the "moods" of this Tibetan Tantric deity relate to the moods of his composition. Then, going beyond the course's mostly 2-dimensional symmetries in geometry, another student did a project titled "Polyhedra and Tessellations of Space". Another example was titled "Symmetry in Human Relationships". This was a very creative and highly interdisciplinary project, supported through the use of graphs and space-time diagrams, which integrated (fairly successfully) ideas from geometry, wave theory, and simple harmonic motion with ones from her own observations of the manner in which humans interact.

In conclusion

Seeing Through Symmetry is designed to help students find pleasurable values in the areas of science and technology through interactions among a variety of disciplines. Because of this it gives students ample opportunity for relating concepts of symmetry to their own disciplines. It thus enables and motivates them also to gain understanding of technical areas outside their major. Such understanding could then feed back to give them a better grasp and appreciation of their own subject.

Even wider: Seeing Through Symmetry is intended to give students an understanding and an appreciation of the many important areas spawned by human creativity. It is to show them that the World is of a piece. And it is to convey the sense that where the human mind journeys there are no barriers.

References

[1] Darvas, György & Nagy, Dénes. 1990-1996. *Symmetry: Culture and Science.*

[2] Gould, Laurence I. 1999. "What is Symmetry, That Educators and Students Should Be Mindful of It"? *Interdisciplinary General Education: Questioning Outside the Lines.* Edited by Marcia Bundy Seabury. New York: College Entrance Examination Board. [This is an earlier essay which can be read if more details are desired.]

[3] Hargittai, István. 1986. *Computers & Mathematics with applications.* Special Issue on Symmetry (Part 1) (January-April), 12B (1/2).

[4] ———. 1986. *Computers & Mathematics with applications.* Special Issue on Symmetry (Part 2) (May-August), 12B (3/4)

[5] Iachello, Francesco 1994. *Symmetries in Science VII*, Edited by B. Gruber and T. Otsuka (Plenum Press, New York).

[6] Kapraff, Jay. 1986. A Course in the Mathematics of Design. *Computers & Mathematics with applications*. Special Issue on Symmetry (Part 2) (May-August), 12B (3/4): 913-948.

[7] ———. 1990. *Connection: The Geometric Bridge between Art and Science*. New York: McGraw-Hill.

[8] Klein, Peter. 1990. "On Symmetry in Science Education". *Symmetry: Culture and Science* 1 (1): 77-91.

[9] Mackay, Alan L. 1986. "But What Is Symmetry"? *Computers & Mathematics with applications*. Special Issue on Symmetry (Part 1) (January-April), 12B (1/2): 19-20.

[10] Van Fraassen, Bas C. 1989. *Laws and Symmetry*. Oxford: Clarendon Press.

SPACE-TIME SYMMETRIES ON CLIFFORD ALGEBRA C_4 BASED REPRESENTATION SPACES OF $SU(4) \sim SO(6)$

B.J. Gruber
Emeritus College of Science
Mailcode 4403
Southern Illinois University at Carbondale
Carbondale, Illinois 62901-4403, USA
and
Fachbereich Physik, Metallurgie und Werkstoffwissenschaften Technische Universitaet Clausthal
D-38678 Clausthal–Zellerfeld
Germany

Abstract Representations of algebras on spaces with operators as basis vectors are discussed. These spaces are constructed using the properties of Clifford algebras. Thus the bases for these representations have, apart from their usual symmetry properties, an "internal" structure determined by the properties of a Clifford algebra. The elements of the Clifford algebra serve both, as operators and basis elements ("operator states").

In ref.[7] it was shown how to construct representation spaces in terms of the "operator states", and moreover how to construct semisimple Lie algebras in terms of the "operator states". In the current article some of the basic concepts of this approach are summarized in the introduction. Then the algebra $su(4) \sim so(6)$, based upon the Clifford algebra C_4, is analyzed from the point of view of "operator state" representations. Assuming space-time properties for $su(4) \sim so(6)$ physical conclusions are drawn, based upon the internal relationships of the various subalgebras which are assigned physical meaning. The Dirac γ matrices are shown to carry, as "operator states", the 4-dimensional vector representation of $so(4)$. Differential operator realizations are obtained on these spaces, as well as differential operator realizations for noncommuting variables on these spaces.

B. Gruber (ed.), Symmetries in Science XI, 239–263.

Introduction

In references [1] and [2] the algebras $su(4) \sim so(6)$ and their symmetry subchains were studied. In [8] boson and fermion realizations were given for these algebras and their semisimple subalgebras, while in [7] "operator states" were used to realize the familiar finite dimensional irreducible representations for these algebras. That is, the elements of the algebra itself (operators) serve as basis vectors (states) for the representations of algebras. The elements of the algebras thus have a dual character: they are operators and states at the same time. Which of the two properties of an "operator state" is to be used depends on the particular situation, namely whether an element of the algebra serves in its function as an operator, or whether it serves as a vector to be acted upon by an operator. It is then clear that, since operator and basis vectors are all elements of the algebra, no elements external to the algebra need to be made use of. For example, the physical vacuum state itself becomes an element of the algebra, and the vectors (states) will have an internal algebraic structure which may be open to physical interpretation.

In [7] this approach was applied to the construction of irreducible finite dimensional representations, to be referred to as irreps. It was demonstrated in [7] that the operators - serving as basis vectors for an irrep of an algebra - can be used to construct larger algebras, which then contain the original algebra as subalgebra. For example, the basis (operator) states for the two four-dimensional irreps [1000] and [1110] of $su(4)$, being operators having definite transformation properties with respect to $su(4)$, can be used to construct the algebra $su(4)$. If the direct product of these operator states is formed, $4 \times 4^* = 15 + 1$, then the set of the 15 operator states of [2110] of $su(4)$ which is obtained in this manner, forms a basis for the selfadjoint 15-dimensional representation of $su(4)$. Moreover, the 15 operator states of this set of operators are the 15 operators of the algebra $su(4)$. It was also shown that the six operator states which form bases for the irreps [1100] and [2211] of $su(4)$ also transform like the two irreps [100000] and [111110] of $su(6)$. It follows then that the operator states of the 35-dimensional irrep contained in $6 \times 6^* = 35 + 1$ are the elements of the algebra $su(6)$.

Since the operator states have this dual property of being both, an operator and a vector (state), the composition law (product) of two elements depends upon whether a given element is considered to act as an operator, or to serve as a vector. Moreover, the direct product of two elements of an algebra depends upon whether these elements are applied in their role as operators or whether they are considered to be vectors. Finally, the choice of a particular ideal may modify a product of an operator acting upon an operator which serves as a state. An operator, taken as a state, has to satisfy the conditions which are imposed by certain ideals which define the space. A familiar example in physics is the condition imposed upon the vacuum state by annihilation operators, $f_i|0>=0$. This condition must hold for all the states of a representation. That is, every state has to act upon the vacuum, and thus any operator Op, considered as a state, becomes Op $|0>$. Two simple examples will be given below in order to briefly demonstrate the concept of the operator state method by means of $su(2)$ as an example.

Before the examples can be discussed it is first necessary to define, and list, the various products (and maps) used in [7] and in this article:

Let $a,\ b,\ c,\ d$ denote elements of a Clifford algebra C. The products (maps) used are

(a) the associative multiplication law $\cdot$ of two operators $a \cdot b = ab$

(b) the Clifford algebra product $(1/2)(ab + ba) = \{a, b\}_+$

(c) the commutator $(ab - ba) = [a, b]$

(d) the direct product $(a \times b) = (a_1 b_2)$

(e) a product $\circ$ (map) $(F^+ \times F) \times (F^+ \times F) \to F^+ \times F$,

$$
\begin{aligned}
&((a \times b)\,,\,(c \times d)) \to (a \times b) \circ (c \times d) = (a(bc) \times d) = (a \times (bc)d)\,,\\
&a(bc) = a \cdot (b \cdot c) \ \text{ the associative multiplication law (a), and}\\
&(a \times b),\ (c \times d),\ (a(bc) \times d) \ \text{ elements of } F^+ \times F
\end{aligned}
$$

with $a,\ c$ elements of F^+ and $b,\ d$ elements of F.

For this map it holds that either

$$(bc) = 0\,, \qquad \text{or} \quad (bc) = p_1 p_2$$

and

$$f_i^+ \cdot p_1 p_2 = f_i^+\,, \qquad p_1 p_2 \cdot f_i = f_i$$

For the definition of $p_1 p_2$ see further below.

(f1) a "derivation law" on $F^+ \times F^+$, $a,\ c,\ d$ elements of F^+, b an element of F

$$(ab) \cdot (c \times d) = ((ab)c \times d) + (c \times (ab)d)$$

(f2) a "commutator" on $F^+ \times F$, $a,\ c$ elements of F^+ and $b,\ d$ elements of F:

$$(ab) \cdot (c \times d) = ((ab)c \times d) + (c \times d(-ab))$$

Applying to (f) the map,

$$\times \to \cdot \quad \text{(dot)}$$

the multiplication laws **(f)** go over into the derivation law (an operator acting upon a direct product state) and the commutator respectively.

(g) independent direct product spaces ("independent particles"):

$$(a \times b)(c \times d) = (ac \times bd), \quad \text{or}$$
$$(a_1 b_2)(c_1 d_2) = (a_1 c_1 \times b_2 d_2)$$

The spaces F^+ and F, referred to above, are quotient spaces of the Clifford algebra C_4 with respect to certain left and right sided ideals. Again a brief summary will be given here, for details see [7].

Consider the set of four elements

$$\{f_1, \ f_1^+, \ f_2, \ f_2^+\}$$

which satisfy the anticommutation rules

$$(1/2)(\{f_i^+, f_k\}_+ = \delta_{ik} 1 \,, \quad \{f_i, f_k\}_+ = 0 \,, \quad \{f_i^+, f_k^+\}_+ = 0 \,.$$

That is, these operators satisfy Fermi particle statistics, $f_i^+ f_i^+ = 0$, and $f_i^+ f_j^+ = -f_j^+ f_i^+$, for i not j. (Later on it will be convenient to absorb a square root 2 into the definition of the f's).

These four elements generate the 16-dimensional Clifford algebra C_4 by forming all possible products with respect to the associative multiplication law (a). This algebra is obviously closed with respect to the associative multiplication law (a), and thus also with respect to the commutator (c). In fact, the space C_4, with respect to the commutator (c), is the Lie algebra $u(4)$. For details see [7].

Defining the two elements

$$p_i = 1 - f_i^+ f_i \,, \qquad p_i p_i = p_i \,, \qquad i = 1, 2$$

the operator $p_1 p_2$ has the property of a vacuum state,

$$f_k \cdot p_1 p_2 = 0 \,, \qquad f_i^+ \cdot p_1 p_2 = f_i^+ \,,$$
$$p_1 p_2 \cdot f_k = f_k \,, \qquad p_1 p_2 \cdot f_i^+ = 0 \,.$$

and thus the "vacuum operator state" $p_1 p_2$ is defined within the algebra C_4 itself.

The two sets

$$\{p_1, \ f_1^+\} \,, \quad \{p_2, \ f_2^+\}$$

and

$$\{p_1, \ f_1\} \,, \quad \{p_2, \ f_2\}$$

generate the two sets

$$F^+ = \{p_1, f_1^+\} \cdot \{p_2, f_2^+\} = \{p_1 p_2\,,\ f_1^+ p_2\,,\ p_1 f_2^+\,,\ f_1^+ f_2^+\},$$

and

$$F = \{p_1, f_1\} \cdot \{p_2, f_2\} = \{p_1 p_2, f_1 p_2\,,\ p_1 f_2\,,\ f_2 f_1\}$$

The sets F^+ and F in turn generate the 16 dimensional Clifford algebra C_4 and $u(4)$in two ways

(1) $F^+ \cdot F = C_4 \sim u(4)$, with the associative multiplication law (a) for the product elements $ab \cdot cd$ in $F^+ \cdot F$, with $a \cdot b = ab$ an element of F^+, and $c \cdot d = cd$ an element of F. The commutator of the elements of $F^+ \cdot F$ closes to $u(4)$.

Or, by using the direct product basis,

(2) $F^+ \times F = C_4 \sim u(4)$, with the multiplication law (e)

$$(ab \times cd) \circ (ef \times gh) = (ab(cd \cdot ef) \times gh)$$

for the elements $(ab \times cd)$ and $(ef \times gh)$ of $F^+ \times F$, with ab and ef in F^+ and cd and gh in F.

The commutator for the elements of $F^+ \times F$ closes with respect to the multiplication law (e) to the algebra $u(4)$.

Case (2) goes over into case (1) by means of the map

$$\begin{array}{ccc} \times & \longrightarrow & \cdot \\ F^+ \times F & \longrightarrow & F^+ \cdot F \\ \text{multiplication (e)} & \longrightarrow & \text{multiplication (a)} \end{array}$$

Two simple example will serve to illustrate some of the comments made above:

The two operators

$$\begin{array}{ll} \{1,\ f^+ 1\} & (= \{p_1\,,\ f^+\} \cdot 1) \\ \text{with } f \cdot 1 = 0\,, & \text{(condition on representation space, generates left ideal)} \\ \text{and with } f^2 = 0\,, & f f^+ + f^+ f = 1 \quad \text{(algebraic properties)} \end{array}$$

form a basis for a two-dimensional spin representation for su(2) with generators

$$\{s_+ = \frac{1}{\sqrt{2}} f^+\,,\quad s_- = \frac{1}{\sqrt{2}} f\,,\quad s_0 = f^+ f - 1/2\}$$

$$[s_0, s_+] = s_+\,,\quad [s_0, s_-] = -s_-\,,\quad [s_+, s_-] = s_0$$

One obtains

$$
\begin{aligned}
&s_+ \cdot 1 = \frac{1}{\sqrt{2}} f^+ \cdot 1 = \frac{1}{\sqrt{2}} f^+ 1, \\
&s_+ \cdot f^+ 1 = \frac{1}{\sqrt{2}} f^+ \cdot f^+ 1 = 0\,, \quad \text{from algebraic property } (f^+)^2 = 0 \\
&s_- \cdot f^+ 1 = \frac{1}{\sqrt{2}} f \cdot f^+ 1 = \frac{1}{\sqrt{2}} (-f^+ f + 1) 1 = \frac{1}{\sqrt{2}} 1 \\
&\text{with } f \cdot 1 = 0\,, \ \text{(defining the space)} \\
&s_- \cdot 1 = \frac{1}{\sqrt{2}} f \cdot 1 = 0\,, \quad \text{with } f \cdot 1 = 0\,, \\
&s_0 \cdot 1 = (f^+ f_- 1/2) \cdot 1 = -(1/2) 1\,, \quad \text{with } f \cdot 1 = 0\,, \\
&s_0 \cdot f^+ 1 = (f^+ f - 1/2) \cdot f^+ 1 = -(1/2) 1\,, \text{ with } f \cdot 1 = 0\,, \ (f^+)^2 = 0
\end{aligned}
$$

In these equations an element of the algebra acts, as operator, upon another element of the algebra, which serves as a basis vector, and maps this basis vector upon another operator which serves as a basis vector. The operators which serve as basis vectors however must satisfy the conditions which define the vector space. That is, the defining relations for the vector space must be applied to them. Therefore the action of an operator A upon an operator B, AB, can be different from the action of the the same two operators if B serves as a vector (the operator which serves as a vector may be mapped into an invariant subspace, and therefore to zero via the relationship which defines the ideal).

A second simple example is the following. The four operators of F^+ form a basis for an the $3 + 1$-dimensional representations of $su(2)$, while the four operators of F form a basis of (in this case equivalent) $3 + 1$ dimensional representation. The bases for the 3-dimensional irreps are obtained from the direct product of two spin representations of $su(2)$,

$$
\begin{aligned}
&\{1,\, f^+ 1\} \times \{1,\, f^+ 1\} \\
&\{1,\, 1f\} \times \{1,\, 1f\}\,,
\end{aligned}
$$

The bases for the 3-dimensional irreps are given by (the operator 1 is now suppressed where the context is clear)

$$
\begin{aligned}
&\{f^+ \times f^+,\ \frac{1}{\sqrt{2}} (f^+ \times 1 + 1 \times f^+),\ (1 \times 1)\}, \quad f \cdot 1 = 0, \\
&\{f \times f,\ \frac{1}{\sqrt{2}} (f \times 1 + 1 \times f),\ (1 \times 1)\}, \quad 1 \cdot f^+ = 0
\end{aligned}
$$

Using the multiplication law **(f1)** one obtains

$$
\begin{aligned}
s_+ \cdot (f^+ \times f^+) &= (\frac{1}{\sqrt{2}} f^+ \cdot f^+) \times f^+ + f^+ \times (\frac{1}{\sqrt{2}} f^+ \cdot f^+) \times f^+ = 0, \\
s_- \cdot (f^+ \times f^+) &= (\frac{1}{\sqrt{2}} f \cdot f^+) \times f^+ + f^+ \times (\frac{1}{\sqrt{2}} f \cdot f^+) \times f^+ = \\
&= \frac{1}{\sqrt{2}}((-f^+ f + 1) \times f^+ + f^+ \times f^+ \times (-f^+ f + 1)) = \\
&= \frac{1}{\sqrt{2}}(f^+ \times 1 + 1 \times f^+) \\
s_- \cdot \frac{1}{\sqrt{2}}&(f^+ \times 1 + 1 \times f^+) = (1 \times 1)
\end{aligned}
$$

etc.

1. Space-Time Symmetries

In reference [1] the algebra $su(4)$ was studied in view of a possible microscopic picture for the $su(6)$ Interacting Boson Model of the Nucleus (IBM). In this article the algebra $su(4)$ is studied in view of applications to space-time symmetries. The methods used in [7] will be applied in this article. In particular representations of $su(4)$ and its subalgebras in terms of algebraic states will be discussed.

There are two $su(4)$ symmetry chains which contain the orbital angular momentum subalgebra $so(3)_l$. These are (see [8])

$$
\begin{array}{ccccccc}
 & & so(3)_l \times so(3)_T & & & & \\
 & / & & & \backslash & & \\
su(4) & & & & & so(3)_l & \\
 & \backslash & & & / & & \\
 & & so(5) & — & so(4) & &
\end{array}
\qquad (1)
$$

For these chains the following algebraic state representations will be obtained:

$$\begin{aligned}
su(4): \quad & [1000] \times [1110] = [2110] + [1111] \\
& \qquad 4 \times 4 = 15 + 1 \\
so(5): \quad & (1/2, 1/2) \times (1/2, 1/2) = ((11) + (10)) + (00) \\
& \qquad 4 \times 4 = 10 + 5 + 1
\end{aligned}$$

$$\begin{aligned}
& so(4) \sim so(3)_S \times so(3)_Q : \\
& \quad (1/2, 1/2) \times (1/2, 1/2) = \\
& \quad = (((11) + (1-1) + (10)_V) + ((10)_V + (00))) + (00) \\
& \quad 4 \times 4 = ((4 + 4 + 3) + (3 + 1)) + 1
\end{aligned}$$

$$\begin{array}{rccc}
so(3)_l \times so(3)_T : & & & \\
(1;0) \times (1;0) = & (1;1) + (1 \times 0) + (0;1) & + & (0;0) \\
(3+1) \times (3+1) = & (3 \times 3) + (3 \times 1) + (1 \times 3) & + & 1
\end{array}$$

$$\begin{array}{rccc}
so(3)_l : & & & \\
(1+0) \times (1+0) = & ((1+1+(1+0)_V) + ((1+0)_V + 0)) & + & 0 \\
(3+1) \times (3+1) = & ((3+3+(3+1)) + ((3+1)+1) & + & 1
\end{array} \tag{2}$$

First the representations will be found which transform according to the chain

$$su(4) \longrightarrow so(5) \longrightarrow so(4) \longrightarrow so(3)_l$$

The operators of the $so(5)$ subalgebra of $su(4) \sim so(6)$ are

$$\begin{aligned}
so(5) \quad = \quad & (H_1' = f_1^+ f_1 - 1/2, \quad H_2' = f_2^+ f_2 - 1/2, \\
& E(-11) = E(-110) = f_2^+ f_1, \\
& E(-10) = \frac{1}{\sqrt{2}}(E(-101) + E(-10-1)) = \frac{1}{\sqrt{2}} f_1, \\
& E(0-1) = \frac{1}{\sqrt{2}}(E(0-11) + E(0-1-1)) = \frac{1}{\sqrt{2}} f_2, \\
& E(-1-1) = E(-1-10) = f_2 f_1, \\
& E(1-1) = E(1-10) = f1 + f2, \\
& E(10) = \frac{1}{\sqrt{2}}(E(10-1) + E(101)) \\
= \quad & f_1^+, \quad E(11) = E(110) = f_1^+ f_2^+, \\
& E(01) = \frac{1}{\sqrt{2}}(E(01-1) + E(011)) = \frac{1}{\sqrt{2}} f_2^+)
\end{aligned} \tag{3}$$

These operators act upon the product states of $F^+ \times F$ as

$$(\mathrm{Op}^{\times})a \times b = ((\mathrm{Op}\, a) \times b - a \times (b\, \mathrm{Op})) \tag{4}$$

It was shown in [7] that the direct product operator state $f_1^+ f_2^+ \times p_1 p_2$ belongs to $su(4)$ weight $(100-1)$, and $so(5)$ weight (110), eq.(3.6) and eq.(3.7) of ref. [7]. In fact, this state is an $so(5)$ extremal state with highest $so(5)$ weight (11). Acting upon this state with the so(5) operators one obtains

$$\begin{aligned}
&H_1^{\times}(f_1 + f_2^+ \times p_1 p_2) = (f_1 + f_1 - 1/2)^{\times}(f_1^+ f_2 + \times p_1 p_2) \\
&\qquad = 1(f_1^+ f_2^+ \times p_1 p_2) \\
&H_2^{\times}(f_1^+ f_2^+ \times p_1 p_2) = (f_2^+ f_2 - 1/2) \times (f_1^+ f_2^+ \times p_1 p_2) \\
&\qquad = 1(f_1^+ f_2^+ \times p_1 p_2) \\
&E(1-1)^{\times}(f_1^+ f_2^+ \times p_1 p_2) = (f_1^+ f_2) \times (f_1^+ f_2^+ \times p_1 p_2) = 0 \\
&E(10)^{\times}(f_1^+ f_2^+ \times p_1 p_2) = ((1/2) f_1^+) \times (f_1^+ f_2^+ \times p_1 p_2) = 0 \\
&E(11)^{\times}(f_1^+ f_2^+ \times p_1 p_2) = (f_1^+ f_2^+) \times (f_1^+ f_2^+ \times p_1 p_2) = 0
\end{aligned}$$

This shows that the operator state $f_1^+ f_2^+ \times p_1 p_2$ belongs indeed to the highest $so(5)$ weight $(1,1)$. Acting upon this state with the lowering operator $E(-10)^{\times}$ of $so(5)$ one obtains

$$\begin{aligned}
&F^+ \times F \longrightarrow F^+ F \\
&E(-10)^{\times}(f_1^+ f_2^+ \times p_1 p_2) \\
&\qquad = \frac{1}{\sqrt{2}}(f_2^+ p_1 \times p_1 p_2 - f_1^+ f_2^+ \times p_2 f_1) \longrightarrow \frac{1}{\sqrt{2}} f_2^+
\end{aligned} \tag{5}$$

The arrow indicates the map from the direct product onto the space of the Clifford algebra, as introduced in [7].

The state given by eq.(5) has $so(5)$ weight (01). Upon restriction of $so(5)$ to $so(4)$ the weight remains the same, while the state becomes extremal. In fact, this state generates one of the two four-vector representations of $so(4)$ which are contained in the direct product $[1000] \times [1110]$.

Making use of the $so(4)$ operators

$$\begin{aligned}
so(4) \;=\; &(H_1' = f_1^+ f_1 - 1/2, \quad H_2' = f_2^+ f_2 - 1/2, \\
&E(-11) = f_2^+ f_1, \quad E(-1-1) = f_2 f_1, \\
&E(1-1) = f_1^+ f_2, \quad E(11) = f_1^+ f_2^+)
\end{aligned} \tag{6}$$

the states of the four-vector representation $V(10)$ of $so(4)$ are obtained as

$$\frac{1}{\sqrt{2}} f_2^+ p_1 \times p_1 p_2 - f_1^+ f_2^+ \times f_1 p_2);$$
$$\frac{1}{\sqrt{2}} f_2^+;\ (01)$$

$$\frac{1}{\sqrt{2}} f_1^+ p_2 \times p_1 p_2 + f_1^+ f_2^+ \times f_2 p_1);$$
$$\frac{1}{\sqrt{2}} f_1^+;\ (10)$$

$$\frac{1}{\sqrt{2}} - f_1^+ p_2 \times f_2^+ f_1^+ p_1 p_2 \times f_2 p_1);$$
$$\frac{1}{\sqrt{2}} f_2;\ (0-1)$$

$$\frac{1}{\sqrt{2}} (f_2^+ p_1 \times f_2 f_1 + p_1 p_2 \times f_1 p_2);$$
$$\frac{1}{\sqrt{2}} f_1;\ (-10) \tag{7}$$

The calculation of the states, as given above for $a \times b \in F^+ \times F$, could have been carried out on the space $F^+F \sim CA$, by using the commutator [Op, ab] for the states $ab \in F^+F$.

The state $\frac{1}{\sqrt{2}}(f_1^+ p_2 \times p_1 p_2 - f_1^+ f_2^+ \times p_1 f_2)$ also belongs to $so(4)$ weight (10), and is orthogonal to the state $\frac{1}{\sqrt{2}}(f_1^+ p_2 \times p_1 p_2 + f_1^+ f_2 + \times p_1 f_2)$. Thus this state generates a second, equivalent four-vector representation of $so(4)$. On the space $F^+F \sim CA$ this state becomes

$$\frac{1}{\sqrt{2}}(f_1^+ p_2 - f_1^+ f_2^+ f_2) = \frac{1}{\sqrt{2}}(f_1^+ e_2) \tag{8}$$

The four operator states which span the representation $V_e(10)$ are, on CA, obtained as

$V_e(10) =$	$(\frac{1}{\sqrt{2}} f_1^+ e_2,$	$\frac{1}{\sqrt{2}} f_2^+ e_1,$	$\frac{1}{\sqrt{2}} f_2 e_1,$	$\frac{1}{\sqrt{2}} f_1 e_2)$
$so(4)$ weights:	(10)	(01)	$(0-1)$	(-10)

(9)

The basis (operator) states for the representation spaces $V(10)$ and $V_e(10)$ can be represented as operators acting on the space F^+. Acting on F^+ yields the matrix representations

$$f_1^+ = \begin{matrix} 0 & 0 & 0 & 0 \\ 1 & 0 & 0 & 0 \\ 0 & 0 & 0 & 0 \\ 0 & 0 & 1 & 0 \end{matrix} \quad f_2^+ = \begin{matrix} 0 & 0 & 0 & 0 \\ 0 & 0 & 0 & 0 \\ 1 & 0 & 0 & 0 \\ 0 & -1 & 0 & 0 \end{matrix}$$

$$f_1 = \begin{matrix} 0 & 1 & 0 & 0 \\ 0 & 0 & 0 & 0 \\ 0 & 0 & 0 & 1 \\ 0 & 0 & 0 & 0 \end{matrix} \quad f_2 = \begin{matrix} 0 & 0 & 1 & 0 \\ 0 & 0 & 0 & -1 \\ 0 & 0 & 0 & 0 \\ 0 & 0 & 0 & 0 \end{matrix} \tag{10}$$

$$f_1^+ e_2 = \begin{matrix} 0 & 0 & 0 & 0 \\ 1 & 0 & 0 & 0 \\ 0 & 0 & 0 & 0 \\ 0 & 0 & -1 & 0 \end{matrix} \quad f_2^+ e_1 = \begin{matrix} 0 & 0 & 0 & 0 \\ 0 & 0 & 0 & 0 \\ 1 & 0 & 0 & 0 \\ 0 & 1 & 0 & 0 \end{matrix}$$

$$f_1 e_2 = \begin{matrix} 0 & 1 & 0 & 0 \\ 0 & 0 & 0 & 0 \\ 0 & 0 & 0 & -1 \\ 0 & 0 & 0 & 0 \end{matrix} \quad f_2 e_1 = \begin{matrix} 0 & 0 & 1 & 0 \\ 0 & 0 & 0 & 1 \\ 0 & 0 & 0 & 0 \\ 0 & 0 & 0 & 0 \end{matrix} \tag{11}$$

The bases for the spaces $V(10)$ and $V_e(10)$ can be re-expressed in terms of the γ matrices. That implies that the γ matrices form bases for the spaces $V(10)$ and $V_e(10)$, a fact which will be made use of later on. One obtains

$$V(10) = (\frac{1}{\sqrt{2}} f_1^+, \ \frac{1}{\sqrt{2}} f_2^+, \ \frac{1}{\sqrt{2}} f_2, \ \frac{1}{\sqrt{2}} f_1)$$

$so(4)$ weights

$$\begin{matrix} & (10) & (01) & (0-1) & (-10) \\ \sim & (\gamma_1 = f_1^+ + f_1, & \gamma_2 = f_2^+ f_2, & \gamma_3 = i(f_1^+ - f_1), & \gamma_4 = f_2^+ - f_2) \end{matrix}$$

$$\gamma_i^2 = 1, \ \ i = 1, 2, 3, \quad \gamma_4^2 = -1, \qquad \gamma_i \gamma_j = -\gamma_i \gamma_j \tag{12}$$

where the γ-matrices represent a Cartesian basis for the space $V(10)$. Moreover, for the second four-vector representation one obtains

$$\begin{array}{llllll} V_e(10) = & (\frac{1}{\sqrt{2}} f_1^+ e_2, & \frac{1}{\sqrt{2}} f_2^+ e_1, & -\frac{1}{\sqrt{2}} f_2 e_1, & \frac{1}{\sqrt{2}} f_1 e_2) & \\ so(4) \text{ weights}: & (10) & (01) & (0-1) & (-10) & \\ & \sim (\gamma_1 e_2, & \gamma_2 e_1, & \gamma_3 e_2, & \gamma_4 e_1), & e_i^2 = 1 \end{array} \tag{13}$$

The spaces $V(10)$ and $V_e(10)$, over the reals R, carry the four dimensional (defining) vector representation of $so(4)$. On the other hand the subset of real unitary (i.e. orthogonal) matrices of $su(4)$ forms the subgroup $so(4)$ of $su(4)$. Under restriction of $su(4)$ to its subgroup $so(4)$ the two inequivalent four dimensional representations $F^+ = [1000]$ and $F = [1110]$ of $su(4)$ go over into the spin representation $(1/2, 1/2)$ of $so(4)$.

The six dimensional (real unitary) representation $[1100]$ of $su(4) \sim so(6)$ is the defining representation of $so(6)$. With respect to the symmetry chain $su(4) \sim so(so6) \to so(5) \to so(4)$ the representation $[1100]$ decomposes as follows,

$$\begin{array}{ccccc} su(4) & & so(5) & & so(4) \\ [1100] & \to & (1,0)+(0,0) & \to & ((1,0)+(0,0))+(0,0) \\ 6 & \to & 5+1 & & (4_V+1)+1 \end{array} \tag{14}$$

This is a consequence of general properties of representation theory. For the special case of direct product operator states, as discussed in this article, the branching of $[1100]$ as shown in eq.(14), can be easily observed from eq.(4.10) and eq.(4.11) of ref. [7]. But it is also observed that, for the case of the operator states, the map from the direct product operator states $F^+ \times F^+$ and $F \times F$ onto the Clifford algebra states $F^+F \sim CA$, does not yield a representation for $su(4) \sim so(6)$, $so(5)$ or $so(4)$. (Instead, the three surviving states span the representation space for the two three dimensional representations of the subalgebra $su(3)$ of $su(4)$ ($[1100] = [100] + [110]$ of $su(3) \to [100], [110]$ of $su(3)$). This implies that for a single species of fermions there exists no six dimensional representation for $so(6)$. There exists however, for a single species of fermions, a five dimensional representation for $so(5)$, as well as two (equivalent) four vector representations $V(10)$ and $V_e(10)$ of $so(4)$. These representations are contained in the 15-dimensional (real orthogonal) representation $[2110]$ of $su(4) \sim so(6)$.

The representation $[2110]$ of $su(4)$ contains in addition to the two $so(4)$ four-vector representations the representations $(1,1)$, $(1,-1)$ and $(0,0)$. The basis operator states for these representations are obtained as

$$
\begin{aligned}
& F^+ \times F \\
& \quad \to F^+F \\
& |(11),(11)> = f_1^+ f_2^+ \times p_1 p_2 \\
& \quad \to f_1^+ f_2^+ = \sqrt{2} Q_+ \\
& |(11),(00)> = \frac{1}{\sqrt{2}}(p_1 p_2 \times p_1 p_2 + f_1^+ f_2^+ \times f_1 f_2) \\
& \quad \to \frac{1}{\sqrt{2}}(p_1 + p_2 - 1) = -\sqrt{2} Q_3 \\
& |(11),(1-1)> = p_1 p_2 \times f_1 f_2 \\
& \quad \to f_1 f_2 = -\sqrt{2} Q_- \\
& |(1-1)(1-1)> = -f_1^+ p_2 \times f_2 p_1 \\
& \quad \to -f_1^+ p_2 = -\sqrt{2} S_+ \\
& |(1-1)(00)> = \frac{1}{\sqrt{2}}(-f_2^+ p_1 \times f_2 p_1 + f_1^+ p_2 \times f_1 p_2) \\
& \quad \to \frac{1}{\sqrt{2}}(f_1^+ f_1 - f_2^+ f_2) = \sqrt{2} S_3 \\
& |(1-1)(1-1)> = -f_2^+ p_1 \times f_1 p_2 \\
& \quad \to f_2^+ f_1 = \sqrt{2} S_- \\
& |(00)(00)> = (1/2)(p_1 p_2 \times p_1 p_2 + f_2^+ p_1 \times f_2 p_1 + \\
& \qquad - f_1^+ p_2 \times f_1 p_2 + f_1^+ f_2^+ \times f_1 f_2) \\
& \quad \to (1/2) e_1
\end{aligned} \tag{15}
$$

The subalgebra $so(3)_l$ of $so(4)$ is of particular importance, since in space-time it corresponds to orbital angular momentum. Its embedding in the symmetry chain is given by

$$
so(3)_l = (l_0 = f_1^+ f_1 - 1/2, \quad l_+ = \frac{1}{\sqrt{2}} f_1^+ \gamma_2, \; l_- = \frac{1}{\sqrt{2}} \gamma_2 f_1) \tag{16}
$$

With respect to $so(3)_l$ the $so(4)$ four vector representation $V(10)$ decomposes into the representations $l = 1$ and $l = 0$. The basis states for these two

$so(3)_l$ representations are obtained as

$$
\begin{aligned}
|l=1, m=1> &= \frac{1}{\sqrt{2}}(f_1^+ p_2 \times p_1 p_2 + f_1^+ f_2^+ \times f_2 p_1) \\
&\rightarrow \frac{1}{\sqrt{2}} f_1^+ \\
|l=1, m=0> &= (1/2)(f_2^+ p_1 \times p_1 p_2 + f_1^+ p_2 \times f_1 f_2 \\
&\qquad + p_1 p_2 \times f_2 p_1 - f_1^+ f_2^+ \times f_1 p_2) \\
&\rightarrow (1/2)(f_2^+ + f_2) \\
|l=1, m=-1> &= \frac{1}{\sqrt{2}}(-f_2^+ p_1 \times f_1 f_2 + p_1 p_2 \times f_1 p_2) \\
&\rightarrow \frac{1}{\sqrt{2}} f_1
\end{aligned}
$$

and

$$
\begin{aligned}
|l=0, m=0> &= (1/2)(-f_2^+ p_1 \times p_1 p_2 - f_1^+ p_2 \times f_1 f_2 \\
&\qquad + p_1 p_2 \times f_2 p_1 + f_1^+ f_2^+ \times f_1 p_2) \\
&\rightarrow -(1/2)(f_2^+ - f_2)
\end{aligned} \tag{17}
$$

Next, the representations will be determined which transform according to the chain

$$su(4) \sim so(6) \rightarrow so(3)_l \times so(3)_T \rightarrow so(3)_l$$

The algebra $so(3)_l \times so(3)_T$ is given as

$$
\begin{aligned}
so(3)_l \times so(3)_T \;\; = \;\; & (l_0 = f_1^+ f_1 - 1/2, \;\; l_+ = \frac{1}{\sqrt{2}} f_1^+ \gamma_2, \;\; l_- = \frac{1}{\sqrt{2}} \gamma_2 f_1, \\
& T_0 = (1/2) e_1 e_2, \;\; T_+ = \frac{1}{\sqrt{2}}(f_2^+ f_1^+ f_1 - f_2 p_1), \\
& T_- = \frac{1}{\sqrt{2}}(f_2 f_1^+ f_1 - f_2^+ p_1))
\end{aligned} \tag{18}
$$

The basis states of the representation [2110] are obtained as:

$l=1, T=1:$	$l_0 = 1,$	0	-1
$T_0 = 1$	$-f_1^+ f_2^+ \times f_2 p_1,$ $-f_1^+ f_2^+ f_2$ $\|11;11>$	$\frac{1}{\sqrt{2}}(-p_1 p_2 \times f_2 p_1 + f_1^+ f_2^+ \times f_1 p_2),$ $-(f_2 p_1 + f_2^+ f_1^+ f_1)$ $\|10;11>$	$p_1 p_2 \times f_1 p_2$ $f_1 p_2$ $\|1-1;11>$
$T_0 = 0$	$\frac{1}{\sqrt{2}}(f_1^+ p_2 \times f_2 p_1 - f_1^+ f_2^+ \times p_1 p_2),$ $-f_1^+ (f_2^+ - f_2^+)$ $\|11;10>$		

$T_0 = 0$ $\quad (1/2)(-f_1^+ p_2 \times f_1 p_2 + f_2^+ p_1 \times f_2 p_1$
$-p_1 p_2 \times p_1 p_2 - f_1^+ f_2^+ \times f_1 f_2)$
$-e_2$
$|10; 10 >$

$T_0 = 0$ $\quad -\frac{1}{\sqrt{2}}(f_2^+ p_1 \times f_1 p_2 + p_1 p_2 \times f_1 f_2)$
$f_1(f_2^+ - f_2^+)$
$|1-1; 10>$

$$\begin{array}{llll} T_0 = -1 & f_1^+ p_2 \times p_1 p_2 & \frac{1}{\sqrt{2}}(f_2^+ p_1 \times p_1 p_2 + f_1^+ p_2 \times f_1 f_2) & -f_2^+ p_1 \times f_1 f_2 \\ & f_1^+ p_2 & f_2^+ p_1 \times f_2 f_1^+ f_1 & -f_2^+ f_1 f_2 \\ & |11; 1-1> & |10; 1-1> & |1-1; 1-1> \end{array} \tag{19}$$

$l = 1, T = 0:$

$$\begin{array}{c} \frac{1}{\sqrt{2}}(f_1^+ p_2 \times f_2 p_1 + f_1^+ f_2^+ \times p_1 p_2) \\ f_1^+(f_2^+ + f_2) \\ |11; 00> \\ \\ (1/2)(f_2^+ p_1 \times f_2 p_1 - f_1^+ p_2 \times f_1 p_2 \\ +p_1 p_2 \times p_1 p_2 - f_1^+ f_2^+ \times f_2 f_1) \\ (1/2)e_1 \\ |10; 00> \\ \\ \frac{1}{\sqrt{2}}(p_1 p_2 \times f_1 f_2 - f_2^+ p_1 \times f_1 p_2) \\ f_1(f_2^+ + f_2) \\ |1-1; 00> \end{array} \tag{20}$$

$l = 0, T = 1:$

$T_0 = 1$ $\quad \frac{1}{\sqrt{2}}(-f_1^+ f_2^+ \times f_1 p_2 - p_1 p_2 \times f_2 p_1)$
$f_2^+ f_1^+ f_1 - f_2 p_1 = 2T_+$
$|00; 11>$

$T_0 = 0$ $\quad -(1/2)(f_1^+ p_2 \times f_1 p_2 + f_2^+ p_1 \times f_2 p_1$
$-p_1 p_2 \times p_1 p_2 - f_1^+ f_2^+ \times f_1 f_2)$
$(1/2)e_1 e_2 = T_0$
$|00; 10>$

$T_0 = -1$ $\quad -\frac{1}{\sqrt{2}}(f_2^+ p_1 \times p_1 p_2 - f_1^+ p_2 \times f_1 f_2)$
$-f_2^+ p_1 + f_2 f_1^+ f_1 = 2T_-$
$|00; 1-1>$

(21)

A change of basis to spherical l states and Cartesian T states yields the three $T = 1$ triplets which have good angular momentum $l = 1$, $l_0 = 1$, $l_0 = 0$ and $l_0 = -1$. These states are

$$
\begin{array}{ll}
 & F^+F \\
 & \\
|11;11> -|11;1-1> & -f_1^+ \in V(10) \\
|11;10> & -f_1^+(f_2^+ - f_2) = -f_1^+\gamma_4 \\
|11;11> +|11;1-1> & f_1^+ e_2 \in V_e(10) \\
 & \\
|10;11> -|10;1-1> & -(f_2^+ + f_2) = -\gamma_2 \in V(10) \\
|10;10> & -e_2 \\
|10;11> +|10;1-1> & (f_2^+ - f_2)e_1 = \gamma_4 e_1 \in V_e(10) \\
 & \\
|1-1;11> -|1-1;1-1> & f_1 \in V(10) \\
|1-1;10> & f_1(f_2^+ - f_2) = f_1\gamma_4 \\
|1-1;11> +|1-1;1-1> & f_1 e_2 \in V_e(10) \\
 & \\
l=0,\ T=1: & \\
|00;11> -|00;1-1> & (f_2^+ - f_2) = \gamma_4 \in V(10) \\
|00;00> & -e_1e_2 \\
|00;11> +|00;1-1> & -(f_2^+ + f_2)e_1 = -\gamma_2 e_1 V_e(10) \\
 & \\
l=1, T=0: & \\
|11;00> & f_1^+(f_2^+ + f_2) = f_1^+\gamma_2 = \sqrt{2}l_+ \\
|10;00> & e_1 = 2l_0 \\
|1-1;00> & -f_1(f_2^+ + f_2) = -f_1\gamma_2 = \sqrt{2}l_-
\end{array}
\tag{22}
$$

In eq.(22) the l-states are given as spherical states, while the T-states are given in Cartesian states. The basis states for the two four-vector representations $V(10)$ and $V_e(10)$ are easily identified. Moreover it is seen that the $so(3)_l$ singlet state $\gamma_4 = |l = 0; m = 0>$ of the basis $V(10)$, eq.(12), transforms like a $T = 1$ triplet state.

At the level of the subalgebras $so(4)$, $so(5)$ and $so(3)_l$ of $so(6)$ the two representations $F^+ = [1000]$ and $F = [1110]$ become equivalent. The following equivalence relations hold for these algebras

$$
\begin{array}{l}
f_1^+ f_2^+ \sim p_1 p_2, \quad f_1^+ p_2 \sim f_2 p_1, \quad f_2^+ p_1 \sim -f_1 p_2, \quad p_1 p_2 \sim f_1 f_2 \\
\phi_1 \sim \phi_4^*, \quad \phi_2 \sim -\phi_3^*, \quad \phi_3 \sim -\phi_2 + *, \quad \phi_4 \sim \phi_1^*
\end{array}
\tag{23}
$$

Thus one can, at the $so(5)$ level, substitute in the representation $[1100]$ of $su(4)$ the states F for the second F^+ of the pair $F^+ \times F^+$ (which will then no longer be a representation of $su(4)$). One obtains, with the states characterized by $\mid so(6); so(3)_l, so(3)_T; l_0, T_0 >$,

$$
\begin{aligned}
& F^+ \times F \\
& \qquad \rightarrow \quad F^+F \quad so(3)_l \\
& |[1100]; 10; 10 >= \frac{1}{\sqrt{2}}(\phi_1\phi_2 - \phi_2\phi_1) \\
& \qquad = \frac{1}{\sqrt{2}}(f_1^+ f_2^+ \times (-f_2 p_1) - f_1^+ p_2 \times p_1 p_2), \\
& \qquad\qquad \rightarrow \quad -\frac{1}{\sqrt{2}} f_1^+, \quad |l = 1, m = 1 > \\
& |[1100]; 10; 00 >= (1/2)(\phi_4\phi_2 - \phi_2\phi_4 + \phi_1\phi_3 - \phi_3\phi_1) \\
& \qquad = (1/2)(p_1 p_2 \times (-f_2 p_1) - f_1^+ p_2 \times f_1 f_2 \\
& \qquad + f_1^+ f_2^+ \times f_1 p_2 - f_2^+ p_1 \times p_1 p_2), \\
& \qquad\qquad \rightarrow \quad -(1/2)(f_2^+ + f_2), \quad |l = 1, m = 0 > \\
& |[1100]; 10; 0-1 >= \frac{1}{\sqrt{2}}(\phi_4\phi_3 - \phi_3\phi_4) \\
& \qquad = \frac{1}{\sqrt{2}}(p_1 p_2 \times f_1 p_2 - f_2^+ p_1 \times f_1 f_2) \\
& \qquad\qquad \rightarrow \quad \frac{1}{\sqrt{2}} f_1, \quad |l = 1, m = 0 > \\
& |[1100]; 01; 01 >= (1/2)(\phi_4\phi_2 - \phi_2\phi_4 - \phi_1\phi_3 + \phi_3\phi_1) \\
& \qquad = (1/2)(p_1 p_2 \times (-f_2 p_1) - f_1^+ p_2 \times f_1 f_2 \\
& \qquad - f_1^+ f_2^+ \times f_1 p_2 + f_2^+ p_1 \times p_1 p_2) \\
& \qquad\qquad \rightarrow \quad (1/2)(f_2^+ - f_2), \quad |l = 0, m = 0 > \\
& |[1100]; 01; 00 >= (1/2)(\phi_2\phi_3 - \phi_3\phi_2 + \phi_1\phi_4 - \phi_4\phi_1) \\
& \qquad = (1/2)(f_1^+ p_2 \times f_1 p_2 - f_2^+ p_1 \times (-f_2 p_1) \\
& \qquad + f_1^+ f_2^+ \times f_1 f_2 - p_1 p_2 \times p_1 p_2) \\
& \qquad\qquad \rightarrow \quad -(1/2) e_1 e_2, \quad |l = 0, m = 0 > \\
& |[1100]; 01; 0-1 >= (1/2)(\phi_2\phi_3 - \phi_3\phi_2 - \phi_1\phi_4 + \phi_4\phi_1) \\
& \qquad = (1/2)(f_1^+ p_2 \times f_1 p_2 - f_2^+ p_1 \times (-f_2 p_1) \\
& \qquad - f_1^+ f_2^+ \times f_1 f_2 + p_1 p_2 \times p_1 p_2) \\
& \qquad\qquad \rightarrow \quad (1/2)1, \quad |l = 0, m = 0 >
\end{aligned}
\tag{24}
$$

The same can be done via the $so(3)_l \times so(3)_T$ chain. The states of this chain are labelled by $| \, so(6); l\, T; l_0 T_0 >$. One obtains, at the $so(3)_l$ level, the states (not a $so(3)_l \times so(3)_T$ representation)

$l = 1:$

$$\begin{array}{lcll} so(3)_l \times so(3)_T & \rightarrow & F^+F & so(3)_l \\ |[1100];10;10> & \rightarrow & -\frac{1}{\sqrt{2}} f_1^+ & |l=1,\, m=1> \\ |[1100];10;00> & \rightarrow & -\frac{1}{\sqrt{2}}(f_2^+ + f_2) & |l=1,\, m=0> \\ |[0110];10;0-1> & \rightarrow & \frac{1}{\sqrt{2}} f_1 & |l=1,\, m=-1> \end{array} \tag{25}$$

and the three singlet states $l = 0:$

$$\begin{array}{lcll} |[1100];01;01> & \rightarrow & (1/2)((e_1e_2+1) & |l=0,\, m=0> \\ |[1100];01;00> & \rightarrow & \frac{1}{\sqrt{2}}(f_2^+ - f_2) & |l=0,\, m=0> \\ |[1100];01;0-1> & \rightarrow & (1/2)((e_1e_2-1) & |l=0,\, m=0> \end{array}$$

Comparing the three singlet states of eq.(25) with the three singlet states of eq.(24) one observes the mixing of the singlets. This is a consequence of constructing these representations of $so(3)_l$ through two different chains.

2. Differential Operator Realizations on Space-Time

In the preceding section it was found that the Dirac γ matrices, eq.(12), and an equivalent set of matrices, eq.(13), form basis elements for the four dimensional vector representation spaces $V(10)$ and $V_e(10)$ of $so(4)$. The spaces $V(10)$ and $V_e(10)$ are assumed to be defined over the field of real numbers R and carry real orthogonal representations of the group $so(4)$. The general element of the space $V(10)$ is given by

$$\gamma = x_1\gamma_1 + x_2\gamma_2 + x_3\gamma_3 + x_4\gamma_4 \in V(10), \quad x_i \in R \tag{26}$$

Later on the variables x_i will be considered as space-time coordinates.

Differential operator realizations on the spaces $V(10)$ and $V_e(10)$ can be found for the elements of the algebras $so(4)$, $so(3)_l$, $so(3)_S$, and $so(3)_Q$. The set of five elements

$$(\gamma_1,\ \gamma_2,\ \gamma_3,\ \gamma_4,\ \gamma_5 = e_1e_2) \tag{27}$$

represents a basis for the 5-dimensional defining representation (10) of $so(5)$, while the linear span of the set of six elements

$$\begin{aligned}&(\gamma_1,\ \gamma_2,\ \gamma_3,\ \gamma_4,\ \gamma_5 = e_1e_2,\ \gamma_6 = 1)\\&\gamma_i\gamma_j = \gamma_j\gamma_i,\ \ i,\, j = -1,\, 2,\, 3,\, 4, \qquad \gamma_1^2 = \gamma_2^2 = \gamma_3^2 = 1\,, \quad \gamma_4^2 = -1\,,\\&\gamma_5^2 = \gamma_6^2 = 1\,, \quad \gamma_i\gamma_6 = \gamma_6\gamma_i\,,\end{aligned} \tag{28}$$

carries the reducible representation $(10) + (00)$ of $so(5)$. All these basis elements represent states obtained from a single fermion species, and none of these spaces carries a representation of either $su(4)$ or $so(3)_l \times so(3)_T$ as can be seen from eq.(4.1) and eq.(4.2) of ref. [7]. It requires a minimum of two fermion species to build the 6-dimensional irreducible representations for these algebras.

Eq.(4.8) of ref. [7] gives the action of the operators with respect to the bases eq.(26-28). This equation should read as

$$((\mathrm{Op}^1)a)b - a(b(\mathrm{Op}^1)) = [(\mathrm{Op}^1),\, ab]$$

which can be obtained by using the associativity of the multiplication law and by adding and subtracting $a(\mathrm{Op}^1 b) = (a\mathrm{Op}^1)b$.

The following differential operators are obtained:

$$so(3)_l: \qquad \begin{aligned}l_0 &= f_1^+ f_1 - 1/2 = \frac{i}{2}\gamma_1\gamma_3 \rightarrow l_0 = i(x_1\partial_2 - x_2\partial_1)\\ l_1 &= \frac{i}{2}\gamma_1\gamma_2, \qquad \rightarrow l_1 = i(x_2\partial_3 - x_3\partial_2)\\ l_2 &= \frac{i}{2}\gamma_3\gamma_2, \qquad \rightarrow l_2 = i(x_3\partial_1 - x_1\partial_3)\\ l_+ &= \frac{1}{\sqrt{2}} f_1^+ \gamma_2\,\gamma_2, \quad I_- = \frac{1}{\sqrt{2}}\gamma_2 f_1, \quad [l_1, l_2] = -il_3\end{aligned} \tag{29}$$

$$so(4): \qquad \begin{aligned}&l_0 = i(x_1\partial_2 - x_2\partial_1), \ \ l_1 = i(x_2\partial_3 - x_3\partial_2),\\ &l_2 = i(x_3\partial_1 - x_1\partial_3),\\ &k_1 = (x_1\partial_4 + x_4\partial_1), \ \ k_2 = (x_2\partial_4 + x_4\partial_2),\\ &k_3 = (x_3\partial_4 + x_4\partial_3))\end{aligned} \tag{30}$$

$$su(2)_S: \qquad \begin{aligned}S_0 &= \frac{i}{2}(x_3\partial_1 - x_1\partial_3) - \frac{1}{2}(x_2\partial_4 + x_4\partial_2) = \frac{1}{2}(l_2 - k_2)\\ S_+ &= \frac{1}{2\sqrt{2}}(il_3 - l_1 + k_1 - ik_3)\\ S_- &= \frac{1}{2\sqrt{2}}(-il_3 - l_1 + k_1 + ik_3)\end{aligned} \tag{31}$$

$$
\begin{aligned}
su(2)_Q: \qquad & Q_0 = \frac{i}{2}(x_3\partial_1 - x_1\partial_3) - \frac{1}{2}(x_2\partial_4 - x_4\partial_2) = \frac{1}{2}(l_2 + k_2) \\
& Q_+ = \frac{1}{2\sqrt{2}}(il_3 - l_1 - k_1 + ik_3) \\
& \qquad Q_- = \frac{1}{2\sqrt{2}}(-il_3 - l_1 - k_1 - ik_3)
\end{aligned}
\tag{32}
$$

$su(2)_{S+Q}$:

$$
\begin{aligned}
S_0 + Q_0 &= i(x_3\partial_1 - x_1\partial_3) = l_2 \\
S_+ + Q_+ &= \frac{1}{\sqrt{2}}((x_2\partial_1 - x_1\partial_2) + i(x_3\partial_2 - x_2\partial_3)) \\
&= -\frac{1}{\sqrt{2}}(l_1 - il_3) \\
S_- + Q_- &= \frac{1}{\sqrt{2}}((x_2\partial_1 - x_1\partial_2) - i(x_3\partial_2 - x_2\partial_3)) \\
&= -\frac{1}{\sqrt{2}}(l_1 + il_3)
\end{aligned}
\tag{33}
$$

$$
\begin{aligned}
l_2 &= S_0 + Q_0, \ l_1 = -\frac{1}{\sqrt{2}}(S_+ + Q_+ + S_- + Q_-), \\
l_3 &= -\frac{1}{\sqrt{2}}(S_+ + Q_+ - S_- - Q_-)
\end{aligned}
\tag{34}
$$

The differential operators obtained on the space $V(10)$ show important relationships among the subalgebras. These relationships hold for the states of a single particle species, as listed by eq.(4.1) of ref. [7]. This analysis shows that there is an intrinsic orbital angular momentum l which is related to spin S and quasispin Q. But since the component of angular momentum l_2 needs to be diagonalized the eigenstates of S_0 and Q_0 are linear combinations of eigenstates of l_2. Spin S and quasispin Q are given as a combination of a rotation and a pure Lorentz transformation.

3. Scalar Products

Let ρ_Λ denote a (infinitesimally) unitary representation of a simple Lie algebra L on a (linear) space V_Λ. Then

$$
(X, Y)_\Lambda = \alpha_\Lambda \,\text{trace}\, (X)^+ \rho_\Lambda(Y)\,, \quad X,\, Y \in L, \tag{35}
$$

defines a scalar product which is positive definite if L is compact. The factor α_Λ depends upon the representation Λ

The 16-dimensional space of the CA carries a realization of the Lie algebra $u(4) \sim su(4) + u(1)$. The elements of $su(4)$ have, in any representation, the property that their trace is zero, while $u(1)$ is a multiple of the identity operator. Thus the scalar product can be extended to the entire CA. From the associativity of the CA multiplication law upon which the anticommutator is based, and the property $\rho(X)^+ = \rho(X^+)$ it follows that

$$(X, Y)_{CA} = (1/4)\ \text{trace}\ \rho_{CA}(X^+ Y)\,, \quad X, Y \in CA \sim su(4) + u(1)\,, \tag{36}$$

defines a scalar product on the CA.

The scalar product can be used to introduce scalar products for algebraic substructures of the CA on appropriately chosen subspaces. For example, the fermion F^+ forms such an algebraic substructure and can be represented on its own space F^+. The scalar product $(X, Y)_{CA}$ induces a scalar product $(X, Y)_{F^+}$ on the space F^+. One obtains

$$\begin{array}{lll} p_1p_2 \cdot p_1p_2 = p_1p_2\,, & f_1^+p_2 \cdot p_1p_2 = f_1^+p_2\,, & \text{etc.} \\ p_1p_2 \cdot f_1^+p_2 = 0\,, & f_1^+p_2 \cdot f_1^+p_2 = 0\,, & \\ p_1p_2 \cdot f_2^+p1 = 0\,, & f_1^+p_2 \cdot f_2^+p_1 = 0\,, & \\ p_1p_2 \cdot f_1^+f_2^+ = 0\,, & f_1^+p_2 \cdot f_1^+f_2^+ = 0\,, & \end{array} \tag{37}$$

where use has been made of the operator-state property of the elements involved, insofar that the operator acts upon an operator, representing a state, and maps this state upon another operator representing a state. It follows

$$\begin{array}{ll} \rho(p_1p_2) = \begin{matrix} 1 & 0 & 0 & 0 \\ 0 & 0 & 0 & 0 \\ 0 & 0 & 0 & 0 \\ 0 & 0 & 0 & 0 \end{matrix} & \rho(f_1^+p_2) = \begin{matrix} 0 & 0 & 0 & 0 \\ 1 & 0 & 0 & 0 \\ 0 & 0 & 0 & 0 \\ 0 & 0 & 0 & 0 \end{matrix} \\ \\ \rho(f_2^+p_1) = \begin{matrix} 0 & 0 & 0 & 0 \\ 0 & 0 & 0 & 0 \\ 1 & 0 & 0 & 0 \\ 0 & 0 & 0 & 0 \end{matrix} & \rho(f_1^+f_2^+) = \begin{matrix} 0 & 0 & 0 & 0 \\ 0 & 0 & 0 & 0 \\ 0 & 0 & 0 & 0 \\ 1 & 0 & 0 & 0 \end{matrix} \end{array} \tag{38}$$

The orthonormality for the basis states of F^+ is easily checked,

$$\begin{aligned} (p_1p_2,\, p_1p_2) &= \text{trace}\ \rho(p_1p_2) = 1 \\ (f_1^+p_2,\, p_1p_2) &= \text{trace}\ \rho(p_2f_1p_1p_2) = \text{trace}\,(0) = 0, \quad \text{etc.}, \end{aligned}$$

with $\alpha_{F^+} = 1$.

For any pair of elements of F^+

$$\begin{aligned} Z &= z_1\, p_1p_2 + z_2 f_1^+ p_2 + z_3 f_2^+ p_1 + z_4 f_1^+ f_2^+ \\ W &= w_1\, p_1p_2 + w_2 f_1^+ p_2 + w_3 f_2^+ p_1 + w_4 f_1^+ f_2^+\,, \qquad z_i,\, w_i \in C, \end{aligned}$$

it holds that

$$(Z, W)_{F^+} = z_1\bar{w}_1 + z_2\bar{w}_2 + z_3\bar{w}_3 + z_4\bar{w}_4 \tag{39}$$

where the bar denotes complex conjugation.

Similarly one obtains a matrix realization of the γ matrices on the space F^+ from the action of the γ upon the operator states F^+,

$$\gamma_1 = \begin{matrix} 0 & 1 & 0 & 0 \\ 1 & 0 & 0 & 0 \\ 0 & 0 & 0 & 1 \\ 0 & 0 & 1 & 0 \end{matrix} \quad \gamma_2 = \begin{matrix} 0 & 0 & 1 & 0 \\ 0 & 0 & 0 & -1 \\ 1 & 0 & 0 & 0 \\ 0 & -1 & 0 & 0 \end{matrix} \quad \gamma_3 = \begin{matrix} 0 & -i & 0 & 0 \\ i & 0 & 0 & 0 \\ 0 & 0 & 0 & -i \\ 0 & 0 & i & 0 \end{matrix}$$

$$\gamma_4 = \begin{matrix} 0 & 0 & -1 & 0 \\ 0 & 0 & 0 & 1 \\ 1 & 0 & 0 & 0 \\ 0 & -1 & 0 & 0 \end{matrix} \quad \gamma_5 = \begin{matrix} 1 & 0 & 0 & 0 \\ 0 & -1 & 0 & 0 \\ 0 & 0 & -1 & 0 \\ 0 & 0 & 0 & 1 \end{matrix} \quad \gamma_6 = \begin{matrix} 1 & 0 & 0 & 0 \\ 0 & 1 & 0 & 0 \\ 0 & 0 & 1 & 0 \\ 0 & 0 & 0 & 1 \end{matrix} \tag{40}$$

The γ_i, $i = 1, 2, 3, 4, 5, 6$, form a basis for the $5 + 1$ dimensional (reducible) representation of $so(5)$. The general element is given as

$$X = x_1\gamma_1 + x_2\gamma_2 + x_3\gamma_3 + x_4\gamma_4 + x_5\gamma_5 + x_6\gamma_6, \qquad x_i \in R. \tag{41}$$

For the symmetric scalar product one obtains

$$(X, Y)_{so(5)} = g_{ij}x_iy_j\,, \qquad g_{ij} = \text{diag}\,(111-11)\,,$$
$$i, j = 1, 2, 3, 4, 5 \qquad x_i, y_j \in R. \tag{42}$$

For the hermitian product $(X, Y)_H$ on the space eq.(40) one obtains,

$$(X, X)_H = \text{ trace } (\rho(X)^+\rho(X)) = x_1^2 + x_2^2 + x_3^2 + x_4^2 + x_5^2\,, \qquad x_i \text{ real}$$

The five dimensional $so(5)$ representation space (10), and the $so(4)$ space $V(10)$, are obvious subspaces of the $5+1$ dimensional representation of $so(5)$.

Once a space-time interpretation is given for the real parameters x_i at the $so(5)$, $so(4)$, and $so(3)_l$ levels, the algebra chains, together with the relationships which hold between the scalar product at the various levels, will yield a physical interpretation for the complex parameters at the $su(4)$ and $so(3)_l \times so(3)_T$ levels.

4. The One Fermion Species Dirac Equation

An operator

$$D = \gamma_1\partial_1 + \gamma_2\partial_2 + \gamma_3\partial_3 + \gamma_4\partial_4 + \gamma_5\partial_5 + \gamma_6\partial_6 = \gamma_i\partial_i \tag{43}$$

is introduced which acts upon plane wave functions of the type

$$\exp(-i(x_1p_1 + x_2p_2 + x_3p_3 - \frac{1}{c}E_1t_1 - \frac{1}{c}E_2t_2 - \frac{1}{c}E_3t_3)) \quad (44)$$

where the x_i and p_i denote ordinary space and linear momentum coordinates, while the E_i and t_i denote three different energies and three different times. The operator D acting upon the wave function eq.(44) yields

$$\gamma_1p_1 + \gamma_2p_2 + \gamma_3p_3 - \gamma_4\frac{1}{c}E_1 - \gamma_5\frac{1}{c}E_2 - \gamma_6\frac{1}{c}E_3 \quad (45)$$

An operator D' is introduced (dual, if t_2 and t_3 are purely imaginary)

$$= \gamma_1\partial_1 + \gamma_2\partial_2 + \gamma_3\partial_3 + \gamma_4\partial_4 - \gamma_5\partial_5 - \gamma_6\partial_6 \quad (46)$$

The operator D' has the property that DD' is obtained as

$$DD' = \partial_1\partial_1 + \partial_2\partial_2 + \partial_3\partial_3 - \partial_4\partial_4 - (\gamma_5\partial_5 + \gamma_6\partial_6)^2 \quad (47)$$

and that the action of DD' upon the plane wave function yields

$$p_1^2\,\mathbb{1} + p_2^2\,\mathbb{1} + p_3^2\,\mathbb{1} - \left(\frac{1}{c}\right)^2 E_1^2\,\mathbb{1} - (\gamma_5\left(\frac{1}{c}\right)^2 E_2^2 + \gamma_6\left(\frac{1}{c}\right)^2 E_3^2)^2 \quad (48)$$

Now, $\gamma_5 = e_1e_2$ is an $so(4)$ singlet while $\gamma_6 = 1$ is an $so(6)$, $so(5)$ and $so(4)$ singlet. This implies, as it has been pointed out earlier, that the basis given by eq.(28) does not carry a representation of $so(6)$, while with respect to $so(5)$ it transforms like $5+1$. And with respect to the Lorentz algebra $so(4)$ the basis vectors transform like $(4+1)+1$. This then appears to indicate that the $so(6)$ symmetry is spontaneously broken, leading to the disappearance of the time coordinate t_3 and the appearance of the mass m_0 in the form of the $so(6)$ invariant $E_3/c^2 = m_0$. At the $so(5)$ level the time t_2 is still not broken, but it is broken at the physical $so(4)$ symmetry level, leading to the disappearance of the time coordinate t_2 and the appearance of the mass breaking term m_Δ in the form $E_2/c^2 = m_\Delta$. Thus eq.(44), (45), and (48) go over into

$$\exp(-i(x_1p_1 + x_2p_2 + x_3p_3 - \frac{1}{c}E_1t_1 - m_\Delta ct_2 - m_0ct_3)) \quad (49)$$

$$\gamma_1p_1 + \gamma_2p_2 + \gamma_3p_3 - \gamma_4\frac{1}{c}E_1 - \gamma_5m_\Delta c - \gamma_6m_0c \quad (50)$$

$$p_1^2\,\mathbb{1} + p_2^2\,\mathbb{1} + p_3^2\,\mathbb{1} - \left(\frac{1}{c}\right)^2 E_1^2\,\mathbb{1} - (\gamma_5m_\Delta c + \gamma_6m_0c)^2 \quad (51)$$

The operators eq.(50) and (51) act upon the spin representation F^+. The γ_5 term has the effect to split the mass of the two 2-spinors $(f_1^+p_2, f_2^+p_1)$ and

$(f_1^+ f_2^+, p_1 p_2)$ of F^+, namely

$$\begin{aligned} \gamma_5 f_1^+ p_2 &= -f_1^+ p_2\,, &\qquad \gamma_5 f_1^+ f_2^+ &= f_1^+ f_2^+\,, \\ \gamma_5 f_2^+ p_1 &= -f_2^+ p_1\,, &\qquad \gamma_5 p_1 p_2 &= p_1 p_2. \end{aligned} \tag{52}$$

Thus, the pair $f_1^+ p_2$, $f_2^+ p_1$ has mass $(m_0 - m_\Delta)$ and the pair $f_1^+ f_2^+$, $p_1 p_2$ has mass $(m_0 + m_\Delta)$.

It is apparent from the description given above that the 4-spinor is not made up from a particle pair and its corresponding antiparticle pair. In fact, these two pairs belong to the two inequivalent spin representations $[1000]$ and $[1110]$ of $su(4)$. However, at the $so(5)$ and $so(4)$ level, these two representations become equivalent, and thus pairs may be substituted.

The two species representation $F^+ \times F^+$ contains the 6-dimensional representation $[1100]$ of $su(4)$. By following the symmetry chains it is seen that this representation transforms like the direct sum $(l = 1,\, T = 0) + (l = 0,\, T = 1)$, at the $so(3)_l \times so(3)_T$ level, and like $l = 1$, $l = 0$, $l = 0$, $l = 0$, at the $so(3)_l$ level. The operator γ_4 is found to transform like $l = 0$ singlet at this level. The coordinate associated with γ_4 is time. The operators γ_5 and γ_6 are also $l = 0$ singlets. Thus it appears that time should be associated with these two operators too, in particular since mass, and mass breaking terms, can be introduced in this manner. The three $l = 0$ singlets form then an $so(3)_T$ time triplet $T = 1$.

5. Differential Operator Realizations for Noncommuting Variables

Non-commuting variables α_i, $i = 1, 2$, are introduced which satisfy the properties

$$\begin{aligned} &\alpha_1 \alpha_2 = -\alpha_2 \alpha_1 \\ &\partial_1 \partial_2 = -\partial_1 \partial_2 \\ &\partial_i \alpha_k = -\alpha_k \partial_i + \partial_{ik} \qquad i,\, k = 1,\, 2 \end{aligned} \tag{53}$$

where the symbol ∂_k denotes the partial derivative with respect to the variable α_k. Note that we do not require

$$\begin{aligned} \alpha_i \alpha_k &= -\alpha_k \alpha_i\,, \qquad i,\, k = 1,\, 2 \\ \partial_i \partial_k &= -\partial_k \partial_i\,, \qquad i,\, k = 1,\, 2 \end{aligned} \tag{54}$$

from which would follow $\alpha_i \alpha_i = 0,\ \ \partial_i \partial_i = 0,\ \ i = 1, 2.$

On the quotient space of the Clifford algebra C_4 with respect to the left ideal l_L

$$\alpha_1^0 \alpha_2^0\, \mathbb{1} + \alpha_1^1 \alpha_2^0 f_1^+ + \alpha_1^0 \alpha_2^1 f_2^+ + \alpha_1^1 \alpha_2^1 f_1^+ f_2^+ \tag{55}$$

with respect to the variables α_i, a differential operator realization is obtained for the fermion creation and annihilation operators,

$$f_i = \partial_i \, , \qquad f_i^+ = \alpha_i(1 - \alpha_i \partial_i) \tag{56}$$

These operators act upon the space spanned by the monomials in the non-commuting variables

$$V(\alpha_1, \, \alpha_2) = (\alpha_1^m \alpha_2^n, \; m, n = 0, \, 1, \, 2, \, 3, \ldots) \tag{57}$$

With respect to the operators eq.(56) the subspace of eq.(57), spanned by the elements

$$V' = (\alpha_1^0 \alpha_2^0, \; \alpha_1^1 \alpha_2^0, \; \alpha_1^0 \alpha_2^1, \; \alpha_1^1 \alpha_2^1) \tag{58}$$

forms an invariant subspace. This is the case since the operators $f_i = \partial_i$ lower the power of the variables α_i, while $f_i^+ \alpha_i = 0$. Thus, the action of the operators $f_i = \partial_i$ and $f_i^+ = \alpha_i(1 - \alpha_i \partial_i)$ on the invariant subspace is equivalent to $\alpha_i \alpha_i = 0$, $\partial_i \partial_i = 0$, $i = 1, 2$.

On the space $V(\alpha_1, \alpha_2)$ the creation and annihilation operators satisfy the following anticommutation relations,

$$\begin{aligned} \{f_i^+ \, , f_k\} &= \partial_{ik} - 2\alpha_i^2 \partial_i \partial_i \delta_{ik} \\ \{f_i^+ \, , f_k^+\} &= 2\alpha_i^4 \partial_i \partial_i \delta_{ik} \\ \{f_i \, , f_k\} &= 2\partial_i \partial_i \delta_{ik} \end{aligned} \tag{59}$$

while, on the space V' the standard anticommutation relations are satisfied.

References

[1] B. Gruber, "Algebraic Shells and the Interacting Boson Model of the Nucleus", in "Symmetries in Science VIII", ed. B. Gruber, Plenum Press, New York, 1995

[2] B. Gruber and M. Ramek, "Boson and Fermion Operator Realizations of su(4) and its Semisimple Subalgebras", in "Symmetries in Science VII", ed. B. Gruber and T. Otsuka, Plenum Press, New York, 1994

FERMIONIC O(8) AND BOSONIC U(36) SYMMETRY SCHEMES FOR HEAVY N=Z NUCLEI

V.K.B. Kota

Physical Research Laboratory,

Ahmedabad 380 009, India

Abstract Isoscalar ($T = 0$) plus isovector ($T = 1$) pairing hamiltonian in LS-coupling, which is important for heavy N=Z nuclei, is solvable in terms of a $O(8)$ algebra for some special values of the mixing parameters that measures the competition between $T = 0$ and $T = 1$ pairing. The $O(8)$ algebra is generated, amongst others, by the $T = 0$ and $T = 1$ pair creation and annihilation operators. Shell model algebras, with only number conserving operators, that are complementary to the $O(8) \supset O(6)$ and $O(8) \supset O(5) \otimes SU(2)$ sub-algebras are discussed in detail. Dyson boson mapping of the $O(8)$ pairing hamiltonian and the addition of quadrupole degrees of freedom give the interacting boson model with the s ($\ell = 0$) and d ($\ell = 2$) bosons carrying spin-isospin degrees of freedom (ST) = $(10) \oplus (01)$. Spectrum generating algebra for this model is $U(36)$. The $U(6) \otimes U(6)$ and $O(36)$ sub-algebras of $U(36)$ are discussed with applications to heavy N=Z nuclei.

Keywords: Shell model, Interacting boson model, Spectrum generating algebras, O(8) algebra, U(36) algebra, N=Z nuclei, Wigner's SU(4), proton drip-line.

1. Introduction

In the last few years study of the structure of heavy odd-odd N=Z nuclei (with A $\geq$ 60) near the proton drip line has become an area of intense research as these nuclei are expected to give new insights into neutron-proton (np) correlations that are hitherto unknown. For example, four particle correlations with possible Wigner's spin-isospin $SU(4)$ symmetry, possible formation of a condensate of $T = 0$ np Cooper pairs, new structures in energy levels due to the competition between isoscalar ($T = 0$) and isovector ($T = 1$) pairing, delay in angular momentum alignments at high spins, an enhanced probability to form an odd-odd nucleus by addition or removal of deuteron-like pair to even-even nuclei etc. With the development of radioactive ion beam (RIB) facilities and large detector arrays, there are now experimental results for the energy spectra

B. Gruber (ed.), Symmetries in Science XI, 265–290.

of ^{62}Ga, ^{66}As, ^{70}Br and ^{74}Rb [1] with many isospin $T = 0$ and $T = 1$ levels identified; in future it is expected that many spectroscopic details of these nuclei will be available. Besides developing models based on shell model and mean-field methods [2] for describing and predicting the spectroscopic properties of these and other N=Z odd-odd nuclei in the A=60-100 region, there are also attempts to develop algebraic models (symmetry schemes) [3-16] as they will give analytical insights into the structure of these nuclei. Given the hamiltonian to be a sum of isoscalar and isovector pairing hamiltonians in LS-coupling for the nucleons, it is solvable in some special situations using O(8) algebra and its subalgebras generated by the pair creation and destruction operators. This is first shown by Flowers and Szpikowski [3] for the case with nucleons in a single ℓ-shell. The 'complementary' (a notion introduced by Moshinsky [17]) number conserving algebras, with several ℓ-orbits, corresponding to the $O(8)$ algebras is a topic of discussion in this article. Going beyond pairs coupled to angular momentum $\ell_{12} = 0$ and including nucleons pairs with $\ell_{12} = 2$ and representing the fermion pairs by ideal bosons, one has the spin-isospin invariant interacting boson model (IBM-ST) [18, 19] with $U(36)$ spectrum generating algebra (SGA); it should be added that the algebraic (group theoretical) interacting boson model (IBM-1 or just IBM) with scalar $s(\ell = 0)$ and quadrupole $d(\ell = 2)$ bosons for quadrupole collective states was introduced in nuclear physics by Arima and Iachello [19]. The $U(36)$ model allows one to study not only pairing but also (quadrupole) deformation effects in heavy N=Z nuclei. The purpose of this article is to give an overview of the developments in the group theoretical aspects of the fermionic $O(8)$ and bosonic $U(36)$ symmetry schemes. Section 2 deals with the fermionic $O(8)$ symmetry schemes. In Section 3 a brief discussion of the Dyson boson mapping of the pairing hamiltonian is given. The boson mappings allows one to go beyond the symmetry limits of the $O(8)$ model and also show the way to the bosonic $U(36)$ model. Section 4 deals with the bosonic $U(36)$ symmetry limits. Finally Section 5 gives conclusions.

2. $O(8)$ symmetry schemes

Let us begin with m nucleons in several ℓ orbits ℓ_1, ℓ_2, ..., then the single nucleon states are $a^{\dagger}_{\ell m_\ell\,;\,\frac{1}{2} m_s\,;\,\frac{1}{2} m_t} |0\rangle$ and they are 4Ω in number where $\Omega = \sum_i (2\ell_i + 1)$. For a single ℓ-orbit, pair states are defined by two nucleon states with orbital angular momentum zero ($L = 0$). Then by antisymmetry, two nucleon pair states have spin(S) and isospin (T) to be $(ST) = (10) \oplus (01)$. With this, the isoscalar and isovector pair creation operators $D^{\dagger}_{\mu}(\ell)$ and $P^{\dagger}_{\mu}(\ell)$

respectively are

$$D^\dagger_\mu(\ell) = \sqrt{\frac{2\ell+1}{2}} \left(a^\dagger_{\ell\frac{1}{2}\frac{1}{2}} a^\dagger_{\ell\frac{1}{2}\frac{1}{2}} \right)^{0,1,0}_{0,\mu,0} , \; P^\dagger_\mu(\ell) = \sqrt{\frac{2\ell+1}{2}} \left(a^\dagger_{\ell\frac{1}{2}\frac{1}{2}} a^\dagger_{\ell\frac{1}{2}\frac{1}{2}} \right)^{0,0,1}_{0,0,\mu} \tag{1}$$

Note that we are using (L, S, T) order in (1). For the multi-orbit case one can define the generalized isoscalar and isovector pair operators $D^\dagger_\mu$ and $P^\dagger_\mu$ as linear combinations of single orbit $D^\dagger_\mu(\ell)$'s and $P^\dagger_\mu(\ell)$'s respectively except for phase factors,

$$D^\dagger_\mu = \sum_\ell \beta_\ell \, D^\dagger_\mu(\ell) \, , \quad P^\dagger_\mu = \sum_\ell \beta_\ell \, P^\dagger_\mu(\ell) \, ; \quad \beta_\ell = +1 \;\; \text{or} \;\; -1 \tag{2}$$

Now it is possible to define the pairing hamiltonian in LS-coupling,

$$H_{pairing} = -(1-x) \sum_\mu P^\dagger_\mu P_\mu \; - \; (1+x) \sum_\mu D^\dagger_\mu D_\mu \tag{3}$$

Note that $P_\mu = (P^\dagger_\mu)^\dagger$ and

$$P_\mu(\ell) = (P_\mu(\ell))^\dagger = (-1)^\mu \sqrt{(2\ell+1)/2} \, (\tilde{a}_{\ell\frac{1}{2}\frac{1}{2}} \tilde{a}_{\ell\frac{1}{2}\frac{1}{2}})^{0,0,1}_{0,0,-\mu}$$

where $\tilde{a}$ is related to a by

$$a_{\ell m_\ell \,;\, \frac{1}{2} m_s \,;\, \frac{1}{2} m_t} = (-1)^{\ell+1+m_\ell-m_s-m_t} \tilde{a}_{\ell -m_\ell \,;\, \frac{1}{2} -m_s \,;\, \frac{1}{2} -m_t} \, .$$

Similarly D_μ and $D_\mu(\ell)$ are defined. At this stage it is also necessary to define spin (S^1_μ), isospin (T^1_μ), Gamow-Teller ($(\sigma\tau)^{1,1}_{\mu,\mu'}$) and number ($\hat{n}$ or the equivalent Q_0) operators,

$$\begin{aligned}
S^1_\mu &= \sum_\ell \sqrt{2\ell+1} \left(a^\dagger_{\ell\frac{1}{2}\frac{1}{2}} \tilde{a}_{\ell\frac{1}{2}\frac{1}{2}} \right)^{0,1,0}_{0,\mu,0} \\
T^1_\mu &= \sum_\ell \sqrt{2\ell+1} \left(a^\dagger_{\ell\frac{1}{2}\frac{1}{2}} \tilde{a}_{\ell\frac{1}{2}\frac{1}{2}} \right)^{0,0,1}_{0,0,\mu} \\
(\sigma\tau)^{1,1}_{\mu,\mu'} &= \sum_\ell \sqrt{2\ell+1} \left(a^\dagger_{\ell\frac{1}{2}\frac{1}{2}} \tilde{a}_{\ell\frac{1}{2}\frac{1}{2}} \right)^{0,1,1}_{0,\mu,\mu'} \\
\hat{n} &= 2 \sum_\ell \sqrt{2\ell+1} \left(a^\dagger_{\ell\frac{1}{2}\frac{1}{2}} \tilde{a}_{\ell\frac{1}{2}\frac{1}{2}} \right)^{0,0,0}_{0,0,0} , \quad Q_0 = \frac{\hat{n}}{2} - \Omega
\end{aligned} \tag{4}$$

By evaluating the commutators it can be shown that the 28 operators $P^\dagger_\mu$, P_μ, $D^\dagger_\mu$, D_μ, S^1_μ, T^1_μ, $(\sigma\tau)^{1,1}_{\mu,\mu'}$ and Q_0 generate the following algebras:

$$\begin{array}{rcl} O(8) & : & P^\dagger_\mu,\ P_\mu,\ D^\dagger_\mu,\ D_\mu,\ S^1_\mu,\ T^1_\mu,\ (\sigma\tau)^{1,1}_{\mu,\mu'},\ Q_0 \\ O(6) & : & S^1_\mu,\ T^1_\mu,\ (\sigma\tau)^{1,1}_{\mu,\mu'} \\ O_S(5) & : & D^\dagger_\mu,\ D_\mu,\ S^1_\mu,\ Q_0 \\ O_T(5) & : & P^\dagger_\mu,\ P_\mu,\ T^1_\mu,\ Q_0 \\ O_S(3) & : & S^1_\mu \\ O_T(3) & : & T^1_\mu \end{array} \tag{5}$$

The $O(6)$ algebra in (5) is nothing but Wigner's spin-isospin supermultiplet $SU(4)$ algebra; its significance will be discussed later. Note that $O_S(3) \sim SU_S(2)$, $O_T(3) \sim SU_T(2)$ and $SU(4) \supset SU_S(2) \otimes SU_T(2)$. Quadratic Casimir operators of the groups in (5) are (besides S^2 for $O_S(3)$ and T^2 for $O_T(3)$),

$$\begin{array}{rcl} C_2(O(8)) & = & 2\left(\sum_\mu P^\dagger_\mu P_\mu + \sum_\mu D^\dagger_\mu D_\mu\right) + C_2(O(6)) + Q_0(Q_0-6) \\ C_2(O(6)) & = & S^2 + T^2 + (\sigma\tau)\cdot(\sigma\tau) \\ C_2(O_S(5)) & = & 2\sum_\mu D^\dagger_\mu D_\mu + S^2 + Q_0(Q_0-3) \\ C_2(O_T(5)) & = & 2\sum_\mu P^\dagger_\mu P_\mu + T^2 + Q_0(Q_0-3) \end{array} \tag{6}$$

In (6), the dot-product is defined by $A^k \cdot B^k = (-1)^k\sqrt{2k+1}(A^kB^k)^0$ and similarly $A^{k_1k_2} \cdot B^{k_1k_2}$ is defined.

Firstly it is seen from (5) that the pairing hamiltonian (3), for any x, has $O(8)$ symmetry as it contains only the generators of $O(8)$. Moreover, by examining the quadratic Casimir invariants in (6), it is seen that the pairing hamiltonian (3) is solvable in the situations $x = 0, 1, -1$ and the corresponding subalgebras (group-subgroup chains) of $O(8)$ are,

$$\begin{array}{rcl} x=0 & : & O(8) \supset O(6) \supset O_S(3) \oplus O_T(3) \\ x=1 & : & O(8) \supset [O_S(5) \supset O_S(3)] \otimes O_T(3) \\ x=-1 & : & O(8) \supset [O_T(5) \supset O_T(3)] \otimes O_S(3) \end{array} \tag{7}$$

All the group-subgroup chains in (7) have number non-conserving operators. We will now consider these chains in more detail in terms of their "complementary" number conserving group chains.

2.1 $O(8) \supset O(6) \supset O_S(3) \oplus O_T(3)$ chain and the complementary $U(4\Omega) \supset [U(\Omega) \supset O(\Omega)] \otimes SU_{ST}(4)$ chain

In the $(\ell_1, \ell_2, \ldots)^m$ space, there is a $U(4\Omega)$ algebra generated by the operators

$$u^{L,S,T}_{\mu_\ell,\mu_S,\mu_T}(\ell_1,\ell_2) = \left(a^\dagger_{\ell_1\frac{1}{2}\frac{1}{2}}\tilde{a}_{\ell_2\frac{1}{2}\frac{1}{2}}\right)^{L,S,T}_{\mu_\ell,\mu_S,\mu_T} \tag{8}$$

and with respect to $U(4\Omega)$, all the m-nucleon states behave as the totally antisymmetric irreducible representation (irrep) $\{1^m\}$. With good (LST), $U(4\Omega)$ algebra can be decomposed into product of space $U(\Omega)$ and spin-isospin $SU_{ST}(4)$ algebras. With this, one has the group chain,

$$U(4\Omega) \supset [U(\Omega) \supset O(\Omega) \supset O_L(3)] \otimes [O_{ST}(6) \supset O_S(3) \oplus O_T(3)] \tag{9}$$

Note that $SU_{ST}(4) \sim O_{ST}(6)$. Following the results in Appendix A and B of [13] it is straight forward to write down the generators and the quadratic Casimir operators (C_2's) of the groups in (9),

$$\begin{aligned}
U(4\Omega) &: \; u^{L,S,T}_{\mu_\ell,\mu_S,\mu_T}(\ell_1,\ell_2) \\
U(\Omega) &: \; 2\,u^{L,0,0}_{\mu_\ell,0,0}(\ell_1,\ell_2) \\
U(4) &: \; X^{S,T} = \sum_\ell \sqrt{2\ell+1}\,u^{0,S,T}_{0,\mu_S,\mu_T}(\ell,\ell) \\
O(6) \sim SU(4) &: \; X^{S,T}\,, \quad (ST) = (10),(01),(11) \\
O(\Omega) &: \; 2\,u^{L=odd,0,0}_{\mu_\ell,0,0}(\ell,\ell),\;\; V^L_\mu(\ell_1,\ell_2) \;\text{ with }\; \ell_1 > \ell_2\,; \\
&\quad\; V^L_\mu(\ell_1,\ell_2) = 2\left[\alpha(\ell_1,\ell_2)(-1)^{\ell_1+\ell_2+L}\right]^{\frac{1}{2}} \times \\
&\quad\; \left\{u^{L,0,0}_{\mu,0,0}(\ell_1,\ell_2) + \alpha(\ell_1,\ell_2)(-1)^L u^{L,0,0}_{\mu,0,0}(\ell_2,\ell_1)\right\}, \\
&\quad\; |\alpha(\ell_1,\ell_2)|^2 = 1,\;\; \alpha(\ell_1,\ell_2)\alpha(\ell_2,\ell_3) = -\alpha(\ell_1,\ell_3) \\
C_2(U(\Omega)) &= 4\sum_{\ell_1,\ell_2,L}(-1)^{\ell_1+\ell_2}u^{L,0,0}(\ell_1,\ell_2)\cdot u^{L,0,0}(\ell_2,\ell_1) \\
C_2(O(\Omega)) &= 8\sum_{\ell,L=odd} u^{L,0,0}(\ell,\ell)\cdot u^{L,0,0}(\ell,\ell) \\
&\quad + \sum_{\ell_1>\ell_2;L} V^L(\ell_1,\ell_2)\cdot V^L(\ell_1,\ell_2) \\
C_2(U(4)) &= \sum_{S,T} X^{S,T}\cdot X^{S,T}
\end{aligned} \tag{10}$$

It should be noted that $O(\Omega)$ is not unique and in the multi-orbit case there are several $O(\Omega)$'s as defined by distinct $\alpha(\ell_1,\ell_2)$'s in (10). Using (10), it can be proved that,

$$\begin{aligned} C_2(U(\Omega)) + C_2(U(4)) &= \hat{n}(4+\Omega) \\ C_2(U(4)) &= C_2(O(6)) + \hat{n}^2/4 \\ C_2(U(\Omega)) - C_2(O(\Omega)) &= 2\left[\sum_\mu P^\dagger_\mu P_\mu + \sum_\mu D^\dagger_\mu D_\mu\right] + \hat{n} \end{aligned} \tag{11}$$

The third equality in (11) is most important and it is valid only when

$$\beta_{\ell_1}\beta_{\ell_2} = -\alpha(\ell_1,\ell_2)\;, \quad \ell_1 \neq \ell_2 \tag{12}$$

The relations in (12) with β's defining the pair operators in multi-orbit case (see Eq. (2)) and α's defining $O(\Omega)$ generators (see Eq. (10)), via $C_2(O(8))$ given in (6), connect $O(8)$ with $O(\Omega)$. In fact using (6,11) it is seen that,

$$C_2(O(8)) = -C_2(O(\Omega)) + \Omega(\Omega+6)\;, \tag{13}$$

$$\begin{aligned} &\sum_\mu P^\dagger_\mu P_\mu + \sum_\mu D^\dagger_\mu D_\mu = \\ &\frac{1}{2}\{-C_2(O(\Omega)) - C_2(O(6)) - Q_0(Q_0-6) + \Omega(\Omega+6)\} \end{aligned} \tag{14}$$

Thus, clearly the chain (9) is equivalent to the $O(8) \supset O(6) \supset O_S(3) \oplus O_T(3)$ chain and it solves the pairing hamiltonian (3) for $x=0$. One important result that follows from (2,10,12) is that in the multi-orbit case, there are multiple definitions of pair operators P and D as given by (2) and for each of these definitions there is a unique $O(\Omega)$ as defined by (10,12). Multiple definitions of pair operators is not possible in the single ℓ case considered in [3, 4, 5]. In all the previous studies involving several orbits, the choice $\beta_\ell = 1$ is made [4-6,8-11]. It can be shown that $D^\dagger_\mu = \sum_\ell \beta_\ell\, D^\dagger_\mu(\ell)$ and $P^\dagger_\mu = \sum_\ell \gamma_\ell\, P^\dagger_\mu(\ell)$ with $\beta_\ell \neq \gamma_\ell$ (except for the choice $\beta_{\ell_1} = \gamma_{\ell_1} = 1$), it is not possible to have $O(8)$ algebra. Thus $O(8)$ will not allow for solving the isovector plus isoscalar pairing hamiltonian with β's different for the isoscalar and isovector parts.

In order to construct the spectra generated by the group chain (9), we will now turn to the irreps of the groups in (9) and their reductions. Throughout this review, we use Wybourne's [20] notations $\{--\}$, $[---]$, $\langle---\rangle$ respectively for denoting $U(N)$, $O(N)$ and $Sp(N)$ irreps. Our starting point is $\{1^m\}$ irrep of $U(4\Omega)$. Its reduction to $U(\Omega)$ irreps is simple as $U(\Omega)$ appears in the direct product subgroup with other group being $SU(4) \sim O(6)$ (or $U(4)$). The $U(4)$ irreps $\{f\} = \{f_1 f_2 f_3 f_4\}$ uniquely define (by transposition) the $U(\Omega)$ irreps

[20],

$$\begin{aligned} U(\Omega) \quad &: \quad \{\tilde{f}\} = \left\{4^{f_4} 3^{f_3-f_4} 2^{f_2-f_3} 1^{f_1-f_2}\right\} ; \\ &\quad \sum_i f_i = m, \quad f_1 \geq f_2 \geq f_3 \geq f_4 \geq 0 \\ O(6) \quad &: \quad [P_1, P_2, P_3] = \\ &\quad \left[\frac{f_1 + f_2 - f_3 - f_4}{2}, \frac{f_1 - f_2 + f_3 - f_4}{2}, \frac{f_1 - f_2 - f_3 + f_4}{2}\right] \end{aligned} \tag{15}$$

Analogous to the $U(\Omega)$ irreps, one can write the $O(\Omega)$ irreps by introducing the quantum numbers v and $[p_1 p_2 p_3]$ as (see [3]),

$$\begin{aligned} O(\Omega) \quad &: \quad \{\tilde{\mu}\} = \left\{4^{\mu_4} 3^{\mu_3-\mu_4} 2^{\mu_2-\mu_3} 1^{\mu_1-\mu_2}\right\} \Leftrightarrow v, [p_1, p_2, p_3] \\ &\quad \sum_i \mu_i = v, \quad \mu_1 \geq \mu_2 \geq \mu_3 \geq \mu_4 \geq 0, \\ &\quad [p_1, p_2, p_3] = \\ &\quad \left[\frac{\mu_1 + \mu_2 - \mu_3 - \mu_4}{2}, \frac{\mu_1 - \mu_2 + \mu_3 - \mu_4}{2}, \frac{\mu_1 - \mu_2 - \mu_3 + \mu_4}{2}\right] \end{aligned} \tag{16}$$

With $\{1^m\}_{U(4\Omega)} \to \{\tilde{f}\}_{U(\Omega)} \otimes [P_1 P_2 P_3]_{O_{ST}(6)}$, the important reduction that is needed is $\{\tilde{f}\}_{U(\Omega)} \to v\,[p_1 p_2 p_3]$ of $O(\Omega)$ (of course $[P_1 P_2 P_3] \to (ST)$ and $v, [p_1 p_2 p_3] \to L$ are also needed). Before addressing this problem, let us examine the matrix elements of the quadratic Casimir invariants. Using the general results

$$\begin{aligned} \langle C_2(U(N))\rangle^{\{F_1 F_2,\ldots\}} &= \sum_i F_i(F_i + N + 1 - 2i) \\ \langle C_2(O(N))\rangle^{[\omega_1 \omega_2,\ldots]} &= \sum_i \omega_i(\omega_i + N - 2i) \end{aligned} \tag{17}$$

it is seen that

$$\begin{aligned} \langle C_2(O(6))\rangle^{[P_1 P_2 P_3]} &= P_1(P_1 + 4) + P_2(P_2 + 2) + P_3^2 \\ \langle C_2(O(\Omega))\rangle^{v,[p_1 p_2 p_3]} &= v(\Omega + 3 - v/4) \\ &\quad - \left[p_1(p_1 + 4) + p_2(p_2 + 2) + p_3^2\right] \\ \Rightarrow \langle C_2(O(8))\rangle^{v,[p_1 p_2 p_3]} &= Q(Q + 6) - \left[p_1(p_1 + 4) + p_2(p_2 + 2) + p_3^2\right] \end{aligned} \tag{18}$$

where $Q = \Omega - v/2$. The last equality follows from (13). From (14,18) it is clear that the states with $v = 0$ will be lowest in energy for the pairing hamiltonian (3) with $x = 0$. Therefore it is meaningful to consider $v = 0$ and $v = 2$ states and workout the allowed $O_{ST}(6)$ irreps. For $\{\lambda\}_{U(\Omega)} \rightarrow [\eta]_{O(\Omega)}$ one has the rule [20]

$$\begin{aligned} \{\lambda\}_{U(\Omega)} &= \sum_{\eta} \Gamma_{\delta\eta\lambda}\; [\eta]_{O(\Omega)} \\ \{\delta\} &= \{0\},\{2\},\{4\},\{22\},\{6\},\{42\},\{222\},\ldots \end{aligned} \tag{19}$$

In (19) $\Gamma_{\delta\eta\lambda}$ is the multiplicity of $\{\lambda\}$ in the reduction $\{\delta\} \otimes \{\eta\} \rightarrow \{\lambda\}$. The form of the $U(\Omega)$ irreps in (15) shows that in (19) the irreps $\{\delta\}$ must be of the type $\{4^r 2^s\}$. Then the corresponding $U(4)$ irreps are of the type $\{r+s, r+s, r, r\}$ and the equivalent $O(6)$ irreps are $[P00]$, $P = s$. This gives via (19) the allowed $O(6)$ irreps for a given $[p_1p_2p_3]$ of $O(\Omega)$ to be $[P] \otimes [p_1p_2p_3]$, a result used in [7]. Applying this for the case with $v = 0$ (then, $[p_1p_2p_3] = [0]$) one has

$$\begin{aligned} m = 4k,\; v = 0 &\;\rightarrow\; [P],\; P = 0,2,4,\ldots,2k \\ m = 4k+2,\; v = 0 &\;\rightarrow\; [P],\; P = 1,3,\ldots,2k+1 \end{aligned} \tag{20}$$

For a symmetric $O(6)$ irrep, it is easy to write down the allowed (ST) values,

$$P = S + T + 2r, \quad r \geq 0 \tag{21}$$

Using (20) and (21) the spectrum can be constructed. The energies are given by

$$\langle H_{pairing}(x=0)\rangle^{m,v=0,P,(ST)} = \frac{1}{2}P(P+4) + \frac{m}{4}\left(\frac{m}{2} - 2\Omega - 6\right) \tag{22}$$

and they depend only on the supermultiplet quantum number P. It is seen from Eqs. (21,22) that in the isospin space the spectrum is soft with $P(P+4)$ spacing for the P multiplets (this is like the $O(6)$ limit of IBM [19] but in isospace). Note that due to $O(6)$ or Wigner's $SU(4)$, in the symmetry limit the ground states exhibit four particle correlations, $|m = 4k, GS\rangle = (D^\dagger.D^\dagger + P^\dagger.P^\dagger)^k\; |0\rangle$ for $m = 4k$ and $|m = 4k+2, GS\rangle = (D^\dagger \cdot D^\dagger + P^\dagger \cdot P^\dagger)^k P^\dagger\; |0\rangle$ for isovector and $|m = 4k+2, GS\rangle = (D^\dagger.D^\dagger + P^\dagger.P^\dagger)^k D^\dagger\; |0\rangle$ for isoscalar ground states with $m = 4k+2$.

For completeness let us consider the classification of $v = 2$ states. In this case $[\tilde{\mu}] = [2] \oplus [11]$ giving $[p] = [p_1p_2p_3] = [1]$ and $[111]$ respectively. Then (19) gives,

$$\begin{array}{ll} [p] = [1]\; : & [p] = [111]\; : \\ \{\tilde{f}\} = \{4^r 2^s\} \otimes \{2\} \Leftrightarrow [s] \otimes [1] & \{\tilde{f}\} = \{4^r 2^s\} \otimes \{11\} \Leftrightarrow [s] \otimes [111] \\ \Rightarrow [P] = [s+1] \oplus [s] \oplus [s-1] & \Rightarrow [P] = [s-1,1,-1] \oplus [s,1] \\ & \qquad \oplus [s+1,1,1] \\ m = 4k\; : \; s = 1,3,5,\ldots,2k-1 & m = 4k\; : \; s = 1,3,5,\ldots,2k-1 \\ m = 4k+2\; : \; s = 0,2,4,\ldots,2k & m = 4k+2\; : \; s = 0,2,4,\ldots,2k \end{array} \tag{23}$$

It should be pointed out that the methods for evaluating $P^\dagger$ and $D^\dagger$ matrix elements in the symmetry defined basis are developed for the $v = 0$ case in [4] and for $v = 2$ (using the so-called vector coherent states) in [7]. The $v = 0$ results are used extensively in numerical studies of the pairing hamiltonian (3) for any strength x (with $\beta_\ell = 1$) in [5, 6]. However the $v = 2$ formalism is not yet applied in numerical studies but, as discussed ahead, is useful in the context of IBM-ST.

2.2 $O(8) \supset [O_S(5) \supset O_S(3)] \otimes O_T(3)$ chain and the complementary $U(4\Omega) \supset [U(2\Omega) \supset Sp(2\Omega) \supset O(\Omega) \otimes SU_T(2)] \otimes SU_S(2)$ chain

In order to identify the number conserving group chain that is complementary to $O(8) \supset [O_S(5) \supset O_S(3)] \otimes O_T(3)$, obviously one has to start with the $U(2\Omega) \otimes SU_S(2)$ subalgebra of $U(4\Omega)$ algebra. As $U(2\Omega)$ contains $O(\Omega) \otimes SU_T(2)$ as a subalgebra, for completing the group chain, one has to find the subalgebras between these two algebras. It is easy to see that $Sp(2\Omega)$ is the subalgebra one is looking for. Therefore the complementary group-subgroup chain is,

$$U(4\Omega) \supset [U(2\Omega) \supset Sp(2\Omega) \supset \{[O(\Omega) \supset O_L(3)] \otimes SU_T(2)\}] \otimes SU_S(2) \tag{24}$$

We will now establish this result.

The generators and the quadratic Casimir operators (C_2's) of the groups in (24) are,

$$\begin{aligned}
U(2\Omega) \quad &: \quad \sqrt{2}\, u^{L,0,T}_{\mu_\ell,0,\mu_T}(\ell_1,\ell_2) \\
U_S(2) \quad &: \quad Y^S = \sum_\ell \sqrt{2(2\ell+1)}\; u^{0,S,0}_{0,\mu_S,0}(\ell,\ell) \\
Sp(2\Omega) \quad &: \quad \sqrt{2}\, u^{L,0,T}_{\mu_\ell,0,\mu_T}(\ell,\ell),\ L+T = odd;\ V^{L,T}_{\mu,\mu_T}(\ell_1,\ell_2),\ \ell_1 > \ell_2 \\
& \quad V^{L,T}_{\mu,\mu_T}(\ell_1,\ell_2) = \sqrt{2}\left[\alpha(\ell_1,\ell_2)(-1)^{\ell_1+\ell_2+L+T}\right]^{\frac{1}{2}} \times \\
& \quad \left\{u^{L,0,T}_{\mu,0,\mu_T}(\ell_1,\ell_2) + \alpha(\ell_1,\ell_2)(-1)^{L+T} u^{L,0,T}_{\mu,0,\mu_T}(\ell_2,\ell_1)\right\} \\
& \quad |\alpha(\ell_1,\ell_2)|^2 = 1,\ \ \alpha(\ell_1,\ell_2)\alpha(\ell_2,\ell_3) = -\alpha(\ell_1,\ell_3) \\
C_2(U(2\Omega)) &= 2 \sum_{\ell_1,\ell_2,L,T} (-1)^{\ell_1+\ell_2} u^{L,0,T}(\ell_1,\ell_2)\cdot u^{L,0,T}(\ell_2,\ell_1) \\
C_2(Sp(2\Omega)) &= 4 \sum_{\ell,L+T=odd} u^{L,0,T}(\ell,\ell)\cdot u^{L,0,T}(\ell,\ell)
\end{aligned}$$

$$+ \sum_{\ell_1 > \ell_2; L,T} V^{L,T}(\ell_1, \ell_2) \cdot V^{L,T}(\ell_1, \ell_2)$$

$$C_2(U_S(2)) = \sum_S Y^S \cdot Y^S \tag{25}$$

It should be noted that $Sp(2\Omega)$ is not unique in the multi-orbit case and just as $O(\Omega)$, it is defined by distinct $\alpha(\ell_1, \ell_2)$'s in (25). Using (25), it can be proved that,

$$\begin{aligned} C_2(U(2\Omega)) + C_2(U_S(2)) &= \hat{n}(2 + 2\Omega) \\ C_2(U_S(2)) &= 2S^2 + \hat{n}^2/2 \\ C_2(U(2\Omega)) - C_2(Sp(2\Omega)) &= 4\left[\sum_\mu D^\dagger_\mu D_\mu\right] - \hat{n} \end{aligned} \tag{26}$$

The third equality in (26) is valid only when (12) is satisfied. Comparing (26) with (6), it is seen that $Sp(2\Omega)$ in (24) is related to $O_S(5)$,

$$C_2(O_S(5)) = -\frac{1}{2}C_2(Sp(2\Omega)) + \Omega(\Omega + 3)\ , \tag{27}$$

$$2\sum_\mu D^\dagger_\mu D_\mu = -\frac{1}{2}C_2(Sp(2\Omega)) - S^2 - Q_0(Q_0 - 3) + \Omega(\Omega + 3) \tag{28}$$

Thus, clearly the chain (24) is equivalent to the $O(8) \supset [O_S(5) \supset O_S(3)] \otimes O_T(3)$ chain and it solves the pairing hamiltonian (3) for $x = 1$.

In order to construct the spectra generated by the group chain (24), we will now turn to the irreps of the groups in (24) and their reductions. Just as before, the starting point is $\{1^m\}$ irrep of $U(4\Omega)$. Its reduction to $U(2\Omega)$ irreps is simple as $U(2\Omega)$ appears in the direct product subgroup with the other group being $SU_S(2)$ (or $U_S(2)$). The $U(2)$ irreps $\{f\} = \{f_1 f_2\}$ uniquely define (by transposition) the $U(2\Omega)$ irreps,

$$U(2\Omega)\ :\ \left\{2^{f_2}1^{f_1 - f_2}\right\}\ ;\quad f_1 + f_2 = m,\ \ f_1 \geq f_2 \geq 0,\ \ S = (f_1 - f_2)/2 \tag{29}$$

Now the $Sp(2\Omega)$ irreps can be written as,

$$Sp(2\Omega)\ :\ \langle 2^{\mu_1} 1^{\mu_2} \rangle\ ;\quad v_S = 2\mu_1 + \mu_2 \quad t = \frac{\mu_2}{2} \tag{30}$$

The v_S and t (reduced isospin) quantum numbers are introduced by examining the eigenvalue expression for $C_2(Sp(2\Omega))$,

$$\begin{aligned}
&\langle C_2(Sp(2\Omega))\rangle^{\langle\lambda_1\lambda_2\ldots\rangle} = \sum_i \lambda_i\left(\lambda_i + 2\Omega + 2 - 2i\right) \\
&\Rightarrow \langle C_2(Sp(2\Omega))\rangle^{\langle 2^{\mu_1}1^{\mu_2}\rangle} = \\
&2\left[\Omega(\Omega+3) - \left(\Omega - \frac{v_s}{2}\right)\left(\Omega - \frac{v_s}{2} + 3\right) - t(t+1)\right]
\end{aligned} \tag{31}$$

General rules for obtaining $U(2\Omega) \supset Sp(2\Omega)$ and $Sp(2\Omega) \supset O(\Omega) \otimes SU_T(2)$ are given in [20]. Here we consider only the $v = 0$ states (note that v labels $O(\Omega)$ irreps). Firstly, it is seen from the results given in [20] that in the reduction $Sp(2\Omega) \supset O(\Omega) \otimes SU_T(2)$, it is possible to get $[0]$ irrep of $O(\Omega)$ only when the $Sp(2\Omega)$ irreps are of the type $\langle 2^r\rangle$ (i.e. for the reduced isospin $t = 0$); note that we are restricting to m even. For $\{\lambda\}_{U(2\Omega)} \to \langle\mu\rangle_{Sp(2\Omega)}$ one has the rule [20]

$$\begin{aligned}
\{\lambda\}_{U(2\Omega)} &= \sum_\mu \Gamma_{\beta\mu\lambda}\ \langle\mu\rangle_{Sp(2\Omega)} \\
\{\beta\} &= \{0\}, \{1^2\}, \{1^4\}, \{2^2\}, \{1^6\}, \{2^2 1^2\}, \{3^2\}, \ldots
\end{aligned} \tag{32}$$

From the form of $U(2\Omega)$ irreps in (29), it is clear that $\{\beta\}$ in (32) must be of the type $\{2^{2\alpha_1}1^{2\alpha_2}\}$. As $\{\mu\}$ is $\{2^r\}$ type (see the discussion just above (32)), the $U(2\Omega)$ irreps follow from $\{2^{2\alpha_1}1^{2\alpha_2}\} \otimes \{2^r\}$. Now using (29) gives,

$$\begin{aligned}
m &\longrightarrow S = 0, 2, 4, \ldots, \frac{m}{2} \\
S &\longrightarrow v_S = (m - 2S) - 4\alpha\ ; \quad \alpha \geq 0
\end{aligned} \tag{33}$$

Finally to obtain $v_S \to [0]_{O(\Omega)}T$ the following procedure can be adopted: (i) consider $U(2\Omega) \supset [U(\Omega) \supset O(\Omega)] \otimes SU_T(2)$ and $S = 0 \Leftrightarrow \{2^\lambda\}_{U(2\Omega)}$; (ii) use the expansion $\{2^\lambda\} = \{1^\lambda\}\{1^\lambda\} - \{1^{\lambda+1}\}\{1^{\lambda-1}\}$; (iii) reduce $\{1^\lambda\}_{U(2\Omega)} \to \{f_1 f_2\}_{U_T(2)}\{\tilde{f}\}_{U(\Omega)} = [T]\{\tilde{f}\}_{U(\Omega)}$; (iv) apply (iii) to each part on the r.h.s of the equation in (ii) and then symbolically,

$$\begin{aligned}
&\{1^\lambda\}\{1^\lambda\} - \{1^{\lambda+1}\}\{1^{\lambda-1}\} = \\
&\sum_{T_1,T_2}\left[[T_1]\{\tilde{f}_1\}\right] \otimes \left[[T_2]\{\tilde{f}_2\}\right] - \sum_{T_3,T_4}\left[[T_3]\{\tilde{f}_3\}\right] \otimes \left[[T_4]\{\tilde{f}_4\}\right]
\end{aligned}$$

(v) say $\{\tilde{f}_1\} \otimes \{\tilde{f}_2\}$ gives a $\{4^r 2^s\}$ type irrep x_{12} times (note that only $\{4^r 2^s\}_{U(\Omega)}$ can give $[0]_{O(\Omega)}$) and similarly define x_{34} and then,

$$\{2^r\}_{U(2\Omega)} \Leftrightarrow S = 0 \longrightarrow \sum_{T_1,T_2} (T_1 \times T_2)^{x_{12}} - (T_3 \times T_4)^{x_{34}}\ .$$

(vi) use, from (33), the reduction $S = 0 \rightarrow v_S = m, m-4, \ldots, 2$ or 0. Starting from $m = 0, 2, 4 \ldots$ and applying (i)-(vi), $v_S \rightarrow [0]_{O(\Omega)}T$ is obtained by successive subtraction. Then one can identify the rule,

$$v_S \rightarrow T = \binom{v_S}{2} - 2\beta\,;\;\; \beta \geq 0 \tag{34}$$

Using Eqs. (33,34) we have $m \rightarrow Sv_ST$ classification for $v = 0$ states and for these $v = 0$ states

$$\langle H_{pairing}(x=1)\rangle^{m,S,v_S,T} = -\frac{1}{4}\,(m - v_S)\,(4\Omega + 6 - m - v_S) + S(S+1) \tag{35}$$

independent of T. For large Ω and a fixed S, the spacing between different v_S multiplets is constant giving vibrational spectrum (with isospin multiplets) in the isospace and $v_S/2$ can be viewed as the phonon number. Even phonon number states appear for $m = 4k$ and odd phonon states appear for $m = 4k+2$. In the symmetry limit, the ground state for $m = 4k$ is $(D^\dagger \cdot D^\dagger)^k\;|0\rangle$.

2.3 $O(8) \supset [O_T(5) \supset O_T(3)] \otimes O_S(3)$ chain and the complementary $U(4\Omega) \supset$ $[U(2\Omega) \supset Sp(2\Omega) \supset O(\Omega) \otimes SU_S(2)] \otimes SU_T(2)$ chain

All the results for the $O(8) \supset [O_T(5) \supset O_T(3)] \otimes O_S(3)$ chain can be obtained from Sect. 2.2 by simply interchanging $S \Leftrightarrow T$ everywhere. For example, for $v = 0$ the quantum numbers and the energy formula are (see also [5]),

$$\begin{gathered} v_T = 2S, 2S+4, \ldots\;,\; v_T \leq m;\;\; T = \frac{m - v_T}{2} - 2\alpha'\;,\; \alpha' \geq 0 \\ \langle H_{pairing}(x=-1)\rangle^{m,S,v_T,T} = \\ -\frac{1}{4}\,(m - v_T)\,(4\Omega + 6 - m - v_T) + T(T+1) \end{gathered} \tag{36}$$

Most significant result that follows from Eq. (36) is that in the symmetry limit one obtains rotational spectra (for a fixed S and fixed v_T) in the isospin space. The ground state for $m = 4k$ is $(P^\dagger \cdot P^\dagger)^k\;|0\rangle$.

In summary, the symmetry chains (7) generate vibrations ($x = 1$), rotations ($x = -1$) and soft ($x = 0$) spectra in isospace. For equal $T = 0$ and $T = 1$ strengths ($x = 0$) there is Wigner's $SU(4)$ symmetry. All these results are valid for the $O(\Omega)$ irrep $[0]$ ($v = 0$) which lies lowest in energy. Using masses calculated from a mass model it is established recently that odd-odd N=Z nuclei with $A > 60$ should exhibit signatures of $SU(4)$ symmetry [21]. Similarly there is also a search for vibrations in isospin space using experimental data for nuclei around ^{56}Ni [22].

3. Dyson boson mapping

Intermediate to the full fermion description of N=Z nuclei in terms of $H_{pairing}$ and its extensions (for general x and/or non degenerate single particle energies only numerical studies as carried out for example in [5, 6] are possible) and IBM-ST boson model (discussed ahead in Sect.4) is the study using Dyson boson mapping of the fermion H's. The Dyson boson mapping is defined by

$$
\begin{aligned}
a_i^\dagger a_j^\dagger &\longrightarrow B_{ij}^\dagger - \sum_{k,l} B_{ik}^\dagger B_{jl}^\dagger B_{kl} \\
a_i a_j &\longrightarrow B_{ji} \\
a_i^\dagger a_j &\longrightarrow \sum_{k} B_{ik}^\dagger B_{jk}
\end{aligned}
\tag{37}
$$

where the boson creation and annihilation operators satisfy the usual boson commutation relations, $[B_{ij}^\dagger B_{kl}^\dagger] = 0$, $[B_{ij} B_{kl}] = 0$ and $[B_{ij} B_{kl}^\dagger] = \delta_{ik}\delta_{jl} - \delta_{il}\delta_{jk}$ and also $B_{ij}^\dagger = -B_{ji}^\dagger$. The Dyson map of $H_{pairing}$ in (3) is obtained by rewriting (37) in angular momentum coupled representation and then following the steps: (i) introducing coupled boson creation (and destruction) operators $B^\dagger_{(\ell_1 \frac{1}{2}\frac{1}{2})(\ell_2 \frac{1}{2}\frac{1}{2})LM_L SM_s TM_T}$; (ii) restricting the bosons B's to $L = 0$ bosons (hereafter called s-bosons); (iii) defining collective s-bosons

$$
s^\dagger_{SM_S;TM_T} = \sum_{\ell} \sqrt{(2\ell+1)/2\Omega}\; \beta_\ell \; B^\dagger_{(\ell\frac{1}{2}\frac{1}{2})(\ell\frac{1}{2}\frac{1}{2})00;SM_S;TM_T}
$$

with $(ST) = (01) \oplus (10)$; (iv) mapping first $D_\mu^\dagger$, $P_\mu^\dagger$, D_μ and P_μ into $B^\dagger$ and B operators and then converting them into $s^\dagger$ and s operators via the replacements

$$
B^\dagger_{(\ell\frac{1}{2}\frac{1}{2})(\ell\frac{1}{2}\frac{1}{2})00,SM_S;TM_T} \rightarrow \sqrt{2(2\ell+1)/\Omega}\; \beta_\ell \; s^\dagger_{SM_S;TM_T}
$$

and similarly $B \rightarrow s$. First results of Dyson mapping of Eq. (3) are due to Dobes and Pittel [8] and more detailed investigation are given in [9, 10, 11]. Applying the steps (i)-(iv), the Dyson mapping for $H_{pairing}$ is (see for example

[9])

$$
\begin{aligned}
H_{pair.} \rightarrow \quad & -(1+x)\left[\Omega\hat{n}_{s:S} - \hat{n}_{s:S}\hat{n}_{s:T} - \frac{3}{2}\left\{\left(s^\dagger_{01}s^\dagger_{01}\right)^{00}\left(\tilde{s}_{10}\tilde{s}_{10}\right)^{00}\right\}^{00}\right. \\
& \left. -9\sum_{S=0,2}\sqrt{2S+1}\,\chi(S)\left\{\left(s^\dagger_{10}s^\dagger_{10}\right)^{S0}\left(\tilde{s}_{10}\tilde{s}_{10}\right)^{S0}\right\}^{00}\right] \\
& -(1-x)\left[\Omega\hat{n}_{s:T} - \hat{n}_{s:S}\hat{n}_{s:T} - \frac{3}{2}\left\{\left(s^\dagger_{10}s^\dagger_{10}\right)^{00}\left(\tilde{s}_{01}\tilde{s}_{01}\right)^{00}\right\}^{00}\right. \\
& \left. -9\sum_{T=0,2}\sqrt{2T+1}\,\chi(T)\left\{\left(s^\dagger_{01}s^\dagger_{01}\right)^{0T}\left(\tilde{s}_{01}\tilde{s}_{01}\right)^{0T}\right\}^{00}\right] \qquad (38)
\end{aligned}
$$

In (38), $\chi(J)$ is the $9-j$ symbol $\left\{\begin{matrix} \frac{1}{2} & \frac{1}{2} & 1 \\ \frac{1}{2} & \frac{1}{2} & 1 \\ 1 & 1 & J \end{matrix}\right\}$. Writing the $S = 2$ term in (38) in terms of the number operator $\hat{n}_{s:S}$ and the $S = 0$ term (similarly for the $T = 2$ term) and converting all the terms into dot-products, Eq. (9) of Ref. [10], which looks different but more appealing, is recovered. Similarly the Dyson boson mapping including spin-orbit term in (3) is given in [11]. Some important results that follow from all the mapping studies are:

i. Naive hermitization (adopted by Van Isacker et al [9]) with $H \rightarrow (H + H^\dagger)/2$, Eq. (38) corresponds to a boson model with s bosons ($\ell = 0$ bosons) carrying $(ST) = (10) \oplus (01)$ and interacting with a one plus two-body force. The SGA for this model is $U_{s:ST}(6)$ and it admits two symmetry limits,

$$
\begin{aligned}
& U_{s:ST}(6) \supset [O_{s:ST}(6) \sim SU_{s:ST}(4)] \supset O_{s:S}(3) \oplus O_{s:T}(3) \quad (I) \\
& U_{s:ST}(6) \supset U_{s:S}(3) \oplus U_{s:T}(3) \supset O_{s:S}(3) \oplus O_{s:T}(3) \quad (II)
\end{aligned} \qquad (39)
$$

For $x = 0$ the boson mapping is hermitian and the boson hamiltonian reduces to a linear combination of the Casimir operators of the algebras in chain (I) in (39). Thus the $SU(4)$ symmetry of the $O(8)$ fermion model is preserved for low-lying states by the interacting s-boson model. The $O(5)$ limits of the $O(8)$ model are similarly recovered by the chain (II) in (39) with isoscalar or isovector boson number is zero (they are generated by the $U(3)$ groups in (39)). Thus the s-boson model keeps (in fact for any x) the physics given by the fermion $O(8)$ model for low-lying states; see [9] for more details.

ii. Palchikov et al [10] used the hermitization procedure with $s^{\dagger}_{10} \to (1+x)^{1/4}s^{\dagger}_{10}$, $s_{10} \to (1+x)^{-1/4}s_{10}$, $s^{\dagger}_{01} \to (1-x)^{1/4}s^{\dagger}_{01}$ and $s_{01} \to (1-x)^{-1/4}s_{01}$. They calculated the overlaps of the exact lowest eigenstates (ground states with $S = T = 0$) for $m = 4k$ (boson number $N = 2k$) systems with α-particle like structure $(A^{\dagger})^k \left|0\right\rangle$ where $A^{\dagger} = (s^{\dagger}_{10} \cdot s^{\dagger}_{10} sin\theta - s^{\dagger}_{01} s^{\dagger}_{01} cos\theta)$. The overlaps are found to be $> 97\%$ for all x values in (3). Note that $\theta = 0$ for $x = -1$, $\pi/4$ for $x = 0$ and $\pi/2$ for $x = 1$. Even for excited states (generated by breaking 1,2, ... α particle structures), the θ are found to be state independent. For odd-odd systems the ground states are close to $(A^{\dagger})^k (s^{\dagger}_{10})^S (s^{\dagger}_{01})^T \left|0\right\rangle$; $(ST) = (10) \oplus (01)$. Moreover the α particle structure of the ground state wavefunctions is also supported by realistic calculations with SDI interaction in a single j-shell. An important consequence of all these results is that it is possible to start with a four particle correlated structure for proton-neutron systems and develop a model for N=Z nuclei similar to the broken-pair model [23] known for identical particle systems.

iii. Juillet and Josse [11] studied the effects of spin-orbit force on properties such as the relative position of $T = 0$ and $T = 1$ states in odd-odd N=Z nuclei. Using Dyson mapping and perturbation theory they showed that $\Delta E = E_{GS}(J = 0, T = 1) - E_{GS}(J = 1, T = 0) = xgf(N) - V^2_{so}\left[h_1(\ell) + xh_2(\ell)\right](N+3)/g$ where f is some function of N, h's are some functions of ℓ, g is the overall strength of $H_{pairing}$ and V_{so} is spin-orbit strength. Then it is clear that, for $x \sim 0$, even a small spin-orbit strength (it will break the $SU(4)$ symmetry) favors isovector ground states as seen in $A \geq 62$ nuclei.

An important outcome of the boson mappings is the recognition that the interacting boson model with s bosons carrying $(ST) = (10) \oplus (01)$ is equivalent to the $O(8)$ pairing model. Then a natural extension is to include d bosons so that quadrupole deformation effects can be included in a group theoretical framework (the corresponding extension of the $O(8)$ model is not available and appear to be more complex than the corresponding sd boson model). The interacting boson model with s ($\ell = 0$) and d ($\ell = 2$) bosons carrying spin-isospin degrees of freedom $(ST) = (10) \oplus (01)$ is called IBM-ST and the SGA for this model is $U(36)$. Now we will consider the symmetry limits of IBM-ST in some detail.

4. IBM-ST and $U(36)$ symmetry schemes

In IBM-ST, the spin-isospin invariant interacting boson model (also called IBM-4), quadrupole collective states are generated by interacting s ($\ell = 0$) and d ($\ell = 2$) bosons with six spin-isospin degrees of freedom (three from

$(ST) = (10)$ and three from $(ST) = (01)$). Given the one boson creation and destruction operators $b^{\dagger}_{\ell,m_\ell;\, S,m_S;\, T,m_T}$ and

$$\tilde{b}_{\ell,m_\ell;\, S,m_S;\, T,m_T} = (-1)^{1+m_\ell+m_S+m_T}\; b_{\ell,-m_\ell;\, S-m_S;\, T,-m_T}\;,$$

the 1296 triple tensors

$$\left(b^{\dagger}_{\ell,s,T}\tilde{b}_{\ell' S',T'}\right)^{L_0,S_0,T_0}_{M_0,M_{S_0},M_{T_0}}$$

generate the model's $U_{sdST}(36)$ SGA. Dynamical symmetry limits ofinteracting boson model IBM-ST correspond to the group chains starting with $U_{sdST}(36)$ generating boson number N and ending with $O_L(3) \otimes [O_S(3) \oplus O_T(3)]$ or $O_J(3) \otimes O_T(3)$ generating states with good LST or JT; $\vec{J} = \vec{L} + \vec{S}$. Before going further it is useful to write down the number, spin, isospin and angular momentum operators,

$$
\begin{aligned}
&\hat{n}_{s;S} = \sqrt{3}\left(s^{\dagger}_{(10)}\tilde{s}_{(10)}\right)^{0,0,0} \;,\quad \hat{n}_{s;T} = \sqrt{3}\left(s^{\dagger}_{(01)}\tilde{s}_{(01)}\right)^{0,0,0}\\
&\hat{n}_{d;S} = \sqrt{15}\left(d^{\dagger}_{(10)}\tilde{d}_{(10)}\right)^{0,0,0} \;,\quad \hat{n}_{d;T} = \sqrt{15}\left(d^{\dagger}_{(01)}\tilde{d}_{(01)}\right)^{0,0,0}\\
&\hat{n}_S = \hat{n}_{s;S} + \hat{n}_{d;S} \;,\quad \hat{n}_T = \hat{n}_{s;T} + \hat{n}_{d;T} \;,\quad \hat{N} = \hat{n}_S + \hat{n}_T\\
&\hat{n}_s = \hat{n}_{s;S} + \hat{n}_{s;T} \;,\quad \hat{n}_d = \hat{n}_{d;S} + \hat{n}_{d;T} \;,\quad \hat{N} \;=\; \hat{n}_s + \hat{n}_d\\
&L^1_{S;\mu} = \sqrt{30}\left(d^{\dagger}_{(10)}\tilde{d}_{(10)}\right)^{1,0,0}_{\mu,0,0} \;,\quad L^1_{T;\mu} = \sqrt{30}\left(d^{\dagger}_{(01)}\tilde{d}_{(01)}\right)^{1,0,0}_{\mu,0,0}\\
&S^1_{s;\mu} = \sqrt{2}\left(s^{\dagger}_{(10)}\tilde{s}_{(10)}\right)^{0,1,0}_{0,\mu,0} \;,\quad S^1_{d;\mu} = \sqrt{10}\left(d^{\dagger}_{(10)}\tilde{d}_{(10)}\right)^{0,1,0}_{0,\mu,0}\\
&T^1_{s;\mu} = \sqrt{2}\left(s^{\dagger}_{(01)}\tilde{s}_{(01)}\right)^{0,0,1}_{0,0,\mu} \;,\quad T^1_{d;\mu} = \sqrt{10}\left(d^{\dagger}_{(01)}\tilde{d}_{(01)}\right)^{0,0,1}_{0,0,\mu}\\
&\vec{L} = \vec{L_S} + \vec{L_T}\;,\;\; \vec{S} = \vec{S}_s + \vec{S}_d\;,\;\; \vec{T} = \vec{T}_s + \vec{T}_d
\end{aligned}
\tag{40}
$$

In (40) n_s, n_d, n_S and n_T give the number of s-bosons, d-bosons, $T = 0$ pairs and $T = 1$ pairs respectively. At the primary level, identified by the first subgroup of $U_{sdST}(36)$, IBM-ST has 4 symmetry limits [12] :

I. Decomposing the $sdST$ space into orbital sd and spin-isospin ST spaces gives $U_{sd}(6) \otimes U_{ST}(6)$ limit chains. The $U_{sd}(6)$ admits the three ($U(5)$, $SU(3)$ and $O(6)$) IBM subgroup chains while $U_{ST}(6)$ admits two chains just as in (39).

II. Decomposing the $sdST$ space into S (with $(ST) = (10)$) and T (with $(ST) = (01)$) spaces gives $U_{sdS}(18) \oplus U_{sdT}(18)$ limit chains and they preserve (n_S, n_T). The group chains admitted by the two $U(18)$ groups are same those of IBM-3 $U(18)$ SGA; see [24]. These subgroups can be combined with each other at various levels to give all the group chains in limit II. All symmetry chains here will break the $SU_{ST}(4)$ symmetry.

III. Decomposing the $sdST$ space into s and d boson spaces gives $U_{sST}(6) \oplus U_{dST}(30)$ limit chains. They preserve (n_s, n_d) (the case with $n_d = 0$ is nothing but the s boson model discussed in Section 3). The $U_{dST}(30)$ admits at the first level $O_{dST}(30)$, $U_d(5) \otimes U_{S_dT_d}(6)$ and $U_{dS}(15) \oplus U_{dT}(15)$ and the subgroups of these groups will follow from I and II and they can be combined at various levels.

IV. The generalized pairing group in the total $sdST$ space is $O_{sdST}(36)$ and this gives the $O_{sdST}(36)$ limit chains. At the first level $O(36)$ subgroups are $O_{sdS}(18) \oplus O_{sdT}(18)$, $O_{sd}(6) \otimes O_{ST}(6)$ and $O_{sST}(6) \oplus O_{dST}(30)$. Further subgroups of these will follow easily from I, II and III and they can be combined at various levels to give the group chains in limit IV; see Fig. 1 in [13]. Some of the group chains here (also in limit III) admit Wigner's $SU_{ST}(4)$ symmetry.

In the last few years some of the group chains in I and IV are studied in detail and the results are briefly discussed here.

4.1 $U_{sd}(6) \otimes U_{ST}(6)$ limit chains

The group-subgroup chains in the $U_{sdST}(36) \supset U_{sd}(6) \otimes U_{ST}(6)$ limit can be written down easily usinginteracting boson model IBM chains and Eq. (39),

$$\begin{array}{lll} U_{sd}(6) \;\; \text{chains:} & : & U_{sd}(6) \supset U(5) \supset O(5) \supset O_L(3) \\ & & U_{sd}(6) \supset SU(3) \supset O_L(3) \\ & & U_{sd}(6) \supset O(6) \supset O(5) \supset O_L(3) \end{array} \tag{41}$$

$$\begin{array}{lll} U_{ST}(6) \;\; \text{chains} & : & U_{ST}(6) \supset U_S(3) \oplus U_T(3) \supset O_S(3) \oplus O_T(3) \\ & & U_{ST}(6) \supset [O_{ST}(6) \sim SU_{ST}(4)] \supset O_S(3) \oplus O_T(3) \end{array} \tag{42}$$

The $U_{sdST}(36)$ irrep being $\{N\}$, the $U_{sd}(6)$ and $U_{ST}(6)$ irreps will have same Young tableaux structure $\{f\} = \{f_1 f_2 \ldots f_6\}$, $\sum_i f_i = N$ and $f_1 \geq f_2 \cdots \geq f_6 \geq 0$. Thus the lowest irrep is $\{N\}$ and the next one is $\{N-1, 1\}$. In the $U_{sd}(6)$ sector they correspond to IBM states and pnIBM (proton-neutron IBM or IBM-2) states with F-spin $F = N/2 - 1$ respectively. Using the chains in (41,42) studied so far are: (i) competition between $T = 0$ and $T = 1$ pairing and the resulting signatures [14]; (ii) Binding energies (BE) of $T = 0$ and $T = 1$ ground states in N=Z nuclei [16]. Here (ii) is discussed in some detail and (i) is differed to Section 4.3.

Considering only $\{N\}$ irrep but otherwise ignoring the space part (assuming ground states have $L = 0$), the spin-isospin structure of the states is governed by the two chains in (42). The allowed $O(6)$ irreps are $\omega_{ST} = N, N-2, \ldots, 1$ or 0 and $\omega_{ST} \to (ST)$ is given by (21). This gives correctly the ground state $(ST) = (00)$ for even-even and $(10) \oplus (01)$ for odd-odd nuclei. Similarly, $SU_S(3)$ generates $T = 0$ pairs $N_{T=0}$ (n_S in (40)) and $SU_T(3)$ generates $T = 1$ pairs $N_{T=1}$ (n_T in (40)) with $N = N_{T=0} + N_{T=1}$; note that $N_S \to S = N_S, N_S - 2, \ldots, 1$ or 0 and similarly $N_T \to T$. The problems (i) and (ii) are addressed by using a hamiltonian that mixes the two chains in (42). The general group theoretical problem need to be solved here is transformation brackets between the chains $U(\mathcal{N}) \supset U(\mathcal{N}_a) \oplus U(\mathcal{N}_b) \supset O(\mathcal{N}_a) \oplus O(\mathcal{N}_b) \supset \ldots$ and $U(\mathcal{N}) \supset O(\mathcal{N}) \supset O(\mathcal{N}_a) \oplus O(\mathcal{N}_b) \supset \ldots$ for symmetric $U(\mathcal{N})$ irreps with $\mathcal{N} = \mathcal{N}_a + \mathcal{N}_b$. For the problem at hand, $(\mathcal{N}_a, \mathcal{N}_b) = (3, 3)$. For the $O(36)$ chain discussed ahead solutions for several other $(\mathcal{N}_a, \mathcal{N}_b)$ are needed (for example (6,30), (18,18) etc.). Before giving a complete solution to this problem let us mention that using the mixing hamiltonian

$$
\begin{aligned}
H_{mix} \;=\;& E_0 + \alpha_1 C_1(U_{ST}(6)) + \alpha_2 C_2(U_{ST}(6)) + \alpha_3 C_2(O_{ST}(6)) \\
& + \alpha_4 C_2(U_S(3)) + \alpha_5 C_2(O_T(3))
\end{aligned}
\tag{43}
$$

the known values of $BE(T=0) - BE(T=1)$ are well described for N=Z nuclei in the mass range ^{56}Ni to ^{100}Sn with predictions for many unknown cases [16]. The Casimirs of $O(6) \sim SU(4)$ and $O_T(3)$ have clear physical meaning and that of $U_S(3)$ has its origin in spin-orbit force. Thus IBM-ST gives a good 'local' mass formula for N=Z nuclei in a region of current experimental interest.

4.2 Transformation brackets between $U(\mathcal{N}) \supset U(\mathcal{N}_a) \oplus U(\mathcal{N}_b) \supset O(\mathcal{N}_a) \oplus O(\mathcal{N}_b)$ and $U(\mathcal{N}) \supset O(\mathcal{N}) \supset O(\mathcal{N}_a) \oplus O(\mathcal{N}_b)$ chains

For symmetric $U(\mathcal{N})$ irreps $\{n\}$ in the $U(\mathcal{N}) \supset U(\mathcal{N}_a) \oplus U(\mathcal{N}_b) \supset O(\mathcal{N}_a) \oplus O(\mathcal{N}_b) \supset K$ chain, the irrep labels for other groups in the chain and their reductions are given by (assuming that $\mathcal{N}_a \geq 3$, $\mathcal{N}_b \geq 3$)

$$
\left| \begin{array}{ccccccccccc}
U(\mathcal{N}) & \supset & U(\mathcal{N}_a) & \oplus & U(\mathcal{N}_b) & \supset & O(\mathcal{N}_a) & \oplus & O(\mathcal{N}_b) & \supset & K \\
\{n\} & & \{n_a\} & & \{n_b\} & & [\omega_a] & & [\omega_b] & & \alpha
\end{array} \right\rangle
$$

$$
\begin{array}{l}
n_a = 0,\, 1,\, 2,\, \cdots,\, n; \quad n_b \;=\; n - n_a \\
\omega_a = n_a, n_a - 2, \ldots, 0 \text{ or } 1, \quad \omega_b = n_b, n_b - 2, \ldots, 0 \text{ or } 1
\end{array}
\tag{44}
$$

In (44) label(s) α for the irreps of K need not be specified as the algebra K do not play any role in the present discussion. On the other hand, for symmetric $U(\mathcal{N})$ irreps $\{n\}$ in the $U(\mathcal{N}) \supset O(\mathcal{N}) \supset O(\mathcal{N}_a) \oplus O(\mathcal{N}_b) \supset K$ chain, the irrep labels for other groups in the chain and their reductions are,

$$
\left| \begin{array}{ccccccccc}
U(\mathcal{N}) & \supset & SO(\mathcal{N}) & \supset & SO(\mathcal{N}_a) & \oplus & SO(\mathcal{N}_b) & \supset & K \\
\{n\} & & [\omega] & & [\omega_a] & & [\omega_b] & & \alpha
\end{array} \right\rangle
$$

$$\omega = n, n-2, \ldots, 0 \text{ or } 1, \quad \omega_a + \omega_b = \omega, \omega - 2, \ldots, 0 \text{ or } 1 \tag{45}$$

As the states in (44) and (45) both form complete set of states, in general it is possible to expand one in terms of the other,

$$|n\omega\,(\omega_a\,\omega_b)\;\alpha\rangle = \sum_{n_a} C_{n_a}^{n,\omega,\omega_a,\omega_b}(\mathcal{N}_a, \mathcal{N}_b)\;\;|n\,(n_a\,n_b)\,(\omega_a\,\omega_b)\;\alpha\rangle \tag{46}$$

In (46) $\sum_{n_a} \mid C_{n_a}^{n,\omega,\omega_a,\omega_b}(\mathcal{N}_a, \mathcal{N}_b) \mid^2 = 1$ and in the summation, $n_a = \omega_a, \omega_a + 2, \cdots, n$ or $n-1$ and $n_b = n - n_a = \omega_b, \omega_b + 2, \cdots, n$ or $n-1$. A realization of the states $|n\;\omega\;\;(\omega_a\,\omega_b)\;\;\alpha\rangle$ is in terms of the eigenstates of $\mathcal{N}$ dimensional harmonic oscillator solved in bispherical co-ordinates in $\mathcal{N} = \mathcal{N}_a + \mathcal{N}_b$ dimensions [25]. This involves Laguerre and Jacobi polynomials. Similarly, the corresponding realization of $|n\;\;(n_a\,n_b)\;\;(\omega_a\,\omega_b)\;\;\alpha\rangle$ basis states is obtained by solving the oscillator equation in $\mathcal{N}_a$ and $\mathcal{N}_b$ co-ordinates separately (each of them involve a Laguerre polynomial). Now the transformation brackets (C_{-}^{---}'s) in (46) are derived using a novel convolution identity for Laguerre polynomials. This identity was derived by Vander Jeugt [26] via $SU(1,1)$ algebra with: (i) using the generators J_0, $J_\pm$ and expanding the simultaneous eigenstates $|kx>$ of the $SU(1,1)$ Casimir operator $\mathcal{C} = J_0^2 - J_0 - J_+ J_-$ and the operator $X = 2J_0 - J_+ - J_-$ in terms of the standard $(\mathcal{C}, J_0)$ eigenstates $|km>$; (ii) defining (Laguerre) polynomials via $<kx|km>\,/<kx|k0>$; (iii) writing $SU(1,1)$ Clebsch - Gordan coefficients in terms of a terminating generalized hypergeometric series; (iv) introducing (Jacobi) polynomials as in (ii) but via the tensor product of two irreps $(k_1) \otimes (k_2)$. Then the final formula is [27],

$$
\begin{array}{l}
|n\;\omega\;\;(\omega_a\,\omega_b)\;\;\alpha\rangle = (-1)^{\phi_r + \phi_{r_{ab}}}\,\mathcal{R}_{n\,\omega\,(\omega_a\,\omega_b)} \quad \times \\[2ex]
\times \displaystyle\sum_{r_a=0}^{r+r_{ab}} (-1)^{\phi_{r_a} + \phi_{r_b}} \frac{C\left(r, r_{ab}, r_a, \omega_a, \omega_b\right)}{\mathcal{M}_{n\,(n_a\,n_b)\,(\omega_a\,\omega_b)}} \quad |n\;\;(n_a\,n_b)\;\;(\omega_a\,\omega_b)\;\;\alpha\rangle\;; \\[2ex]
\mathcal{R}_{n\,\omega\,(\omega_a\,\omega_b)} = \\[2ex]
= \left[\dfrac{4(r!)\,(r_{ab}!)\,\left(\omega + \frac{\mathcal{N}}{2} - 1\right)\Gamma\left(\omega + \frac{\mathcal{N}}{2} - r_{ab} - 1\right)}{\Gamma\left(\omega + \frac{\mathcal{N}}{2} + r\right)\Gamma\left(\omega_a + \frac{\mathcal{N}_a}{2} + r_{ab}\right)\Gamma\left(\omega_b + \frac{\mathcal{N}_b}{2} + r_{ab}\right)}\right]^{1/2}
\end{array}
$$

$$
\mathcal{M}_{n\,(n_a\,n_b)\,(\omega_a\,\omega_b)} = \left[\frac{4\,(r_a!)\,(r_b!)}{\Gamma(\omega_a + r_a + \frac{\mathcal{N}_a}{2})\,\Gamma(\omega_b + r_b + \frac{\mathcal{N}_b}{2})}\right]^{1/2}
$$

$$
C\,(r, r_{ab}, r_a, \omega_a, \omega_b) = \frac{(-1)^{r_{ab}}\,\Gamma\left(\omega_a + r_{ab} + \frac{\mathcal{N}_a}{2}\right)\,(r + r_{ab})!}{r!\,r_{ab}!\,\Gamma\left(\omega_a + \frac{\mathcal{N}_a}{2}\right)} \times
$$

$$
\times\ {}_3F_2\left(\begin{matrix}\omega_a + \omega_b + r_{ab} + \frac{\mathcal{N}}{2} - 1,\ -r_a,\ -r_{ab} \\ \omega_a + \frac{\mathcal{N}_a}{2},\ -r - r_{ab}\end{matrix}\ ;\ 1\right)
$$

$$
\begin{aligned}
& n = 2r + \omega,\ \omega = 2r_{ab} + \omega_a + \omega_b, \\
& r_a + r_b = r + r_{ab},\ n_a = 2r_a + \omega_a,\ n_b = 2r_b + \omega_b
\end{aligned} \tag{47}
$$

Standard phase convention is to use $\phi_k = k$, $k = r, r_{ab}, r_a, r_b$. It should be noted that the final formula given by (47) for the C^{--}_{--} coefficients in (46) is independent of the specific realizations used in the derivation. The formula involves a terminating ${}_3F_2(1)$ generalized hypergeometric series. More importantly, though it is derived assuming $\mathcal{N}_a \geq 3$, $\mathcal{N}_b \geq 3$, in fact it is proved to be applicable for all $\mathcal{N}_a, \mathcal{N}_b$ [27]. The result in (47) takes much simpler form for the important situation with $\omega = n$ and this is used in many applications [13, 24].

4.3 $O_{sdST}(36) \supset O_{S_sT_s}(6) \oplus O_{dST}(30) \supset O_L(3) \otimes O_{ST}(6)$ limit

In the $O_{sdST}(36) \supset O_{S_sT_s}(6) \oplus O_{dST}(30) \supset O_L(3) \otimes O_{ST}(6)$ limit the group chain and the irrep labels for the basis states are [13],

$$
\begin{array}{l}
\left| \begin{array}{ccccccccc} U_{sdST}(36) & \supset & O_{sdST}(36) & \supset & O_{S_sT_s}(6) & \oplus & [O_{dST}(30) & \supset & \{O_d(5) \\ \{N\} & & [\omega] & & [\omega_s] & & [\omega_d] & & [\omega_1\omega_2] \end{array} \right. \\
\quad \begin{array}{cccccccc} \supset & O_L(3)\} & \otimes & O_{S_dT_d}(6)] & \supset & O_L(3) & \otimes & O_{ST}(6) \\ & L & & [\sigma_1\sigma_2\sigma_3] & & L & & [\sigma_a\sigma_b\sigma_c] \end{array} \\
\left. \begin{array}{ccccccccccc} \supset & O_L(3) & \otimes & [O_S(3) & \oplus & O_T(3)] & \supset & O_J(3) & \otimes & O_T(3) \\ & L & & S & & T & & \vec{J} = \vec{L} + \vec{S} & & T \end{array} \right\rangle
\end{array} \tag{48}
$$

The reduction of $\{N\} \to [\omega]$ and $[\omega] \to [\omega_s]\,[\omega_d]$ is simple (see (21),(44)), $N \to \omega = N, N-2, \ldots, 0$ or 1 and $\omega = 2r + \omega_s + \omega_d$; $r = 0, 1, 2, \ldots, \frac{\omega}{2}$. Other irreps in (48) follow from the results given in [13]. Table I gives the irrep labels for low-lying states with $\omega = N$ for odd-odd nuclei. It is seen that ω_d acts as a phonon number and the symmetry limit generates several extra levels for each phonon excitation as compared to the vibrational nuclei. In the $O_{sdST}(36) \supset O_{S_sT_s}(6) \oplus O_{dST}(30) \supset O_L(3) \otimes O_{ST}(6)$ limit studied are [12, 13, 15]: (i) number of $T = 0$ pairs in the ground states of even-even and odd-odd nuclei; (ii) some of the spectroscopic properties of ^{74}Rb; (iii) $B(E2)$

values for low-lying states and also for the yrast band in N=Z odd-odd nuclei with $(ST) = (01)$; (iv) two nucleon transfer strengths for exciting the ground states of N=Z even-even nuclei to states of N=Z odd-odd nuclei. Let us discuss (i) and (iii) briefly.

A quantity of physical interest which is much discussed in recent literature (see for example [14]) is number of $T = 0$ (or pn) pairs ($N_{T=0}$) in the ground states (GS's) of heavy N$\approx$Z nuclei as it will determine the role of pn pairing near the proton drip line. With T_z =(N$-$Z)/2 and the assumption of good Wigner spin-isospin $SU(4)$ will determine the ground state $O_{ST}(6) \sim SU_{ST}(4)$ irreps,

$$\begin{array}{l}
\text{Even-Even Nuclei: N=Z or N} \neq \text{Z} \\
O_{ST}(6)\ :\ [T]\, S=0, T=|T_z|, \quad N+T= \text{ even} \\
\text{Odd-odd Nuclei: N=Z} \\
O_{ST}(6)\ :\ [1]\, (ST)=(10)\oplus(01), \quad N= \text{ odd, } T=0 \text{ or } 1 \\
\text{Odd-odd Nuclei: N} \neq \text{Z} \\
O_{ST}(6)\ :\ [T,1]\ (ST)=(0,|T_z|) \text{ or } (1,|T_z|), \quad N+T= \text{ odd}
\end{array} \tag{49}$$

As in the $O(6)$ limit of IBM it is expected that GS's have $[\omega]_{O_{sdST}(36)} = [N]$ and ω_d of $O_{dST}(30)$ takes smallest possible value. These and (49) will determine uniquely, for any nucleus, the structure of GS's in the $O_{sdST}(36) \supset O_{S_sT_s}(6) \oplus O_{dST}(30) \supset O_L(3) \otimes O_{ST}(6)$ limit. The operator $\hat{N}_{T=0}$ whose expectation value $\left\langle \hat{N}_{T=0} \right\rangle$ gives number of $T = 0$ pairs is already given in (40) (there it is denoted as $\hat{n}_S$). Using the C-coefficients of Sect. 4.2 with $(\mathcal{N}_a, \mathcal{N}_b)$ = (6,30), (3,3) and (15,15) one can derive the analytical formulas [13],

$$\begin{array}{l}
\left\langle \hat{N}_{T=0} \right\rangle^{N,[T](0,T)}_{\text{even-even}} = \dfrac{(N-T)}{4(T+3)(N+16)}\ [T(N+T+4)+6(N+16)] \\
\left\langle \hat{N}_{T=0} \right\rangle^{N,[1](1,0)}_{\text{odd-odd, N=Z}} = (9N^2+162N+101)/16(N+16)\ , \\
\left\langle \hat{N}_{T=0} \right\rangle^{N,[1](0,1)}_{\text{odd-odd, N=Z}} = (7N^2+94N-101)/16(N+16) \\
\left\langle \hat{N}_{T=0} \right\rangle^{N,[T,1](S,T)}_{\text{odd-odd, N}\neq\text{Z}} \\
\quad = \delta_{S,1} + \left(3+\delta_{S,1}\dfrac{2}{T+1}\right) \dfrac{(N-T+29)(N-T-1)}{4(T+3)(N+16)} \\
\quad + \left(15+\delta_{S,1}\dfrac{2T}{T+1}\right) \dfrac{(N+T+3)(N-T-1)}{64(N+16)}, \quad S=0,1
\end{array} \tag{50}$$

Eq. (50) show that there is even-even to odd-odd staggering in the symmetry limit in the number of $T = 0$ pairs. One can go beyond the symmetry limit and carry out mixing calculations using,

$$H_{mix} = \alpha\, C_2\left(O_{S_sT_s}(6)\right) + \beta\, C_2\left(SU_{sS}(3)\right) + \gamma\, \left[C_2\left(SU_{dST}(30)\right)\right] \quad (51)$$

In (51), β/α measures the competition between $T = 0$ and $T = 1$ pairing; for s-bosons the energy for $T = 0$ pairs is $\epsilon(T_s = 0)/\alpha = 5 + 4(\beta/\alpha)$ and for $T = 1$ pairs $\epsilon(T_s = 1)/\alpha = 5$. Results of the mixing calculations are given in detail in [15, 13]. Here it suffices to state that the C-coefficients of Section 4.2 allow the calculations to be performed. It should be added that the results of $O_{sdST}(36) \supset O_{S_sT_s}(6) \oplus O_{dST}(30) \supset O_L(3) \otimes O_{ST}(6)$ limit are similar to the results obtained [14] using $U_{sd}(6) \otimes U_{ST}(6)$ limit.

For detailed spectroscopy, a quantity of great interest is $B(E2)$'s along the Yrast band. The $O_{sdST}(36) \supset O_{S_sT_s}(6) \oplus O_{dST}(30) \supset O_L(3) \otimes O_{ST}(6)$ limit gives, for odd-odd N=Z nuclei for the $(ST) = (01)$ band (note that for $A > 60$ nuclei ground states have $T = 1$), for the ratio $R(N, L) = B(E2; N, L \to N, L - 2)/B(E2; N, 2^+ \to N, 0^+)$, which is parameter free, the expression [13],

$$R(N, L) \overset{N\to\infty,\, N>>L}{\longrightarrow} \frac{5(L + 5 - 5(-1)^{L/2})}{2(L + 28)} \quad ; L = 2, 4, 6, 8, \ldots \quad (52)$$

Thus in the symmetry limit, for odd-odd nuclei, $B(E2)$'s exhibit a isospin (see Table I for s and d boson isospins for yrast levels, they alternate between 0 and 1) dictated $\Delta L = \Delta J = 4$ staggering for the yrast levels with $T = 1$. For example $R(9, L)$ = 1, 0.21, 1.06, 0.29, 1.04, 0.27, 0.95, 0.17, 0.79 for L = 2, 4, 6, 8, 10, 12, 14, 16, 18 respectively.

5. Conclusions

In this article given is a short review of the developments in the fermionic $O(8)$ and bosonic $U(36)$ symmetry schemes for heavy N=Z nuclei. Briefly discussed is also their applications. Clearly, more detailed study of $v = 2, 4$ states in the $O(8)$ model are needed and they will be useful in shell model mapping of IBM-ST model as they will give one and two d boson states respectively. It should be added that already there are attempts (see [28]) in this direction by exploiting the $SU(4)$ algebra present in the $O(8)$ and IBM-ST symmetry schemes. Similarly a detailed study of the various symmetry schemes of the $U(36)$ IBM-ST model are needed. Towards this end it is necessary to solve the problem of transformation brackets between $U(\mathcal{N} \times \mathcal{M}) \supset U(\mathcal{N}) \otimes U(\mathcal{M}) \supset O(\mathcal{N}) \otimes O(\mathcal{M}) \supset \ldots$ and $U(\mathcal{N} \times \mathcal{M}) \supset O(\mathcal{N} \times \mathcal{M}) \supset O(\mathcal{N}) \otimes O(\mathcal{M}) \supset \ldots$. Finally we add that the identification of $U(36) \supset O(36)$ limits led to the identification of a new $U(12) \supset O(12)$ symmetry limit of pnIBM and with this complete classification of symmetry chains for the pnIBM model is obtained recently; see Table II.

$O_{ST}(6)$ irrep	$[\omega_d]$	$[\omega_1\omega_2]$	L^π	$[\sigma_1\sigma_2\sigma_3]$	$[\omega_s]$
[0]	[0]	[0]	0^+	[0]	[0]
	[1]	[1]	2^+	[1]	[1]
	[2]	[2]	4^+	[0]	[0]
				[2]	[2]
			2^+	[0]	[0]
				[2]	[2]
		[0]	0^+	[2]	[2]
	[3]	[3]	6^+	[1]	[1]
				[3]	[3]
	−−	−−	−−	−−	−−
[1]	[0]	[0]	0^+	[0]	[1]
	[1]	[1]	2^+	[1]	[0]
				[1]	[2]
	[2]	[2]	4^+	[0]	[1]
				[2]	[1]
				[2]	[3]
			2^+	[0]	[1]
				[2]	[1]
				[2]	[3]
		[11]	3^+	[11]	[1]
			1^+	[11]	[1]
		[0]	0^+	[2]	[1]
				[2]	[3]
	[3]	[3]	6^+	[1]	[0]
				[1]	[2]
				[3]	[2]
				[3]	[4]

Table 1. Quantum numbers for low-lying states in the $O_{sdST}(36) \supset O_{S_sT_s}(6) \oplus O_{dST}(30) \supset O_L(3) \otimes O_{ST}(6)$ limit with $\omega = N$. Note that $(ST) = (00)$ for even-even nuclei with $[0]_{O_{ST}(6)}$ and $(ST) = (10) \oplus (01)$ for odd-odd nuclei with $[1]_{O_{ST}(6)}$.

Complete Classification of pn-IBM Symmetry Schemes

proton bosons π and neutron bosons ν

with $\ell = 0$ $(s_\pi \, , \, s_\nu)$ and $\ell = 2$ $(d_\pi \, , \, d_\nu)$

SGA: U(12)

good quantum numbers: N, L, $M_{F_z} = (N_\pi - N_\nu)/2$

$$
\begin{array}{lll}
 & & \nearrow \; U_{sd}(6) \otimes SU(2) \\
 & & \nearrow \; U_\pi(6) \otimes U_\nu(6) \\
 & U_{sd\pi\nu}(12) & \\
 & & \searrow \; U_s(2) \oplus U_d(10) \\
 & & \searrow \; O_{sd\pi\nu}(12)
\end{array}
$$

$$
\begin{array}{lllll}
 & \nearrow \; O(6) \otimes O(2) & & & \\
O(12) & & & \nearrow & U(5) \otimes SU(2) \\
 & & U(10) & & \\
 & & & \searrow & O(10) \supset O(5) \otimes O(2) \\
 & \searrow \; O(2) \oplus O(10) & & &
\end{array}
$$

U(6)'s generate U(5), SU(3) and O(6) symmetries

SU(2)'s generate F-spin, O(2)'s generate F_z

$U(6) \otimes SU(2)$ generates mixed symmetry states,

(scissors 1^+ states in rotational nuclei)

seen in ^{156}Gd, ^{196}Pt, ^{134}Ba, ^{94}Mo

$U_\pi(6) \otimes U_\nu(6) \supset SU_\pi(3) \otimes \overline{SU_\nu(3)} \rightarrow$ triaxial shapes

O(12) breaks F-spin symmetry and it is generated

by the pairs: $\left[s_\pi^\dagger s_\nu^\dagger + \sqrt{5}\beta \left(d_\pi^\dagger d_\nu^\dagger \right)^0 \right]$, $\beta = \pm 1$

V.K.B. Kota, Pramana-J. Phys. **60**, 59 (2003).

Table 2. Overview of pnIBM symmetry limits.

Acknowledgments

Thanks are due to Ms V. Sreevidya, Ms K. Madhuparna and Mr S. Sabareesan for their help in deriving some of the results presented in Sections 2 and 3. This article is dedicated to Prof. Francesco Iachello on his 60th birthday.

References

[1] D. Rudolph et al, *Phys. Rev. Lett.* **76**, 376 (1996); S.M. Vincent et al, *Phys. Lett.* **B437**, 264 (1998); R. Grzywacz et al, *Phys. Lett.* **B429**, 247 (1998); D.G. Jenkins et al, *Phys. Rev.* **C 65**, 064307 (2002).

[2] D. Dean, S. Koonin, K. Langanke, and P. Radha, *Phys. Lett.* **B399**, 1 (1997); A. Poves and G. Martinez-Pinedo, *Phys. Lett.* **B430**, 203 (1998); T. Otsuka, T. Mizusaki, and M. Honma, *J. Phys. G* **25**, 699 (1999); R. Sahu and V.K.B. Kota, *Phys. Rev. C* **66**, 024301 (2002); A. Petrovici, K.W. Schmid, and A. Faessler, *Nucl. Phys.* **A708**, 190 (2002).

[3] B.H. Flowers and S. Szpikowski, *Proc. Phys. Soc.* **84**, 673 (1964).

[4] S.C. Pang, *Nucl. Phys.* **A128**, 497 (1969).

[5] J.A. Evans, G.G. Dussel, E.E. Maqueda and R.P.J. Perazzo, *Nucl. Phys.* **A367**, 77 (1981).

[6] G.G. Dussel, E.E. Maqueda, R.P.J. Perazzo and J.A. Evans, *Nucl. Phys.* **A450**, 164 (1986).

[7] K.T. Hecht, *Nucl. Phys.* **A444**, 189 (1985).

[8] J. Dobes and S. Pittel, *Phys. Rev. C* **57**, 688 (1998).

[9] P. Van Isacker, J. Dukelsky, S. Pittel and O. Juillet, *J. Phys. G* **24**, 1261 (1998).

[10] Yu. V. Palchikov, J. Dobes and R.V. Jolos, *Phys. Rev. C* **63**, 034320 (2001).

[11] O. Juillet and S. Josse, *Eur. Phys. J.* **A8**, 291 (2000).

[12] V.K.B. Kota, *J. Phys. G* **24**, 1535 (1998).

[13] V.K.B. Kota, *Ann. Phys.* (N.Y.) **280**, 1 (2000).

[14] P. Van Isacker and D.D. Warner, *Phys. Rev. Lett.* **78**, 3266 (1997).

[15] V.K.B. Kota, *Pramana-J. Phys.* **51**, 727 (1998).

[16] E. Baldini-Neto, C.L. Lima and P. Van Isacker, *Phys. Rev. C* **65**, 064303 (2002).

[17] M. Moshinsky and C. Quesne, *Phys. Lett.* **B29**, 484 (1969).

[18] J.P. Elliott and J.A. Evans, *Phys. Lett.* **B101**, 216 (1981).

[19] F. Iachello and A. Arima, "The Interacting Boson Model", Cambridge University Press, Cambridge, 1987.

[20] B.G. Wybourne, "Symmetry Principles and Atomic Spectroscopy", Wiley Interscience, New York, 1970).

[21] R.C. Nayak and V.K.B. Kota, *Phys. Rev. C* **64**, 057303 (2001).

[22] A.O. Macchiavelli et al, *Phys. Lett.* **B480**, 1 (2000).

[23] K. Allaart, E. Boeker, G. Bonsignori, M. Savoia and Y.K. Gambhir, *Phys. Rep.* **169**, 209 (1988).

[24] V.K.B. Kota, *Ann. Phys.* (N.Y.) **265**, 101 (1998); J.E. Garcia-Ramos and P. Van Isacker, *Ann. Phys.* (N.Y.) **274**, 45 (1999).

[25] N. Ya. Vilenkin, "Special Functions and the Theory of Group representations", AMS Translations (Providence, RI, 1968), Vol.22.

[26] J. Van der Jeugt, *J. Math. Phys.* **38**, 2728 (1997); H.T. Koelink and J. Van der Jeugt, *SIAM (Soc. Ind. Appl. Math) J. Math. Anal.* **29**, 794 (1998); S. Lievens and J. Van der Jeugt, *J. Math. Phys.* **43**, 3824 (2002).

[27] V.K.B. Kota, *J. Math. Phys.* **38**, 6639 (1997).

[28] O. Juillet, P. Van Isacker and D.D. Warner, *Phys. Rev. C* **63**, 054312 (2001).

INTERACTION AND FUSION OF ELEMENTARY SYSTEMS

P. Kramer
Institut für Theoretische Physik der Universität
72076 Tübingen, Germany
peter.kramer@uni-tuebingen.de

Abstract The notion due to R.G. Newton and E.P. Wigner (1949) of an elementary system ES is sharpened to a system on a Lie group G-manifold as configuration space and the unitary irreducible representations (IR) of G as states. We study pairs of elementary systems with configuration space taken as the direct product $(\mathbf{G} \times \mathbf{G})$-manifold and with an interaction invariant under the right action of the subgroup $\text{diag}(G \times G)$. The $(\mathbf{G} \times \mathbf{G})$-manifold is split into a new external group manifold $\langle X \rangle$ transformed by left action, and a new internal group manifold $\langle x \rangle$ unchanged under $\text{diag}(G \times G)$. By use of Kronecker products we transform IR pair states to external/internal coordinates. The general concept of fusion due to de Broglie (1932-34) is expressed in the new coordinates as the limit where x goes to the identity element. For elementary Poincaré systems, the distinction between massive Mackey and covariant fields becomes crucial. The presence of a full Poincaré-manifold and of corresponding observables are illuminated by position operators. The space translation parameters of the Poincaré group are related to the relativistic position operators of Newton and Wigner. For two Dirac elementary systems of equal mass m we recover by fusion the field of Bargmann and Wigner (1948) of spin $S = 1$ which can be rewritten in terms of a massive vector field. The total mass of the Bargmann-Wigner field is shown to be minimal, $M = 2m$. Interaction schemes are sketched for pairs of Euclidean and Poincaré-manifolds and ES. By Frobenius reciprocity, the process of fusion allows for a counterpart termed scission. Scission is constructed by use of the theory of induced representations.

Keywords: Elementary systems, Lie group $\mathbf{G}$-manifolds, external/internal coordinates, de Broglie-fusion.

1. Introduction

Our aim is to sharpen the classical theory of elementary systems due to E. Schrödinger [19], V. Bargmann, E.P. Wigner [2], T.D. Newton [16], A.S. Wightman [22], and others, extend it to several elementary systems, get insight into the concept of fusion as proposed by de Broglie, compare [4], and describe

B. Gruber (ed.), Symmetries in Science XI, 291–315.

interaction schemes. We use standard tools of a Lie group G and its irreducible representations and exemplify some concepts with the familiar group $G = SU(2)$.

(i) *Groups and irreducible representations*: Our first observation is that the group **G** provides a geometric G-manifold. Once the manifold is equipped with an invariant measure, there is a scalar product and a Hilbert space $\mathcal{H}^G$ of square integrable functions on G. The set of unitary irreducible representations (IR) of **G** provide an orthogonal and complete basis of this Hilbert space, compare Appendix A. The Lie algebra $\mathcal{L}G$ provides operators on this Hilbert space with definite Hermitian properties and so allows to construct observables for the notion of a physical system associated with the group.

(ii) *Left and right group actions*: The multiplication law of a Lie group, $g_1, g_2 \in G \rightarrow g_1 g_2 \in G$, provides the left and right G-actions $g' : g \rightarrow (g')^{-1} g$ and $g' : g \rightarrow g g'$ which when applied to functions $f = f(g)$ yield operator representations. Moreover the left and right actions on G commute with one another.

(iii) *Kronecker products*: An important role in Lie group representations applied to physics is played by the Kronecker product. This product arises naturally from the direct product group $(G \times G)$ with elements (g_1, g_2). The direct product group has a subgroup $\mathrm{diag}(G \times G)$ with elements (g_1, g_1). The corresponding group/subgroup reduction of representations yields the standard Kronecker product. In examples we shall often assume for simplicity that G be simply reducible (Wigner), that is, that any IR of G appears with multiplicity 1 or 0 in the Kronecker product.

2. Elementary systems on G-manifolds.

Consider a Lie group G and D^j, one of its irreducible representations.

Elementary Systems of general type:

A first standard definition of an elementary system ES associated with (G, D^j) according to Newton and Wigner [16] is a (set of) state(s) $\psi = \psi(x)$ in a Hilbert space $\mathcal{H}^{G/G_0}$ of square integrable functions such that

(ESa): ES admits a geometric unitary action of G on $\mathcal{H}^{G/G_0}$ over a coset space G/G_0 of the form $(T_g \psi)(x) = \psi(g^{-1} x)$,

(ESb): the states of ES under G transform irreducibly with the representation D^j.

In physics one says that the system ES represents a geometric symmetry group G. The irreducibility (ESb) implies that the state space of ES does not admit a subspace invariant under G. This novel description of elementary systems by irreducibility abandons the classical atomistic notion of an indivisible piece of matter and replaces it by a notion in terms of fields, states, and groups of symmetry. Actually Newton and Wigner [16] p. 400 distinguish between an elementary system and an elementary particle. For an elementary system they require only - that it should not be useful to consider the particle as a union of other particles-.

What then are the possible elementary systems for a given group? Given a Hilbert space for the system S, assumption (Ia) allows to decompose it into orbits under G. By standard theorems, any orbit can be characterized by a stability group G_0 as a homogeneous or coset space G/G_0. Therefore the geometric G-action on functions over a fixed orbit may be traced back to a representation of G explicitly reduced w.r.t the subgroup G_0. Applying this to all irreducible representations of G we obtain all possible irreducible states according to assumption ESb. The G-action on G/G_0 is an instance of a Lie transformation group, with all the implications on the Lie theorems as discussed in Gilmore [7] pp. 87-119. We also refer to Wigner for the attempt to partially rewrite the states of a system in terms of irreducible representations [23] pp. 210-219.

Elementary Systems on G-manifolds:
There is a sharper notion of elementary systems implicit in the work of Newton and Wigner [16] and Wightman [22]. Here we **identify** an IR D^j of G with an elementary system or elementary particle on a configuration space taken as a **G**-manifold and quantum numbers j, m, k. An elementary particle is then a square integrable state on $\mathcal{H}^G$ whose domain of definition or configuration space is the full group **G**.

To keep a clear distinction between different appearances of the same groups, from now on we shall denote by boldface **G** a group manifold serving as part of configuration space for ES's, and by G a group acting on such manifolds.

Compared to the previous notion, the sharper notion of an elementary system arises as a special case where we restrict to a single orbit $G/I = G$ with I the identity subgroup of G. The orbit G/I as a manifold has maximal dimension and so we shall have to interpret the additional parameters which may appear in it, see below. For fixed G, the possible elementary systems on the **G**-manifold are in one-to-one correspondence with the set of its irreducible representations.

For the case of relativistic elementary systems with **G** the Poincaré group, there is a subtle interplay between these notions and the group parameters which we shall analyze in a later section.

3. Two elementary systems: The $(\mathrm{G} \times \mathrm{G})$-manifold and its submanifold splitting under $\mathrm{diag}(G \times G)$.

We shall adopt the sharper notion of elementarity and develop some of its consequences.
(i) We adopt the direct product manifold $(\mathbf{G} \times \mathbf{G})$ as the configuration space of two interacting elementary particles. This approach extends to more than two

elementary particles, see remark 2 at the end of this section.
(ii) For the interaction we assume as symmetry group the right G-action of the diagonal group $\mathrm{diag}(G \times G) \sim G$ on $(\mathbf{G} \times \mathbf{G})$.
(iii) We introduce on $(\mathbf{G} \times \mathbf{G})$ in **Prop 1**, **Def 2** the new external/internal coordinates $\langle X, x\rangle := \langle g_1, g_1(g_2)^{-1}\rangle$. The symmetry group acts on X from the right and keeps x. There is a second G-action on $(\mathbf{G} \times \mathbf{G})$, **Prop 3**, which commutes with the symmetry group, keeps X, and acts on x from the right. This group allows the construction of interaction operators.

First we summarize some group properties. The direct product $(G \times G)$ is the group with elements (g_1, g_2) and multiplication rule $(h_1, h_2)(g_1, g_2) := (h_1 g_1, h_2 g_2)$. The multiplication rule allows to define new left and right actions on $(G \times G)$. In the spirit of elementary systems on $\mathbf{G}$-manifolds, we assume as the configuration space of two elementary particles the direct product manifold $(\mathbf{G} \times \mathbf{G})$ with coordinates (g_1, g_2) corresponding to pairs of group elements. There are natural left and right actions $(G \times G)$ on the manifold $(\mathbf{G} \times \mathbf{G})$.

The right G-action of the diagonal subgroup on $(\mathbf{G} \times \mathbf{G})$ reads

$$(g, g) : (g_1, g_2) \rightarrow (g_1 g, g_2 g). \tag{1}$$

In our interpretation, the diagonal subgroup with this right action

on the product manifold with elements (g_1, g_2) becomes the symmetry group of two interacting elementary systems. The free particles belong to a single Kronecker product IR of $(\mathbf{G} \times \mathbf{G})$. The interaction we assume to commute not with the two independent right actions of $(G \times G)$ but only with the right action eq. 17 of the diagonal subgroup which transforms the two particle coordinates in the same way. It follows that the symmetry group in going from non-interacting to interacting particles subduces to a subgroup. The restriction of the symmetry group for an interacting system reflects the view of standard quantum theory: When we speak of a relativistic invariant interaction between relativistic particles, we consider transformations which affect the coordinates of these particles in the same way. We illuminate this view by two examples.

Example 1: **Action of** $\mathrm{diag}(\mathbf{SU(2)} \times \mathbf{SU(2)})$.

Consider a two-electron atom (without spin). Without the electron-electron interaction, the 2-electron states correspond to irreducible representations D^{l_1}, D^{l_2} of $(SU(2) \times SU(2))$. The Coulomb repulsion of the two electrons commutes only with the diagonal subgroup $\mathrm{diag}(SU(2) \times SU(2))$, with IRs determined by the total orbital angular momentum L. In example 3 below we describe simple interactions schemes for a pair of $SU(2)$-manifolds.

Our aim is to find new coordinates on the product manifold $(\mathbf{G} \times \mathbf{G})$ with simple transformation properties under the action eq. 17 of the symmetry group $\mathrm{diag}(G \times G)$. We also wish to characterize interaction operators commuting with this group.

Example 2: **Separation of dynamics in external and relative coordinates.**
Consider a quantum system of two non-relativistic particles of coordinate vectors x^1, x^2 and equal mass m and a Hamiltonian

$$H = \frac{1}{2m}((p^1)^2 + (p^2)^2) + V(x^1 - x^2). \tag{2}$$

The Hamiltonian is invariant under those (diagonal) translations which act in the same way on both particles. In the new external and internal coordinates and momenta

$$\begin{aligned} X: &= \frac{1}{2}(x^1 + x^2),\ P := p^1 + p^2, \\ x: &= x^1 - x^2,\ p := \frac{1}{2}(p^1 - p^2) \end{aligned} \tag{3}$$

the Hamiltonian becomes

$$H = \frac{1}{4m}P^2 + \frac{1}{4m}p^2 + V(x). \tag{4}$$

Clearly the Hamiltonian separates in terms of the new coordinates. The solution of the eigenvalue equations may be written as

$$\psi(X, x) = \exp(iK \cdot X)\phi(x). \tag{5}$$

The first factor dependent on X is a plane wave. It corresponds to a single IR characterized by K of the diagonal translation group, which is a symmetry group of the system.

Our aim is now to generalize the notion of external and internal coordinates to $(\mathbf{G} \times \mathbf{G})$ manifolds for non-abelian groups. Once we have achieved this goal we can take up the notion [16] of a composite elementary system by saying that it can be described exclusively in terms of the external coordinates. Moreover by the process of fusion we can remove the internal coordinates and return to a single elementary particle.

1 Prop: **Internal and external coordinates on** $(\mathbf{G} \times \mathbf{G})$.
For $(g_1, g_2) \in (G \times G)$, the group element $x := g_2 g_1^{-1}$ is easily shown from $(g_1 g)(g_2 g)^{-1} = g_1 g_2^{-1}$ to be invariant under the right action eq. 17 of diag$(G \times G)$.

2 Def: Replace the pair (g_1, g_2) of group coordinates on the $(\mathbf{G} \times \mathbf{G})$-manifold by the new pair

$$\langle X, x\rangle := \langle g_1, g_1 g_2^{-1}\rangle, \tag{6}$$

call X the external coordinate and x the internal or relative coordinate of the pair $(1, 2)$ of systems. Both (X, x) by themselves are G-manifolds, i.e. admit as manifolds the parameters of G. We use pointed brackets for the external/internal coordinates. Note that their transformation properties under left or right group actions must be inferred from their expression eq. 6 in terms of the original product coordinates (g_1, g_2)!

The right action eq. 17 of the diagonal group on the new pair of coordinates is

$$\begin{aligned}(g,g) \;\in\; & \mathrm{diag}(G \times G) \\ & (g_1, g_2) \to (g_1 g, g_2 g), \\ & \langle X, x\rangle \to \langle Xg, x\rangle \end{aligned} \tag{7}$$

So under this right action of $\mathrm{diag}(G \times G)$, the product manifold $(\mathbf{G} \times \mathbf{G})$ splits into the external manifold X, transformed by right G-action, and the internal manifold x which is left unchanged. Since we shall adopt $\mathrm{diag}(G \times G)$ as the symmetry group of the interacting system, any state of this system must correspond to a single IR of this group. From the transformation properties of $\langle X, x\rangle$ under $\mathrm{diag}(G \times G)$ we shall derive below that the external $\mathbf{G}$-submanifold X in the two-particle state carries this overall IR of the composite system.

We turn to the interaction and to the role played by the internal coordinate x. Interaction operators commuting with the right action of $\mathrm{diag}\ (G \times G)$ must essentially act on x. To find such operators we now construct complementary G-actions which
(i) commute with the right action of $diag(G \times G)$ and
(ii) keep X but transform x.

Condition (i) is fulfilled by the general left action of $(G \times G)$ on $(\mathbf{G} \times \mathbf{G})$. We use inverse group elements for left actions and so anticipate the homomorphism properties under the geometric action on states. Expressed in terms of $\langle X, x\rangle$ the left action yields

$$\begin{aligned}(g_1, g_2) &\to (l_1^{-1}, l_2^{-1})(g_1, g_2) = (l_1^{-1} g_1, l_2^{-1} g_2) \\ \langle X, x\rangle &\to \langle l_1^{-1} g_1, l_1^{-1} g_1 g_2^{-1} l_2\rangle = \langle l_1^{-1} X, l_1^{-1} x l_2\rangle. \end{aligned} \tag{8}$$

To fulfill in addition condition (ii) we must restrict the left action in eq. 7 as $(l_1^{-1}, l_2^{-1}) \to (e, l_2^{-1})$ and obtain from eq. 7 under this restricted left G-action a corresponding right G-action on x,

$$(e, l_2^{-1}) : \langle X, x\rangle \to \langle X, x l_2\rangle. \tag{9}$$

3 Prop: **Interaction operators commuting with the right action of** $\mathrm{diag}\ (G \times G)$ on $(\mathbf{G} \times \mathbf{G})$:

Any operator built from the left action of $(I \times G)$ on $(\mathbf{G} \times \mathbf{G})$ commutes with the right action of the symmetry group $diag(G \times G)$, leaves X unchanged, and transforms $x \to x l_2$ by right G-action. These interaction operators include all operators from the Lie and enveloping algebra of the right action on x.

Remark 1: It is possible to devise a second independent G-action beyond eq. 9 which keeps X and acts on the internal coordinate from the left according to $p \in G : \langle X, x \rangle \to \langle X, px \rangle$. By use of the transforms $g_1 = X, g_2 = x^{-1}X$ inverse to eq. 6 one deduces in terms of the initial coordinates on the product manifold the G-action $p : (g_1, g_2) \to (g_1, g_2(g_1^{-1}p^{-1}g_1))$. This left G-action on x can be shown to commute both with the right G-action of the symmetry group $\text{diag}(G \times G)$ and with the right G-action on x described in **Prop 3**.

Remark 2: We indicate the generalization to more than two particles: For three particles, the configuration space is taken as the direct product $(\mathbf{G} \times \mathbf{G} \times \mathbf{G})$-manifold. As the symmetry group we assume the right action of the diagonal subgroup $\text{diag}(G \times G \times G) \sim G$. After applying the coordinate transformation of the form
$(g_1, g_2, g_3) \to \langle X, x_2, x_3 \rangle := \langle g_1, g_1 g_2^{-1}, g_2 g_3^{-1} \rangle$,
the symmetry group acts exclusively on $X = g_1$. G-actions only on $x_2 = g_1 g_2^{-1}, x_3 = g_2 g_3^{-1}$ could be constructed which would provide interaction operators commuting with the symmetry group.

Remark 3: In classical relativistic phase-space mechanics there is a no-interaction theorem [5], [21] 535-545: From assuming standard Poisson relations between the Poincaré group generators and the individual particle coordinates and momenta, and from a relativistic world-line condition it follows that no interaction is allowed. A unified approach to the classical dynamics of N interacting relativistic particles is given in [1]. The authors start from a configuration space built as the semidirect product of the Poincaré group with the product of N 4-vectors. The full Poincaré group enters this configuration space. Its action as given in section IV of [1] transforms all 4-vectors in the same way.

A one-time quantum approach to the relativistic many-body problem is developed and applied in [15]. The authors introduce relative coordinates and interactions depending on them.

These approaches share certain geometric and group relations with the present approach through G-manifolds, but the treatment of classical versus quantum mechanics, the role of symmetry and dynamical groups, and the setting of time coordinates requires further comparative studies.

4. Examples of internal coordinates on $(\mathrm{G} \times \mathrm{G})$. Translation groups.

For the continuous commutative and additive translation group $T = (a)$ we get, upon writing group multiplication by addition, for the internal coordinate, $x = g_1(g_2)^{-1} \to (a_1 - a_2)$ which is the relative coordinate as expected.

This result is in line with Example 2 considered earlier.

Inhomogeneous matrix groups.

The Euclidean and Poincaré group are semidirect products of a normal translation group with a matrix group $T \times_s G$ with elements (a, g) and product $(a_1, g_1)(a_2, g_2) = (a_1 + g_1 a_2, g_1 g_2)$. For the present purpose which involves right actions it is convenient to rewrite the product form of the group elements according to

$$(a, g) = (0, g)(a', e) = (ga', g),\ (a, g)^{-1} = (-a', g^{-1}),\ a' := g^{-1}a. \tag{10}$$

We get from eq. 6 by group multiplication and use of eq. 10 the internal group coordinate

$$x = (a_1, g_1)(a_2, g_2)^{-1} = (g_1 a_1', g_1)(-a_2', g_2^{-1}) = (0, g_1)(a_1' - a_2', e)(0, g_2^{-1}) \tag{11}$$

Therefore, up to left and right homogeneous transformations, the internal group coordinate x always is the difference of two translation vectors. This result is in line with intuitive expectations about a two-particle relative coordinate.

5. Kronecker products and two-particle state decompositions on $(G \times G)$.

Kronecker product representations allow to analyze a class of two-particle states and interactions on $(G \times G)$.

The $(\mathbf{G} \times \mathbf{G})$-manifold supports a Hilbert space $\mathcal{H}^G \times \mathcal{H}^G$ with a basis given by products of IR of $\mathbf{G}$.

We first give results from representation theory, use a notation familiar from $SU(2)$ and disregard multiplicity. Group/subgroup relations imply subduction rules for the IR. Under right action eq. 17 of the diagonal subgroup diag$(G \times G)$ on $(G \times G)$, the IR $D^{j_1} \times D^{j_2}$ of $(G \times G)$ can be decomposed by Wigner coefficients into IR D^{j_3} with respect to this action,

$$\begin{aligned}
&D^{((j_1,j_2)j_3)}_{m_3 k_3}(g_1, g_2)\\
&= \sum_{m_1 k_1 m_2 k_2} \langle j_1 m_1 j_2 m_2 | j_3 m_3 \rangle D^{j_1}_{m_1 k_1}(g_1) D^{j_2}_{m_2 k_2}(g_2) \langle j_1 k_1 j_2 k_2 | j_3 k_3 \rangle,\\
&\sum_{k_1 k_2} D^{j_1}_{m_1 k_1}(g_1) D^{j_2}_{m_2 k_2}(g_2) \langle j_1 k_1 j_2 k_2 | j_3 k_3 \rangle\\
&= \sum_{m_3} \langle j_1 m_1 j_2 m_2 | j_3 m_3 \rangle D^{((j_1,j_2)j_3)}_{m_3 k_3}(g_1, g_2).
\end{aligned} \tag{12}$$

The second line arises from the first one by use of the orthogonality relations of the Wigner coefficients. Next we subduce IR from $(G \times G)$ to diag$(G \times G)$

and find

$$\left[D^{j_1}_{m_1 k_1}(g_1) D^{j_2}_{m_2 k_2}(g_2)\right]_{g_1=g_2}$$
$$= \sum_{j_3 m_3 k_3} \langle j_1 m_1 j_2 m_2 | j_3 m_3 \rangle D^{j_3}_{m_3 k_3}(g_1) \langle j_1 k_1 j_2 k_2 | j_3 k_3 \rangle \quad (13)$$

In the frame of Newton and Wigner [16] extended to two elementary particles, we now interpret the second line of eq. 11 as an IR coupled two-particle state ψ on $(\mathbf{G} \times \mathbf{G})$ since it transforms irreducibly under the left action of $\text{diag}(G \times G)$. Next we transform this state by eq. 6 to the group manifolds $\langle X, x \rangle$.

Upon insertion of $g_1 = X, g_2 = x^{-1}X$ into eq. 11 we find with eq. 12 and by use of the representation property

$$\psi(X, x, (j_1, m_1), (j_2, m_2), (j_3, k_3))$$
$$:= \sum_{k_1, k_2} D^{j_1}_{m_1 k_1}(X) D^{j_2}_{m_2 k_2}(x^{-1}X) \langle j_1 k_1 j_2 k_2 | j_3 k_3 \rangle$$
$$= \sum_{s_2 s_3} D^{j_3}_{s_3 k_3}(X) D^{j_2}_{m_2 s_2}(x^{-1}) \langle j_1 m_1 j_2 s_2 | j_3 s_3 \rangle . \quad (14)$$

In eq. 13, the IR two-particle state on $(\mathbf{G} \times \mathbf{G})$ eq. 11 with right-hand Wigner coupling is written in terms of the new external and internal $\mathbf{G}$-manifolds. Under the right action of the symmetry group $\text{diag}(G \times G)$, the total IR $D^{j_3}(X)$ is carried entirely by the external manifold X. The label k_3 of D^{j_3} is fixed and transforms under the right action of $\text{diag}(G \times G)$. The sum over the label s_3 reflects the remaining coupling between the external and internal G-manifolds. The IR $D^{j_2}(x^{-1})$ is linked to j_3, j_1 by a triangle condition $\Delta(j_1, j_2, j_3)$ between the IR.

Example 3: **Casimir two-particle interactions.**

It follows from **Prop 3** in particular that the second order Casimir operator of G acting from the left (or right, since x by itself is a G-manifold) on x, denoted by $C_2(x)$, is invariant under the symmetry group $\text{diag}(G \times G)$.

The state eq. 11, first line is an eigenstate of the second-order Casimir operators $C_2(g_1), C_2(g_2)$, and, eq. 13, an eigenstate of $C_2(X), C_2(x)$. Applying $C_2(x)$ to the expression in eq. 13 for the example $G = SU(2)$ we find the eigenvalue $j_2(j_2+1)$. For fixed representation $D^{j_3}(X)$ we can even find the spectrum of the Casimir operator $C_2(x)$: Its states of eigenvalue $j_2(j_2+1)$ are degenerate and labelled by j_1. The degeneracy ranges over all j_2 such that $\exists j_1 : \Delta(j_1, j_2, j_3)$ where Δ denotes the triangle condition of IR. In any such case we can construct a state of type eq. 13. The possible values of j_1 for fixed j_2, j_3 label the degeneracy which therefore is given by

$$\deg(j_2)|_{j_3} = j_2 + j_3 - |j_2 - j_3| . \quad (15)$$

For example for $j_3 = 1, j_2 = 1/2$ we get $\deg(1/2)_1 = 1$. -

Example 4: Hamiltonian two-particle interactions.

It is natural to extend the Casimir interaction into an interaction operator

$$\mathcal{O} = C_2(x) + V(x) \tag{16}$$

whose action on the internal group coordinates x resembles the Hamiltonian in a two-body Schroedinger equation. The eigenstates of the general operators in eq. 16 still have the term $D^{j_3}(X)$ as functions of X but will have a more general dependence on x than the state eq. 13. In particular we cannot expect a single IR of the commuting group eq. 9.

6. Fusion of two elementary systems on $(G \times G)$.

De Broglie's theory of fusion is formulated in terms of wave equations, see [4]. It has no relation to nuclear fusion. We approach the geometry of de Broglie fusion with the groups $(G \times G) > \mathrm{diag}(G \times G)$, the external/internal coordinates on $(\mathbf{G} \times \mathbf{G})$, and the Kronecker products of IR.

Consider the diagonal restriction $(g_1, g_2) \to (g_1, g_1)$. In the external/internal coordinates on $(G \times G)$ eq. 6 we have $\langle X, x\rangle \to \langle X, e\rangle$.

4 Def: Fusion of two elementary systems is the limit $x \to e$ taken for Kronecker coupled two particle states on $(\mathbf{G} \times \mathbf{G})$.

Under this fusion two elementary particles go into a single elementary particle while the internal submanifold shrinks to a point. The irreducible two-particle state eq. 13 under fusion $x \to e$ in agreement with eqs. 11, 12 reduces to $D^{j_3}_{m_3k_3}(X)$.

Fusion may be achieved by the action of an operator. We can construct a two-body operator, invariant under the right action of $\mathrm{diag}(G \times G)$, which fuses the two particles into a single one. Define the fusion integral operator $\mathcal{F}$ acting on a function $\phi(x)$ as

$$(\mathcal{F} * \phi)(x) := \int d\mu(\tilde{x})\delta(e, \tilde{x})\phi(\tilde{x}) = \phi(e). \tag{17}$$

The kernel of this operator may be rewritten by use of Appendix A in terms of characters,

$$\delta(e, \tilde{x}) = \sum_j (|j|/|G|)\chi^j(\tilde{x}). \tag{18}$$

Then clearly the application of this operator to a two-particle product state,

$$\begin{aligned} \mathcal{F} \quad &* \quad \left[D^{j_1}_{m_1k_1}(g_1)D^{j_2}_{m_2k_2}(g_2)\right] \\ &= \sum_{j_3m_3k_3} \langle j_1m_1j_2m_2|j_3m_3\rangle D^{j_3}_{m_3k_3}(X)\langle j_1k_1j_2k_2|j_3k_3\rangle, \end{aligned} \tag{19}$$

produces a sum of fused irreducible single-particle states as functions of X with corresponding algebraic probability amplitudes. An additional projector $P^{j_3 m_3 k_3}$ would reduce this sum to the components of a single irreducible state. By construction, two elementary particles can fuse only into a state j_3 contained in the Kronecker product $(j_1 \times j_2)$ of their individual IR.

7. Elementary systems on the Poincaré-manifold.

We turn to the group theory of massive relativistic fields. From the covariant fields defined on massive orbits of Minkowski momentum space we pass to irreducible Mackey representations of the Poincaré group. As was stressed by Bargmann and Wigner [3] p.212, two descriptions which are equivalent as IR may be quite different with respect to appearance, observables and possible interactions. The most important relations between different forms of IR of the Poincaré group are discussed in [17].

7.1 Mackey and covariant fields.

Following [17] I p. 112-114, we write down the construction of the induced Dirac representations for $S = 1/2$ and mass m. The full group is the Poincaré semidirect product group $P = T \times_s G$, with T the group of space-time translations and $G = Sl(2, C)$. For the little group $K = SU(2) < G$ of massive fields we choose a finite unitary IR $d(k)$.

The standard Mackey functions $f(g) = f^{Ma}(g)$ have (i) the induction property, (ii) the transformation under right action, and (iii) the scalar product

$$\begin{aligned}
(i) \qquad & k \in K < G: \; f(kg) = d(k) f(g), \\
(ii) \qquad & (T_{g'}^{Ma} f)(g) =: \tilde{f}(g) = f(gg'), \\
(iii) \qquad & (f, f') = \int d\mu(c)(f(g), f'(g))_K, \; d\mu(cg) = d\mu(g)
\end{aligned} \tag{20}$$

The operators $T_{g'}^{Ma}$ have the homomorphism property

$$T_{g'_1}^{Ma} T_{g'_2}^{Ma} = T_{(g'_1 g'_2)}^{Ma}. \tag{21}$$

The covariant fields according to [17] are constructed by using instead of K the larger subgroup $G = Sl(2, C)$ and a finite (non-unitary) representation $D(g)$ which when restricted to K is required to be unitary. The covariant field is determined from the Mackey field as

$$\phi(g) := D^{-1}(g) f(g) \tag{22}$$

The properties (i,ii,iii) eq. 19 for $\phi(g)$ are then derived (!) from eqs. 19 and 21 and read

$$\begin{aligned} (i) &\quad k \in K < G: \ \phi(kg) = \phi(g), \\ (ii) &\quad (T_{g'}^{Co}\phi)(g) =: \tilde{\phi}(g) = D(g')\phi(gg'), \\ (iii) &\quad (\phi, \phi') = \int d\mu(c)(\phi(g), D^{+}(g)D(g)\phi'(g)). \end{aligned} \tag{23}$$

The group operators $T_{g'}^{Co}$ were redefined in this equation. In [17] it is shown that the standard wave equations of covariant fields are projection equations which assure the irreducibility in the reduction of $D(g)$ under $Sl(2,C) \downarrow SU(2)$. Note from (i) eq. 22 that $\phi(g)$ becomes a function exclusively on the cosets $K\backslash G$. Compared to a function on the full group, the covariant field is truncated with respect to the parameters of K. The new transformation operators T_g^{Co} (ii) resemble the group action on a tensor field. The typical fields used for elementary particles are of covariant form. We particularize these equations to the Poincaré group, the semidirect product $\mathcal{T} \times_s Sl(2,C)$ of the space-time translation group $\mathcal{T}$ and the group $Sl(2,C)$ with elements $(a,g), a \in \mathcal{T},\ g \in Sl(2,C)$. The Lorentz representation is $L(g),\ g \in Sl(2,C)$. For the covariant form we extend $SU(2)$ to $Sl(2,C)$ and choose $D(g)$ as the Dirac representation. The cosets G/H can be labelled by $(0,c),\ c \in Sl(2,C)/SU(2)$ which is a manifold of dimension 3, see Appendix B. A crucial observation is that these cosets are in one-to-one correspondence to the points of the mass shell: We can reinterpret each coset c by a vector k pointing to a fixed point of the mass shell in Minkowski momentum space, $k = L(c^{-1})\ \overset{0}{k},\ \overset{0}{k} = (mc/h, 0, 0, 0)$, with L being the Lorentz representation of c^{-1}. The orbit has the structure $SL(2,C)/SU(2)$ and the representative point $L(e)\ \overset{0}{k} = (mc/h, 0, 0, 0)$ and so applies to any massive IR of the Poincaré group with little group $SU(2)$. The covariant relativistic fields apparently represent elementary systems of general type. They appear as vector-valued functions on the massive part of Minkowski momentum space.

7.2 From covariant to Mackey fields.

The covariant massive fields were derived in eq. 22 from the Mackey fields. Now we wish, starting from the irreducible covariant fields as input, to reconstruct the Mackey fields. For the converse to eq. 22 we find

5 Prop: Mackey fields from covariant fields.

Given a covariant relativistic field with the properties $(i - iii)$ eq. 22 and a representation $D(g)$ of K, define a Mackey field and its transformation by

$$
\begin{aligned}
f^{Ma}(g) \quad &:= \quad D(g)f^{Co}(g) \\
(ii) \qquad & (T^{Ma}_{g'} f^{Ma})(g) := f^{Ma}(gg') \\
(iii) \qquad & (f^{Ma}, f'^{Ma}) = \int d\mu(c)(f^{Ma}(g), f'^{Ma}(g))_K, \qquad (24)
\end{aligned}
$$

Then we can prove from eq. 22 $(i - iii)$ the Mackey left transformation rule eq. 19,

$$
\begin{aligned}
(i) \qquad & k \in K < G: \ f^{Ma}(kg) = D(kg)f^{Co}(kg) \\
= \quad & D(k)f^{Ma}(g). \qquad (25)
\end{aligned}
$$

7.3 Obstruction of Poincaré-manifolds by covariant fields.

We turn to the interpretation of the fields in terms of G-manifolds. The Mackey fields are proper induced representations of the Poincaré group, they are defined on any element $(a, g),\ a \in T,\ g \in Sl(2, C)$ of the 10-parameter Poincaré G-manifold.

This is not the case for covariant fields. Due to the property (i) eq. 22, these fields are functions on the cosets $SU(2)\backslash Sl(2, C)$. They are truncated w.r.t a three-parametric left subgroup $SU(2)$ of $Sl(2, C)$ and therefore depend on 3 real parameters out of 6 parameters of $Sl(2, C)$. By this property they are adapted and can be interpreted on massive orbits or mass shells in Minkowski momentum space of type $Sl(2, C)/SU(2)$. The group G acts on the covariant fields as a Lie transformation group.

The covariant fields then fail to meet the crucial assumptions of being full irreducible representations of the Poincaré group made in section 2 for elementary systems on this G-manifold. This failure also obstructs the approach of section 3 with left and right actions and external/internal coordinates. By the reconstruction of eq. 23, we can recover the Mackey fields only up to this subgroup $SU(2)$ in the form $f^{Ma}(c)$.

A way to overcome these obstructions is found as follows:

6 Prop: All 10 Poincaré group parameters can be found for reconstructed Mackey fields.

In the reconstructed fields, we reintroduce the missing $SU(2)$ subgroup by a right action of $Sl(2, C)$. For $c \in SU(2)\backslash Sl(2, C)$ from a set of coset generators and $g \in Sl(2, C)$ we have a unique pair u', c' with $cg = u'c',\ u' \in SU(2), c' \in SU(2)\backslash Sl(2, C)$.

Conversely, since the products $u'c', u' \in SU(2), c' \in SU(2)\backslash Sl(2, C)$ cover $Sl(2, C)$, for any product $u'c',\ u' \in SU(2), c' \in SU(2)\backslash Sl(2, C)$ we can find a pair c, g such that $u'c' = cg$, as is shown in the lemma of Appendix

B. Applied to the reconstructed Mackey field we get

$$(T_g^{Ma} f^{Ma})(c) = f^{Ma}(cg) = f^{Ma}(u'c') = D(u') f^{Ma}(c'). \qquad (26)$$

Written out in more detail, the elements of the Poincaré group are $(a, g), a \in T, g \in Sl(2, C)$. The reconstructed Mackey fields depend on cosets $(0, c), c \in SU(2)\backslash Sl(2, C)$. Acting on $(0, c)$ from the right first with the translation (a, e) and then with the element $g \in Sl(2, C)$ we find with the Lorentz action $k = L(c^{-1}) \overset{0}{k}$

$$\begin{aligned} (0, c)(a, e)(0, g) &= (ca, u'c'), \\ (T_{(a,e)(0,g')} f^{Ma})(0, c) &= f^{Ma}(c\, a, u'c') \\ &= \exp(ik^{\mu} a_{\mu}) D(u') f^{Ma}(0, c') \end{aligned} \qquad (27)$$

The Mackey fields by eq. 26 are now augmented by 4 translation parameters a_{μ} and 3 parameters of $u' \in SU(2)$. Together with the 3 coset parameters of c' we get all 10 Poincaré parameters. Part of the translation parameters $(c\, a, e)$ in eq. 26 will be interpreted in section 8 as position variables.

7 Prop : Right actions on covariant and reconstructed Mackey fields.

For the right group action on covariant and Mackey fields we have the following results: Applied from the right, $Sl(2, C)$ and its Lie generators act on covariant fields as a Lie transformation group and form a Lie algebra $\mathcal{L}Sl(2, C)$. Here the Lie generators involves only the " group parameters of the coset $SU(2)\backslash Sl(2, C)$.

The group action of $Sl(2, C)$ applied from the right to (reconstructed) Mackey fields and its Lie generators can be found by standard methods from the structure function as first-order differential operators in the group parameters. The generators form the Lie algebra $\mathcal{L}Sl(2, C)$ and in general involve all the 10 parameters of the group, see [7].

Example 5: Generators of $SU(2)$ on coset spaces.

We illustrate the difference between actions on $G/I = G$ and on $G/H, H \neq I$ with $G = SU(2)$. We adopt the standard Euler angle parameters α, β, γ from Edmonds [6]. The generators for the left action of $SU(2)$ on $SU(2)$ from [6] p. 64 take the form

$$\begin{aligned} L_1 &= -i \left[-\cos(\alpha) \cot(\beta) \frac{\partial}{\partial \alpha} - \sin(\alpha) \frac{\partial}{\partial \beta} + \frac{\cos(\alpha)}{\sin(\beta)} \frac{\partial}{\partial \gamma} \right], \\ L_2 &= -i \left[-\sin(\alpha) \cot(\beta) \frac{\partial}{\partial \alpha} + \cos(\alpha) \frac{\partial}{\partial \beta} + \frac{\sin(\alpha)}{\sin(\beta)} \frac{\partial}{\partial \gamma} \right], \\ L_3 &= -i \frac{\partial}{\partial \alpha}. \end{aligned} \qquad (28)$$

and fulfill the standard commutation rules of $\mathcal{L}SU(2)$. For the coset space $SU(2)/U(1)$ of dimension 2, the coset representatives depend only on the two

Euler angles α, β. The generators for the left action of $SU(2)$ on the coset space $SU(2)/U(1)$ are obtained from the expressions of eq. 27 by dropping the derivatives with respect to the parameter γ. With the replacement $(\alpha, \beta) \to (\phi, \theta)$, the new generators take the form familiar from the action of $SU(2)$ on R^3 with the same orbit structure, [6] p. 11. This is another Lie group action, and so the commutator relations must be the same as before.

8. Relativistic position operators and coordinates.

Position operators for relativistic fields were considered first by Newton and Wigner in [16]. Wightman [22] discusses localizability and Mackey imprimitivity in terms of representations of the Euclidean subgroup of the Poincaré group. Lorente and Roman [13] analyze position and spin operators for a wide variety of relativistic fields. With the present analysis we wish to demonstrate the presence of the Poincaré manifold beyond Minkowski space and the distinction between Mackey and covariant analysis.

8.1 Position operators for relativistic Mackey fields.

First we consider in line with [16] the Klein-Gordon field $\psi(k)$. The essential condition on the localized eigenstates of the position operators is eq. (6a) in [16]. It demands that the scalar product between a localized state and its Euclidean translate vanishes. The localized Klein-Gordon state is then found as

$$\psi^{loc}(k^l, k^0) = \exp(ik^r a_r)\,\sqrt{k^0},\ (k^0)^2 = \sum_j (k^j)^2 + (\frac{mc}{h})^2. \tag{29}$$

It is an eigenstate of the Newton-Wigner position operators [16] defined as

$$Q^{Wi,r} := (-i)\left[\frac{\partial}{\partial k_r} + \frac{k^r}{2(k^0)^2}\right],\ r = 1, 2, 3. \tag{30}$$

So far we did not distinguish between covariant and Mackey fields. For the Klein-Gordon field this distinction collapses and so our previous analysis is correct. Since we have to consider differential operators with respect to k^μ, in general we expect a distinction between position operators acting on Mackey and on covariant fields respectively. It turns out that the position operators are simpler for Mackey fields.

The massive fields depend on four momenta, but these in turn are constrained to a mass shell. We shall often write the fields as functions $\psi(k_l, k_0)$ of four momentum variables but keep in mind that $k^0 = k^0(k^1, k^2, k^3)$.

Our present approach to Mackey position operators runs as follows: Consider a shift $k^s = -k_s \to k^s(v^s) = k^s + v^s$ of the momenta, applied to a massive Mackey field. Since the mass shell is conserved, the component $k^0 = \sqrt{\sum_j (k^j)^2 + (mc/h)^2}$ must change according to $k^0 \to k^0(v^s) =$

$\sqrt{\sum_j (k^j)^2 + 2k^s v^s + (k^s)^2 + (mc/h)^2}$. The standard generator on the transformed fields depending on the momentum shift is obtained by differentiation of the fields first w.r.t k^s, k^0, then w.r.t $v^s = -v_s$ and subsequent evaluation at $v^s = 0$,

$$Q^{Ma,s} = (-i)(\frac{\partial}{\partial k_s})^{\text{tot}} = (-i)\left[\frac{\partial}{\partial k_s} + \frac{k^s}{k_0}\frac{\partial}{\partial k_0}\right], \tag{31}$$

By the upper index tot we indicate that we mean the total differentiation of the field w.r.t k_s. This expression differs from the one given by Newton and Wigner. For the general form of a massive Mackey field

$$\psi = \exp(i\sum_j k^j a_j)\psi^0(k_l, k_0) \tag{32}$$

this operator yields

$$Q^{Ma,s}\psi = \exp(i\sum_j k^j a_j)\left[a^s - i\frac{\partial}{\partial k_s} - i\frac{k^s}{k_0}\frac{\partial}{\partial k_0}\right]\psi^0(k_l, k_0). \tag{33}$$

In particular if $\psi^{0,\text{loc}}(k_l, k_0) = (k^0)^\alpha$, we find

$$Q^{Ma,s}\exp(i\sum_j k^j a_j)(k^0)^\alpha = \left[a^s - i\alpha\frac{k^s}{(k_0)^2}\right]\exp(i\sum_j k^j a_j)(k^0)^\alpha. \tag{34}$$

so that this field is an eigenfunction of the new position operator. Applied to a localized Mackey field with $\alpha = 1/2$, the Mackey position operators eq. 34 give the same eigenvalues as the ones of Newton and Wigner.

We find that the eigenvalues of the position operators yield essentially the three parameters of the Euclidean translation group. This is to be expected since the Mackey fields live on the Poincaré G-manifold and therefore must depend on all parameters of this group. The momenta stand for the cosets c of $Sl(2,C)/SU(2)$ and therefore are functions of the group parameters.

8.2 Position operators for fields with spin.

Position operators for relativistic fields with spin were analyzed by Newton and Wigner [16]. The fields are constructed by the technique of Bargmann and Wigner [3]. The Bargmann-Wigner field equations are copies of the Dirac equation. Since the Dirac field and equation is constructed in covariant form, the Bargmann-Wigner fields too belong to this form.

Newton and Wigner consider massive relativistic Bargmann-Wigner fields with spin s. The scalar product of these fields from [3] p. 89 and [16] eq.(15)

is

$$\langle \psi^{\rm Wi}, \psi^{\rm Wi} \rangle$$
$$= \int d^3k \, k_0^{-2\alpha} \, (\psi^{\rm Ma}(k)^+ \psi^{\rm Ma}(k)),$$
$$k_0 = \sqrt{\sum (k_j)^2 + (mc/h)^2}, \tag{35}$$

with $2\alpha = 2s + 1$ and k_μ on the mass shell. The position operators of [16] are essentially given by momentum derivatives as in eq. 31. Modifications arise first of all by making these derivatives hermitian with respect to the scalar product eq. 34. Newton and Wigner construct completely symmetric states of spin s in terms of pure spin variables. To these they apply projection operators which assure that the projected fields obey the general Bargmann-Wigner equations given in [3]. In [16] eq. (22) they construct the position operators and in [16] eq. (21b) their eigenstates in momentum space. These Wigner position operators refer to covariant fields and their transformation properties.

We have argued above that the covariant fields should be replaced by Mackey fields in order to describe them as genuine states on the full Poincaré G-manifold. It would be of interest to find corresponding Mackey position operators which refer to the Mackey form of fields with spin s and their transformation properties, and to relate them to the Wigner type. This detailed elaboration and comparison cannot be given here.

The main point for considering position operators in relation to the presence of Poincaré G-manifolds can be seen without detailed analysis: Clearly the Mackey fields become explicit functions of the Euclidean translation parameters of the Poincareé G-manifold. Moreover the construction of position operators shows that these translation parameters are associated with observables of the fields on the full Poincaré G-manifold. More work is necessary in order to display the three $SU(2)$ parameters which are missing in the covariant fields.

9. From Dirac fields to Bargmann-Wigner fields by fusion.

We now wish to derive the Bargmann-Wigner field equations from the construction of certain Kronecker products of Dirac spinors. As a result we shall reinterpret the mass of the Bargmann-Wigner field. We shall use covariant fields in order to keep in line with the standard references. The Dirac IR of the Poincaré group we denote by $D^{(m,1/2)}$. Each Dirac spinor will be described by a four-component object depending on momentum variables. The Dirac equation we take as

$$(P\chi)(k) = (h/mc) \sum_\mu \gamma^\mu k_\mu \chi(k) = \chi(k) \tag{36}$$

According to Schrödinger [19] and to [17] this equation is a projection on a single IR of $SU(2)$.

The spinor $\chi(k)$ for general k on the mass shell is determined from its value $\chi(\overset{0}{k})$ by the Dirac and Lorentz representations $D(c), L(c)$ of an element $c \in Sl(2,C)$. We still have the freedom of choosing the two first Dirac spinor components as functions of $\overset{0}{k}$. Consider two such spinors and the Kronecker product of their transformations. These Kronecker products fall into the symmetric and the antisymmetric part. Construct three basis functions for the symmetric Kronecker product according to the spin coupling rules $\frac{1}{2} \times \frac{1}{2} \rightarrow 1$,

$$\begin{aligned} M=1 &: \psi(1,2,\overset{0}{k}) = \chi_1(1,\overset{0}{k})\chi_1(2,\overset{0}{k}), \\ M=0 &: \psi(1,2,\overset{0}{k}) = \sqrt{\frac{1}{2}}(\chi_1(1,\overset{0}{k})\chi_2(2,\overset{0}{k}) + \chi_2(1,\overset{0}{k})\chi_1(2,\overset{0}{k})), \\ M=-1 &: \psi(1,2,\overset{0}{k}) = \chi_2(1,\overset{0}{k})\chi_2(2,\overset{0}{k}). \end{aligned} \tag{37}$$

Application of $D(\kappa)$ to the products yields 10 symmetric basis functions of a symmetric bilinear field with components

$$\psi_{ij}(1,2,k) = \frac{1}{2}\left[\chi_i(1,k)\chi_j(2,k) + \chi_j(2,k)\chi_i(1,k)\right] \tag{38}$$

To write the spinors as explicit functions on the Poincaré group with elements (a,g) we apply to both the same translation operator and introduce

$$(T_{(a,e)}\chi)(s,k) = \exp i(k^{\mu}a_{\mu})\chi(s,k), \;\; s=1,2. \tag{39}$$

Since the Poincaré group elements for both spinors are identified, (a_1, g_1) $=(a_2, g_2)$, we speak of the fusion of the representations. We could even identify the components of the two spinors at $k = \overset{0}{k}$ without getting a trivial result in eq. 36.

Both spinors obey the same Dirac projection equation. Application to ψ from eq. 38 of the matrix operator $P = P(k)$ from the left yields

$$\begin{aligned} (P\psi)_{i,j} &= \frac{1}{2}\left[(P\chi(1,k))_i\chi(2,k)_j + \chi(1,k)_i(P(\chi(2,k))_j\right] \\ &= \psi_{i,j}. \end{aligned} \tag{40}$$

Similarly one finds under application from the right

$$(\psi P^T)_{i,j} = \psi_{i,j}. \tag{41}$$

These two equations are the Bargmann-Wigner equations for ψ_{ij}.

8 Prop: Symmetric Bargmann-Wigner field from two Dirac fields. The Bargmann-Wigner field and equations can be obtained by the fusion of two symmetrized Dirac spinor representations $D^{(m,1/2)}$. It can be shown that the Bargmann-Wigner equations imply that the field ψ belongs to the single irreducible massive vector field and representation $D^{(2m,1)}$.

The mass $M = 2m$ of the Bargmann-Wigner field can be directly verified from the transformation properties of the product bases eq. 36 under translations. One finds in particular

$$\begin{aligned}(T^{Co}_{(0,a')}(T_{(a,g)}\chi))(k) &= \exp i(k^\mu a'_\mu + (L(g^{-1})k)^\mu a_\mu)\chi(L(g^{-1})k),\\ (T^{Co}_{(0,a')}\psi)(k) &= \exp i\, 2(k^\mu a'_\mu + (L(g^{-1})k)^\mu a_\mu)\psi(L(g^{-1})k)\end{aligned} \tag{42}$$

The second line results from the bilinear form eq. 38 of ψ and implies that the wave vector for the Bargmann-Wigner field is $K_\mu = 2k_\mu$ and hence the mass is $M = 2m$. This is the minimal mass in the Kronecker product of two Dirac representations of equal mass according to Joos [9]. Again we emphasize [17] that a single IR $D^{(M,S)}$ of the Poincaré group has different realizations in terms of fields and field equations.

By a detailed analysis of the Dirac representation of $Sl(2, C)$ and its relation to the Lorentz group, the Bargmann-Wigner field can be related to a massive vector field. This analysis is given for example in [14] pp. 30-32 in a non-standard metric, compare also [11]. In this way one finally arrives at a relativistic massive vector field which resembles the construction by fusion of the photon suggested by de Broglie.

There exists an antisymmetric counterpart of the Bargmann-Wigner construction eqs. 39, 41. The corresponding antisymmetric field corresponds to $D^{(2m,0)}$. This antisymmetric field can be rewritten in terms of a scalar Klein-Gordon field as shown in [11].

10. Elementary systems in interaction.

We sketch two interacting elementary systems were use can be made of the external and internal submanifolds $\langle X, x\rangle$.

10.1 Euclidean invariant interactions.

Consider two elementary systems on the Euclidean group $G = ISO(3, R)$. The submanifolds $\langle X, x\rangle$ are as given in eq. 11. Let $C(X), C(x)$ denote the Casimir operators corresponding to the squares of the momenta and hence related to the kinetic energy. An interaction operator of type

$$H = C(x)/(2\mu) + V(|a'_1 - a'_2|^2) \tag{43}$$

commutes with the symmetry group diag$(G \times G)$ given as a function of group elements and generators of the dynamical group. But, since the second potential part V is not expressible by the right G-action of the dynamical group G, we can no longer expect a single IR of G w.r.t x. Since V and hence H are invariant under the rotation subgroup $SO(3,R)$, the eigenvalues of H belong to fixed IR of the rotation group.

10.2 Interacting Dirac spinor fields.

As an example of a non-trivial interacting relativistic system we consider two Dirac fields in interaction. We denote the single-particle observables of these particles by upper indices $1,2$. The Poincaré group parameters for mass, translation parameters and momenta are then $(m_i\ a^i, g_i)\ i = 1,2$ with $m_1 = m_2$. We have a correspondence between $g \in Sl(2,C)$, coset generators c and momenta by $k(g) = L(g^{-1}) \overset{0}{k} = L(c^{-1}) \overset{0}{k}$ where $g = uc, u \in SU(2)$. We pass from two group elements (a^1, g_1),(a^2, g_2) with the help of the relations of eqs. 10, 11 and Appendix C to the external and internal group elements $(A, G) = (a^1, g_1)$, $(a, g) = (0, g_1)(a'^1 - a'^2, g_2^{-1})$. We have extracted from (a, g) two pure Lorentz transformation to the left and to the right. Since the interaction operator will be covariant under Lorentz transformations, these factor does not affect the interaction and hence may be dropped.

Reduced masses and some relevant scalar products are treated in Appendix C.

For the part depending on the external G-manifold we may choose for example the IR $D^{(M',0)}$, a Klein-Gordon field $\psi(K)$ of spin $S = 0$ and obeying a Klein-Gordon equation

$$(\sum_\nu K^\nu K_\nu - (M'c/h)^2)\psi(K) = 0. \tag{44}$$

The alternative IR $D^{(M',1)}$ for the external G-manifold requires extra consideration on the spin. Note that the mass M' in both cases obeys $M' \geq 2m$. We can use a plane wave solution for the external manifold.

The interaction may be constructed as an operator acting on the internal coordinates (a, g) or rather $(a, k(g))$. We can take this interaction operator as an extension of the free Dirac operator w.r.t. (a, g) with reduced mass μ, see Appendix C, minimally coupled to an electromagnetic field in the form

$$(h/\mu c) \sum_\nu \gamma^\nu (k_\nu - \frac{e\mu}{h} A_\nu(k))\chi(k) = \chi(k). \tag{45}$$

This equation is Lorentz invariant and therefore allows to drop the two Lorentz transformations appearing in the internal coordinates (a, g). It is customary to consider eq. 45 in position rather than momentum space. The transform to position space involves a Fourier transform. The mass shell condition in position space is expressed by the usual second order differential equation. A new feature in position space is that the distinction between the forward and backward light cone in momentum space has no local counterpart [17] p.146-148. This leads to the doubling of orbits and hence IR of the Poincaré group, and to the appearance of positive and negative energy solutions. Moreover the equation may be simplified by a Foldy-Wouthuysen transformation of the type reviewed in [20].

Eq. 45, with an appropriate choice of the vector potential A_μ, serves as the basis for a description of the relativistic hydrogen atom or of positronium. We refer to the classical treatment by Pauli [18] pp. 155-166 and to [12]. From field theory [8] one can construct improved versions of the vector potential.

11. Scission of an elementary system.

Fusion was analyzed in section 3 in terms of reduction of representations for the group/subgroup scheme $(G \times G) > \mathrm{diag}(G \times G)$. Frobenius reciprocity for the converse induction of representations in the present interpretation leads to the notion of scission of elementary systems of general type. Again we stress that scission here is a concept of group theory and has no relation to nuclear scission. By the process of scission, a single ES is decomposed into a pair of ES. The IR of the three ES are again linked by the same triangle condition as in fusion.

Frobenius reciprocity applies to the group/subgroup pair under consideration. It relates two reductions of reducible representations: (1) An IR of the group under restriction to the subgroup can be reduced into IRs of the subgroup. (2) A representation of the group, induced from an IR of the subgroup, can be reduced into the IRs of the full group. Both processes are reciprocal to one another.

We now induce from an irreducible representation of $H = \mathrm{diag}(G \times G)$ a reducible representation of $(G \times G)$. The coset generators of $H < (G \times G)$ and their transformation under $(G \times G)$ we choose as

$$\begin{aligned} \tilde{c}_i &:= (e, c_i),\ c_i \in G, \\ (g_1, g_2)(e, c_i) &= (e, c_j)(g_1, g_1), \\ g_2 c_i &= c_j g_1 \end{aligned} \tag{46}$$

The elements c_i, c_j run over G. We choose the irreducible representation $D^{j_3}_{m_3 k_3}(g_1)$ of $\mathrm{diag}(G \times G)$ and get the induced representation

$$D^{j_3 \uparrow}_{l m_3, i k_3}(g_1, g_2) = \delta(c_l^{-1} g_2 c_i, g_1) D^{j_3}_{m_3 k_3}(g_1) \tag{47}$$

This representation is reducible into the IR of $(G \times G)$. The explicit reduction may be obtained by use of standard Young operators and reads

$$D^{j_3 \uparrow}_{lm_3, ik_3}(g_1, g_2)$$
$$= \sum_{j_1 m_1 k_1 j_2 m_2 k_2} C(lj_3m_3, j_1m_1j_2m_2)$$
$$D^{j_1}_{m_1k_1}(g_1)D^{j_2}_{m_2k_2}(g_2)C(j_1k_1j_2k_2, ij_3k_3),$$
$$C(lj_3m_3, j_1m_1j_2m_2) = \left[\frac{|j_1||j_2|}{|G||j_3|}\right]^{1/2} \sum_{\mu_2} D^{j_2}_{\mu_2 m_2}(c_l^{-1})\langle j_1m_1j_2\mu_2|j_3m_3\rangle,$$
$$C(j_1k_1j_2k_2, ij_3k_3) = \left[\frac{|j_1||j_2|}{|G||j_3|}\right]^{1/2} \sum_{\kappa_2} D^{j_2}_{k_2\kappa_2}(c_i)\langle j_1k_1j_2\kappa_2|j_3k_3\rangle. \quad (48)$$

The reduction coefficients C obey the two unitary relations

$$\sum_{j_1 m_1 j_2 m_2} C(lj_3m_3, j_1m_1j_2m_2)C(j_1m_1j_2m_2, ij_3k_3) = \delta(c_l, c_i)\delta_{m_3,k_3},$$
$$\sum_{ij_3m_3} C(j_1'k_1j_2'k_2, ij_3m_3)C(ij_3m_3, j_1m_1j_2m_2) = \delta_{j_1'j_1}\delta_{j_2'j_2}\delta_{k_1m_1}\delta_{k_2m_2}. \quad (49)$$

Frobenius reciprocity assures that the multiplicities obey

$$m(j_3 \uparrow (j_1 \times j_2)) = m((j_1 \times j_2) \downarrow j_3) \quad (50)$$

The multiplicities take the values $1, 0$ due to the assumed simple reducibility.

9 Prop : **Scission of elementary systems**. Now we give the scission interpretation of the induced representation in terms of elementary particles. The induced representation eq. 47 is a two-particle state. Particle 1 is in an elementary irreducible state D^{j_3} whereas particle 2 is in a reducible state located at $g_2 = c_l g_1 c_i^{-1}$. The reduction process yields the possible irreducible product states $D^{j_1} \times D^{j_2}$ of two elementary particles. For given j_3, the possible IRs of the pair are restricted through reciprocity such that j_3 must be contained in the Kronecker product $(j_1 \times j_2)$. We can construct a scission operator $\mathcal{S}$. It would create from the irreducible state D^{j_3} of the first particle a second particle in a reducible state localized at g_2. Next we could project with an operator $P^{j_1 m_1 k_1 j_2 m_2 k_2}$ the reducible two-particle state into an irreducible one. Products of the coefficients C would yield the corresponding amplitudes.

The product of the scission and projection operators are the counterparts of the fusion and projection operator introduced in the previous section. The rules for the possible triples (j_1, j_2, j_3) in scission by Frobenius reciprocity eq. 50 are identical to the rules in fusion.

12. Conclusion.

A novel approach to the geometry and dynamics of elementary systems is based on direct products of G-manifolds. External and internal coordinates split under the right action of the diagonal symmetry group G. Various directions of future research open up and deserve investigation in detail.

13. Appendix.

13.1 A: Orthogonality and completeness of unitary representations.

We use the notation familiar from $SU(2)$. $|G|$ denotes the invariant integral over G, $|j|$ the dimension of the IR D^j.

The orthogonality and completeness relations of the IR of a compact Lie group G, [3] pp. 134-158, read

$$\begin{aligned}\int D^{j_1}_{k_1 m_1}(g^{-1}) D^{j_2}_{m_2 k_2}(g) d\mu(g) &= \delta(j_1 j_2)\delta(m_1 m_2)\delta(k_1 k_2)\frac{|G|}{|j_1|},\\ \sum_{jmk} \frac{|j|}{|G|} D^j_{mk}(g') D^j_{km}(g^{-1}) &= \delta(g', g) \end{aligned} \tag{51}$$

13.2 B: Parameters, cosets and multiplication rules for $Sl(2, C)$.

We start from a double coset decomposition of a general element $g \in Sl(2, C)$ into left and right factors $(u_1, u_2) \in SU(2)$ and a double coset representative $q(\lambda)$,

$$\begin{aligned} g &= u_1 q(\lambda) u_2,\\ q(\lambda) &= \begin{bmatrix} \lambda & 0 \\ 0 & \lambda^{-1} \end{bmatrix}, \ \lambda > 0 \end{aligned} \tag{52}$$

This equation is easily shown by diagonalizing the positive definite hermitian matrices gg^+ and g^+g respectively and taking the square roots of the eigenvalues. For (u_1, u_2) we use a factorization into three Euler angles $(\alpha_i, \beta_i, \gamma_i)$. The Lorentz transformation corresponding to $q(\lambda)$ is a boost between the coordinates x^0, x^3 so that the rotation $u_3(\gamma_1)$ commutes with it. Therefore we must eliminate one parameter say α_2 in the product eq. 51 to get a unique parametrization in terms of 6 real Lorentz parameters $(\alpha_1, \beta_1, \gamma_1, \lambda, \beta_2, \gamma_2)$.

The coset representatives of $SU(2)\backslash Sl(2, C)$ are now given by

$$c = c(\lambda, \beta_2, \gamma_2) = q(\lambda) u_2(\beta_2, \gamma_2). \tag{53}$$

Given the set $\langle c \rangle$ of coset generators from eq. 53, the right action of g has a coset decomposition,

$$cg = u'c'. \tag{54}$$

For **Prop 5** in section 5 we wish to prove the stronger converse
Lemma: For any pair $u' \in SU(2), c' \in SU(2)\backslash Sl(2,C)$ and any $c \in SU(2)\backslash Sl(2,C)$ there exists a unique $g \in Sl(2,C)$ such that eq. 54 holds.
Proof: For any chosen fixed u', c', we choose c and $g := (c)^{-1}u'c'$. Clearly eq. 54 is fulfilled.

13.3 C: Observables in the relativistic 2-body system.

Consider two relativistic particles with rest masses m_1, m_2, momenta k^1, k^2, and translation parameters a^1, a^2. We introduce new external and internal momenta and translation parameters by

$$\begin{aligned} \begin{bmatrix} K \\ k \end{bmatrix} &= \begin{bmatrix} 1 & 1 \\ \frac{m_2}{M} & -\frac{m_1}{M} \end{bmatrix} \begin{bmatrix} k^1 \\ k^2 \end{bmatrix}, \begin{bmatrix} k^1 \\ k^2 \end{bmatrix} = \begin{bmatrix} \frac{m_1}{M} & 1 \\ \frac{m_2}{M} & -1 \end{bmatrix} \begin{bmatrix} K \\ k \end{bmatrix}, \\ \begin{bmatrix} A \\ a \end{bmatrix} &= \begin{bmatrix} \frac{m_1}{M} & \frac{m_2}{M} \\ 1 & -1 \end{bmatrix} \begin{bmatrix} a^1 \\ a^2 \end{bmatrix}, \begin{bmatrix} a^1 \\ a^2 \end{bmatrix} = \begin{bmatrix} 1 & \frac{m_2}{M} \\ 1 & -\frac{m_1}{M} \end{bmatrix} \begin{bmatrix} A \\ a \end{bmatrix}. \end{aligned} \tag{55}$$

We define the masses $M := m_1 + m_2,\ \mu := m_1 m_2/(m_1 + m_2)$. But the mass M must be distinguished from the mass M' of the external state. This mass M' is determined by the square of the momentum observable $K = k^1 + k^2$ as

$$\langle K, K \rangle = (\frac{M'c}{h})^2 \geq (\frac{Mc}{h})^2. \tag{56}$$

One can verify the following relations between relativistic scalar products of 4-vectors referring to the two particles on one hand and to the external and internal vectors on the other:

$$\begin{aligned} \frac{1}{2m_1}\langle k^1, k^1 \rangle + \frac{1}{2m_2}\langle k^2, k^2 \rangle &= \frac{1}{2M}\langle K, K \rangle + \frac{1}{2\mu}\langle k, k \rangle, \\ \langle k^1, a^1 \rangle + \langle k^2, a^2 \rangle &= \langle K, A \rangle + \langle k, a \rangle \end{aligned} \tag{57}$$

The first row relates the mass Casimir operators and the second one the expressions appearing in the unitary space-time translation operators.

A similar relation allows to rewrite the sum of the generators of Lorentz transformations for the two particles as a sum of generators of Lorentz transformations with respect to the external and internal coordinates.

References

[1] Balachandran, A.P., Marmo, G., Mukunda, N., Nilsson, J.S., Simoni, A., Sudarshan, E.C.G., and Zaccaria, F. (1984) *Unified geometrical approach to relativistic particle dynamics*, J. Math. Phys. **25**, 167-176

[2] Bargmann, V. and Wigner, E.P. (1948), *Group Theoretical Discussion of Relativistic Wave Equations*, Proc. Nat. Acad. Sci. USA **34**, 211-223

[3] Barut, A.O. and Raczka, R. (1977), *Theory of group representations and applications*, Polish Scientific Publishers, Warszawa

[4] Borne, Th., Lochak, G., and Stumpf, H. (2001), *Nonperturbative Quantum Field Theory and the Structure of Matter*, Kluwer, Dordrecht

[5] Currie, D.G., Jordan, T.F., and Sudarshan, E.C.G. (1963), *Relativistic invariance and Hamiltonian theories of interacting particles* Rev. Mod. Phys. **35**, 350-375

[6] Edmonds, A.R. (1957), *Angular Momentum in,Quantum Mechanics*, Princeton University Press, Princeton

[7] Gilmore, R. (1974), *Lie groups, Lie algebras, and some of their applications*, Wiley, New York

[8] Itzykson, C. and Zuber, J. B. (1980), *Quantum field theory*, McGraw-Hill, New York

[9] Joos, H., *Zur Darstellungstheorie der inhomogenen Lorentzgruppe als Grundlage quantenmechanischer Kinematik*, Fortschritte der Physik **10**, 65-146

[10] Kramer, P., John, G., and Schenzle, D. (1981), *Group theory and the interaction of composite nucleon systems*, Vieweg, Braunschweig

[11] Kramer, P. (2003), *Dirac fields and their Kronecker products*, in preparation

[12] Landau, L.D., Lifschitz, E.M., and Pitajevski, L.P., *Relativistic Quantum Theory*,

[13] Lorente, M. and Roman P. (1974), *General expressions for the position and spin operators of relativistic fields*, J. Math. Phys. **15**, 70-74

[14] Lurie, D. (1968), *Particles and Fields*, Wiley, New York

[15] Moshinsky, M. and Riquer, V. (2003), *The relativistic many-body problem and application to bottomonium*, J. Phys. A: Math. Gen. **36**, 2163-2174

[16] Newton, T.D. and Wigner, E.P. (1949), *Localized states for elementary systems*, Rev. Mod. Phys. **21**, 400-406

[17] Niederer, U.H. and O'Raifeartaigh, L. (1974), *Realizations of the unitary representations of the inhomogeneous space-time groups I,II*, Fortschritte der Physik **22**, I: 111-129, II: 131-157

[18] Pauli W. (1958), *Die allgemeinen Prinzipien der Wellenmechanik*, in: Handbuch der Physik V Teil 1, Springer, Berlin

[19] Schrödinger, E. (1930), *quoted in [16]*, Berl. Ber. 418, (1930); 63, (1931)

[20] Silenko, A.J. (2003), *Foldy-Wouthuysen transformation for relativistic particles in external fields*, J. Math. Phys. **44**, 2952-2966

[21] Sudarshan, E.C.G. and Mukunda, N. (1974), *Classical dynamics: a modern perspective* Wiley, New York

[22] Wightman, A.S. (1962), *On the localizability of quantum mechanical systems*, Rev. Mod. Phys. **34**, 845-872

[23] Wigner, E.P., *Group theory and its applications to the quantum theory of atomic spectra*, Academic Press, New York

PROPAGATION IN CROSSED ELECTRIC AND MAGNETIC FIELDS: THE QUANTUM SOURCE APPROACH

T. Kramer
Physik-Department T30,
Technische Universität München
James-Franck-Straße, 85747 Garching, Germany
tkramer@ph.tum.de

C. Bracher
Department of Physics and Atmospheric Science,
Dalhousie University
Halifax, N.S. B3H 3J5, Canada
cbracher@fizz.phys.dal.ca

Abstract The propagation of electrons in static and uniform electromagnetic fields is a standard topic of classical electrodynamics. The Hamilton function is given by a quadratic polynomial in the positions and momenta. The corresponding quantum-mechanical problem has been analyzed in great detail and the eigenfunctions and time evolution operators are well-known. Surprisingly, the energy-dependent counterpart of the time-evolution operator, the Green function, is not easily accessible. However in many situations one is interested in the evolution of a system that started with emitted particles that carry a specific energy. In the following we present a suitable approach to study this type of matter waves arising from a localized region in space. Two applications are discussed, the photodetachment current in external fields and the quantum Hall effect in a fermionic electron gas.

Keywords: Green function. Electric and magnetic fields. Hall effect.

1. Introduction

In quantum mechanics, static and uniform electric and magnetic fields are represented by a quadratic Hamiltonian (i.e. a second order polynomial in the canonical coordinates r_i and momenta p_i). Moshinsky and Winternitz carried

B. Gruber (ed.), Symmetries in Science XI, 317–353.

out a detailed group-theoretical analysis of this class of Hamiltonians and their eigenfunctions [40]. Quadratic Hamiltonians are connected to linear canonical transformations [39] and therefore a general phase-space approach gives valuable information for their classification in different dimensions. Nieto used the Moyal phase-space representation to develop a general method for deriving the time-evolution operator for quadratic Hamiltonians [41]. However, its energy-dependent counterpart, the Green function, withstands such a systematic analysis and is not available in analytic form for many physical relevant potentials. Also other methods, like the Feynman path-integral approach, are not capable to derive the exact energy-dependent Green function.

In experiments, the energy of particles is often easier controlled than the time of travel. Under these circumstances, the energy-dependent Green function is relevant for the description of the system. Monochromatic particle sources arise in numerous applications of quantum mechanics. In accelerator physics sources located far away from the scattering region lead to boundary conditions in the form of incoming plane waves. In this contribution, we study the behaviour of spatially localized electron sources in perpendicular, homogeneous electric and magnetic fields. Our discussion is based on the framework of quantum source theory, a variant of the scattering formalism that is especially suited to describe scattering events restricted to a region of finite volume [9, 11, 36, 34]. The idea of quantum sources was first promoted by Schwinger [47] but has not found widespread attention. Therefore, we briefly introduce the concept and some basic results derived from it, and stress its connection to the propagator approach to quantum mechanics [18]. In fact, stationary elastic scattering at pointlike sources is fully described in terms of the energy Green function. Several fundamental properties of the quantum system, like the scattering wave function, current density distribution, cross section, and local density of states, immediately follow from this functional. Here, we explore in detail isotropic point sources in crossed external static fields both in two- and three-dimensional configuration space. The results are in agreement with experimental findings in a recent photodetachment experiment, and offer an alternative viewpoint towards the anomalous Hall effect observed in low-dimensional semiconductor devices.

2. Elastic scattering and quantum sources

In preparation for our later discussion, we illustrate the quantum source formalism using potential scattering in external static fields as an example. We assume that the potential $V(\mathbf{r})$ represents a localized disturbance, and that the charged quantum particles otherwise move in the external electromagnetic potentials $\mathbf{A}(\mathbf{r})$, $\Phi(\mathbf{r})$:

$$\mathbf{H} = \mathbf{H}_0 + V(\mathbf{r}) = \frac{1}{2m}\left(\mathbf{p} - q\mathbf{A}(\mathbf{r})/c\right)^2 + q\Phi(\mathbf{r}) + V(\mathbf{r}) \,. \tag{1}$$

The scattering solutions $\psi(\mathbf{r})$, which are eigenfunctions of the Hamiltonian $\mathbf{H}$ with energy E, then usually are decomposed into two parts, $\psi(\mathbf{r}) = \psi_{\text{in}}(\mathbf{r}) + \psi_{\text{sc}}(\mathbf{r})$, where the incident wave is a solution for the unperturbed system $\mathbf{H}_0$: $\mathbf{H}_0\psi_{\text{in}}(\mathbf{r}) = E\psi_{\text{in}}(\mathbf{r})$, whereas the remainder $\psi_{\text{sc}}(\mathbf{r})$ represents the scattering wave. By comparison with (1), we find that $\psi_{\text{sc}}(\mathbf{r})$ obeys:

$$\left[E - \mathbf{H}_0 - V(\mathbf{r})\right]\psi_{\text{sc}}(\mathbf{r}) = V(\mathbf{r})\psi_{\text{in}}(\mathbf{r}) \,. \tag{2}$$

Hence, $\psi_{\text{sc}}(\mathbf{r})$ is a solution to the inhomogeneous Schrödinger equation of the full Hamiltonian $\mathbf{H} = \mathbf{H}_0 + V(\mathbf{r})$, where we denote the right-hand side in (2) as the source term $\sigma(\mathbf{r})$:

$$\sigma(\mathbf{r}) := V(\mathbf{r})\,\psi_{\text{in}}(\mathbf{r}) \,. \tag{3}$$

Equation (2) suggests the following physical interpretation: The incoming wave $\psi_{\text{in}}(\mathbf{r})$, via the perturbation $V(\mathbf{r})$, feeds particles into the scattering wave $\psi_{\text{sc}}(\mathbf{r})$ that is governed by the Hamiltonian $\mathbf{H}$. While not commonly seen in standard quantum theory, inhomogeneous partial differential equations are familiar from other branches of physics, the heat conduction equation and Maxwell's equations being examples for the introduction of sources. For these problems, a sophisticated mathematical framework in the form of Green functions has been developed. Accordingly, we introduce the energy-dependent Green function $G(\mathbf{r}, \mathbf{r}'; E)$ for the Hamiltonian $\mathbf{H}$ defined via [16]

$$\left[E - \mathbf{H}_0 - V(\mathbf{r})\right] G(\mathbf{r}, \mathbf{r}'; E) = \delta(\mathbf{r} - \mathbf{r}'). \tag{4}$$

Formally, the solution to equation (2) is given by a convolution integral comprising the source term and the Green function

$$\psi_{\text{sc}}(\mathbf{r}) = \int \mathrm{d}^3r'\, G(\mathbf{r}, \mathbf{r}'; E)\sigma(\mathbf{r}'). \tag{5}$$

We infer that the scattering wave generated by the source $\sigma(\mathbf{r})$ allows for an interpretation as the linear superposition of "fundamental" waves $G(\mathbf{r}, \mathbf{r}'; E)$ emitted from point sources $C\delta(\mathbf{r} - \mathbf{r}')$ located at $\mathbf{r}'$.

2.1 Connection to the propagator

In the continuous spectrum of $\mathbf{H}$, the Green function is not uniquely defined. Depending on our choice for $G(\mathbf{r}, \mathbf{r}'; E)$, we obtain a set of wave functions $\psi_{\text{sc}}(\mathbf{r})$ that differ only by eigenfunctions $\psi_{\text{hom}}(\mathbf{r})$ of $\mathbf{H}$. This ambiguity is resolved by the demand that $G(\mathbf{r}, \mathbf{r}'; E)$ presents a retarded solution that enforces outgoing-wave behaviour of the scattering wave $\psi_{\rm sc}(\mathbf{r})$ at large distances from the source. The representation of $G(\mathbf{r},\mathbf{r}';E)$ as a Laplace transform of the quantum propagator $K(\mathbf{r},t|\mathbf{r}',t_0)$ [18] guarantees the proper choice of boundary conditions for the Green function [16]:

$$G(\mathbf{r},\mathbf{r}';E) = -\frac{\mathrm{i}}{\hbar}\lim_{\eta\to 0_+}\int_0^\infty \mathrm{d}T\ \mathrm{e}^{\mathrm{i}ET/\hbar-\eta T/\hbar}\ K(\mathbf{r},T|\mathbf{r}',0). \tag{6}$$

where $K(\mathbf{r},t|\mathbf{r}',t_0)$ denotes the coordinate space representation of the time evolution operator $\mathbf{U}(t,t_0)$

$$K(\mathbf{r},t|\mathbf{r}',t_0) := \langle\mathbf{r}|\mathbf{U}(t,t_0)|\mathbf{r}'\rangle. \tag{7}$$

Since for a conservative system, $\mathbf{U}(T,0) = \exp(-\mathrm{i}\mathbf{H}T/\hbar)$ holds, we may formally integrate (6) to obtain:

$$G(\mathbf{r},\mathbf{r}';E) = \lim_{\eta\to 0_+}\left\langle\mathbf{r}\left|\frac{1}{E-\mathbf{H}+\mathrm{i}\eta}\right|\mathbf{r}'\right\rangle. \tag{8}$$

Therefore, the Green function represents the resolvent assigned to the Hamiltonian $\mathbf{H}$ in configuration space. According to (8), $G(\mathbf{r},\mathbf{r}';E)$ indeed acts as an "inverse" to the operator $E-\mathbf{H}$.

At least in principle, knowledge of the full Green function permits the exact evaluation of the scattering wave $\psi_{\rm sc}(\mathbf{r})$ (5). In general, however, $G(\mathbf{r},\mathbf{r}';E)$ is not available in analytic form. In the favourable situation that we can find an expression for the Green function $G_0(\mathbf{r},\mathbf{r}';E)$ associated with the unperturbed Hamiltonian $\mathbf{H}_0$, $G(\mathbf{r},\mathbf{r}';E)$ formally may be expanded into a series via the Dyson equation:

$$\frac{1}{E-\mathbf{H}} = \frac{1}{E-\mathbf{H}_0}\left[1+\mathbf{V}\frac{1}{E-\mathbf{H}}\right]. \tag{9}$$

Replacing $G(\mathbf{r}\mathbf{r}';E)$ by $G_0(\mathbf{r},\mathbf{r}';E)$, i. e., neglect of the rescattering terms that involve the perturbation $\mathbf{V}$, is equivalent to the leading order of perturbation theory in the conventional scattering formalism, which we will endorse in the following.

While the quantum propagators $K(\mathbf{r},t|\mathbf{r}',t_0)$ are tabulated for a fairly extensive set of potentials [21, 32], few energy Green functions are available in closed form for problems in more than one spatial dimension. This list includes free particles in two and three dimensions, as well as the Coulomb problem [26, 27], uniformly accelerated quantum motion [10, 13, 19, 49], the isotropic harmonic oscillator [3], motion in a homogeneous magnetic field [15, 20], and parallel electric and magnetic fields [17, 35], all in three-dimensional configuration space.

2.2 Currents generated by quantum sources

A first obvious quantity of interest are the currents associated with the scattering wave $\psi_{sc}(\mathbf{r})$ generated by the source $\sigma(\mathbf{r})$ (5). The current density distribution $\mathbf{j}(\mathbf{r})$ is defined in the usual fashion via

$$\mathbf{j}(\mathbf{r}) = \frac{\hbar}{m}\Im[\psi_{sc}(\mathbf{r})^* \boldsymbol{\nabla}\psi_{sc}(\mathbf{r})] - \frac{q\mathbf{A}(\mathbf{r})}{m}|\psi_{sc}(\mathbf{r})|^2, \tag{10}$$

where $\mathbf{A}(\mathbf{r})$ denotes the vector potential, and displays the spatial distribution of the quanta in the scattering wave, i. e., is directly related to the differential cross section of the scattering process. Integration of $\mathbf{j}(\mathbf{r})$ over a surface enclosing $\sigma(\mathbf{r})$ will yield the total current $J(E)$ emitted by the source which, in turn, is a measure of the total scattering rate. For a concise expression, we first note that the inhomogeneous Schrödinger equation (2) gives rise to a modified equation of continuity. Instead of $\nabla \cdot \mathbf{j}(\mathbf{r}) = 0$, valid for a stationary system in the absence of sources, we now find:

$$\boldsymbol{\nabla} \cdot \mathbf{j}(\mathbf{r}) = -\frac{2}{\hbar}\Im\left[\sigma(\mathbf{r})^* \psi_{sc}(\mathbf{r})\right]. \tag{11}$$

Thus, the inhomogeneity $\sigma(\mathbf{r})$ acts also as a source for the particle current $\mathbf{j}(\mathbf{r})$. Since the current is conserved outside the source region, the surface integral may be replaced by a spatial integration over $\boldsymbol{\nabla} \cdot \mathbf{j}(\mathbf{r})$ covering the source volume, and upon insertion of (5) for the scattering wave in (11), we obtain a bilinear expression for the total flux $J(E)$:

$$J(E) = -\frac{2}{\hbar}\Im\left[\int \mathrm{d}^3r \int \mathrm{d}^3r' \sigma(\mathbf{r})^* G(\mathbf{r},\mathbf{r}';E)\sigma(\mathbf{r}')\right]. \tag{12}$$

For pointlike sources $\sigma(\mathbf{r}) = C\delta(\mathbf{r}-\mathbf{r}')$, where C is a measure for the source strength, the calculation of the scattering currents simplifies considerably. In this case, the scattering wave is a multiple of the Green function, $\psi_{sc}(\mathbf{r}) = CG(\mathbf{r},\mathbf{r}';E)$, and the pattern of the current distribution follows from (10). Point sources yield a particularly simple expression for the total cross section:

$$J(\mathbf{r}';E) = -\frac{2|C|^2}{\hbar}\lim_{\mathbf{r}\to\mathbf{r}'}\Im\left\{G(\mathbf{r},\mathbf{r}';E)\right\}. \tag{13}$$

In passing, we remark that for $\mathbf{r} \to \mathbf{r}'$, the Green function $G(\mathbf{r},\mathbf{r}';E)$ diverges in more than one spatial dimension, while its imaginary part remains well-defined in the limit and is proportional to the scattering rate. The statement (13) is closely related to the optical theorem of conventional scattering theory [45].

For reference, we list the total currents emitted by a free-particle point source of unit strength ($C = 1$) in one-, two-, and three-di - men - sio - nal configuration space. For $E > 0$, they read:

$$J^{(1D)}_{\rm free}(E) = \frac{2m}{\hbar^3 k}\,, \qquad J^{(2D)}_{\rm free}(E) = \frac{m}{\hbar^3}\,, \qquad J^{(3D)}_{\rm free}(E) = \frac{mk}{\pi\hbar^3}\,, \tag{14}$$

where $k = \sqrt{2mE}/\hbar$ denotes the wave number of the particles.

2.3 Density of States

Somewhat surprisingly, the local density of states (LDOS) $n(\mathbf{r}'; E)$ of the quantum system, i. e., the accumulated density $|\psi(\mathbf{r})|^2$ of the eigenfunctions of the system with energy E, evaluated at $\mathbf{r}'$, is, apart from a prefactor, identical to the total current $J(\mathbf{r}'; E)$ emitted by a point source located at the same position. Formally, this equivalence is established from equation (8) by setting $\mathbf{r} = \mathbf{r}'$ and using the distribution relation [24] $\Im[(z + i\eta)^{-1}] = -\pi \,\mathrm{sgn}\,\eta \cdot \delta(z)$ that holds in the limit $\eta \to 0$. Thus, we formally obtain:

$$\Im[G(\mathbf{r}'\mathbf{r}'; E)] = -\pi \left\langle \mathbf{r}' \left|\delta(E - \mathbf{H})\right| \mathbf{r}' \right\rangle \,. \tag{15}$$

The right-hand side of this relation formally contains the spatial representation of the density of states operator $\delta(E - \mathbf{H})$, and we conclude that the LDOS is linked to the imaginary part of the Green function. In conjunction with (13), this implies that the density of states is directly proportional to the previously defined total current $J(\mathbf{r}'; E)$:

$$J(\mathbf{r}'; E) = \frac{2\pi}{\hbar}|C|^2 n(\mathbf{r}'; E). \tag{16}$$

We note here that the localized eigenstates of $\mathbf{H}$ that make up the discrete spectrum of the Hamiltonian are irrelevant for the imaginary part of the Green function, and thus do not contribute to the current. It is the unbounded solutions in the continuous spectrum of $\mathbf{H}$ that are entirely responsible for the outgoing wave character of $G(\mathbf{r}, \mathbf{r}'; E)$ and constitute $J(\mathbf{r}'; E)$. From (16), we conclude that a non-vanishing density of states is therefore directly related to an extended flow pattern in position space. We will show examples of this behaviour in Section 5.6.

2.4 Construction of the Green function

In the following, we will briefly discuss some techniques that are useful in establishing the energy Green function for simple systems.

Matching of solutions For one-dimensional systems, the inhomogeneous Schrödinger equation (2) reduces to a linear ordinary differential equation of second order, and if it exists at all, the Green function $G(z, z'; \epsilon)$ is always available as a product of solutions $\psi_{\epsilon,<}(z_<)$ and $\psi_{\epsilon,>}(z_>)$ that behave regularly in the sectors $z \to \pm\infty$, respectively, and are matched at the source position $z = z'$:

$$G(z, z'; \epsilon) = \frac{2m}{\hbar^2} \frac{\psi_{\epsilon,<}(z_<)\psi_{\epsilon,>}(z_>)}{W[\psi_{\epsilon,<}, \psi_{\epsilon,>}]} \, . \tag{17}$$

Here we introduced the symbols $z_< = \min(z, z')$ and $z_> = \max(z, z')$, and $W[\psi_{\epsilon,<}, \psi_{\epsilon,>}]$ denotes the Wronskian of the two solutions. The basic example for this strategy is the free particle problem in one spatial dimension, where for $E > 0$ ($E < 0$) $\psi_{\epsilon,<}(z_<)$ and $\psi_{\epsilon,>}(z_>)$ are outgoing (evanescent) waves in either direction:

$$G^{(1D)}_{\text{free}}(z, z'; E) = \begin{cases} -\frac{m}{\hbar^2 \kappa} \exp(-\kappa|z - z'|) & (E < 0) \, , \\ \\ -\frac{im}{\hbar^2 k} \exp(ik|z - z'|) & (E > 0) \, . \end{cases} \tag{18}$$

Here, $k = \sqrt{2mE}/\hbar$, $\kappa = \sqrt{-2mE}/\hbar$ again denote the wave number of the particle. In passing, we remark that the few higher-dimensional Green functions that can be found in analytical form usually have been determined by formal extensions of this technique [3, 10, 26, 49].

Eigenfunction expansion In section 2.1, we found a formal position space representation for the retarded energy Green function as a special resolvent of the Hamiltonian operator $\mathbf{H}$. Expanding (8) into a complete set of eigenstates $|\psi_\epsilon\rangle$ of $\mathbf{H}$ (where $\mathbf{H}|\psi_\epsilon\rangle = \epsilon|\psi_\epsilon\rangle$), we may alternatively express the Green function as a sum over all properly normalized eigenfunctions $\psi_\epsilon(\mathbf{r}) = \langle \mathbf{r}|\psi_\epsilon\rangle$ of the system:

$$G(\mathbf{r}, \mathbf{r}'; E) = \lim_{\eta \to 0_+} \sum_{|\psi_\epsilon\rangle} \frac{\psi_\epsilon(\mathbf{r}')^* \psi_\epsilon(\mathbf{r})}{E - \epsilon + i\eta} \, . \tag{19}$$

We will encounter an example of this decomposition in Section 3.3.1.

Complex convolution

We noted before that the quantum propagator $K(\mathbf{r}, t|\mathbf{r}', t_0)$ is generally more easily available than the Green function $G(\mathbf{r}, \mathbf{r}'; E)$ [21, 32]. In part, this situation is the consequence of the simple composition properties of $K(\mathbf{r}, t|\mathbf{r}', t_0)$. Assume that the (conservative) Hamiltonian $\mathbf{H}$ of the system can be written as the sum of commuting, lower-dimensional parts: $\mathbf{H} = \mathbf{H}_1 + \mathbf{H}_2$, where $\mathbf{H}_1\mathbf{H}_2 = \mathbf{H}_2\mathbf{H}_1$. Then, the corresponding evolution operator obeys $\mathbf{U}(T, 0) = e^{i\mathbf{H}T/\hbar} = e^{i\mathbf{H}_1 T/\hbar} e^{i\mathbf{H}_2 T/\hbar} = \mathbf{U}_1(T, 0)\mathbf{U}_2(T, 0)$, and thus reduces to a product of its constituents. This property is transferred to their spatial representations, the propagators:

$$K(\mathbf{r}, t|\mathbf{r}', t_0) = K_1(\mathbf{r}_1, t|\mathbf{r}'_1, t_0) K_2(\mathbf{r}_2, t|\mathbf{r}'_2, t_0) \, , \tag{20}$$

where $\mathbf{r}_1, \mathbf{r}_2$ are the projections of $\mathbf{r}$ onto the subspaces of $\mathbf{H}_1$, $\mathbf{H}_2$.

Unfortunately, the simple multiplicative property (20) does not extend to the energy domain. We may, however, exploit it to derive a corresponding statement for the Green functions $G_\nu(\mathbf{r}_\nu, \mathbf{r}'_\nu; E)$. Equation (6) shows that the evolution operator $\mathbf{U}(T, 0)$ and the resolvent operator $[E - \mathbf{H}]^{-1}$, which yields the Green function in configuration space (8), are linked through a Laplace transform. Since the image of a product of Laplace transforms is represented by the convolution integral of the images of the factors, we obtain $[E-\mathbf{H}]^{-1} = \frac{\mathrm{i}}{2\pi} \int \mathrm{d}E' \, [E' - \mathbf{H}_1]^{-1} [E - E' - \mathbf{H}_2]^{-1}$, or, in position representation:

$$G(\mathbf{r}, \mathbf{r}'; E) = \frac{\mathrm{i}}{2\pi} \int_{-\infty}^{\infty} \mathrm{d}E' \, G_1(\mathbf{r}_1, \mathbf{r}'_1; E') \, G_2(\mathbf{r}_2, \mathbf{r}'_2; E - E'). \tag{21}$$

Due to the generally complicated form of the energy-dependent Green function, the practical value of this relation is limited. We will, however, present an application in the following section.

3. Matter waves in crossed electric and magnetic fields

As our example of interest, we study quantum sources of charged particles in an environment of homogeneous, static electric and magnetic fields $\boldsymbol{\mathcal{E}}$, $\boldsymbol{\mathcal{B}}$. The Hamiltonian $\mathbf{H}$ in this case may be written as the sum of commuting parts $\mathbf{H}_\parallel(r_\parallel)$ and $\mathbf{H}_\perp(\mathbf{r}_\perp)$ in the sense of Section 2.4 and reads:

$$\mathbf{H}_\parallel(r_\parallel) = -\frac{\hbar^2}{2m} \frac{\partial^2}{\partial r_\parallel^2} - q r_\parallel \mathcal{E}_\parallel \,, \tag{22}$$

$$\mathbf{H}_\perp(\mathbf{r}_\perp) = \frac{1}{2m} \left[-\mathrm{i}\hbar \boldsymbol{\nabla}_\perp - \frac{q}{2} (\boldsymbol{\mathcal{B}} \times \mathbf{r}_\perp) \right]^2 - q \mathbf{r}_\perp \cdot \boldsymbol{\mathcal{E}}_\perp \,. \tag{23}$$

Here, the subscripts in $r_\parallel$ and $\mathbf{r}_\perp$ denote the directions parallel and perpendicular to the magnetic field $\boldsymbol{\mathcal{B}}$, respectively. We chose the electromagnetic potentials $\mathbf{A}(\mathbf{r}) = \frac{1}{2}(\boldsymbol{\mathcal{B}} \times \mathbf{r})$ and $\Phi(\mathbf{r}) = -\mathbf{r} \cdot \boldsymbol{\mathcal{E}}$ as particular gauge in $\mathbf{H} = \mathbf{H}_\parallel(r_\parallel) + \mathbf{H}_\perp(\mathbf{r}_\perp)$ (22), (23), but all observable quantities, e. g. the currents $\mathbf{j}(\mathbf{r})$ and $J(E)$ (10), (12), are invariant under gauge transformations, unlike the propagator and Green function. (We note that under a change of gauge field $\chi(\mathbf{r}, t)$, the source term (3) must be correspondingly modified.)

3.1 The quantum propagator

According to the composition properties outlined in Section 2.4, the propagator for a particle in homogeneous fields $\boldsymbol{\mathcal{E}}$, $\boldsymbol{\mathcal{B}}$ at arbitrary angle may be written as a product (20):

$$K_{\boldsymbol{\mathcal{E}},\boldsymbol{\mathcal{B}}}(\mathbf{r}, t | \mathbf{r}', 0) = K_\parallel(r_\parallel, t | r'_\parallel, 0) K_\perp(\mathbf{r}_\perp, t | \mathbf{r}'_\perp, 0) \,, \tag{24}$$

Here, $K_\parallel(r_\parallel, t | r'_\parallel, 0)$ is the propagator for a uniformly accelerated particle in one dimension that has been known from the beginnings of quantum mechanics

[10, 18, 31, 32]:

$$K_{\|}(r_{\|},t|r'_{\|},0) = \sqrt{\frac{m}{2\pi \mathrm{i}\hbar t}} \quad (25)$$
$$\times \exp\left\{\frac{\mathrm{i}}{\hbar}\left[\frac{m}{2t}(r_{\|}-r'_{\|})^2 + \frac{qt}{2}\mathcal{E}_{\|}(r_{\|}+r'_{\|}) - \frac{q^2\mathcal{E}_{\|}^2 t^3}{24m}\right]\right\} ,$$

whereas $K_{\perp}(\mathbf{r}_{\perp},t|\mathbf{r}'_{\perp},0)$, the propagator for a charge moving in two dimensions subject to perpendicular electric and magnetic fields, was unraveled much later [25, 41, 14]:

$$K_{\perp}(\mathbf{r}_{\perp},t|\mathbf{r}'_{\perp},0) = \frac{m\omega_L}{2\pi\mathrm{i}\hbar\sin(\omega_L t)} \exp\left\{\frac{\mathrm{i}}{\hbar}\left[\frac{q}{2}\boldsymbol{\mathcal{B}}\cdot(\mathbf{r}'_{\perp}\times\mathbf{r}_{\perp})\right.\right. \quad (26)$$
$$+\frac{qt}{2}\boldsymbol{\mathcal{E}}_{\perp}\cdot(\mathbf{r}_{\perp}+\mathbf{r}'_{\perp}) + 2m\mathbf{v}_D\cdot(\mathbf{r}_{\perp}-\mathbf{r}'_{\perp}) - \frac{m}{2}v_D^2 t$$
$$\left.\left.+\frac{m\omega_L}{2}\cot(\omega_L t)\left[(\mathbf{r}_{\perp}-\mathbf{r}'_{\perp})^2 - 4\mathbf{v}_D\cdot(\mathbf{r}_{\perp}-\mathbf{r}'_{\perp})t + v_D^2 t^2\right]\right]\right\} .$$

In this expression, we introduced the Larmor frequency ω_L and the drift velocity $\mathbf{v}_D$:

$$\omega_L = q\mathcal{B}/(2m) , \qquad \mathbf{v}_D = (\boldsymbol{\mathcal{E}}\times\boldsymbol{\mathcal{B}})/\mathcal{B}^2 . \quad (27)$$

Interestingly, Schwinger derived the relativistic propagator much earlier [46], but apparently no simple transition to the non-relativistic case exists.

The equations (25) and (26) reveal a simple symmetry property of the propagator under translations of the coordinate origin:

$$K_{\boldsymbol{\mathcal{E}},\boldsymbol{\mathcal{B}}}(\mathbf{r},t|\mathbf{r}',0) = \exp\left\{\frac{\mathrm{i}q}{\hbar}\left[\frac{1}{2}\boldsymbol{\mathcal{B}}\cdot(\mathbf{r}'\times\mathbf{r}) + \mathbf{r}'\cdot\boldsymbol{\mathcal{E}}t\right]\right\} K_{\boldsymbol{\mathcal{E}},\boldsymbol{\mathcal{B}}}(\mathbf{r}-\mathbf{r}',t|\mathbf{o},0) . \quad (28)$$

(Alternatively, the symmetry (28) may be viewed as the effect of a gauge transform that shifts the origin of the potentials [34].) Of primary interest in our study is the corresponding energy Green function $G_{\boldsymbol{\mathcal{E}},\boldsymbol{\mathcal{B}}}(\mathbf{r},\mathbf{r}';E)$ (4). Its analytical expression is unknown, however. Hence, the results displayed in subsequent figures were obtained by numerical evaluation of the integral representation (6). Fortunately, the symmetry (28) of the propagator, together with (6), permits to predict the behaviour of $G_{\boldsymbol{\mathcal{E}},\boldsymbol{\mathcal{B}}}(\mathbf{r},\mathbf{r}';E)$ under coordinate transformations:

$$G_{\boldsymbol{\mathcal{E}},\boldsymbol{\mathcal{B}}}(\mathbf{r},\mathbf{r}';E) = \exp\left\{\frac{\mathrm{i}q}{2\hbar}\boldsymbol{\mathcal{B}}\cdot(\mathbf{r}'\times\mathbf{r})\right\} G_{\boldsymbol{\mathcal{E}},\boldsymbol{\mathcal{B}}}(\mathbf{r}-\mathbf{r}',\mathbf{o};E+q\mathbf{r}'\cdot\boldsymbol{\mathcal{E}}) . \quad (29)$$

This relation immediately extends any result obtained for $\mathbf{r}' = \mathbf{o}$ to general source locations $\mathbf{r}' \neq \mathbf{o}$.

3.2 Purely magnetic field

As a simple example, we first inquire into the dynamics of a charge in a purely magnetic field ($\boldsymbol{\mathcal{E}} = \mathbf{o}$). In two spatial dimensions, the relevant propagator $K_\perp(\mathbf{r}_\perp, t|\mathbf{o}, 0)$ (26) reduces to:

$$K_{\boldsymbol{\mathcal{B}}}^{(2D)}(\mathbf{r}_\perp, t|\mathbf{o}, 0) = \frac{m\omega_L}{2\pi \mathrm{i}\hbar \sin(\omega_L t)} \exp\left\{\frac{\mathrm{i}m\omega_L}{2\hbar}\mathbf{r}_\perp^2 \cot(\omega_L t)\right\} . \tag{30}$$

As the propagator is periodic in t, we may expand it into a series using the generating function of the Laguerre polynomials $\mathrm{L}_k(z)$ [1]:

$$\frac{1}{1-z} \exp\left\{\frac{xz}{z-1}\right\} = \sum_{n=0}^{\infty} \mathrm{L}_n(x)\, z^n , \tag{31}$$

where we set $z = \mathrm{e}^{2\mathrm{i}\omega_L t}$ and $x = m\omega_L r_\perp^2/\hbar$. This procedure yields:

$$K_{\boldsymbol{\mathcal{B}}}^{(2D)}(\mathbf{r}_\perp, t|\mathbf{o}, 0) = \frac{m\omega_L}{\pi\hbar} \mathrm{e}^{-m\omega_L r_\perp^2/2\hbar} \sum_{n=0}^{\infty} \mathrm{L}_n\left(\frac{m\omega_L r_\perp^2}{\hbar}\right) \mathrm{e}^{-\mathrm{i}(2n+1)\omega_L t} . \tag{32}$$

This form is easily recognized as the decomposition of the time evolution operator $\mathbf{U}_\perp(t, 0)$ into the eigenfunctions of $\mathbf{H}_\perp$ (23) populating the Landau levels $E_n = (2n+1)\hbar\omega_L$.

The two-dimensional energy Green function $G_{\boldsymbol{\mathcal{B}}}^{(2D)}(\mathbf{r}_\perp, \mathbf{o}; E)$ follows from (32) after the Laplace transform (6) which immediately yields an infinite series expression:

$$G_{\boldsymbol{\mathcal{B}}}^{(2D)}(\mathbf{r}_\perp, \mathbf{o}; E) = \frac{m\omega_L}{\pi\hbar} \mathrm{e}^{-m\omega_L r_\perp^2/2\hbar} \lim_{\eta\to 0} \sum_{n=0}^{\infty} \frac{\mathrm{L}_n\left(m\omega_L r_\perp^2/\hbar\right)}{E - \hbar\omega_L(2n+1) + \mathrm{i}\eta} , \tag{33}$$

which clearly resembles the formal eigenfunction expansion (19). In the important case $\mathbf{r}_\perp \to \mathbf{o}$, we extract the density of states for a two-dimensional gas of charges subject to a magnetic field:

$$n_{\boldsymbol{\mathcal{B}}}^{(2D)}(E) = -\frac{1}{\pi}\Im\{G_{\boldsymbol{\mathcal{B}}}^{(2D)}(\mathbf{o}, \mathbf{o}, E)\} = \frac{m\omega_L}{\pi\hbar} \sum_{n=0}^{\infty} \delta\big[E - \hbar\omega_L(2n+1)\big] . \tag{34}$$

Here, we again made use of the distribution relation $\Im[(z+\mathrm{i}\eta)^{-1}] = -\pi \operatorname{sgn}\eta \cdot \delta(z)$ [24] that holds for $\eta \to 0$. The resulting discrete δ–array is indicated in Figure 1. Equation (34) expresses the fact that the eigenstates take on only the discrete energy values E_n.

In three spatial dimensions, we have to multiply the propagator $K_{\boldsymbol{\mathcal{B}}}^{(2D)}$ $(\mathbf{r}_\perp, t|\mathbf{o}, 0)$ with the free-particle propagator in one dimension $K_{\text{free}}^{(1D)}$

$$(r_\parallel, t|0, 0) = \sqrt{m/(2\pi \mathrm{i}\hbar t)} \exp\left[\mathrm{i}m r_\parallel^2/(2\hbar t)\right]$$

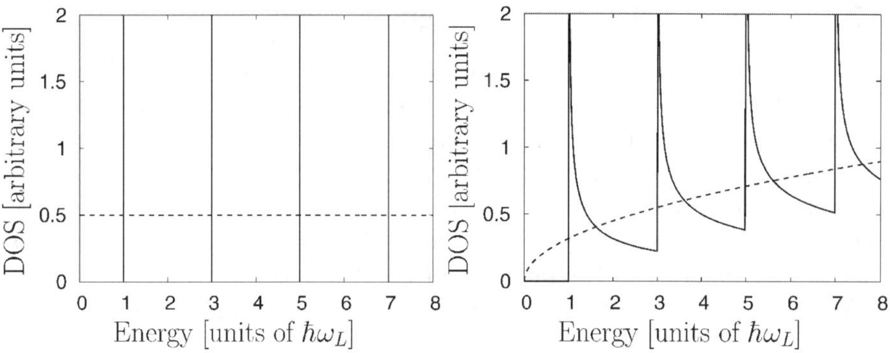

Figure 1. Electronic density of states in two (left panel) and three dimensions (right panel) in a purely magnetic field. The dashed line denotes Wigner's threshold law (14), valid for free particles.

that follows from (25) once we set $\mathcal{E}_\| = 0$. Its Laplace transform (6) is the free-particle energy Green function $G^{(1D)}_{\text{free}}(r_\|, 0; E)$ (18) that we derived in the preceding section. Similarly, we may transform the product of $K^{(1D)}_{\text{free}}(r_\|, t|0, 0)$ with the series expansion of $K^{(2D)}_{\boldsymbol{\mathcal{B}}}(\mathbf{r}_\perp, t|\mathbf{o}, 0)$ (32) to determine the three-dimensional Green function of a charge in a homogeneous magnetic field:

$$G^{(3D)}_{\boldsymbol{\mathcal{B}}}(\mathbf{r}, \mathbf{o}; E) = \frac{m\omega_L}{\pi\hbar}\, \mathrm{e}^{-m\omega_{\mathcal{B}} r_\perp^2/2\hbar} \times \sum_{n=0}^{\infty} \mathrm{L}_n\left(m\omega_L r_\perp^2/\hbar\right) G^{(1D)}_{\text{free}}\left[r_\|, 0; E - (2n+1)\hbar\omega_L\right] . \tag{35}$$

(A representation in closed form is stated in [15].) Only the terms with positive effective energy ($E - (2n+1)\hbar\omega_L > 0$) contribute to the density of states $n^{(3D)}_{\boldsymbol{\mathcal{B}}}(E) = -\frac{1}{\pi}\Im[G^{(3D)}_{\boldsymbol{\mathcal{B}}}(\mathbf{o}, \mathbf{o}; E)]$, as comparison with (18) shows. With the help of (14), (18) we find:

$$n^{(3D)}_{\boldsymbol{\mathcal{B}}}(E) = \frac{m^{3/2}\omega_L}{\sqrt{2}\,\pi^2\hbar^2} \sum_{n=0}^{\infty} \frac{\Theta\left[E - (2n+1)\hbar\omega_L\right]}{\sqrt{E - (2n+1)\hbar\omega_L}} . \tag{36}$$

This superposition of effectively one-dimensional free-particle sources is displayed in Figure 1. In passing, we point out that the limes $\mathcal{B} \to 0$ in (36) is not well-defined; only after averaging over a small energy range δE, the Wigner free-particle law $n^{(3D)}_{\text{free}}(E) = mk/2\pi^2\hbar^2$ (14) will emerge. Finally, we note that the result (35) extends to the case of parallel electric and magnetic fields $\boldsymbol{\mathcal{E}} \parallel \boldsymbol{\mathcal{B}}$, once the free particle Green function in the sum is replaced by the one-dimensional Green function for a uniformly accelerated particle [17, 35].

3.3 Crossed electric and magnetic fields

Unlike the case of a purely magnetic or parallel fields, the energy-dependent Green function $G_{\boldsymbol{\mathcal{E}},\boldsymbol{\mathcal{B}}}(\mathbf{r},\mathbf{o};E)$ for a particle in crossed electric and magnetic fields $\boldsymbol{\mathcal{E}}$, $\boldsymbol{\mathcal{B}}$ is not available in closed form. Thus, in this section we limit our considerations to the current emitted by a point source, or equivalently, the density of states $n(\mathbf{o};E)$. In a purely magnetic field, the degeneracy of the energy spectrum leads to peculiar shapes of the density of states functionals $n_{\boldsymbol{\mathcal{B}}}(E)$ (see Figure 1). The presence of an additional perpendicular electric field $\boldsymbol{\mathcal{E}}$ lifts these degeneracies and renders a broadened Landau level structure.

3.3.1 Density of states in two dimensions. We first examine the two-dimensional case. Here, the Green function $G^{(2D)}_{\boldsymbol{\mathcal{E}}\times\boldsymbol{\mathcal{B}}}(\mathbf{r}_\perp,\mathbf{o};E)$ in perpendicular fields is formally given by the Laplace transform (6) of the propagator $K_\perp(\mathbf{r}_\perp,t|\mathbf{o},0)$ (26). Since we are only interested in the imaginary part of the Green function at $\mathbf{r}_\perp=\mathbf{o}$ (13), (16) we may rewrite this relation and express the density of states $n^{(2D)}_{\boldsymbol{\mathcal{E}}\times\boldsymbol{\mathcal{B}}}(\mathbf{o};E)$ as the Fourier transform of the propagator:

$$n^{(2D)}_{\boldsymbol{\mathcal{E}}\times\boldsymbol{\mathcal{B}}}(\mathbf{o};E)=\frac{1}{2\pi\hbar}\int_{-\infty}^{\infty}\mathrm{d}T\,\mathrm{e}^{\mathrm{i}ET/\hbar}\,K_\perp(\mathbf{o},T|\mathbf{o},0)\,. \tag{37}$$

(Formally, the density of states operator $\delta(E-\mathbf{H})$ (15) is the Fourier transform of the time evolution operator $\mathbf{U}(T,0)=\exp(-\mathrm{i}\mathbf{H}T/\hbar)$, and the identity (37) follows in configuration space representation.) Alternatively, we may determine $n^{(2D)}_{\boldsymbol{\mathcal{E}}\times\boldsymbol{\mathcal{B}}}(\mathbf{o};E)$ by direct summation over a complete set of eigenstates of $\mathbf{H}_\perp(\mathbf{r}_\perp)$. We will explore both routes below.

Eigenfunction method A complete set of eigenfunctions for a charge in perpendicular fields is conveniently determined in the Landau gauge $\mathbf{A}=(-\mathcal{B}y,0,0)$ [29]. Here, we assume that the magnetic field points into the z direction while the electric field component $\boldsymbol{\mathcal{E}}_\perp$ is aligned to the y–axis, so the charges drift along the x–axis. The corresponding Hamiltonian $\mathbf{H}'_\perp(x,y)$ is given by:

$$\mathbf{H}'_\perp(x,y)=\frac{1}{2m}\left(-\mathrm{i}\hbar\partial_x+q\mathcal{B}y\right)^2-\frac{1}{2m}\partial_y^2-q\mathcal{E}_\perp y\,. \tag{38}$$

The eigenfunctions are products of shifted oscillator functions with a plane wave in drift direction, and read properly normalized:

$$\psi_{n,y_c}(x,y)=\left(\frac{q\mathcal{B}}{2\pi\hbar}\right)^{1/2}\exp\left[\frac{\mathrm{i}}{\hbar}\left(mv_D-q\mathcal{B}y_c\right)x\right]\frac{1}{\sqrt{l}}\,u_n\left(\frac{y-y_c}{l}\right), \tag{39}$$

where

$$u_n(\xi) = \left(\frac{1}{2^n n!\,\sqrt{\pi}}\right)^{1/2} \mathrm{e}^{-\xi^2/2}\mathrm{H}_n(\xi) \ . \tag{40}$$

Here, the magnetic length $l = \sqrt{\hbar/(q\mathcal{B})}$ determines the extension of the wave function in the direction of $\boldsymbol{\mathcal{E}}_\perp$, while the continuous variable y_c denotes its centroid. $\mathrm{H}_n(\xi)$ is a Hermite polynomial of order n [1]. The terms in the corresponding eigenenergy $E_n(y_c)$:

$$E_n(y_c) = (2n+1)\hbar\omega_L + mv_D^2/2 - q\mathcal{E}_\perp y_c \ , \tag{41}$$

reflect the Landau level, the kinetic energy of the drift motion and the potential energy in the electric field, respectively. Summation over all eigenstates (39) yields the density of states (37):

$$\begin{aligned} n^{(2D)}_{\boldsymbol{\mathcal{E}}\times\boldsymbol{\mathcal{B}}}(\mathbf{o};E) &= \sum_{n=0}^{\infty}\int_{-\infty}^{\infty} \mathrm{d}y_c\,\delta\big[E - E_n(y_c)\big]\,\big|\psi_{n,y_c}(\mathbf{o})\big|^2 \\ &= \frac{qB}{2\pi\hbar}\sum_{n=0}^{\infty}\frac{1}{2^n n!\sqrt{\pi}\,\Gamma}\,\mathrm{e}^{-E_n^2/\Gamma^2}\,\big[\mathrm{H}_n(E_n/\Gamma)\big]^2 \ . \end{aligned} \tag{42}$$

Comparison with (40) shows that the density of states is itself a sum over squares of regularly spaced oscillator functions, albeit in energy space; their width Γ and shifts E_n are given by:

$$\Gamma = q\mathcal{E}_\perp l = \mathcal{E}_\perp\sqrt{\frac{q\hbar}{\mathcal{B}}}\ , \qquad E_n = E - (2n+1)\hbar\omega_L - mv_D^2/2 \ . \tag{43}$$

Note that the centers of these oscillator states coincide with the Landau levels, apart from a constant shift due to the drift motion. As $\mathcal{E}_\perp \to 0$, the width Γ tends towards zero, and the discrete energy levels familiar from a purely magnetic field emerge (34).

Propagator method To obtain the density of states (42) using the propagator transform (37), we must first extract the function $K_\perp(\mathbf{o},t|\mathbf{o},0)$ from (26):

$$K_\perp(\mathbf{o},t|\mathbf{o},0) = \frac{m\omega_L}{2\pi\mathrm{i}\hbar\sin(\omega_L t)}\exp\left\{\frac{\mathrm{i}mv_D^2 t}{2\hbar}\left[\omega_L t\cot(\omega_L t) - 1\right]\right\} \ . \tag{44}$$

This expression formally resembles the propagator in a purely magnetic field (30), and the generating function expansion (31) again yields the series expansion of $K_\perp(\mathbf{o},t|\mathbf{o},0)$. As a function of the width parameter Γ defined above

(43), it reads:

$$K_\perp(\mathbf{o},t|\mathbf{o},0) = \frac{q\mathcal{B}}{2\pi\hbar}\,\mathrm{e}^{-\Gamma^2t^2/(4\hbar^2)}\sum_{n=0}^{\infty}\mathrm{L}_n\left(\frac{\Gamma^2t^2}{2\hbar^2}\right) \times \exp\left\{-\frac{\mathrm{i}t}{\hbar}\left[(2n+1)\hbar\omega_L+\frac{mv_D^2}{2}\right]\right\} . \quad (45)$$

Indeed, the density of states in crossed fields (42) follows after term-by-term integration of this sum in (37), as can be shown using the Fourier transform [1]:

$$\int_{-\infty}^{\infty}\mathrm{d}t\,\mathrm{e}^{-(t-\mathrm{i}x)^2}\left[\mathrm{H}_n(t)\right]^2 = 2^n n!\sqrt{\pi}\,\mathrm{L}_n(2x^2)\,. \quad (46)$$

For a more detailed discussion of this approach, see [37]. While the method appears unnecessarily complicated for the determination of $n^{(2D)}_{\mathcal{E}\times\mathcal{B}}(\mathbf{o};E)$, the propagator formalism clearly offers an advantage when evaluating the Green function $G^{(2D)}_{\mathcal{E}\times\mathcal{B}}(\mathbf{r}_\perp,\mathbf{o};E)$ for $\mathbf{r}_\perp\neq\mathbf{o}$, since a single numerical integration (6) will suffice to obtain the complete Green function. We will show examples below.

Canonical transformation method The result gained by the two previous methods is also consistent with the mapping of the original Hamiltonian in crossed fields in equation (38) to the Hamiltonian of a shifted harmonic oscillator. Details of the corresponding canonical transformation and its unitary representation are discussed in [33, 34].

Properties of the density of states functional The functional form of the two-dimensional density of states (42) in crossed fields is a major topic in Refs. [34, 37]. Here we content ourselves with a short summary of the main features. As noted above, the density of states consists of a sum over equally weighted harmonic oscillator eigenstates that appear not in configuration space but as functions of the energy E. As the eigenfunctions $u_n(\xi)$ (40) form an orthonormal set, the total contribution of each sum term, i. e., each Landau level is given by:

$$\int_{-\infty}^{\infty}\mathrm{d}E\;n^{(2D)}_{n,\mathcal{E}\times\mathcal{B}}(E) = \frac{e\mathcal{B}}{2\pi\hbar}\,. \quad (47)$$

This result is in accordance with the quantization of the Landau levels in a purely magnetic field (34). For each Landau level n, the density of states has a Gaussian envelope with width Γ that is split into $n+1$ intervals by the n simple zeroes $\xi_{n,j}$ ($j=1,\ldots,n$) of the polynomial $\mathrm{H}_n(\xi)$. In Fig. 2, we plot the resulting density of states for various electric field strengths $\mathcal{E}_\perp$. For small $\mathcal{E}_\perp$, the overlap between adjacent Landau levels is negligible, as the

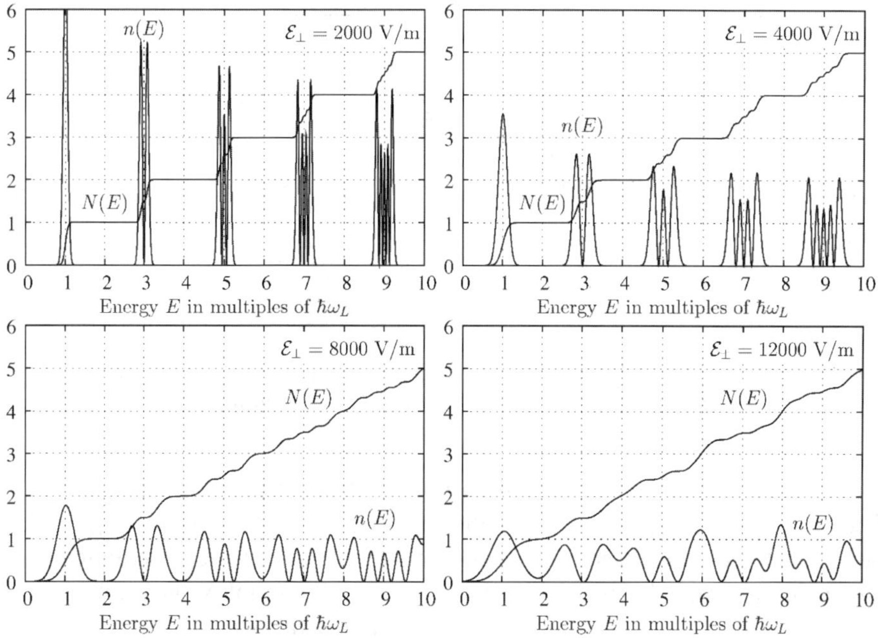

Figure 2. Two-dimensional local density of states (LDOS) $n(\mathbf{o}; E)$ (in units of $m/(\pi\hbar^2)$) and integrated LDOS $N(\mathbf{o}; E)$ (in units of $q\mathcal{B}/(2\pi\hbar)$) at four different electric fields $\mathcal{E} = 2000, 4000, 8000, 12000$ V/m and for a magnetic field $\mathcal{B} = 5$ T as a function of the scaled energy $E/(\hbar\omega_L)$ according to equation (42). Near the nth Landau level at $E = (2n+1)\hbar\omega_L$, the DOS renders the probability distribution of a one-dimensional harmonic oscillator in the nth eigenstate.

DOS drops off exponentially between them. With increasing electric field, the Landau levels broaden and finally coalesce. We infer from equation (40) that the classical turning point of harmonic motion, $\xi_n^{\text{tp}} = \sqrt{2n+1}$, provides a practical measure for the width of the partial density of states $n^{(2D)}_{n,\boldsymbol{\mathcal{E}}\times\boldsymbol{\mathcal{B}}}(E)$. The populated region in energy between adjacent Landau levels $n-1$, n is then approximately given by the ratio:

$$\frac{\text{combined half widths of levels } \Gamma(\xi_{n-1}^{\text{tp}} + \xi_n^{\text{tp}})}{\text{level spacing } 2\hbar\omega_L} \sim 2\sqrt{2n}\,\frac{m\mathcal{E}_\perp}{\sqrt{q\hbar\mathcal{B}^3}}\,. \qquad (48)$$

The overall extension of the modulated Landau levels increases with $n^{1/2}$ for fixed fields. Note that all features of $n^{(2D)}_{n,\boldsymbol{\mathcal{E}}\times\boldsymbol{\mathcal{B}}}(E)$, including the nodes, scale linearly in width with the electric field $\mathcal{E}_\perp$. While for small ratios in (48) individual levels remain well separated, with increasing overlap the density of states becomes a smooth function of the energy E. This transition is clearly visible in Figure 2.

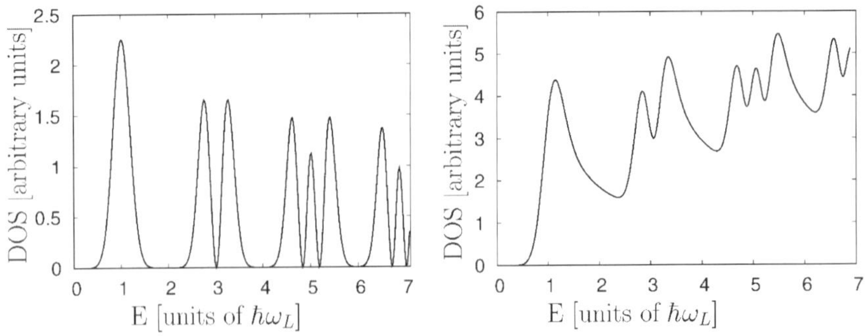

Figure 3. Electronic density of states in two (left panel) and three dimensions (right panel) in perpendicular electric and magnetic fields ($\mathcal{B} = 0.5$ T, $\mathcal{E}_\perp = 200$ V/m).

3.3.2 Extension to three dimensions. In three dimensions, besides the Green function $G_{\boldsymbol{\mathcal{E}},\boldsymbol{\mathcal{B}}}(\mathbf{r},\mathbf{o};E)$ even the density of states functional $n^{(3D)}_{\boldsymbol{\mathcal{E}},\boldsymbol{\mathcal{B}}}(\mathbf{o};E)$ defies evaluation in closed form. However, simple integral representations are available. Starting from the identity (37), we may employ the composition property (24) in order to obtain an integral representation:

$$n^{(3D)}_{\boldsymbol{\mathcal{E}},\boldsymbol{\mathcal{B}}}(\mathbf{o};E) = \frac{1}{2\pi\hbar}\int_{-\infty}^{\infty} \mathrm{d}T\, \mathrm{e}^{\mathrm{i}ET/\hbar}\, K_\perp(\mathbf{o},T|\mathbf{o},0)K_\parallel(0,T|0,0)\;. \qquad (49)$$

Alternatively, we may formally perform the integration in (49), which leads to a convolution integral of the individual transforms, similar to (21). This approach yields a simple composition theorem for the density of states:

$$n^{(3D)}_{\boldsymbol{\mathcal{E}},\boldsymbol{\mathcal{B}}}(\mathbf{o};E) = \int_{-\infty}^{\infty} \mathrm{d}E'\, n^{(2D)}_{\boldsymbol{\mathcal{E}}\times\boldsymbol{\mathcal{B}}}(\mathbf{o};E')\, n^{(1D)}_{\mathcal{E}_\parallel}(0;E-E')\;. \qquad (50)$$

While the former expression is better suited for numerical calculations, (50) yields more physical insight: Inserting the series expansion (42) into (50),we infer that the three-dimensional density of states can again be interpreted as a sum over individual Landau levels n, where their actual contribution $n^{(3D)}_{n,\boldsymbol{\mathcal{E}},\boldsymbol{\mathcal{B}}}(\mathbf{o};E)$ follows from convolution of the oscillator function $\left[u_n(E_n/\Gamma)\right]^2$ (40) with the one-dimensional density of states $n^{(1D)}_{\boldsymbol{\mathcal{E}}_\parallel}(0;E-E')$. (In the case of perpendicular fields ($\mathcal{E}_\parallel = 0$), these integrals can be expanded into series of parabolic cylinder functions [7], but we will not elaborate this point further.) In Figure 3 we compare the analytic two-dimensional solution (42) and the corresponding three-dimensional density of states (49) in orthogonal fields. The close relation between both functionals is clearly displayed, as well as the separation of $n^{(3D)}_{\boldsymbol{\mathcal{E}},\boldsymbol{\mathcal{B}}}(\mathbf{o};E)$ into individual Landau levels.

3.4 Spin

A slight complication occurs if the motion of charges with spin, like electrons, is considered, since the spin interacts with the magnetic field $\mathcal{B}$. However, for uniform magnetic field, this interaction merely causes a constant effective energy shift $\Delta E = \pm\frac{1}{2}\, g\hbar\omega_L$ if we select the magnetic field direction as axis of quantization. Thus, the Green functions for each spin component follow from its scalar counterpart by adjusting their energy, $G_{\uparrow,\downarrow}(\mathbf{r},\mathbf{r}';E) = G(\mathbf{r},\mathbf{r}';E\pm\Delta E)$. Similarly, the spin dependent densities of states become

$$n_{\uparrow,\downarrow}(E) = n\left(E \pm \frac{1}{2} g\hbar\omega_L\right) , \qquad (51)$$

and the total density of states including spin can be mapped back to the scalar quantity: $n_{\uparrow\downarrow}(E) = n_\uparrow(E) + n_\downarrow(E)$. Hence we defer the inclusion of spin for the moment.

4. Application: Photodetachment

In a photodetachment experiment, electrons are detached from negatively charged ions A^- due to the interaction with a laser field:

$$A^- + h\nu \rightarrow A + e^- . \qquad (52)$$

The detached electron is emitted with a definite energy E given by the difference between its binding energy (or affinity) and the photon energy. This process allows a description in terms of quantum sources. In near-threshold detachment ($E \rightarrow 0$), it is reasonable to model the ion as a point source because its size is small compared to the de Broglie wavelength of the emitted electron. (Here, we consider only the generation of s–waves. For the general case of multipole emission, see Ref. [11].) The photodetachment current in external fields is then linked to the relevant energy-dependent Green function: For a laser beam illuminating the ions for the duration T, their survival probability is given by

$$R(E) \propto \exp[-J(E)T], \qquad (53)$$

where $J(E)$ denotes the total current defined in equation (13). In practice, an external electric field can provide a virtual double-slit environment that allows to probe the energy of the emitted electron (and thereby the electron affinity of the ion) with extreme accuracy [5, 6, 11].

The combination of electric and magnetic fields imprints a non-trivial structure on the detachment rate and allows to identify features of the underlying energy-dependent Green function. Unfortunately a direct experimental observation of these features is obscured by several effects. Typically, the negative ions are confined in an ion trap, where they still have a large kinetic energy.

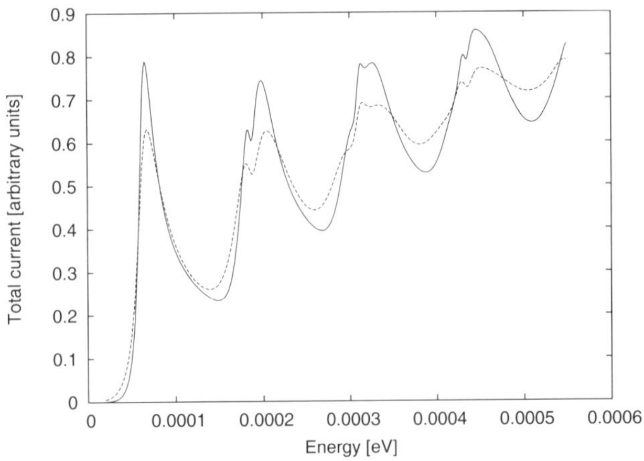

Figure 4. Thermally averaged curves for the total photocurrent as a function of the electron energy. Magnetic field: $\mathcal{B} = 1.07$ T, ion mass: $m = 32$ u. Solid line: $T = 400$ K, dashed line $T = 950$ K (see also [7], Figure 2). The substructure and broadening of the Landau levels due to the perpendicular electric field is visible. However, the features are washed out (compared to Figure 3) due to the averaging over a wide range of electric field values.

In a thermal ion cloud the momentum distribution $P(\mathbf{p})$ is given by Maxwell's expression:

$$P(\mathbf{p}) = \frac{1}{(2\pi m k_B T)^{3/2}} \exp\left(-p^2/(2mk_B T)\right). \tag{54}$$

In an external magnetic field, the charges will experience an electric field in their rest frame due to the transformation of the fields [28] that accounts for the Lorentz force,

$$\mathcal{E}_\perp(\mathbf{p}) = \frac{1}{m}\mathbf{p} \times \boldsymbol{\mathcal{B}}. \tag{55}$$

This electric field is exactly perpendicular to the momentum and the external magnetic field. A stationary source will only emerge if we consider the photon-electron interaction in the rest frame of the ion. Hence, we employ the three-dimensional Green function for crossed electromagnetic fields to describe the photodetachment of moving ions in a purely magnetic field.

The averaging effect of varying electric fields due to the thermal motion is displayed in Figure 4. A comparison with the plot for a single value of the electric field (right panel in Figure 3) shows that the substructure of the Landau levels changes. Some features are still visible, like the division of the first level due to the zero of the first Hermite polynomial. Another complication stems from the Zeeman splitting of the ionic energy levels in an external magnetic field. Experimentally, usually a superposition of many allowed transitions is

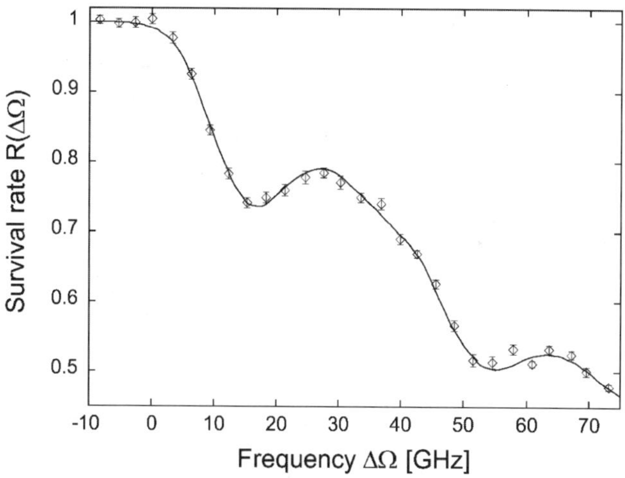

Figure 5. Ratio of surviving ions $R(\Delta\Omega)$ in photodetachment of S^- in an external magnetic field as a function of the laser detuning $\Delta\Omega$. The solid line is the theoretical prediction, the circles represent experimental data [56]. Parameters used: $\mathcal{B} = 1$ T, ion mass $m = 32$ amu, $T = 2100$ K.

observed. A recent experiment is compared to the theory sketched here in Ref. [56]. As shown in Figure 5 the agreement is excellent and underlines the validity of the quantum source approach. An alternative theoretical description, together with earlier experimental data is put forward in [7, 8].

5. Application: Quantum Hall effect

Another application of the Green function in crossed fields is the quantum Hall effect in a two-dimensional electron gas (2DEG). In Figure 6 we show the basic geometry of the sample. In the system, a constant current is sent along the x-axis of the sample. Perpendicular to the surface of the electron gas a strong magnetic field is applied. In a classical picture, initially electrons entering the sample are deflected to one edge, and a potential across the sample builds up until the Lorentz force is compensated by the induced electric field. The electrons then drift in the crossed fields with the constant velocity $\mathbf{v}_D$ (27) perpendicular to both fields. (Note that this mechanism leads to a loss-free stationary current in the presence of an electric field, unlike conventional transport theory, where the current is limited by inelastic scattering instead.) The linear relation between electric field and current density in two dimensions is expressed by the conductivity tensor $\boldsymbol{\sigma}$:

$$\begin{pmatrix} j_x \\ j_y \end{pmatrix} = \begin{pmatrix} \sigma_{xx} & \sigma_{xy} \\ \sigma_{yx} & \sigma_{yy} \end{pmatrix} \begin{pmatrix} \mathcal{E}_x \\ \mathcal{E}_y \end{pmatrix}. \tag{56}$$

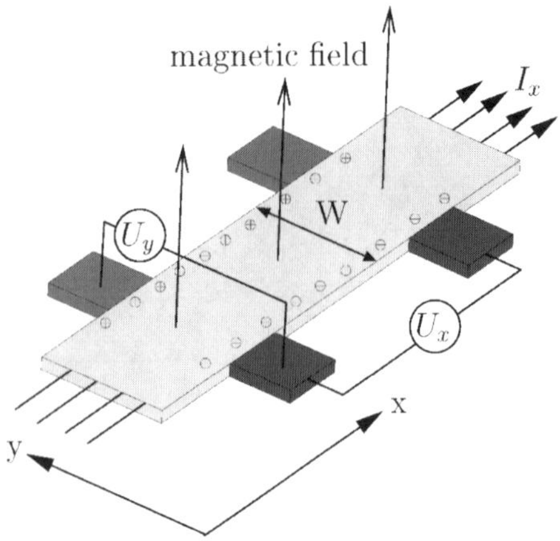

Figure 6. Schematic picture of a Hall bar. A constant current I_x is flowing along the x-axis. Perpendicular to the current and an external magnetic field, the Hall field is established along the y-axis to counterbalance the deflection of the electrons. Experiments record the Hall potential U_y and the longitudinal potential U_x.

Its inverse, the resistivity tensor $\boldsymbol{\rho}$, is related to the conductivity via

$$\rho_{xy} = \frac{\sigma_{yx}}{\sigma_{xx}^2 + \sigma_{xy}^2}, \qquad \rho_{xx} = \frac{\sigma_{xx}}{\sigma_{xx}^2 + \sigma_{xy}^2}. \tag{57}$$

We remark that for $\sigma_{xy} \neq 0$, vanishing resistivity ρ_{xx} implies vanishing conductivity σ_{xx}. In the setup shown in Figure 6, for stationary current density j_x and transverse electric field $\mathcal{E}_y$, the current j_y has to be zero. We will now explore different models for the conductivity in the two-dimensional Hall effect.

5.1 Drift transport of electrons

5.1.1 Classical transport. In this section we will review the classical transport of electrons in a Hall sample. We will take a modest level of scattering of the conduction electrons into account. In a simple Drude-like model the dynamics of the electrons is governed by the Lorentz force, amended for a term that incorporates friction via a relaxation time τ:

$$m\frac{\mathrm{d}\mathbf{v}}{\mathrm{d}t} = e\boldsymbol{\mathcal{E}} + e\mathbf{v} \times \boldsymbol{\mathcal{B}} - \frac{m}{\tau}\mathbf{v}. \tag{58}$$

Under stationary conditions $\mathbf{v}$ is constant and together with the current density $\mathbf{j} = Ne\mathbf{v}$ the components of the conductivity tensor in equation (56) become

$$\boldsymbol{\sigma} = \frac{e^2 N\tau}{m} \frac{1}{1+\omega_C^2\tau^2} \begin{pmatrix} 1 & -\omega_C\tau \\ \omega_C\tau & 1 \end{pmatrix}, \tag{59}$$

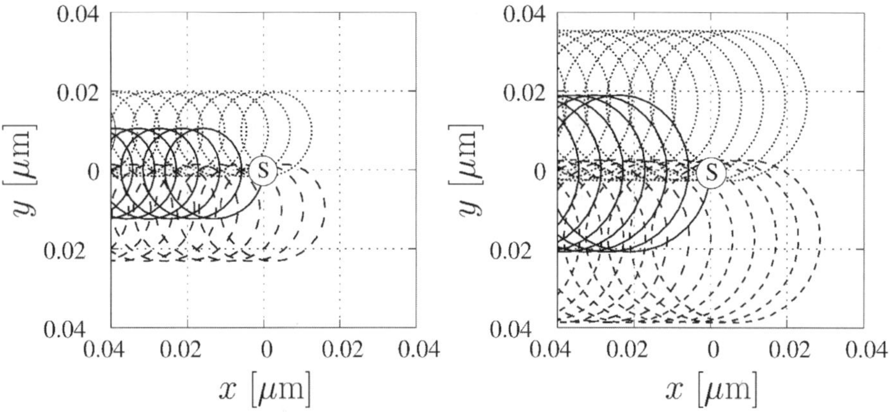

Figure 7. Classical trajectories from a point source ⓈS located at $x = y = 0$. The magnetic field $\mathcal{B} = 5$ T is oriented perpendicular to the plane, the electric field $\mathcal{E} = 4000$ V/m along $\mathbf{e}_y$. Left panel: Energy $E = \hbar\omega_L$, right panel: $E = 3\hbar\omega_L$. While the radius of cyclotron motion varies, the average drift velocity $v_D = \mathcal{E}/\mathcal{B}$ is the same in both panels.

where $\omega_C = 2\omega_L = e\mathcal{B}/m$. Inverting this matrix we extract the resistivity components

$$\rho_{xy} = \frac{\mathcal{B}}{Ne}, \qquad \rho_{xx} = \frac{m}{Ne^2\tau}. \tag{60}$$

In order to connect this picture to the classical drift of electrons in crossed fields we use the relation

$$\mathcal{E}_y = \rho_{xy} j_x = \frac{\mathcal{B}}{Ne} j_x, \qquad j_y = 0. \tag{61}$$

Solving for $v_x^{cl} = j_x/(eN)$ yields $v_x^{cl} = v_D = \mathcal{E}_y/\mathcal{B}$: The current density along the x-direction is given by the electron density N, multiplied by the drift velocity v_D. Some classical electron trajectories are shown in Figure 7. As we will see in the next section, the quantum mechanical picture radically diverts from these results.

If we assume a field independent carrier density N, equation (60) leads to the classical Hall effect: The Hall resistivity ρ_{xy} is linearly dependent on the magnetic field $\mathcal{B}$, and the constant of proportionality renders the number of carriers that participate in the transport.

5.1.2 Quantum mechanical drift. What do the quantum current and transport look like in our quantum source model of electronic matter waves? Equations (10) and (13) yield quantum mechanical expressions for the currents originating from a point source. In Figures 8 and 9 we plot the spatial current distribution generated in a magnetic field of $\mathcal{B} = 5$ T and an electric field of $\mathcal{E}_y = 4000$ V/m. In the first plot we chose $E = \hbar\omega_L$ which corresponds to

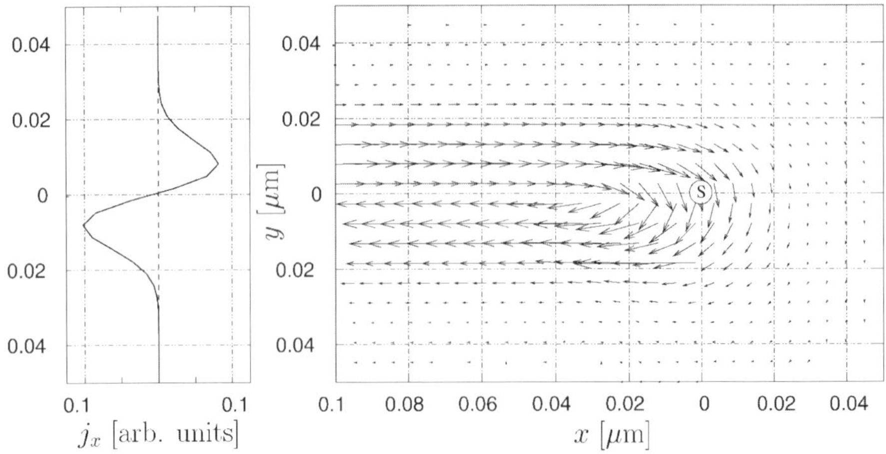

Figure 8. Current density from a point source ⓈS located at $x = y = 0$ in crossed fields $\mathcal{B} = 5$ T, $\mathcal{E}_y = 4000$ V/m (cf. Figure 7). The electron energy is $E = \hbar\omega_L$. Left panel: The component $j_x(x, y)$ at $x = -0.1$ μm. Right panel: Spatial current flow. The arrows indicate the direction of the current and their length is proportional to $|j|^{1/4}$.

the first maximum in the density of states (42) (see Figure 3), whereas in the second plot the energy $E = 3\hbar\omega_L$ is close to a minimum in the total current. Some corresponding classical trajectories are shown in Figure 7. The quantum mechanical current distribution shows some intriguing features: In the vicinity of the source (located at the origin), a complicated flow pattern emerges. At some distance from the source, the current follows the classical drift direction, but is split into two stripes with anti-parallel current vectors. We will discuss the implications of these oppositely flowing currents again in Section 5.6.

The current distribution for stronger Hall fields is depicted in Figure 10. Here the drift velocity is $v_D = 4000$ m/s. Three classical trajectories are included in the plot. The nearly circular orbits of Figure 7 are distorted to trochoidal shapes [43], and also the quantum mechanical current profile changes considerably.

Total current vs. current density The equation of continuity is valid for any surface enclosing the point source, and therefore the spatial current density integrated over such a closed surface must yield the total current, which by (16) is proportional to the density of states (42). Since the total current is available in analytic form, we may use this relation to cross-check our numerical evaluation of the spatial current density. As a function of E, the functional (42) repeatedly virtually drops to zero. This almost vanishing total current does not imply vanishing current density, however, as Figure 9 demonstrates: Circular flow patterns lead to a very small net-flow from the source region. Further away the contributions of oppositely directed flows cancel each other almost perfectly.

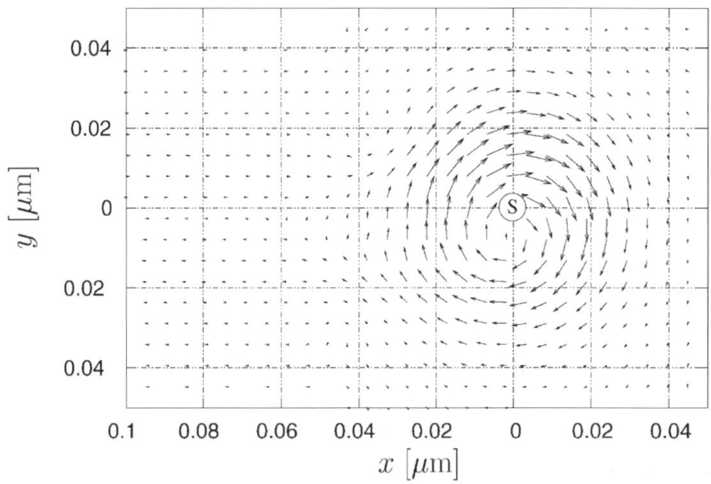

Figure 9. Current density from a point source ⓈS located at $x = y = 0$ in crossed fields $\mathcal{B} = 5$ T, $\mathcal{E}_y = 4000$ V/m (cf. Figure 7). The electron energy is $E = 3\hbar\omega_L$. The arrows indicate the direction of the spatial current flow and their length is proportional to $|j|^{1/4}$.

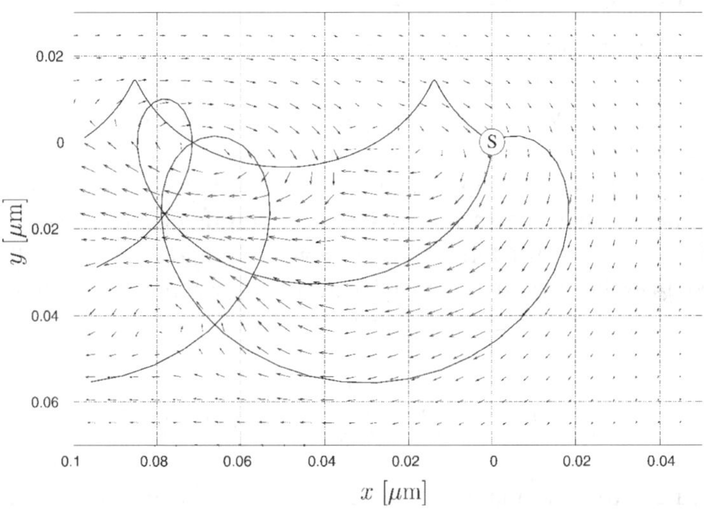

Figure 10. Current density from a point source ⓈS located at $x = y = 0$ in perpendicular fields $\mathcal{B} = 2$ T, $\mathcal{E}_y = 8000$ V/m. The electron energy is $E = \hbar\omega_L$. The arrows indicate the direction of the current flow and their length is proportional to $|j|^{1/4}$. Three classical trajectories of electrons with the same energy are also shown.

Recovering the drift velocity In order to define an average velocity along the drift direction we proceed as follows: For a slice along the y-axis at some fixed distance x from the source we calculate the integrated density $\varrho(x; E)$

using

$$\varrho(x;E) = \int_{-\infty}^{+\infty} \mathrm{d}y \, |G(\mathbf{r} = (x,y), \mathbf{o}; E)|^2 \tag{62}$$

and then define the ratio

$$v_x^{\mathrm{av}}(x;E) = J(\mathbf{o};E)/\varrho(x;E) \tag{63}$$

as the average velocity. This procedure yields values very close to the classical drift velocity $v_D = 800$ m/s in both cases illustrated. However, $J(\mathbf{o};E)$ and $\varrho(x;E)$ are drastically different for the two energies chosen in Figures 8 and 9. (We note that the local velocity field $\mathbf{v}(\mathbf{r}) = \mathbf{j}(\mathbf{r})/|G(\mathbf{r} = (x,y), \mathbf{o}; E)|^2$ greatly varies with $\mathbf{r}$. Only the integrated quantity reproduces the drift.) Furthermore, it is important to realize that the intensity of the current $J(\mathbf{o};E)$ is exponentially suppressed at certain energies as shown in Figure 3. This is in sharp contrast to the classical picture, where a constant drift transport occurs for all energy values of the injected electrons.

5.2 Fermionic matter waves

The density of states is a single-particle quantity. In a solid, many electrons take part in the conduction process. For a non-interacting system, the available single-particle energy levels are occupied according to Fermi-Dirac statistics. Taking spin into account, two electrons may share each state. For a system that exhibits a point spectrum of the energy levels (e.g. atoms, or a purely magnetic field in two dimensions (34)), the resulting electronic configuration is similar to the shell structure of atoms. For a continuous spectrum, the Fermi energy controls the integrated carrier density of the system, which at temperature $T \to 0$ is given by the integrated density of states $N(\mathbf{o};E_F)$:

$$N(\mathbf{o}; E_F, \mathcal{E}_y, \mathcal{B}) = \int_{-\infty}^{E_F} \mathrm{d}E \; n(\mathbf{o}; E, \mathcal{E}_y, \mathcal{B}). \tag{64}$$

In crossed electric and magnetic fields, (64) is available in closed form; see Ref. [34], Appendix C. Clearly, $N(\mathbf{o};E_F)$ depends on the external magnetic and electric fields.

Here, we should point out that in multi-electron systems Coulomb-type interactions will occur. In the following we will ignore these interactions. The Pauli principle (which requires antisymmetric wave-functions for fermions) ensures that two electrons sharing the same spin will not be at the same position. Moreover scattering events that redistribute the electrons to different energy states take place only if the process involves filled initial and empty final states. At $T \to 0$, these states are available only close to the Fermi level of the system.

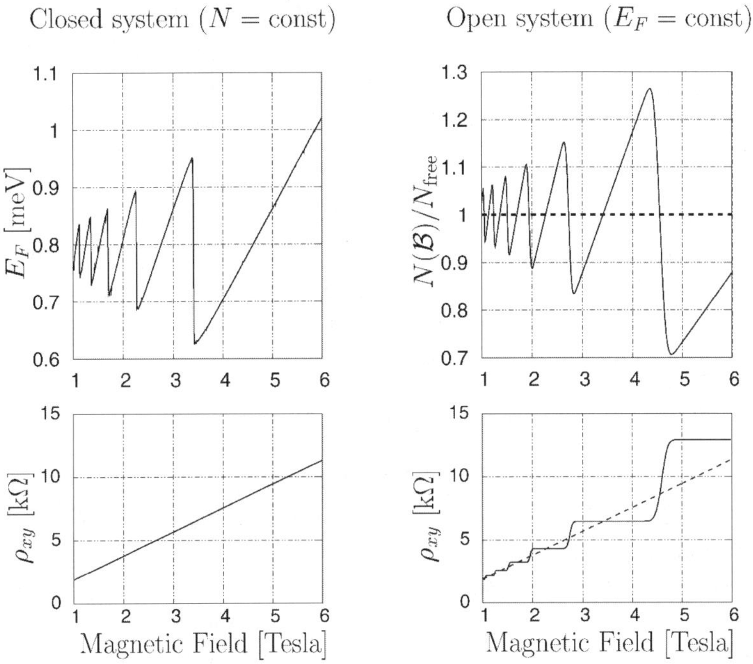

Figure 11. [

Top left: Fluctuation of the Fermi energy as a function of the magnetic field for fixed carrier concentration. Top right: Fluctuation of the current carrier concentration as a function of the magnetic field for fixed Fermi energy. Lower panels: Corresponding resistivity plots (the dashed line denotes the average number of carriers).

5.3 Fermi energy in open and closed system

Closed system If we treat the two-dimensional electron gas as a system that is closed and decoupled from reservoirs, the number of electrons is a fixed quantity. The Fermi energy is determined from the relation

$$N = \text{const} = N(\mathbf{o}; E_F, \mathcal{E}_y, \mathcal{B}) = \int_{-\infty}^{E_F} \mathrm{d}E\, n(\mathbf{o}; E, \mathcal{E}_y, \mathcal{B}). \tag{65}$$

Changing the external fields (and therefore the density of states $n(\mathbf{o}; E)$) leads to jumps in the Fermi energy as depicted in Figure 11 (top left panel): The upper limit of the integral has to be adjusted in order to keep the number of carriers N constant. Another implication of constant carrier density is a linear relationship between ρ_{xy} and $\mathcal{B}$ (60). Thus different mechanisms have to be invoked in order to explain the existence of finite Hall "plateaus" of constant conductivity in the closed system. Proposals include the formation of one-dimensional conduction channels along the edges of the sample, and disorder. In a one-dimensional device Landauer quantization gives rise to discrete values of the conductivity. Disorder is supposed to lead to localized states populated by electrons which do not participate in the transport but nevertheless allow to adjust the Fermi energy smoothly [22].

Open system In the following we consider the implications of an open Hall system, where electrons can enter and leave the system through the contacts. In this picture the Fermi energy is fixed, while the number of particles fluctuates around the average free-particle value observed for $\mathcal{B} = 0$. In Figure 11 (upper right panel) we plot the oscillations of

$$N(\mathbf{o}; E_F, \mathcal{E}, \mathcal{B})/N_{\text{free}}^{(2D)}(E_F) \tag{66}$$

as a function of the magnetic field $\mathcal{B}$. For comparison, we also show the resulting Hall resistivity $\rho_{xy} = \mathcal{B}/(Ne)$ in Figure 11 (lower right panel) together with its classical counterpart obtained by using $N = N_{\text{free}}^{(2D)}(E_F)$ (14). The following section is devoted to a detailed discussion of this curve. Here, we merely note that the difference in carrier density vanishes at the intersection points of both resistivity curves. Otherwise, excess charges will be present whose electrostatic interaction will lead to potentials that subsequently alter the Fermi energy of the system. In the present discussion we will neglect this feedback mechanism.

Experimental evidence for fluctuations Experiments show two types of fluctuations in quantum Hall systems as a function of the external magnetic field:

Density fluctuations are directly observed in [44] and fit well into the picture of an open system. These results contradict the basic theoretical assumptions for quantum Hall systems in Ref. [22, 55], where $N = \text{const}$ is used to determine the currents.

According to another experiment [53, 54] the electrostatic potential measured atop the two-dimensional quantum Hall system fluctuates as a function of the magnetic field. These changes are interpreted by the authors as variations of the (local) chemical potential. However, according to the basic ideas of most quantum Hall theories [22], disorder should buffer these oscillations and lead to a smooth variation of the Fermi-energy. The implications of these observations are profound, since, in the view of a closed system, jumps of the chemical potential would prevent the formation of Hall plateaus [22] (see Figure 11). Disorder must be invoked to allow smooth variation of the Fermi energy. From the viewpoint of an open system, these fluctuations are correlated with the intersections of the classical and quantum mechanical Hall resistivities. The excess charges lead also to a varying electrostatic potential above the Hall system. In an open system, disorder is not an essential ingredient of plateau

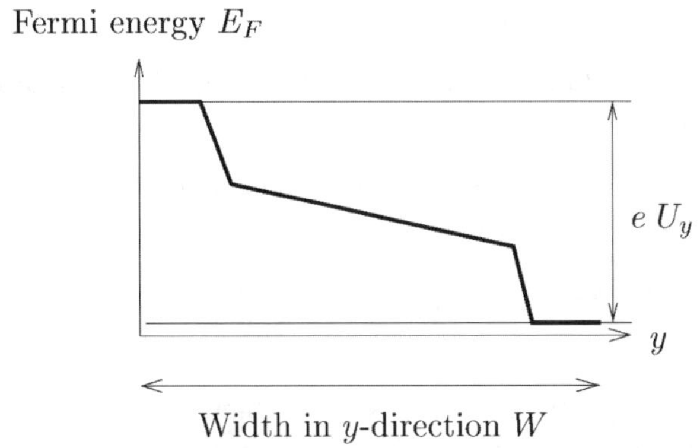

Figure 12. Schematic sketch of a possible variation of the Hall potential and the Fermi energy throughout the sample according to [2].

formation. A careful analysis of these intersection points and their matching to the observed fluctuations is crucial to investigate this issue.

5.4 Fermi energy and Hall potential variations

Up to now we considered only the injection of electrons from a single point source. We will now extend the formalism to cover a continuous "wire" of point sources along the current injecting contacts. Also, we will introduce the possibility of a local variation of the Fermi energy $E_F(y)$and the Hall field $\mathcal{E}_y(y)$, as sketched in Figure 12. The current of a macroscopic device with width W is given by integrating the current density over the width of the device:

$$I_x = \int_0^W \mathrm{d}y\; j_x(x,y) = \int_0^W \mathrm{d}y\; \sigma_{xy}\mathcal{E}_y(y)\;. \tag{67}$$

For a given form of the Hall potential and the Fermi energy we are now in a position to calculate the resistivity ρ.

5.5 Calculation of the Hall resistivity and the current flow

For the realistic calculation of a Hall resistivity curve we have to incorporate some material parameters in the theory:

a Electrons in solids are characterized by an effective mass m^*.

b Similarly the magnetic g-factor of the electron depends on the material and possibly on the magnetic field $\mathcal{B}$.

c In some materials additional degeneracies appear (e.g. the "valley splitting" in silicon).

d All observations are made at a finite temperature T.

e The electric field and Fermi energy may vary along the direction of the Hall field $\mathcal{E}_y$.

f In a multi-electron system Coulomb interactions between the electrons and the positive background charges occur.

g In a non-perfect sample disorder and electron-phonon interactions are present.

5.5.1 A new expression for the Hall conductivity. The combination of the quantum source model with the Pauli principle allows us to obtain a purely quantum mechanical expression for the current along the drift direction. Working in the eigenfunction expansion for the Green function (see Section 3.3.1), each eigenstate supports the current

$$j_x^{n,y_c}(\mathbf{r}) = \frac{\hbar}{m}\Im\left\{\psi_{n,y_c}(\mathbf{r})^*\partial_x\psi_{n,y_c}(\mathbf{r})\right\} + \frac{eA_x}{m}|\psi_{n,y_c}(\mathbf{r})|^2. \tag{68}$$

The properly weighted current is given by

$$j_x(\mathbf{r};E) = \sum_{n=0}^{\infty}\int \mathrm{d}y_c\ \delta[E - E_n(y_c)]j_x^{n,y_c}(\mathbf{r}). \tag{69}$$

A short calculation yields

$$\begin{aligned} j_x(\mathbf{r};E) &= v_D\sum_{n=0}^{\infty}\int \mathrm{d}y_c\ \delta[E - E_n(y_c)]\ |\psi_{n,y_c}(\mathbf{r})|^2 \\ &+\frac{e\mathcal{B}}{m}\sum_{n=0}^{\infty}\int \mathrm{d}y_c\ \delta[E - E_n(y_c)]\ (y_c - y)\ |\psi_{n,y_c}(\mathbf{r})|^2. \end{aligned} \tag{70}$$

We already evaluated the first term in (42). A (macroscopic) conductivity is obtained by integrating over y. Since the second integral runs over a function antisymmetric in $(y - y_c)$, the second term vanishes. Inserting $v_D = \mathcal{E}/\mathcal{B}$ gives

$$j_x(\mathbf{r};E) = \frac{e\mathcal{E}}{\mathcal{B}}n^{(2D)}_{\boldsymbol{\mathcal{E}}\times\boldsymbol{\mathcal{B}}}(\mathbf{r};E). \tag{71}$$

Without integration, the second term is responsible for the complicated flow pattern seen in the current pictures analyzed in Section 5.1.2. In a many electron system with Fermi energy E_F the conductivity in Ohms law $\mathbf{j} = \boldsymbol{\sigma}\cdot\mathbf{E}$ becomes

$$\sigma_{xy} = \frac{e}{\mathcal{B}}\int_{-\infty}^{E_F}\mathrm{d}E\ n^{(2D)}_{\boldsymbol{\mathcal{E}}\times\boldsymbol{\mathcal{B}}}(\mathbf{r};E) = \frac{e}{\mathcal{B}}N(\mathbf{r};E_F), \tag{72}$$

$$\sigma_{xx} = 0. \tag{73}$$

We emphasize that this expression couples the specific form of the density of states in crossed electromagnetic fields with the drift velocity. It is not possible to separate the quantity $N(\mathbf{r}; E_F)$ from the drift velocity and introduce it as an independent classical parameter.

5.5.2 A simple model for the quantum Hall effect including scattering.

The previous model is not complete, because the longitudinal resistance is always zero. Experiments show a non-vanishing σ_{xx} if E_F coincidences with a Landau-level. A natural extension of the model is the incorporation of scattering. A simple, yet instructive model for the Hall effect that incorporates effects 1–4 is presented in [34, 37]. We start from a conductivity tensor similar to the classical expression (59). However, we take into account the external field and energy dependence of all quantities:

$$\boldsymbol{\sigma}(E) = \frac{e^2 n(\mathbf{o}; E)\tau(E)}{m} \frac{1}{1+\omega_C^2 \tau(E)^2} \begin{pmatrix} 1 & -\omega_C \tau(E) \\ \omega_C \tau(E) & 1 \end{pmatrix}. \quad (74)$$

We will not consider a locally varying electric field. The discussion also assumes that the electrons are injected at a point-contact located at $\mathbf{o}$. We can lift this restriction by introducing a position dependent Fermi-level [23] in the system

$$E_F(\mathbf{r}) = E_F(\mathbf{o}) + e\, \mathbf{r} \cdot \mathcal{E}. \quad (75)$$

Usually, unequal Fermi levels result in a current. In the Hall geometry this current along the electric field is absent as the electrons only drift perpendicular to the magnetic and electric fields. From the symmetry relation (29) we obtain a translational invariance in the sense that

$$n(\mathbf{r}; E_F(\mathbf{r})) = n(\mathbf{o}; E_F(\mathbf{o})). \quad (76)$$

Therefore we can regard the local density of states as the global density of states in the system. A more sophisticated model, which incorporates a (slow) variation of the electric fields and the Fermi energy, is sketched in Section 5.6. In the Appendix we show, that the Lorentz-force model for the conductivity may be replaced by the expression for the probability current defined in (13).

For $T \to 0$, the total conductivity is obtained by integrating over the occupied energy range

$$\boldsymbol{\sigma} = \int_{-\infty}^{E_F} \mathrm{d}E\, \boldsymbol{\sigma}(E). \quad (77)$$

For strong magnetic fields, the energy-dependent relaxation time $\tau(E)$ satisfies $[\omega_C \tau(E)]^2 \gg 1$, except in the vicinity of E_F, and the transversal component σ_{xy} thus mirrors the integrated density of states $N(\mathbf{o}; E_F)$ (64)

$$\sigma_{xy} = \frac{e}{\mathcal{B}} \int_{-\infty}^{E_F} \mathrm{d}E\, \frac{n(\mathbf{o}; E)}{1+[\omega_C \tau(E)]^{-2}} = \frac{e}{\mathcal{B}} N(\mathbf{o}; E_F) \quad (78)$$

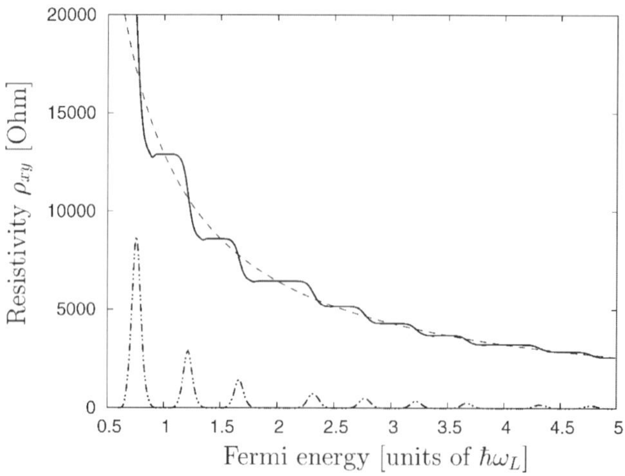

Figure 13. Quantum Hall effect for a constant magnetic field $\mathcal{B} = 19$ T. Effective mass $m^* = 0.2m_e$, $T = 1.5$ K. Notice that any substructure of the Landau levels is washed out by thermal averaging. The dashed line shows the classical Hall line. The lower dash-dotted line is proportional to the longitudinal resistance ρ_{xx}. Corresponding experimental data is shown in [51].

[cf. (60)]. The last expression is already known from our first model: The quantization of the plateaus in the Hall effect does not depend on the scattering.

The longitudinal component is more difficult to evaluate since it involves assumptions about the scattering events. If we assume that only electrons with energies close to the Fermi energy contribute significantly to σ_{xx}, we obtain:

$$\sigma_{xx}(E_F) = \frac{e}{\mathcal{B}} \int_{-\infty}^{E_F} \mathrm{d}E\, n(\mathbf{o}; E) \frac{\omega_C \tau(E)}{1 + \omega_C^2 \tau(E)^2} \approx D n(\mathbf{o}; E_F). \tag{79}$$

Here D denotes some constant that may depend on the material parameters and the fields. For finite temperature, significant scattering may also take place in an energy range of several $k_B T$ around E_F. The value of $\sigma_{xx}(T)$ then follows after suitable averaging [52]

$$\sigma_{xx}(T) = \int \mathrm{d}E \left(-\frac{\partial f(E,T)}{\partial E} \right) \sigma_{xx}(E), \tag{80}$$

where $f(E,T)$ denotes the Fermi-Dirac distribution

$$f(E,T) = \frac{1}{\mathrm{e}^{(E-E_F)/(k_B T)} + 1}. \tag{81}$$

The Hall resistivity for a fixed magnetic field In early experiments on the quantum Hall effect, the resistivity was measured under the condition of a fixed current along the x-axis ($I_x = \text{const}$) and a fixed magnetic field [51]. By varying the gate voltage in the experimental Si–MOSFET system, the Fermi energy is adjusted. We will assume a linear relationship between the gate voltage and the Fermi energy. Then, the value of the Hall field $\mathcal{E}_y$ is a solution of the implicit equation

$$\mathcal{E}_y = \rho_{xy}(E_F, \mathcal{E}_y, \mathcal{B})\, j_x \tag{82}$$

for given $E_F, \mathcal{B}, j_x$. Here, the resistivity ρ_{xy} is related to the conductivity components (78), (80) via (57). For the interpretation of data, we have to include an additional degeneracy besides spin that occurs in silicon, the "valley splitting" which effectively doubles each level, leading to a total of four repetitions of each Landau level. Introducing the additional valley quantum number $v = \pm\frac{1}{2}$ and the corresponding energy shift E_v [50], the density of states given in equation (51) becomes

$$n_{\uparrow,\downarrow,\mathrm{valley}}(\mathbf{o}; E) = n\,(\mathbf{o}; E \pm g\hbar\omega_L/2 \pm vE_v)\,. \tag{83}$$

Temperature dependence If the ratio $k_B T/(\hbar\omega_L)$ becomes close to unity, the Hall plateaus disappear, since $\sigma_{xx}(T)$ as given by equation (80) is no longer approaching zero between two Landau levels. However, the introduction of an effective mass m^* can lead to large modifications of $\omega_L^* = eB/(2m^*)$. Since the width $k_B T$ is independent of material parameters, it can be used as an independent energy scale to access the values of m^*, g^*. In Figure 13 we assume an effective mass $m^*/m_e = 0.2$. A higher effective mass would be inconsistent with the reported temperature of $T = 1.5$ K, since a smaller energy range $\hbar\omega_L^*$ cannot accommodate four separated peaks of individual width $k_B T$. Thus, the temperature dependence of ρ_{xy} may be used to determine some of the material parameters of the system.

The Hall resistivity as a function of the magnetic field Nowadays, GaAs–heterostructures are commonly used to provide the two-di - men - sional electron gas for the quantum Hall effect. Advantages are cleaner samples with very high mobilities and the absence of valley splittings. However, in these samples the number of electrons for $\mathcal{B} = 0$ is virtually constant and largely independent of the gate voltage. Therefore the magnetic field is varied while keeping the gate voltage fixed. In Figure 14 we show a typical plot of the resulting resistivity. As mentioned before, the intersection points of the averaged or classical Hall resistivity with the quantum mechanical result for ρ_{xy} deliver important information: They allow to analyze the spin splitting and other parameters of the system.

5.5.3 Fractional effects. As shown before, the electric field leads to additional zeroes in the density of states and consequently to subdivided Landau levels. Their fractional values of the filling factor are analyzed in [37] and

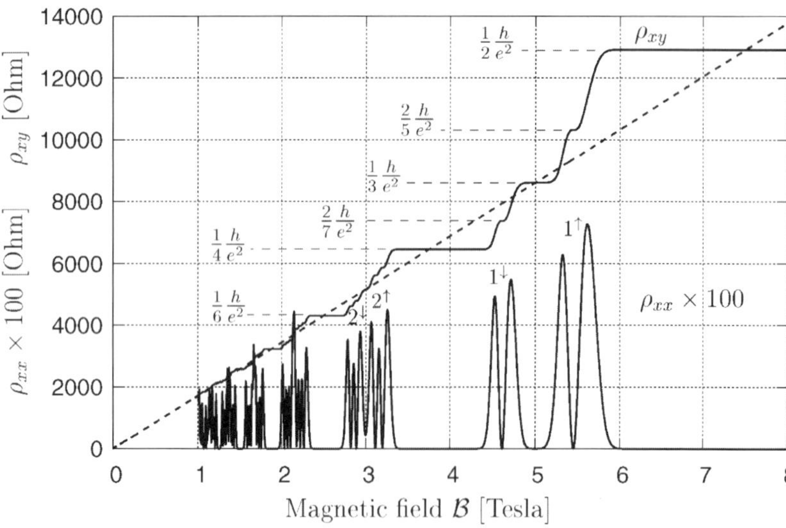

Figure 14. Quantum Hall effect at strong magnetic fields ($B > 1$ Tesla) for a non-interacting two-dimensional electron gas. The plot shows the Hall resistance ρ_{xy} and longitudinal resistance ρ_{xx} as a function of the magnetic field $\mathcal{B}$ for fixed Fermi energy ($E_F = 0.868$ meV). Effective mass $m^* = m_e$, effective g-factor $g = \frac{1}{2}$, current density $j_x = 0.2$ A/m, $\tau(E_F) = 10^{-11}$ s, $T = 0.1$ K. The dashed line represents the classical Hall resistance ρ_{xy} with a constant level density. Experimental results are shown in [42]. Note that j_x is chosen fairly large in order to show the substructure of the Landau levels.

displayed in Figure 14. Experimentally, the appearance of the fractional filling factors arising from the electric field might be difficult to detect, since uniform and high current densities are required at very low temperatures. Furthermore, the fractional quantum Hall effect may overshadow the single-particle structure if Coulomb interactions dominate the Hall field contribution. Experiments show plateaus in ρ_{xy} for simple fractions of the filling factor in the first ($n = 0$) Landau level. In the presented model such features are not explained. The effect is attributed to collective modes of the system that are caused by interactions between the electrons [55]. We explicitly did not include Coulomb interactions in our model system.

The best candidates for the detection of field-induced fractional filling factors are exactly half-filled odd Landau levels, since they are left largely unaffected by the averaging caused by a non-uniform electric field. (A similar effect prevails in three dimensions, where the dip associated with the second Landau level is still visible in Figure 4 despite extensive averaging over different field strengths.) In contrast, according to the standard theory of the fractional quantum Hall effect half-filled levels do not induce plateau formation.

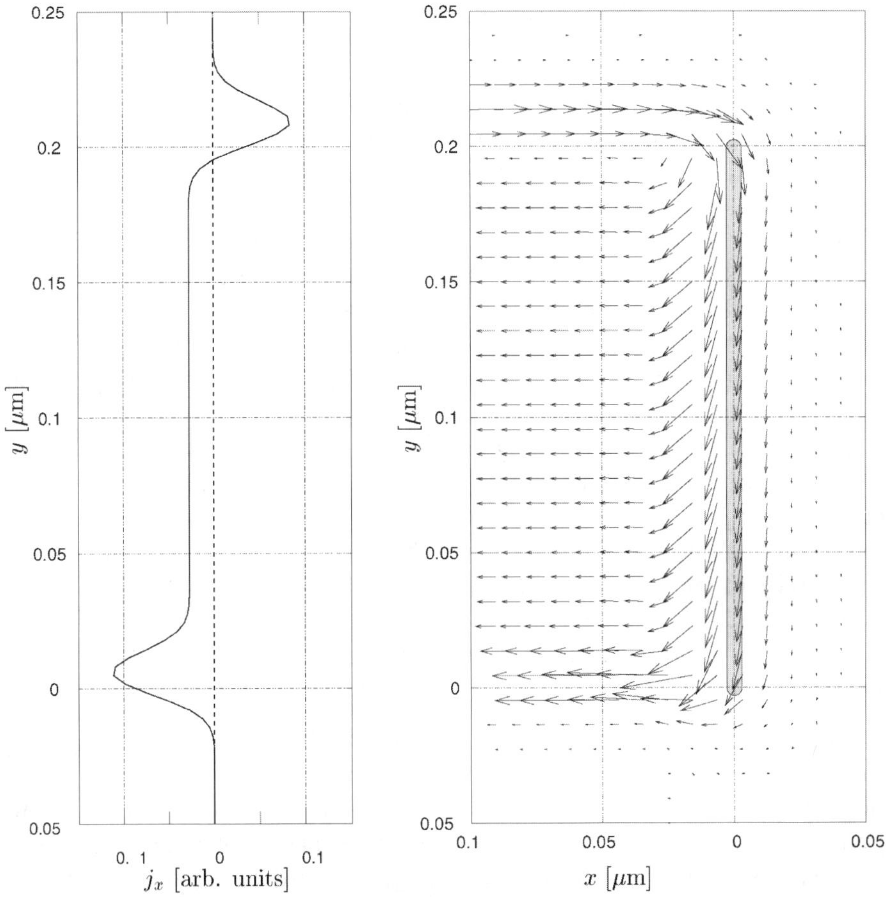

Figure 15. Quantum mechanical current density from an extended "wire" (shaded region). The electron emitting region extends from $y = 0$ to 0.2 μm. Left panel: The component $j_x(x, y)$ at $x = -0.1$ μm. Right panel: Spatial current flow. The arrows indicate the direction of the current and their length is proportional to $|j|^{1/4}$. In this example, an effective edge current forms with unequal magnitudes at each edge. The bulk also carries a constant current flow.

5.5.4 Hall-field dependence of the plateau width. Another signature of the presence of an electric field dependent broadening of Landau levels is the breakdown behavior for high currents. Kawaji and co-workers conducted extensive experiments on the characteristics of the breakdown and find experimentally a dependency that is exactly the same as the one obtained in equation (48): The width of the plateaus decreases linearly with increasing current. The experimental observation of this behaviour is reported in [4, 30, 48].

5.6 Current distribution

The form of the Hall potential is actually experimentally accessible [2], and a schematic result is sketched in Figure 12. We should note that the theory presented here could be applied for any given variation of the Hall potential and the Fermi energy. A starting point for models that provide this input could be the self-consistent potentials obtained in Refs. [12, 38]. For the sake of simplicity, let us discuss here only a straightforward extension in which we treat the emission of independent electrons along a constant Hall field. We already calculated the current distribution for a point source. For the "wire" described above, we obtain the global current profile by summing over the current contributions of the point sources. Here, we will use eq. (75) and assume that the Hall field is constant across the probe.

Figure 15 displays the resulting flow pattern. The previously described oppositely flowing currents are shifted to the edges, while in the bulk a uniform current emerges. In this way effective "edge" currents are established in a model of an electron emitting contact of finite width. We should note that the magnitudes of the oppositely directed edge currents differ, as already seen for point sources.

6. Conclusions

The quantum source formalism provides an excellent basis for the analysis of the propagation of matter waves in external fields. While some classical properties of the motion of particles prevail in the quantum mechanical case, a smooth transition from quantum to classical mechanics is generally not observed. The Landau quantization due to the magnetic field and the combination with an electric field have profound implications for the spatial current distribution and intensity.

Present-day nanotechnological devices, operated at very low temperatures, can actually provide experimental data for the propagation of electronic matter waves and allow to test theoretical predictions. We analyzed a simple model of the quantum Hall effect as one example. Even this non-interacting electron picture already shows a wealth of interesting features and gives access to important parameters of the system.

Acknowledgments

T. K. would like to thank Prof. Bruno Gruber for the invitation to present this paper at the "International symposium symmetries in science XIII". This work was financially supported by the Deutsche Forschungsgemeinschaft (project number Kl315/6-1, T. K.), and the Alexander von Humboldt foundation and the Killam trust (C. B.).

References

[1] Abramowitz, M. and I. Stegun: 1965, *Handbook of Mathematical Functions*. New York: Dover.

[2] Ahlswede, E.: 2002, 'Potential- und Stromverteilung beim Quanten-Hall-Effekt bestimmt mittels Rasterkraftmikroskopie'. Ph.D. thesis, Max-Planck-Institut für Festkörperforschung, Stuttgart. Online: http://elib.uni-stuttgart.de/opus/volltexte/2002/1187/.

[3] Bakhrakh, V. and S. Vetchinkin: 1971, 'Green's functions of the Schrödinger equation for the simplest systems'. *Theoret. Math. Phys.* **6**, 283. [Teoret. Mat. Fiz. 6, 392–402 (1971)].

[4] Bliek, L., E. Braun, G. Hein, V. Kose, J. Niemeyer, G. Weimann, and W. Schlapp: 1986, 'Critical current density for the dissipationless quantum Hall effect'. *Semicond. Sci. Technol.* **1**, 110.

[5] Blondel, C., C. Delsart, and F. Dulieu: 1996, 'The photodetachment microscope'. *Phys. Rev. Lett.* **77**, 3755.

[6] Blondel, C., C. Delsart, F. Dulieu, and C. Valli: 1999, 'Photodetachment microscopy of O^-'. *Eur. Phys. J. D* **5**, 207.

[7] Blumberg, W.A.M., W.M. Itano, and D.J. Larson: 1979, 'Theory of the photodetachment of negative ions in a magnetic field'. *Phys. Rev. A* **19**, 139–148.

[8] Blumberg, W.A.M., R.M. Jopson, and D. J. Larson: 1978, 'Precision laser photodetachment spectroscopy in magnetic fields'. *Phys. Rev. Lett.* **40**, 1320–1323.

[9] Bracher, C.: 1999, 'Quantum Ballistic Motion and its Applications'. Ph.D. thesis, Technische Universität München. Unpublished.

[10] Bracher, C., W. Becker, S. Gurvitz, M. Kleber, and M. Marinov: 1998, 'Three-dimensional tunneling in quantum ballistic motion'. *Am. J. Phys.* **66**, 38.

[11] Bracher, C., T. Kramer, and M. Kleber: 2003, 'Ballistic matter waves with angular momentum: Exact solutions and applications'. *Phys. Rev. A* **67**, 043601–1.

[12] Chklovskii, D.B., B.I. Shklovskii, and L.I. Glazman: 1992, 'Electrostatics of edge channels'. *Phys. Rev. B* **46**, 4026. Erratum: ibidem, p. 15606.

[13] Dalidchik, F. and V. Slonim: 1976, 'Strong exchange interaction effects in a homogeneous electric field'. *Sov. Phys. JETP* **43**, 25. [Zh. Eksp. Teor. Fiz. 70, 47–60 (1976)].

[14] de Souza, C. and A. de Souza Dutra: 1988, 'Galilean transformation and the path integral propagator for a crossed electric and magnetic field'. *Am. J. Phys.* **57**, 330.

[15] Dodonov, V., I. Malkin, and V. Man'ko: 1975, 'The Green function of the stationary Schrödinger equation for a particle in a uniform magnetic field'. *Phys. Lett. A* **51**, 133.

[16] Economou, E.: 1983, *Green's Functions in Quantum Physics (Solid–State Sciences 7)*. Berlin: Springer.

[17] Fabrikant, I.: 1991, 'Near-threshold photodetachment of H^- in parallel and crossed electric and magnetic fields'. *Phys. Rev. A* **43**, 258.

[18] Feynman, R. and A. Hibbs: 1965, *Quantum Mechanics and Path Integrals*. New York: McGraw-Hill.

[19] Gottlieb, B., M. Kleber, and J. Krause: 1991, 'Tunneling from a 3–dimensional quantum well in an electric field: An analytic solution'. *Z. Phys. A – Hadrons and Nuclei* **339**, 201.

[20] Gountaroulis, G.: 1972, 'Green-function of the free electron in a uniform magnetic field'. *Phys. Lett. A* **40**, 132.

[21] Grosche, C. and F. Steiner: 1998, *Handbook of Feynman Path Integrals*, Vol. 145 of *Springer Tracts in Modern Physics*. Berlin: Springer.

[22] Hajdu, J. (ed.): 1994, *Introduction to the Theory of the Integer Quantum Hall Effect*. Weinheim: VCH.

[23] Halperin, B.: 1986, 'The quantized Hall Effect'. *Sci. Am.* **254**, 40.

[24] Halperin, I. and L. Schwartz: 1952, *Introduction to the Theory of Distributions*. Toronto: University of Toronto Press.

[25] Horing, N., H. Cui, and G. Fiorenza: 1986, 'Nonrelativistic Schrödinger Green's function for crossed time-dependent electric and magnetic fields'. *Phys. Rev. A* **34**, 612.

[26] Hostler, L.: 1963, 'Coulomb Green's functions and the Furry approximation'. *J. Math. Phys.* **5**, 591.

[27] Hostler, L. and R. Pratt: 1964, 'Coulomb Green's function in closed form'. *Phys. Rev. Lett.* **10**, 469.

[28] Jackson, J.: 1975, *Classical Electrodynamics*. New York: Wiley, 2nd edition.

[29] Johnson, B., J. Hirschfelder, and K.-H. Yang: 1983, 'Interaction of atoms, molecules, and ions with constant electric and magnetic fields'. *Rev. Mod. Phys.* **55**, 109.

[30] Kawaji, S., K. Hirakawa, and M. Nagata: 1993, 'Device-width dependence of plateau width in quantum Hall states'. *Physica B* **184**, 17–20.

[31] Kennard, E.: 1927, 'Zur Quantenmechanik einfacher Bewegungstypen'. *Z. Phys.* **44**, 326–352.

[32] Kleber, M.: 1994, 'Exact solutions for time–dependent phenomena in quantum mechanics'. *Phys. Rep.* **236**(6), 331.

[33] Kramer, T.: 2000, 'Quantum ballistic motion in uniform electric and magnetic fields'. Diploma thesis, Technische Universität München. Unpublished.

[34] Kramer, T.: 2003, 'Matter waves from localized sources in homogeneous force fields'. Ph.D. thesis, Technische Universität München. Online: http://tumb1.biblio.tu-muenchen.de/publ/diss/ph/2003/kramer.pdf.

[35] Kramer, T., C. Bracher, and M. Kleber: 2001, 'Four-path interference and uncertainty principle in photodetachment microscopy'. *Europhys. Lett.* **56**, 471.

[36] Kramer, T., C. Bracher, and M. Kleber: 2002, 'Matter waves from quantum sources in a force field'. *J. Phys. A: Math. Gen.* **35**, 8361.

[37] Kramer, T., C. Bracher, and M. Kleber: 2003, 'Electron propagation in crossed magnetic and electric fields'. *submitted for publication*. Online: http://arxiv.org/abs/quant-ph/0307228.

[38] Lier, K. and R. Gerhardts: 1994, 'Self-consistent calculations of edge channels in laterally confined two-dimensional electron systems'. *Phys. Rev. B* **50**, 7757.

[39] Moshinsky, M. and C. Quesne: 1971, 'Linear Canonical Transformations and Their Unitary Representations'. *J. Math. Phys.* **12**, 1772–1780.

[40] Moshinsky, M. and P. Winternitz: 1980, 'Quadratic Hamiltonians in phase space and their eigenstates'. *J. Math. Phys.* **21**, 1667–1682.

[41] Nieto, L.M.: 1992, 'Green's function for crossed time-dependent electric and magnetic fields. Phase-space quantum mechanics approach'. *J. Math. Phys.* **33**(10), 3402–3409.

[42] Paalanen, M., D. Tsui, and A. Gossard: 1982, 'Quantized Hall effect at low temperatures'. *Phys. Rev. B* **25**, 5566.

[43] Peters, A.D. and J.B. Delos: 1993, 'Photodetachment cross-section of H^- in crossed electric and magnetic-fields. 1. Closed-orbit theory'. *Phys. Rev. A* **47**, 3020–3035.

[44] Raymond, A., S. Juillaguet, I. Elmezouar, W. Zawadzki, M. Sadowski, M. Kamel-Saadi, and B. Etienne: 1999, 'Oscillations of 2D electron density in GaAs/$Ga_{0.67}Al_{0.33}$As heterostructures in the QHE regime'. *Semicond. Sci. Technol.* **14**, 915.

[45] Sakurai, J.: 1994, *Modern Quantum Mechanics*. Addison-Wesley.

[46] Schwinger, J.: 1951, 'On Gauge Invariance and Vacuum Polarization'. *Phys. Rev.* **82**, 664.

[47] Schwinger, J.: 1973, *Particles, Sources, and Fields*, Vol. 2. Addison-Wesley.

[48] Shimada, T., T. Okamoto, and S. Kawaji: 1998, 'Hall electric field-dependent broadening of extended state bands in Landau levels and breakdown of the quantum Hall effect'. *Physica B* **249-251**, 107–110.

[49] Slonim, V. and F. Dalidchik: 1976, 'Impurity electroabsorption and photodestruction of negative ions in an electric field'. *Sov. Phys. JETP* **44**, 1081. [Zh. Eksp. Teor. Fiz. 71, 2057–2067 (1976)].

[50] v. Klitzing, K.: 1981, 'The fine structure constant α. A contribution of semiconductor physics to the determination of α.'. *Adv. Solid State Phys.* **21**, 1.

[51] v. Klitzing, K., G. Dorda, and M. Pepper: 1980, 'New method for high-accuracy determination of the fine-structure constant based on quantized Hall resistance'. *Phys. Rev. Lett.* **45**, 494.

[52] Wei, H., D. Tsui, and A. Pruisken: 1985, 'Localization and scaling in the quantum Hall regime'. *Phys. Rev. B* **33**, 1488.

[53] Wei, Y., J. Weis, K. v. Klitzing, and K. Eberl: 1997, 'Single-electron transistor as an electrometer measuring chemical potential variations'. *Appl. Phys. Lett.* **71**, 2514.

[54] Weitz, P., E. Ahlswede, J. Weis, K. v. Klitzing, and K. Eberl: 2000, 'A low-temperature scanning force microscope for investigating buried two-dimensional electron systems under quantum Hall conditions'. *Appl. Surf. Sci.* **157**, 349.

[55] Yoshioka, D. (ed.): 2002, *The Quantum Hall Effect*. Berlin: Springer.

[56] Yukich, J., T. Kramer, and C. Bracher: 2003, 'Observed photodetachment in parallel electric and magnetic fields'. *Phys. Rev. A*. accepted for publication, online: http://arxiv.org/abs/physics/0304039.

ON GROUP THEORETICAL ASPECTS, HYPERGEOMETRIC TRANSFORMATIONS AND SYMMETRIES OF ANGULAR MOMENTUM COEFFICIENTS

C. Krattenthaler
Institut Girard Desargues,
Université Claude Bernard Lyon-I,
21, avenue Claude Bernard,
F-69622 Villeurbanne, Cedex.
kratt@euler.univ-lyon1.fr www: http://euler.univ-lyon1.fr/home/kratt

K. Srinivasa Rao
The Institute of Mathematical Sciences,
Chennai-600 113, India
rao@imsc.res.in

Keywords: beta integral, generalized hypergeometric series, Kampé de Fériet functions

Abstract In this article we present a review of the work on Group theory of ordinary and basic Hypergeometric transformations and the application of the transformations of ordinary hypergeometric series in Quantum Theory of Angular Momentum.

1. Introduction

The fact that the coupling / recoupling coefficients of Quantum Theory of Angular Momentum (QTAM) were related to the hypergeometric functions [1], [2] and the use of a balanced (or Saalschützian) transformation formula to obtain what was claimed to be a *new* symmetry [3] for the Racah (or 6-j) coefficient was the starting point for a study of the symmetries of these angular momentum coefficients in terms of generalized hypergeometric functions of unit argument. Early studies [4] in this area of QTAM established that if the 3-j coefficient is to be written as a ${}_3F_2(1)$, a set of six ${}_3F_2(1)$s is necessary and sufficient to account for the 72 symmetries of the 3-j coefficient [5]. In the case of the 6-j coefficient it was shown that there exist a set of three ${}_4F_3(1)$s [6] and an

B. Gruber (ed.), Symmetries in Science XI, 355–375.

equivalent set of four ${}_4F_3(1)$s [7]. These two sets were shown to be related to each other through the *reversal* of hypergeometric series [8] and this completed an understanding of the symmetries of the 6-j coefficient in terms of the sets of ${}_4F_3(1)$s demonstrating a confirmation of Askey's remark [9] that the 3-j and 6-j coefficients that arise in QTAM are *hypergeometric functions and many of their elementary properties are best understood when considered as such.*

Wu observed [10] that while the 3-j and the 6-j coefficient can be related to a ${}_3F_2(1)$ and ${}_4F_3(1)$, respectively, the 9-j coefficient cannot be related to a ${}_7F_6(1)$ but that it is related to a *new* hypergeometric function [11] was yet another starting point. This led to the simplest known formula for the 9-j coefficient, a triple sum series due to Alisauskas, Jucys and Bandzaitis [12] being related [13] to a triple hypergeometric series, with unit arguments. From a novel way of looking at the symmetries of the 9-j coefficient, the Bailey transform for a Saalschützian ${}_4F_3(1)$ and a transformation of a Kampé de Fériet function into a Saalschützian ${}_4F_3(1)$ or its Bailey transform were derived [14].

The above results were reviewed in an earlier contribution to the *Symmetries in Science* series symposium [15]. Section 2 reviews the Group theoretical aspects of hypergeometric transformations and the 24 Kummer solutions of the Gauss second order ordinary differential equation being related to the symmetries of the cube. In section 3, the beta integral method to generate new transformations from old is reviewed and more results obtained recently are presented. Section 4 relates the ${}_7F_6(1)$ forms for the 6-j coefficient with the sets of ${}_4F_3(1)$s using a Whipple transformation and presents an understanding of the symmetries of the 6-j coefficient in terms of these sets. In section 5, a q-generalization of the new summation theorem obtained for the ${}_3F_2(a, a, x; 1+a, 1+a+N; 1)$, where N is a non-negative integer is derived.

2. Group theoretical aspects of hypergeometric transformations

Pfaff's transformation ([16], p.68) also referred to as Saalschütz's theorem [17]:

$$ {}_2F_1(a,b;c;x) = (1-x)^{-a}\, {}_2F_1(a, c-b; c; \frac{x}{x-1}) \tag{1} $$

when iterated generates the well-known Euler transformation formulae [8]:

$$ \begin{aligned} {}_2F_1(a,b;c;x) &= (1-x)^{-a}\, {}_2F_1(a, c-b; c; \frac{x}{x-1}) && (2) \\ &= (1-x)^{-b}\, {}_2F_1(c-a, b; c; \frac{x}{x-1}) && (3) \\ &= (1-x)^{c-a-b}\, {}_2F_1(c-a, c-b; c; x), && (4) \end{aligned} $$

on each of which is superposed the trivial numerator parameter permutation symmetry:

$$ {}_2F_1(a,b;c;x) = {}_2F_1(b,a;c;x), \tag{5} $$

so that we have a set of eight transformations. If we denote the Pfaff transformation by g_1, and the numerator parameter permutation of the ${}_2F_1$ by g_2, then these two elementary transformations satisfy the properties:

$$ g_1^2 = 1, \qquad g_2^2 = 1, \qquad (g_1 g_2)^4 = 1 $$

and the group generated by the two elements subjected to these relations is the dihedral group D_8, also known as the group of symmetries of the square. It is a subgroup of the symmetric group S_4, the permutation group on four elements.

Erdélyi and Weber [18] stated that the recursive use of the Thomae [19] transformation for a ${}_3F_2(1)$ resulted in a new transformation. This was the starting point for a study of the group theory of transformations. Bayer, Louck and Stein [20] showed that the group of transformations of the non-terminating ${}_3F_2(1)$ transformation is S_5 and that the group of transformations of the terminating balanced (or Saalschützian) ${}_4F_3(1)$ series is S_6. The study of the group theory of the terminating ${}_3F_2(1)$ transformations, obtained by a recursive application of the terminating version of the Thomae transformation was done by Srinivasa Rao et al. [21].

The remark of Hardy [22] that the Thomae transformation for the non-terminating ${}_3F_2(1)$:

$$ {}_3F_2\left(\begin{array}{c} a,\ b,\ c\,;1 \\ d,\ e \end{array}\right) = \frac{\Gamma(d,e,s)}{\Gamma(a,s+b,s+c)}\, {}_3F_2\left(\begin{array}{c} d-a,\ e-a,\ s\,;1 \\ s+b,\ s+c \end{array}\right) \tag{6} $$

where $\Gamma(x,y,\cdots) = \Gamma(x)\Gamma(y)\cdots$ and $s = d+e-a-b-c$ is the parameter excess,

is an expression of the theorem that

$$ \frac{1}{\Gamma(\beta_1)\Gamma(\beta_2)\Gamma(\beta_1+\beta_2-\alpha_1-\alpha_2-\alpha_3)} F\left(\begin{array}{c} \alpha_1,\alpha_2,\alpha_3 \\ \beta_1,\beta_2 \end{array}\right) $$

is a symmetric function of the five arguments

$$ \beta_1, \beta_2, \beta_1+\beta_2-\alpha_2-\alpha_3, \beta_1+\beta_2-\alpha_3-\alpha_1, \beta_1+\beta_2-\alpha_1-\alpha_2, $$

clearly implies a group theoretical interpretation for the Thomae formula. As in the case of the terminating ${}_3F_2$ series, a recursive use of this transformation results in our obtaining a set of **ten** non-terminating Thomae transformations ([23], Appendix 1). In [24], the following function was constructed:

$$ \begin{aligned} &f(x_1,x_2,x_3,x_4,x_5) \\ &\quad = \frac{1}{\Gamma(s,2x_4,2x_5)}\, {}_3F_2\left(\begin{array}{c} 2x_1-s,\ 2x_2-s,\ 2x_3-s\,;1 \\ 2x_4,\quad 2x_5 \end{array}\right) \end{aligned} \tag{7} $$

where $s = x_1+x_2+x_3-x_4-x_5$. That this function is symmetric in all the five variables is a consequence of the Thomae transformation for non-terminating ${}_3F_2(1)$ has been proved as follows: the function $f(\vec{x}) \equiv f(x_1, x_2, x_3, x_4, x_5)$ is manifestly invariant for permutations of (x_1, x_2, x_3) and (x_4, x_5). Consider the permutation $p: \; x_1 \to x_2 \to x_3 \to x_4 \to x_5 \to x_1$, which is a permutation of order 5. Upon relabeling the parameters of the ${}_3F_2(1)$ in $f(\vec{x})$ as ${}_3F_2\left(\begin{matrix} a,\ b,\ c\,;1 \\ d,\ e \end{matrix}\right)$, it is straight forward to see that corresponding to

$$f(\vec{x}) \;=\; f(p.\vec{x}) \tag{8}$$

we get:

$${}_3F_2\left(\begin{matrix} a,\ b,\ c\,;1 \\ d,\ e \end{matrix}\right) = \frac{\Gamma(d, s)}{\Gamma(d-a, s-a)}\, {}_3F_2\left(\begin{matrix} e-c,\ e-b,\ a\,;1 \\ e,\ s+a \end{matrix}\right), \tag{9}$$

belonging to the set of ten Thomae transformations. Since $f(\vec{x})$ is invariant under this permutation p of order 5, and under the transposition, $x_4 \to x_5$, manifestly, the group generated by these two generators (viz. p and the transposition) is the complete group of permutations on 5 elements, i.e. the symmetric group S_5. This result is, obviously, *a succinct, quintessential one-line statement* for the group of Thomae transformations.

The two-term relation for the terminating balanced (or Saalschützian) ${}_4F_3$ of unit argument is:

$$\begin{aligned} {}_4F_3\left(\begin{matrix} A, B, C, -n;\ 1 \\ E, F, G \end{matrix}\right) &= \frac{(F-C)_n (G-C)_n}{(F)_n (G)_n} \\ &\times {}_4F_3\left(\begin{matrix} E-A, & E-B, & C & ,-n;\ 1 \\ E, E+F-A-B, E+G-A-B \end{matrix}\right) \end{aligned} \tag{10}$$

where the Pochhammer symbol is defined for $n \geq 1$ as:

$$(a)_n = \frac{\Gamma(a+n)}{\Gamma(a)} = a(a+1)(a+2)\cdots(a+n-1), \quad \text{with} \quad (a)_0 = 1. \tag{11}$$

A recursive use of this terminating ${}_4F_3$ transformation results in a set of **twenty** transformations ([23], Appendix 2). In this case, the function:

$$\begin{aligned} & f(x_1, x_2, x_3, x_4, x_5, x_6) \\ &= (x_1+x_2+x_3+x_4, x_1+x_2+x_3+x_5, x_1+x_2+x_3+x_6)_n \\ &\times\ {}_4F_3\left(\begin{matrix} x_1+x_2, x_2+x_3, x_3+x_1, -n;\ 1 \\ x_1+x_2+x_3+x_4, x_1+x_2+x_3+x_5, x_1+x_2+x_3+x_6 \end{matrix}\right) \end{aligned} \tag{12}$$

with $(x, y, \cdots)_n = (x)_n (y)_n \cdots$ and $x_1 + x_2 + x_3 + x_4 + x_5 + x_6 = 1 - n$, for some non-negative integer n, being symmetric in all the six variables $x_1, x_2, x_3, x_4, x_5, x_6$ is a consequence of the ${}_4F_3(1)$ transformation.

Hardy's clue enabled us [25] to look for the invariance groups for all the known transformations of basic hypergeometric series. A summary of these results is presented below:

The Heine transformation, in the standard notation [26]:

$$ {}_2\phi_1(a, b; c; q, x) = \frac{(a, bz; q)_\infty}{(c, z; q)_\infty} {}_2\phi_1(c/a, z; bz; q, a), \tag{13} $$

when iterated yields a set of 12 transformations [27]. The invariance group is the dihedral group D_{12} – the group of symmetries of the hexagon (sometimes also denoted as D_6). We construct the function:

$$ \begin{aligned} & f(x_1, x_2, x_3, x_4, x_5, x_6) \\ &= \left(x_1 x_4, \frac{x_2 x_6}{x_1}; q \right)_\infty {}_2\phi_1 \left(\frac{x_1 x_3}{x_2}, \frac{x_1 x_5}{x_6}; x_1 x_4; q, \frac{x_2 x_6}{x_1} \right), \end{aligned} \tag{14} $$

whose invariance under the dihedral group acting on the six variables $x_1, \cdots, x_6$ generates the set of 12 Heine transformations.

The function:

$$ \begin{aligned} f(x) &= f(x_1, x_2, x_3, x_4, x_5) \\ &= \left(\frac{x_1 x_2 x_3}{x_4 x_5}, x_4^2, x_5^2; q \right)_\infty {}_3\phi_2 \left(\begin{matrix} \frac{x_1 x_4 x_5}{x_2 x_3}, \frac{x_2 x_4 x_5}{x_1 x_3}, \frac{x_3 x_4 x_5}{x_1 x_2} \\ x_4^2, x_5^2 \end{matrix} ; q, \frac{x_1 x_2 x_3}{x_4 x_5} \right) \end{aligned} \tag{15} $$

is symmetric in the five variables $x_1, \cdots, x_5$. For the cyclic permutation: $p = (14325)$ – by this we mean the permutation $x_1 \to x_4 \to x_3 \to x_2 \to x_5 \to x_1$:

$$ p.(x_1, x_2, x_3, x_4, x_5) = (x_4, x_5, x_2, x_3, x_1) \tag{16} $$

On relabeling the parameters of the ${}_3\phi_2$s for the equation: $f(x) = f(p.x)$, where x is the five component vector, we get the transformation:

$$ {}_3\phi_2 \left(\begin{matrix} a, b, c \\ d, e \end{matrix} ; q, \frac{de}{abc} \right) = \left(b, \frac{de}{ab}, \frac{de}{ac} \right)_\infty {}_3\phi_2 \left(\begin{matrix} \frac{d}{b}, \frac{e}{b}, s \\ \frac{de}{ab}, \frac{de}{ac} \end{matrix} ; q, b \right), \tag{17} $$

which is the q- analogue of the non-terminating Thomae transformation. The defined function is manifestly invariant under the permutations (x_1, x_2, x_3) and (x_4, x_5). $f(x)$ is also invariant under the transposition $x_4 \leftrightarrow x_5$, as well as the above cyclic permutation p of order 5. Hence, the group generated by p and the transposition is S_5 which is thus the invariance group of the q-analogue of the Thomae non-terminating transformation.

Let $x_1, \cdots, x_6$ satisfy the condition: $x_1x_2x_3x_4x_5x_6 = q^{1-n}$ for some non-negative integer n. Then the function:

$$\begin{aligned} f(x) &= f(x_1, x_2, x_3, x_4, x_5, x_6) \\ &= q^{n(n-1)/2}(x_1x_2x_3x_4, x_1x_2x_3x_5, x_1x_2x_3x_6; q)_n \frac{1}{(x_1x_2x_3)^n} \\ &\times {}_4\phi_3\left(\begin{matrix} q^{-n}, x_2x_3, x_1x_3, x_1x_2 \\ x_1x_2x_3x_4, x_1x_2x_3x_5, x_1x_2x_3x_6 \end{matrix}; q, q\right), \end{aligned} \tag{18}$$

being symmetric in the variables $x_1, \cdots, x_6$ is a manifestation of the q-analogue of the terminating balanced (Saalschützian) ${}_4F_3(1)$ transformation. The invariance group is S_6 and the proof is along the same lines as that given above.

The above are the q-analogues of the results [24] of Bayer, Louck and Stein [20] for ordinary hypergeometric series transformations.

For terminating ${}_3\phi_2$ series, the q-analogue of the Whipple transformation obtained by Sears [27] is:

$${}_3\phi_2\left(\begin{matrix} q^{-n}, b, c \\ d, e \end{matrix}; q, q\right) = \frac{\left(c, \frac{de}{bc}; q\right)_n}{(d, e; q)_n} (b)^n \, {}_3\phi_2\left(\begin{matrix} q^{-n}, \frac{d}{c}, \frac{b}{c} \\ \frac{de}{bc}, \frac{q^{1-n}}{c} \end{matrix}; q, \frac{q}{b}\right), \tag{19}$$

which when recursively applied yields 72 transformations, as in the case of the terminating ${}_3F_2(1)$ investigated in [23]. This 72-element group cannot be a subgroup of S_5 by Cayley's and Lagrange's theorems (since 72 is not a factor of 120). Thus, it is shown in [24] that the invariance group of the terminating ${}_3\phi_2$ series transformations is a 72-element subgroup of S_6 generated by the permutations: (24) and (123456).

It was also shown in [24] that the transformations of the very-well-poised ${}_8\phi_7$ series belong to the invariance group (of order 1920), which is a subgroup of signed permutations on 5 elements coincides with the Weyl group of a root system of type D_5.

In 1836, Kummer published a set of six distinct solutions for the second oder ordinary Gauss differential equation characterized by three regular singular points. Each of these six solutions has four forms which are related to each other by Euler's transformations (2)–(4). Thus, there are in all 24 solutions of the Gauss hypergeometric equation which can be found in most classical text books [28]. Recently, Posser [29] related these solutions of Kummer to one another through a finite group (of order 24, or 48 if one includes the mirror symmetries) of transformations. It is remarkable that these 24 solutions have been related [25] to the (rotational) group of symmetries of the cube. This is achieved through the function:

$$f(x) = {}_2F_1(\frac{1}{2} + x_1 + x_2 + x_3, \frac{1}{2} + x_1 + x_2 + x_4; \frac{1}{2} + x_1 - x_6; -\frac{x_1 + x_6}{x_3 + x_4}), \tag{20}$$

where the six variables $x_1, \cdots, x_6$ (identified with the six faces of a cube labeled as $1, \cdots, 6$ such that labels of the opposite faces of the cube add to 7) satisfy the constraint:

$$\sum_{i=1}^{6} x_i = 0.$$

The symmetries of the cube can then be associated with the 24 permutations of a group $\epsilon\, G$ (a subgroup of the 720 permutations belonging to S_6) of the indices labeling its faces. Identifying the four arguments of the above ${}_2F_1$ with a, b, c, z, if we denote by $g \,\epsilon\, G$ one of these symmetries then $f(g.x)$ will correspond to one of the Kummer solutions. In [25] it is shown that the 24 complete Kummer solutions (up to a constant) can be related to the permutations which generate the symmetries of the cube and that the mirror symmetries, or the reflection symmetries of the cube, are the ones which correspond to the interchange of the numerator parameters of the ${}_2F_1$ function. The intimate connection between the symmetries of the cube and the 24 Kummer solutions of the Gauss differential equation, is an aesthetically beautiful result discovered 90 years after the discovery of the Gauss equation and 66 years after Kummer established its complete set of solutions.

3. Beta Integral Method

In [30] it is shown how from known identities for hypergeometric series, with lesser number of numerator and denominator parameters, involving the argument z, $1-z$ or combinations of their powers, identities for hypergeometric series with more number of numerator and denominator parameters but at some fixed argument (usually 1) can be derived. The basic idea is to multiply the known hypergeometric series identity by $z^{a-1}(1-z)^{b-1}$, integrate term-wise, use the beta integral representation for the hypergeometric function and rewrite the result in terms of a new hypergeometric series. This beta integral method has been automated using computer algebra and the software package HYP of Krattenthaler [31] to generate some old and some new results. In this section we present more results which add to the large collection of known hypergeometric identities.

The Chapter 11 in the second Notebook of Ramanujan [33] – systematically studied and edited by Bruce C. Berndt [34], over a period of more than two decades (1975 - 1997) – contains many results on quadratic transformations of hypergeometric series and many theorems on products of hypergeometric series. Most of these transformations can be used in our beta integral method and we get many interesting results, many of them new.

To illustrate the procedure in this beta integral method, consider Entry 2 of Ramanujan in Chapter 11 of his second Notebook ([34], p.49), which after $x \to -x$ and relabeling of the parameters reads:

$$ {}_2F_1(a,b;2b;\frac{-2x}{1-x}) = (1-x)^a {}_2F_1(\frac{a}{2},\frac{a+1}{2};\frac{2b+1}{2};x^2). \tag{21} $$

The ${}_2F_1$ functions are written as series, both sides are multiplied by the factor $x^{d-1}(1-x)^{e-d-1}$ and integrated term-by-term with respect to the variable x from the limits 0 to 1 and using the standard properties of the beta and gamma functions, it is straight forward to obtain the following result after minor simplifications:

$$ \begin{aligned} {}_3F_2\left(\begin{array}{c} a,\ b,\ d\,;2 \\ 2b,\ 1+d-e \end{array}\right) &= \frac{\Gamma(e,a-d+e)}{\Gamma(a+e,-d+e)} \\ &\times\ {}_4F_3\left(\begin{array}{c} \frac{a}{2},\ \frac{a}{2}+\frac{1}{2},\ \frac{d}{2},\ \frac{d}{2}+\frac{1}{2}\,;1 \\ b+\frac{1}{2},\ \frac{a}{2}+\frac{e}{2},\ \frac{a}{2}+\frac{e}{2}+\frac{1}{2} \end{array}\right), \end{aligned} \tag{22} $$

where a, b or d must be a negative integer. This result is similar to one which will be obtained from the transformation T2136 of [31] and Bailey's result ([32], (4.18)).

Entry 3 of Ramanujan ([33], Ch. XI, Vol. 2), due to Gauss [35] and (4.10) in [32] is:

$$ {}_2F_1(r,m;2m;\frac{4x}{(1+x)^2}) = (a+x)^{2r} {}_2F_1(r,r-m+\frac{1}{2};m+\frac{1}{2};x^2), \tag{23} $$

after replacing $x \to -x$, yields by the beta integral method, after multiplying by $x^{\delta-1}(1-x)^{\varepsilon-\delta-1}$ and integrating with respect to the variable x:

$$ \begin{aligned} {}_4F_3\left(\begin{array}{c} r,m,\delta,1-\varepsilon \\ 2m,\frac{1+\delta-\varepsilon}{2},1+\frac{\delta-\varepsilon}{2} \end{array};1\right) &= \frac{\Gamma(\varepsilon,2r+\varepsilon-\delta)}{\Gamma(\varepsilon-\delta,2r+\varepsilon)} \\ &\times {}_4F_3\left(\begin{array}{c} r,r-m+\frac{1}{2},\frac{\delta}{2},\frac{\delta+1}{2} \\ m+\frac{1}{2},r+\frac{\varepsilon}{2},r+\frac{\varepsilon+1}{2} \end{array};1\right), \end{aligned} \tag{24} $$

which is a combination of the two transformations T2136 and T2140 in [31].

Entry 15 of Ramanujan ([33], Ch. XI, Vol. 2) in terms of hypergeometric series ([34], Part II, p.59) is:

$$ {}_0F_1(\gamma;x){}_0F_1(\delta;x) = {}_2F_3\left(\begin{array}{c} \frac{1}{2}(\gamma+\delta),\frac{1}{2}(\gamma+\delta-1);4x \\ \gamma,\delta,\gamma+\delta-1 \end{array}\right), \tag{25} $$

after multiplication by $x^{\alpha-1}(1-x)^{\beta-\alpha-1}$ and integration results in:

$$F^{1:0;1}_{1:1;1}\left(\begin{array}{l}\alpha\ :\ -;-;\\ \beta\ :\ \gamma\ ;\ \delta\end{array};1,1\right)$$
$$={}_3F_4\left(\begin{array}{l}\alpha,\frac{1}{2}(\gamma+\delta),\frac{1}{2}(\gamma+\delta-1)\\ \beta,\gamma,\delta,\gamma+\delta-1\end{array};4\right), \tag{26}$$

where α being a negative integer assures the convergence of the hypergeometric series, since it has 4 as its argument.

If x is arbitrary, Entry 16 of Ramanujan ([33], Ch. XI, Vol. 2) in terms of hypergeometric series, after a brief calculation ([34], Part II, p.59) can be written as:

$$\begin{aligned}&{}_0F_2(m+1,n+1;x){}_0F_2(m+1,n+1;-x)\\ &={}_3F_8\left(\begin{array}{l}\frac{1}{3}(m+n+1),\frac{1}{3}(m+n+2),\\ \frac{1}{2}(m+n+1),\frac{1}{2}(m+n+2),\end{array}\right.\\ &\left.\begin{array}{l}\frac{1}{3}(m+n+3)\\ m+1,n+1,\frac{1}{2}(m+1),\frac{1}{2}(m+2),\frac{1}{2}(n+1),\frac{1}{2}(n+2)\end{array};-\frac{27}{64}x^2\right),\end{aligned} \tag{27}$$

which by the beta integration method results in the new transformation:

$$\begin{aligned}&F^{1:0;0}_{1:2;2}\left(\begin{array}{l}\alpha\ :\ --\ ;\ --\\ \beta\ :\ m+1,n+1;\ m+1,n+1\end{array};1,-1\right)\\ &={}_5F_{10}\left(\begin{array}{l}\frac{1}{2}\alpha,\frac{1}{2}\alpha+\frac{1}{2},\frac{1}{3}(m+n+1),\frac{1}{3}(m+n+2),\\ \frac{1}{2}(m+n+1),\frac{1}{2}(m+n+2),m+1,n+1,\frac{1}{2}(m+1),\end{array}\right.\\ &\left.\begin{array}{l}\frac{1}{3}(m+n+3)\\ \frac{1}{2}(m+2),\frac{1}{2}(n+1),\frac{1}{2}(n+2),\frac{1}{2}\beta,\frac{1}{2}\beta+\frac{1}{2}\end{array};-\frac{27}{64}\right).\end{aligned} \tag{28}$$

Entry 18 of Ramanujan ([33], Ch. XI, Vol. 2) and ([34], Part II, p.61):

$${}_1F_1(-\beta;\gamma;-x){}_1F_1(-\beta;\gamma;x)\ =\ {}_2F_3\left(\begin{array}{l}-\beta,\beta+\gamma\\ \gamma,\frac{1}{2}\gamma,\frac{1}{2}(\gamma+1)\end{array};\frac{x^2}{4}\right), \tag{29}$$

results in:

$$F^{1:1;1}_{1:1;1}\left(\begin{array}{l}\delta\ :\ -\beta;-\beta\\ \varepsilon\ :\ \gamma;\ \gamma\end{array};-1,1\right)$$
$$={}_4F_5\left(\begin{array}{l}-\beta,\beta+\gamma,\frac{1}{2}\delta,\frac{1}{2}(\delta+1)\\ \gamma,\frac{1}{2}\gamma,\frac{1}{2}(\gamma+1),\frac{1}{2}\varepsilon,\frac{1}{2}(\varepsilon+1)\end{array};\frac{1}{4}\right). \tag{30}$$

Corresponding to Example 7 after Entry 20 ([34], part II, p.63), which is a special case of Entry 18 of Ramanujan with $\beta=-\frac{1}{2}$ and $\gamma=1$:

$${}_1F_1(\frac{1}{2};1;x){}_1F_1(\frac{1}{2};1;-x)\ =\ {}_1F_2(\frac{1}{2};1,1;\frac{x^2}{4}), \tag{31}$$

we get the result:

$$F^{1:1;1}_{1:1;1}\left(\begin{matrix}\delta & : \frac{1}{2}; \frac{1}{2} \\ \varepsilon & : 1; 1\end{matrix}; -1, 1\right) = {}_3F_4\left(\begin{matrix}\frac{1}{2}, \frac{1}{2}\delta, \frac{1}{2}(\delta+1) \\ 1, 1, \frac{1}{2}\varepsilon, \frac{1}{2}(\varepsilon+1)\end{matrix}; \frac{1}{4}\right), \quad (32)$$

which is the same as the result (30) with $\beta = -\frac{1}{2}$ and $\gamma = 1$, except for the ${}_4F_5$ reducing to a ${}_3F_4$, due to a numerator and a denominator parameter in the former having the same value $\frac{1}{2}$.

However, though Example 8 after Entry 20 has been shown ([34], part II, p.64), to be a special case of Entry 18 of Ramanujan with $\beta = -1$, $\gamma = \frac{3}{2}$ and x replaced by $\frac{x}{2}$, viz.:

$${}_1F_1(1; \frac{3}{2}; \frac{x}{2})\,{}_1F_1(1; \frac{3}{2}; -\frac{x}{2}) = {}_2F_3(1, \frac{1}{2}; \frac{3}{2}, \frac{3}{4}, \frac{5}{4}; \frac{x^2}{16}), \quad (33)$$

the result we derive from it by our beta integral method:

$$F^{1:1;1}_{1:1;1}\left(\begin{matrix}\delta & : 1; 1 \\ \varepsilon & : \frac{3}{2}; \frac{3}{2}\end{matrix}; \frac{1}{2}, -\frac{1}{2}\right) = {}_4F_5\left(\begin{matrix}1, \frac{1}{2}, \frac{1}{2}\delta, \frac{1}{2}(\delta+1) \\ \frac{3}{2}, \frac{3}{4}, \frac{5}{4}, \frac{1}{2}\varepsilon, \frac{1}{2}(\varepsilon+1)\end{matrix}; \frac{1}{16}\right), \quad (34)$$

is interestingly **not** a special case of (30) with $\beta = -1$, $\gamma = \frac{3}{2}$ and x replaced by $\frac{x}{2}$.

Example 9, after Entry 20 of Ramanujan ([33], Ch. XI, Vol.2) is obtained by setting $\beta = -1$ and $\gamma = n + 1$ in Entry 18 as pointed out by ([34], p. 64, Part II). In this case the product of ${}_1F_1$s related to a ${}_2F_3$ transformation yields the derived result which is (30) when the parameters are set to $\beta = -1$ and $\gamma = n + 1$.

Entry 21 of Ramanujan in his second Notebook ([33], Ch. XI), a three-term relation, is originally due to Kummer ([37], p.82) and in hypergeometric notation ([34], p.64, Part II) it is:

$$\begin{aligned}&{}_2F_1(m, n; \frac{m+n+1}{2}; \frac{1+x}{2}) \\ &= \Gamma\left(\begin{matrix}\frac{1}{2}, \frac{m+n+1}{2} \\ \frac{m+1}{2}, \frac{n+1}{2}\end{matrix}\right) {}_2F_1(\frac{m}{2}, \frac{n}{2}; \frac{1}{2}; x^2) \\ &+ \Gamma\left(\begin{matrix}-\frac{1}{2}, \frac{m+n+1}{2} \\ \frac{m}{2}, \frac{n}{2}\end{matrix}\right) {}_2F_1(\frac{m+1}{2}, \frac{n+1}{2}; \frac{3}{2}; x^2)\end{aligned}$$

and this results, in our method, in the transformation:

$$
\begin{aligned}
{}_3F_2\left(\begin{matrix} m, n, -r+s \\ \frac{m+n+1}{2}, s \end{matrix} ; \frac{1}{2}\right) &= \Gamma\left(\begin{matrix} \frac{1}{2}, \frac{m+n+1}{2} \\ \frac{m+1}{2}, \frac{n+1}{2} \end{matrix}\right) {}_4F_3\left(\begin{matrix} \frac{m}{2}, \frac{n}{2}, \frac{r}{2}, \frac{r+1}{2} \\ \frac{1}{2}, \frac{s}{2}, \frac{s+1}{2} \end{matrix} ; 1\right) \\
&+ \Gamma\left(\begin{matrix} -\frac{1}{2}, \frac{m+n+1}{2}, r+1, s \\ \frac{m}{2}, \frac{n}{2}, r, s+1 \end{matrix}\right) {}_4F_3\left(\begin{matrix} \frac{m+1}{2}, \frac{n+1}{2}, \frac{r+1}{2}, r+1 \\ \frac{3}{2}, \frac{s+1}{2}, \frac{s}{2}+1 \end{matrix} ; 1\right)
\end{aligned}
\tag{35}
$$

We have seen how several results can be derived from known linear and quad - ra - tic transformations of ordinary hypergeometric series, using the beta integral me - thod.

4. 6-j coefficient in terms of sets of ${}_7V_6$

In [38], it was shown that the highly symmetric form of the Racah or 6-j angular momentum recoupling coefficient, due to Regge [39], which exhibits the 144 symmetries of the coefficient, can be formally cast into a ${}_5F_4(1)$ form. Though the ${}_5F_4(1)$ exhibits the 144 symmetries, it has the property that for real integral and half integral values of the angular momenta, the numerator and denominator parameters are integers. The termination of the series, which occurs due to the numerator parameter being a negative integer, unfortunately, occurs after the zero due to the denominator parameter! Hence, this formal expansion represents a divergent series and is not useful.

The claim that a new symmetry was found by Minton [40] for the 6-j coefficient which did not satisfy even the triangle inequalities [41] led us to the set I [6] of ${}_4F_3(1)$s for the 6-j coefficient. Racah's [42] achievement was to show that the recoupling coefficient for three angular momenta, $W(abcd; ef)$, can be written as a single sum series (independent of the 3-j coefficients and hence of the projection quantum numbers of angular momenta), viz:

$$
\left\{\begin{matrix} a & b & e \\ d & c & f \end{matrix}\right\} = (-1)^{a+b+c+d}\, W(abcd; ef) \tag{36}
$$

$$
= N \sum_P (-1)^P\, (P+1)!\, \{\prod_{i=1}^{4} (P-\alpha_i)! \prod_{j=1}^{3} (P-\beta_j)!\}^{-1}, \tag{37}
$$

with

$$N = (-1)^{a+b+c+d} \triangle(a,b,e) \triangle(c,d,e) \triangle(a,c,f) \triangle(b,d,f), \quad (38)$$

$$\triangle(x,y,z) = \left[\frac{(-x+y+z)!(x-y+z)!(x+y-z)!}{(x+y+z+1)!}\right]^{1/2}, \quad (39)$$

$$\alpha_1 = a+b+e,\ \alpha_2 = c+d+e,\ \alpha_3 = a+c+f,\ \alpha_4 = b+d+f,$$
$$\beta_1 = a+b+c+d,\ \beta_2 = a+d+e+f,\ \beta_3 = b+c+e+f \quad (40)$$

and

$$P_{min} \leq P \leq P_{max}, \quad (41)$$
$$P_{min} = max(\alpha_1, \alpha_2, \alpha_3, \alpha_4), \quad P_{max} = min(\beta_1, \beta_2, \beta_3). \quad (42)$$

By setting in (37), $s = \beta_k - P, k = 1,2,3$, in succession, a set I of three series expansions, and in turn, the set I of the following three ${}_4F_3(1)$s has been obtained in [6]:

$$\begin{Bmatrix} a & b & e \\ d & c & f \end{Bmatrix} = (-1)^{E+1} N\, \Gamma(1-E) \times [\Gamma(1-A, 1-B, 1-C, 1-D, F, G)]^{-1} \times {}_4F_3(ABCD\,;\ EFG\,;\ 1), \quad (43)$$

where

$$A = e-a-b,\ B = e-c-d,\ C = f-a-c,\ D = f-b-d,$$
$$E = -a-b-c-d-1,\ F = e+f-b-c+1,$$
$$G = e+f-a-d+1, \quad (44)$$

for $k = 1$ and for $k = 2$ and 3, the numerator and denominator parameter sets are:

$$A = a-b-e,\ B = d-c-e,\ C = a-c-f,\ D = d-b-f,$$
$$E = -b-c-e-f-1,\ F = a+d-b-c+1,$$
$$G = a+d-e-f+1, \quad (45)$$

and

$$A = b-a-e,\ B = c-d-e,\ C = c-a-f,\ D = b-d-f,$$
$$E = -a-d-e-f-a,\ F = b+c-a-d+1,$$
$$G = b+c-e-f+1. \quad (46)$$

The Minton procedure was to apply a Saalschützian ${}_4F_3(1)$ transformation to the ${}_4F_3(1)$ in (43) (with the parameters given by (44)) and identify the transformed

${}_4F_3(1)$ with a 6-j coefficient using again (43). We have shown in [5] that this procedure will, *at best*, result in a $j \to -j-1$ substitution for one or more of the six angular momenta in the 6-j coefficient, which violates triangle inequalities [41], though it is a mathematically valid symmetry in quantum theory of angular momentum.

A hypergeometric series is called very well-poised if the numerator (a_i) and denominator (b_i) parameters of the hypergeometric series are such that:

$$a_i + b_i = 1 + a_0, \quad \text{for} \quad i = 1, 2, \cdots n,$$

and amongst the parameters a_i occurs $1 + a_0/2$. The standard abbreviated notation for a very well-poised hypergeometric series uses its numerator parameters only and it is:

$$\begin{aligned} &{}_{r+1}V_r(a_0;\ a_2, a_3, \cdots, a_r;\ z) = \\ &= {}_{r+1}F_r\left(\begin{matrix} a_0, 1+\frac{a_0}{2}, & a_2, & a_3, & \cdots, & a_r \\ \frac{a_0}{2}, & 1+a_0-a_2, & 1+a_0-a_3, & \cdots, & 1+a_0-a_r \end{matrix} ; z \right). \end{aligned} \tag{47}$$

The transformation of a balanced or Saalschützian ${}_4F_3(1)$ series into a well-poised ${}_7F_6(1)$ series, with the special form of the second parameter, was given by Whipple [43]:

$$\begin{aligned} &{}_4F_3\left(\begin{matrix} x, y, z, -n \\ u, v, w \end{matrix} ; 1 \right) \\ &= \Gamma\left(\begin{matrix} v+w-x, 1+y-u, 1+z-u, 1-n-u \\ 1+y-n-u, 1+z-n-u, 1+y+z-u, 1-u \end{matrix} \right) \times \\ &\quad {}_7F_6\left(\begin{matrix} a, 1+\frac{a}{2}, w-x, v-x, & y, \\ \frac{a}{2}, \quad v, & w, 1+z-u-n, \end{matrix} \right), \\ &\quad \left. \begin{matrix} z, & -n \\ 1+y-u-n, & 1+y+z-u \end{matrix} ; 1 \right) \end{aligned} \tag{48}$$

in which $a = y+z-u-n = w+v-x-1$ (and $x+y+z-n+1 = u+v+w$) and the ${}_7F_6(1)$ in the expression above can be written in the shortened notation for the very well-poised hypergeometric series as:

$${}_7F_6(-;-;1) \equiv {}_7V_6(a;\ w-x, v-x, y, z, -n). \tag{49}$$

Recently, Alisauskas [44] has claimed that he has found *a new expression for the 6-j coefficient of SU(2)* and it is (with minor changes in the notation):

$$\left\{ \begin{matrix} a & b & e \\ d & c & f \end{matrix} \right\} = \nabla(a,c,f)\nabla(b,d,f)\nabla(e,d,c) \bigtriangleup (a,b,e)$$

$$\times \sum_j \frac{(-1)^{d+c+f-(a+b+e)/2+j}\,(2j+1)}{\nabla^2(\frac{1}{2}(b+e-a),d,j)\nabla^2(\frac{1}{2}(a-b+e),c,j)\nabla^2(\frac{1}{2}(a+b-e),f,j)}, \tag{50}$$

where

$$\nabla(x,y,z) = \triangle(x,y,z)\frac{(x+y+z+1)!}{(-x+y+z)!}$$

and he states that the sum is the well-poised ${}_7F_6(1)$. He also obtains another expression for the 6-j coefficient by substituting $a \to -a-1, b \to -b-1$, which is also a well-poised ${}_7F_6(1)$.

We use of the Whipple transformation for the ${}_4F_3(1)$ forms of the 6-j coefficient, and since in this set of ${}_4F_3(1)$s all the four numerator parameters are negative integers, there are four different ways in which the given transformation can be used to obtain well-poised ${}_7V_6(1)$ forms for the 6-j coefficient. In the table given below are the four different permutations and the corresponding ${}_7V_6(1)$ forms we obtain for the 6-j coefficient.

Parameter permutation	${}_7V_6$ (parameters)
(4231);(231)	${}_7V_6$(A+B+C-E; G-D,F-D,A,B,C)
(3241);(231)	${}_7V_6$(A+B+D-E; G-C,F-C,A,B,D)
(2341);(231)	${}_7V_6$(A+C+D-E; G-B,F-B,A,C,D)
(1234);(231)	${}_7V_6$(B+C+D-E; G-A,F-A,B,C,D)

These are the only independent ${}_7V_6$ forms for the 6-j coefficient. Of these, the one corresponding to the second of the three ${}_4F_3(1)$ forms (45) for the 6-j coefficient corresponds to the 'new' expression reported by Alisauskas.

In our approach we have shown [5] that there are two sets of ${}_4F_3(1)$s which are related to each other by 'reversal' of series, which completely maps the set I of three ${}_4F_3(1)$s – explicitly given here in this section – onto the set II of four ${}_4F_3(1)$s). From the table given above, it is clear that we get for each of the ${}_4F_3(1)$s belonging to set I of hypergeometric functions, a set of four ${}_7V_6(1)$s. It is straight forward to then observe that each of these ${}_7V_6(1)$s will account for only 12 of the 144 symmetries of the 6-j coefficient. For instance, the first entry in the table above clearly shows that the symmetries exhibited are due to the permutations of the parameters A, B, C (S_3) and the parameters F, G (S_2). It is also clear from an examination of the table that the 12 symmetries corresponding to each of the four ${}_7V_6(1)$s are distinctly different. Thus, we can conclude, that if we want to express the 6-j coefficient as a ${}_7V_6(1)$, then **a set of 12 of ${}_7V_6(1)$s is necessary and sufficient to account for the 144 symmetries** of the 6-j-coefficient.

Equivalently, corresponding to the set II of four ${}_4F_3(1)$s, one can write down a set of 16 ${}_7V_6(1)$s, each of which will account for only 9 of the 144 symmetries of the 6-j coefficient, enabling us to draw the conclusion that if the 6-j coefficient is represented as a ${}_7V_6(1)$ derived by the use of the Whipple transformation on the ${}_4F_3(1)$ belonging to set II of ${}_4F_3(1)$s for the 6-j coefficient, then a set of 16 ${}_7V_6(1)$s is necessary and sufficient to account for the 144 symmetries of the 6-j coefficient.

This completes our understanding of the 144 symmetries of the 6-j coefficient in terms of the equivalent sets of 12 or 16 ${}_7V_6(1)$s. The 'new' expression obtained by Alisauskas is just one member of the set of 12 ${}_7V_6(1)$s.

5. A q-generalization of a new ${}_3F_2$ summation theorem

Ramanujan's Example 7, after Entry 43, in Chapter XII, of the first Notebook reads is:

$$\frac{\pi}{\tan(\pi x)} \frac{\lfloor 2x}{\left(2^x \lfloor x\right)^2 (1-2x)} \left(\sum \frac{1}{2x} - \frac{1}{2}\sum \frac{1}{x} + \frac{1}{1-2x} - \frac{\pi}{2}\tan(\pi x)\right)$$
$$= \frac{1}{1^2} + \frac{x}{\lfloor 1} \cdot \frac{1}{3^2} + \frac{x(x+1)}{\lfloor 2} \cdot \frac{1}{5^2} + \&\text{c.(XII, 43, Ex.7)}$$

where, $\lfloor x$ is his notation for the *gamma function* $\Gamma(x+1)$. In [45], we have stated and proved a new summation theorem for the ${}_3F_2(x, a, a; 1+a, 1+a+N; 1)$, and showed that the above entry is a very special case of that theorem. Also, for $x = 1, a = 1/2$ and $N = 0$, our theorem reduces to Dixon's theorem for a well-posed ${}_3F_2(1)$. In this section, we obtain a q-generalization of the Krattenthaler-Srinivasa Rao Theorem [30].

Let us define the following q-analogue of the digamma function (see [46, (2.10)]), with q being a fixed complex number with $|q| < 1$,

$$\tilde{\psi}_q(x) := -(1-q)\sum_{n=0}^{\infty} \frac{xq^n}{1-xq^n}. \tag{51}$$

Then we have the following theorem:

Theorem 1. *Let N be a non-negative integer and a be a complex number which is not of the form q^{-m}, where m is a non-negative integer. If $|x| > |q^{N+2}|$, then*

$$
\begin{aligned}
&{}_3\phi_2\left[\begin{matrix} a,\, a,\, x \\ aq,\, aq^{N+1} \end{matrix}\, ;\, q,\, \frac{q^{N+2}}{x}\right] \\
&\qquad = \frac{(1-a)}{(1-q)} \frac{(q^{N+1};q)_\infty\,(aq/x;q)_\infty}{a^{N+1}\,(aq^{N+1};q)_\infty\,(q/x;q)_\infty} \\
&\qquad \times \Bigl(\tilde{\psi}_q\left(aq^{N+1}/x\right) - \tilde{\psi}_q(a) - \tilde{\psi}_q(q^{N+1}) + \tilde{\psi}_q(q) \\
&\qquad - \sum_{k=1}^{N} \frac{(1-q)}{(1-q^k)} \frac{(q/a;q)_k\,(q^{-N};q)_k}{(q;q)_k\,(x/aq^N;q)_k} x^k \Bigr) . \qquad (52)
\end{aligned}
$$

Proof. We start with the ${}_3\phi_2$-series

$$
{}_3\phi_2\left[\begin{matrix} a, x, a\delta \\ aq, aq^{1+N} \end{matrix}\, ;\, q,\, \frac{q^{N+2}}{\delta x}\right].
$$

Clearly, for $\delta = 1$ this series reduces to the ${}_3\phi_2$-series on the left-hand side of 52. To the above ${}_3\phi_2$-series we apply the three-term transformation formula (see [47, (3.3.3); Appendix (III.33)])

$$
\begin{aligned}
&{}_3\phi_2\left[\begin{matrix} A, B, C \\ D, E \end{matrix}\, ;\, q,\, \frac{DE}{ABC}\right] \\
&= \frac{(\frac{E}{B}, \frac{E}{C}, \frac{Cq}{A}, \frac{q}{D}; q)_\infty}{(C, \frac{Cq}{D}, \frac{q}{A}, \frac{E}{BC}; q)_\infty}\, {}_3\phi_2\left[\begin{matrix} C, \frac{D}{A}, \frac{Cq}{E} \\ \frac{Cq}{A}, \frac{BCq}{E} \end{matrix}\, ;\, q,\, \frac{Bq}{D}\right] \\
&\quad - \frac{(\frac{q}{D}, \frac{Eq}{D}, B, C, \frac{D}{A}, \frac{DE}{BCq}, \frac{BCq^2}{DE}; q)_\infty}{(\frac{D}{q}, E, \frac{Bq}{D}, \frac{Cq}{D}, \frac{q}{A}, \frac{E}{BC}, \frac{BCq}{E}; q)_\infty} \\
&\quad \times {}_3\phi_2\left[\begin{matrix} \frac{Aq}{D}, \frac{Bq}{D}, \frac{Cq}{D} \\ \frac{q^2}{D}, \frac{Eq}{D} \end{matrix}\, ;\, q,\, \frac{DE}{ABC}\right], \qquad (53)
\end{aligned}
$$

with $A = a$, $B = x$, $C = a\delta$, $D = aq$, $E = aq^{N+1}$. Since for this choice of parameters we have $Aq/D = 1$, the second ${}_3\phi_2$-series on the right-hand side of 53 reduces to 1, as a result of the application of 53 to our ${}_3\phi_2$-series we obtain the expression

$$
\begin{aligned}
&\frac{(aq^{N+1}/x, q^{N+1}/\delta, \delta q, \frac{1}{a}; q)_\infty}{(aq^{N+1}, \delta, \frac{q}{a}, q^{N+1}/\delta; q)_\infty}\, {}_3\phi_2\left[\begin{matrix} \frac{\delta}{q^N}, a\delta, q \\ \delta q, \frac{\delta x}{q^N} \end{matrix}\, ;\, q,\, \frac{x}{a}\right] \\
&- \frac{(\frac{1}{a}, q^{N+1}, x, a\delta, q, \frac{aq^{N+1}}{\delta x}, \frac{\delta x}{aq^N}; q)_\infty}{(a, aq^{1+N}, \frac{x}{a}, \delta, \frac{q}{a}, \frac{q^{N+1}}{\delta x}, \frac{\delta x}{q^N}; q)_\infty}.
\end{aligned}
$$

To ensure convergence of the latter ${}_3\phi_2$-series we must have $|x| < |a|$, which in turn requires that $|a| > |q^{N+2}|$. So, let us for the moment assume these two restrictions.

We continue by applying the following ${}_3\phi_2$ transformation formula (see [47, (3.2.7); Appendix (III.9)])

$$ {}_3\phi_2\left[\begin{matrix} a, b, c \\ d, e \end{matrix} ; q, \frac{de}{abc}\right] = \frac{(e/a, de/bc; q)_\infty}{(e, de/abc; q)_\infty} {}_3\phi_2\left[\begin{matrix} a, d/b, d/c \\ d, de/bc \end{matrix} ; q, \frac{e}{a}\right] $$

to our series. Thus we obtain the expression

$$ \frac{1}{1-\delta}\left(\frac{(x, \delta x/aq^N, aq^{N+1}/x, q^{N+1}/\delta, 1/a; q)_\infty}{(\delta x/q^N, x/a, aq^{N+1}, q/a, q^{N+1}/\delta x; q)_\infty} {}_3\phi_2 \left[\begin{matrix} \delta/q^N, q/a, \delta \\ \delta q, \delta x/aq^N \end{matrix} ; q, x\right] \right. $$
$$ \left. - \frac{(1/a, q^{N+1}, x, a\delta, q, aq^{N+1}/\delta x, \delta x/aq^N; q)_\infty}{(a, aq^{N+1}, x/a, \delta q, q/a, q^{1+N}/\delta x, \delta x/q^N; q)_\infty}\right). \quad (54) $$

For convergence of the latter ${}_3\phi_2$-series, we assume that, in addition to the restrictions that we already imposed, we have $|x| < 1$.

We now compute the limit of 54 as $\delta \to 1$, by using de l'Hôpital's rule. A straight-forward calculation shows that this limit is exactly equal to the right-hand side of 52.

As it stands, the assertion is only demonstrated for $|q^{N+1}| < |x| < \min\{1, |a|\}$. However, by analytic continuation, Equation 52 is true for any values of x and a for which the ${}_3\phi_2$-series on the left-hand side converges, i.e., for $|x| < |q^{N+1}|$.
□

In [46, (2.12)] it was shown that, although the limit $\lim_{q\uparrow 1} \tilde{\psi}_q(q^A)$ does not even exist, we have

$$ \lim_{q\uparrow 1}\left(\tilde{\psi}_q(q^A) - \tilde{\psi}_q(q^B)\right) = \psi(A) - \psi(B), $$

where $\psi(z)$ is the ordinary digamma function, $\psi(z) := \Gamma'(z)/\Gamma(z)$. Therefore, if in 52 we replace a by q^a and x by q^x, and then let $q \to 1$, we obtain the formula

$$
\begin{aligned}
{}_3F_2\left[\begin{matrix} a, a, x \\ a+1, a+N+1 \end{matrix}; 1\right] &= \frac{a\,\Gamma(a+N+1)\,\Gamma(1-x)}{N!\,\Gamma(a-x+1)} \\
&\times \Big(\psi(a-x+N+1) - \psi(a) - \psi(N+1) + \psi(1) \\
&- \sum_{k=1}^{N} \frac{(1-a)_k(-N)_k}{k\cdot k!\,(a-x+1)_k}\Big). \qquad (55)
\end{aligned}
$$

In view of $\psi(1) = -\gamma$, where γ is the Euler–Mascheroni constant, this formula agrees almost with Theorem 1 from [48],

$$
\begin{aligned}
{}_3F_2\left[\begin{matrix} a, a, x \\ a+1, a+N+1 \end{matrix}; 1\right] &= \frac{a\,\Gamma(a+N+1)\,\Gamma(1-x)}{N!\,\Gamma(a-x+1)} \\
&\times \Big(\psi(a-x+1) - \psi(a) - \psi(N+1) - \gamma \\
&- \sum_{k=1}^{N} \frac{(a)_k\,(-N)_k}{k\cdot k!(a-x+1)_k}\Big). \qquad (56)
\end{aligned}
$$

However, it is not too difficult to see directly that these two forms agree. In fact, we are going to derive a q-analogue of the form 56, by starting from 52. As it will turn out, this q-analogue will not be as elegant as 56. In this q-analogue, a simplification takes place only if $q = 1$.

It is sufficient to just consider the sum on the right-hand side of 52,

$$
\sum_{k=1}^{N} \frac{(1-q)}{(1-q^k)} \frac{(q/a;q)_k\,(q^{-N};q)_k}{(q;q)_k\,(x/aq^N;q)_k} x^k.
$$

For our purposes it will be necessary to write this sum as a limit,

$$
\begin{aligned}
&\lim_{\delta\to 1} \sum_{k=1}^{N} \frac{(1-q)}{(1-q^k\delta)} \frac{(q/a;q)_k\,(q^{-N};q)_k}{(q;q)_k\,(x/aq^N;q)_k} x^k \\
&= \lim_{\delta\to 1} \left(\sum_{k=0}^{N} \frac{(1-q)}{(1-q^k\delta)} \frac{(q/a;q)_k\,(q^{-N};q)_k}{(q;q)_k\,(x/aq^N;q)_k} x^k - \frac{1-q}{1-\delta} \right).
\end{aligned}
$$

In hypergeometric notation this is

$$
\frac{1-q}{1-\delta} \left({}_3\phi_2\left[\begin{matrix} q/a, \delta, q^{-N} \\ \delta q, x/aq^N \end{matrix}; q, x\right] - 1 \right).
$$

We transform this ${}_3\phi_2$-series by using the transformation formula (see [47, (3.2.5); Appendix (III.13)])

$$
\begin{aligned}
&{}_3\phi_2\left[\begin{matrix} A, B, q^{-N} \\ D, E \end{matrix}; q, \frac{DEq^N}{AB}\right] \\
&= \frac{(E/B;q)_N}{(E;q)_N}\,{}_3\phi_2\left[\begin{matrix} q^{-N}, B, D/A \\ D, Bq^{1-N}/E \end{matrix}; q, q\right],
\end{aligned}
$$

where N is a nonnegative integer. Thus we arrive at the expression

$$
\lim_{\delta\to 1}\frac{1-q}{1-\delta}\left(\frac{(x/a\delta q^N;q)_N}{(x/aq^N;q)_N}\,{}_3\phi_2\left[\begin{matrix} q^{-N}, \delta, a\delta \\ \delta q, a\delta q/x \end{matrix}; q, q\right] - 1\right).
$$

This limit can be easily evaluated by means of de l'Hôpital's rule. The result is

$$
-(1-q)\sum_{k=1}^{N}\frac{\frac{x}{aq^k}}{1-\frac{x}{aq^k}} + \sum_{k=1}^{N}\frac{(1-q)}{(1-q^k)}\frac{(a;q)_k\,(q^{-N};q)_k}{(q;q)_k\,(aq/x;q)_k}q^k.
$$

We have thus derived the following corollary.

Corollary 2. *Let N be a non-negative integer and a be a complex number which is not of the form q^{-m}, where m is a non-negative integer. If $|x| > |q^{N+2}|$, then*

$$
\begin{aligned}
&{}_3\phi_2\left[\begin{matrix} a, a, x \\ aq, aq^{N+1} \end{matrix}; q, \frac{q^{N+2}}{x}\right] = \frac{(1-a)}{(1-q)}\frac{(q^{N+1};q)_\infty\,(\frac{aq}{x};q_\infty}{a^{N+1}\,(aq^{N+1};q)_\infty\,(\frac{q}{x};q)_\infty} \\
&\times\left(\tilde{\psi}_q\left(aq^{N+1}/x\right) + (1-q)\sum_{k=1}^{N}\frac{\frac{x}{aq^k}}{1-\frac{x}{aq^k}} - \tilde{\psi}_q(a) - \tilde{\psi}_q(q^{N+1}) + \tilde{\psi}_q(q)\right. \\
&\left.-\sum_{k=1}^{N}\frac{(1-q)}{(1-q^k)}\frac{(a;q)_k\,(q^{-N};q)_k}{(q;q)_k\,(aq/x;q)_k}q^k\right). \qquad (57)
\end{aligned}
$$

The first terms in parentheses on the right-hand side,

$$
\tilde{\psi}_q\left(aq^{N+1}/x\right) + (1-q)\sum_{k=1}^{N}\frac{\frac{x}{aq^k}}{1-\frac{x}{aq^k}} = \tilde{\psi}_q\left(aq^{N+1}/x\right) - (1-q)\sum_{k=1}^{N}\frac{1}{1-\frac{aq^k}{x}},
$$

do not fit together to add up to $\tilde{\psi}_q\,(aq/x)$. However, in the limit $q \uparrow 1$ (where we have replaced a by q^a and x by q^x), this expression tends to

$$
\psi\,(a-x+N+1) - \sum_{k=1}^{N}\frac{1}{a+k-x},
$$

which does indeed tend to $\psi(a-x+1)$.

We now list some an interesting special case:

• If we choose $N=0$ in 52, then we obtain the formula

$$
{}_3\phi_2\left[\begin{array}{c} a, a, x \\ aq, aq \end{array}; q, \frac{q^2}{x}\right]
= -\frac{\left(1-\frac{1}{a}\right)}{(1-q)} \frac{(q;q)_\infty\,(aq/x;q)_\infty}{(aq;q)_\infty\,(q/x;q)_\infty} \left(\tilde{\psi}_q\,(aq/x) - \tilde{\psi}_q(a)\right). \quad (58)
$$

This is a q-analogue of Example 5 of Section 10 of Chapter 10 of Ramanujan's second Notebook ([34], p. 26, Part II).

Acknowledgments

One of us (KSR) wishes to thank Professor Bruno Gruber for the kind invitation to the XIII Symmetries in Science Symposium and for the excellent hospitality. It was the second pleasurable experience he had and it gave him an opportunity to meet again friends of Gruber and also spend a week in Vienna en route, during which period a part of this work was done.

References

[1] M.E. Rose, *Multipole Fields*, Wiley, New York (1955).

[2] Yu.A. Smorodinskii and L.A. Shelepin, Soviet Phys. Uspekhi **15** (1972) 1.

[3] B.M. Minton, J. Math. Phys. **11** (1970) 1375.

[4] K. Srinivasa Rao and V. Rajeswari, *Quantum Theory of Angular Momentum: Selected Topics*, Springer-Verlag, Heidelberg and Narosa Publishing House, New Delhi (1993).

[5] K. Srinivasa Rao, J. Phys. **A 11** (1978) L69.

[6] K. Srinivasa Rao, T.S. Santhanam and K. Venkatesh, J. Math. Phys. **16** (1975) 1528.

[7] K. Srinivasa Rao and K. Venkatesh, in *Group Theoretical Methods in Physics*, eds. R.T. Sharp and B. Kolman, Academic Press (1977) p. 649.

[8] Lucy J. Slater, *Generalized Hypergeometric Functions*, Cambridge Univ. Press (1966).

[9] R.A. Askey, in Preface to *Special Functions*, George E. Andrews, Richard Askey and Ranjan Roy, Encycl. of Maths. and its Applns., Cambridge Univ. Press, Vol. **71** (1991).

[10] A.C.T. Wu, J. Math. Phys.**13** (1972) 84.

[11] A.C.T. Wu, J. Math. Phys.**14** (1973) 1222.

[12] S.J. Alisauskas, A.P. Jucys, J. Math. Phys. **12** (1971) 594; A.P. Jucys and A.A. Bandzaitis, *Angular Momentum in Quantum Physics*, Mokslas, Vilnius (1977).

[13] K. Srinivasa Rao and V. Rajeswari, J. Math. Phys. **30** (1989) 1016.

[14] K. Srinivasa Rao and J. Van der Jeugt, Int. J. Theor. Phys. **37** (1998) 891.

[15] K. Srinivasa Rao, Symmetries in Science, **XI**, ed. B. Gruber and M. Ramek, (1998) 383.

[16] G.E. Andrews, R. Askey and R. Roy, *Special Functions*, Encycl. of Maths. and its Applns., **71**, Cambridge Univ. Press (1999).

[17] L. Saalschütz, Zeitschr. für Math. und physik **35** (1890) 186.

[18] M. Weber and A. Erdélyi, Am. Math. Monthly **159** (1952) 163.

[19] J. Thomae, J. Reine Agnew. Math.**87** (1879) 26.

[20] W.A. Beyer, J.D. Louck and P.R. Stein, Group theoretical basis of some identities for the generalized hypergeometric series. J. Math. Phys. 28 (1987), no. 3, 497–508.

[21] K. Srinivasa Rao, J. Van der Jeugt, J. Raynal, R. Jagannathan and V. Rajeswari, J. Phys. A: Math. and Gen. **25** (1992) 861.

[22] G.H. Hardy, in *Ramanujan: Twelve Lectures on Subjects suggested by his Life and Work*, AMS Chelsea Pub., Providence, Rhode Island (1999) p.111.

[23] K. Srinivasa Rao, H.-D. Doebner and P. Natterman, in *Number theoretic methods – Future trends*, ed. S. Kanemitsu and C. Jia, Kluwer Acad. Publs., Dodrecht (2002) 381.

[24] J. Van der Jeugt and K. Srinivasa Rao, J. Math. Phys. **40** (1999) 6692.

[25] S. Lievens, K. Srinivasa Rao and J. Van der Jeugt, Integral Transforms and Special Functions (2003) to appear.

[26] G. Gasper and M. Rahman, *Basic Hypergeometric Series*, Encycl. Maths. and Applns., Ed. G.-C. Rota, **35**, Cambridge Univ. Press (1990).

[27] D.B. Sears, Proc. London Math. Soc. (2) **52** (1950) 14; ibid (1953) 138, 158 and 181.

[28] E.T. Whittaker and G.N. Watson, *A Course of modern analysis*, Cambridge Univ. Press (1965).

[29] R.T. Posser, Am. Math. Monthly **101** (1994) 535.

[30] C. Krattenthaler and K. Srinivasa Rao, J. Comput. and Applied Maths. **160** (2003) 159 – 173.

[31] C. Krattenthaler, *HYP*, Manual for a *Mathematica* package for handling hypergeometric series, available from http://www.mat.univie.ac.at/People/kratt

[32] W.N. Bailey, Proc. London Math. Soc. (2), **28** (1928) 242.

[33] *Notebooks of Srinivasa Ramanujan*, (facsimile edition) 2 Volumes, Tata Institute of Fundamental Research, Bombay, 1957; also, Narosa (1987).

[34] Bruce C. Berndt, *Ramanujan Notebooks*, Springer- Verlag, New York, Part I (1985), Part II (1989), Part III (1991), Part IV (1994), Part V (1997).

[35] C.F. Gauss, *Disquisitiones generales circa seriem infinitam* $1 + \frac{\alpha\beta}{1\cdot\gamma}x + \frac{\alpha(\alpha+1)\beta(\beta+1)}{1\cdot2\cdot\gamma(\gamma+1)}xx + \frac{\alpha(\alpha+1)(\alpha+2)\beta(\beta+1)()\beta+2}{1\cdot2\cdot3\cdot\gamma(\gamma+1)(\gamma+2)}x^3 + etc.$, Pars prior, Comm. soc. regiae sci. Gottingensis rec. 2 (1812), reprinted in C.F. Gauss's *Werke*, band 3, Königlichen Gesellschaft der Wissenschaften, Göttingen, 1876, pp. 123–162.

[36] G.H. Hardy, Proc. Camb. Phil. Soc. **21** (1923) 492 – 503.

[37] E.E. Kummer, J. Reine Angew. Math. **15** (1836) 39 – 83, 127 – 172.

[38] K. Srinivasa Rao, *Hypergeometric series and Quantum Theory of Angular Momentum*, in *Selected Topics in Special Functions*, Eds.: R.P. Agarwal, H.L. Manocha and K. Srinivasa Rao, Allied Publishers Ltd. (2001) 93 – 134.

[39] T. Regge, Nuo. Cim. **10** (1958) 544; *ibid* **11** (1959) 116.

[40] B.M. Minton, J. Math. Phys. **11** (1970) 3061.

[41] Vinaya Joshi, J. Math. Phys. **12** (1971) 1134

[42] G. Racah, Phys. Rev. **61** (1942) 186; **62** (1942) 438; **63** (1943) 367; **76** (1949) 1352.

[43] F.J.W. Whipple, Proc. Lon. Math. Soc. (2) **25** (1926) 525 – 544.

[44] S. Alisauskas, J. Math. Phys. **33** (1992) p. 2203 (Appendix C).

[45] K. Srinivasa Rao, G. Vanden Berghe and C. Krattenthaler, arXiv:math.CA/0304317v1 dated April 22, 2003 (submitted for publication).

[46] C. Krattenthaler and H.M. Srivastava, *Summations for basic hypergeometric series involving a q-analogue of the digamma function*, Comput. Math. Appl. **32** (1996), 73–91.

[47] G. Gasper and M. Rahman, *Basic Hypergeometric Series*, Encyclopedia of Mathematics And Its Applications 35, Cambridge University Press, Cambridge, 1990.

[48] K. Srinivasa Rao, G. Vanden Berghe and C. Krattenthaler, *An entry of Ramanujan on hypergeometric series in his notebooks*, preprint.

TENSOR AND SPIN REPRESENTATIONS OF SO(4) AND DISCRETE QUANTUM GRAVITY

M. Lorente
Departamento de Física,
Universidad de Oviedo,
33007 Oviedo, Spain

P. Kramer
Institut für theoretische Physik
Universität Tübingen,
72076 Tübingen, Germany

Abstract Starting from the defining transformations of complex matrices for the $SO(4)$ group, we construct the fundamental representation and the tensor and spinor representations of the group $SO(4)$. Given the commutation relations for the corresponding algebra, the unitary representations of the group in terms of the generalized Euler angles are constructed. These mathematical results help us to a more complete description of the Barret-Crane model in Quantum Gravity. In particular a complete realization of the weight function for the partition function is given and a new geometrical interpretation of the asymptotic limit for the Regge action is presented.

Keywords: SO(4) group, tensor representation, spin representation, quantum gravity, spin networks.

1. Discrete models in quantum gravity

The use of discrete models in Physics has become very popular, mainly for two reasons. It helps to find the solutions of some differential equations by numerical methods, which would not be possible to solve by analytic methods. Besides that, the introduction of a lattice is equivalent to the introduction of a cut-off in the momentum variable for the field in order to achieve the finite limit of the solution. In the case of relativistic field equations -like the Dirac,

B. Gruber (ed.), Symmetries in Science XI, 377–394.

Klein-Gordon, and the electromagnetic interactions- we have worked out some particular cases [1].

There is an other motivation for the discrete models and it is based in some philosophical presuppositions that the space-time structure is discrete. This is more attractive in the case of general relativity and quantum gravity because it makes more transparent the connection between the discrete properties of the intrinsic curvature and the background independent gravitational field.

This last approach was started rigorously by Regge in the early sixties [2]. He introduces some triangulation in a Riemannian manifold, out of which he constructs local curvature, coordinate independent, on the polyhedra. With the help of the total curvature on the vertices of the discrete manifold he constructs a finite action which, in the continuous limit, becomes the standard Hilbert-Einstein action of general relativity.

Regge himself applied his method ("Regge calculus") to quantum gravity in three dimensions [3]. In this work he assigns some representation of the $SU(2)$ group to the edges of the triangles. To be more precise, to every tetrahedron appearing in the discrete triangulation of the manifold he associates a 6j-symbol in such a way that the spin eigenvalues of the corresponding representation satisfy sum rules described by the edges and vertices of the tetrahedra. Since the value of the 6j-symbol has a continuous limit when some edges of the tetrahedra become very large, he could calculate the sum of this limit for all the 6j-symbols attached to the tetrahedra, and in this way he could compare it with the continuous Hilbert-Einstein action corresponding to an Euclidean non planar manifold.

A different approach to the discretization of space and time was taken by Penrose [4]. Given some graph representing the interaction of elementary units satisfying the rules of angular momentum without an underlying space, he constructs out of this network ("spin network") the properties of total angular momentum as a derived concept. Later this model was applied to quantum gravity in the sense of Ponzano and Regge. In general, a spin network is a triple (γ, ρ, i) where γ is a graph with a finite set of edges e, a finite set of vertices v, ρ_e is the representation of a group G attached to an edge, and i_v is an intertwiner attached to each vertex. If we take the product of the amplitudes corresponding to all the edges and vertices (given in terms of the representations and intertwiners) we obtain the particular diagram of some quantum state.

Although the physical consequences of Penrose's ideas were soon considered to be equivalent to the Ponzano-Regge approach to quantum gravity, the last method was taken as guiding rule in the calculation of partition functions. We can mention a few results. Turaev and Viro [6] calculated the state sum for a 3d-triangulated manifold with tetrahedra described by 6j-symbols using the $SU(2)_q$ group. This model was enlarged to 4-dimensional triangulations and was proved by Turaev, Oguri, Crane and Yetter [7] to be independent of the triangulation (the "TOCY model").

A different approach was introduced by Boulatov [8] that led to the same partition function as the TOCY model, but with the advantage that the terms corresponding to the kinematics and the interaction could be distinguished. For this purpose he introduced some fields defined over the elements of the groups $SO(3)$, invariant under the action of the group, and attached to the edges of the tetrahedra. The kine - ma - ti - cal term corresponds to the self interacting field over each edge and the interaction term corresponds to the fields defined in different edges and coupled among themselves. This method (the Boulatov matrix model) was very soon enlarged to 4-dimensional triangulations by Ooguri [9]. In both models the fields over the matrix elements of the group are expanded in terms of the representations of the group and then integrated out, with the result of a partition function extended to the amplitudes over all tetrahedra, all edges and vertices of the triangulation.

A more abstract approach was taken by Barret and Crane, generalizing Penrose's spin networks to 4 dimensions. The novelty of this models consists in the association of representation of the $SO(4)$ groups to the faces of the tetrahedra. We will come back to this model in section 5.

Because we are interested in the physical and mathematical properties of the Barret-Crane model, we mention briefly some recent work about this model combined with the matrix model approach of Boulatov and Ooguri [10]. In this work the 2d– quantum space-time emerges as a Feynman graph, in the manner of the 4d– matrix models. In this way a spin foam model is connected to the Feynman diagram of quantum gravity.

In this paper we have tried to implement all the mathematical consequences of Barret-Crane model using the group theoretical properties of $SO(4)$ applied to the 4d-triangulation of manifolds in terms of 4-simplices. It turns out that when we take into account the description of spin representations of $SO(4)$ the weight function given by Barret and Williams is incomplete; besides the values for the areas in the Regge action can be calculated in our paper directly from geometrical considerations.

2. The groups $\mathrm{SO}(4, \mathrm{R})$ and $\mathrm{SU}(2) \times \mathrm{SU}(2)$

The rotation group in 4 dimensions is the group of linear transformations that leaves the quadratic form $x_1^2 + x_2^2 + x_3^2 + x_4^2$ invariant. The well known fact that this group is locally isomorphic to $SU(2) \times SU(2)$ enables one to decompose the group action in the following way:

Take a complex matrix (not necessarily unimodular)

$$w = \begin{pmatrix} y & z \\ -\bar{z} & \bar{y} \end{pmatrix} \quad , \quad y = x_1 + ix_2, -\bar{z} = x_3 + ix_4, \tag{1}$$

where w satisfies $w\,w^+ = \det(w)$.

We define the *complete* group action

$$w \to w' = u_1 w u_2, \tag{2}$$

where $u_1, u_2 \in SU(2)$ correspond to the left, right action, respectively,

$$\begin{aligned} u_1 &= \begin{pmatrix} \alpha & \beta \\ -\bar\beta & \bar\alpha \end{pmatrix} \in SU(2)^L \ , \quad \alpha\bar\alpha + \beta\bar\beta = 1, \\ u_2 &= \begin{pmatrix} \gamma & \delta \\ -\bar\delta & \bar\gamma \end{pmatrix} \in SU(2)^R \ , \quad \gamma\bar\gamma + \delta\bar\delta = 1. \end{aligned}$$

The complete group action satisfies:

$$w'\,w'^+ = \det(w') = w\,w^+ = \det(w), \tag{3}$$

or ${x'_1}^2 + {x'_2}^2 + {x'_3}^2 + {x'_4}^2 = {x_1}^2 + {x_2}^2 + {x_3}^2 + {x_4}^2$, which corresponds to the defining relation for $SO(4, R)$.

In order to make connection with R^4, we take only the *left* action $w' = u_1 w$ and express the matrix elements of w as a 4-vector

$$\begin{pmatrix} y' \\ -\bar z' \\ z' \\ \bar y' \end{pmatrix} = \begin{pmatrix} \alpha & \beta & 0 & 0 \\ -\bar\beta & \bar\alpha & 0 & 0 \\ 0 & 0 & \alpha & \beta \\ 0 & 0 & -\bar\beta & \bar\alpha \end{pmatrix} \begin{pmatrix} y \\ -\bar z \\ z \\ \bar y \end{pmatrix}. \tag{4}$$

Substituting $y = x_1 + i x_2$, $-\bar z = x_3 + i x_4$, and $\alpha = \alpha_1 + i\alpha_2$, $\beta = \beta_1 + i\beta_2$, we get

$$\begin{pmatrix} x'_1 \\ x'_2 \\ x'_3 \\ x'_4 \end{pmatrix} = \begin{pmatrix} \alpha_1 & -\alpha_2 & \beta_1 & -\beta_2 \\ \alpha_2 & \alpha_1 & \beta_2 & \beta_1 \\ -\beta_1 & -\beta_2 & \alpha_1 & \alpha_2 \\ \beta_2 & -\beta_1 & -\alpha_2 & \alpha_1 \end{pmatrix} \begin{pmatrix} x_1 \\ x_2 \\ x_3 \\ x_4 \end{pmatrix}. \tag{5}$$

Obviously, the transformation matrix is orthogonal. Similarly for the right action $w' = w u_2^+$ we get

$$\begin{pmatrix} y' \\ -\bar z' \\ z' \\ \bar y' \end{pmatrix} = \begin{pmatrix} \bar\gamma & 0 & \bar\delta & 0 \\ 0 & \bar\gamma & 0 & \bar\delta \\ -\delta & 0 & \gamma & 0 \\ 0 & -\delta & 0 & \gamma \end{pmatrix} \begin{pmatrix} y \\ -\bar z \\ z \\ \bar y \end{pmatrix}, \tag{6}$$

and after substituting $\gamma = \gamma_1 + i\gamma_2$, $\delta = \delta_1 + i\delta_2$, we get

$$\begin{pmatrix} x'_1 \\ x'_2 \\ x'_3 \\ x'_4 \end{pmatrix} = \begin{pmatrix} \gamma_1 & \gamma_2 & -\delta_1 & \delta_2 \\ -\gamma_2 & \gamma_1 & \delta_2 & \delta_1 \\ \delta_1 & -\delta_2 & \gamma_1 & \gamma_2 \\ -\delta_2 & -\delta_1 & -\gamma_2 & \gamma_1 \end{pmatrix} \begin{pmatrix} x_1 \\ x_2 \\ x_3 \\ x_4 \end{pmatrix}, \tag{7}$$

where the transformation matrix is orthogonal.

If we take the complete action

$$\begin{pmatrix} y' & z' \\ -\bar{z}' & \bar{y}' \end{pmatrix} = \begin{pmatrix} \alpha & \beta \\ -\bar{\beta} & \bar{\alpha} \end{pmatrix} \begin{pmatrix} y & z \\ -\bar{z} & \bar{y} \end{pmatrix} \begin{pmatrix} \bar{\gamma} & -\delta \\ \bar{\delta} & \gamma \end{pmatrix},$$

we get

$$\begin{pmatrix} y' \\ -\bar{z}' \\ z' \\ \bar{y}' \end{pmatrix} = \begin{pmatrix} \alpha\bar{\gamma} & \beta\bar{\gamma} & \alpha\bar{\delta} & \beta\bar{\delta} \\ -\bar{\beta}\bar{\gamma} & \bar{\alpha}\bar{\gamma} & -\bar{\beta}\bar{\delta} & \bar{\alpha}\bar{\delta} \\ -\alpha\delta & -\beta\delta & \alpha\gamma & \beta\gamma \\ \bar{\beta}\delta & -\bar{\alpha}\delta & -\bar{\beta}\gamma & \bar{\alpha}\gamma \end{pmatrix} \begin{pmatrix} y \\ -\bar{z} \\ z \\ \bar{y} \end{pmatrix} =$$

$$= \begin{pmatrix} \alpha & \beta & 0 & 0 \\ -\bar{\beta} & \bar{\alpha} & 0 & 0 \\ 0 & 0 & \alpha & \beta \\ 0 & 0 & -\bar{\beta} & \bar{\alpha} \end{pmatrix} \begin{pmatrix} \bar{\gamma} & 0 & \bar{\delta} & 0 \\ 0 & \bar{\gamma} & 0 & \bar{\delta} \\ -\delta & 0 & \gamma & 0 \\ 0 & -\delta & 0 & \gamma \end{pmatrix} \begin{pmatrix} y \\ -\bar{z} \\ z \\ \bar{y} \end{pmatrix} \tag{8}$$

and taking $y = x_1 + ix_2$, $-\bar{z} = x_3 + ix_4$ we get the general transformation matrix for the 4-dimensional vector in R^4 under the group $SO(4, R)$ as

$$\begin{pmatrix} x'_1 \\ x'_2 \\ x'_3 \\ x'_4 \end{pmatrix} =$$

$$= \begin{pmatrix} \alpha_1 & -\alpha_2 & \beta_1 & -\beta_2 \\ \alpha_2 & \alpha_1 & \beta_2 & \beta_1 \\ -\beta_1 & -\beta_2 & \alpha_1 & \alpha_2 \\ \beta_2 & -\beta_1 & -\alpha_2 & \alpha_1 \end{pmatrix} \begin{pmatrix} \gamma_1 & \gamma_2 & -\delta_1 & \delta_2 \\ -\gamma_2 & \gamma_1 & \delta_2 & \delta_1 \\ \delta_1 & -\delta_2 & \gamma_1 & \gamma_2 \\ -\delta_2 & -\delta_1 & -\gamma_2 & \gamma_1 \end{pmatrix} \begin{pmatrix} x_1 \\ x_2 \\ x_3 \\ x_4 \end{pmatrix}. \tag{9}$$

Notice that the eight parameters $\alpha_1, \alpha_2, \beta_1, \beta_2, \gamma_1, \gamma_2, \delta_1, \delta_2$ with the constraints $\alpha_1^2 + \alpha_2^2 + \beta_1^2 + \beta_2^2 = 1$, $\gamma_1^2 + \gamma_2^2 + \delta_1^2 + \delta_2^2 = 1$, can be considered the Cayley parameters for the $SO(4)$ group [11].

3. Tensor and spinor representations of $\mathrm{SO}(4, \mathrm{R})$

Given the fundamental 4-dimensional representation of $SO(4, R)$ in terms of the parameters $\alpha, \beta, \gamma, \delta$, as given in (9),

$$x'_\mu = g_{\mu\nu} x_\nu, \tag{10}$$

the tensor representations are defined in the usual way

$$T_{k'_1 k'_2 \ldots k'_n} = g_{k'_1 k_1} \quad \cdots \quad g_{k'_n k_n} T_{k_1 k_2 \ldots k_n}, \quad (k'_i, k_i = 1, 2, 3, 4)\,. \tag{11}$$

For the sake of simplicity we take the second rank tensors. We can decompose them into totally symmetric and antisymmetric tensors, namely,

$$\begin{array}{rclr} S_{ij} & \equiv & x_i y_j + x_j y_i & \text{(totally symmetric)}, \\ A_{ij} & \equiv & x_i y_j - x_j y_i & \text{(antisymmetric)}. \end{array}$$

If we subtract the trace from S_{ij} we get a tensor that transforms under an irreducible representation. For the antisymmetric tensor the situation is more delicate. In general we have

$$A'_{ij} \equiv x'_i y'_j - x'_j y'_i = (g_{i\ell} g_{jm} - g_{j\ell} g_{im})\, A_{\ell m}. \tag{12}$$

This representation of dimension 6 is still reducible. For simplicity take the left action of the group given in (5). The linear combination of the antisymmetric tensor components are transformed among themselves in the following way:

$$\begin{pmatrix} A'_{12} + A'_{34} \\ A'_{31} + A'_{24} \\ A'_{23} + A'_{14} \end{pmatrix} = \begin{pmatrix} A_{12} + A_{34} \\ A_{31} + A_{24} \\ A_{23} + A_{14} \end{pmatrix}, \tag{13}$$

$$\begin{aligned} & \begin{pmatrix} A'_{12} - A'_{34} \\ A'_{31} - A'_{24} \\ A'_{23} - A'_{14} \end{pmatrix} = \\ = & \begin{pmatrix} \alpha_1^2 + \alpha_2^2 - \beta_1^2 - \beta_2^2 & -2\,(\alpha_1\beta_2 - \alpha_2\beta_1) & -2\,(\alpha_1\beta_1 + \alpha_2\beta_2) \\ 2\,(\alpha_1\beta_2 + \alpha_2\beta_1) & \alpha_1^2 - \alpha_2^2 + \beta_1^2 - \beta_2^2 & 2\,(\alpha_1\alpha_2 - \beta_1\beta_2) \\ 2\,(\alpha_1\beta_1 - \alpha_2\beta_2) & -2\,(\alpha_1\alpha_2 + \beta_1\beta_2) & \alpha_1^2 - \alpha_2^2 - \beta_1^2 + \beta_2^2 \end{pmatrix} \\ \times & \begin{pmatrix} A_{12} - A_{34} \\ A_{31} - A_{24} \\ A_{23} - A_{14} \end{pmatrix}. \end{aligned} \tag{14}$$

In the case of the right action given by (7) the 6-dimensional representation for the antisymmetric second rank tensor decomposes into two irreducible 3-dimensional representation of $SO(4,R)$. For this purpose one takes the linear combination of the components of the antisymmetric tensor as before:

$$\begin{pmatrix} A'_{23} - A'_{14} \\ A'_{31} - A'_{24} \\ A'_{12} - A'_{34} \end{pmatrix} = \begin{pmatrix} A_{23} - A_{14} \\ A_{31} - A_{24} \\ A_{12} - A_{34} \end{pmatrix}, \tag{15}$$

$$\begin{pmatrix} A'_{23}+A'_{14} \\ A'_{31}+A'_{24} \\ A'_{12}+A'_{34} \end{pmatrix} =$$

$$= \begin{pmatrix} \gamma_1^2-\gamma_2^2-\delta_1^2+\delta_2^2 & 2\,(\gamma_1\gamma_2+\delta_1\delta_2) & -2\,(\gamma_1\delta_1-\gamma_2\delta_2) \\ -2\,(\gamma_1\gamma_2-\delta_1\delta_2) & \gamma_1^2-\gamma_2^2+\delta_1^2-\delta_2^2 & 2\,(\gamma_1\delta_2+\gamma_2\delta_1) \\ 2\,(\gamma_1\delta_1+\gamma_2\delta_2) & -2\,(\gamma_1\delta_2-\gamma_2\delta_1) & \gamma_1^2+\gamma_2^2-\delta_1^2-\delta_2^2 \end{pmatrix} \times$$

$$\times \begin{pmatrix} A_{23}+A_{14} \\ A_{31}+A_{24} \\ A_{12}+A_{34} \end{pmatrix}. \tag{16}$$

Therefore the 6-dimensional representation for the antisymmetric tensor decomposes into two irreducible 3-dimensional irreducible representation of the $SO(4, R)$ group.

For the spinor representation of $SU(2)^L$ we take

$$\begin{pmatrix} a'_1 \\ a'_2 \end{pmatrix} = \begin{pmatrix} \alpha & \beta \\ -\bar{\beta} & \bar{\alpha} \end{pmatrix} \begin{pmatrix} a_1 \\ a_2 \end{pmatrix}, \quad a_1, a_2 \in \not\subset \tag{17}$$

Let $a^{i_1 i_2 \ldots i_k}$, $(i_1, i_2, \ldots i_k = 1, 2)$ be a set of complex numbers of dimension 2^k which transform under the $SU(2)^L$ group as follows:

$$a^{i'_1 \ldots i'_k} = u_{i'_1 i_1} \ldots u_{i'_k i_k} a^{i_1 \ldots i_k}, \tag{18}$$

where $u_{i'_1 i_1}, u_{i'_2 i_2} \ldots$ are the components of $u \in SO(2)^L$. If $a^{i_1 \ldots i_k}$ is totally symmetric in the indices $i_1 \ldots i_k$ the representation of dimension $(k+1)$ is irreducible. In an analogous way we can define an irreducible representation of $SU(2)^R$ with respect to the totally symmetric multispinor of dimension $(\ell+1)$.

For the general group $SU(4, R) \sim SU(2)^L \otimes SU(2)^R$ we can take a set of totally symmetric multispinors that transform under the $SO(4)$ group as

$$a^{i'_1 \ldots i'_k\, j'_1 \ldots j'_\ell} = u_{i'_1 i_1} \ldots u_{i'_k i_k} \bar{v}_{j'_1 j_1} \ldots \bar{v}_{j'_\ell i_\ell} a^{i_1 \ldots i_k j_1 \ldots j_\ell} \tag{19}$$

where $u_{i'_1 i_1} \ldots$ are the components of a general element of $SU(2)^L$ and $\bar{v}_{j'_\ell i_\ell}$ are the components of a general element of $SU(2)^R$. They define an irreducible representation of $SO(4, R)$ of dimension $(k+1)(\ell+1)$ and with labels (see next section)

$$\ell_0 = \frac{k-\ell}{2}, \quad \ell_1 = \frac{k+\ell}{2} + 1. \tag{20}$$

4. Representations of the algebra SO(4, R)

Let J_1, J_2, J_3 be the generators corresponding to the rotations in the planes (x_2, x_3), (x_3, x_1), and (x_1, x_2) respectively, and K_1, K_2, K_3 the generators corresponding to the rotations (boost) in the planes (x_1, x_4), (x_2, x_4) and (x_3, x_4)

respectively. They satisfy the following conmutation relations:

$$\begin{aligned} &[J_p, J_q] = i\varepsilon_{pqr}J_r \quad , \quad p,q,r = 1,2,3, \\ &[J_p, K_q] = i\varepsilon_{pqr}K_r, \\ &[K_p, K_q] = i\varepsilon_{pqr}J_r. \end{aligned} \tag{21}$$

If one defines $\bar{A} = \frac{1}{2}\left(\bar{J} + \bar{K}\right) \quad , \quad \bar{B} = \frac{1}{2}\left(\bar{J} - \bar{K}\right),$

with $\quad \bar{J} = (J_1, J_2, J_3) \quad , \quad \bar{K} = (K_1, K_2, K_3)$, then

$$\begin{aligned} &[A_p, A_q] = i\varepsilon_{pqr}A_r \quad , \quad p,q,r = 1,2,3, \\ &[B_p, B_q] = i\varepsilon_{pqr}B_r, \\ &[A_p, B_q] = 0, \end{aligned} \tag{22}$$

that is to say, the algebra so(4) decomposes into two simple algebras su(2) × su(2)

Let $\phi_{m_1 m_2}$ be a basis where $\bar{A}^2$, A_3 and $\bar{B}^2$, B_3 are diagonal. Then a unitary irreducible representation for the sets $\{A_\pm \equiv A_1 \pm iA_2, A_3\}$ and $\{B_\pm \equiv B_1 \pm iB_2, B_3\}$ is given by

$$\begin{aligned} A_\pm \phi_{m_1 m_2} &= \sqrt{(j_1 \mp m_1)(j_1 \pm m_1 + 1)}\phi_{m_1 \pm 1, m_2}, \\ A_3 \phi_{m_1 m_2} &= m_1 \phi_{m_1 m_2} \quad , \quad -j_1 \leq m_1 \leq j_1, \end{aligned} \tag{23}$$

$$\begin{aligned} B_\pm \phi_{m_1 m_2} &= \sqrt{(j_2 \mp m_2)(j_2 \pm m_2 + 1)}\phi_{m_1 m_2 \pm 1}, \\ B_3 \phi_{m_1 m_2} &= m_2 \phi_{m_1 m_2} \quad , \quad -j_2 \leq m_2 \leq j_2. \end{aligned}$$

We change now to a new basis

$$\psi_{jm} = \sum_{m_1 + m_2 = m} \langle j_1 m_1 j_2 m_2 \mid jm \rangle \, \phi_{m_1 m_2} \tag{24}$$

that corresponds to the Gelfand-Zetlin basis for so(4),

$$\psi_{jm} = \left| \begin{matrix} j_1 + j_2 & , & j_1 - j_2 \\ & j & \\ & m & \end{matrix} \right\rangle .$$

In this basis the representation for the generators $\bar{J}$, $\bar{K}$ of so(4) are given by [12]

$$\begin{aligned} J_\pm \psi_{jm} &= \sqrt{(j \mp m)(j \pm m + 1)}\psi_{jm\pm 1}, \\ J_3 \psi_{jm} &= m\psi_{jm}, \\ K_3 \psi_{jm} &= a_{jm}\psi_{j-1,m} + b_{jm}\psi_{jm} + a_{j+1,m}\psi_{j+1,m}, \end{aligned} \tag{25}$$

where

$$a_{jm} \equiv \left(\frac{\left(j^2 - m^2\right)\left(j^2 - \ell_0^2\right)\left(\ell_1^2 - j^2\right)}{\left(2j-1\right) j^2 \left(2j+1\right)} \right)^{1/2} \quad , \quad b_{jm} = \frac{m \ell_0 \ell_1}{j\left(j+1\right)},$$

with $\ell_0 = j_1 - j_2$, $\ell_1 = j_1 + j_2 + 1$ the labels of the representations.

The representation for K_1, K_2 are obtained with the help of the commutation relations.

The Casimir operators are

$$\left(\bar{J}^2 + \bar{K}^2\right)\psi_{jm} = \left(\ell_0^2 + \ell_1^2 - 1\right)\psi_{jm}, \tag{26}$$

$$\bar{J} \cdot \bar{K}\psi_{jm} = \ell_0 \ell_1 \psi_{jm}. \tag{27}$$

The representations in the bases ψ_{jm} are irreducible in the following cases

$$\begin{aligned} \ell_0 &= j_1 - j_2 = 0, \pm\frac{1}{2}, \pm 1, \pm\frac{3}{2}, \pm 2, \ldots, \\ \ell_1 &= j_1 + j_2 - 1 = \left|\ell_0\right| + 1, \left|\ell_0\right| + 2, \ldots, \\ j &= \left|j_1 - j_2\right|, \ldots, j_1 + j_2. \end{aligned}$$

If we exponentiate the infinitesimal generators we obtain the finite representations of $SO(4, R)$ given in terms of the rotation angles. An element U of $SO(4, R)$ is given as [13]

$$U\left(\varphi, \theta, \tau, \alpha, \beta, \gamma\right) = R_3\left(\varphi\right) R_2\left(\theta\right) S_3\left(\tau\right) R_3\left(\alpha\right) R_2\left(\beta\right) R_3\left(\gamma\right), \tag{28}$$

where R_2 is the rotation matrix in the $(x_1 x_3)$ plane, R_3 the rotation matrix in the $(x_1 x_2)$ plane and S_3 the rotation ("boost") in the $(x_3 x_4)$ plane, and

$$0 \leq \beta, \psi, \theta \leq \pi \quad , \quad 0 \leq \alpha, \varphi, \gamma \leq 2\pi.$$

In the basis ψ_{jm} the action of S_3 is as follows:

$$S_3\left(\tau\right)\psi_{jm} = \sum_{j'} d^{j_1 j_2}_{j' jm}\left(\tau\right)\psi_{j'm}, \tag{29}$$

where

$$d^{j_1 j_2}_{j' jm}\left(\tau\right) = \sum_{m_1 m_2} \langle j_1 j_2 m_1 m_2 \mid jm\rangle e^{-i(m_1 - m_2)\tau} \left\langle j_1 j_2 m_1 m_2 \mid j' m\right\rangle \tag{30}$$

is the Biedenharn-Dolginov function [14].

From this function the general irreducible representations of the operator U in terms of rotation angles is [13]:

$$U\left(\varphi, \theta, \tau, \alpha, \beta, \gamma\right)\psi_{jm} = \sum_{j'm'} D^{j_1 j_2}_{j'm' jm}\left(\varphi, \theta, \tau, \alpha, \beta, \gamma\right)\psi_{j'm'}, \tag{31}$$

where

$$D^{j_1 j_2}_{j'm'jm}(\varphi,\theta,\tau,\alpha,\beta,\gamma) = \sum_{m''} D^{j'}_{m'm''}(\varphi,\theta,0) d^{j_1 j_2}_{j'jm''}(\tau) D^{j}_{m''m}(\alpha,\beta,\gamma). \tag{32}$$

We now give some particular values of these representations. In the case of spin $j = 1/2$ we know

$$R_3(\alpha) R_2(\beta) R_3(\gamma) = \begin{pmatrix} \cos\frac{\beta}{2} e^{i\frac{\alpha+\gamma}{2}} & i\sin\frac{\beta}{2} e^{-i\left(\frac{\gamma-\alpha}{2}\right)} \\ i\sin\frac{\beta}{2} e^{i\frac{\gamma-\alpha}{2}} & \cos\frac{\beta}{2} e^{-i\left(\frac{\alpha+\gamma}{2}\right)} \end{pmatrix}$$

Introducing the new parameters $\alpha+\gamma=\delta$, $\gamma-\alpha=\eta$ and the variables

$$x_1 = \cos\frac{\beta}{2}\cos\frac{\delta}{2} \quad , \quad x_2 = \cos\frac{\beta}{2}\sin\frac{\delta}{2},$$
$$x_3 = \sin\frac{\beta}{2}\sin\frac{\eta}{2} \quad , \quad x_4 = \sin\frac{\beta}{2}\cos\frac{\eta}{2},$$

we have

$$R_3(\alpha) R_2(\beta) R_3(\gamma) = \begin{pmatrix} x_1 + ix_2 & x_3 + ix_4 \\ -x_3 + ix_4 & x_1 - ix_2 \end{pmatrix}. \tag{33}$$

Similarly we have

$$R_3(\varphi) R_2(\theta) S_3(\tau) = \begin{pmatrix} y_1 + iy_2 & y_3 + iy_4 \\ -y_3 + iy_4 & y_1 - iy_2 \end{pmatrix}, \tag{34}$$

with

$$y_1 = \cos\frac{\theta}{2}\cos\frac{\varphi+\tau}{2} \quad , \quad y_2 = \cos\frac{\theta}{2}\sin\frac{\varphi+\tau}{2},$$
$$y_3 = \sin\frac{\theta}{2}\sin\frac{\tau-\varphi}{2} \quad , \quad y_4 = \sin\frac{\theta}{2}\cos\frac{\tau-\varphi}{2}.$$

For the Biedenharn-Dolginov function we have some particular values [15]

$$d^{[j_+,0]}_{jmm}(\tau) = i^{j-m} 2^j \sqrt{2j+1}\,\Gamma(j+1) \times$$
$$\times \left(\frac{\Gamma\left(m+\frac{3}{2}\right)\Gamma(j_+-m+1)\Gamma(j_+-j+1)\Gamma(j+m+1)}{\Gamma\left(\frac{3}{2}\right)\Gamma(j_++m+2)\Gamma(j_++j+2)\Gamma(j-m+1)\Gamma(m+1)} \right)^{\frac{1}{2}}$$
$$\times (\sin\tau)^{j-m} C^{j+1}_{j_+-j}(\cos\tau), \tag{35}$$

where $j_+ \equiv j_1 + j_2$, $j_- = j_1 - j_2 = 0$, and $C^{\nu}_n(\cos\tau)$ are the Gegenbauer (ultraspherical) polynomials which are related to the Jacobi polynomials by

$$C^{\nu}_n(\cos\tau) = \frac{\Gamma\left(\nu+\frac{3}{2}\right)\Gamma(2\nu+n)}{\Gamma(2\nu)\Gamma\left(\nu+n+\frac{1}{2}\right)} P_n^{\left(\nu-\frac{1}{2},\nu-\frac{1}{2}\right)}(\cos\tau),$$

from which it can be deduced that

$$d_{000}^{[j_+,0]}(\tau) = \frac{1}{j_+ + 1} \frac{\sin((j_+ + 1)\tau)}{\sin\tau}. \tag{36}$$

From the asymptotic relations of $C_n^{\nu}(\cos\tau)$ it can be proved

$$d_{jmm}^{[j_+,0]}(\tau) \underset{j_+\to\infty}{\longrightarrow} \frac{i^{j-m}}{j_+^{m+1}} \frac{\cos\left[(j_+ + 1)\tau - \frac{1}{2}(j+1)\pi\right]}{(\sin\tau)^{m+1}}. \tag{37}$$

5. Relativistic spin network in 4-dimensions

We address ourselves to the Barret-Crane model that generalized Penrose's spin networks from three dimensions to four dimensions [16]. They characterize the geometrical properties of 4-simplices, out of which the tesselation of the 4-dimensional manifold is made, and then attach to them the representations of $SO(4)$.

A geometric 4-simplex in Euclidean space is given by the embedding of an ordered set of 5 points in $R^4(0, x, y, z, t)$ which is required to be non-degenerate (the points should not lie in any hyperplane). Each triangle in it determines a bivector constructed out of the vectors for the edges. Barret and Crane proved that classically, a geometric 4-simplex in Euclidean space is completely characterized (up to parallel translation an inversion through the origin) by a set of 10 bivectors b_i, each corresponding to a triangle in the 4-simplex and satisfying the following properties:

i) the bivector changes sign if the orientation of the triangle is changed;

ii) each bivector is simple, i.e. is given by the wedge product of two vectors for the edges;

iii) if two triangles share a common edge, the sum of the two bivector is simple;

iv) the sum (considering orientation) of the 4 bivectors corresponding to the faces of a tetrahedron is zero;

v) for six triangles sharing the same vertex, the six corresponding bivectors are linearly independent;

vi) the bivectors (thought of as operators) corresponding to triangles meeting at a vertex of a tetrahedron satisfy tr $b_1[b_2, b_3] > 0$ i.e. the tetrahedron has non-zero volume.

Then Barret and Crane define the quantum 4-simplex with the help of bivectors thought as elements of the Lie algebra $SO(4)$, associating a representation to each triangle and a tensor to each tetrahedron. The representations chosen should satisfy the following conditions corresponding to the geometrical ones:

i) different orientations of a triangle correspond to dual representations;

ii) the representations of the triangles are "simple" representations of $SO(4)$, i.e. $j_1 = j_2$;

iii) given two triangles, if we decompose the pair of representations into its Clebsch-Gordan series, the tensor for the tetrahedron is decomposed into summands which are non-zero only for simple representations;

iv) the tensor for the tetrahedron is invariant under $SO(4)$.

Now it is easy to construct an amplitude for the quantum 4-simplex. The graph for a relativistic spin network is the 1-complex, dual to the boundary of the 4-simplex, having five 4-valent vertices (corresponding to the five tetrahedra), with each of the ten edges connecting two different vertices (corresponding to the ten triangles of the 4-simples each shared by two tetrahedra). Now we associate to each triangle (the dual of which is an edge) a simple representation of the algebra $SO(4)$ and to each tetrahedra (the dual of which is a vertex) a intertwiner; and to a 4-simplex the product of the five intertwiner with the indices suitable contracted, and the sum for all possible representations. The proposed state sum suitable for quantum gravity for a given triangulation (decomposed into 4-simplices) is

$$Z_{BC} = \sum_{J} \prod_{\text{triang.}} A_{\text{tr}} \prod_{\text{tetrahedra}} A_{\text{tetr.}} \prod_{4-\text{simplies}} A_{\text{simp.}} \tag{38}$$

where the sum extends to all possible values of the representations J.

6. The triple product in R^4

Before we apply the representation theory developed in previous sections to the Barret-Crane model we introduce some geometrical properties based in the triple product that generalizes the vector (cross) product in R^3. Given three vectors in R^4, we define the triple product:

$$\begin{aligned} u \wedge v \wedge w &= -v \wedge u \wedge w = -u \wedge w \wedge v = -w \wedge v \wedge u = v \wedge w \wedge u = \\ &= w \wedge u \wedge v, \\ u \wedge u \wedge v &= u \wedge v \wedge u = v \wedge u \wedge u = 0. \end{aligned} \tag{39}$$

If the vectors in R^4 have cartesian coordinates

$$u = (u_1, u_2, u_3, u_4)\ ,\ v = (v_1, v_2, v_3, v_4)\ ,\ w = (w_1, w_2, w_3, w_4)\,,$$

we define an orthonormal basis in R^4

$$\hat{\imath} = (1,0,0,0)\ ,\ \hat{\jmath} = (0,1,0,0)\ ,\ \hat{k} = (0,0,1,0)\ ,\ \hat{\ell} = (0,0,0,1)\,.$$

The triple product of these vectors satisfies

$$\hat{\imath}\wedge\hat{\jmath}\wedge\hat{k} = -\hat{\ell}\,,\ \hat{\jmath}\wedge\hat{k}\wedge\hat{\ell} = \hat{\imath}\,,\ \hat{k}\wedge\hat{\ell}\wedge\hat{\imath} = -\hat{\jmath}\,,\ \hat{\imath}\wedge\hat{\jmath}\wedge\hat{\ell} = \hat{k}$$

.

In coordinates the triple product is given by the determinant

$$u\wedge v\wedge w = \begin{vmatrix} \hat{\imath} & \hat{\jmath} & \hat{k} & \hat{\ell} \\ u_1 & u_2 & u_3 & u_4 \\ v_1 & v_2 & v_3 & v_4 \\ w_1 & w_2 & w_3 & w_4 \end{vmatrix}\,. \tag{40}$$

The scalar quadruple product is defined by

$$\begin{aligned} a\cdot(b\wedge c\wedge d) &= \begin{vmatrix} a_1 & a_2 & a_3 & a_4 \\ b_1 & b_2 & b_3 & b_4 \\ c_1 & c_2 & c_3 & c_4 \\ d_1 & d_2 & d_3 & d_4 \end{vmatrix} = [abcd] = -\,[abdc] = \\ &= -\,[acbd] = [acdb] \text{ and so on.} \end{aligned} \tag{41}$$

It follows: $a\cdot a\wedge b\wedge c = b\cdot a\wedge b\wedge c = c\cdot a\wedge b\wedge c = 0$.

We can use the properties of the three vector for the description of the 4-simplex. Let $\{0,x,y,z,t\}$ be the 4-simplex in R^4. Two tetrahedra have a common face

$\{0,x,y,z\}\cap\{0,x,y,t\} = \{0,x,y\}\,.$

Each tetrahedron is embedded in an hyperplane characterized by a vector perpendicular to all the vectors forming the tetrahedron. For instance,

$\{0,x,y,z\}$ is characterized by $a = x\wedge y\wedge z$,

$\{0,x,y,t\}$ is characterized by $b = x\wedge y\wedge t$.

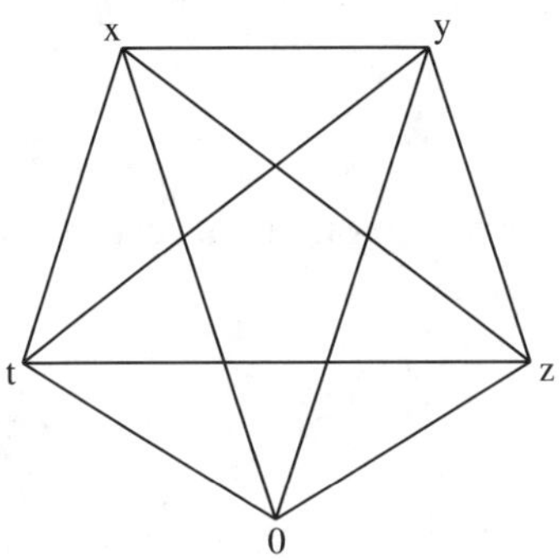

The vector a satisfies $a \cdot x = a \cdot y = a \cdot z = 0$,
the vector b satisfies $b \cdot x = b \cdot y = b \cdot t = 0$.

The triangle $\{0, x, y\}$ shared by the two tetrahedra is characterized by the bivector $x \wedge y$. The plane where the triangle is embedded is defined by the two vectors a, b, forming the angle ϕ, given by

$$\cos \phi = a \cdot b.$$

The bivector $a \wedge b$ can be calculated with the help of trivectors as

$$a \wedge b = [x\, y\, z\, t]^* \, (x \wedge y) \, .$$

Obviously $a \wedge b$ is perpendicular to $x \wedge y$

$$\langle a \wedge b, x \wedge y \rangle = (a \cdot x)\,(b \cdot y) - (a \cdot y)\,(b \cdot x) = 0. \tag{42}$$

For completeness we add some useful properties of bivectors in R^4. The six components of a bivector can be written as

$$\begin{array}{lll} B_{\mu\nu} = x_\mu y_\nu - x_\nu y_\mu \, , & \mu, \nu = 1, 2, 3, 4 \, , & B = \left(\bar{J}, \bar{K}\right) , \\ J_1 = (x_2 y_3 - x_3 y_2) \, , & J_2 = (x_3 y_1 - x_1 y_3) \, , & J_3 = (x_1 y_2 - x_2 y_1) \, , \\ K_1 = (x_1 y_4 - x_4 y_1) \, , & K_2 = (x_2 y_4 - x_3 y_1) \, , & K_3 = (x_3 y_4 - x_4 y_1) \, . \end{array}$$

The six components of the dual of a bivector are
$^*B = \left(\bar{K}, \bar{J}\right) \, , \quad ^*B_{\alpha\beta} = \frac{1}{2} b_{\mu\nu} \, \varepsilon_{\mu\nu\alpha\beta}$.

We take the linear combinations of $\bar{J}, \bar{K}$

$$\bar{M} = \frac{1}{2}\left(\bar{J} + \bar{K}\right), \qquad \bar{N} = \frac{1}{2}\left(\bar{J} - \bar{K}\right). \tag{43}$$

They form the bivector $\left(\bar{M}, \bar{N}\right)$, whose dual is:

$$^*\,(M, N) = (M, -N)\, , \tag{44}$$

therefore $\bar{M}$ can be considered the self-dual part, $\bar{N}$ the antiselfdual part of the bivector $\bar{M}, \bar{N}$ $\bar{M}$ and $\bar{N}$ coincides with the basis for the irreducible tensor representations of section 3. The norm of the bivectors can be explicitly calculated.

$$\begin{aligned} \|B\|^2 &= \langle B, B \rangle = J^2 + K^2 = \|x\|^2 \, \|y\|^2 - |x, y|^2 = \\ &= \|x\|^2 \, \|y\|^2 \sin^2 \phi(x, y) = 4Area^2 \, \{0, x, y\} \, , \end{aligned} \tag{45}$$

$$\|^*B\|^2 = \langle ^*B, ^*B \rangle = J^2 + K^2 = \|B\|^2 \, . \tag{46}$$

Finally, the scalar product of two vectors in R^4 can be expressed in terms of the corresponding $U\,(2, \mathbb{C})$ matrices

Let $X \Leftrightarrow \begin{pmatrix} x_1 + i x_2 & x_3 + i x_4 \\ -x_3 + i x_4 & x_1 - i x_2 \end{pmatrix}, \; Y \Leftrightarrow \begin{pmatrix} y_1 + i y_2 & y_3 + i y_4 \\ -y_3 + i y_4 & y_1 - i y_2 \end{pmatrix}$.

Then

$$Tr\left(X^+ Y\right) = x_1 y_1 + x_2 y_2 + x_3 y_3 + x_4 y_4. \tag{47}$$

7. Evaluation of the spin sum for the relativistic spin network

The five tetrahedra in the 4–simplex are numbered by $k = 1, 2, 3, 4, 5$ and the triangles are indexed by the pair k, l of tetrahedra which intersect on the triangle kl. To each triangle we associate a simple representation of $SO(4)$ labelled by j_{kl}, that corresponds to the same spin for each part of the $SU(2) \otimes SU(2)$ group. The matrix representing an element $g \in SU(2)$ in the irreducible representation of spin j_{kl} belonging to a triangle is denoted by $\rho_{kl}(g)$. An element $h_k \in SU(2)$ is assigned to each tetrahedron k. The invariant I is defined by integrating a function of these variables over each copy of $SU(2)$:

$$I = (-1)^{\sum\limits_{k<\ell} 2j_{k\ell}} \int\limits_{h \in SU(2)^5} \prod \mathrm{Tr}\rho_{k\ell}\left(h_k h_\ell^{-1}\right) \tag{48}$$

The geometrical interpretation of this formula given by Barret [17] is that since the manifold $SU(2)$ is isomorphic to S^3, each variable $h \in SU(2)$ can be regarded as a unit vector in R^4. This unit vector can be identified with the vector perpendicular to the hyperplane where the tetrahedron is embedded. The two variables h_k, h_l correspond in this picture to the two vectors a, b that were defined in the last section.

In our opinion there is some disagreement between the conditions given in Ref. [16] and the application of formula (2.1) in Ref.[17]. In the former paper an irreducible representation of $SO(4)$ with two labels $j_1 = j_2$ is assigned to each triangle in the 4–simplex. In the last paper, a representation of $SU(2)$ is assigned to each triangle. Therefore we have the standard values for the trace of a general representation of the group $SU(2)$ with spin j, namely, $\sin{(2j+1)}\,\phi / \sin\phi$ (Formula 4.1 of Ref. 17).

The disagreement can be avoided if one takes the trace with respect to the irreducible representation of $SO(4)$ as described in Sections 3 and 4, where the parameters of the group $SO(4)$ are the 3+3 cartesian independent coordinates of the two unit vectors h_k, h_ℓ, as defined before, or the 6 rotation angles of formula (28). In the last case we choose a system of reference for R_4 such that one unit vector corresponding to h_k, say a, has components $(0, 0, 0, 1)$ and the other one h_ℓ, say b, is located in the plane $(x_3 x_4)$ forming an angle τ with the first vector. In this particular situation all the rotation angles $\alpha = \beta = \gamma = \vartheta = \varphi = 0$ and the representation is restricted to $S_3(\tau)$.

From (31) and (32) the general element representation of $SO(4)$ is restricted to

$$D^{j_1 j_2}_{j'm'jm}(0, 0, \tau, 0, 0, 0) = d^{j_1 j_2}_{j'jm}(\tau) \equiv d^{[j_+, j_-]}_{j'jm}(\tau). \tag{49}$$

In the case of a simple representations of $SO(4)$ $j_- = j_1 - j_2 = 0$, and the trace becomes

$$\mathrm{tr} D^{[j_+,0]}_{j'm'jm}(\tau) = \sum_{j=0}^{j_+} \sum_{m=-j}^{j} d^{[j_+,0]}_{jjm}(\tau). \tag{50}$$

Obviously this expression does not coincide with formula (4.1) of Ref. [17] except in the term

$$d^{[j_+,0]}_{000}(\tau) = \frac{1}{j_+ + 1} \frac{\sin\left(\left(\frac{1}{j_+ + 1}\right)\tau\right)}{\sin\tau}, \text{ Ref [15], (IV.2.9).}$$

For other values of the Biedenharn-Dolginov function we can use the asymptotic expression (37) for $m = j$. With this formula it is still possible to give an geometrical interpretation of the probability amplitude encompassed in the trace. In fact, the spin dependent factor appearing in the exponential of (37)

$$e^{i(2j_{k\ell}+1)\tau_{k\ell}}, \tag{51}$$

corresponding to two tetrahedra $k\ell$ intersecting the triangle $k\ell$, can be interpreted as the product of the angle between the two vectors h_k, h_ℓ perpendicular to the triangle and the area $A_{k\ell}$ of the intersecting triangle.

For the proof we identify the component of the antisymmetric tensor $(\bar{J}, \bar{K})$ with the components of the infinitesimal generators of the $SO(4)$ group

$$J_{\mu\nu} \equiv i\left(x_\mu \frac{\partial}{\partial x_\nu} - x_\nu \frac{\partial}{\partial x_\mu}\right).$$

From (43) and (45) we have $\|B\|^2 = 4\,(A_{k\ell})^2 = 2\left(\bar{M}^2 + \bar{N}^2\right)$

But $\bar{M}^2$ and $\bar{N}^2$ are the Casimir operators of the $SU(2) \otimes SU(2)$ group with eigenvalues $j_1\,(j_1+1)$ and $j_2\,(j_2+1)$.

For large values of $j_1 = j_2 = j_{k\ell}$ we have

$$2\left(\bar{M}^2 + \bar{N}^2\right) \cong 4j_{k\ell}^2 + 4j_{k\ell} + 1 = (2j_{k\ell}+1)^2,$$

therefore $\frac{1}{2}\,(2j_{k\ell}+1) = A_{k\ell}$ where $A_{k\ell}$ is the area of the triangle characterized by the two vectors h_k and h_ℓ and $j_{k\ell}$ is the spin corresponding to the representation $\rho_{k\ell}$ associated to the triangle kl. Substituting this result in (51) we obtain the asymptotic value of the amplitude given by Barret and Williams [18]

Acknowledgments

The authors would like to express their gratitude to Professor Bruno Gruber for inviting them to the Colloquium "Symmetries in Science XIII". One of the authors (M. L.) wants to extend his gratitude to the Director of the Institut für theoretische Physik of the University of Tübingen, where part of this work was done, and also to the Ministerio de Ciencia y Tecnología (Spain) for the financial support under grant BFM-2000–0357.

References

[1] Lorente, M. "Basis for a discrete special relativity". *Int. J. Theor. Phys.*, 15:927–947, 1976.

Lorente, M. "A new scheme for the Klein-Gordon and Dirac Fields on the lattice with Axial Anomaly". *Int. J. Group Theor. in Phys.*, 1:105–121, 1993.

Lorente M. and P. Kramer. "Representations of the discrete inhomogeneous Lorentz group and Dirac wave equation on the lattice". *J. Phys. A, Math. Gen.*, 32:2481–2497, 1999.

[2] Regge, T. "General Relativity without coordinates". *Il Nuovo Cimento*, 19:558–571, 1961.

[3] Ponzano, G. and T. Regge. "Semiclassical limit of Racah coefficients". In *Spectroscopy and group theoretical methods in Physics* (F. Bloch et al. ed.), North Holland, Amsterdam, 1968.

[4] Penrose, R. "Angular momentum: an approach to combinatorial space-time". In *Quantum theory and Beyond* (T. Bastin, ed.), C.U.P. Cambridge, 1971.

[5] Hasslacher B. and M. Perry. "Spin networks are simplicial quantum gravity". *Phys. Lett*, 103 B:21–24, 1981.

[6] Turaev V.G. and O. Viro. *Topology*, 31:865–902, 1992.

[7] Crane, L. and D. Yetter "A categorial construction of 4d-topological quantum field theories". In *Quantum Topology* (L. Kaufmann, ed.) World Scientific, Singapore, 1993.

[8] Boulatov, D. "A model of three-dimensional lattice gravity". *Mod. Phys. Lett. A*, 7:1629–1646, 1992.

[9] Ooguri, H. "Topological lattice models in 4-dimensions". *Mod. Phys. Lett. A*, 7:2799–2810, 1992.

[10] De Pietri, R., L. Freidel, K. Krasnov and C. Rovelli "Barret-Crame model from a Boulatov-Ooguri field theory over a homogeneous space". *Nucl. Phys. B*, 574:785–806, 2000. M. Reisenberg, C. Rovelli, "Spin foams as Feynman diagrams", gr–qc/0002083.

[11] Lorente, M. "Cayley parametrization of semisimple Lie groups and its Application to Physical Laws in a (3+1)-Dimensional Cubic Lattice". *Int. J. Theor. Phys.*, 11:213–247, 1974.

[12] Nikiforov, A.F., S.K. Suslov and V.B. Uvarov. *Classical Orthogonal Polynomials of a Discrete Variable*. Springer, Berlin, 1991, p. 271–273.

[13] Frazer, W.R., F.R. Halpern, H.M. Lipinski and R.S. Snider "O(4) Expansion of Off-Shell Scattering Amplitudes". *Phys. Rev.*, 176:2047–2053, 1968.

[14] Biedenharn, L.C. "Wigner Coefficients for the R_4 Group and Some Appli - cations". *J. Math. Phys.*, 2:433–441, 1961.

[15] Barut, A.O. and R. Wilson "Some new identities of Clebsch-Gordan coefficients and representation functions of SO(2,1) and SO(4)". *J. Math. Phys.*, 17:900–915, 1976.

[16] Barret, J. and L. Crane "Relativistic spin networks and Quantum gravity". *J. Math. Phys.*, 39:3296, 1998.

[17] Barret, J. "The Classical Evaluation of Relativistic Spin Networks". *Adv. Theor. Math. Phys.*, 2:593–600, 1998.

Oriti, D. "Space time geometry from Algebra: spin foam models for non-perturbative quantum gravity". *Rep. Progr. Phys.*, 64:1489–1544, 2001.

[18] Barret, J.W. and R.M. Williams. "The asymptotics of an Amplitude for the 4-simplex". *Adv. Theor. Math. Phys.*, 3:209–215, 1999.

THE GEOMETRY OF DENSITY STATES, POSITIVE MAPS AND TOMOGRAMS

V.I. Man'ko[†], G. Marmo[⋆], E.C.G. Sudarshan[‡] and F. Zaccaria[⋆]
[†] *P.N. Lebedev Physical Institute,*
Leninskii Prospect, 53,
Moscow 119991 Russia
manko@sci.lebedev.ru
[⋆]*Dipartimento di Scienze Fisiche,*
Università "Federico II" di Napoli
and Istituto Nazionale di Fisica Nucleare, Sezione di Napoli,
Complesso Universitario di Monte Sant'Angelo,
Via Cintia, I-80126 Napoli, Italy
marmo@na.infn.it
zaccaria@na.infn.it
[‡] *Physics Department,*
Center for Particle Physics,
University of Texas,
78712 Austin, Texas, USA
sudarshan@physics.utexas.edu

Abstract The positive and not completely positive maps of density matrices, which are contractive maps, are discussed as elements of a semigroup. A new kind of positive map (the purification map), which is nonlinear map, is introduced. The density matrices are considered as vectors, linear maps among matrices are represented by superoperators given in the form of higher dimensional matrices. Probability representation of spin states (spin tomography) is reviewed and $U(N)$-tomogram of spin states is presented. Properties of the tomograms as probability distribution functions are studied. Notion of tomographic purity of spin states is introduced. Entanglement and separability of density matrices are expressed in terms of properties of the tomographic joint probability distributions of random spin projections which depend also on unitary group parameters. A new positivity criterion for hermitian matrices is formulated. An entanglement criterion is given in terms of a function depending on unitary group parameters and semigroup of positive map parameters. The function is constructed as sum of moduli of $U(N)$-tomographic symbols of the hermitian matrix obtained after action on the density matrix of composite system by a positive but not completely positive map of the subsystem density matrix. Some two-qubit and two-qutritt states are considered

B. Gruber (ed.), Symmetries in Science XI, 395–443.

as examples of entangled states. The connection with the star-product quantisation is discussed. The structure of the set of density matrices and their relation to unitary group and Lie algebra of the unitary group are studied. Nonlinear quantum evolution of state vector obtained by means of applying purification rule of density matrices evolving via dynamical maps is considered. Some connection of positive maps and entanglement with random matrices is discussed and used.

Keywords: Unitary group, entanglement, adjoint representation, tomogram, operator symbol, random matrix.

1. Introduction

The states in quantum mechanics are associated with vectors in Hilbert space [1] (it is better to say with rays) in the case of pure states. For mixed state, one associates the state with density matrix [2, 3]. In classical mechanics (statistical mechanics), the states are associated with joint probability distributions in phase space. There is an essential difference in the concept of states in classical and quantum mechanics. This difference is clearly pointed out by the phenomenon of entanglement. The notion of entanglement [4] is related to the quantum superposition rule of the states of subsystems for a given multipartite system. For pure states, the notion of entanglement and separability can be given as follows.

If the wave function [5] of a state of a bipartite system is represented as the product of two wave functions depending on coordinates of the subsystems, the state is simply separable; otherwise, the state is simply entangled. An intrinsic approach to the entanglement measure was suggested in [6]. The measure was introduced as the distance between the system density matrix and the tensor product of the associated states. For the subsystems, the association being realized via partial traces. There are several other different characteristics and measures of entanglement considered by several authors [7–13]. For example, there are measures related to entropy (see, [14–24]). Also linear entropy of entanglement was used in [25–27], "concurrences" in [28, 29] and "covariance entanglement measure" in [30]. Each of the entanglement measures describes some degree of correlation between the subsystems' properties.

The notion of entanglement is not an absolute notion for a given system but depends on the decomposition into subsystems. The same quantum state can be considered as entangled, if one kind of division of the system into subsystems is given, or as completely disentangled, if another decomposition of the system into subsystems is considered.

For instance, the state of two continuous quadratures can be entangled in Cartesian coordinates and disentangled in polar coordinates. Coordinates are considered as measurable observables labelling the subsystems of the given system. The choice of different subsystems mathematically implies the existence of two different sets of the subsystems' characteristics (we focus on bipartite case). We may consider the Hilbert space of states $H(1,2)$ or $H(1',2')$. The Hilbert space for the total system is, of course, the same but the index $(1,2)$ means that there are two sets of operators P_1 and P_2, which select subsystem states 1 and 2. The index $(1',2')$ means that there are two other sets of operators P_1' and P_2', which select subsystem states $1'$ and $2'$. The operators $P_{1,2}$ and $P'_{1',2'}$ have specific properties. They are represented as tensor products of operators acting in the space of states of the subsystem 1 (or 2) and unit operators acting in the subsystem 2 (or 1). In other words, we consider the space H, which can be treated as the tensor product of spaces $H(1)$ and $H(2)$ or $H(1')$ and $H(2')$. In the subsystems 1 and 2, there are basis vectors $\mid n_1\rangle$ and $\mid m_2\rangle$, and in the subsystems $1'$ and $2'$ there are basis vectors $\mid n_1'\rangle$ and $\mid m_2'\rangle$. The vectors $\mid n_1\rangle \mid m_2\rangle$ and $\mid n_1'\rangle \mid m_2'\rangle$ form the sets of basis vectors in the composite Hilbert space, respectively. These two sets are related by means of unitary transformation. An example of such a composite system is a bipartite spin system.

If one has spin-j_1 [the space $H(1)$] and spin-j_2 [the space $H(2)$] systems, the combined system can be treated as having basis
$\mid j_1m_1\rangle \mid j_2m_2\rangle$.

Another basis in the composite-system-state space can be considered in the form $\mid jm\rangle$, where j is one of the numbers $|j_1-j_2|, |j_1-j_2|+1,\ldots,j_1+j_2$ and $m=m_1+m_2$. The basis $\mid jm\rangle$ is related to the basis $\mid j_1m_1\rangle \mid j_2m_2\rangle$ by means of the unitary transform given by Clebsch–Gordon coefficients C $(j_1m_1j_2m_2|jm)$. From the viewpoint of the given definition, the states $\mid jm\rangle$ are entangled states in the original basis. Another example is the separation of the hydrogen atom in terms of parabolic coordinates used while discussing the Stark effect.

The spin states can be described by means of the tomographic map [31–33]. For bipartite spin systems, the states were described by the tomographic probabilities in [34, 35]. Some properties of the tomographic spin description were studied in [36]. In the tomographic approach, the problems of the quantum state entanglement can be cast into the form of some relations among the probability distribution functions. On the other hand, to have a clear picture of entanglement, one needs a mathematical formulation of the properties of the density matrix of the composite system, a description of the linear space of the composite system states. Since a density matrix is hermitian, the space of states may be embedded as a subset of the Lie algebra of the unitary group, carrying the adjoint representation of $U(n^2)$, where $n^2=(2j+1)^2$ is the dimension of the spin states of two spinning particles. Thus one may try to characterize the entanglement phenomena by using various structures present in the space of the adjoint representation of the $U(n^2)$ group.

The aim of this paper is to give a review of different aspects of density matrices and positive map and connect entanglement problems with the properties of tomographic probability distributions and discuss the properties of the convex set of positive states for composite system by taking into account the subsystem structures. We used [6] the Hilbert–Schmidt distance to calculate the measure of entanglement as the distance between a given state and the tensor product of the partial traces of the density matrix of the given state. In [37] another measure of entanglement as a characteristic of subsystem correlations was introduced. This measure is determined via the covariance matrix of some observables. A review of different approaches to the entanglement notion and entanglement measures is given in [38] where the approach to describe entanglement and separability of composite systems is based, e.g., on entropy methods.

Due to a variety of approaches to the entanglement problem, one needs to understand better what in reality the word "entanglement" describes. Is it a synonym of the word "correlation" between two subsystems or does it have to capture some specific correlations attributed completely and only to the quantum domain?

The paper is organized as follows.

In section 2 we discuss division of composite systems onto subsystems and relation of the density matrix to adjoint representation of unitary group in generic terms of vector representation of matrices; we study also completely positive maps of density matrices. In section 3 we consider a vector representation of probability distribution functions and notion of distance between the probability distributions and density matrices. In section 4 we present definition of separable quantum state of a composite system and criterion of separability. In section 5 the entanglement is considered in terms of operator symbols. Particular tomographic probability representation of quantum states and tomographic symbols are reviewed in section 6. Symbols of multipartite states are studied in section 7. In section 8 spin tomography is reviewed. An example of qubit state is done in section 9. The unitary spin tomogram is introduced in section 10 while in section 11 dynamical map and corresponding quantum evolution equations are discussed as well as examples of concrete positive maps. Conclusions and perspectives are presented in section 12.

2. Composite system

In this section, we review the meaning and notion of composite system in terms of additional structures on the linear space of state for the composite system.

2.1 Difference of states and observables

In quantum mechanics, there are two basic aspects, which are associated with linear operators acting in a Hilbert space. The first one is related to the concept of quantum state and the second one, to the concept of observable. These two concepts of state and observable are paired via a map with values in probability measures on the real line. Often states are described by Hermitian nonnegative, trace-class, matrices. The observables are described by Hermitian operators. Though both states and observables are identified with Hermitian objects, there is an essential difference between the corresponding objects. The observables have an additional product structure. Thus we may consider the product of two linear operators corresponding to observables.

For the states, the notion of product is redundant. The product of two states is not a state. For states, one keeps only the linear structure of vector space. For finite n-dimensional system, the Hermitian states and the Hermitian observables may be mapped into the Lie algebra of the unitary group $U(n)$. But the states correspond to nonnegative Hermitian operators. Observables can be associated with both types of operators, including nonnegative and nonpositive ones. The space of states is therefore a convex-linear space which, in principle, is not equipped with a product structure. Due to this, transformations in the linear space of states need not preserve any product structure. In the set of observables, one has to be concerned with what is happening with the product of operators when some transformations are performed. State vectors can be transformed into other state vectors. Density operators also can be transformed. We will consider linear transformations of the density operators. The density operator has nonnegative eigenvalues. In any representation, diagonal elements of density matrix have physical meaning of probability distribution function.

Density operator can be decomposed as a sum of eigenprojectors with coefficients equal to its eigenvalues. Each one of the projectors defines a pure state. There exists a basis in which every eigenprojector of rank one is represented by a diagonal matrix of rank one with only one matrix element equal to one and all other matrix elements equal to zero. Other density matrices with similar properties belong to the orbit of the unitary group on the starting eigenprojector. Depending on the number of distinct nonzero values determines the class of the orbit. Since density matrices of higher rank belong to an appropriate orbit of a convex sum of the different diagonal eigenprojectors (in special basis), we may say that generic density matrices belong to the orbits of the unitary group acting on the diagonal density matrices which belong to the Cartan subalgebra of the Lie algebra of the unitary group. Any convex sum of density matrices can be treated as a mean value of a random density matrix. The positive coefficients of the convex sum can be interpreted as a probability distribution function which makes the averaging providing the final value of the convex sum. The set of density matrices may be identified with the union of the orbits of the unitary group acting on diagonal density matrices considered as elements of the Cartan subalgebra.

2.2 Matrices as vectors, density operators and superoperators

When matrices represent states it may be convenient to identify them with vectors. In this case, a density matrix can be considered as a vector with additional properties of its components. If the identifications are done elegantly, we can see the real Hilbert space of density matrices in terms of vectors with real components. In this case, linear transforms of the matrix can be interpreted as matrices called superoperators. It means that density matrices–vectors undergoing real linear transformations are acted on by the matrices representing the action of the superoperators of the linear map. This construction can be continued. Thus we can get a chain of vector spaces of higher and higher dimensions. Let us first introduce some extra constructions of the map of a matrix onto a vector. Given a rectangular matrix M with elements M_{id}, where $i = 1, 2, \ldots, n$ and $d = 1, 2, \ldots, m$, one can consider the matrix as a vector $\vec{\mathcal{M}}$ with $N = nm$ components constructed by the following rule:

$$\mathcal{M}_1 = M_{11}, \qquad \mathcal{M}_2 = M_{12}, \qquad \mathcal{M}_m = M_{1m}, \tag{1}$$
$$\mathcal{M}_{m+1} = M_{21}, \ldots \mathcal{M}_N = M_{nm}.$$

Thus we construct the map $M \to \vec{\mathcal{M}} = \hat{t}_{\vec{\mathcal{M}}M} M$.

We have introduced the linear operator $\hat{t}_{\vec{\mathcal{M}}M}$ which maps the matrix M onto a vector $\vec{\mathcal{M}}$. Now we introduce the inverse operator $\hat{p}_{\vec{\mathcal{M}}M}$ which maps a given column vector in the space with dimension $N = mn$ onto a rectangular matrix. This means that given a vector $\vec{\mathcal{M}} = \mathcal{M}_1, \ldots, \mathcal{M}_N$, we relabel its components by introducing two indices $i = 1, \ldots, n$ and $d = 1, \ldots, m$. The relabeling is accomplished according to (0). Then we collect the relabeled components into a matrix table. Eventually we get the map

$$\hat{p}_{\vec{\mathcal{M}}M} \vec{\mathcal{M}} = M. \tag{2}$$

The composition of these two maps

$$\hat{t}_{\vec{\mathcal{M}}M} \hat{p}_{\vec{\mathcal{M}}M} \vec{\mathcal{M}} = 1 \cdot \vec{\mathcal{M}} \tag{3}$$

acts as the unit operator in the linear space of vectors.

Given a $n \times n$ matrix the map considered can also be applied. The matrix can be treated as an n^2-dimensional vector and, vice versa, the vector of dimension n^2 may be mapped by this procedure onto the $n \times n$ matrix.

Let us consider a linear operator acting on the vector $\vec{\mathcal{M}}$ and related to a linear transform of the matrix M. First, we study the correspondence of the linear transform of the form

$$M \to gM = M_g^l \tag{4}$$

to the transform of the vector

$$\vec{\mathcal{M}} \to \vec{\mathcal{M}}_g^l = \mathcal{L}_g^l \vec{\mathcal{M}}. \tag{5}$$

One can show that the $n^2 \times n^2$ matrix $\mathcal{L}_g^l$ is determined by the tensor product of the $n \times n$ matrix g and $n \times n$ unit matrix, i.e.,

$$\mathcal{L}_g^l = g \otimes 1. \tag{6}$$

Analogously, the linear transform of the matrix M of the form

$$M \to Mg = M_g^r \tag{7}$$

induces the linear transform of the vector $\vec{\mathcal{M}}$ of the form

$$\vec{\mathcal{M}} \to \vec{\mathcal{M}}_g^r = \hat{t}_{\vec{\mathcal{M}}\mathcal{M}} M_g^r = \mathcal{L}_g^r \vec{\mathcal{M}}, \tag{8}$$

where the $n^2 \times n^2$ matrix $\mathcal{L}_g^r$ reads

$$\mathcal{L}_g^r = 1 \otimes g^{\mathrm{tr}}. \tag{9}$$

Similarity transformation of the matrix M of the form

$$M \to gMg^{-1} \tag{10}$$

induces the corresponding linear transform of the vector $\vec{\mathcal{M}}$ of the form

$$\vec{\mathcal{M}} \to \vec{\mathcal{M}}_s = \mathcal{L}_g^s \vec{\mathcal{M}}, \tag{11}$$

where the $n^2 \times n^2$ matrix $\mathcal{L}_g^s$ reads

$$\mathcal{L}_g^s = g \otimes (g^{-1})^{\mathrm{tr}}. \tag{12}$$

Starting with vectors, one may ask how to identify on them a product structure which would make $\hat{p}_{\vec{\mathcal{M}}\mathcal{N}}$ into an algebra homomorphism. An associative algebraic structure on the vector space may be defined by imitating the procedure one uses to define star-products on the space of functions on phase space. One can define the associative product of two N-vectors $\vec{\mathcal{M}}_1$ and $\vec{\mathcal{M}}_2$ using the rule

$$\vec{\mathcal{M}} = \vec{\mathcal{M}}_1 \star \vec{\mathcal{M}}_2, \tag{13}$$

where

$$\vec{\mathcal{M}}_k = \sum_{l,s=1}^{N} K^k_{ls}(\vec{\mathcal{M}}_1)_l(\vec{\mathcal{M}}_2)_s. \tag{14}$$

If one applies a linear transform to the vectors $\vec{\mathcal{M}}_1$, $\vec{\mathcal{M}}_2$, $\vec{\mathcal{M}}$ of the form

$$\vec{\mathcal{M}}_1 \to \vec{\mathcal{M}}'_1 = \mathcal{L}\vec{\mathcal{M}}_1, \qquad \vec{\mathcal{M}}_2 \to \vec{\mathcal{M}}'_2 = \mathcal{L}\vec{\mathcal{M}}_2, \qquad \vec{\mathcal{M}} \to \vec{\mathcal{M}}' = \mathcal{L}\vec{\mathcal{M}},$$

and requires the invariance of the star-product kernel, one finds

$$\vec{\mathcal{M}}'_1 \star \vec{\mathcal{M}}'_2 = \vec{\mathcal{M}}', \qquad \text{if} \qquad \mathcal{L} = G \otimes G^{-1\mathrm{tr}}, \quad G \in GL(n).$$

The kernel K^k_{ls} (structure constants) which determines the associative star-product satisfies a quadratic equation. Thus if one wants to make the correspondence of the vector star-product to the standard matrix product (row by column), the matrix M must be constructed appropriately. For example, if the vector star-product is commutative, the matrix M corresponding to the N-vector $\vec{\mathcal{M}}$ can be chosen as a diagonal $N \times N$ matrix. This consideration shows that the map of matrices on the vectors provides the star-product of the vectors (defining the structure constants or the kernel of the star-product) and, conversely, if one starts with vectors and uses matrices with the standard multiplication rule, it will be the map to be determined by the structure constants (or by the kernel of the vector star-product).

The constructed space of matrices associated with vectors enables one to enlarge the dimensionality of the group acting in the linear space of matrices in comparison with the standard one, i.e., we may relax the requirement of invariance of the product structure. In general, given a $n \times n$ matrix M the left action, the right action, and the similarity transformation of the matrix are related to the complex group $GL(n)$. On the other hand, the linear transformations in the linear space of n^2-vectors $\vec{\mathcal{M}}$ obtained by using the introduced map are determined by the matrices belonging to the group $GL(n^2)$. There are transformations on the vectors which cannot be simply represented on matrices. If $M \to \Phi(M)$ is a linear homogeneous function of the matrix M, we may represent it by

$$\Phi_{ab} = B_{aa',\,bb'} M_{a'b'}.$$

Under rather clear conditions, $B_{aa',bb'}$ can be expressed in terms of its nonnormalized left and right eigenvectors:

$$B_{aa',bb'} = \sum_{\nu} x_{aa'}(\nu) y^{\dagger}_{bb'}(\nu),$$

being an index for eigenvalues, which corresponds to

$$\Phi(M) = xMy^{\dagger} = \sum_{\nu=1}^{n^2} x(\nu) M y^{\dagger}(\nu).$$

There are possible linear transforms on the matrices and corresponding linear transforms on the induced vector space which do not give rise to a group structure but possess only the structure of algebra. One can describe the map of $n \times n$ matrices M (source space) onto vectors $\vec{\mathcal{M}}$ (target space) using specific basis in the space of the matrices. The basis is given by the matrices E_{jk} $(j, k = 1, 2, \ldots, n)$ with all matrix elements equal to zero except the element in the jth row and kth column which is equal to unity. One has the obvious property

$$M_{jk} = \mathrm{Tr}\,(ME_{jk})\,. \tag{15}$$

In our procedure, the basis matrix E_{jk} is mapped onto the basis column-vector $\vec{\mathcal{E}}_{jk}$, which has all components equal to zero except the unity component related to the position in the matrix determined by the numbers j and k. Then one has

$$\vec{\mathcal{M}} = \sum_{j,k=1}^{n} \mathrm{Tr}\,(ME_{jk})\,\vec{\mathcal{E}}_{jk}. \tag{16}$$

For example, for similarity transformation of the finite matrix M, one has

$$\vec{\mathcal{M}}_g^s = \sum_{j,k=1}^{N} \mathrm{Tr}\left(gMg^{-1}E_{jk}\right)\vec{\mathcal{E}}_{jk}. \tag{17}$$

Now we will define the notion of 'composite' vector which corresponds to dividing a quantum system into subsystems.

We will use the following terminology.

In general, the given linear space of dimensionality $N = mn$ has a structure of a bipartite system, if the space is equipped with the operator $\hat{p}_{\vec{\mathcal{M}}M}$ and the matrix M (obtained by means of the map) has matrix elements in factorizable form

$$M_{id} \to x_i y_d. \tag{18}$$

This $M = x \otimes y$ corresponds to the special case of nonentangled states. Otherwise, one needs

$$M = \sum_{\nu} x(\nu) \otimes y(\nu).$$

In fact, to consider in detail the entanglement phenomenon, in the bipartite system of spin-1/2, one has to introduce a hierarchy of three linear spaces. The first space of pure spin states is the two-dimensional linear space of complex vectors

$$\mid \vec{x}\rangle = \begin{pmatrix} x_1 \\ x_2 \end{pmatrix}. \tag{19}$$

In this space, the scalar product is defined as follows:

$$\langle \vec{x} \mid \vec{y}\rangle = x_1^* y_1 + x_2^* y_2. \tag{20}$$

So it is a two-dimensional Hilbert space. We do not equip this space with a vector star-product structure. In the primary linear space, one introduces linear operators $\hat{M}$ which are described by 2×2 matrices M. Due to the map discussed in the previous section, the matrices are represented by 4-vectors $\vec{\mathcal{M}}$ belonging to the second complex 4-dimensional space. The star-product of the vectors $\vec{\mathcal{M}}$ determined by the kernel $\mathcal{K}_{ls}^{k}$ is defined in such a manner in order to correspond to the standard rule of multiplication of the matrices.

In addition to the star-product structure, we introduce the scalar product of the vectors $\vec{\mathcal{M}}_1$ and $\vec{\mathcal{M}}_2$, in view of the definition

$$\langle \vec{\mathcal{M}}_1 \mid \vec{\mathcal{M}}_2 \rangle = \mathrm{Tr}\,(M_1^{\dagger} M_2), \tag{21}$$

which is the trace formula for the scalar product of matrices.

This means introducing the real metric $g^{\alpha\beta}$ in the standard notation for scalar product

$$\langle \vec{\mathcal{M}}_1 \mid \vec{\mathcal{M}}_2 \rangle = \sum_{\alpha,\beta=1}^{4} (\mathcal{M}_1)^*_{\alpha} g^{\alpha\beta} (\mathcal{M}_2)_{\beta}, \tag{22}$$

where the matrix $g^{\alpha\beta}$ is of the form

$$g^{\alpha\beta} = \begin{pmatrix} 1 & 0 & 0 & 0 \\ 0 & 0 & 1 & 0 \\ 0 & 1 & 0 & 0 \\ 0 & 0 & 0 & 1 \end{pmatrix}, \qquad g^{\alpha j} g^{j\beta} = \delta^{\alpha\beta}. \tag{23}$$

The scalar product is invariant under the action of the group of nonsingular 4×4 matrices ℓ, which satisfy the condition

$$\ell^{-1} = \ell^{\dagger} g \ell. \tag{24}$$

The product of matrices ℓ satisfies the same condition since $g^2 = 1$.

Thus, the space of operators $\hat{M}$ in the primary two-dimensional space of spin states is mapped onto the linear space which is equipped with a scalar product (metric Hilbert space structure) and an associative star-product (kernel satisfying the quadratic associativity equation). In the linear space of the 4-vectors $\vec{\mathcal{M}}$, we introduce linear operators (superoperators), which can be associated with the algebra of 4×4 complex matrices.

Let us now focus on density matrices. This means that our matrix M is considered as a density matrix ρ which describes a quantum state. We consider here the action of the unitary transformation $U(n)$ of the density matrices and corresponding transformations on the vector space. If one has the structure of a bipartite system, we also consider the action of local gauge transformation both in the "source space" of density matrices and in the "target space" of the corresponding vectors.

The $n\times n$ density matrix ρ has matrix elements

$$\rho_{ik} = \rho^*_{ki}, \qquad \operatorname{Tr}\rho = 1, \qquad \langle\psi|\rho|\psi\rangle \geq 0. \tag{25}$$

Since the density matrix is hermitian, it can always be identified as an element of the convex subset of the linear space associated with the Lie algebra of $U(n)$ group, on which the group $U(n)$ acts with the adjoint representation

$$\rho \to \rho_U = U\rho U^\dagger. \tag{26}$$

The system is said to be bipartite if the space of representation is equipped with an additional structure. This means that for

$$n^2 = n_1 \cdot n_2,$$

where, for simplicity, $n_1 = n_2 = n$, one can make first the map of $n\times n$ matrix ρ onto n^2-dimensional vector $\vec{\rho}$ according to the previous procedure, i.e., one equips the space by an operator $\hat{t}_{\vec{\rho}\rho}$. Given this vector one makes a relabeling of the vector $\vec{\rho}$ components according to the rule

$$\vec{\rho} \to \rho_{id,ke}, \quad i,k = 1,2,\ldots,n_1, \quad d,e = 1,2,\ldots,n_2, \tag{27}$$

i.e., obtaining again the quadratic matrix

$$\rho_q = \hat{p}_{\rho_q\vec{\rho}}\vec{\rho}. \tag{28}$$

The unitary transform (26) of the density matrix induces a linear transform of the vector $\vec{\rho}$ of the form

$$\vec{\rho} \to \vec{\rho}_U = (U \otimes U^*)\vec{\rho}. \tag{29}$$

There exist linear transforms (called positive maps) of the density matrix, which preserve its trace, hermicity, and positivity. In some cases, they have the following form introduced in [39]

$$\rho_0 \to \rho_U = L_U\rho_0 = \sum_k p_k U_k \rho_0 U_k^\dagger, \qquad \sum_k p_k = 1, \tag{30}$$

where U_k are unitary matrices and p_k are positive numbers.

If the initial density matrix is diagonal, i.e., it belongs to the Cartan subalgebra of the Lie algebra of the unitary group, the diagonal elements of the obtained matrix give a "smoother" probability distribution than the initial one. A generic transformation preserving previously stated properties may be given in the form (see [39, 40])

$$\rho_0 \to \rho_V = L_V\rho_0 = \sum_k V_k \rho_0 V_k^\dagger, \qquad \sum_k V_k^\dagger V_k = 1. \tag{31}$$

For example, if V_k $(k = 1, 2, \ldots, N)$ are taken as square roots of orthogonal projectors onto complete set of N state, the map provides the map of the density matrix ρ_0 onto diagonal density matrix ρ_{0d} which has the same diagonal elements as ρ_0 has. In this case, the matrices V_k have the only nonzero matrix element which is equal to one. Such a map may be called "decoherence map" because it removes all nondiagonal elements of the density matrix ρ_0 killing any phase relations. In quantum information terminology, one uses also the name "phase damping channel." More general map may be given if one takes V_k as N generic diagonal density matrices, in which eigenvalues are obtained by N circular permutations from the initial one. Due to this map, one has a new matrix with the same diagonal matrix elements but with changed nondiagonal elements. The purity of this matrix is smaller then the purity of the initial one. This means that the map is contractive. All matrices with the same diagonal elements up to permutations belong to a given orbit of the unitary group.

For a large number of terms with randomly chosen matrices V_k in the sum in (31), the above map gives the most stochastic density matrix

$$\rho_0 \to \rho_s = L_1 \rho_0 = (n)^{-1} 1.$$

Its four-dimensional matrix L_1 for the qubit case has four matrix elements different from zero. These matrix elements are equal to one. They have the labels L_{11}, L_{14}, L_{41}, L_{44}. The map with two nonzero matrix elements $L_{41} = L_{44} = 0$ provides pure-state density matrix from any ρ_0. The transform (30) is the partial case of the transform (31). We discuss the transforms separately since they are used in the literature in the presented form.

One can see that the constructed map of density matrices onto vectors provides the corresponding transforms of the vectors, i.e.,

$$\vec{\rho}_0 \to \vec{\rho}_U = \sum_k p_k (U_k \otimes U_k^*) \vec{\rho}_0 \tag{32}$$

and

$$\vec{\rho}_0 \to \vec{\rho}_V = \sum_k (V_k \otimes V_k^*) \vec{\rho}_0. \tag{33}$$

It is obvious that the set of linear transforms of vectors, which preserve their properties of being image of density matrices, is essentially larger than the standard unitary transform of the density matrices.

Formulae (32) and (33) mean that the positive map superoperators acting on the density matrix in the vector representation are described by $n^2 \times n^2$ matrices

$$\mathcal{L}_U = \sum_k p_k (U_k \otimes U_k^*) \tag{34}$$

and

$$\mathcal{L}_V = \sum_k V_k \otimes V_k^*, \tag{35}$$

respectively.

The positive map is called "noncompletely positive" if

$$\mathcal{L}=\sum_k V_k\otimes V_k^*-\sum_s v_s\otimes v_s^*, \qquad \sum_k V_k^\dagger V_k-\sum_s v_s^\dagger v_s=1.$$

This map is related to a possible "nonphysical" evolution of a subsystem.

Formula (34) can be considered in the context of random matrix representation. In fact, the matrix $\mathcal{L}_U$ can be interpreted as the weighted mean value of the random matrix $U_k\otimes U_k^*$. The dependence of matrix elements and positive numbers p_k on index k means that we have a probability distribution function p_k and averaging of the random matrix $U_k\otimes U_k^*$ by means of the distribution function. So the matrix $\mathcal{L}_U$ reads

$$\mathcal{L}_U=\langle U\otimes U^*\rangle. \tag{36}$$

Let us consider an example of a 2×2 unitary matrix. We can consider a matrix of the $SU(2)$ group of the form

$$u=\begin{pmatrix}\alpha & \beta\\ -\beta^* & \alpha^*\end{pmatrix}, \qquad |\alpha|^2+|\beta|^2=1. \tag{37}$$

The 4×4 matrix $\mathcal{L}_U$ takes the form

$$\mathcal{L}_U=\begin{pmatrix}\ell & m & m^* & 1-\ell\\ -n & s & -q & n\\ -n^* & -q^* & s^* & n^*\\ 1-\ell & -m & -m^* & \ell\end{pmatrix}. \tag{38}$$

The matrix elements of the matrix $\mathcal{L}_U$ are the means

$$\begin{aligned} m &= \langle\alpha\beta^*\rangle,\\ \ell &= \langle\alpha\alpha^*\rangle,\\ n &= \langle\alpha\beta\rangle,\\ s &= \langle\alpha^2\rangle,\\ q &= \langle\beta^2\rangle. \end{aligned} \tag{39}$$

The moduli of these matrix elements are smaller than unity.

The determinant of the matrix $\mathcal{L}_U$ reads

$$\det\mathcal{L}_U=(1-2\ell)\left(|q|^2-|s|^2\right)+4\,\mathrm{Re}\left[q^*m^*n+mns^*\right]. \tag{40}$$

If one represents the matrix $\mathcal{L}_U$ in block form

$$\mathcal{L}_U=\begin{pmatrix}A & B\\ C & D\end{pmatrix}, \tag{41}$$

then

$$A = \begin{pmatrix} \ell & m \\ -n & s \end{pmatrix}, \qquad B = \begin{pmatrix} m^* & 1-\ell \\ -q & n \end{pmatrix}, \tag{42}$$

and

$$D = \sigma_2 A^* \sigma_2, \qquad C = -\sigma_2 B^* \sigma_2, \tag{43}$$

where σ_2 is the Pauli matrix.

One can check that the product of two different matrices $\mathcal{L}_U$ can be cast in the same form. This means that the matrices $\mathcal{L}_U$ form a 9-parameter compact semigroup. It means that the product of two matrices from the set (semigroup) belongs to the same set. It means that composition is inner like the one for groups. There is a unity element in the semigroup, however, there exist elements which have no inverse. In our case, these elements are described, e.g., by the matrices with zero determinant. Also the elements, which are matrices with nonzero determinants, have no inverse elements in this set, since the map corresponding to the inverse of these matrices is not positive one. For example, in the case $\ell = 1/2$ and $m = 0$, one has the matrices

$$A = \begin{pmatrix} 1/2 & 0 \\ -n & s \end{pmatrix}, \qquad B = \begin{pmatrix} 0 & 1/2 \\ -q & n \end{pmatrix}. \tag{44}$$

The determinant of the matrix $\mathcal{L}_U$ in this case is equal to zero. All the matrices $\mathcal{L}_U$ have the eigenvector

$$\vec{\rho}_0 = \begin{pmatrix} 1/2 \\ 0 \\ 0 \\ 1/2 \end{pmatrix}, \tag{45}$$

i.e.,

$$\mathcal{L}_U \vec{\rho}_0 = \vec{\rho}_0. \tag{46}$$

This eigenvector corresponds to the density matrix

$$\rho_1 = \begin{pmatrix} 1/2 & 0 \\ 0 & 1/2 \end{pmatrix}, \tag{47}$$

which is obviously invariant of the positive map.

For random matrix, one has correlations of the random matrix elements, e.g., $\langle \alpha\alpha^* \rangle \neq \langle \alpha \rangle \langle \alpha^* \rangle$.

The matrix $\mathcal{L}_p$

$$\mathcal{L}_p = \begin{pmatrix} 1 & 0 & 0 & 0 \\ 0 & 0 & 1 & 0 \\ 0 & 1 & 0 & 0 \\ 0 & 0 & 0 & 1 \end{pmatrix} \tag{48}$$

maps the vector

$$\vec{\rho}_{\rm in} = \begin{pmatrix} \rho_{11} \\ \rho_{12} \\ \rho_{21} \\ \rho_{22} \end{pmatrix} \qquad (49)$$

onto the vector

$$\vec{\rho}_{\rm t} = \begin{pmatrix} \rho_{11} \\ \rho_{21} \\ \rho_{12} \\ \rho_{22} \end{pmatrix}. \qquad (50)$$

This means that the positive map (48) connects the positive density matrix with its transpose (or complex conjugate). This map can be presented as the connection of the matrix ρ with its transpose of the form

$$\rho \to \rho^{\rm tr} = \rho^* = \frac{1}{2}\Big(\rho + \sigma_1\rho\sigma_1 - \sigma_2\rho\sigma_2 + \sigma_3\rho\sigma_3\Big).$$

There is no unitary transform connecting these matrices.

There is noncompletely positive map in the N-dimensional case, which is given by the generalized formula (for some ε)

$$\rho \to \rho_s = -\varepsilon\rho + \frac{1+\varepsilon}{N}\,1_N.$$

In quantum information terminology, it is called "depolarizing channel".

For the qubit case, matrix form of this map reads

$$L = \begin{pmatrix} \frac{1-\varepsilon}{2} & 0 & 0 & \frac{1+\varepsilon}{2} \\ 0 & -\varepsilon & 0 & 0 \\ 0 & 0 & -\varepsilon & 0 \\ \frac{1+\varepsilon}{2} & 0 & 0 & \frac{1-\varepsilon}{2} \end{pmatrix}. \qquad (51)$$

Thus we constructed the matrix representation of the positive map of density operators of the spin-1/2 system. This particular set of matrices realize the representation of the semigroup of real numbers $-1 \le \varepsilon \le 1$. If one considers the product $\varepsilon_1\varepsilon_2 = \varepsilon_3$, the result ε_3 belongs to the semigroup. Only two elements 1 and -1 of the semigroup have the inverse. These two elements form the finite subgroup of the semigroup. The semigroup itself without element $\varepsilon = 0$ can be embedded into the group of real numbers with natural multiplication rule. Each matrix L has an inverse element in this group but all the parameters of the inverse elements η live out of the segment $-1, 1$. The group of the real numbers is commutative. The matrices of the nonunitary representation of this group commute too. It means that we have nonunitary reducible representation of the semigroup which is also commutative. To construct this representation, one needs to use the map of matrices on the vectors discussed in the previous section. Formulae (31) and (35) can be interpreted also in the context of the random matrix representation, but we use the uniform distribution for averaging in this case. So one has equality (35) in the form

$$\mathcal{L}_V = \langle V \otimes V^* \rangle \tag{52}$$

and the equality

$$\langle V^\dagger V \rangle = 1, \tag{53}$$

which provides constraints for the random matrices V used.

Using the random matrix formalism, the positive (but not completely positive) maps can be presented in the form

$$\mathcal{L} = \langle V \otimes V^* \rangle - \langle v \otimes v^* \rangle, \qquad \langle V^\dagger V \rangle - \langle v^\dagger v \rangle = 1.$$

One can characterize the action of positive map on a density matrix ρ by the parameter

$$\kappa = \frac{\mathrm{Tr}\,(\mathcal{L}\rho)^2}{\mathrm{Tr}\,\rho^2} = \frac{\mu_{\mathcal{L}\rho}}{\mu_\rho} \leq 1.$$

As a remark we note that in [39] the positive maps (30) and (31) were used to describe the non-Hamiltonian evolution of quantum states for open systems.

We have to point out that, in general, such evolution is not described by first-order-in-time differential equation. As in the previous case, if there are additional structures for the matrix in the form

$$\rho_{id,ke} \rightarrow x_i y_d z_k t_e, \tag{54}$$

which means associating with the initial linear space two extra linear spaces where x_i, z_k are considered as vector components in the n_1-dimensional linear space and y_d, t_e are vector components in n_2-dimensional vector space, we see that one has bipartite structure of the initial space of state [bipartite structure of the space of adjoint representations of the group $U(n)$].

Usually the adjoint representation of any group is defined per se without any reference to possible substructures. Here we introduce the space with extra structure. In addition to being the space of the adjoint representation of the group $U(n)$, it has the structure of a bipartite system. The generalization to multipartite (N-partite) structure is straightforward. One needs only the representation of positive integer n^2 in the form

$$n^2 = \prod_{k=1}^{N} n_k^2. \tag{55}$$

If one considers the more general map given by superoperator (35) rewritten in the form

$$\mathcal{L}_V = \langle V \otimes V^* \rangle, \qquad \langle V^\dagger V \rangle = 1,$$

the number of parameters determining the matrix $\mathcal{L}_V$ can be easily evaluated. For example, for $n = 2$,

$$V = \begin{pmatrix} a & b \\ c & d \end{pmatrix}, \qquad V^* = \begin{pmatrix} a^* & b^* \\ c^* & d^* \end{pmatrix},$$

where the matrix elements are complex numbers, the normalization condition provides four constraints for the real and imaginary parts of the matrix elements of the following matrix:

$$\mathcal{L}_V = \begin{pmatrix} \langle |a|^2 \rangle & \langle ab^* \rangle & \langle ba^* \rangle & \langle bb^* \rangle \\ \langle ac^* \rangle & \langle ad^* \rangle & \langle bc^* \rangle & \langle bd^* \rangle \\ \langle ca^* \rangle & \langle cb^* \rangle & \langle da^* \rangle & \langle db^* \rangle \\ \langle cc^* \rangle & \langle cd^* \rangle & \langle dc^* \rangle & \langle dd^* \rangle \end{pmatrix},$$

namely,

$$\langle |a|^2 \rangle + \langle |c|^2 \rangle = 1, \qquad \langle |b|^2 \rangle + \langle |d|^2 \rangle = 1, \qquad \langle a^* b \rangle + \langle c^* d \rangle = 0.$$

Due to the structure of the matrix $\mathcal{L}_V$, there are six complex parameters

$$\langle ab^* \rangle, \quad \langle ac^* \rangle, \quad \langle ad^* \rangle, \quad \langle bc^* \rangle, \quad \langle bd^* \rangle, \quad \langle cd^* \rangle$$

or 12 real parameters.

The geometrical picture of the positive map can be clarified if one considers the transform of the positive density matrix into another density matrix as the transform of an ellipsoid into another ellipsoid. A generic positive transform means a generic transform of the ellipsoid, which changes its orientation, values of semiaxis, and position in the space. But the transform does not change the ellipsoidal surface into a hyperboloidal or paraboloidal surface. For pure states, the positive density matrix defines the quadratic form which is maximally degenerated. In this sense, the "ellipsoid" includes all its degenerate forms corresponding to the density matrix of rank less than n (in n-dimensional case). The number of parameters defining the map $\langle V \otimes V^* \rangle$ in the n-dimensional case is equal to $n^2(n^2 - 1)$.

The linear space of Hermitian matrices is also equipped with the commutator structure defining the Lie algebra of the group $U(n)$. The kernel that defines this structure (Lie product structure) is determined by the kernel that determines the star-product.

3. Distributions as vectors

In quantum mechanics, one needs the concept of distance between the quantum states. In this section, we consider the notion of distance between the quantum states in terms of vectors. First, let us discuss the notion of distance between conventional probability distributions. This notion is well known in the classical probability theory.

Given the probability distribution $P(k)$, $k = 1, 2, \dots N$, one can introduce the vector $\vec{P}$ in the form of a column with components $P_1 = P(1)$, $P_2 = P(2), \dots, P_N = P(N)$. The vector satisfies the condition

$$\sum_{k=1}^{N} P_k = 1. \tag{56}$$

This set of vectors does not form a linear space but only a convex subset. Nevertheless, in this set one can introduce a distance between two distributions by using the one suggested by the vector space structure of the ambient space:

$$D^2 = \left(\vec{P}_1 - \vec{P}_2\right)^2 = \sum_k P_{1k} P_{1k} + \sum_k P_{2k} P_{2k} - 2 \sum_k P_{1k} P_{2k}. \tag{57}$$

Of course, one may use other identifications of distributions with vectors.

Since all $P(k) \geq 0$, one can introduce $\mathcal{P}_k = \sqrt{P(k)}$ as components of the vector $\vec{\mathcal{P}}$. The $\vec{\mathcal{P}}$ can be thought of as a column with nonnegative components. Then the distance between the two distributions takes the form

$$\mathcal{D}^2 = \left(\vec{\mathcal{P}}_1 - \vec{\mathcal{P}}_2\right)^2 = 2 - 2 \sum_k \sqrt{P_1(k) P_2(k)}. \tag{58}$$

The two different definitions (56) and (57) can be used as distances between the distributions.

Let us discuss now the notion of distance between the quantum states determined by density matrices. In the density-matrix space (in the set of linear space of the adjoint $U(n)$ representation), one can introduce distances analogously. The first case is

$$\mathrm{Tr}\,(\rho_1 - \rho_2)^2 = D^2 \tag{59}$$

and the second case is

$$\mathrm{Tr}\,(\sqrt{\rho_1} - \sqrt{\rho_2})^2 = \mathcal{D}^2. \tag{60}$$

In fact, the distances introduced can be written naturally as norms of vectors associated to density matrices

$$D^2 = |\vec{\rho}_1 - \vec{\rho}_2|^2 \tag{61}$$

and

$$\mathcal{D}^2 = \left((\vec{\sqrt{\rho}}_1) - (\vec{\sqrt{\rho}}_2)\right)^2, \tag{62}$$

respectively.

In the above expressions, we use scalar product of vectors $\vec{\rho}_1$ and $\vec{\rho}_2$ as well as scalar products of vectors $(\vec{\sqrt{\rho_1}})$ and $(\vec{\sqrt{\rho_2}})$, respectively.

Both definitions immediately follow by identification of either matrices ρ_1 and ρ_2 with vectors according to the map of the previous sections or matrices $\sqrt{\rho_1}$ and $\sqrt{\rho_2}$ with vectors. Since the density matrices ρ_1 and ρ_2 have nonnegative eigenvalues, the matrices $\sqrt{\rho_1}$ and $\sqrt{\rho_2}$ are defined without ambiguity. This means that the vectors $(\vec{\sqrt{\rho_1}})$ and $(\vec{\sqrt{\rho_2}})$ are also defined without ambiguity. It is obvious that using this construction and introducing linear map of positive vectors $\vec{\sqrt{\rho}}$, one induces nonlinear map of density matrices. Other analogous functions, in addition to square root function, can be used to create other nonlinear positive maps.

4. Separable systems and separability criterion

According to the definition, the system density matrix is called separable (for composite system) but not simply separable, if there is decomposition of the form

$$\rho_{AB} = \sum_k p_k \left(\rho_A^{(k)} \otimes \rho_B^{(k)}\right), \qquad \sum_k p_k = 1, \qquad 1 \geq p_k \geq 0. \tag{63}$$

This is Hilbert's problem of biquadrates. Is a positive biquadratic the positive sum of products of positive quadratics? In this formula, one may use also sum over two different indices. Using another labelling in such sum over two different indices, this sum can be always represented as the sum over only one index. The formula does not demand orthogonality of the density operators $\rho_A^{(k)}$ and $\rho_B^{(k)}$ for different k. Since every density matrix is a convex sum of pure density matrices, one could demand that $\rho_A^{(k)}$ and $\rho_B^{(k)}$ be pure. This formula can be interpreted in the context of random matrix representation reading

$$\rho_{AB} = \langle \rho_A \otimes \rho_B \rangle, \tag{64}$$

where ρ_A and ρ_B are considered as random density matrices of the subsystems A and B, respectively. One can use the clarified structure of the density matrix set as the union of orbits obtained by action of the unitary group on projectors of rank one with matrix form containing only one nonzero matrix element. Then the separable density matrix of bipartite composite system can be always written as the sum of $n_1 n_2$ tensor products (or corresponding mean tensor product), i.e., in (64) the factors are state projectors. Each of tensor products contains random unitary matrices of local transforms of the fixed local projector for one subsystem and for the second subsystem. It means that an arbitrary projector of rank one of a subsystem can be always presented in the product form $\rho_A^{(k)} = u_A^{(k)} \rho_A u_A^{(k)\dagger}$, where $u_A^{(k)}$ is a unitary local transform and ρ_A is a fixed projector.

There are several criteria for the system to be separable. We suggest in the next sections a new approach to the problem of separability and entanglement based on the tomographic probability description of quantum states. The states which cannot be represented in the form (63) by definition are called entangled states [38]. Thus the states are entangled if in formula (63) at least one coefficient (or more) p_i is negative which means that the positive ones can take values greater than unity.

Let us discuss the condition for the system state to be separable. According to the partial transpose criterion [41], the system is separable if the partial transpose of the matrix ρ_{AB} (63) gives a positive density matrix. This condition is necessary but not sufficient. Let us discuss this condition within the framework of positive-map matrix representation. For example, for a spin-1/2 bipartite system, we have shown that the map of a density matrix onto its transpose belongs to the matrix semigroup of matrices $\mathcal{L}$. One should point out that this map cannot be obtained by means of averaging with all positive probability distributions p_k. On the other hand, it is obvious that the generic criterion, which contains the Peres criterion as a partial case, can be formulated as follows.

Let us map the density matrix ρ_{AB} of a bipartite system onto vector $\vec{\rho}_{AB}$. Let the vector $\vec{\rho}_{AB}$ be acted upon by an arbitrary matrix, which represents the positive maps in subsystems A and B. Thus we get a new vector

$$\vec{\rho}_{AB}^{(p)} = \left(\mathcal{L}_A \otimes \mathcal{L}_B\right)\vec{\rho}_{AB}. \tag{65}$$

Let us construct the density matrix $\rho_{AB}^{(p)}$ using the inverse map of the vectors onto matrices. If the initial density matrix is separable, the new density matrix $\rho_{AB}^{(p)}$ must be positive (and separable).

In the case of the bipartite spin-1/2 system, by choosing $\mathcal{L}_A = 1$ and with $\mathcal{L}_B$ being the matrix coinciding with the matrix $g^{\alpha\beta}$, we obtain the Peres criterion as a partial case of the criterion of separability formulated above. Thus, our criterion means that the separable matrix keeps positivity under the action of the tensor product of two semigroups. In the case of the bipartite spin-1/2 system, the 16×16 matrix of the semigroup tensor product of positive contractive maps (52) is determined by 24 parameters. Among these parameters, one can have some correlations.

Let us discuss the positive map (52) which is determined by the semigroup for the n-dimensional system. It can be realized also as follows.

The $n\times n$ Hermitian generic matrix ρ can be mapped onto essentially real n^2-vector $\vec{\rho}$ by the map described above. The complex vector $\vec{\rho}$ is mapped onto the real vector $\vec{\rho}_{\mathrm{r}}$ by multiplying by the unitary matrix S, i.e.,

$$\vec{\rho}_{\mathrm{r}} = S\vec{\rho}, \qquad \vec{\rho} = S^{-1}\vec{\rho}_{\mathrm{r}}. \tag{66}$$

The matrix S is composed from n unity blocks and the blocks

$$S_b^{(jk)} = \frac{1}{\sqrt{2}} \begin{pmatrix} 1 & 1 \\ -i & i \end{pmatrix}, \tag{67}$$

where j corresponds to a column and k corresponds to a row in the matrix ρ.

For example, in the case $n = 2$, one has the vector $\vec{\rho}_{\mathrm{r}}$ of the form

$$\vec{\rho}_{\mathrm{r}} = \begin{pmatrix} \rho_{11} \\ \sqrt{2}\,\mathrm{Re}\,\rho_{12} \\ \sqrt{2}\,\mathrm{Im}\,\rho_{12} \\ \rho_{22} \end{pmatrix}. \tag{68}$$

One has the equalities

$$\vec{\rho}_{\mathrm{r}}^{\,2} = \vec{\rho}^{\,2} = \mathrm{Tr}\,\rho^2. \tag{69}$$

The semigroup preserves the trace of the density matrix. Also the discrete transforms, which are described by the matrix with diagonal matrix blocks of the form

$$\mathcal{D} = \begin{pmatrix} 1 & 0 & 0 & 0 \\ 0 & 1 & 0 & 0 \\ 0 & 0 & -1 & 0 \\ 0 & 0 & 0 & 1 \end{pmatrix}, \tag{70}$$

preserve positivity of the density matrix.

For the spin case, the semigroup contains 12 parameters.

Thus, the direct product of the semigroup (52) and the discrete group of the transform D defines positive map preserving positivity of the density operator. One can include also all the matrices which correspond to other not completely positive maps. The considered representation contains only real vectors and their real positive transforms. This means that one can construct representation of semigroup of positive maps by real matrices.

5. Symbols, star-product and entanglement

In this section, we describe how entangled states and separable states can be studied using properties of symbols and density operators of different kinds, e.g., from the viewpoint of the Wigner function or tomogram. The general scheme of constructing the operator symbols is as follows [36].

Given a Hilbert space H and an operator $\hat{A}$ acting on this space, let us suppose that we have a set of operators $\hat{U}(\mathbf{x})$ acting transitively on H parametrized by n-dimensional vectors $\mathbf{x} = (x_1, x_2, \ldots, x_n)$. We construct the c-number function $f_{\hat{A}}(\mathbf{x})$ (we call it the symbol of the operator $\hat{A}$) using the definition

$$f_{\hat{A}}(\mathbf{x}) = \mathrm{Tr}\left[\hat{A}\hat{U}(\mathbf{x})\right]. \tag{71}$$

Let us suppose that relation (71) has an inverse, i.e., there exists a set of operators $\hat{D}(\mathbf{x})$ acting on the Hilbert space such that

$$\hat{A} = \int f_{\hat{A}}(\mathbf{x})\hat{D}(\mathbf{x})\,d\mathbf{x}, \qquad \mathrm{Tr}\,\hat{A} = \int f_{\hat{A}}(\mathbf{x})\,\mathrm{Tr}\,\hat{D}(\mathbf{x})\,d\mathbf{x}. \tag{72}$$

One needs a measure in $\mathbf{x}$ to define the integral in above formulae. Then, we will consider relations (71) and (72) as relations determining the invertible map from the operator $\hat{A}$ onto the function $f_{\hat{A}}(\mathbf{x})$. Multiplying both sides of Eq. (2) by the operator $\hat{U}(\mathbf{x}')$ and taking the trace, one can satisfy the consistency condition for the operators $\hat{U}(\mathbf{x}')$ and $\hat{D}(\mathbf{x})$

$$\mathrm{Tr}\left[\hat{U}(\mathbf{x}')\hat{D}(\mathbf{x})\right] = \delta\left(\mathbf{x}' - \mathbf{x}\right). \tag{73}$$

The consistency condition (73) follows from the relation

$$f_{\hat{A}}(\mathbf{x}) = \int K(\mathbf{x}, \mathbf{x}')f_{\hat{A}}(\mathbf{x}')\,d\mathbf{x}'. \tag{74}$$

The kernel in (74) is equal to the standard Dirac delta-function, if the set of functions $f_{\hat{A}}(\mathbf{x})$ is a complete set.

In fact, we could consider relations of the form

$$\hat{A} \rightarrow f_{\hat{A}}(\mathbf{x}) \tag{75}$$

and

$$f_{\hat{A}}(\mathbf{x}) \rightarrow \hat{A}. \tag{76}$$

The most important property of the map is the existence of the associative product (star-product) of the functions.

We introduce the product (star-product) of two functions $f_{\hat{A}}(\mathbf{x})$ and $f_{\hat{B}}(\mathbf{x})$ corresponding to two operators $\hat{A}$ and $\hat{B}$ by the relationships

$$f_{\hat{A}\hat{B}}(\mathbf{x}) = f_{\hat{A}}(\mathbf{x}) * f_{\hat{B}}(\mathbf{x}) := \mathrm{Tr}\left[\hat{A}\hat{B}\hat{U}(\mathbf{x})\right]. \tag{77}$$

Since the standard product of operators on a Hilbert space is an associative product, i.e., $\hat{A}(\hat{B}\hat{C}) = (\hat{A}\hat{B})\hat{C}$, it is obvious that formula (77) defines an associative product for the functions $f_{\hat{A}}(\mathbf{x})$, i.e.,

$$f_{\hat{A}}(\mathbf{x}) * \left(f_{\hat{B}}(\mathbf{x}) * f_{\hat{C}}(\mathbf{x})\right) = \left(f_{\hat{A}}(\mathbf{x}) * f_{\hat{B}}(\mathbf{x})\right) * f_{\hat{C}}(\mathbf{x}). \tag{78}$$

Using formulae (71) and (72), one can write down a composition rule for two symbols $f_{\hat{A}}(\mathbf{x})$ and $f_{\hat{B}}(\mathbf{x})$, which determines the star-product of these symbols. The composition rule is described by the formula

$$f_{\hat{A}}(\mathbf{x}) * f_{\hat{B}}(\mathbf{x}) = \int f_{\hat{A}}(\mathbf{x}'')f_{\hat{B}}(\mathbf{x}')K(\mathbf{x}'', \mathbf{x}', \mathbf{x})\,d\mathbf{x}'\,d\mathbf{x}''. \tag{79}$$

The kernel in the integral of (79) is determined by the trace of the product of the basic operators, which we use to construct the map

$$K(\mathbf{x}'', \mathbf{x}', \mathbf{x}) = \mathrm{Tr}\left[\hat{D}(\mathbf{x}'')\hat{D}(\mathbf{x}')\hat{U}(\mathbf{x})\right]. \tag{80}$$

The kernel function satisfies the composition property $K * K = K$.

6. Tomographic representation

In this section, we will consider an example of the probability representation of quantum mechanics [42]. In the probability representation of quantum mechanics, the state is described by a family of probabilities [43–45]. According to the general scheme, one can introduce for the operator $\hat{A}$ the function $f_{\hat{A}}(\mathbf{x})$, where

$$\mathbf{x} = (x_1, x_2, x_3) \equiv (X, \mu, \nu),$$

which we denote here as $w_{\hat{A}}(X, \mu, \nu)$ depending on the position X and the parameters μ and ν of the reference frame

$$w_{\hat{A}}(X, \mu, \nu) = \mathrm{Tr}\left[\hat{A}\hat{U}(\mathbf{x})\right]. \tag{81}$$

We call the function $w_{\hat{A}}(X, \mu, \nu)$ the tomographic symbol of the operator $\hat{A}$. The operator $\hat{U}(\mathbf{x})$ is given by

$$\begin{aligned}\hat{U}(\mathbf{x}) \equiv \hat{U}(X, \mu, \nu) &= \exp\left(\frac{i\lambda}{2}(\hat{q}\hat{p} + \hat{p}\hat{q})\right)\exp\left(\frac{i\theta}{2}(\hat{q}^2 + \hat{p}^2)\right) \mid X\rangle\langle X \mid \\ &\quad \times \exp\left(-\frac{i\theta}{2}(\hat{q}^2 + \hat{p}^2)\right)\exp\left(-\frac{i\lambda}{2}(\hat{q}\hat{p} + \hat{p}\hat{q})\right) \\ &= \hat{U}_{\mu\nu} \mid X\rangle\langle X \mid \hat{U}^{\dagger}_{\mu\nu}.\end{aligned} \tag{82}$$

The tomographic symbol is the homogeneous version of the Moyal phase-space density. The angle θ and parameter λ in terms of the reference phase-space frame parameters are given by

$$\mu = e^{\lambda}\cos\theta, \qquad \nu = e^{-\lambda}\sin\theta,$$

that is, $\hat{q}$ and $\hat{p}$ are position and momentum operators

$$\hat{q} \mid X\rangle = X \mid X\rangle \tag{83}$$

and $\mid X\rangle\langle X \mid$ is the projection density. One has the canonical transform of quadratures

$$\hat{X} = \hat{U}_{\mu\nu}\,\hat{q}\,\hat{U}^{\dagger}_{\mu\nu} = \mu\hat{q} + \nu\hat{p},$$

$$\hat{P} = \hat{U}_{\mu\nu}\,\hat{p}\,\hat{U}^{\dagger}_{\mu\nu} = \frac{1+\sqrt{1-4\mu^2\nu^2}}{2\mu}\,\hat{p} - \frac{1-\sqrt{1-4\mu^2\nu^2}}{2\nu}\,\hat{q}.$$

Using the approach of [46] one obtains the relationship

$$\hat{U}(X,\mu,\nu) = \delta(X - \mu\hat{q} - \nu\hat{p}).$$

In the case we are considering, the inverse transform determining the operator in terms of the tomogram [see Eq. (72)] will be of the form

$$\hat{A} = \int w_{\hat{A}}(X,\mu,\nu)\hat{D}(X,\mu,\nu)\,dX\,d\mu\,d\nu, \tag{84}$$

where

$$\hat{D}(\mathbf{x}) \equiv \hat{D}(X,\mu,\nu) = \frac{1}{2\pi}\exp\left(iX - i\nu\hat{p} - i\mu\hat{q}\right). \tag{85}$$

The trace of the above operator, which provides the kernel determining the trace of an arbitrary operator in the tomographic representation, reads

$$\mathrm{Tr}\,\hat{D}(\mathbf{x}) = e^{iX}\delta(\mu)\delta(\nu).$$

The function $w_{\hat{A}}(X,\mu,\nu)$ satisfies the relation

$$w_{\hat{A}}(\lambda X,\lambda\mu,\lambda\nu) = \frac{1}{|\lambda|}\,w_{\hat{A}}(X,\mu,\nu). \tag{86}$$

This means that the tomographic symbols of operators are homogeneous functions of three variables.

If one takes two operators $\hat{A}_1$ and $\hat{A}_2$, which are expressed through the corresponding functions by the formulas

$$\begin{aligned} \hat{A}_1 &= \int w_{\hat{A}_1}(X',\mu',\nu')\hat{D}(X',\mu',\nu')\,dX'\,d\mu'\,d\nu', \\ \hat{A}_2 &= \int w_{\hat{A}_2}(X'',\mu'',\nu'')\hat{D}(X'',\mu'',\nu'')dX''\,d\mu''\,d\nu'', \end{aligned} \tag{87}$$

and $\hat{A}$ denotes the product of $\hat{A}_1$ and $\hat{A}_2$, then the function $w_{\hat{A}}(X,\mu,\nu)$, which corresponds to $\hat{A}$, is the star-product of the functions $w_{\hat{A}_1}(X,\mu,\nu)$ and $w_{\hat{A}_2}(X,\mu,\nu)$. Thus this product

$$w_{\hat{A}}(X,\mu,\nu) = w_{\hat{A}_1}(X,\mu,\nu) * w_{\hat{A}_2}(X,\mu,\nu)$$

reads

$$w_{\hat{A}}(X,\mu,\nu) = \int w_{\hat{A}_1}(\mathbf{x}'')w_{\hat{A}_2}(\mathbf{x}')K(\mathbf{x}'',\mathbf{x}',\mathbf{x})\,d\mathbf{x}''\,d\mathbf{x}', \tag{88}$$

with kernel given by

$$K(\mathbf{x}'',\mathbf{x}',\mathbf{x}) = \mathrm{Tr}\left[\hat{D}(X'',\mu'',\nu'')\hat{D}(X',\mu',\nu')\hat{U}(X,\mu,\nu)\right]. \tag{89}$$

The explicit form of the kernel reads

$$\begin{aligned} &K(X_1,\mu_1,\nu_1,X_2,\mu_2,\nu_2,X\mu,\nu)\\ &= \frac{\delta\Big(\mu(\nu_1+\nu_2)-\nu(\mu_1+\mu_2)\Big)}{4\pi^2}\exp\left(\frac{i}{2}\Big\{(\nu_1\mu_2-\nu_2\mu_1)+2X_1+2X_2\right.\\ &\left.-\left[\frac{1-\sqrt{1-4\mu^2\nu^2}}{\nu}(\nu_1+\nu_2)+\frac{1+\sqrt{1-4\nu^2\mu^2}}{\mu}(\mu_1+\mu_2)\right]X\Big\}\right). \end{aligned} \tag{90}$$

7. Multipartite systems

Let us assume that for multimode (N-mode) system one has

$$\hat{U}(\vec{y}) = \prod_{k=1}^{N}\otimes\hat{U}\left(\vec{x}^{(k)}\right), \tag{91}$$

$$\hat{D}(\vec{y}) = \prod_{k=1}^{N}\otimes\hat{D}\left(\vec{x}^{(k)}\right), \tag{92}$$

where

$$\vec{y} = \left(x_1^{(1)},x_2^{(1)},\ldots,x_m^{(1)},x_1^{(2)},x_2^{(2)},\ldots,x_m^{(N)}\right). \tag{93}$$

This means that the symbol of the density operator of the composite system reads

$$f_\rho(\vec{y}) = \mathrm{Tr}\left[\hat{\rho}\prod_{k=1}^{N}\otimes\hat{U}(\vec{x}^{(k)})\right]. \tag{94}$$

The inverse transform reads

$$\hat{\rho} = \int d\vec{y}\, f_\rho(\vec{y})\prod_{k=1}^{N}\otimes\hat{D}(\vec{x}^{(k)}),\qquad d\vec{y} = \prod_{k=1}^{N}\prod_{s=1}^{m}dx_s^{(k)}. \tag{95}$$

Now we formulate the properties of the symbols in the case of entangled and separable states, respectively.

Given a composite m-partite system with density operator $\hat{\rho}$.

If the nonnegative operator can be presented in the form of a "probabilistic sum"

$$\hat{\rho} = \sum_{\vec{z}}\mathcal{P}(\vec{z})\hat{\rho}_{\vec{z}}^{(a_1)}\otimes\hat{\rho}_{\vec{z}}^{(a_2)}\otimes\cdots\otimes\hat{\rho}_{\vec{z}}^{(a_m)}, \tag{96}$$

with positive probability distribution function $\mathcal{P}(\vec{z})$, where the components of $\vec{z}$ can be either discrete or continuous, we call the state a "separable state". Without loss of generality, all factors in the tensor products can be considered as projectors of rank one. This means that the symbol of the state can be presented in the form

$$f_\rho(\vec{y}) = \sum_{\vec{z}} \mathcal{P}(\vec{z}) \prod_{k=1}^{m} f_\rho^{(a_k)}(\vec{x}_k, \vec{z}). \tag{97}$$

Analogous formula can be written for the tomogram of separable state. We point out that in the multipartite case one can introduce random symbols and represent the symbol of separable density matrix of composite system as mean value of pointwise products of symbols of subsystem density operators. As in the bipartite case, one can use sum over different indices but this sum can be always reduced to the sum over only one index common for all the subsystems. It is important that for separable state its symbol always can be represented as the sum containing number of summants which is equal to dimensionality of composite system. Each term in the sum is equal to mean value of random projector. The random projector is constructed as the product of diagonal projectors of rank one in each subsystem considered in random local basis obtained by means of random unitary local transforms.

8. Spin tomography

Below we concentrate on bipartite spin systems.

The tomographic probability (spin tomogram) completely determines the density matrix of a spin state. It has been introduced in [31, 32, 36]. The tomographic probability for the spin-j state is defined via the density matrix by the formula

$$\langle jm \mid D^\dagger(g)\rho D(g) \mid jm\rangle = W^{(j)}(m, \vec{0}), \qquad m = -j, -j+1, \ldots, j, \tag{98}$$

where $D(g)$ is the matrix of $SU(2)$-group representation depending on the group element g determined by three Euler angles. It is useful to generalize the construction of spin tomogram.

One can introduce unitary spin tomograms $w(m, u)$ by replacing in above formula (98) the matrix $D(g)$ by generic unitary matrix u. For the case of higher spins $j = 1, 3/2, 2, \ldots$, the $n \times n$ projector matrix

$$\rho_1 = \begin{pmatrix} 1 & 0 & \cdots & 0 \\ 0 & 0 & \cdots & 0 \\ \cdot & \cdot & \cdots & \cdot \\ \cdot & \cdot & \cdots & \cdot \\ 0 & 0 & \cdots & 0 \end{pmatrix}, \qquad n = 2j+1 \tag{99}$$

has the unitary spin tomogram denoted as

$$w_1(j,u) = |u_{11}|^2, \quad w_1(j-1,u) = |u_{12}|^2, \quad \dots \quad w_1(-j,u) = |u_{1n}|^2. \tag{100}$$

Other projectors

$$\rho_k = \begin{pmatrix} 0 & 0 & \cdots & \cdots & 0 \\ \cdot & \cdot & \cdots & \cdot & \cdot \\ 0 & \cdots & 1 & \cdots & 0 \\ \cdot & \cdot & \cdot & \cdots & \cdot \\ 0 & 0 & \cdot & \cdots & 0 \end{pmatrix}, \tag{101}$$

in which unity is located in kth column, have the tomogram $w_k(m,u)$ of the form

$$w_k(j,u) = |u_{k1}|^2, \quad w_k(j-1,u) = |u_{k2}|^2, \quad \dots \quad w_k(-j,u) = |u_{kn}|^2. \tag{102}$$

In connection with the decomposition of any density matrix in the form

$$\rho = \sum_{jk} \rho_{jk} E_{jk}, \tag{103}$$

the unitary spin tomogram can be presented in form of the decomposition

$$w_\rho(m,u) = \sum_{jk} \rho_{jk} w_{jk}(m,u), \tag{104}$$

where $w_{jk}(m,u)$ are basic unitary spin symbols of transition operators E_{jk} of the form

$$w_{jk}(m,u) = \langle jm \mid u^\dagger E_{jk} u \mid jm \rangle. \tag{105}$$

If one uses a map

$$\rho \rightarrow \rho', \tag{106}$$

the unitary spin tomogram is transformed as

$$w_\rho(m,u) \rightarrow w'_\rho(m,u) = \sum_{jk} \rho'_{jk} w_{jk}(m,u). \tag{107}$$

If the transform (106) is a linear one

$$\rho_{jk} \rightarrow \rho'_{jk} = L_{jk,ps} \rho_{ps}, \tag{108}$$

the transform reads

$$w'_\rho(m,u) = \sum_{ps} \rho_{ps} w'_{ps}(m,u). \tag{109}$$

Here

$$w'_{ps}(m,u) = \sum_{jk} L_{jk,ps} w_{jk}(m,u) \tag{110}$$

is the linear transform of the basic tomographic symbols of the operators E_{jk}.

Let us now discuss some properties of usual spin tomograms.

The set of the tomogram values for each $\vec{0}$ is an overcomplete set. We need only a finite number of independent locations which will give information on the density matrix of the spin state. Due to the structure of the formula, there are only two Euler angles involved. They are combined into the unit vector

$$\vec{0} = (\cos\phi \sin\vartheta, \sin\phi \sin\vartheta, \cos\vartheta). \tag{111}$$

This is the Hopf map from S^3 to S^2.

The physical meaning of the probability $W(m, \vec{0})$ is the following.

It is the probability to find, in the state with the density matrix ρ, the spin projection on direction $\vec{0}$ equal to m. For a bipartite system, the spin tomogram is defined as follows:

$$W(m_1 m_2 \vec{0}_1 \vec{0}_2) = \langle j_1 m_1 j_2 m_2 \mid D^\dagger(g_1) D^\dagger(g_2) \rho D(g_1) D(g_2) \mid j_1 m_1 j_2 m_2 \rangle. \tag{112}$$

It completely determines the density matrix ρ. It has the meaning of the joint probability distribution for spin j_1 and j_2 projections m_1 and m_2 on directions $\vec{0}_1$ and $\vec{0}_2$. Since the map $\rho \rightleftharpoons W$ is linear and invertible, the definition of separable system can be rewritten in the following form for the decomposition of the joint probability into a sum of products (of factorized probabilities):

$$W(m_1 m_2 \vec{0}_1 \vec{0}_2) = \sum_k p_k W^{(k)}(m_1 \vec{0}_1) \tilde{W}^{(k)}(m_2 \vec{0}_2). \tag{113}$$

This form can be considered to formulate the criterion of separability of the two-spin state.

One can present this formula in the form

$$W(m_1 m_2 \vec{0}_1 \vec{0}_2) = \langle W(m_1 \vec{0}_1) \tilde{W}(m_2 \vec{0}_2) \rangle, \tag{114}$$

where we interpret the positive numbers p_k as probability distributions. Thus separability means the possibility to represent joint probability distribution in the form of average product of two random probability distributions.

The state is separable iff the tomogram can be written in the form (113) with $\sum_k p_k = 1$, $p_k \geq 0$. It seems that we simply use the definition but, in fact, we cast the problem of separability into the form of the property of the positive joint probability distribution of two random variables. This is an area of probability theory and one can use the results and theorems on joint probability distributions. If one does not use any theorem, one has to study the solvability of relation (113) considered as the equation for unknown probability distribution p_k and unknown probability functions $W^{(k)}(m_1 \vec{0}_1)$ and $W^{(k)}(m_2 \vec{0}_2)$.

9. Example of spin-1/2 bipartite system

For the spin-1/2 state, the generic density matrix can be presented in the form

$$\rho = \frac{1}{2}\left(1 + \vec{\sigma}\cdot\vec{n}\right), \qquad \vec{n} = (n_1, n_2, n_3), \tag{115}$$

where $\vec{\sigma}$ are Pauli matrices and $\vec{n}^2 \leq 1$, with the vector $\vec{n}$ for a pure state being the unit vector. This decomposition means that we use as basis in 4-dimensional vector space the vectors corresponding to the Pauli matrices and the unit matrix, i.e.,

$$\vec{\sigma}_1 = \begin{pmatrix} 0 \\ 1 \\ 1 \\ 0 \end{pmatrix}, \qquad \vec{\sigma}_2 = \begin{pmatrix} 0 \\ -i \\ i \\ 0 \end{pmatrix}, \qquad \vec{\sigma}_3 = \begin{pmatrix} 1 \\ 0 \\ 0 \\ -1 \end{pmatrix}, \qquad \vec{1} = \begin{pmatrix} 1 \\ 0 \\ 0 \\ 1 \end{pmatrix}. \tag{116}$$

The density matrix vector

$$\vec{\rho} = \begin{pmatrix} \rho_{11} \\ \rho_{12} \\ \rho_{21} \\ \rho_{22} \end{pmatrix} \tag{117}$$

is decomposed in terms of the basis vectors

$$\vec{\rho} = \frac{1}{2}\left(\vec{1} + n_1\vec{\sigma}_1 + n_2\vec{\sigma}_2 + n_3\vec{\sigma}_3\right). \tag{118}$$

This means that the spin tomogram of the spin-1/2 state can be given in the form

$$W\left(\frac{1}{2}, \vec{0}\right) = \frac{1}{2} + \frac{\vec{n}\cdot\vec{0}}{2}, \qquad W\left(-\frac{1}{2}, \vec{0}\right) = \frac{1}{2} - \frac{\vec{n}\cdot\vec{0}}{2}. \tag{119}$$

We can consider tomograms of specific spin state. If the state is pure state with density matrix

$$\rho_+ = \begin{pmatrix} 1 & 0 \\ 0 & 0 \end{pmatrix}, \tag{120}$$

the spin tomogram $W(m, \vec{0})$, where

$$m = \pm\frac{1}{2}, \qquad \vec{0} = (\sin\theta\cos\varphi, \sin\theta\sin\varphi, \cos\theta)$$

has the values

$$W_+\left(\frac{1}{2}, \vec{0}\right) = \cos^2\frac{\theta}{2}, \quad W_+\left(-\frac{1}{2}, \vec{0}\right) = \sin^2\frac{\theta}{2}. \tag{121}$$

The tomogram of the pure state

$$\rho_- = \begin{pmatrix} 0 & 0 \\ 0 & 1 \end{pmatrix}, \tag{122}$$

has the values

$$\begin{aligned} W_-\left(\frac{1}{2}, \vec{0}\right) &= \sin^2\frac{\theta}{2} = \cos^2\frac{\pi-\theta}{2}, \\ W_-\left(-\frac{1}{2}, \vec{0}\right) &= \cos^2\frac{\theta}{2} = \sin^2\frac{\pi-\theta}{2}. \end{aligned} \tag{123}$$

The spin tomogram of the diagonal density matrix

$$\rho_d = \begin{pmatrix} \rho_{11} & 0 \\ 0 & \rho_{22} \end{pmatrix} \tag{124}$$

equals

$$W_d(m, \vec{0}) = \rho_{11} W_+(m, \vec{0}_+) + \rho_{22} W_-(m, \vec{0}_-). \tag{125}$$

The generic density matrix which has eigenvalues ρ_{11} and ρ_{22} can be presented in the form

$$u_0 \rho_d u_0^\dagger, \tag{126}$$

where the unitary matrix u_0 has columns containing components of normalized eigenvectors of the density matrix ρ.

This means that the tomogram of the state with the matrix ρ reads

$$W_\rho(m, \vec{0}) = \langle m \mid u^\dagger u_0 \rho_d u_0^\dagger u \mid m\rangle. \tag{127}$$

The elements of the group can be combined

$$u^\dagger u_0 = \tilde{u}. \tag{128}$$

Thus the tomogram becomes

$$W_\rho(m, \vec{0}) = W_d(m, \vec{0}'), \tag{129}$$

where the angle $\vec{0}'$ corresponds to the Euler angle calculated from the product of two unitary matrices $u_0^\dagger u$.

One can use the property of numbers ρ_{11} and ρ_{22} to interpret formula (125) as averaging

$$W_d(m, \vec{0}) = \langle W(m, \vec{0}')\rangle, \tag{130}$$

where one interprets two functions $W_+(m, \vec{0})$ and $W_-(m, \vec{0}')$ as the realization of "random" probability distribution function $W_\pm(m, \vec{0})$. Then one has

$$W_\rho(m, \vec{0}) = \langle W(m, \vec{0}')\rangle. \tag{131}$$

The new vector $\vec{0}'$ has the parameter θ' obtained from the initial parameter θ by action of the unitary matrix on the initial unitary matrix u.

Inserting these probability values into relation (113) for each value of k, we get the relationships

$$\begin{aligned} W\left(\frac{1}{2},\frac{1}{2},\vec{0}_1,\vec{0}_2\right) &= \frac{1}{4}+\frac{1}{2}\left(\sum_k p_k\vec{n}_k\right)\cdot\vec{0}_1+\frac{1}{2}\left(\sum_k p_k\vec{n}_k^*\right)\cdot\vec{0}_2 \\ &\quad +\sum_k p_k\left(\vec{n}_k\cdot\vec{0}_1\right)\left(\vec{n}_k^*\cdot\vec{0}_2\right), \end{aligned} \tag{132}$$

$$\begin{aligned} W\left(\frac{1}{2},-\frac{1}{2},\vec{0}_1,\vec{0}_2\right) &= \frac{1}{4}+\frac{1}{2}\left(\sum_k p_k\vec{n}_k\right)\cdot\vec{0}_1-\frac{1}{2}\left(\sum_k p_k\vec{n}_k^*\right)\cdot\vec{0}_2 \\ &\quad -\sum_k p_k\left(\vec{n}_k\cdot\vec{0}_1\right)\left(\vec{n}_k^*\cdot\vec{0}_2\right), \end{aligned} \tag{133}$$

$$\begin{aligned} W\left(-\frac{1}{2},\frac{1}{2},\vec{0}_1,\vec{0}_2\right) &= \frac{1}{4}-\frac{1}{2}\left(\sum_k p_k\vec{n}_k\right)\cdot\vec{0}_1+\frac{1}{2}\left(\sum_k p_k\vec{n}_k^*\right)\cdot\vec{0}_2 \\ &\quad -\sum_k p_k\left(\vec{n}_k\cdot\vec{0}_1\right)\left(\vec{n}_k^*\cdot\vec{0}_2\right). \end{aligned} \tag{134}$$

One has the normalization property

$$\sum_{m_1,m_2=-1/2}^{1/2} W(m_1 m_2\vec{0}_1\vec{0}_2)=1. \tag{135}$$

One easily gets

$$W\left(\frac{1}{2},\frac{1}{2},\vec{0}_1,\vec{0}_2\right)+W\left(\frac{1}{2},-\frac{1}{2},\vec{0}_1,\vec{0}_2\right)=\frac{1}{2}+\left(\sum_k p_k\vec{n}_k\right)\cdot\vec{0}_1. \tag{136}$$

This means that the derivative in $\vec{0}_1$ on the left-hand side gives

$$\frac{\partial}{\partial\vec{0}_1}\left[W\left(\frac{1}{2},\frac{1}{2},\vec{0}_1,\vec{0}_2\right)+W\left(\frac{1}{2},-\frac{1}{2},\vec{0}_1,\vec{0}_2\right)\right]=\left(\sum_k p_k\vec{n}_k\right). \tag{137}$$

Analogously

$$\frac{\partial}{\partial\vec{0}_2}\left[W\left(\frac{1}{2},\frac{1}{2},\vec{0}_1,\vec{0}_2\right)+W\left(-\frac{1}{2},\frac{1}{2},\vec{0}_1,\vec{0}_2\right)\right]=\left(\sum_k p_k\vec{n}_k^{(\star)}\right). \tag{138}$$

Taking the sum of (132) and (133)) one sees that

$$\frac{1}{2}\frac{\partial}{\partial \vec{0}_i}\frac{\partial}{\partial \vec{0}_j}\left[W\left(\frac{1}{2},-\frac{1}{2},\vec{0}_1,\vec{0}_2\right)+W\left(-\frac{1}{2},\frac{1}{2},\vec{0}_1,\vec{0}_2\right)\right]$$
$$=-\sum_k p_k(n_k)_i(n_k^{(\star)})_j. \tag{139}$$

Since we look for the solution where $p_k \geq 0$, we can introduce

$$\vec{N}_k = \sqrt{p_k}\vec{n}_k, \qquad \vec{N}_k^{(\star)} = \sqrt{p_k}\vec{n}_k^{(\star)}. \tag{140}$$

This means that the derivative in (138) can be presented as a tensor

$$-T_{ij} = \sum_k (N_k)_i(N_k^{(\star)})_j. \tag{141}$$

One has

$$\sum_k p_k\vec{n}_k = \sum_k \sqrt{p_k}\vec{N}_k, \tag{142}$$

$$\sum_k p_k\vec{n}_k^{\star} = \sum_k \sqrt{p_k}\vec{N}_k^{(\star)}. \tag{143}$$

The conditions of solvability of the obtained equations is a criterion for separability or entanglement of a bipartite quantum spin state. Using the arguments on the representation of the tomogram (tomographic symbol) as sum of random basic projector symbols we get that for two qubits the separable state has the tomogram with following properties. All four values of joint probability distribution function are equal to mean values of product of two cosine of two different angles squared, product of sine of two different angles squared and product of sine and cosine squared, respectively. The entangled matrix does not provide such structure.

As an example, we consider the Werner state. For the Werner state (see, e.g., [47]) with the density matrix

$$\rho_{AB} = \begin{pmatrix} \frac{1+p}{4} & 0 & 0 & \frac{p}{2} \\ 0 & \frac{1-p}{4} & 0 & 0 \\ 0 & 0 & \frac{1-p}{4} & 0 \\ \frac{p}{2} & 0 & 0 & \frac{1+p}{4} \end{pmatrix},$$
$$\rho_A = \rho_B = \frac{1}{2}\begin{pmatrix} 1 & 0 \\ 0 & 1 \end{pmatrix}, \tag{144}$$

one can reconstruct the known results that for $p < 1/3$ the state is separable and for $p > 1/3$ the state is entangled, since in the decomposition of the density operator in the form (113) the state

$$\rho_0 = \frac{1}{4}\begin{pmatrix} 1 & 0 & 0 & 0 \\ 0 & 1 & 0 & 0 \\ 0 & 0 & 1 & 0 \\ 0 & 0 & 0 & 1 \end{pmatrix} \tag{145}$$

has the weight $p_0 = (1 - 3p)/4$.

For $p > 1/3$, the coefficient p_o becomes negative.

There is some extension of the presented consideration.

Let us consider the state with the density matrix (nonnegative and Hermitian)

$$\rho = \begin{pmatrix} R_{11} & 0 & 0 & R_{12} \\ 0 & \rho_{11} & \rho_{12} & 0 \\ 0 & \rho_{21} & \rho_{22} & 0 \\ R_{21} & 0 & 0 & R_{22} \end{pmatrix}, \qquad \operatorname{Tr}\rho = 1. \tag{146}$$

Using the procedure of mapping the matrix onto vector $\vec{\rho}$ and applying to the vector the nonlocal linear transform corresponding to the Peres partial transpose and making the inverse map of the transformed vector onto the matrix, we obtain

$$\rho^m = \begin{pmatrix} R_{11} & 0 & 0 & \rho_{12} \\ 0 & \rho_{11} & R_{12} & 0 \\ 0 & R_{21} & \rho_{22} & 0 \\ \rho_{21} & 0 & 0 & R_{22} \end{pmatrix}. \tag{147}$$

In the case of separable matrix ρ, the matrix ρ^m is a nonnegative matrix. Calculating the eigenvalues of ρ^m and applying the condition of their positivity, we get

$$R_{11}R_{22} \geq |\rho_{12}|^2, \qquad \rho_{11}\rho_{22} \geq |R_{12}|^2. \tag{148}$$

Violation of these inequalities gives a signal that ρ is entangled. For Werner state (143), Eq. (148) means

$$1 + p > 0, \qquad 1 - p > 2p, \tag{149}$$

which recovers the condition of separability $p < 1/3$ mentioned above.

The joint probability distribution (112) of separable state is positive after making the local and nonlocal (partial transpose-like) transforms connected with positive map semigroup. But for entangled state, function (112) can take negative values after making this map in the function and replacing on the right-hand side of this equality the product of two matrices $D(g)$ by generic unitary transform u. This is a criterion of entanglement in terms of unitary spin tomogram of the state of multiparticle system.

A simpler and more transparent case is the generalized Werner model with density matrix

$$\rho = \frac{1}{4}\left(1 + \mu_1\sigma_1 \otimes \tau_1 + \mu_2\sigma_2 \otimes \tau_2 + \mu_3\sigma_3 \otimes \tau_3\right). \tag{150}$$

Here the density matrix is expressed in terms of tensor products of two sets of Pauli matrices σ_k and τ_k $(k = 1, 2, 3)$, which are chosen in the standard form.

Its eigenvalues are

$$1-\mu_1-\mu_2-\mu_3, \quad 1+\mu_1+\mu_2-\mu_3, \quad 1+\mu_1-\mu_2+\mu_3, \quad 1-\mu_1+\mu_2+\mu_3.$$

These eigenvalues are related to the vertices of a regular tetrahedron. The partially time-reversed density matrix is

$$\widetilde{\rho} = \frac{1}{4}\left(1 - \mu_1\sigma_1 \otimes \tau_1 - \mu_2\sigma_2 \otimes \tau_2 - \mu_3\sigma_3 \otimes \tau_3\right), \tag{151}$$

which may be viewed as

$$\widetilde{\rho} = L^{(1)} \otimes L^{(2)}\rho \qquad \text{with} \qquad L^{(1)}\rho^{(1)} = \rho^{(1)}, \quad L^{(2)}\rho^{(2)} = 1-\rho^{(2)}. \tag{152}$$

The eigenvalues of this are

$$1+\mu_1+\mu_2+\mu_3, \quad 1+\mu_1-\mu_2-\mu_3, \quad 1-\mu_1+\mu_2-\mu_3, \quad 1-\mu_1-\mu_2+\mu_3.$$

These form an inverted tetrahedron and they have the common domain which is a regular octahedron. The unitary spin tomograms can be written down by inspection and we may verify that all the relations required by the separability criterion (see the next section for details) are satisfied by any point inside the octahedron for ρ and for $L^{(1)} \otimes L^{(2)}\rho$ but the relations connected with positivity condition expressed in terms of positivity of unitary spin tomogram fail when it lies outside.

10. Tomogram of the group $U(n)$

In this section we discuss in more detail the separability criterion using introduced notion of unitary spin tomogram.

In order to formulate a criterion of separability for a with spin j_1 and j_2, we introduce the tomogram $w(\vec{l}, \vec{m}, g^{(n)})$ for the group $U(n)$, where

$$n = n_1 n_2, \qquad n_1 = 2j_1 + 1, \qquad n_2 = 2j_2 + 1,$$

and $g^{(n)}$ are parameters of the group element. Vectors $\vec{l}$ and $\vec{m}$ label a basis $\mid \vec{l}, \vec{m}\rangle$ of the fundamental representation of the group $U(n)$. For example, since this representation is irreducible, being reduced to the representation of the $U(n_1) \otimes U(n_2)$ subgroup of the group $U(n)$, the basis can be chosen as the product of basis vectors:

$$\mid j_1, m_1\rangle \mid j_2, m_2\rangle = \mid j_1, j_2, m_1, m_2\rangle. \tag{153}$$

Due to the irreducibility of this representation of the group $U(n)$ and its subgroup, there exists a unitary transform $u^{\vec{l}\vec{m}}_{j_1 j_2 m_1 m_2} \mid \vec{l}, \vec{m}\rangle$ such that

$$\mid j_1, j_2, m_1, m_2\rangle = \sum_{\vec{l}\vec{m}} u^{\vec{l}\vec{m}}_{j_1 j_2 m_1 m_2} \mid \vec{l}, \vec{m}\rangle, \tag{154}$$

$$\mid \vec{l}\vec{m}\rangle = \sum_{m_1 m_2} (u^{-1})^{\vec{l}\vec{m}}_{j_1 j_2 m_1 m_2} \mid j_l, j_2, m_1, m_2\rangle. \tag{155}$$

One can define the $U(n)$ tomogram for a Hermitian nonnegative $n \times n$ density matrix ρ, which belongs to the Lie algebra of the group $U(n)$, by a generic formula

$$w(\vec{l}, \vec{m}, g^{(n)}) = \langle \vec{l}, \vec{m} \mid U^{\dagger}(g^{(n)}) \rho U(g^{(n)}) \mid \vec{l}, \vec{m}\rangle. \tag{156}$$

Formula (156) defines the tomogram in the basis $\mid \vec{l}, \vec{m}\rangle$ for arbitrary irreducible representation of the unitary group. But below we focus only on tomograms connected with spins.

Let us define the $U(n)$ tomogram using the basis $\mid j_1, j_2, m_1, m_2\rangle$ namely for fundamental representation, i.e.,

$$\begin{aligned} & w^{(j_1,j_2)}(m_1, m_2, g^{(n)}) \\ & = \langle j_1, j_2, m_1, m_2 \mid U^{\dagger}(g^{(n)}) \rho U(g^{(n)}) \mid j_1, j_2, m_1, m_2\rangle. \end{aligned} \tag{157}$$

This unitary spin tomogram becomes the spin-tomogram [34] for the $g^{(n)} \in U(2) \otimes U(2)$ subgroup of the group $U(n)$. The properties of this tomogram follow from its definition as the joint probability distribution of two random spin projections m_1, m_2 depending on $g^{(n)}$ parameters.

One has the normalization condition

$$\sum_{m_1, m_2} w^{(j_1,j_2)}(m_1, m_2, g^{(n)}) = 1. \tag{158}$$

Also all the probabilities are nonnegative, i.e.,

$$w^{(j_1,j_2)}(m_l, m_2, g^{(n)}) \geq 0. \tag{159}$$

Due to this, one has

$$\sum_{m_1, m_2} |w^{(j_1,j_2)}(m_l, m_2, g^{(n)})| = 1. \tag{160}$$

For the spin-tomogram,

$$g^{(n)} \rightarrow \left(\vec{O}_1, \vec{O}_2\right) \tag{161}$$

and

$$w^{(j_1,j_2)}(m_l, m_2, g^{(n)}) \rightarrow w(m_1, m_2, \vec{O}_1, \vec{O}_2). \tag{162}$$

The separability and entanglement condition discussed in the previous section for a bipartite spin-tomogram can be considered also from the viewpoint of the properties of a $U(n)$ tomogram. If the two-spin $n\times n$ density matrix ρ is separable, it remains separable under the action of the generic positive map of the subsystem density matrices. This map can be described as follows.

Let ρ be mapped onto vector $\vec{\rho}$ with n^2 components. The components are simply ordered rows of the matrix ρ, i.e.,

$$\vec{\rho} = \Big(\rho_{11}, \rho_{12}, \ldots, \rho_{1n}, \rho_{21}, \rho_{22}, \ldots, \rho_{nn},\Big). \tag{163}$$

Let the $n^2\times n^2$ matrix L be taken in the form

$$L = \sum_s p_s L_s^{(j_1)} \otimes L_s^{(j_2)}, \qquad p_s \geq 0, \quad \sum_s p_s = 1, \tag{164}$$

where the $n_1\times n_1$ matrix $L_s^{(j_1)}$ and the $n_2\times n_2$ matrix $L_s^{(j_2)}$ describe the positive maps of density matrices of spin-j_1 and spin-j_2 subsystems, respectively. We map vector $\vec{\rho}$ onto vector $\vec{\rho}_L$

$$\vec{\rho}_L = L\vec{\rho} \tag{165}$$

and construct the $n\times n$ matrix ρ_L, which corresponds to the vector $\vec{\rho}_L$. Then we consider the $U(n)$ tomogram of the matrix ρ_L, i.e.,

$$\begin{aligned} &w_L^{(j_1,j_2)}(m_l, m_2, g^{(n)}) \\ &= \langle j_1, j_2, m_1, m_2 \mid U^\dagger(g^{(n)})\rho_L U(g^{(n)}) \mid j_1, j_2, m_l, m_2\rangle. \end{aligned} \tag{166}$$

Using this tomogram we introduce the function

$$F(g^{(n)}, L) = \sum_{m_1,m_2} \left| w_L^{(j_1,j_2)}(m_1, m_2, g^{(n)})\right|. \tag{167}$$

For separable states, this function does not depend on the $U(n)$-group parameter $g^{(n)}$ and positive-map matrix elements of the matrix L.

For the normalized density matrix ρ of the bipartite spin system, this function reads

$$F(g^{(n)}, L) = 1. \tag{168}$$

For entangled states, this function depends on $g^{(n)}$ and L and is not equal to unity. This property can be chosen as a necessary and sufficient condition for separability of bipartite spin-states. We introduce also tomographic purity parameter μ_k of kth order by the formula

$$\mu_k(g^{(n)}, L) = \sum_{m_1 m_2} \left| w_L^{(j_1,j_2)}(m_1, m_2, g^{(h)})\right|^k.$$

For identity semigroup element L and specific $g_0^{(n)}$ unitary transform diagonalizing the density matrix, the tomographic purity μ_2 is identical to purity parameter of the state ρ. The parameters for $k = 2, 3, \ldots$, correspond to $\operatorname{Tr}\rho^{k+1}$.

In fact, the formulated approach can be extended to multipartite systems too. The generalization is as follows.

Given N spin-systems with spins $j_1, j_2, \ldots, j_N$, let us consider the group $U(n)$ with

$$n = \prod_{k=1}^{N} n_k, \qquad n_k = 2j_k + 1. \tag{169}$$

Let us introduce the basis

$$\mid \vec{m}\rangle = \prod_{k=1}^{N} \mid j_k m_k\rangle \tag{170}$$

in the linear space of the fundamental representation of the group $U(n)$. We define now the $U(n)$ tomogram of a state with the $n\times n$ matrix ρ:

$$w_\rho(\vec{m}, g^{(n)}) = \langle \vec{m} \mid U^\dagger(g^{(n)})\rho U(g^{(n)}) \mid \vec{m}\rangle. \tag{171}$$

For a positive Hermitian matrix ρ with $\operatorname{Tr}\rho = 1$, we formulate the criterion of separability as follows.

Let the map matrix L be of the form

$$L = \sum_s p_s\Big(\prod_{k=1}^{N} \otimes L_s^{(k)}\Big), \qquad p_s \geq 0, \quad \sum_s p_s = 1, \tag{172}$$

where $L_s^{(k)}$ is the positive-map matrix of the density matrix of the kth spin subsystem. We construct the matrix ρ_L as in the case of the bipartite system using the matrix L. The function

$$F(g^{(n)}, L) = \sum_{\vec{m}} |w_{\rho_L}(\vec{m}, g^{(n)})| \geq 1 \tag{173}$$

is equal to unity for separable state and depends on the matrix L and $U(n)$-parameters $g^{(n)}$ for entangled states.

This criterion can be applied also in the case of continuous variables, e.g., for Gaussian states of photons. Function (173) can provide the measure of entanglement. Thus one can use the maximum value (or a mean value) of this function as a characteristic of entanglement. In the previous section, we considered the generalized Werner states. Using the above criterion, one can get the domain of values of the parameters of the states for which one has separability or entanglement. In fact, the separability criterion is related to the following positivity criterion of finite or infinite (trace class) matrix A. The matrix A is positive iff the sum of moduli of diagonal matrix elements of the matrix $UAU^\dagger$ is equal to a positive trace of the matrix A for an arbitrary unitary matrix U.

11. Dynamical map and purification

In this section, we consider the connection of positive maps with purification procedure. In fact, formula

$$\rho \to \rho' = \sum_k p_k U_k \rho U_k^\dagger, \tag{174}$$

where U_k are unitary operators, can be considered in the form

$$\rho \to \rho' = \sum_k p_k \rho_k, \qquad p_k \geq 0, \qquad \sum_k p_k = 1. \tag{175}$$

Here the density operators ρ_k read

$$\rho_k = U_k \rho U_k^\dagger, \tag{176}$$

and the maps which are not sufficiently general keep the most degenerate density matrix fixed. This form is the form of probabilistic addition. This mixture of density operators can be purified with the help of a fiducial projector P_0

$$\rho' \to \rho'' = N \left[\sum_{kj} \sqrt{p_k p_j} \frac{\rho_k P_0 \rho_j}{\sqrt{\mathrm{Tr}\, \rho_k P_0 \rho_j P_0}} \right], \tag{177}$$

where P_0 and N is a normalization constant

$$N^{-1} = \mathrm{Tr} \left(\sum_{kj} \sqrt{p_k p_j} \frac{\rho_k P_0 \rho_j}{\sqrt{\mathrm{Tr}\, \rho_k P_0 \rho_j P_0}} \right). \tag{178}$$

The normalization is unnecessary if all ρ_k are mutually orthogonal. We call this map a purification map. It maps the density matrix of mixed state on the density matrix of pure state.

The map (174) could be interpreted as the evolution in time of the initial matrix ρ_0 considering unitary operators $U_k(t)$ depending on time. Thus one has

$$\rho_0 \to \rho(t) = \sum_k p_k U_k(t) \rho_0 U_k^\dagger(t). \tag{179}$$

In this case, the purification procedure provides the dynamical map of a pure state

$$\mid \psi_0\rangle\langle\psi_0 \mid\rightarrow\mid \psi(t)\rangle\langle\psi(t) \mid, \tag{180}$$

where $\mid \psi(t)\rangle$ obeys a nonlinear equation and, in the general case, this map does not define a one parameter group of transformations not even locally.

For some specific cases, the evolution (179) can be described by a semigroup. The density matrix (179) obeys then a first-order differential equation in time for this case [27–29].

More specifically, the reason why there is no differential equation in time for the generic case is due to the absence of the property

$$\rho_{ij}(t_2) = \sum_{mn} K_{ij}^{mn}(t_2, t_1)\rho_{mn}(t_1), \tag{181}$$

where the kernel of evolution operator satisfies

$$K_{ij}^{mn}(t_3, t_2)K_{mn}^{pq}(t_2, t_1) = K_{ij}^{pg}(t_3, t_1). \tag{182}$$

Thus, via a purification procedure and a dynamical map applied to a density matrix we get a pure state (nonlinear dynamical map). This map can be used in nonlinear models of quantum evolution. All the linear positive maps both completely positive and not completely positive are contractive. This means, for example, that purity parameter $\mu = \mathrm{Tr}\,\rho^2$ after performing the positive map generically becomes smaller. There are maps for which the purity parameter is preserved, for example,

$$\rho \rightarrow \rho^{\mathrm{tr}}, \qquad \rho \rightarrow -\rho + \frac{2}{N}\,1. \tag{183}$$

These linear maps include also unitary transform

$$\rho \rightarrow u\rho u^{\dagger}. \tag{184}$$

All the maps obtained by means of convex addition of density matrices are contractive. There are no linear maps which provide dilation. The existence of such maps would mean that there are matrices $\widetilde{L}_{jk,ps}$ of positive maps satisfying the condition

$$\widetilde{L}L = 1, \qquad \widetilde{L} = L^{-1}. \tag{185}$$

For example, for the matrices L_ε the inverse matrices exist for $\varepsilon \neq 0$. But these inverse matrices do not provide positive trace preserving maps. Since the purification procedure provides a positive map, which increases the purity parameter, the composition of linear map with the purification map provides the possibility to recover the initial density matrix ρ which was the object of action of a positive linear map. It means that the purification map $\hat{L}_p$ can give

$$\hat{L}_{P_0}(L\rho) = 1\rho \tag{186}$$

for any density matrix ρ but the choice of fiducial projector depends on ρ (the initial condition).

Thus one has also for completely positive maps

$$\rho \to \rho' = \sum_k \rho'_k, \qquad \rho_k = V_k \rho V_k^\dagger, \qquad \sum_k V_k^\dagger V_k = 1. \tag{187}$$

Making polar decomposition

$$\rho_k = \sqrt{\rho_{0k}} U_k, \qquad U_k U_k^\dagger = 1, \qquad \rho_{0k} \geq 0$$

and introducing the positive numbers $p_k = \text{Tr}\,\rho_{0k}$, we construct the map

$$\rho' \to \rho'' = \left\{ \sum_{kj} \sqrt{p_k p_j} \frac{\tilde{\rho}_k P_0 \tilde{\rho}_j + \tilde{\rho}_j P_0 \tilde{\rho}_k}{\sqrt{T \tilde{\rho}_k P_0 \tilde{\rho}_j P_0}} \right\}, \quad \sum_k p_k = 1, \quad p_k \tilde{\rho}_k = \rho_k. \tag{188}$$

The matrix ρ'' is a matrix of rank one for projector P_0. The projector P_0 is not orthogonal to matrix ρ_k. Taking N orthogonal projectors $P_0^{(s)}$ $(s = 1, 2, \ldots, N)$ and obtaining N projectors ρ''_s, one can combine them in order to get the initial matrix ρ. It means that one can take convex sum of the N pure states ρ''_s to recover the initial mixed state ρ. Another way to make the state with higher purity was demonstrated using the modified purification procedure in [48]. For qubit state, one has

$$\rho = p_1 \rho_1 + p_2 \rho_2 + \kappa \sqrt{p_1 p_2} \frac{\rho_1 P_0 \rho_2 + \rho_2 P_0 \rho_1}{\sqrt{\text{Tr}\,\rho_1 P_0 \rho_2 P_0}}, \quad p_1 + p_2 = 1, \tag{189}$$

where the decoherence parameter $0 \leq \kappa \leq 1$ is used. If $\kappa \sim 1$, we increase purity.

Let us discuss the map (187) using its matrix form, i.e.,

$$\rho_{\alpha\beta} \to \rho'_{\alpha\beta} = \sum_{ij} \mathcal{L}_{\alpha\beta,ij} \rho_{ij}. \tag{190}$$

The matrix $\mathcal{L}_{\alpha\beta,ij}$ is expressed in terms of the matrices V_k as

$$\mathcal{L}_{\alpha\beta,ij} = \sum_k (V_k)_{\alpha i} (V_k^*)_{\beta j}. \tag{191}$$

One can construct another positive map [49]

$$\rho \to \rho' = \sum_k r_k \text{Tr}\,(R_k \rho), \tag{192}$$

where r_k are density matrices and R_k are positive operators satisfying the normalization condition

$$\sum_k R_k = 1. \tag{193}$$

The matrix corresponding to this map (called entanglement breaking map [50]) reads

$$\mathcal{L}^{b}_{\alpha\beta,ij} = \sum_k (r_k)_{\alpha\beta}(R_k^*)_{ij}. \tag{194}$$

The entanglement breaking map is contractive positive map. There exist some special cases of completely positive maps. For example,

$$\rho \to -\varepsilon\rho + \frac{1+\varepsilon}{N}\,\rho \tag{195}$$

differs from the depolarizing map by replacing the unity operator by the density operator. Another map reads

$$\rho \to \frac{1 - \operatorname{diag}\rho}{N}\,. \tag{196}$$

The decoherence map (phase damping map) of the kind

$$\rho_{ij} \to \begin{cases} \rho_{ij}, & i = j \\ \lambda\rho_{ij}, & i \neq j, \end{cases} \tag{197}$$

where $|\lambda| < 1$ provides contractive map with uniform change of off-diagonal matrix elements of the density matrix.

Let us discuss the property of tomogram of bipartite system with density matrix ρ_{12}. If the density matrix is separable, than the depolarizing map of the second subsystem provides the following density matrix

$$\rho_{12} \to \rho_\varepsilon = -\varepsilon\rho_{12} + \frac{1+\varepsilon}{N_2}\,\underline{\rho^{(1)}} \otimes 1_2, \tag{198}$$

where

$$\underline{\rho^{(1)}} = \mathrm{Tr}_2(\rho_{12}) \tag{199}$$

and 1_2 is the N_2-dimensional unity matrix. Then one has the property of unitary spin tomogram

$$w_\varepsilon(m_1, m_2, g^{(n)}) = -\varepsilon w_{12}(m_1, m_2, g^{(n)}) + \frac{1+\varepsilon}{N_2}\,\underline{w}(m_1, m_2, g^{(n)}), \tag{200}$$

where $g^{(n)}$ is matrix of $U\Big((2j_1+1)(2j_2+1)\Big)$ unitary transform;
$w_\varepsilon(m_1, m_2, g^{(n)})$ is the tomogram of transformed density matrix of bipartite system;
$\underline{w}(m_1, m_2, g^{(n)})$ is the unitary spin tomogram of tensor product of partial trace $\underline{\rho^{(1)}}$ over the second subsystem's coordinates of the density matrix ρ_{12} and unity operator 1_2;

$w_{12}(m_1, m_2, g^{(n)})$ is the unitary spin tomogram of the state with density matrix ρ_{12}.

The criterion of separability means

$$\sum_{m_1=-j_1}^{j_1} \sum_{m_2=-j_2}^{j_2} \left| \frac{1+\varepsilon}{2j_2+1} w(m_1, m_2, g^{(n)}) - \varepsilon w_{12}(m_1, m_2, g^{(n)}) \right| = 1 \quad (201)$$

for arbitrary $g^{(n)}$ and ε.

For Werner states ρ_W, the tomogram of transformed state (in this case, it means that $p \to -\varepsilon p$) is related to the initial-state tomogram w_W

$$w_\varepsilon(m_1, m_2, g^{(n)}) = -\varepsilon w_{12}(m_1, m_2, g^{(n)}) + \frac{1+\varepsilon}{4}. \quad (202)$$

The criterion of separability yields

$$\sum_{m_1, m_2=-1/2}^{1/2} \left| \frac{1+\varepsilon}{4} - \varepsilon w_W(m_1, m_2, g^{(n)}) \right| = 1. \quad (203)$$

Equality (203) takes place for arbitrary $g^{(n)}$ and ε only for $|p| \leq 1/3$. For $p > 1/3$, the above sum depends on $g^{(n)}$ and ε and it is larger than one.

It is obvious if one calculates the tomogram using the element of the unitary group of the form

$$g_0^{(n)} = \begin{pmatrix} 0 & 0 & 0 & 1 \\ 0 & 1 & 0 & 0 \\ 0 & 0 & 1 & 0 \\ 1 & 0 & 0 & 0 \end{pmatrix}. \quad (204)$$

At this point, the sum (203) reads

$$\sum_{m_1, m_2=-1/2}^{1/2} \left| \frac{1+\varepsilon}{4} - \varepsilon w_W(m_1, m_2, g^{(n)}) \right| = 3 \left| \frac{1+\varepsilon p}{4} \right| + \left| \frac{1-3p\varepsilon}{4} \right|. \quad (205)$$

One can see that this sum equals to one independently on the value of parameter $|\varepsilon| \leq 1$ only for values $|p| \leq 1/3$. For $p = 1$, the maximum value of the sum equals $2 = (1+3\varepsilon)/2$ $(\varepsilon = 1)$. This value can characterize the degree of entanglement of Werner state.

We have introduced positive nonlinear map of density matrix which is purification map. The purification map can be combined with contractive maps discussed. The tomographic-probability distributions under discussion can be completely described by their characteristic functions. This means that the relation of tomogram property to entanglement can be formulated in terms of the properties of characteristic functions.

One can also check the criterion using example of two-qutrite pure entangled state with wave function

$$| \psi\rangle = \frac{1}{\sqrt{3}} \sum_{m=-1}^{1} | u_m\rangle \, | v_m\rangle. \tag{206}$$

The sum defining the criterion of separability for specific $U(9)$ transform $g_0^{(n)}$ which is diagonalizing the hermitian matrix $L_\varepsilon \mid \psi\rangle\langle\psi \mid$ reads

$$F(\varepsilon, g_0^{(n)}) = 8 \left| \frac{1+\varepsilon}{9} \right| + \left| \frac{1-8\varepsilon}{9} \right|. \tag{207}$$

For $1/2 > \varepsilon > 1/8$, this sum is larger than one, that means that the state is entangled. For $\varepsilon = 1/2$, the function has maximum and it is equal to 5/3.

The entanglement of the considered state can be detected using partial transposition criterion too.

For the case of pure entangled state of two-qutrite system with the wave function

$$| \Phi\rangle = \frac{1}{\sqrt{2}} \Big(| u_1\rangle \, | v_1\rangle + | u_0\rangle \, | v_0\rangle \Big), \tag{208}$$

in which the states with spin projections $m = -1$ do not participate, the partial transpose criterion does not detect entanglement. But our criterion yields for specific $U(9)$ transform $g_0^{(n)}$, which diagonalizes the hermitian matrix $L_\varepsilon \mid \Phi\rangle\langle\Phi \mid$ the following expression for the function $F(\varepsilon, g_0^{(n)})$, which reads

$$F(\varepsilon, g_0^{(n)}) = 5 \, \frac{|1+\varepsilon|}{6} + \frac{|1-5\varepsilon|}{6}. \tag{209}$$

The function takes maximum value for $\varepsilon = 1/2$ that equals to 3/2. This value is smaller than 5/3 of the previous case. It corresponds to our intuition that the superposition of three product states of two qutrite system is more entangled than the superposition of only two such product states.

The criterion can be extended to multipartite spin system.

We have to apply for n-partite system the transform of the density matrix ρ of the form

$$L_{\vec{\varepsilon}} = L_{\varepsilon_1}^{(1)} \otimes L_{\varepsilon_2}^{(2)} \otimes \ldots \otimes L_{\varepsilon_n}^{(n)}, \tag{210}$$

where the transform $L_{\varepsilon_k}^{(k)}$ acts as depolarizing map on the kth subsystem. If the state is separable

$$\rho = \sum_k p_k \rho_k^{(1)} \otimes \rho_k^{(2)} \otimes \ldots \otimes \rho_k^{(n)}, \quad \sum_k p_k = 1, \quad p_k \geq 0, \tag{211}$$

each of the terms $\rho_k^{(j)}$ $(j = 1, 2, \ldots, n)$ in the tensor product is replaced by the term

$$\rho_k^{(j)} \to -\varepsilon_j \rho_k^{(j)} + \frac{1+\varepsilon_j}{N_j}\, 1_j. \tag{212}$$

This means that the transformed density matrix reads

$$\rho \to L_{\vec{\varepsilon}}\rho = \sum_k p_k \left[\prod_{j=1}^{n} \otimes \left(-\varepsilon \rho_k^{(j)} + \frac{1+\varepsilon_j}{N_j}\, 1_j \right) \right]. \tag{213}$$

The unitary spin tomogram of the transformed density matrix takes the form $(\vec{\varepsilon} = \varepsilon_1, \varepsilon_2, \ldots, \varepsilon_n)$

$$w_{\vec{\varepsilon}}(m_1, m_2, \ldots, m_n, g^{(N)}) = \sum_k p_k w_{pr}^{(k)}(m_1, m_2, \ldots, m_n, g^{(N)}, \vec{\varepsilon}), \tag{214}$$

where $N = \prod_{s=1}^{n}(2j_s + 1)$ and element $g^{(N)}$ is the unitary matrix in N-dimensional space. The tomogram $w_{pr}^{(k)}(m_1, m_2, \ldots, m_n, g^{(N)}, \vec{\varepsilon})$ is the joint probability distribution of spin projections $m_s = -j_s, -j_s + 1, \ldots, j_s$, which depends on the unitary transform $g^{(N)}$ in the state with density matrix

$$\rho_k = \prod_{s=1}^{n} \otimes \left(-\varepsilon_s \rho_k^{(s)} + \frac{1+\varepsilon_s}{N_s}\, 1_s \right). \tag{215}$$

For the elements

$$g_{pr}^{(N)} = \prod_{s=1}^{n} \otimes u_s(2j_s + 1),$$

where $u_s(2j_s + 1)$ is unitary matrix, the tomogram (214) takes the form of sum of the products

$$w_{\vec{\varepsilon}}(m_1, m_2, \ldots, m_n, g_{pr}^{(N)}) = \sum_k p_k \prod_{s=1}^{n} w_k\Big(m_s, u_s(2j_s + 1), \varepsilon_s\Big), \tag{216}$$

with $w_k(m_s, u_s(2j_s + 1), \varepsilon_s)$ being the unitary spin tomograms of the sth spin subsystem with transformed density matrix $L_{\varepsilon_s}\rho_k^{(s)}$. If one uses as the matrix $u_s(2j_s+1)$, the matrix of unitary irreducible representation of the $SU(2)$ group, the tomogram w_k depends only on the two parameters defining the point on the sphere S^2.

For a separable state of the multipartite system, one has

$$\sum_{m_1, \ldots, m_n} \left| w_{\vec{\varepsilon}}(m_1, m_2, \ldots, m_n, g^{(N)}) \right| = 1 \tag{217}$$

for all elements $g^{(N)}$ and all parameters $\vec{\varepsilon}$.

For entangled state, there can be some values of parameters $\vec{\varepsilon}$ and group elements $g^{(N)}$ for which the sum is larger than one.

12. Conclusions

We summarize the results of the paper.

The notion of entangled states (first discussed by Schrödinger [4, 51]) has attracted a lot of efforts to find a criterion and quantitative characteristics of entanglement. A criterion based on partial transpose transform of subsystem density matrix (complex conjugation of the subsystem density matrix or its time reverse) provides the necessary and sufficient condition of separability of the system of two qubits and qubit-qutritt system [52]. The phase-space representation of the quantum states and time reverse transform (change of the signs of the subsystem momenta) of the Wigner function in the case of Gaussian state was applied to study the separability and entanglement of photon states in [13]. Recently it was pointed out that the tomographic approach of reconstructing the Wigner function of quantum state [43–45] can be developed to consider the positive probability distribution (tomogram) as an alternative to density matrix (or wave function) because the complete set of tomograms contains the complete information on the quantum state [42]. This representation (called probability representation) was constructed also for spin states including a bipartite system of two spins. Up to now the problem of entanglement was not discussed in the tomographic representation. Some remarks on tomograms and entanglement of photon states in the process of Raman scattering were done in [53]. The tomographic approach has the advantage of dealing with positive probabilities and one deals with standard probability distributions which are positive and normalized.

We studied the properties of separable and entangled state of multipartite system using the tomographic probability distributions. The positive and completely positive maps of density matrices [39, 54] induce specific properties of the tomograms. The properties of the positive maps were studied in [55]. We formulated necessary and sufficient conditions of separability and entanglement of multipartite systems in terms of properties of the quantum tomogram. Since the tomograms were shown [36] to be related to the star-product quantization procedure [56], we discuss entanglement and separability properties in terms of generic operator symbols. The tomographic symbols of generic spin operators were studied in [36]. Then we focused on properties of entanglement and separability of a bipartite system using spin tomograms ($SU(2)$-tomograms) and tomograms of the $U(N)$-group.

The idea of the approach suggested can be summarized as following.

The positive but not completely positive linear maps of a subsystem density matrix do preserve the positivity of separable density matrix of the composite system. These maps contain also maps which do not preserve the positivity of the initial density matrix of an entangled state for the composite system. It means that the set of all linear positive maps of the subsystem density matrix (this set is semigroup) creates from the initial entangled positive density matrix of composite system a set of hermitian matrices including the matrices with negative eigenvalues. To detect the entanglement we use the tomographic symbols of the obtained hermitian matrices. The tomographic symbols of state density matrices (state tomograms) are standard probabilities. In view of this the tomographic symbols of the obtained hermitian matrices corresponding to initial separable state preserve all the properties of the probability representation including positivity and normalization. But in case of entangled state the tomographic symbols of the obtained hermitian matrices can take negative values. The different behaviour of tomograms of separable and entangled states of composite systems under action of the semigroup of positive maps of the subsystem density matrix provides the tomographic criterion of the separability.

To conclude, we point out the main result of the work.

We found the criterion of separability which is given by equation (173). The criterion is valid for multiparticle spin system. The criterion can be called "tomographic criterion" of separability. The tomographic criterion can be considered also for symplectic tomograms of multimode photon states. The condition of separability is sufficient because there always exists a unitary group element by means of which any hermitian matrix can be diagonalized. It means that tomographic symbol of nonpositive hermitian matrix has nonpositive values for some unitary group parameters. The suggested criterion is connected with properties of the constructed function (173) which for given density matrix depends on unitary group parameters g and the parameters of positive map semigroup L. For separable density matrix the dependence on unitary group parameters and the semigroup parameters disappears and the function becomes constant equal to unity. For entangled states the function differs from unity and depends on both group and semigroup parameters. The suggested criterion can be considered as some complementary test of separability together with other criteria available in the literature (see, for example, [38, 52]). We point out that suggested criterion differs from available usual ones by the kind of the necessary numerical calculations. To apply this criterion one needs to calculate the sum of moduli of diagonal matrix elements of product of three matrices. One of the matrices is hermitian and two others are unitary ones. This procedure does not need the calculation of the eigenvalues of a matrix. The structure of positive (including not completely positive) map semigroup with elements L needs extra investigation (see, for example, [57]). We found also a test of entanglement based on the property of unitary spin tomogram.

The discussed purification map can be applied to find new quantum evolution equations in addition to known ones [58–61]. The application of different forms of positive maps [55, 62] and supermatrix representation of the maps [63, 64] are useful for better understanding of the computations. Entanglement phenomena can be considered using symbols of density matrix of different kinds, e.g., particular quasidistributions [65] as well as tomographic symbols [36]. The difference of symbols of entangled and separable density operators for different schemes of the star-product quantization needs further investigations as well as test of entanglement of some generalizations of Werner state [66, 40] in multipartite case. A relation of tomographic approach to different positive maps [67] should be investigated. The tomographic symbols are analytic in group parameters. This can be used to find extrema of tomograms which give information on degree of entanglement.

Acknowledgments

V.I. M. and E.C.G. S. thank Dipartimento di Scienze Fisiche, Università "Federico II" di Napoli and Istituto Nazionale di Fisica Nucleare, Sezione di Napoli for kind hospitality. V.I. M. is grateful to the Russian Foundation for Basic Research for partial support under Project No. 01-02-17745.

References

[1] P.A.M. Dirac 1958 "The Principles of Quantum Mechanics" (Oxford: Pergamon)

[2] L.D. Landau 1927 *Z. Phys* **45** 430

[3] J. von Neumann 1932 "Mathematische Grudlagen der Quantenmechanik" (Berlin: Springer); Nov. 1927 *Göttingenische Nachrichten* **11** S245

[4] E. Schrödinger 1935 *Naturwissenschaften* **23** 807; 823; 844

[5] E. Schrödinger 1926 *Ann. d. Phys.* Lpz **79** 489

[6] V.I. Man'ko, G. Marmo, E.C.G. Sudarshan and F. Zaccaria 1999 *J. Russ. Laser Res.* **20** 421; 2002 *J. Phys. A: Math. Gen.* **35** 7173

[7] M. Horodecki, P. Horodecki and R. Horodecki 1996 *Phys. Lett.* A **223** 1

[8] S. Hill and W.K. Wootters 1997 *Phys. Rev. Lett.* **78** 5022
W.K. Wootters 1998 *Phys. Rev. Lett.* **80** 2245

[9] K. Zyczkowski, P. Horodecki, A. Sanpera and M. Lewenstein 1998 *Phys. Rev.* A **58** 883

[10] S. Popescu and D. Rohrlich 1997 *Phys. Rev.* A **56** R3319

[11] S. Abe and A.K. Rajagopal 2002 *Physica* A **289** 157

[12] C.H. Bennett, D.P. Di Vincenzo, J.A. Smolin and W.L. Wootters 1996 *Phys. Rev.* A **54** 3824

[13] R. Simon 2002 *Phys. Rev. Lett.* **84** 2726

[14] S.M. Barnett and S.J.D. Phoenix 1989 *Phys. Rev.* A **40** 2404; 1991 ibid **44** 535

[15] A. Mann, B.C. Sanders and W.J. Munro 1995 *Phys. Rev.* A **51** 989

[16] C.H. Bennet, H.J. Bernstein, S. Popescu and B. Schumacher 1996 *Phys. Rev.* A **53** 2046

[17] V. Vedral and M.B. Plenio 1998 *Phys. Rev.* A **57** 1619

[18] B.-G. Englert, M. Löffler, O. Benson, B. Varcoe, M. Weidinger and H. Walter 1998 *Fortschr. Phys.* **46** 897

[19] M. Horodecki, P. Horodecki and R. Horodecki 2000 *Phys. Rev. Lett.* **84** 2014

[20] S. Parker, S. Bose and M.B. Plenio 2000 *Phys. Rev.* A **61** 032305

[21] C.H. Bennet, S. Popescu, D. Rohrlich, J.A. Smolin and A.V. Thapliyal 2001 *Phys. Rev.* A **63** 012307

[22] K. Piątek and W. Leoński 2001 *J. Phys. A: Math. Gen.* **34** 4951

[23] K. Audenaert, J. Esert, E. Jané, M.B. Plenio, S. Virmani and R.B. De Moor 2001 *Phys. Rev. Lett.* **87** 217902

[24] K. Furuya, M.C. Nemes and G.Q. Pellegrino 1998 *Phys. Rev. Lett.* **80** 5524

[25] K. Życzkowski, P. Horofecki, A. Savpera and M. Lewenstein 1998 *Phys. Rev.* A **58** 883

[26] W.J. Munro, D.F.V. James, A.G. White and P.G. Kwait 2001 *Phys. Rev.* A **64** 030302

[27] G. Vidal and R.F. Werner 2002 *Phys. Rev.* A **65** 032314

[28] F. Coffman, J. Kundu and W.K. Wootters 2000 *Phys. Rev.* A **61** 052306

[29] P. Badziag, P. Deuar, M. Horodecki, P. Horodecki and R. Horodecki 2002 *J. Mod. Opt.* **49** 1289

[30] M.A. Andreata, A.V. Dodonov and V.V. Dodonov 2002 *J. Russ. Laser Res.* **23** 531

[31] V.V. Dodonov and V.I. Man'ko 1997 *Phys. Lett.* A **229** 335

[32] O. Man'ko and V.I. Man'ko 1997 *JETP* **85** 430

[33] A.B. Klimov, O.V. Man'ko, V.I. Man'ko, Yu. F. Smirnov and V.N. Tolstoy 2002 *J. Phys. A: Math. Gen.* **35** 6101

[34] V.A. Andreev and V.I. Man'ko 1998 *JETP* **87** 239

[35] V.I. Man'ko and S.S. Safonov 1998 *Yad. Fiz.* **61** 658

[36] O.V. Man'ko, V.I. Man'ko and G. Marmo 2000 *Phys Scr.* **62** 446; 2002 *J. Phys. A: Math. Gen.* **35** 699

[37] V.V. Dodonov, A.S.M. De Castro and S.S. Misrahi 2002 *Phys. Lett.* A **296** 73
A.S.M. De Castro and V.V. Dodonov 2003 *J. Russ. Laser Res.* **23** 93; 2003 *J. Opt. B: Quantum Semiclass. Opt.* **5** S593

[38] Special Issue on Entanglement 2002 *J. Math. Phys.* **43** No. 9

[39] E.C.G. Sudarshan, P.M. Mathews and J. Rau 1961 *Phys. Rev.* **121** 920

[40] E.C.G. Sudarshan and A. Shaji 2003 "Structure and parametrization of stochastic maps of density matrix" ArXiv quant-ph/0205051 v2; 2003 *J. Phys. A: Math. Gen.* **36** (in press)

[41] A. Peres 1996 *Phys. Rev. Lett.* **77** 1413

[42] S. Mancini, V.I. Man'ko and P. Tombesi 1996 *Phys. Lett.* A **213** 1; 1997 *Found. Phys.* **27** 801

[43] J. Bertrand and P. Bertrand 1987 *Found. Phys.* **17** 397

[44] K. Vogel and H. Risken 1989 *Phys. Rev.* A **40** 2847

[45] S. Mancini, V.I. Man'ko and P. Tombesi 1995 *Quantum Semiclass. Opt.* **7** 615
G.M. D'Ariano, S. Mancini, V.I. Man'ko and P. Tombesi 1996 *Quantum Semiclass. Opt.* **8** 1017

[46] M.A. Man'ko, V.I. Man'ko and R.V. Mendes 2001 *J. Phys. A: Math. Gen.* **24** 8321

[47] V.I. Man'ko, G. Marmo, E.C.G. Sudarshan and F. Zaccaria 2003 "Entanglement in probability representation of quantum states and tomographic criterion of separability" *J. Opt. B: Quantum Semiclass. Opt.* (in press); 2003 *J. Russ. Laser Res.* **24** 507

[48] V.I. Man'ko, G. Marmo, E.C.G. Sudarshan and F. Zaccaria 2000 "Inner composition law of pure-spin states" in "Spin-Statistics Connection and Commutation Relations" R.C. Hilborn and G.M. Tino (eds.) *AIP Conference Proceedings* **545** 92

[49] A.S. Holevo 1999 *Russ. Math. Surveys* **53** 1295

[50] P.W. Shor 2002 *J. Math. Phys.* **43** 4334

[51] E. Schrödinger 1935 *Proc. Cambridge Philos. Soc.* **31** 555

[52] P.B. Slater 2003 *J. Opt. B: Quantum Semiclass. Opt.* **5** S691

[53] S.V. Kuznetsov, O.V. Man'ko and N.V. Tcherniega 2003 *J. Opt. B: Quantum Semiclass. Opt.* **6** S503

[54] W.F. Stinespring 1955 *Proc. Amer. Math. Soc.* **6** 211

[55] L.C. Woronowicz 1976 *Rep. Math. Phys.* **10** 165

[56] F. Bayen, M. Flato, C. Fronsdal, A. Lichnerovicz and D. Sternheimer 1975 *Lett. Math. Phys.* **1** 521

[57] C. King 2002 "The capacity of the quantum depolarizing channel" ArXiv quant-ph/0204172 v2

[58] J.E. Moyal 1949 *Proc. Cambridge Philos. Soc.* **45** 99

[59] A. Kossakovski 1972 *Rep. Math. Phys.* **3** 247

[60] G. Lindblad 1976 *Comm. Math. Phys.* **48** 119

[61] V. Gorini, A. Kossakovski and E.C.G. Sudarshan 1978 *Rep. Math. Phys.* **18** 149

[62] K. Kraus 1973 *Ann. Phys.* NY **64** 311

[63] M.D. Choi 1975 *Lin. Alg. Appl.* **10** 285; 1970 *Canadian J. Math.* **24** 520; 1976 *Illinois J. Math.* **48** 119

[64] T.F. Havel 2003 *J. Math. Phys.* **44** 534

[65] E.C.G. Sudarshan 1963 *Phys. Rev. Lett.* **10** 277
C.L. Mehta and E.C.G. Sudarshan 1965 *Phys. Rev.* B **138** 274

[66] R. Werner 1989 *Phys. Rev.* A **40** 4277

[67] A. Kossakowski 2003 "A class of linear positive maps in matrix algebras" ArXiv quant-ph/0307132 v1

OBJECTIVE EXISTENCE AND RELATIVITY GROUPS

G. Marmo[*,†,a)] and B. Preziosi[*,‡,b)]

[*] *Dipartimento di Scienze Fisiche*

Università di Napoli "Federico II"

Complesso Universitario di Monte S. Angelo

80126 - Napoli (Italy)

[†] *Istituto Nazionale di Fisica Nucleare (Italy)*

[‡] *Istituto Nazionale di Fisica della Materia (Italy*

[a)] giuseppe.marmo@na.infn.it

[a)] bruno.preziosi@na.infn.it

Abstract A minimal requirement for different inertial observers to be equivalent is that each one is perceived by the other as existing always in the past and in the future. This aspect is formalized in the notion of mutual objective existence. Using just this notion, we show that, in the bidimensional case (x_0, x_1), the linear transformations $A(x_0, x_1)$ connecting two different frames form a Lorentz group (or its contractions, Galilei and Carroll). In three dimensions (one time) x_0 and two space variables x_1 and x_2) the transformations compatible with the mutual objective existence are the product of $A_1(x_0, x_1)A_2(x_0, x_2)$, where both A_1 and A_2 are one of the previous transformations, times a space transformation $R(x_1, x_2)$, which is obliged to be Euclidean when both A_1 and A_2 are Lorentzian.

Keywords: PACS: 03.30 - Special Relativity

In memory of Ruggiero de Ritis

1. Introduction

In 1905 Einstein published his theory of electrodynamics of moving bodies [7], accepted in the body of physical science under the name of the special theory of relativity. Since, thanks to the formulation of Minkowski [8], the carrier manifold for the perception of the external world became space-time. In the words of Minkowski:

B. Gruber (ed.), Symmetries in Science XI, 445–458.

> *"Henceforth space by itself, and time by itself are doomed to fade away into mere shadows, and only a kind of union of the two will preserve an independent reality".*

Only a "kind of union" of space and time is to be allowed. Our aim is to discuss and investigate this kind of union.

Usually space and time are distinguishable in a relativity world model due to the indefinite nature of the associated pseudo Riemannian structure of the space-time manifold. This is the property that Reichenbach [9] calls the singular nature of time.

In our approach, however, we do consider a four dimensional continuum as the stage for the events of the external world, but we do not assume any preexisting (pseudo)Riemannian structure (this approach is more close to the spirit of the General Relativity). We have to rely on a more deep distinction between space and time. Indeed we may rely on Weyl's [10] analysis of space, time and matter:

> *"Time is the primitive form of the stream of consciousness. It is a fact, however obscure and perplexing to our minds, that the contents of consciousness do not present themselves simply as being, but as being now filling the form of the enduring present with a varying content.*
>
> *So that one does not say this is but this is now, yet now no more. It we project ourselves outside the stream of consciousness and represent its contents as an object, it becomes an event happening in time, the separate stages of which stand to one another in the relation of earlier and later.*
>
> *Just as time is the form of the stream of consciousness, so one may justifiably assert that space is the form of external material reality".*

Existence is therefore perceived as here and now.

A minimalist translation of Weyl's assertions into a mathematical object is provided by the notion of reference frame, i.e. a rank one $(1-1)$ tensor field T defined on the four dimensional continuum $\mathcal{M}$ with the property $T \cdot T \propto T$. By writing T as the tensor product of a column vector (identifying a concept of time) and a row vector (identifying a concept of space) we give $\mathcal{M}$ a separation into space and time. The identification of an equivalence class of reference frames will allow a "kind of union" of the individual separations into space and time in such a way that existence is shared objectively by all of them.

2. Reference frames: space and time

A reference frame imparts a natural temporal grain of the space-time manifold and a natural transverse spatial location. A suggestive intuitive model of this separation is obtained by imagining that at each instant of time one could take a photograph of the entire cosmos. Imagine these photographs bond into a book. The book stands for the whole space-time and the leaves of the book stand for the spatial location. The temporal grain is determined by one-dimensional line transverse to the spatial leaves. A curve in the book which intersects any leaf just once is a world line.

To give our presentation a tutorial character, we make the assumption that $\mathcal{M}$ is actually a vector space. Therefore T can be written as a matrix, tensor product of a column vector

$$\varepsilon \equiv \begin{pmatrix} e_0 \\ e_1 \\ e_2 \\ e_3 \end{pmatrix}$$

and the row vector $\alpha \equiv (a^0, a^1, a^2, a^3)$ to obtain $T = \varepsilon \otimes \alpha$, i.e. $(T)^k_j = e_j a^k$, with the requirement that $Tr\,T = a^0 e_0 + a^1 e_1 + a^2 e_2 + a^3 e_3 = \alpha(\varepsilon) \neq 0$. If we normalize T by requiring $Tr\,T = 1$, we get $T^2 = T$.

We may associate an ordinary differential equation with ε by setting

$$\frac{dx^0}{ds} = a^0\,, \quad \frac{dx^1}{ds} = a^1\,, \quad \frac{dx^2}{ds} = a^2\,, \quad \frac{dx^3}{ds} = a^3.$$

Any solution of these equations defines a world-line and will be called an observer associated with the reference frame T. A representative observer will be the solution defined by the initial conditions

$$x^0(s=0) = 0\,, \quad x^1(s=0) = 0\,, \quad x^2(s=0) = 0 \quad x^3(s=0) = 0\,.$$

Spatial leaves associated with the reference frame T are defined by the vectors $y \equiv (y^0, y^1, y^2, y^3)$ which satisfy $\alpha(y) = a^0 y_0 + a^1 y_1 + a^2 y_2 + a^3 y_3 = c$; a representative leaf will be the one defined by $c = 0$. It is possible to choose a mutual orientation between space and time by requiring $\alpha(\varepsilon) > 0$.

It is now clear that existing objects for the reference frame T should have a world line with tangent vector v_m at the point m of space-time such that $\alpha(v_m) \neq 0$. Those world-lines for which $\alpha(v_m) = 0$ will be perceived as existing only at a given instant of time, not here but all over the line, and will not be perceived neither in the past nor in the future. Therefore, we shall say that two reference frames are mutually compatible if $T_1 \cdot T_2 \neq 0$ and $T_2 \cdot T_1 \neq 0$, i.e. observers of either one are perceived as existing by all other observers of the other reference frame.

This property, for the two frames of reference, will be called the *mutual objective existence (m.o.e.)* condition.

It is not difficult to show that the *m.o.e.* condition does not define an equivalence relation, i.e. it does not satisfy the transitivity property. On the other hand, if we want to define an objective existence, the notion should not privilege one frame with respect to the others. To determine equivalence classes of frames, the best efficient way is to act with a group of linear transformations on a fiducial one, say T_0, and require that all frames obtained from T_0 pairwise satisfy the *m.o.e.* condition.

Therefore we get an equation for the permissible transformations φ_a by requiring that

$$(\varphi_a^* T_0) \cdot (\varphi_b^* T_0) \neq 0\,, \qquad \varphi_b^* T_0) \cdot (\varphi_a^* T_0) \neq 0\,,$$

It is convenient to search for one-parameter groups of permissible transformations and look for their permissibility (i.e. the composition of any two is still permissible); this amounts to look for one parameter subgroups in the connected component of the identity for $GL(4, R)$.

If the transformation φ_a is identified with the matrix A, in terms of the row vector α and the column vector ε of T_0, we have the permissibility condition in the form

$$\alpha(A \cdot \varepsilon) > 0\,, \quad (A^T \alpha) \cdot \varepsilon > 0\,,$$

where A^T stands for the transposed matrix of A.

To solve this inequality equation for the unknown matrix A will be the main aim of this paper and the group of transformations satisfying the inequality will be the relativity group determined by the *m.o.e.* condition.

3. The two-dimensional transformation group

In this section we consider $T_0 = \varepsilon \otimes \alpha$ with $\alpha = (1, 0)$ and $\varepsilon = \begin{pmatrix} x_0 \\ x_1 \end{pmatrix}$.

Neglecting translations, a generic 2-dimensional linear finite transformation with positive determinant, when written in the form

$$\begin{aligned} \begin{pmatrix} x_0' \\ x_1' \end{pmatrix} &= A(\lambda_0, \lambda_1) \begin{pmatrix} x_0 \\ x_1 \end{pmatrix} \equiv \\ &\equiv \begin{pmatrix} \dfrac{c_1 e^{\lambda_0} - c_0 e^{\lambda_1}}{c_1 - c_0} & \dfrac{e^{\lambda_1} - e^{\lambda_0}}{c_1 - c_0} \\ -c_0 c_1 \dfrac{e^{\lambda_1} - e^{\lambda_0}}{c_1 - c_0} & \dfrac{c_1 e^{\lambda_1} - c_0 e^{\lambda_0}}{c_1 - c_0} \end{pmatrix} \begin{pmatrix} x_0 \\ x_1 \end{pmatrix} = \\ &= B^{-1} \begin{pmatrix} e^{\lambda_0} & 0 \\ 0 & e^{\lambda_1} \end{pmatrix} B, \quad \text{where} \quad B = \begin{pmatrix} c_1 & 1 \\ c_0 & 1 \end{pmatrix} \end{aligned}$$

exhibits, for fixed c's, the following useful properties:

$$A(\lambda_0, \lambda_1) A(\lambda_0', \lambda_1') = A(\lambda_0 + \lambda_0', \lambda_1 + \lambda_1'), \quad \det A = e^{\lambda_0 + \lambda_1}.$$

The λ's play the role of evolution parameters.

A simultaneous rescaling of x_0 and x_1 by the factor $e^{-(\lambda_0 + \lambda_1)/2}$, which is in any case real and positive, does not modify the physical descriptions and gives

$$A(\lambda) = \frac{1}{c_1 - c_0} \begin{pmatrix} c_1 e^{\lambda} - c_0 e^{-\lambda} & e^{-\lambda} - e^{\lambda} \\ -c_0 c_1 \left(e^{-\lambda} - e^{\lambda}\right) & c_1 e^{-\lambda} - c_0 e^{\lambda} \end{pmatrix} \tag{1}$$

where $\lambda = (\lambda_0 - \lambda_1)/2$.

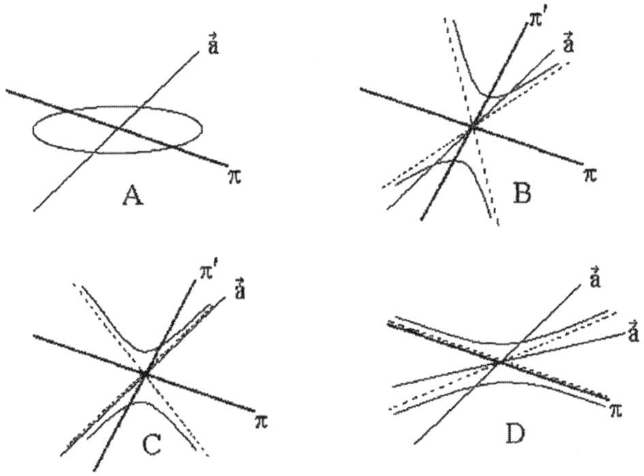

Figure 1. On two-dimensional transformation group.

i) If λ is imaginary ($\lambda \rightarrow i\mu$), the reality of A obliges $c_1 = -c_0 = ic$ and $(A^n)_{00} = \cos n\mu$: the *m.o.e.* condition is violated only if μ is a rational number, but in any case matter transforms continuously in antimatter. Following Feynman [11] we are here identifying particles travelling backwards in time with antiparticles.

Here and in the following we are dealing with discrete transformations; in the next, we will find other situations in which $(A^n)_{00}$ not necessarily becomes zero for some n. This implies that the *m.o.e.* condition is not necessarily violated, but the time direction of an observer is inverted for a different observer and what is matter for the first one becomes antimatter for the second one. For this reason we will discard them.

ii) If λ is real, the c's are real and the quadratic form $(c_1x_0 - x_1)(c_0x_0 - x_1)$ is invariant. The world line associated with the column vector $(x_0, 0)^T$, which is representative of an observer who is always in the origin of his frame, is transformed by A^n in a world line $(x'_{0n}, x'_{1n})^T$ which, when $n \rightarrow \infty$ or $n \rightarrow -\infty$, tends to the line $c_1x'_0 - x'_1 = 0$ or to the line $c_0x'_0 - x'_1 = 0$. If c_0 and c_1 have equal sign, $(x'_{0n}, x'_{1n})^T$ crosses the $(0, x')^T$ line and must be discarded (see fig. 1B), while, if they have opposite sign, $(x'_{0n}, x'_{1n})^T$ moves between the previous lines, which may be considered asymptotic limits (see fig. 1B), and never passes through the x' line. The *m.o.e.* condition is also satisfied in two interesting limit cases: the Carroll case, $c_1c_0 = 0$, in which one of the asymptotes coincides with the x_0 axis (see fig. 1C) and the Galilei case, $c_1c_0 \to \infty$, in which one of the asymptotes coincides with the x_1 axis (see fig. 1D)

The previous quadratic form may be written in normal form. This choice follows the well known Reichenbach analysis [9], who showed that the synchronization procedure which is symmetric and transitive is the one in which two observers are synchronized in such a way that light signals take the same time in going from one to the other. This implies the synchronization identity $\frac{\partial x'_0}{\partial x_0} = \frac{\partial x_0}{\partial x'_0}$, that is to say $A_{00}(\lambda) = A_{00}(-\lambda)$, or $c_0 = -c_1 = c$.

We must remark that other authors [12] - [15] have derived in the past the Lorentz transformations without introducing the constance of the light speed, under conditions compatible with space-time rotations.

4. The three-dimensional transformation group

In this section we consider $T_0 = \varepsilon \otimes \alpha$ with $\alpha = (1, 0, 0)$ and $\varepsilon = (x_0, x_1, x_2)^T$.

As in the bi-dimensional case, it is convenient to express the 3-dimensio - nal transformation matrix A with positive determinant in the form:

$$\begin{pmatrix} x'_0 \\ x'_1 \\ x'_2 \end{pmatrix} = A \begin{pmatrix} x_0 \\ x_1 \\ x_2 \end{pmatrix}$$

where

$$A = \frac{1}{\det C} \begin{pmatrix} c_{0i}\widetilde{c}_{0i}e^{\lambda_i} & c_{0i}\widetilde{c}_{1i}e^{\lambda_i} & c_{0i}\widetilde{c}_{2i}e^{\lambda_i} \\ c_{1i}\widetilde{c}_{0i}e^{\lambda_i} & c_{1i}\widetilde{c}_{1i}e^{\lambda_i} & c_{1i}\widetilde{c}_{2i}e^{\lambda_i} \\ c_{2i}\widetilde{c}_{0i}e^{\lambda_i} & c_{2i}\widetilde{c}_{1i}e^{\lambda_i} & c_{2i}\widetilde{c}_{2i}e^{\lambda_i} \end{pmatrix}$$

and $\widetilde{c}_{ji}$ are the minors with sign of the elements of the matrix

$$C = \begin{pmatrix} 1 & 1 & 1 \\ c_{10} & c_{11} & c_{12} \\ c_{20} & c_{21} & c_{22} \end{pmatrix}, \qquad \det C = \widetilde{c}_{00} + \widetilde{c}_{01} + \widetilde{c}_{02} \quad \widetilde{\widetilde{c}}_{ik} = c_{ik} \det C.$$

Notice that

$$\begin{aligned} A(\lambda_0, \lambda_1, \lambda_2)\, A(\lambda'_0, \lambda'_1, \lambda'_2) &= A(\lambda_0 + \lambda'_0, \lambda_1 + \lambda'_1, \lambda_2 + \lambda'_2) \\ A^n(\lambda_0, \lambda_1, \lambda_2) = A(n\lambda_0, n\lambda_1, n\lambda_2)\,, &\qquad \det A = e^{\lambda_0 + \lambda_1 + \lambda_2}\,. \end{aligned}$$

The matrix A is generic among those with positive determinant; its nine elements are in fact depending on nine parameters (3 λ's and 6 c's) which may be also complex, A being in any case real. Then, at least one of the λ's, e.g.

λ_0, must be real; a simultaneous rescaling of the x variables by the real factor $e^{\frac{\lambda_1+\lambda_2}{2}}$ gives the following form for A_{00}:

$$A_{00} = \frac{1}{\det C}(\widetilde{c}_{00}e^{\mu_0} + \widetilde{c}_{01}e^{\mu} + \widetilde{c}_{02}e^{-\mu}),$$
$$\text{where } \mu_0 = \lambda_0 - \frac{\lambda_1+\lambda_2}{2} \quad \text{and} \quad \mu = \frac{\lambda_1-\lambda_2}{2}. \tag{2}$$

All c's real if μ is real, while, if μ is imaginary, c_{10} and c_{20} are real, $c_{11} = c^*_{12}, c_{21} = c^*_{22}$.
It will be useful in the next to use the following identities:

$$\begin{aligned}(\widetilde{c}_{00}x'_0 + \widetilde{c}_{10}x'_1 + \widetilde{c}_{20}x'_2) &= e^{\mu_0}(\widetilde{c}_{00}x_0 + \widetilde{c}_{10}x_1 + \widetilde{c}_{20}x_2)\,,\\ (\widetilde{c}_{01}x'_0 + \widetilde{c}_{11}x'_1 + \widetilde{c}_{21}x'_2) &= e^{\mu}(\widetilde{c}_{01}x_0 + \widetilde{c}_{11}x_1 + \widetilde{c}_{21}x_2)\,,\\ (\widetilde{c}_{02}x'_0 + \widetilde{c}_{12}x'_1 + \widetilde{c}_{22}x'_2) &= e^{-\mu}(\widetilde{c}_{02}x_0 + \widetilde{c}_{12}x_1 + \widetilde{c}_{22}x_2).\end{aligned} \tag{3}$$

We treat in the next paragraph the real case; in the subsequent one we will consider the complex one.

4.1 Real eigenvalues

4.1.1 Carroll-like or Galilei-like possibilities. It is clearly very useful to look for connections between the 3-dimensional case and the 2-dimensional one discussed in the previous section. It is of particular interest to analyze the possibility of factorizing the matrix A in terms of matrices, acting non trivially only on a coordinate plane; we denote them by $G(x_0, x_1)$, $F(x_0, x_2)$ and $R(x_1, x_2)$, where both F and G are matrices met in section 3 and R is any transformation in the plane (x_1, x_2). As the *m.o.e.* condition is heavily dependent on the behaviour of the element A_{00}, it is convenient to look for those factorizations which assure that, if F_{00} and G_{00} satisfy that condition also A_{00} does it. Then we have four possibilities: FGR, GFR, RFG and RGF.

The first two decompositions have the form

$$A = FGR = \begin{pmatrix} \frac{A_{00}}{G_{00}} & 0 & \frac{-\widetilde{A}_{11}\widetilde{A}_{20}F_{22}}{\mathcal{A}_{20}R_{22}} \\ 0 & 1 & 0 \\ \frac{A_{20}}{G_{00}} & 0 & \frac{-\widetilde{A}_{11}\widetilde{A}_{00}}{\mathcal{A}_{20}R_{22}} \end{pmatrix} \begin{pmatrix} G_{00} & \frac{-\widetilde{A}_{10}G_{00}G_{11}}{\mathcal{A}_{20}} & 0 \\ A_{10} & G_{11} & 0 \\ 0 & 0 & 1 \end{pmatrix} \cdot$$

$$\cdot \begin{pmatrix} 1 & 0 & 0 \\ 0 & \frac{A_{11}}{G_{11}} & \frac{A_{12}}{G_{11}} \\ 0 & -\frac{\widetilde{A}_{12}R_{22}}{\widetilde{A}_{11}} & R_{22} \end{pmatrix},$$

$$A = GFR = \begin{pmatrix} \frac{A_{00}}{F_{00}} & \frac{-\widetilde{A}_{10}G_{11}}{\widetilde{A}_{00}} & 0 \\ \frac{A_{10}}{F_{00}} & G_{11} & 0 \\ 0 & 0 & 1 \end{pmatrix} \begin{pmatrix} F_{00} & 0 & \frac{-\widetilde{A}_{20}F_{00}F_{22}}{\mathcal{A}_{10}} \\ 0 & 1 & 0 \\ A_{20} & 0 & F_{22} \end{pmatrix} \cdot$$

$$\cdot \begin{pmatrix} 1 & 0 & 0 \\ 0 & \frac{\widetilde{A}_{22}\widetilde{A}_{00}}{\mathcal{A}_{10}G_{11}} & \frac{-\widetilde{A}_{21}\widetilde{A}_{00}}{\mathcal{A}_{10}G_{11}} \\ 0 & \frac{A_{21}}{F_{22}} & \frac{A_{22}}{F_{22}} \end{pmatrix},$$

respectively, where $\mathcal{A}_{10} = A_{00}\widetilde{A}_{00} + A_{10}\widetilde{A}_{10}$ and $\mathcal{A}_{20} = A_{00}\widetilde{A}_{00} + A_{20}\widetilde{A}_{20}$.

Notice that the first decomposition is meaningless if $\mathcal{A}_{20} = 0$, while the second one is not allowed when $\mathcal{A}_{10} = 0$. If these two conditions were verified simultaneously, we would have $\det A + A_{00}\widetilde{A}_{00} = 0$, which is against the *m.o.e.* condition. So, at least one of the previous decompositions is allowed. In both cases, if either A_{10} or A_{20} are null, the matrix A is the product of a Carroll transformation times another transformation, which may be Lorentz or Galilei or Carroll (if both A_{01} and A_{02} are null) times a generic space transformation.

If we look for the possibility of factorizing A in the forms RFG or RGF, we meet the symmetrical situation in which A_{10} and A_{20} are substituted by A_{01} and A_{02}. As before, at least one of the decompositions is allowed.

4.1.2 General case. Relations (3) suggest the following transformation for the space variables:

$$X_1 = \frac{\widetilde{c}_{11}x_1 + \widetilde{c}_{21}x_2}{\det C}, \qquad X_2 = \frac{\widetilde{c}_{12}x_1 + \widetilde{c}_{22}x_2}{\det C},$$

$$x_1 = \widetilde{c}_{22}X_1 - \widetilde{c}_{21}X_2\,, \qquad x_2 = -\widetilde{c}_{12}X_1 + \widetilde{c}_{11}X_2\,.$$

In terms of these new variables the transformation depends only on the three variables $\widetilde{c}_{00}$, $\widetilde{c}_{01}$, $\widetilde{c}_{02}$. In fact

$$\left(\, x_0' \quad X_1' \quad X_2' \,\right)^T = A\left(\, x_0 \quad X_1 \quad X_2 \,\right)^T,$$

where

$$A = \begin{pmatrix} A_{00} & e^{\mu} - \dfrac{\widetilde{c}_{01}}{\det C}e^{\mu_0} \\ \dfrac{\widetilde{c}_{01}}{\det C}(e^{\mu} - A_{00}) & (\dfrac{\widetilde{c}_{01}}{\det C})^2 e^{\mu_0} + (1 - \dfrac{\widetilde{c}_{01}}{\det C})e^{\mu} \\ \dfrac{\widetilde{c}_{02}}{\det C}(e^{-\mu} - A_{00}) & \dfrac{\widetilde{c}_{02}}{\det C}(\dfrac{\widetilde{c}_{01}}{\det C}e^{\mu_0} - e^{\mu}) \end{pmatrix.$$

$$\begin{matrix} e^{-\mu} - \dfrac{\widetilde{c}_{02}}{\det C}e^{\mu_0} \\ \dfrac{\widetilde{c}_{01}}{\det C}(\dfrac{\widetilde{c}_{02}}{\det C}e^{\mu_0} - e^{-\mu}) \\ (\dfrac{\widetilde{c}_{02}}{\det C})^2 e^{\mu_0} + (1 - \dfrac{\widetilde{c}_{02}}{\det C})e^{-\mu} \end{matrix}\Bigg)\,,$$

and A_{00} is given by (2).

Relations (3) become:

$$\begin{aligned} \widetilde{c}_{00}x_0' - X_1' - X_2' &= e^{\mu_0}(\widetilde{c}_{00}x_0 - X_1 - X_2)\,, \\ \widetilde{c}_{01}x_0' + X_1' &= e^{\mu}(\widetilde{c}_{01}x_0 + X_1) \\ \widetilde{c}_{02}x_0' + X_2' &= e^{-\mu}(\widetilde{c}_{02}x_0 + X_2)\,. \end{aligned} \tag{4}$$

We may now look at the consequences of the *m.o.e.* condition, which formally states that $(x_0, 0, 0)$ cannot be transformed in (x_0', X_1', X_2'), where $x_0' = 0$, which implies that $(A^n)_{00} = \widetilde{c}_{00}e^{n\mu_0} + \widetilde{c}_{01}e^{n\mu} + \widetilde{c}_{02}e^{-n\mu}$, when n is supposed to be a continuous variable, cannot never be zero; in the discrete case this condition implies that $(A^n)_{00}$ cannot never change sign when going from n to $n+1$.

Coming back to the possible decompositions described in the previous subsection, we can apply to F and G the Reichenbach criterion, which implies $F_{00} = (F^{-1})_{00}$, $G_{00} = (G^{-1})_{00}$. Then

$$\widetilde{c}_{00}e^{n\mu_0} + \widetilde{c}_{01}e^{n\mu} + \widetilde{c}_{02}e^{-n\mu} = \widetilde{c}_{00}e^{-n\mu_0} + \widetilde{c}_{01}e^{-n\mu} + \widetilde{c}_{02}e^{n\mu}\,,$$

or

$$\widetilde{c}_{00}\sinh n\mu_0 + (\widetilde{c}_{01} - \widetilde{c}_{02})\sinh n\mu = 0$$

If $\mu = 0$, the transformation becomes trivial. So we must have:

a) either $\widetilde{c}_{01} = \widetilde{c}_{02}, \widetilde{c}_{00} = 0$

b) or $\widetilde{c}_{01} = \widetilde{c}_{02}, \mu_0 = 0$.

Let us now discuss these two cases separately:

CASE a): $\widetilde{c}_{01} = \widetilde{c}_{02}, \widetilde{c}_{00} = 0$.

The transformation A reduces to

$$A = \begin{pmatrix} \cosh\mu & e^{\mu} - \frac{1}{2}e^{\mu_0} & e^{-\mu} - \frac{1}{2}e^{\mu_0} \\ \frac{1}{2}\sinh\mu & \frac{1}{4}e^{\mu_0} + \frac{1}{2}e^{\mu} & \frac{1}{4}e^{\mu_0} - \frac{1}{2}e^{\mu} \\ -\frac{1}{2}\sinh\mu & \frac{1}{4}e^{\mu_0} - \frac{1}{2}e^{\mu} & \frac{1}{4}e^{\mu_0} + \frac{1}{2}e^{-\mu} \end{pmatrix}$$

This transformation implies

$$X_1' + X_2' = \frac{1}{2}e^{\mu_0}(X_1 + X_2),$$
$$\left(X_0' + 2X_1'\right)\left(X_0' + 2X_2'\right) = (X_0 + 2X_1)(X_0 + 2X_2)$$

The vector $(1, 0, 0)$ transforms in a vector such that $X_1' = -X_2'$, $X_0'^2 - 4X_1'^2 = X_0^2 - 4X_1^2$. The transformation is in fact bidimensional; it couples a Lorentz with another time-space transformation which obey the *m.o.e.* condition.

CASE b): $\widetilde{c}_{01} = \widetilde{c}_{02}, \mu_0 = 0$.

$$(A^n)_{00} = \frac{1 + 2\frac{\widetilde{c}_{01}}{\widetilde{c}_{00}}\cosh n\mu}{1 + 2\frac{\widetilde{c}_{01}}{\widetilde{c}_{00}}}$$

$$X_0' = \frac{\widetilde{c}_{0i}e^{\mu_i}}{\det C}X_0, \qquad X_1' = \frac{\widetilde{c}_{01}}{\det C}(e^{\mu}X_0 - X_0'),$$
$$X_2' = \frac{\widetilde{c}_{01}}{\det C}(e^{-\mu}X_0 - X_0')-$$

The *m.o.e.* condition implies either

i) $\frac{\widetilde{c}_{01}}{\widetilde{c}_{00}} > 0$ (and then $1 + \frac{\widetilde{c}_{00}}{\widetilde{c}_{01}} > 1, 1 + 2\frac{\widetilde{c}_{01}}{\widetilde{c}_{00}} > 1$) or

ii) $\frac{\widetilde{c}_{01}}{\widetilde{c}_{00}} < -\frac{1}{2}$ (and then $1 + \frac{\widetilde{c}_{00}}{\widetilde{c}_{01}} > -1, 1 + 2\frac{\widetilde{c}_{01}}{\widetilde{c}_{00}} < 0$)

$$\begin{aligned} X_0' - \frac{\det C}{\widetilde{c}_{00}}(X_1' + X_2') &= X_0, \\ X_0' + \frac{\det C}{\widetilde{c}_{01}} X_1' &= e^{\mu} X_0, \\ X_0' + \frac{\det C}{\widetilde{c}_{01}} X_2' &= e^{-\mu} X_0. \end{aligned} \tag{5}$$

If we combine the square of the first relation with the product of the last two, eliminating the mixed term in $X_0'(X_1' + X_2')$, we obtain the following invariant:

$$X_0'^2 + \left(1 + 2\frac{\widetilde{c}_{00}}{\widetilde{c}_{01}}\right)\left(X_1' - \left(1 + \frac{\widetilde{c}_{00}}{\widetilde{c}_{01}}\right) X_2'\right)^2 +$$
$$-X_2'^2 \left(\frac{\widetilde{c}_{00}}{\widetilde{c}_{01}}\right)^2 \left(1 + 2\frac{\widetilde{c}_{00}}{\widetilde{c}_{01}}\right) = inv$$

In the case i) we have $\left(1 + 2\frac{\widetilde{c}_{00}}{\widetilde{c}_{01}}\right)$ >0; if this is the case, the bidimensional transformation in the plane $\left(X_0, X_1 - (1 + \frac{\widetilde{c}_{00}}{\widetilde{c}_{01}})X_2\right)$ corresponds to a transformation which is incompatible with the *m.o.e.* condition, as shown in the previous section. Only the case ii), in which the space structure is Euclidean, is then admissible.

4.2 Complex eigenvalues ($\mu_1 \to i\mu\,,\quad \mu_2 \to -i\mu$)

The reality of the transformation A obliges C and $\widetilde{C}$ to have the following form:

$$C = \begin{pmatrix} 1 & 1 & 1 \\ c_1 k \sin(\nu - \alpha_1) & c_1 e^{i\alpha_1} & c_1 e^{-i\alpha_1} \\ c_2 k \sin(\nu - \alpha_2) & c_2 e^{i\alpha_2} & c_2 e^{-i\alpha_2} \end{pmatrix}$$
$$\widetilde{C} = \begin{pmatrix} 2i\delta & k\delta e^{i\nu} & -k\delta e^{-i\nu} \\ 2ih_2 & g_2 - ih_2 & -g_2 - ih_2 \\ -2ih_1 & -g_1 + ih_1 & g_1 + ih_1 \end{pmatrix},$$

where

$$g_i = c_i(\cos\alpha_i - k\sin(\nu - \alpha_i)), \quad h_i = c_i \sin\alpha_i, \quad \delta = c_1 c_2 \sin(\alpha_1 - \alpha_2)\,.$$

As

$$(A^n)_{00} = \frac{\widetilde{c}_{00} e^{n\mu_0} + \widetilde{c}_{01} e^{in\mu} + \widetilde{c}_{02} e^{-in\mu}}{\widetilde{c}_{00} + \widetilde{c}_{01} + \widetilde{c}_{02}} = \frac{e^{n\mu_0} + k\cos(n\mu + \nu)}{1 + k\cos\nu},$$

the *m.o.e.* existence is violated if $\mu_0 \neq 0$ either for $n \to +\infty$ or for $n \to -\infty$. Then $\mu_0 = 0$; moreover $(A^n)_{00}$ is always positive if $|k| > 1$ and change continuously sign if $|k| < 1$.
If we put

$$X_0 = k\delta x_0\,, \quad X_1 = g_2 x_1 - g_1 x_2\,, \quad X_2 = h_2 x_1 - h_1 x_2\,,$$

relations (10) imply

$$\begin{aligned} &X_0' + kX_2' = inv\,, \\ &(X_0' + (\cos\nu X_1' - \sin\nu X_2'))^2 + (\sin\nu X_1' - \cos\nu X_2')^2 = inv. \end{aligned}$$

The last invariant implies situations which are not compatible with the *m.o.e.* existence (e.g., the case $|k| < 1$).

In conclusion complex eigenvalues are not compatible with the *m.o.e.* existence.

4.3 Remark

So far we have derived the possible forms of finite linear transformations, which satisfy the condition that different instants of time x_0, going from $-\infty$ to $+\infty$, of an observer are transformed into different instants of time x_0', going from $-\infty$ to $+\infty$, of another observer.

In the bidimensional case we have found that the unique transformations which obey this rule have Lorentz form, which may restrict, for particular values of the parameters, to Galilei or Carrol ones.

In the three-dimensional case, the transformations, which are compatible with the previous condition, may be combinations of two transformations A (in (x_0, x_1)) and B (in (x_0, x_2)), where A and B may be any of the previous three transformations. If A and B are not both Lorentzian the space structure is not specified. Viceversa, if they are both Lorentzian, the space structure must necessarily be Euclidean.

The analysis for the four dimensional space-time can be repeated along the same lines; it becomes more cumbersome and therefore will not be pursued here. However we will make few comments on how to generalize this picture to the context of general relativity.

5. On the transition to General Relativity and conclusions

Having translated Weyl's assertion into a tensorial object, i.e. a rank one $(1,1)-$tensor field T with the property $T \cdot T \propto T$, it is not difficult to extend our approach to a framework were Minkowski space-time is replaced by a four-dimensional differential manifold $\mathcal{M}$. In this more general setting the column vector ε is replaced by a vector field and the row vector is replaced by a 1-form

$\alpha = a^0 dx_0 + a^1 dx_1 + a^2 dx_2 + a^3 dx_3$. These two geometrical objects define $T = \varepsilon \otimes \alpha$ and require that $\alpha(\varepsilon) > 0$ all over $\mathcal{M}$, therefore we may choose a normalization such that $Tr\,(T) = \alpha(\varepsilon) = 1$ and get $T \cdot T = T$. For simplicity we still denote the normalized tensor T by $\varepsilon \otimes \alpha$, with the understanding that now $\alpha(\varepsilon) = 1$. We notice that to have a splitting of $\mathcal{M}$ into space and time, we have to require that α admits an "integrating factor", i.e. $\alpha \wedge d\alpha = 0$ and that ε is a complete vector field, i.e., for any initial condition, integral curves for ε exist from $-\infty$ to $+\infty$ in the evolution parameter.

By selecting a fiducial T_0, we may consider a subgroup of the diffeomorphism group for $\mathcal{M}$, which satisfy $(\varphi_a^* T_0) \cdot (\varphi_b^* T_0) \neq 0$ and $(\varphi_b^* T_0) \cdot (\varphi_a^* T_0) \neq 0$.

We may again write the equation for the transformation φ in the form

$$\alpha(\varphi_* \varepsilon) > 0\,, \qquad (\varphi^* \alpha)(\varepsilon) > 0\,.$$

Now we have to solve this inequality for the diffeomorphism φ.

When T is such that $\alpha = d\tau$ and ε defines a congruence of world-limes (a family of observers) which are solutions of a second order differential equation associated with a flat generalized connection [16] (observers will be *inertial observers*), we are back to the situation we have already described in the previous sections. A generalized solution of the previous inequalities in terms of the allowed diffeomorphisms will be considered elsewhere, here we close by stressing that the *mutual objective existence* condition is able to capture the notion of causality (earlier and later) along with an objective meaning of a "kind of union" of space and time for a class of equivalent observers more general than inertial ones.

Acknowledgments

A preliminary account of these results was presented in the Symposium Symmetries XIII at Bregenz (July 2003). We would like to thank the participants for their comment and B. Gruber for his kind invitation.

References

[1] Einstein, A. 1905 *Ann. Phys.*, Lpz., **17**, 891-921

[2] Minkowski, H. 1923, *Raum, Zeit, Materie* (Berlin, J. Springer);
English translation, Brose, H.L. 1952 *(Space, time and matter)* (Dover, New York)

[3] Reichenbach, H. 1928, *Philosophie der Raum-Zeit-Lehre* (Walter de Gruipter, Berlin, Leipzig);
English transl., Reichenbach M. and Freund J. 1958 *The Philosophy of Space and Time* (Dover, New York)

[4] Weyl, H. 1923, *Raum, Zeit, Materie* (Berlin, J. Springer);
English translation, Brose, H.L. 1952 *(Space, time and matter)* (Dover, New York)

[5] Feynman, R.P., Weinberg, S. *Elementary particles and the laws of physics* (Cambridge University Press, Cambridge 1987)

[6] Ignatowski, W. von, 1910 *Arch. Math. Phys.*, Lpz, **17**, 1

[7] Levy-Leblond, J.M. 1976 *Am. J. Phys.* **44**, 271; id. 1977 Nuovo Cimento **7**, 187; Levy-Leblond, J.M. and Prevost, J.P. 1979, *Am. J. Phys.* **47**, 12

[8] Mermin, N.D. 1984 *Am. J. Phys.* **52**, 119

[9] Ross, A.W. *Am. J. Phys.* **55,** 174

[10] de Ritis, R., Marmo, G. and Preziosi, B. 1999 *General Relativity and Gravitation* **31**, 1501

SURVIVAL OF QUASI-SPIN STRUCTURE IN ISOMERS OF $N \sim 82$ NUCLEI

H. Nakada, T. Matsuzawa and K. Ogawa
Department of Physics,
Chiba University,
Chiba 263-8522, JAPAN

Abstract The structure of isomers in $N \sim 82$ nuclei is reinvestigated. It is pointed out that the quasi-spin structure survives in the isomers and the final states of their decays, as a partial dynamical symmetry. Using this group structure, an extended seniority reduction formula is derived and the presence of pair excitations out of the $Z = 64$ core is revealed. By taking into account the core excitation, an anomaly in neutron effective charges for $N = 81$ and $N = 83$ nuclei is greatly reduced.

Keywords: Quasi-spin, isomer, shell structure, $E2$ transition

1. Introduction

As is well-known, atomic nuclei have magic numbers, which originate from the shell structure formed by spherical single-particle orbits. However, recent experiments show that the magic numbers are not rigorous and somewhat depend on proton and neutron numbers (Z and N, respectively). It is now an important and contemporary question how stiff individual shells are.

The subshell closure at $Z = 64$ in $N \sim 82$ nuclei was established about 25 years ago [10]. High excitation energy (E_x) of the 2^+ state and relatively low E_x of the 3^- state in ^{146}Gd, as well as a kink in the two-proton separation energy (S_{2p}), give the indication for the subshell closure. Moreover, $E2$ transition strengths of isomers in $Z > 64$, $N = 82$ nuclei were measured [7, 8, 14], and their Z-dependence was successfully described by a $(0h_{11/2})^n$ shell model assuming the ^{146}Gd core [11]. At a glance, this seems to indicate that the $Z = 64$ core is very stiff. On the other hand, the kink in S_{2p} is not so conspicuous as in S_{2n} at $N = 82$. In addition, systematic calculations for $E_x(10^+)$ suggest the presence of sizable pair excitations out of the $Z = 64$ core [2].

B. Gruber (ed.), Symmetries in Science XI, 459–470.

We have reinvestigated the structure of the isomers in the $Z > 64$, $N \sim 82$ nuclei [12, 13]. Aided by the survival of quasi-spin structure in the isomers, some of the problems have been solved.

2. Brief survey of quasi-spin

For each spherical single-particle orbit j, the creation and annihilation operators $(a^\dagger_{jm}, a_{jm})$ form an $\mathrm{O}(4\Omega_j)$ algebra, where $\Omega_j = (2j+1)/2$. There is an $\mathrm{SU}(2)$ subalgebra known as the quasi-spin (QS), whose generators are

$$S_{j+} = \sqrt{\frac{\Omega_j}{2}}[a^\dagger_j a^\dagger_j]^{(0)}, \quad S_{j-} = (S_{j+})^\dagger \quad S_{j0} = \frac{1}{2}(N_j - \Omega_j), \tag{1}$$

where N_j denotes the number operator. The quadratic Casimir operator of the QS defines the so-called seniority ν_j,

$$\mathbf{S}_j = \frac{1}{2}(S_{j+}S_{j-} + S_{j-}S_{j+}) + S_{j0}^2 = S_j(S_j+1), \quad S_j = \frac{1}{2}(\Omega_j - \nu_j). \tag{2}$$

We can apply well-known results of the $\mathrm{SU}(2)$ algebra such as the Clebsch-Gordan coefficients and the Wigner-Eckart theorem. There is a unitary symplectic algebra dual to the QS, $\mathrm{USp}(2\Omega_j)$ [19], whose generators are $[a^\dagger_j \tilde{a}_j]^{(\lambda)}$ ($\lambda =$ odd). Note that the angular momentum operators are proportional to the generators with $\lambda = 1$.

For a shell comprised of multiple orbits $(j_1, j_2, \cdots, j_k)$, we may consider the product of the QS group, $\mathrm{SU}(2)_{j_1} \otimes \mathrm{SU}(2)_{j_2} \otimes \cdots \otimes \mathrm{SU}(2)_{j_k}$. While this product group structure is usually broken via mixing among different representations, the total seniority $\nu = \sum_j \nu_j$ remains a good quantum number in spherical nuclei. It is commented that if and only if the single-particle energies are degenerate, we again have an $\mathrm{SU}(2)$ structure [9, 3],

$$\mathrm{SU}(2)_{j_1} \otimes \mathrm{SU}(2)_{j_2} \otimes \cdots \otimes \mathrm{SU}(2)_{j_k} \supset \mathrm{SU}(2)_{j_1 \oplus j_2 \oplus \cdots \oplus j_k}. \tag{3}$$

3. Previous studies of $Z > 64$, $N = 82$ nuclei

The $Z > 64$, $N = 82$ nuclei systematically have isomers with $J^\pi = 10^+$ (for even-Z) or $27/2^-$ (for odd-Z). Decay out of these isomers is highly dominated by the $E2$ transition. The $E2$ transition strengths from the isomers have been measured [7, 8, 14]. Lawson investigated the structure of the isomers, assuming the $(0h_{11/2})^n$ configuration on top of the ^{146}Gd inert core ($n = Z - 64$). The two-body interaction matrix elements are determined from the observed levels in ^{148}Dy, which show that the QS $\mathrm{SU}(2)_{0h_{11/2}}$ is a good dynamical symmetry. The Wigner-Eckart theorem for $\mathrm{SU}(2)_{0h_{11/2}}$ derives the seniority reduction formulae (SRF). For the $E2$ transition of the 10^+ and $27/2^-$ isomers, the SRF gives

$$\langle (0h_{11/2})^n J_f^\pi || T(E2) || (0h_{11/2})^n J_i^\pi \rangle$$
$$= \frac{\Omega - n}{\Omega - \nu} \langle (0h_{11/2})^\nu J_f^\pi || T(E2) || (0h_{11/2})^\nu J_i^\pi \rangle , \tag{4}$$

since $T(E2)$ is a QS-vector. This leads to $B(E2) \propto (\Omega - n)^2$, indicating that the $E2$ transition is vanishingly weak at $Z \simeq 70$. This nature, called seniority isomerism, seems to be observed in the experiments. Owing to the success of Lawson's model, the $(0h_{11/2})^n$ configuration ($n = Z - 64$) has often been assumed for proton degrees of freedom, when analyzing data of $Z > 64$, $N \sim 82$ nuclei.

However, Lawson's model has a problem. The $1d_{3/2}$ and $2s_{1/2}$ orbits lie very closely to $0h_{11/2}$ in this region, indicating significant mixing due to the pairing interaction. If this mixing is taken into consideration, the minimum of $B(E2)$ with respect to Z is displaced, in disagreement with the data. Noticing this problem, Blomqvist gave a conjecture based on the BCS approximation [4],

$$\langle (\xi r)^n J_f^\pi || T(E2) || (\xi r)^n J_i^\pi \rangle = \frac{\Omega - \langle N_\xi \rangle}{\Omega - \nu} \langle \xi^\nu J_f^\pi || T(E2) || \xi^\nu J_i^\pi \rangle , \tag{5}$$

where ξ denotes $0h_{11/2}$ in this case and r represents all the other orbits. From this conjecture, Blomqvist argued that the suppression of $\langle N_{0h_{11/2}} \rangle$ at $Z \simeq 70$ due to the mixing to $1d_{3/2}$ and $2s_{1/2}$ should be compensated by the excitation across the $Z = 64$ gap. However, the particle number conservation $\mathrm{U}(1)_N$ is broken in the BCS approximation. Assessment of the influence of the $\mathrm{U}(1)_N$ symmetry could be important.

4. Survival of quasi-spin structure in $N = 82$ nuclei

In solving the problem for the isomers of the $Z > 64$, $N = 82$ nuclei, a key ingredient is the survival of the quasi-spin (QS) structure. The total seniority ν should be a good quantum number in the $Z > 64$, $N = 82$ nuclei, because they are spherical. Both the 10^+ $(27/2^-)$ isomers and the 8^+ $(23/2^-)$ final states of their decays have $\nu = 2$ $(\nu = 3)$. Since the j values of the neighboring orbits are not so high, this seniority should be solely carried by $0h_{11/2}$. This means that, although $\xi = 0h_{11/2}$ is not isolated, the isomers and the decay final states belong to a specific representation of the product group $\mathrm{SU}(2)_{j_1} \otimes \mathrm{SU}(2)_{j_2} \otimes \cdots \otimes \mathrm{SU}(2)_{j_k} = \mathrm{SU}(2)_\xi \otimes [\mathrm{SU}(2)_r]^{k-1}$. Namely, for the isomer $|i\rangle$ and the final state $|f\rangle$, we have

$$[H, \mathbf{S}_j^2] |i\rangle \simeq [H, \mathbf{S}_j^2] |f\rangle \simeq 0 , \tag{6}$$

and the product QS group gives a partial dynamical symmetry [1]. It is noted that the dual group $\mathrm{USp}(2\Omega_\xi) \otimes [\mathrm{USp}(2\Omega_{j_r})]^{k-1}$ is also a good partial symmetry.

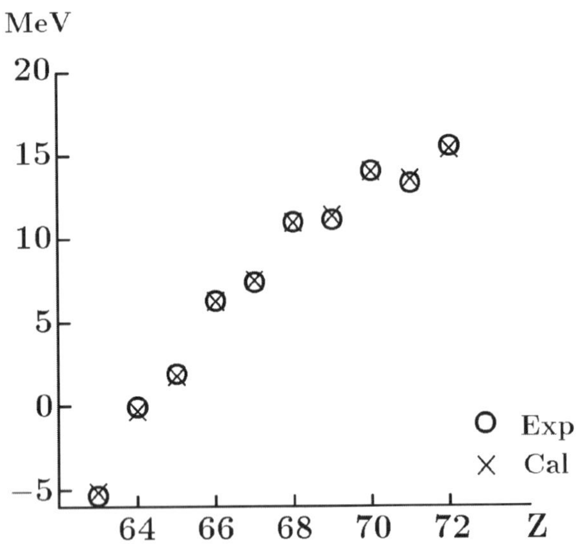

Figure 1. Comparison of calculated and experimental binding energies for $N = 82$ nuclei.

Based on this QS structure, we have proven Eq. (5) as an extension of the SRF [12], maintaining the number conservation $\mathrm{U}(1)_N$. An additional assumption is needed for the exact derivation: a similarity in the pairing part of the wave-functions between the initial and final states. It is also proven [12] that this assumption is fulfilled if V_ξ (two-body interaction between nucleons in the ξ orbit) in H holds the $\mathrm{SU}(2) \times \mathrm{USp}(2\Omega_\xi)$ symmetry, which is connected to the short-range character of the residual interaction among nucleons.

Thus Blomqvist's conjecture should be correct. As already mentioned, approximate degeneracy among $0h_{11/2}$, $1d_{3/2}$ and $2s_{1/2}$ necessarily leads to significant mixing of the configurations involving $1d_{3/2}$ and $2s_{1/2}$ via the pairing interaction. The observed Z-dependence of $B(E2)$ showing a minimum at $Z \simeq 70$ indicates that the depletion of $\langle N_{0h_{11/2}} \rangle$ due to the mixing has to be compensated by the core excitation. Therefore, contrary to the naive expectation from the success of Lawson's model, the $Z = 64$ core should not be very stiff.

In order to confirm quantitatively the above mechanism resulting from the survival of the QS structure, we have carried out a shell model calculation. We take a $(0g_{7/2}1d_{5/2}1d_{3/2}2s_{1/2}0h_{11/2})^n$ model space on top of the ^{132}Sn core $(n = Z - 50)$, with the truncation $N_{0g_{7/2}} + N_{1d_{5/2}} \geq 8$ and $\nu \leq 3$. The effective Hamiltonian consists of the single-particle energies and the modified surface-delta interaction with the parameters adjusted to data of energies in

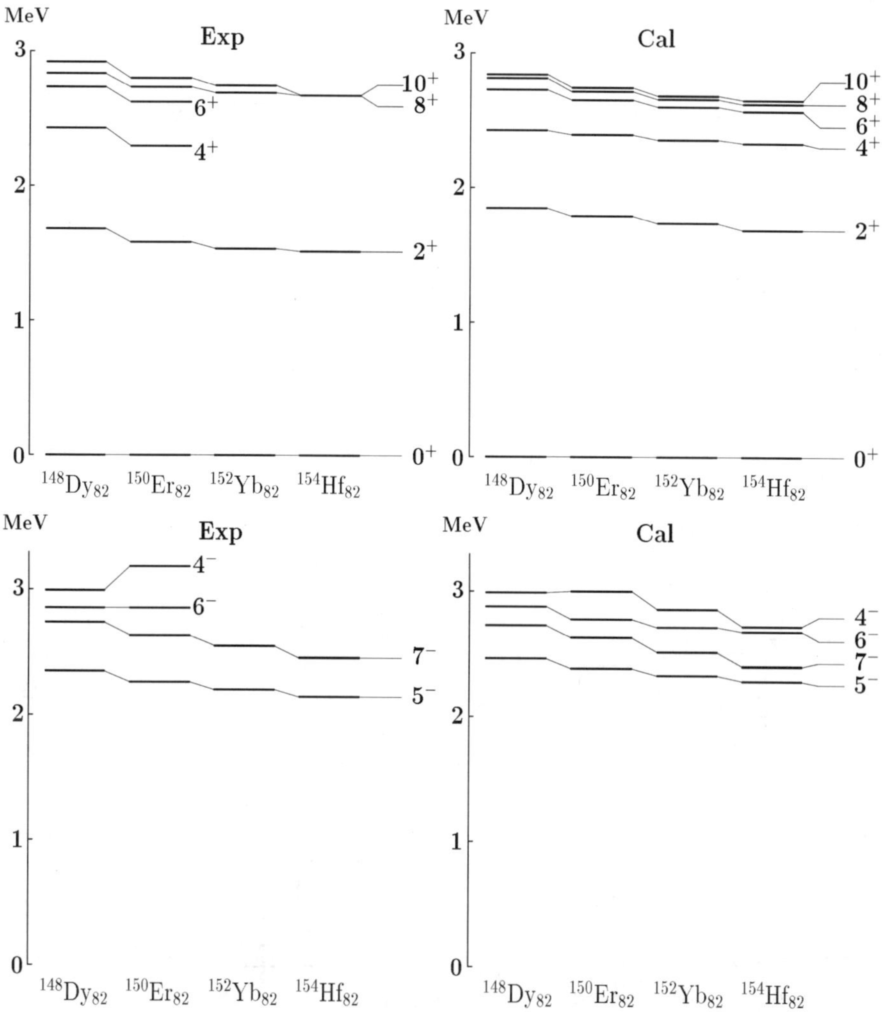

Figure 2. Comparison of calculated and experimental energy levels for even-Z, $N = 82$ nuclei.

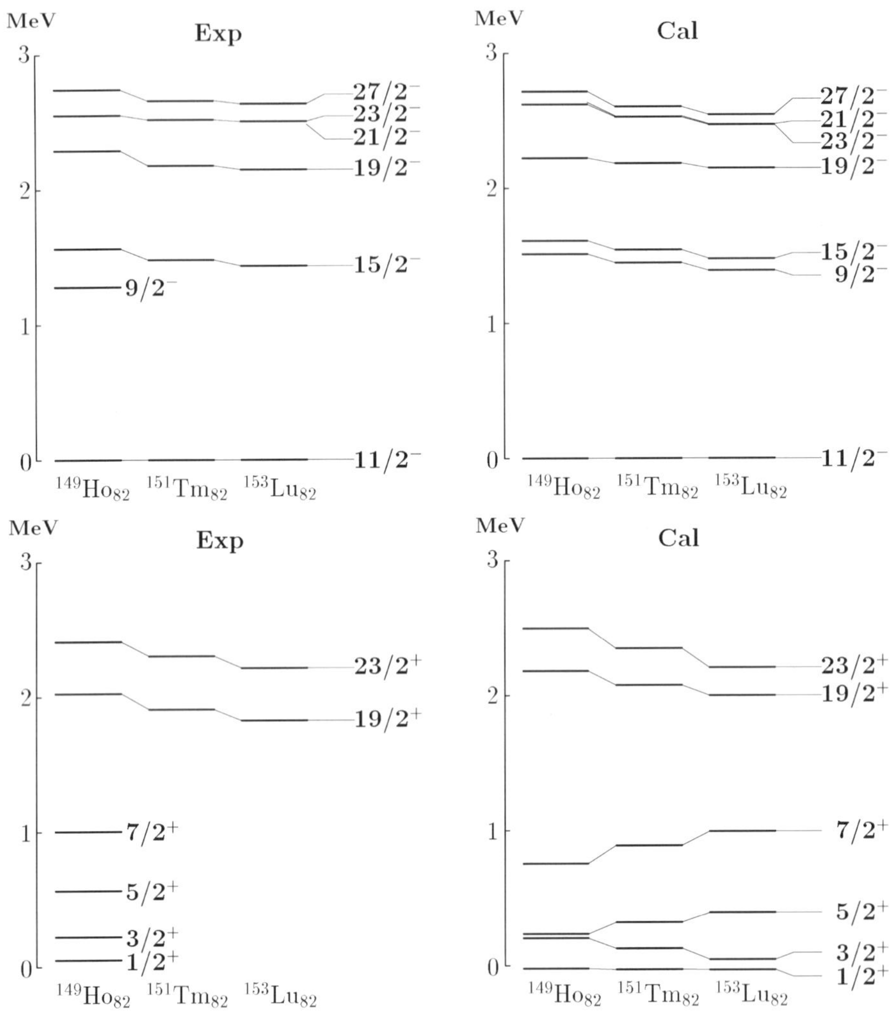

Figure 3. Comparison of calculated and experimental energy levels for odd-Z, $N = 82$ nuclei.

^{146}Gd and neighboring nuclei. In Fig. 1, we compare the calculated binding energies, in relative to the binding energy of ^{146}Gd, with the experimental data. Energy levels of even-Z, $N = 82$ nuclei are presented in Fig. 2, both for $\pi = +$ and $\pi = -$ levels. Those of odd-Z, $N = 82$ nuclei are shown in Fig. 3. It is remarked that the $\pi = -$ ($\pi = +$) levels of the even-Z (odd-Z) nuclei, which are out of the model space in Lawson's model, are also reproduced.

By using the wave-functions obtained above, we have calculated the $E2$ transition strengths from the isomers. The calculated $B(E2)$ values and their Z-dependence are shown in Fig. 4, in comparison with the experimental data and the results of Lawson's model. In reproducing the $E2$ strengths, we have introduced a single additional adjustable parameter, the effective charge for protons. Apart from it, it is clear that the Z-dependence of the $E2$ strengths is reproduced very well.

In the present calculation, Eq. (5) is exact for the decays of the 10^+ isomers, because of the seniority truncation and the short-range character of the residual interaction. Although this is not the case for the decays of the $27/2^-$ isomers, Eq. (5) still holds to an excellent approximation. As discussed above, the minimum of $B(E2)$ at $Z \simeq 70$ is reproduced in the balance between the coupling of $0h_{11/2}$ to the $(1d_{3/2}2s_{1/2})$ orbits and a sizable pair excitation from $(0g_{7/2}1d_{5/2})$ to $0h_{11/2}$. In Fig. 5, the calculated number expectation values in the isomers $\langle N_{0h_{11/2}}\rangle$ and $\langle N_{1d_{3/2}} + N_{2s_{1/2}}\rangle$ are depicted. In comparison, the number expectation values in the model without the $Z = 64$ core excitation (*i.e.* the $(1d_{3/2}2s_{1/2}0h_{11/2})^n$ model) are also presented. This figure clarifies the significance of the $Z = 64$ core excitation.

5. $N = 81$ and 83 nuclei

In $Z > 64$, $N = 83$ nuclei with odd-Z, the 17^+ isomers have systematically been observed. It is considered, in the first approximation, that these isomers are formed by a neutron in the $1f_{7/2}$ orbit which is coupled to the $27/2^-$ states of the $N = 82$ nuclei. In $Z > 64$, $N = 81$ nuclei with even-Z, $27/2^-$ isomers have been observed. These are considered to have a $0h_{11/2}$ neutron-hole coupled to the 10^+ states of the $N = 82$ nuclei. In both cases the isomers decay via $E2$ transitions. The $E2$ strengths have been measured via the lifetimes of the isomers [15, 16, 5, 6, 17, 18].

The systematics of the $E2$ strengths from the isomers in the $N = 81$ and 83 nuclei was analyzed by using simple models, in which Lawson's model (*i.e.* the $(0h_{11/2})^n$ configuration) was assumed for proton degrees of freedom, and a $1f_{7/2}$ neutron (a $0h_{11/2}$ neutron-hole) is added for the $N = 83$ ($N = 81$) case. The contribution of a neutron (or a neutron-hole) may shift the minimum with respect to Z in the $B(E2)$ values. From this analysis, quite anomalous values were obtained for the neutron effective charges; $e_\nu^{\text{eff}} \simeq 3.6e$ for the $N = 83$

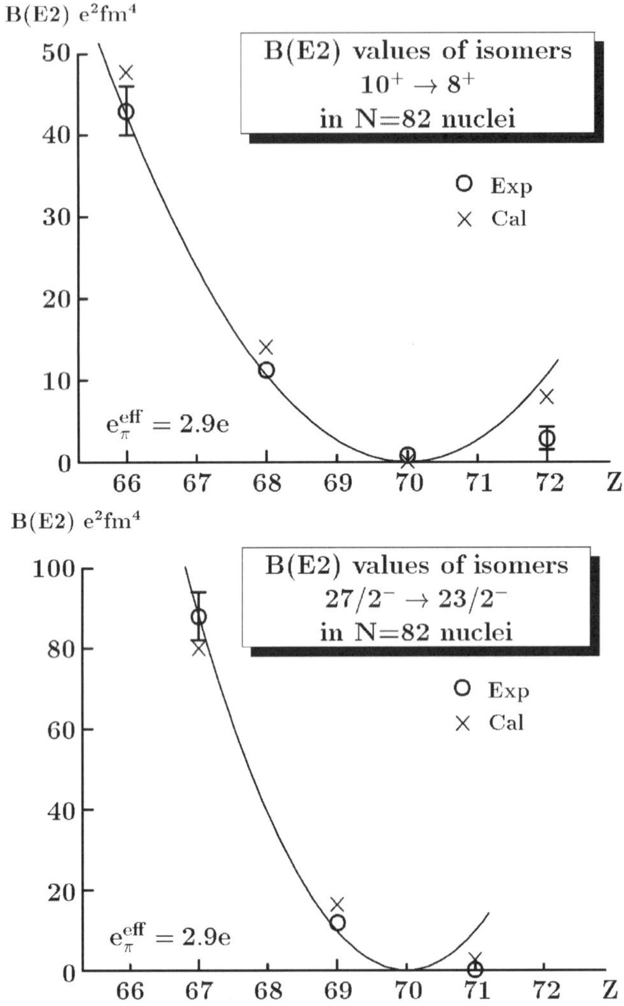

Figure 4. $B(E2)$ values from 10^+ and $27/2^-$ isomers in $Z > 64$, $N = 82$ nuclei. The lines show the results of Lawson's model.

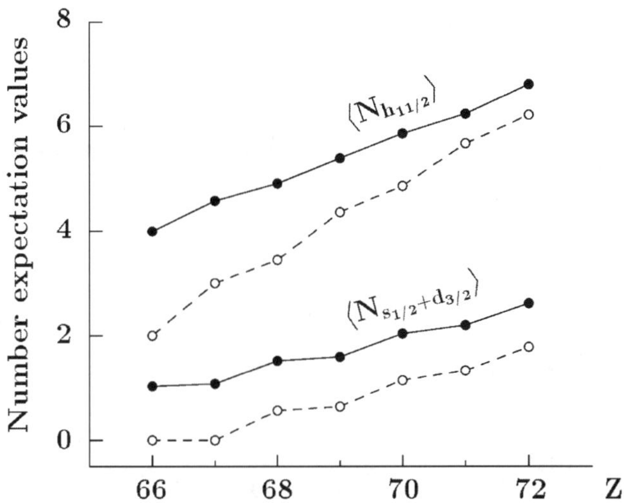

Figure 5. Calculated $\langle N_{0h_{11/2}} \rangle$ and $\langle N_{1d_{3/2}} + N_{2s_{1/2}} \rangle$ in the 10^+ and $27/2^-$ isomers. Solid lines (dotted lines) are the results with (without) the $Z = 64$ core excitation.

nuclei, which is remarkably larger than $e_\pi^{\rm eff}$, and vanishing $e_\nu^{\rm eff}$ for the $N = 81$ nuclei.

Since we now know that the pair excitation out of the $Z = 64$ core is important in the $N = 82$ nuclei, it will be interesting to see how the core excitation affects the structure of the $N = 81$ and 83 isomers. For this purpose, we have carried out shell model calculations in relatively large model spaces. For the $N = 83$ nuclei, we take the $[\pi(0g_{7/2}1d_{5/2}1d_{3/2}2s_{1/2}0h_{11/2})]^n \times [\nu(0h_{9/2}1f_{7/2}0i_{13/2})]^1$ model space ($n = Z - 64$). For the $N = 81$ nuclei, the $[\pi(0g_{7/2}1d_{5/2}1d_{3/2}2s_{1/2}0h_{11/2})]^n \times [\nu(0g_{7/2}1d_{5/2}1d_{3/2}2s_{1/2}0h_{11/2})]^{-1}$ model space is adopted. In both cases, truncations by the total seniority and by the particle number excited out of $\pi(0g_{7/2}1d_{5/2})$ are introduced, in a consistent manner with the $N = 82$ nuclei. For the proton-neutron interaction, we use a Yukawa form with the range of one-pion exchange. The strength of the interaction, as well as the single-neutron energies, are adjusted to the experimental energies of the relevant nuclei. We have confirmed that, in the 17^+ isomers and the 15^+ final states of their decays in the $N = 83$ nuclei, the mixings of $\nu(0h_{9/2})^1$ and $\nu(0i_{13/2})^1$ configurations are negligibly small. In the $N = 81$ nuclei, the $\nu(0h_{11/2})^{-1}$ configuration necessarily dominates in the $27/2^-$ isomers and the $23/2^-$ final states, due to the high J value and the conservation of the total seniority.

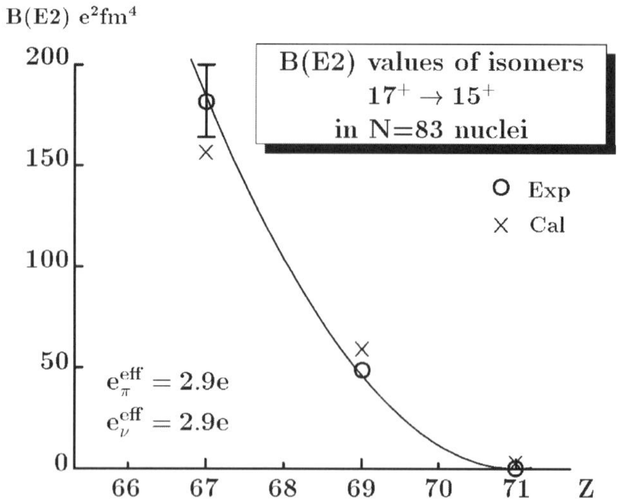

Figure 6. $B(E2)$ values from 17^+ isomers in $Z > 64$, $N = 83$ nuclei.

Nevertheless, the proton states do not necessarily remain pure in the $N = 81$ and 83 nuclei. For instance, in the $27/2^-$ isomer of the $N = 81$ nuclei, $[|\pi\, 10^+\rangle \otimes |\nu\, 0h^{-1}_{11/2}\rangle]^{(27/2)}$ and $[|\pi\, 8^+\rangle \otimes |\nu\, 0h^{-1}_{11/2}\rangle]^{(27/2)}$ may couple to each other. However, even if we take into account this coupling of proton configurations, we can derive a formula both for the 17^+ isomers of the $N = 83$ nuclei and for the $27/2^-$ isomers of the $N = 81$ nuclei [13], such as

$$\langle Z\, J^\pi_f || T(E2) || Z\, J^\pi_i \rangle = T_p\, \frac{\Omega_{0h_{11/2}} - \langle N_{\pi 0h_{11/2}} \rangle}{\Omega_{0h_{11/2}} - \nu} + T_n\,. \tag{7}$$

Here we have assumed that the orbit occupied by the last neutron (or neutron-hole) is unique. T_p and T_n denote appropriate proton and neutron matrix elements which are almost independent of Z. In practice, T_p and T_n are sums of several terms, including the mixing amplitudes of the proton configurations. Eq. (7) clarifies the Z-dependence of the $E2$ strengths from the isomers. The minimum of the $B(E2)$ values with respect to Z may be shifted from $Z \simeq 70$ in the $N = 82$ nuclei, because of the almost constant neutron contribution. However, the amount of the shift depends on the mixing amplitudes of the proton configurations. This point was not taken into consideration in the previous analysis.

In Fig. 6, we compare the shell model results of the $B(E2)$ values from the 17^+ isomers of the $N = 83$ nuclei with the experimental data. We present the results for the $27/2^-$ isomers of the $N = 81$ nuclei in Fig. 7. We have fixed

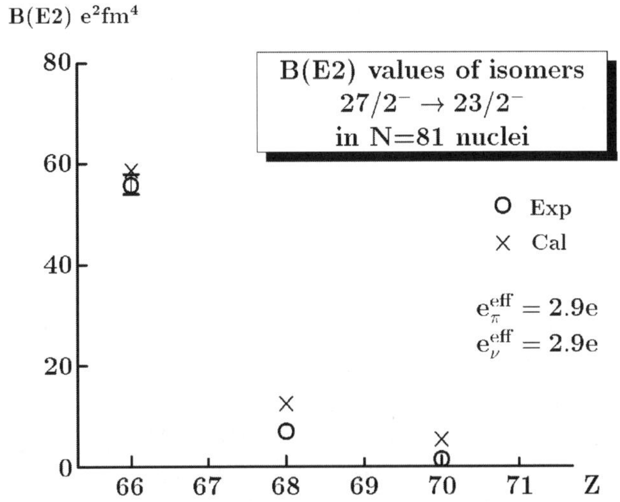

Figure 7. $B(E2)$ values from $27/2^-$ isomers in $Z > 64$, $N = 81$ nuclei.

the proton effective charge $e_\pi^{\rm eff}$ to the value obtained in the $N = 82$ nuclei, and adjusted $e_\nu^{\rm eff}$ to the measured $B(E2)$ values of these isomers in the $N = 81$ and 83 nuclei. The observed Z-dependence of the $B(E2)$ values is reproduced for both cases. It is remarked that $e_\nu^{\rm eff}$ is the same for the $N = 81$ and 83 cases. Although $e_\nu^{\rm eff}$ looks relatively large and its origin is still an open problem, the anomaly in $e_\nu^{\rm eff}$ is greatly reduced.

6. Summary

We have reinvestigated the structure of isomers in $N \sim 82$ nuclei. The quasi-spin structure survives in the isomers and the final states of their decays as a partial dynamical symmetry. This symmetry is quite useful to clarify the physics relevant to the isomers and their decays. Via the quasi-spin group structure, an extended seniority reduction formula has been derived. This reveals the presence of pair excitations out of the $Z = 64$ core, contrary to the earlier model assuming a single configuration. Effects of the core excitation in the isomers of the $N = 81$ and 83 nuclei have also been investigated. The anomaly in neutron effective charges in their $E2$ decays is greatly reduced by the effect of the core excitation.

Acknowledgments

The authors are grateful to W. Bentz for careful reading of the manuscript. This work is financially supported as Grant-in-Aid for Scientific Research (C), No. 13640263, by the Ministry of Education, Culture, Sports, Science and Technology, Japan.

References

[1] Alhassid, Y., and A. Leviatan: 1992, *Journal of Physics*, A **25**, p. L1265

[2] Andreozzi, F., A. Covello, A. Gargano, and A. Porrino: 1992, *Physical Review*, C **45**, p. 2008

[3] Arima, A. and M. Ichimura: 1966, *Progress of Theoretical Physics*, **36**, p. 296

[4] Blomqvist, J.: *International review of nuclear physics*, vol. 1, p. 1. World Scientific, Singapore, 1984.

[5] Broda, R. *et al.*: 1985, *Zeitschrift für Physik*, A **321**, p. 287

[6] Broda, R. *et al.*: 1987, *Zeitschrift für Physik*, A **327**, p. 403

[7] Daly, P.J. *et al.*: 1980, *Zeitschrift für Physik*, A **298**, p. 173

[8] Helppi, H. *et al.*: 1982, *Physics Letters*, B **115**, p. 11

[9] Kerman, A.K.: 1961 *Annals of Physics*, **12**, p. 300

[10] Kleinheinz, P.K. *et al.*: 1979, *Zeitschrift für Physik*, A **290**, p. 279

[11] Lawson, R.D.: 1981, *Zeitschrift für Physik*, A **303**, p. 51

[12] Matsuzawa, T., H. Nakada, K. Ogawa, and G. Momoki: 2000, *Physical Review*, C **62**, 054304

[13] Matsuzawa, T. *Shell structure in proton-rich nuclei around* $Z = 64$. Ph.D. thesis, Chiba University, Japan, 2002.

[14] McNeill, J.H. *et al.*: 1993, *Zeitschrift für Physik*, A **344**, p. 369

[15] McNeill, J.H. *et al.*: 1986, *Zeitschrift für Physik*, A **325**, p. 27

[16] McNeill, J.H. *et al.*: 1990, *Zeitschrift für Physik*, A **335**, p. 241

[17] Nisius, D. *et al.*: 1993, *Physical Review*, C **47**, p. 1929

[18] Nisius, D. *et al.*: 1995, *Physical Review*, C **52**, p. 1355

[19] Rosensteel, G. and D.J. Rowe: 2003, *Physical Review*, C **67**, 014303

IRREVERSIBLE WAVE PACKET DYNAMICS IN A MAGNETIC FIELD

D. Schuch
Institut für Theoretische Physik,
Robert-Mayer-Strasse 8-10,
D-60054 Frankfurt am Main, Germany
Schuch@em.uni-frankfurt.de

Abstract It has been shown that the irreversible dynamics of a system in a dissipative environment can be described by an effective one-particle Hamiltonian (in the classical case) or an effective one-particle Schrödinger equation (in the quantum mechanical case) without taking into account the environmental degrees of freedom explicitly. In the quantum mechanical case, the description can be achieved either by a nonlinear modification of the Schrödinger equation or an explicitly time-dependent Hamiltonian operator - where both forms are connected via a non-unitary transformation, corresponding to a non-canonical transformation on the classical level. Taking into account an external magnetic field, in both formalisms, an additional unphysical gauge-dependent term occurs in the equations of motion. This shortcoming can be eliminated if the Lagrangian and Hamiltonian that provide Maxwell's equations of electrodynamics undergo, in the dissipative case, the same kind of transformation as the mechanical degrees of freedom of the system. As a result, a consistent description in terms of exact analytical wave packet-solutions for the dissipative motion in a magnetic field can be obtained that will be discussed in detail.

Keywords: Dissipation, magnetic field, dissipative Hamiltonians, nonlinear Schrödinger equations, wave packet dynamics

Abbreviations: SE - Schrödinger equation;
NLSE - Nonlinear Schrödinger equation
WP - Wave packet

1. Introduction

Classical mechanics in its Lagrangian and Hamiltonian forms is a very fundamental theory - not only because also other essential physical theories, e.g., electrodynamics, can be formulated in Lagrangian and Hamiltonian ways - but,

B. Gruber (ed.), Symmetries in Science XI, 471–490.

also, because the underlying equations of quantum mechanics can be obtained from this form of classical mechanics. However, classical Hamiltonian mechanics, as well as quantum mechanics, are reversible theories, i.e., the direction of time does not matter; whereas the world surrounding us obviously prefers one direction of time: everything - including ourselves - is ageing; we observe an "arrow of time". Therefore, the question of how to resolve this obvious discrepancy between the observable, macroscopic world and fundamental theories governing the behaviour of the microscopic world has attracted the interest of scientists for a long time and many different answers describing classical and quantum mechanical systems with broken time-reversal symmetry have been given - some more satisfactory, others less.

In this paper, the irreversible dynamics of a system interacting with a dissipative environment that exerts a frictional force proportional to velocity shall be considered. Further, the description of the dissipative system shall be given by an effective one-body Hamiltonian (classical) or one-body Schrödinger equation (SE) (quantum mechanical) for the system alone, including the frictional effect of the environment, but not its details.

It has been shown that it is possible to reach this goal with explicitly time-dependent Hamiltonians [7, 8] or nonlinear modifications of the SE [9, 10], where both formalisms are uniquely connected via a non-canonical (classical) or non-unitary (quantum mechanical) transformation [11, 12, 13].

New problems emerge when a magnetic field is included. In both formalisms, apparently unphysical, gauge-dependent, additional terms occur in the equations of motion or in the potential. This caused, e.g., Wagner [14] to reject the method using time-dependent Hamiltonians on physical grounds. It will be shown in the following how this shortcoming can be eliminated and a consistent description of dissipative systems can be reached - also with the presence of a magnetic field - what the corresponding quantum mechanical solutions look like and what properties they possess.

The paper is arranged as follows: in Section 2, a short description of the two effective formalisms for the description of dissipative systems without magnetic field is given; in Section 3, a constant magnetic field is included, the corresponding changes in the formalisms and the resulting problems are discussed; in Section 4, it is shown how the problems can be solved giving also the electromagnetic field a proper and consistent treatment; the correct quantum mechanical solutions for the motion in a magnetic field are given in Section 5 and their properties discussed in Section 6. Finally, some conclusions and perspective are presented.

2. Effective Descriptions of Dissipative Systems without Electromagnetic Fields

2.1 Modified Hamiltonian Formalism

The most frequently-used, explicitly time-dependent approach for the description of dissipative, frictionally-damped systems is the one by Caldirola [7] and Kanai [8]. They start with the Lagrangian

$$\bar{L}_{CK} = e^{\gamma t}\left[\frac{m}{2}\dot{x}^2 - V(x)\right] = e^{\gamma t}\, L \tag{1}$$

which, via the Euler-Lagrange equation, yields an equation of motion including a linear velocity-dependent frictional force,

$$m\ddot{x} + m\gamma\dot{x} + \frac{\partial}{\partial x}V(x) = 0\ , \tag{2}$$

with the friction coefficient γ.

From this Lagrangian, the canonical momentum $\bar{p}$ is obtained in the usual way via

$$\frac{\partial}{\partial \dot{x}}\bar{L}_{CK} = m\dot{x}e^{\gamma t} = pe^{\gamma t} = \bar{p}\ . \tag{3}$$

It is important to realize that the canonical momentum $\bar{p}$ is not only different from the physical (kinetic) momentum $p = m\dot{x}$, but, in particular, the transition from the physical variables x and $p = m\dot{x}$ to the canonical variables $\bar{x} = x$ and $\bar{p} = pe^{\gamma t}$ represents a **non − canonical** transformation.

With the canonical momentum, the Hamiltonian corresponding to $\bar{L}_{CK}$ can be obtained in the usual way as

$$\bar{H}_{CK} = \frac{1}{2m}e^{-\gamma t}\,\bar{p}^2 + e^{\gamma t}V(x)\ , \tag{4}$$

which yields the proper equations of motion equivalent to the Newton-type equation (2).

The transition to quantum mechanics is achieved - in position space - replacing the **canonical** momentum by a differential operator according to

$$\bar{p} \to \bar{p}_{op} = \frac{\hbar}{i}\frac{\partial}{\partial x}. \tag{5}$$

Substituting this into the Hamiltonian $\bar{H}_{CK}$, finally leads to the modified SE

$$\begin{aligned} i\hbar\frac{\partial}{\partial t}\bar{\Psi}_{CK}(x,t) &= \bar{H}_{CK,op}\,\bar{\Psi}_{CK}(x,t) \\ &= \left\{e^{-\gamma t}\left(-\frac{\hbar^2}{2m}\frac{\partial^2}{\partial x^2}\right) + e^{\gamma t}V\right\}\bar{\Psi}_{CK}(x,t)\ . \end{aligned} \tag{6}$$

An apparent violation of the uncertainty principle - connected with the form of the **kinetic** momentum - has been criticized by several authors [15, 16, 17], but, it was possible to show [11] that the unphysical result is due to a wrong interpretation of the wave function $\bar{\Psi}$ which neglects the fact that the aforementioned non-canonical transformation of the variables is connected with a non-unitary transformation of the wave functions.

2.2 Modified Method of Madelung and Mrowka

In order to circumvent possible problems connected with Hamiltonians for dissipative systems and their quantization, several attempts were made adding a friction-term directly to the Hamiltonian operator [18, 19, 20, 21]. Most of these approaches use - as a guideline to finding the proper form of the additional term - Ehrenfest's Theorem, in the sense that this term should provide the above-mentioned frictional force in the equation of motion for the mean values. However, this does not allow for a unique definition of the the friction-term; therefore, several different ones are proposed in the literature, each with some physical shortcoming.

We tried to avoid this ambiguity by breaking the time-symmetry not on the level of the averaged equation of motion, but on the level of the density function [1] $\varrho(\mathbf{r},t) = \Psi^*(\mathbf{r},t)\Psi(\mathbf{r},t)$, modifying a method used by Madelung and Mrowka [22, 23] to 'rederive' the time-dependent SE without making use of the classical Hamiltonian formalism but only using Newton's equation of motion.

Madelung und Mrowka's basic idea is to separate the (real) continuity equation for the probability density $\varrho(\mathbf{r},t)$,

$$\frac{\partial}{\partial t}\varrho + \nabla\mathbf{j} = \frac{\partial}{\partial t}\varrho + \nabla(\varrho\,\mathbf{v}) = 0 \,, \tag{7}$$

into two (complex) equations for the wave function $\Psi(\mathbf{r},t)$ and its complex conjugate $\Psi^*(\mathbf{r},t)$.

The density ϱ, as well as the current density $\mathbf{j}= \varrho\mathbf{v}$, were assumed - in analogy to optics - to be bilinear forms of some complex field amplitudes $\alpha(\mathbf{r},t)$ and $\beta(\mathbf{r},t)$ (with $\beta = \alpha^*$),

$$\begin{aligned} \varrho &= \alpha\cdot\beta \\ \mathbf{j} &= C(\beta\nabla\alpha - \alpha\nabla\beta). \end{aligned} \tag{8}$$

Separation can be achieved if a separation function $f(\mathbf{r},t)$ is introduced,

[1] Bold face quantities denote vectors.

$$\dot{\alpha} + C\Delta\alpha + f\alpha = 0 \qquad (9)$$
$$\dot{\beta} - C\Delta\beta - f\beta = 0.$$

The physical meaning of the separation function becomes obvious when Eqs.(9) are used to determine the mean value of Newton's equation of motion,

$$\begin{aligned} < \mathbf{F} > &= m\frac{d}{dt} < \mathbf{v} >= m\frac{d}{dt}\int 2C\left(\dot{\beta}\nabla\alpha - \dot{\alpha}\nabla\beta\right) d\mathbf{r} \\ &= m\int \varrho\left(-\nabla(2Cf)\right) d\mathbf{r}. \end{aligned} \qquad (10)$$

This yields $f = \frac{1}{2mC}V$, where V is the potential of the problem and the constant C can be determined by comparison with, e.g., spectroscopic experiments to be $C = \hbar/2mi$. Inserting all of this into Eqs.(9) and replacing α by Ψ, the first of Eqs.(9) is just the time-dependent SE,

$$\begin{aligned} i\hbar\frac{\partial}{\partial t}\Psi(\mathbf{r},t) &= \left(-\frac{\hbar^2}{2m}\Delta + V(\mathbf{r})\right)\Psi(\mathbf{r},t) \\ &= \tilde{H}_{CK,op}\bar{\Psi}(\mathbf{r},t) = H_L\,\Psi(\mathbf{r},t)\ . \end{aligned} \qquad (11)$$

Our modification consists of an additional diffusion term to the continuity equation, thus arriving at the Fokker-Planck-type equation

$$\frac{\partial}{\partial t}\varrho + \nabla(\varrho\,\mathbf{v}) - D\Delta\varrho = 0 \qquad (12)$$

with diffusion coeffecient D.

Due to the diffusion term, a separation, in general, is no longer possible; however, it can be achieved if the additional condition

$$-D\frac{\Delta\varrho}{\varrho} = \gamma\left(\ell n\varrho - < \ell n\varrho >\right) \qquad (13)$$

where $< \ldots >$ denote mean values calculated according to $< \ldots >= \int \Psi^* \ldots \Psi d\mathbf{r}$, is fulfilled. After separation, the SE contains an additional logarithmic nonlinear term, (the logarithmic NLSE),

$$i\hbar\frac{\partial}{\partial t}\Psi = \left\{H_L + \gamma\frac{\hbar}{i}\left(\ell n\Psi - < \ell n\Psi >\right)\right\}\Psi\ . \qquad (14)$$

When Ehrenfest's theorem is considered, the physical meaning of the nonlinear term becomes obvious as the logarithmic term gives rise to the additional friction-term $m\gamma <\dot{\mathbf{r}}>$ in the classical equation of motion.

The connection with the afore-mentioned time-dependent approach of Cal - di - ro - la and Kanai can be found using Schrödinger's original definition of action [24],

$$S_c = \frac{\hbar}{i} \ell n \Psi, \tag{15}$$

(note that S_c is complex if Ψ is complex) which allows for writing the SE (divided by Ψ) as a Hamilton-Jacobi-Equation

$$\frac{\partial}{\partial t} S_c + H\left(\mathbf{r}, \nabla S_c, t\right) \;=\; 0\,. \tag{16}$$

The same form can be achieved for the NLSE (14), if the transformed action and Hamiltonian

$$\bar{S}_c \;=\; e^{\gamma t} S_c \quad \text{and} \quad \bar{H} \;=\; e^{\gamma t} H \tag{17}$$

are introduced,

$$\frac{\partial}{\partial t} \bar{S}_c + \bar{H} \;=\; 0\;. \tag{18}$$

The operator corresponding to $\bar{H}$ is then just the Caldirola-Kanai operator, $\bar{H}_{op}$, and the above-mentioned non-unitary connection between the wave-functions is given by

$$\bar{S}_c \;=\; \ell n \bar{\Psi} \;=\; e^{\gamma t} \ell n \Psi \;=\; e^{\gamma t} S_c\;; \tag{19}$$

for further details, see [11, 12, 13].

3. Motion of a Charged Particle in a Constant Magnetic Field, without and with Dissipation

3.1 Problematic Aspects of the Modified Hamiltonian

The classical Lagrangian function for the motion in a magnetic field is given by

$$L = \frac{m}{2} \mathbf{v}^2 + \frac{q}{c} \mathbf{v} \cdot \mathbf{A} - V(\mathbf{r}) \tag{20}$$

with the position vector $\mathbf{r} = (x_1, x_2, x_3)$, the corresponding velocity $\mathbf{v} = (\dot{x}_1, \dot{x}_2, \dot{x}_3)$, the charge q of the particle and the velocity of light c. We assume the direction of the magnetic field is that of x_3, $\mathbf{B} = (0, 0, B)$ and the vector potential has the form $\mathbf{A} = \frac{1}{2}(\mathbf{B} \times \mathbf{r})$. The motion in the x_3-direction will be that of a free particle and, therefore, suppressed in the further treatment of the problem; in this case $\mathbf{r}$ will have the meaning $\mathbf{r}_\perp = (x_1, x_2)$.

With the canonical momentum

$$\frac{\partial L}{\partial x_i} = p_i = m\dot{x}_i + \frac{q}{c} A_i, \tag{21}$$

the corresponding Hamiltonian can be written as

$$H = \frac{1}{2m}(\mathbf{p} - \frac{q}{c}\mathbf{A})^2 + V(\mathbf{r}) \quad , \tag{22}$$

which yields the equations

$$\ddot{x}_1 - \omega_c \dot{x}_2 = 0, \qquad \ddot{x}_2 + \omega_c \dot{x}_1 = 0 \tag{23}$$

for the motion in the (x_1, x_2)-plane, where $\omega_c = qB/mc$ is the cyclotron frequency. Eqs.(23) describe the motion on a circle with constant radius and constant angular velocity $-\omega_c$.

Including the frictional force, we assume the same transformed Hamil - ton-Jacobi-equation to be valid as in the case without magnetic field but, now, with the transformed Hamiltonian function

$$\bar{H} = \frac{1}{2m} e^{-\gamma t}(\bar{\mathbf{p}} - e^{\gamma t}\frac{q}{c}\mathbf{A})^2 + e^{\gamma t}V(\mathbf{r}) = \bar{T} + \bar{V}, \tag{24}$$

again with $\bar{p_i} = e^{\gamma t} p_i$.

The corresponding equations of motion (with $V = q\Phi$; Φ: electric potential) are:

$$\ddot{x}_1 - \omega_c \dot{x}_2 + \gamma \dot{x}_1 + \gamma \frac{q}{mc} A_1 + \frac{q}{m}\frac{\partial}{\partial x_1}\Phi = 0, \tag{25}$$

$$\ddot{x}_2 + \omega_c \dot{x}_1 + \gamma \dot{x}_2 + \gamma \frac{q}{mc} A_2 + \frac{q}{m}\frac{\partial}{\partial x_2}\Phi = 0.$$

These equations depend explicitly on the vector potential $\mathbf{A}$ and are, therefore, not gauge-invariant - which led to the aforementioned rejection of this approach in the presence of magnetic fields.

3.2 Problematic Aspects of the Modified Madelung-Mrowka Me - thod

Without dissipation, the inclusion of a magnetic field in the Madelung-Mrowka method can be achieved by simply changing the definition of the probability current into

$$\mathbf{j} = \varrho \mathbf{v} = C(\beta \nabla \alpha - \alpha \nabla \beta) - \mathbf{a}\varrho \tag{26}$$

with a constant vector $\mathbf{a}$ still being a bilinear form in α and β. Separation of the continuity equation now leads to the equation

$$\dot{\alpha} + C\Delta\alpha - (\mathbf{a}\nabla\alpha) - \frac{\alpha}{2} div\mathbf{a} + f\alpha = 0 \tag{27}$$

for α, which, inserted into the averaged equation of motion, yields for the mean value of the force

$$< \mathbf{F} >= m \int \varrho \left\{ -\nabla(2Cf - \frac{1}{2}a^2) - \dot{\mathbf{a}} + \mathbf{v} \times rot\mathbf{a} \right\} d\mathbf{r}. \tag{28}$$

In this case, the separation constant has the form

$$f = \frac{1}{2mC} q\Phi(\mathbf{r}) + \frac{a^2}{4C} = \frac{i}{\hbar} \left(q\Phi + \frac{1}{2m} \frac{q^2}{c^2} \mathbf{A}^2(\mathbf{r}) \right), \tag{29}$$

where C has the same form as before and $\mathbf{a}$ is proportional to the vector potential $\mathbf{A}$ according to $\mathbf{a} = \frac{q}{mc}\mathbf{A}(\mathbf{r})$.

Using the relation

$$\frac{1}{c}\frac{\partial}{\partial t}\mathbf{A} = -\mathbf{E} - \nabla\Phi, \tag{30}$$

equivalent to one of Maxwell's homogeneous equations (with **E**=electric field vector), Eq.(28) can be written as

$$< \mathbf{F} >= \int \varrho \left\{ q\mathbf{E} + \frac{q}{c}(\mathbf{v} \times \mathbf{B}) \right\} d\mathbf{r} \tag{31}$$

which is equivalent to Eqs.(23); and the SE corresponding to Eq.(27) is given by

$$i\hbar \frac{\partial}{\partial t}\Psi = \left\{ \frac{1}{2m} \left(\frac{\hbar}{i}\nabla - \frac{q}{c}\mathbf{A} \right)^2 + q\Phi \right\} \Psi = H_{L,M}\Psi \quad . \tag{32}$$

Considering the dissipative case, the change for the current is the same as in (26), only the diffusion term in the Fokker-Planck equation leads - together with the condition of separability (13) - to the additional terms $\gamma(\ell n\alpha + Z)\alpha$ on the lhs of Eq.(27). Without magnetic field, the choice $Z = - < \ell n\alpha >$ was sufficient to guarantee conservation of probability and fulfilment of all physical requirements. Here, the complex quantity $Z = Z_R + iZ_I$, or, more precisely, the gradient of its imaginary part, must fulfil an additional condition since the mean value of the force is now given by

$$< \mathbf{F} >= \int \varrho \left\{ q\mathbf{E} + \frac{q}{c}(\mathbf{v} \times \mathbf{B}) - m\gamma\mathbf{v} \right\} d\mathbf{r} - \gamma \int \varrho \left\{ \frac{q}{c}\mathbf{A} + \hbar\nabla Z_I \right\} d\mathbf{r}. \tag{33}$$

The first integral on the rhs of Eq.(33) provides the desired equation of motion but the second integral must vanish! Without magnetic field, $\nabla Z_I = 0$ is valid; with magnetic field, however, this must be modified. The choice $\nabla Z_I = -\frac{q}{\hbar c}\mathbf{A}(\langle \mathbf{r} \rangle)$ with $< \mathbf{r} >=$ classical trajectory allows for obtaining the correct classical equations of motion but introduces an additional term into the NLSE that now reads

$$i\hbar\frac{\partial}{\partial t}\Psi = \left\{ H_{L,M} + \gamma\frac{\hbar}{i}\left(\ell n\Psi - \langle \ell n\Psi\rangle\right) - \gamma\frac{q}{c}\left(\mathbf{A}(\langle\mathbf{r}\rangle)\cdot\mathbf{r}\right)\right\}\Psi \ . \qquad (34)$$

What is the origin and physical meaning of this additional term $-\gamma\frac{q}{c}\mathbf{A}(\langle\mathbf{r}\rangle)\cdot\mathbf{r}$, apart from the fact that it is necessary to guarantee the validity of the averaged equations of motion, including a frictional force proportional to velocity? For $\nabla Z_I = 0$, the frictional force would be proportional to momentum, $\mathbf{p} = m\mathbf{v} + \frac{q}{c}\mathbf{A}$, instead of velocity, thus depending on the vector potential $\mathbf{A}$ and is, therefore, not gauge-invariant, facing the same problem as described in Section 3.1 for the other approach. The answer to the aforementioned questions will be given in the next section. NB: the mean value of the additional term in the NLSE (34) vanishes according to

$$\begin{aligned} <\mathbf{A}(\langle\mathbf{r}\rangle)\cdot\mathbf{r}> &= \mathbf{A}(\langle\mathbf{r}\rangle)<\mathbf{r}> \\ &= \frac{B}{2}(-<x_2><x_1>+<x_1><x_2>) = 0. \qquad (35)\end{aligned}$$

4. Inclusion of the Electromagnetic Field Aspect

The energy of a system in a magnetic field - and thus its Hamiltonian - has not only a contribution from the mechanical aspect but, also, from the electric and magnetic fields. These can be taken into account by

$$H_{field} = \int d\mathbf{r} \quad \mathcal{H}_{field} \qquad (36)$$

with the Hamiltonian density

$$\mathcal{H}_{field} = \frac{1}{8\pi}(\mathbf{E}^2 + \mathbf{B}^2) \qquad (37)$$

and the total Hamiltonian is given by

$$H_{tot} = H + H_{field}. \qquad (38)$$

From the Lagrangian corresponding to the field contribution,

$$L_{field} = \int d\mathbf{r}\mathcal{L}_{field}$$

with $\mathcal{L}_{field} = \frac{1}{8\pi}(\mathbf{E}^2 - \mathbf{B}^2)$, Maxwell's equations can be derived [25], where the homogeneous ones are equivalent to

$$\frac{1}{c}\frac{\partial}{\partial t}\mathbf{A} = -\mathbf{E} - \nabla\Phi, \quad \mathbf{B} = \nabla \times \mathbf{A}. \tag{39}$$

Transition to the canonical level, including dissipation, should also change, according to (17), $\mathcal{H}_{field}$ into $\bar{\mathcal{H}}_{field}$,

$$\bar{\mathcal{H}}_{field} = e^{\gamma t}\,\mathcal{H}_{field}. \tag{40}$$

On the canonical level, the form of the homogeneous Maxwell's equations (39) should be unchanged, i.e.,

$$\frac{1}{c}\frac{\partial}{\partial t}\bar{\mathbf{A}} = -\bar{\mathbf{E}} - \nabla\bar{\Phi}, \quad \bar{\mathbf{B}} = \nabla \times \bar{\mathbf{A}} \tag{41}$$

should be valid.

Since - apart from the electric charge q - the electric potential Φ can be identified with the potential V if no external potentials are present, it follows from Eq.(24) that

$$q\bar{\Phi} = \bar{V} = e^{\gamma t}V = e^{\gamma t}q\Phi. \tag{42}$$

Inserting this into the first of Eqs. (41), a way to keep these equations consistent is, therefore, to also define $\bar{\mathbf{E}} = e^{\gamma t}\mathbf{E}$, $\bar{\mathbf{B}} = e^{\gamma t}\mathbf{B}$ and $\bar{\mathbf{A}} = e^{\gamma t}\mathbf{A}$. This implies that

$$\bar{\mathcal{L}}_{field} = \frac{1}{8\pi}e^{\gamma t}(\mathbf{E}^2 - \mathbf{B}^2) = e^{\gamma t}\mathcal{L}_{field} \tag{43}$$

which is also in agreement with the Caldirola-Kanai form for a dissipative Lagrangian of a mechanical system.

The Hamiltonian density $\bar{\mathcal{H}}_{field}$ can then be written in a form similar to $\bar{T}$, namely

$$\bar{\mathcal{H}}_{field} = \frac{1}{8\pi}e^{-\gamma t}(\bar{\mathbf{E}}^2 + \bar{\mathbf{B}}^2) \tag{44}$$

and the mechanical Hamiltonian takes the form

$$\bar{H} = \frac{1}{2m}e^{-\gamma t}(\bar{\mathbf{p}} - \frac{q}{c}\bar{\mathbf{A}})^2 + q\bar{\Phi}. \tag{45}$$

This Hamiltonian is identical with the one in Eq.(24) and would yield the same problematic equations of motion (25), including the terms depending on the vector potential. However, from Eq.(41), follows now (in the absence of an external electric field $\mathbf{E}$) that

$$\nabla\bar{\Phi} = e^{\gamma t}\nabla\Phi = -\frac{1}{c}\frac{\partial}{\partial t}\bar{\mathbf{A}} = -\frac{1}{c}\frac{\partial}{\partial t}(e^{\gamma t}\mathbf{A}). \tag{46}$$

For a time-independent vector potential **A** on the physical level, i.e. $\frac{\partial}{\partial t}\mathbf{A} = 0$, it follows with

$$\frac{\partial}{\partial t}\bar{\mathbf{A}} = \gamma\bar{\mathbf{A}} = \gamma e^{\gamma t}\mathbf{A} \tag{47}$$

that

$$\frac{\partial}{\partial x_i}\Phi = -\frac{1}{c}\gamma A_i. \tag{48}$$

Therefore, the last two terms on the lhs of Eqs.(25) cancel and we obtain the desired equations independent of the vector potential,

$$\ddot{x}_1 - \omega_c\dot{x}_2 + \gamma\dot{x}_1 = 0, \qquad \ddot{x}_2 + \omega_c\dot{x}_1 + \gamma\dot{x}_2 = 0. \tag{49}$$

Note that, without dissipation, i.e. for $\gamma = 0$, the term $\nabla\bar{\Phi}$ and, consequently, $\nabla\Phi$ disappears, i.e. Φ and, thus, V on the physical level is just a constant (with respect to spatial variables) and will be chosen to be zero in the following.

From Eq.(48) follows, with $V = q\Phi$, that agreement with the **classical** equations for the scalar and vector potential can be achieved, if the potential $\bar{V}$ fulfils

$$-\frac{\partial}{\partial x_i}\bar{V} = -e^{\gamma t}\frac{\partial}{\partial x_i}V = \gamma\frac{q}{c}\bar{A}_i = \gamma e^{\gamma t}\frac{q}{c}A_i. \tag{50}$$

Since, so far, the electromagnetic contribution is treated only on a classical level, the occurring variables are represented quantum mechanically by their respective mean values. So, the potential $\bar{V}$ that enters the quantum mechanical description on the **canonical level** can be given (up to a purely time-dependent additive term) as

$$\bar{V} = -\gamma\frac{q}{c}\mathbf{A}(<\mathbf{r}>)\ \cdot\ \mathbf{r}e^{\gamma t}, \tag{51}$$

which corresponds on the **physical level** to the additional potential-term

$$V = -\gamma\frac{q}{c}\mathbf{A}(<\mathbf{r}>)\ \cdot\ \mathbf{r}. \tag{52}$$

This is, however, exactly the term that had to be added in the log NLSE (34) in order to obtain the proper equations of motion for the mean values!

Therefore, after the aforementioned non-unitary transformation, the Caldirola-Kanai Hamiltonian (including $\bar{V}$) and the log NLSE (including V) describe the same physical problem of a frictionally-damped motion - also in the presence of a constant magnetic field.

5. Quantum Mechanical Solutions in a Magnetic Field, without and with Dissipation

5.1 Solutions without Dissipation

From the time-dependent quantum mechanical problem of the two-di - men - sio - nal motion in a magnetic field without dissipation, as described by the SE (32), the corresponding time-independent problem can be obtained via the product ansatz $\Psi(\mathbf{r},t) = \psi_N(\mathbf{r})\exp(-\frac{i}{\hbar}E_N t)$. The eigenfunctions $\psi_N(\mathbf{r})$, belonging to the energy eigenvalues $E_N = (N+\frac{1}{2})\hbar\omega_c$ with integer quantum numbers N, were called Schrauben functions by Jannussis [26] and are of the form

$$\begin{aligned}\psi_N(\mathbf{r}) &= \left(\frac{b}{2\pi}\left(\frac{b}{2}\right)^N\frac{1}{N!}\right)^{\frac{1}{2}} \\ &\exp\left\{\frac{iq}{\hbar c}\mathbf{A}(\mathbf{r}_0)\mathbf{r} - \frac{b}{4}(\mathbf{r}-\mathbf{r}_0)^2\right\} \mid \mathbf{r}-\mathbf{r}_0\mid^N e^{-iN\phi} \quad (53)\end{aligned}$$

where $\phi = arctan(x_2 - x_{02}/x_1 - x_{01})$, $b = qB/\hbar c = m\omega_c/\hbar$ with cyclotron frequency ω_c.

An elegant way of obtaining these eigenfunctions has been shown by Moshinsky [21]. Rewriting the Hamiltonian of Eq.(32) (in components), using atomic units $(m = q = \hbar = 1)$ and $b' = \frac{\hbar}{2m}b = \frac{\omega_c}{2}$, yields

$$H_\perp = \frac{1}{2}[(p_1^2+p_2^2) + b'^2(x_1^2+x_2^2)] + b'[-(x_1p_2 - x_2p_1)] \quad . \quad (54)$$

This $H_\perp$ can be expressed in terms of circular creation and annihilation operators,

$$\eta_\pm = \frac{1}{\sqrt{2}}\left(\sqrt{b'}x_\pm - \frac{1}{\sqrt{b'}}\frac{\partial}{\partial x_\mp}\right), \quad \xi_\pm = \frac{1}{\sqrt{2}}(\sqrt{b'}x_\mp + \frac{1}{\sqrt{b'}}\frac{\partial}{\partial x_\pm}) , \quad (55)$$

using circular coordinates $x_\pm = \frac{1}{\sqrt{2}}(x_1 \pm ix_2)$, as

$$H_\perp = b'[(\eta_+\xi_+ + \eta_-\xi_- + 1) - (\eta_+\xi_+ - \eta_-\xi_-)] = b'(2\eta_-\xi_- + 1) . \quad (56)$$

With the help of $\xi_\pm \mid 0 >= 0$, one can obtain the ground state solution $\mid 0 >$ and, applying the creation operators, also all the excited states as:

$$\begin{aligned}|N, z_0 >= b'^{\frac{1}{2}}(\pi N!)^{-\frac{1}{2}}b'^{\frac{N}{2}}[(x_1-x_{01}) + i(x_2-x_{02})]^N \\ \exp\{-(b'/2)[(x_1-x_{01})^2 + (x_2-x_{02})^2]\}\exp[ib'(x_{02}x_1 - x_{01}x_2)] , \quad (57)\end{aligned}$$

the so-called **coherent states**, centered at (x_{01}, x_{02}), where

$$z_0 = \sqrt{b'}(x_{01} + ix_{02}).$$

For $b' = \frac{\hbar}{2m}b$, they are (up to the sign of the phase factor) identical with the Schrauben functions.

By superpositioning the Schrauben functions or coherent states, solutions in the form of Gaussian wave packets (WPs) can be obtained - if the initial state is chosen as a Gaussian [22], i.e.,

$$\begin{aligned} \Psi_{WP}(t) &= e^{-i\frac{\omega_c}{2}t} \sum_{N=0}^{\infty} A_N |N, z_0 > e^{-iN\omega_c t} \\ &= \left(\frac{b}{2\pi}\right)^{\frac{1}{2}} \exp\left\{-\frac{b}{4}(\mathbf{R} - \mathbf{a}(t))^2 \right. \\ &\quad \left. + \frac{i}{\hbar}\left[\left(m\langle\dot{\mathbf{r}}\rangle - \frac{q}{c}\mathbf{A}(\langle\mathbf{r}\rangle)\right)(\mathbf{R} - \mathbf{a}(t)) + K(t)\right]\right\} \end{aligned} \tag{58}$$

where $\mathbf{R} = \mathbf{r} - \mathbf{r}_0$, $\mathbf{R} - \mathbf{a} = \mathbf{r} - < \mathbf{r} >$, see also Fig.1.

Using the orthonormality of the coherent states and appropriate initial conditions, the coefficients A_N can be determined as

$$A_N = \left(\left(\frac{b}{2}\right)^N \frac{1}{N!}\right)^{1/2} |\mathbf{a}_0|^N \exp\left\{-\frac{b}{4}\mathbf{a}_0^2 - \frac{iq}{\hbar c}\mathbf{A}(\mathbf{r}) \cdot \mathbf{a}_0\right\}. \tag{59}$$

The corresponding density $\varrho_{WP} = \Psi^*_{WP}\Psi_{WP}$ is given by

$$\varrho_{WP}(\mathbf{r}, t) = \frac{b}{2\pi} \exp\left\{-\frac{b}{2}(\mathbf{R} - \mathbf{a}(t))^2\right\}, \tag{60}$$

i.e., it attains its maximum value along a circle with the classical radius $\mid \mathbf{a} \mid$.

5.2 Solutions with Dissipation

On the **canonical level**, without time-dependence, the Caldirola-Kanai Hamiltonian

$$\bar{H} = \frac{1}{2m} e^{-\gamma t} (\bar{\mathbf{p}} - \bar{\mathbf{A}})^2 \tag{61}$$

can be canonically quantized (i.e. $\bar{\mathbf{p}} \rightarrow \frac{\hbar}{i}\nabla$) and expressed in terms of modified creation/annihilation operators. This allows one to obtain coherent states modified accordingly. For details, see [23].

Considering the time-dependent situation on the canonical level, the potential $\bar{V} = -\gamma\mathbf{A}(< \mathbf{r} >)\cdot\mathbf{r}e^{\gamma t}$ must be included, and again time-dependent Gaussian

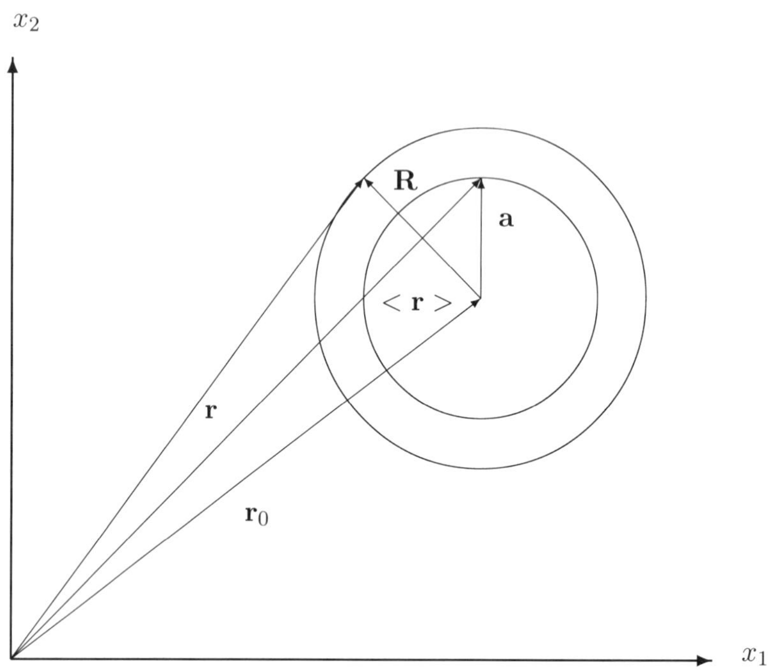

Figure 1. $\mathbf{r}$: vector of position in the (x_1, x_2) plane; $< \mathbf{r} >$: classical trajectory; $\mathbf{r}_0$: center of circular motion; $\mathbf{R} = \mathbf{r} - \mathbf{r}_0$ with $| \mathbf{R} |$ = radius of circular motion; $\mathbf{a} =< \mathbf{r} > -\mathbf{r}_0$ with $| \mathbf{a} |$ = classical radius.

WPs can be obtained as solutions; however, with complex coefficient of the quadratic term in the exponent,

$$\Psi_{WP,CK} = \bar{N}_{CK} \exp\left\{ e^{\gamma t} \left[-\frac{\bar{b}+i\gamma}{4} (\mathbf{R}-\mathbf{a}(t))^2 + \frac{i}{\hbar} \left[<\mathbf{p}> (\mathbf{R}-\mathbf{a}(t)) + e^{-\gamma t} \bar{K}(t) \right] \right] \right\} \quad (62)$$

where $\mathbf{a}(t)$ follows the classical damped trajectory, $\bar{N}_{CK} = (\bar{b}' e^{\gamma t}/\pi)^{1/2}$ with $\bar{b}' = (b'^2 - \frac{\gamma^2}{4})^{1/2} = \frac{1}{2}(b^2-\gamma^2)^{1/2} = \frac{\bar{b}}{2}$ and $\bar{K}(t)$ is again a purely time-dependent phase factor - not relevant for the following discussions.

The corresponding density is given by

$$\bar{\varrho}_{CK} = |\bar{\Psi}_{WP,CK}(\mathbf{r},t)|^2 = \bar{N}_{CK}\bar{N}^*_{CK} \exp\left\{ e^{\gamma t} \left[-\frac{\bar{b}}{2} (\mathbf{R}-\mathbf{a}(t))^2 \right] \right\} . \quad (63)$$

The WP-solution on the **physical level** can be obtained either using the non-unitary transformation

$$\ell n \Psi_{WP,NL} = e^{-\gamma t} \ell n \bar{\Psi}_{WP,CK} \quad (64)$$

with subsequent normalization of the WP $\Psi_{WP,NL}$, or by direct solution of the log NLSE

$$i\hbar \frac{\partial}{\partial t} \Psi_{WP,NL} = \left\{ \frac{1}{2m} \left(\frac{\hbar}{i} \nabla - \frac{q}{c} \mathbf{A} \right)^2 + \gamma \frac{\hbar}{i} \left(\ell n \Psi_{WP} - \langle \ell n \Psi_{WP} \rangle \right) \right\} \Psi_{WP,NL} - \left\{ \gamma \frac{q}{c} \left(\mathbf{A}(\langle \mathbf{r} \rangle) \cdot \mathbf{r} \right) \right\} \Psi_{WP,NL} \quad . \quad (65)$$

One obtains, with $b = 2b'$ and $\bar{b} = (b^2 - \gamma^2)^{1/2}$, the WP as

$$\Psi_{WP,NL} = N_{NL} \exp\left\{ -\frac{\bar{b}+i\gamma}{4} (\mathbf{R}-\mathbf{a}(t))^2 + \frac{i}{\hbar} \left[<\mathbf{p}> (\mathbf{R}-\mathbf{a}(t)) + e^{-\gamma t} \bar{K}(t) \right] \right\} \quad (66)$$

with the corresponding density

$$\varrho_{NL}(t) = |\Psi_{WP,NL}(\mathbf{r},t)|^2 = \frac{\bar{b}}{2\pi}\exp\left\{-\frac{\bar{b}}{2}\left(\mathbf{R}-\mathbf{a}(t)\right)^2\right\}, \qquad (67)$$

where $\bar{b} = \frac{m}{\hbar}(\omega_c^2 - \gamma^2)^{1/2}$.

From here, it follows that the maximum of the WP on the physical level spirals to the origin of the circular motion, $\mathbf{r}_0$, with exponentially-decreasing radius, $\mathbf{a}\,(t) = \mathbf{a}_0 e^{-\gamma t}$.

6. Properties of the Dissipative Wave Packet Solutions in a Magnetic Field

The WP-solutions of the NLSE allow a direct physical interpretation of its properties. Gaussian WPs are characterized by two parameters: their maximum and their width. As has been shown, the maximum follows the classical circular motion with exponentially-decaying radius.

The width in position space is proportional to a quantity, α, that my be time-dependent and is proportional to the mean square deviation of position according to

$$\alpha_i^2 = \frac{2m}{\hbar} < \tilde{x}_i{}^2 >= \frac{2m}{\hbar}\left(< x_i^2 > - < x_i >^2\right). \qquad (68)$$

This quantity enters the quantum mechanical contribution to the energy (if $< H >= E = E_{class} + \tilde{E}$) as

$$\tilde{E}_{NL} = \sum_{i=1,2} \frac{\hbar}{4}\left[\left(\dot{\alpha}_i - \frac{\gamma}{2}\alpha_i\right)^2 + \frac{1}{\alpha_i^2} + \left(\frac{\omega_c}{2}\right)^2 \alpha_i^2\right] \qquad (69)$$

and can also contribute to the probability current

$$\mathbf{j}_{NL} = \varrho\mathbf{v} - D\nabla\varrho = \begin{pmatrix} \dot{\eta}_1 + \left(\frac{\dot{\alpha}_1}{\alpha_1}\right)\tilde{x}_1 + \frac{\omega_c}{2}\tilde{x}_2 \\ \\ \dot{\eta}_2 + \left(\frac{\dot{\alpha}_2}{\alpha_2}\right)\tilde{x}_2 - \frac{\omega_c}{2}\tilde{x}_1 \end{pmatrix} \cdot \varrho \qquad (70)$$

(with $\tilde{x}_i = x_i - < x_i >$) if α is time-dependent.

The above-mentioned WP (66) is a particular solution of the NLSE with constant width, $\alpha^2 = (\frac{\omega^2}{4} - \frac{\gamma^2}{4}) = const.$

General WP-solutions have time-dependent coefficients of the quadratic term in the exponent. This coefficient fulfils a complex nonlinear Riccati equation that can be transformed into a (real) nonlinear Ermakov equation for α (see, e.g., [24]).

The log NLSE including magnetic field provides for each direction x_i an equation of the form

$$\ddot{\alpha}_i + \left(\frac{\omega_c^2 - \gamma^2}{4}\right)\alpha_i = \frac{1}{\alpha_i^3} \ . \tag{71}$$

Three qualitatively-different cases can be distinguished:
1) $\omega_c > \gamma$: undercritical damping (strong magnetic field);
2) $\omega_c = \gamma$: aperiodic limit ('resonance' case);
3) $\omega_c < \gamma$: overcritical damping (weak magnetic field).

In each case, different contributions to $\tilde{E}_{NL}$ and $\mathbf{j_{NL}}$ occur.

6.1 Undercritical Damping, $\omega_c > \gamma$

In this case, WP-solutions with constant width exist. Therefore, they do not provide any contribution to the current (that would be proportional to $\dot{\alpha}/\alpha$). The contribution to the quantum mechanical (ground state) energy is constant and higher than in the case without dissipative environment,

$$\tilde{E}_{NL} = 2\frac{\hbar}{4}\frac{\omega_c^2}{\Omega} = \frac{\hbar}{2}\omega_c\left(\frac{\omega_c}{\Omega}\right) > \frac{\hbar}{2}\omega_c = const \tag{72}$$

with $\Omega = (\omega_c^2 - \gamma^2)^{1/2}$. This could be interpreted as a partial backtransfer of energy from the bath to the system - just by being in contact with the environment.

But there exist also WP-solutions that show an oscillatory behaviour of the WP-width according to

$$\alpha_i^2 = \alpha_{i0}^2 \left\{ \beta_{i0}^2 \left[\frac{\sin(\frac{\Omega}{2}t)}{\frac{\Omega}{2}}\right]^2 + \cos^2(\frac{\Omega}{2}t) \right\} \ , \tag{73}$$

with the relative change in time of the width also oscillating,

$$\frac{\dot{\alpha}_i}{\alpha_i} = \frac{\left(\frac{\beta_{i0}^2}{\frac{\Omega}{2}} - \frac{\Omega}{2}\right)\cos(\frac{\Omega}{2}t)\sin(\frac{\Omega}{2}t)}{\left\{\beta_{i0}^2\left[\frac{\sin(\frac{\Omega}{2}t)}{\frac{\Omega}{2}}\right]^2 + \cos^2(\frac{\Omega}{2}t)\right\}} \ , \tag{74}$$

where $\beta_{i0} = \frac{1}{\alpha_{i0}^2}$.

This causes oscillatory contributions to the quantum energy and current. The current oscillations should lead to emission of radiation in the mm range (estimated power radiated from a small sample of solid: approx. $P = 0.03W$; this should be measurable with today's experimental methods). Further details will be published elsewhere (see [25]).

6.2 Aperiodic Limit, $\omega_c = \gamma$

Here, the Ermakov equation has the same form as that of the WP for the free motion without friction and magnetic fields, so the time-dependence of the width has the form

$$\alpha_i^2 = \alpha_{i0}^2 \left\{1 + (\beta_{i0} t)^2\right\} \tag{75}$$

and the contribution to the current, given by

$$\frac{\dot{\alpha}_i}{\alpha_i} = \beta_{i0} \frac{\beta_{i0} t}{1 + (\beta_{i0} t)^2} \;, \tag{76}$$

is Lorentzian-shaped in time, i.e., after an initial increase, it decreases till it finally approaches zero asymptotically.

Most interesting, however, is the behaviour of the quantum mechanical energy contribution - since it grows proportional to t^2, $\tilde{E}_{NL} \propto t^2$. This means, while the classical contribution of the energy connected with the motion of the WP-maximum is exponentially decreasing, there is a backtransfer of energy from the environment to the quantum mechanical degrees of the system in a resonance-type fashion. This does not contradict classical thermodynamics since this energy does not cause an acceleration of the motion of the WP-maximum. A similar situation was also obtained in the case of the damped harmonic oscillator (for details, see [26]).

6.3 Overcritical Damping, $\omega_c < \gamma$

In this case, the WP-width and the contribution to the current can be expressed with $A = \pm\sqrt{\gamma^2 - \omega_c^2}$ as

$$\alpha_i^2 = \alpha_{i0}^2 \left\{ e^{At} + \left(\frac{\beta_{i0}}{\frac{A}{2}}\right)^2 \sinh^2(\frac{A}{2} t) \right\} \tag{77}$$

and

$$\frac{\dot{\alpha}_i}{\alpha_i} = \frac{A}{2} \frac{e^{At} + \left(\frac{\beta_{i0}}{\frac{A}{2}}\right)^2 \sinh(\frac{A}{2}t)\cosh(\frac{A}{2}t)}{\left\{ e^{At} + \left(\frac{\beta_{i0}}{\frac{A}{2}}\right)^2 \sinh^2(\frac{A}{2}t) \right\}} \;. \tag{78}$$

After a steep initial increase, the contribution to the current approaches, asymptotically, a finite value that is identical with its initial value.

Also here a backtransfer of energy from the environment to the quantum mechanical degrees of freedom is possible. The time-dependence is more complicated than in the case with resonance.

7. Conclusions and Perspectives

In this paper, it has been shown that it is possible to find a consistent effective description - classical, as well as quantum mechanical - of a system interacting with a dissipative environment in a magnetic field without violating any physical principles.

Further aims are:

i) experiments to measure the predicted quantum mechanical effects (e.g. oscillating currents, backtransfer of energy from the bath, etc.). An advantage of the systems considered here (in comparison with the harmonic oscillator) is that the magnetic field, and hence ω_c, can be better and more accurately manipulated than the frequency of an harmonic oscillator;

ii) a further transformation of the canonical Hamiltonian $\bar{H}(t)$, that is explicitly time-dependent, should be found to obtain - on the canonical level - a Hamiltonian that not only provides the correct equations of motion, but, also is a constant of motion (similar to the case of the damped harmonic oscillator, see [12, 13]);

iii) an extension of the formalism that works for mechanical systems, without and with magnetic fields, to other Lagrangian/Hamiltonian systems - e.g., relativistic problems, field theories, etc. - following the same scheme: a) finding a non-canonical/non-unitary transformation of L, H to obtain the correct equations of motion and to achieve canonical quantization; b) further canonical/unitary transformation to obtain a constant of motion; c) back-transformation to the physical level to obtain the corresponding NLSE that allows for a direct physical interpretation of the quantum mechanical results.

Acknowledgments

The author would like to thank Professor Marcos Moshinsky for his suggestion to look deeper into the problem of dissipation in a magnetic field and for many stimulating and fruitful discussions during his stay with him in Mexico and would also like to express his gratitude as a recipient of the CONACYT Project 32421E that made the visit possible.

References

[1] Caldirola, P.: 1941, *Nuovo Cimento* **18**, 393

[2] Kanai, E.: 1948, *Prog. Theor. Phys.* **3**, 440

[3] Schuch, D., Chung, K.-M., and Hartmann, H.: 1983, *J. Math. Phys.* **24**, 1652

[4] Schuch, D. and Chung, K.-M.: 1986, *Int. J. Quant. Chem.* **29**, 1561

[5] Schuch, D.: 1997, *Phys. Rev. A* **55**, 935

[6] Schuch, D.: 1999, *Int. J. Quant. Chem.* **72**, 537

[7] Schuch, D.: *Symmetries in Science X*, Gruber, B. and Ramek, M. Eds., pp. 345-355. Plenum Press, New York, USA, 1998

[8] Wagner, H.-J.: 1994, *Z. Phys. B* **95**, 261

[9] Brittin, W.E.: 1950, *Phys. Rev.* **77**, 396

[10] Ray, J.R.: 1979, *Lett. Nuovo Cim.* **25**, 47

[11] Greenberger, D.M.: 1979, *J. Math. Phys.* **20**, 762

[12] Kostin, M.D.: 1972, *J. Chem. Phys.* **57**, 3589

[13] Hasse, R.W.: 1975, *J. Math. Phys.* **16**, 2005

[14] Albrecht, K.: 1975, *Phys. Lett. B* **56**, 127

[15] Gisin, N.: 1981, *J. Phys. A* **14**, 2259

[16] Madelung, E.: *Die mathematischen Hilfsmittel des Physikers*, p. 432. Springer, Berlin, 1950

[17] Mrowka, B.: 1951, *Z. Phys.* **130**, 164

[18] Schrödinger, E.: 1926, *Ann. Phys., Lpz.* **79**, 361

[19] Schwinger, J., DeRaad Jr., L.L., Milton, K.A., and Tsai, W.-Y.: *Classical Electrodynamics*, Sect. 9. Perseus Books, Reading, MA, USA, 1998

[20] Jannussis, A.: 1964, *Phys. Status Solidi* **6**, 217

[21] Loyola, G., Moshinsky M., and Szepaniak, A.: 1989, *Am. J. Phys.* **57**, 811

[22] Hartmann, H. and Chung, K.-M.: *Lecture Notes in Chemistry 7, Ion Cyclotron Resonance Spectrometry*, pp. 17-32. Springer, Berlin, 1978

[23] Schuch, D. and Moshinsky, M.: 2003, *J. Phys. A: Math Gen.* **36**, 6571

[24] Schuch, D.: 1989, *Int. J. Quant. Chem., Quant. Chem. Symp.* **23**, 59

[25] Nuñez, M., Hess, P.O., and Schuch, D.: 2003, submitted for publication

[26] Schuch, D.: 2002, *J. Phys. A.: Math. Gen.* **35**, 8615

FROM QUANTUM GROUPS TO GENETIC MUTATIONS

A. Sciarrino
Dipartimento di Scienze Fisiche, Università di Napoli "Federico II"
I.N.F.N., Sezione di Napoli - Italy
sciarrino@na.infn.it

Abstract In the framework of the crystal basis model of the genetic code, where each codon is assigned to an irreducible representation of $U_{q\to 0}(sl(2) \oplus sl(2))$, single base mutation matrices are introduced. The strength of the mutation is assumed to depend on the "distance" between the codons. Preliminary general predictions of the model are compared with experimental data, with a satisfactory agreement.

Keywords: Quantum groups, genetic code, codon, crystal basis model

1. Introduction

Among the numerous and important questions offered to the theoretical physicist by the sciences of life, the ones relative to the genetic code present a particular interest. The DNA structure and the mechanism of polypeptid fixation from codons possess appealing aspects for the theorist and, indeed, the first proposal of genetic code may be ascribed to G. Gamow [1] in 1954, less than year after the discovery of DNA by Watson and Crick. Let us briefly recall some essential features, see e.g. [2]. First the DNA macromolecule is constituted by two linear chains of nucleotides in a double helix shape. There are four different nucleotides, characterised by their bases: adenine (A) and guanine (G) (purines family), cytosine (C) and thymine (T) (pyrimidines family). Note also that an A (resp. T) base in one strand is connected, with two hydrogen bonds, to a T (resp. A) base in the other strand, while a C (resp. G) base is related to a G (resp. C) base, with three hydrogen bonds. The genetic information is transmitted via the messenger ribonucleic acid or mRNA. During this operation, called transcription, the A, G, C, T bases in one strand of the DNA are associated respectively to the U, C, G, A bases, (U denoting the uracile base) of

B. Gruber (ed.), Symmetries in Science XI, 491–503.

Table 1. The eukaryotic or standard code code. Upper labels denote different irreps.

codon	amino acid	J_H	J_V	$J_{3,H}$	$J_{3,V}$	codon	amino acid	J_H	J_V	$J_{H,3}$	$J_{V,3}$
CCC	Pro P	$3/2$	$3/2$	$3/2$	$3/2$	UCC	Ser S	$3/2$	$3/2$	$1/2$	$3/2$
CCU	Pro P	$(1/2$	$3/2)^1$	$1/2$	$3/2$	UCU	Ser S	$(1/2$	$3/2)^1$	$-1/2$	$3/2$
CCG	Pro P	$(3/2$	$1/2)^1$	$3/2$	$1/2$	UCG	Ser S	$(3/2$	$1/2)^1$	$1/2$	$1/2$
CCA	Pro P	$(1/2$	$1/2)^1$	$1/2$	$1/2$	UCA	Ser S	$(1/2$	$1/2)^1$	$-1/2$	$1/2$
CUC	Leu L	$(1/2$	$3/2)^2$	$1/2$	$3/2$	UUC	Phe F	$3/2$	$3/2$	$-1/2$	$3/2$
CUU	Leu L	$(1/2$	$3/2)^2$	$-1/2$	$3/2$	UUU	Phe F	$3/2$	$3/2$	$-3/2$	$3/2$
CUG	Leu L	$(1/2$	$1/2)^3$	$1/2$	$1/2$	UUG	Leu L	$(3/2$	$1/2)^1$	$-1/2$	$1/2$
CUA	Leu L	$(1/2$	$1/2)^3$	$-1/2$	$1/2$	UUA	Leu L	$(3/2$	$1/2)^1$	$-3/2$	$1/2$
CGC	Arg R	$(3/2$	$1/2)^2$	$3/2$	$1/2$	UGC	Cys C	$(3/2$	$1/2)^2$	$1/2$	$1/2$
CGU	Arg R	$(1/2$	$1/2)^2$	$1/2$	$1/2$	UGU	Cys C	$(1/2$	$1/2)^2$	$-1/2$	$1/2$
CGG	Arg R	$(3/2$	$1/2)^2$	$3/2$	$-1/2$	UGG	Trp W	$(3/2$	$1/2)^2$	$1/2$	$-1/2$
CGA	Arg R	$(1/2$	$1/2)^2$	$1/2$	$-1/2$	UGA	Ter	$(1/2$	$1/2)^2$	$-1/2$	$-1/2$
CAC	His H	$(1/2$	$1/2)^4$	$1/2$	$1/2$	UAC	Tyr Y	$(3/2$	$1/2)^2$	$-1/2$	$1/2$
CAU	His H	$(1/2$	$1/2)^4$	$-1/2$	$1/2$	UAU	Tyr Y	$(3/2$	$1/2)^2$	$-3/2$	$1/2$
CAG	Gln Q	$(1/2$	$1/2)^4$	$1/2$	$-1/2$	UAG	Ter	$(3/2$	$1/2)^2$	$-1/2$	$-1/2$
CAA	Gln Q	$(1/2$	$1/2)^4$	$-1/2$	$-1/2$	UAA	Ter	$(3/2$	$1/2)^2$	$-3/2$	$-1/2$
GCC	Ala A	$3/2$	$3/2$	$3/2$	$1/2$	ACC	Thr T	$3/2$	$3/2$	$1/2$	$1/2$
GCU	Ala A	$(1/2$	$3/2)^1$	$1/2$	$1/2$	ACU	Thr T	$(1/2$	$3/2)^1$	$-1/2$	$1/2$
GCG	Ala A	$(3/2$	$1/2)^1$	$3/2$	$-1/2$	ACG	Thr T	$(3/2$	$1/2)^1$	$1/2$	$-1/2$
GCA	Ala A	$(1/2$	$1/2)^1$	$1/2$	$-1/2$	ACA	Thr T	$(1/2$	$1/2)^1$	$-1/2$	$-1/2$
GUC	Val V	$(1/2$	$3/2)^2$	$1/2$	$1/2$	AUC	Ile I	$3/2$	$3/2$	$-1/2$	$1/2$
GUU	Val V	$(1/2$	$3/2)^2$	$-1/2$	$1/2$	AUU	Ile I	$3/2$	$3/2$	$-3/2$	$1/2$
GUG	Val V	$(1/2$	$1/2)^3$	$1/2$	$-1/2$	AUG	Met M	$(3/2$	$1/2)^1$	$-1/2$	$-1/2$
GUA	Val V	$(1/2$	$1/2)^3$	$-1/2$	$-1/2$	AUA	Ile I	$(3/2$	$1/2)^1$	$-3/2$	$-1/2$
GGC	Gly G	$3/2$	$3/2$	$3/2$	$-1/2$	AGC	Ser S	$3/2$	$3/2$	$1/2$	$-1/2$
GGU	Gly G	$(1/2$	$3/2)^1$	$1/2$	$-1/2$	AGU	Ser S	$(1/2$	$3/2)^1$	$-1/2$	$-1/2$
GGG	Gly G	$3/2$	$3/2$	$3/2$	$-3/2$	AGG	Arg R	$3/2$	$3/2$	$1/2$	$-3/2$
GGA	Gly G	$(1/2$	$3/2)^1$	$1/2$	$-3/2$	AGA	Arg R	$(1/2$	$3/2)^1$	$-1/2$	$-3/2$
GAC	Asp D	$(1/2$	$3/2)^2$	$1/2$	$-1/2$	AAC	Asn N	$3/2$	$3/2$	$-1/2$	$-1/2$
GAU	Asp D	$(1/2$	$3/2)^2$	$-1/2$	$-1/2$	AAU	Asn N	$3/2$	$3/2$	$-3/2$	$-1/2$
GAG	Glu E	$(1/2$	$3/2)^2$	$1/2$	$-3/2$	AAG	Lys K	$3/2$	$3/2$	$-1/2$	$-3/2$
GAA	Glu E	$(1/2$	$3/2)^2$	$-1/2$	$-3/2$	AAA	Lys K	$3/2$	$3/2$	$-3/2$	$-3/2$

RNA. Then, a triplet of nucleotides or codon will be related to an amino-acid. More precisely, a codon is defined as an ordered sequence of three nucleotides, e.g. AAG, AGA and GAA, and one enumerates in this way $4 \times 4 \times 4 = 64$ different codons.

In the universal eukariotic code (see Table 1), 61 of such triplets encode the amino-acids, while the three codons UAA, UAG and UGA, which are called

non-sense or stop-codons, play the role to stop the biosynthesis process. Indeed, the genetic code is the association between codons and amino-acids. But since one distinguishes only 20 amino-acids [1] related to the 61 codons, it follows that the genetic code is degenerate. From Table 1, one remarks the presence of 3 sextets, 5 quadruplets, 1 triplet, 9 doublets and 2 singlets of codons, each multiplet corresponding to a specific amino-acid. Since its appearance on the earth life has been characterized by its continuous change. Spontaneous **genetic mutations**, i.e. modifications of the DNA genomic sequences, play a fundamental role in the evolution. In the present paper I only deal with point mutation, that is with single base (single nucleotide) changes. More generally, mutations include changes of more than one nucleotide, insertions and deletions of nucleotides, frame-shifts and inversions. The point mutations are usually modelled by stationary, homogeneous Markov process, which assume:
1) the nucleotide positions are stochastically independent one from another, which is clearly not true in functional sequences;
2) the mutation is not depending on the site and is constant in time, which ignores the existence of "hot spots" for mutations as well as the probable existence of evolutionary spurts;
3) the nucleotide frequencies are equilibrium frequencies.

Moreover, phenomenologically, a change of the 3rd nucleotide is more frequent than the change of the 1st nucleotide, the latter being more frequent than the change of the second one.

In the following the labels i, j run in the set analysed, e.g. $i, j \in \{C, T, G, A\}$ (T being replaced by U in RNA) for single nucleotides changes or i, j run in a 20-dim set for the amino-acids substitution matrix or in a 64-dim set for for the codon substitution matrix. The transition matrix $\mathbf{Q}$, where $Q_{ij} > 0 \; (i \neq j)$ represents the transition rate between the j state and the i state, in the chosen unit of "time", and it is normalised to

$$0 > Q_{ii} = 1 - \sum_{j \neq i} Q_{ij} \tag{1}$$

The evolution matrix $\mathbf{P}$, where $P_{ij}(t)$ gives the probability that the j state at time $t = 0$, will be replaced, at time t, by the i state, satisfies the differential equation

$$\frac{dP_{ij}(t)}{dt} = \sum_k P_{ik}(t)\, Q_{kj} \quad \longrightarrow \quad \mathbf{P}(t) = \mathbf{P}(0) \exp \mathbf{Q}\, t \quad \mathbf{P}(0) = \mathbf{1} \tag{2}$$

[1] Alanine (Ala), Arginine (Arg), Asparagine (Asn), Aspartic acid (Asp), Cysteine (Cys), Glutamine (Gln), Glutamic acid (Glu), Glycine (Gly), Histidine (His), Isoleucine (Ile), Leucine (Leu), Lysine (Lys), Methionine (Met), Phenylalanine (Phe), Proline (Pro), Serine (Ser), Threonine (Thr), Tryptophane (Trp), Tyrosine (Tyr), Valine (Val).

In the Markov model, with discretized time τ, we have

$$\mathbf{P}((n+1)\tau) = \mathbf{Q}\,\mathbf{P}(n\tau) \tag{3}$$

The most simple reversible model describing single nucleotide changes depends on 1 parameter and the most complex not reversible model depends on 12 parameters [3] [2]. These models consider the DNA sequences as set of nucleotides each nucleotide evolving independently of the others. They are not able to make, a priori, any prediction on the reversibility of a mutation and naturally predict that a nucleotide change happens at the same rate independently of which codon it belongs to and are indeed unable to explain:
i) the dependence of mutations on the nature of the neighbouring nucleotides [5]. These features can of course be accounted introducing more new unknown parameters or new type of models, see [6];
ii) the fact that mutations occur more frequently between amino acids with similar physical-chemical properties, which generally have similar functional roles. Generally in the literature it is stated that the nature of the 2nd nucleotide strongly determines the physical-chemical properties. In the seventies Konopolchenko and Rumer [7] have remarked that amino acids with similar physical-chemical properties can be described by assigning a suitable charge Q to the first dinucleotide (called "root" by the authors) of the codon, in particular "strong roots" ("weak roots"), corresponding to multiplets of codons of dimension 4 (≤ 3), have $Q > 0$ ($Q < 0$). Note that sextets appear as the sum of a quartet and of a doublet.

The aim of this paper is to propose a model in which the strength of the mutation depends on a suitably defined **distance** between codons. This model reduces to the Markov model if the distance dependence is assumed constant, but it is able, in principle, to take into some account the points i)-ii). The first requirement to build such a model is to identify codons as mathematical objects, in particular as vectors in a suitable space. This will be done in the framework of the *crystal basis model* of the genetic code [8]. In this model the 4 nucleotides are assigned to the (4-dim fundamental) irreducible representation (irrep.) $(1/2, 1/2)$ of $U_{q\to 0}(sl(2) \oplus sl(2))$ with the following assignment for the values of the third component of $\vec{J}$ for the two $sl(2)$ which in the following will be denoted as $sl_H(2)$ and $sl_V(2)$:

$$\mathrm{C} \equiv (+\tfrac{1}{2}, +\tfrac{1}{2}) \quad \mathrm{T/U} \equiv (-\tfrac{1}{2}, +\tfrac{1}{2}) \quad \mathrm{G} \equiv (+\tfrac{1}{2}, -\tfrac{1}{2}) \quad \mathrm{A} \equiv (-\tfrac{1}{2}, -\tfrac{1}{2}) \tag{4}$$

and the codons, triple of nucleotides, to the 3-fold tensor product of $(1/2, 1/2)$. The assignment of the codons to the different irreps. and the correspondence with the encoded amino acid in the eukaryotic code is provided in Table 1.

[2]For a review of the different Markov models with a large list of the original papers, see Cap. 3 of [4]

Let us emphasize that the assignment of the codons to the different irreps. is a straightforward consequence of the assumed labelling of the nucleotides eq.(4) and of the Kashiwara's theorem on the tensor product of irreps. in the crystal basis [9]. In the following we call **nearest codon** codons differing by only one nucleotide. The effects of a single nucleotide mutation in the codons are represented, neglecting the mutations into or from the three stop codons which are not detectable in the considered set of experimental data, by a $61x61$ (symmetric) matrix, whose elements, in first approximation, will be assumed vanishing if non connecting nearest codons.

In [10] it has been shown that amino acids with similar properties can grouped together looking to the content of the irrep. of the first dinucleotide (or "root"), in particular to the values of the charge Q and of the third generator of $sl_V(2)$. The charge Q can be expressed as[3]

$$Q = 4\, J_{3,H} \; + \; C_V(J_{3,V} + 1) \; - \; 1 \qquad (5)$$

In that paper the analysis has been performed for 10 physico-chemical properties: the Chou–Fasman conformational parameters, which give a measure of the probability of the amino acid to form respectively a helix, a sheet and a turn; the Grantham polarity; the relative hydrophilicity; the thermodynamic activation parameters at 298 K: ΔH (enthalpy, in kJ/mol), ΔG (free energy, in kJ/mol) and ΔS (entropy, in J/mole/K); the dissociation constants at 298 K; the isoelectronic point, i.e. the pH value at which no electrophoresis occurs. The strength of the mutation inducing operator is assumed to depend on the distance between the initial codon and the final codon, i.e. the codon appearing as result of the mutation. In the literature many attempts to define distance between codons exist based on the similarity of their physico-chemical properties or of those of the encoded amino-acid. Sometimes the distance between amino acids is defined by the strength of their mutation. Here I follow a completely different approach as I define a priori a distance and then I try to derive the strength of their mutation.

2. The mutation matrix

In order to be able to define the distance we make a correspondence between a codon and a point in n-dim. Euclidean space. For sake of simplicity, presently we assume a 1-dim space.[4] The correspondence between codons and real numbers is realized through the eigenvalues of the following operator

$$\hat{X} = [\alpha\, Q^1 \; - \; \beta\, J^1_{3,V}(J^1_{3,V} - 1) \; + \; 4\gamma\, (C_H + C_V)]\; 2\, (J_{3,H} + \eta J_{3,V}) \qquad (6)$$

[3]Note that the numerical values of eq.(5) are slightly different from those of [7].

[4]The use of a 2-dim space, related to the roots of the two commuting $sl(2)$, may seem the most naturale choice.

Table 2. Dinucleotides representation content and charge Q.

dimer	J_H	J_V	J_{3H}	J_{3V}	Q	dimer	J_H	J_V	J_{3H}	J_{3V}	Q
CC	1	1	1	1	7	GC	1	1	1	0	5
CU	0	1	0	1	1	GU	0	1	0	0	1
CG	1	0	1	0	3	GG	1	1	1	-1	3
CA	0	0	0	0	-1	GA	0	1	0	-1	-1
UC	1	1	0	1	3	AC	1	1	0	0	1
UU	1	1	-1	1	-1	AU	1	1	-1	0	-3
UG	1	0	0	0	-1	AG	1	1	0	-1	-1
UA	1	0	-1	0	-5	AA	1	1	-1	-1	-5

where α, β, γ and η are real positive parameters ($\eta > 1$ as mutations between pyrimidines and purines (transversions, $\Delta J_{3,V} \neq 0$) occur less frequently than mutations between pyrimidines or purines (transitions $\Delta J_{3,H} \neq 0$)); Q^1 and $J^1_{3,V}$ are, respectively, the "charge", given by eq.(5), and the third generators of $sl_V(2)$ of the first dinucleotide of the codon XYZ, that is XY, and $C_H, J_{3,H}$ (resp. $C_V, J_{3,V}$) are the Casimir operator and the third generator of $sl_H(2)$ (resp. $sl_V(2)$) for the trinucleotide state or codon,

$$\hat{X}\ \psi(XYZ) = r(XYZ)\ \psi(XYZ) \tag{7}$$

where $\psi(XYZ)$ is the state $\in V$, V being the space of the irreps. of

$$U_{q \to 0}(sl_H(2) \oplus sl_V(2)) \,,$$

corresponding to the XYZ codon and, using the same notation for the operators and for their eigenvalues,

$$r = [\alpha\, Q^1 \; - \; \beta\, J^1_{3,V}(J^1_{3,V} - 1) \; + \; 4\gamma\,(C_H + C_V)]\; 2\,(J_{3,H} + \eta J_{3,V}) \tag{8}$$

the values of the quantities appearing in eq.(8) are given in Table 2 and Table 1. The transition matrix between the codon $i = XYZ$ and the codon $j = X'Y'Z'$ is

$$Q_{ji} = F(d_{ji})\, q_{ji} \qquad\qquad j \neq i \tag{9}$$

where $F(d_{ji})$, the strength of the transition, is a **decreasing** function of the argument and d_{ji} is the distance between the initial and final codon

$$d_{ji} = |r(X'Y'Z') \; - \; r(XYZ)| \tag{10}$$

and q_{ji} is the element of a matrix $\mathbf{q}$ such that

$$q_{ji} = 1 \quad \text{i,j nearest codons} \qquad q_{ji} = 0 \quad \text{otherwise} \tag{11}$$

If the strength are considered as constants, our model is essentially equivalent to a reversible Markov model with constant parameters. Of course there is an arbitrary infinite way of defining the correspondence between a codon and a point of an Euclidean space. Our choice is such that to a larger variation of the charge, i.e. to a larger variation of the physical-chemical properties, corresponds a larger distance and that the distance between codons in the same irrep. is lower that codon in different irreps.. Generally, from eq.(8), the distance between two codons, differing by a nucleotide in the middle position or in the first position, is larger, due to the change of the value of the charge, than the distance between two codons, differing by a nucleotide in the third position. At this stage our model can be considered as a markovian model with neighbours depending parameters.

3. Amino acid substitution matrices

In this section we recall the definition and the differences between the experimentally determined mutation matrix. The sequences alignment of proteins is a most powerful tool to get insights on the protein functions and to compute substitution rates due to evolutionary processes. The first scheme was proposed in the seventies by M. Dayhoff [11] and it is generally considered as the standard scheme. It is based on the alignments of protein sequences that are at least 85% identical. The evolutionary distance in measured in "accepted point mutation" (PAM). Two sequences are said to be 1 PAM distant if they differ on average by one accepted-point mutation per 100 amino acids. The term "accepted" means that the mutation of the amino acid has been incorporated into the protein's progeny, i.e. the mutation has not produced harmful consequences. The original Dayhoff matrix, by construction, was biased by the sample of proteins available at that time, mainly small globular proteins, and emphasized the rate of mutation in the highly mutable amino acids. A matrix, taking into account substitutions poorly represented in the original Dayhoff's analysis and making use of a statistics about 35 times higher, was computed in ref. [12] and it is known as PET91. [5] We make a comparison between our data and the 1-PAM PET91 matrix, see Table II of [12]. In that table the data are referred to the substitution of the amino acids, so we cannot compare them directly with our predictions, which refer to the codon mutations. We have to consider for each amino acid the multiplet of codons encoding it and then to consider only the one-nucleotide mutation. In this process we have to take into account the preferred codon usages, which depend on the biological species and on the type of gene analysed. In this preliminary analysis we make the simple (and definitely incorrect) assumption of an uniform codon usage. The experimental Dayhoff

[5] To study the relations for distant sequences a more reliable model has been proposed in 1992 [13], which is presently known as block substitution matrix (BLOSUM).

matrix entries between the amino acids a and b are identified as

$$M_{ab} = \sum_{i,j} f_i^a \, M_{ab}^{ij} \tag{12}$$

where f_i^a is the frequency of the i codon in the amino acid a,M_{ab}^{ij} is the substitution rate matrix for the codons and the sum is over all the codons encoding the amino acids a and b and differing by only one nucleotide. The comparison with experimental data requires one more assumption. We have to compare the matrix

$$\mathbf{P}(t) = \exp \mathbf{Q}\, t \tag{13}$$

with the $x - PAM$ mutation matrix $\mathbf{M}_{x-PAM}$ which is computed at a x distance between the amino acids sequences. Commonly $1 - PAM$ evolutionary distance is considered to correspond to a time interval of $\approx 1x10^7$ years and the correspondence between the PAM matrix and the instantaneous rate matrix is

$$\mathbf{M}_{1-PAM} = \exp \mathbf{Q}\, t \;\approx\; \mathbf{1} + 0.1\mathbf{Q} \tag{14}$$

i.e. the unit of time is chosen $\tau_0 = 1x10^8$ years. It should be remarked that the above matrices, by construction, are really divergence matrices, that is they provide the probability that the j state in the first sequence, will be replaced by the i state in the second $x - PAM$ distant sequence. Moreover these matrices have been build up assuming a symmetric probability of mutation between two amino acids and, consequently, the estimated rate is lower for the amino acid which has a larger frequency. Therefore, strictly speaking, a direct comparison between the rate matrix eq.(9) and the amino acid substitution matrices is uncorrect. However, as in the present work we present only semiquantitative comparison, our conclusions should not be sensibly affected by the above remarks.

4. Predictions of model

4.1 Stability

From the assignment of the codons to the different irreps., see Table 1, and the assumed distance, see eq.(10), we can make a set of general predictions, independently of the structure of the F function and of the detailed values of α, β, γ and η. Considering a single-nucleotide mutation, each codon can make transition in the (9) nearest codons. Some of these codons can be synonymous (silent mutations) or stop codons (nonsense mutations), both being unobservable in the framework of the substitution matrices. However, without a thorough analysis of their physical-chemical properties and/or their functional functions, we should expect amino acids encoded by multiplets of the same dimension to be approximately equally stable, i.e. the diagonal entries of the mutation matrix

Table 3. Relative mutability for the 20 amino acids with respect to Ala, arbitrarily fixed to 100, from Table III of [12].

amino acid	$PET91$	Dayhoff	amino acid	$PET91$	Dayhoff
Ala	100	100	Leu	54	40
Arg	83	65	Lys	72	56
Asn	104	134	Met	93	94
Asp	86	106	Phe	51	41
Cys	44	20	Pro	58	56
Gln	84	93	Ser	117	120
Glu	77	102	Thr	107	97
Gly	50	49	Trp	25	18
His	91	66	Tyr	50	41
Ile	103	96	Val	98	74

M should be of the same order. In the crystal basis model, see Table 1, not all the codons are on the same foot as they belong to different irreps. spaces. We indeed expect that mutations between codons in the same irrep. to occur more frequently than mutations between different irreps., provided that the values of $J^1_{3,V}$ are close and the signs of their charge Q are the same. This requires that we have to compare respectively long multiplets and short multiplets between them. Moreover in each fixed space, the codons represented by highest or lowest weight are "surrounded" by a smaller number of nearest codons. From an analysis of the positions of the codons in the different irreps., we derive, from eq.(8), a hierarchy in the stability.

$$\begin{gathered} Gly > Pro > Ala > Thr > Ser^* \\ Phe > Lys > Ile^{**} > Asn \\ Leu^* > Val \qquad Glu > Asp \\ His \approx Gln \qquad Trp >> Met \end{gathered} \tag{15}$$

where the * (**) is written to recall that we are dealing with a sextet (triplet), so our analysis is less reliable. A comparison with the experimental data from the PET91 and Dayhoff matrices for the average mutability, see Table 3, shows a remarkably satisfactory agreement (higher stability implies lower mutability). Note that the comparison between His and Gln which, at first sight, is not satisfactory with the Dayhoff data, should be analyzed on the light of the wide range of variation of the values of the average relative mutability for the doublets (between 20 and 134). A more detailed analysis should require an evaluation of the form of the F functions and of the values of the constants appearing in eq.(8).

4.2 Relation between rates

In the following we use the standard notation Y = C, U (pyrimidines) and R = G, A (purines) and N for any nucleotide. First we look for qualitative prediction for the rate of transition between two amino acids a and b (R($a \Leftrightarrow b$)) which follow directly from eq.(9) and from the assumed behaviour of the F function, without any information of the values of α, β, γ. From an inspection of eqs.(10), (8), (5) and Tables 2, 1, we can write a set of inequalities between the rates for several amino acids. The results of our analysis are reported in Table 4 where for any couple of amino acids we write the experimental values (Exp) taken from PET matrix [12]. Of course we cannot make any more precise statement on the range of the inequalities, due to the yet undefined F function.

Table 4. Theoretical inequalities for the rate mutations between two couples of amino acids. In the last two columns the experimental rate, from [12], for each couple.

Theor: Rate(I)	$<$	Rate(II)	Exp-I	Exp-II
$R(Asp \Leftrightarrow Ala)$	$<$	$R(Glu \Leftrightarrow Ala)$	63	82
$R(His \Leftrightarrow Pro)$	$<$	R$(Gln \Leftrightarrow Pro)$	58	81
$R(Gly \Leftrightarrow Arg)$	$<$	$R(Gly \Leftrightarrow Ser)$	70	129
$R(Gly \Leftrightarrow Asp)$	$\lesssim$	$R(Gly \Leftrightarrow Glu)$	66	70
$R(Trp \Leftrightarrow Arg)$	$\lesssim$	$R(Met \Leftrightarrow Thr)$	7	123
$R(Gly \Leftrightarrow Arg)$	$\lesssim$	$R(Gly \Leftrightarrow Glu)$	70	70
$R(Gln \Leftrightarrow Arg)$	$<$	$R(His \Leftrightarrow Arg)$	154	164
$R(Asn \Leftrightarrow Asp)$	$<$	$R(Asn \Leftrightarrow Ser)$	284	344
$R(Lys \Leftrightarrow Gln)$	$<$	$R(Asn \Leftrightarrow His)$	122	150
$R(Lys \Leftrightarrow Arg)$	$<$	$R(Asn \Leftrightarrow Ser)$	334	344
$R(Ala \Leftrightarrow Thr)$	$<$	$R(Ala \Leftrightarrow Ser)$	267	284
$R(Met \Leftrightarrow Thr)$	$<$	$R(Met \Leftrightarrow Val)$	123	201
$R(Tyr \Leftrightarrow Asp)$	$<$	$R(Tyr \Leftrightarrow Ser)$	23	43
$R(Tyr \Leftrightarrow Ser)$	$<$	$R(Tyr \Leftrightarrow His)$	43	134
$R(Val \Leftrightarrow Leu)$	$<$	$R(Val \Leftrightarrow Ala)$	161	226
$R(Val \Leftrightarrow Ala)$	$<$	$R(Val \Leftrightarrow Ile)$	226	504
$R(Ser \Leftrightarrow Thr)$	$<$	$R(Ser \Leftrightarrow Ala)$	278	297
$R(Pro \Leftrightarrow Thr)$	$<$	$R(Pro \Leftrightarrow Leu)$	69	97
$R(Pro \Leftrightarrow Thr)$	$<$	$R(Ser \Leftrightarrow Ala)$	69	297
$R(Pro \Leftrightarrow Ala)$	$<$	$R(His \Leftrightarrow Arg)$	150	164
$R(Ile \Leftrightarrow Thr)$	$<$	$R(His \Leftrightarrow Arg)$	149	164
$R(His \Leftrightarrow Arg)$	$<$	$R(Pro \Leftrightarrow Ser)$	164	190
$R(Thr \Leftrightarrow Ile)$	$<$	$R(Thr \Leftrightarrow Ser)$	134	325

From the experimental data that $R(Phe \Leftrightarrow Leu) > R(Phe \Leftrightarrow Tyr)$ (Exp.: 230 | 179) we derive $\eta > 2$. Then we expect

$$R(Ala \Leftrightarrow Pro) \;<\; R(Ala \Leftrightarrow Val) \qquad \text{Exp:} \quad 23\,|\,193 \tag{16}$$

Let we remark that the following mutations between doublets: $Asn \Leftrightarrow Lys$ (AAY $\Leftrightarrow$ AAR), $Asp \Leftrightarrow Glu$ (GAY $\Leftrightarrow$ GAR), $His \Leftrightarrow Gln$ (CAY $\Leftrightarrow$ CAR), share the common features to involve a mutation in the 3rd nucleotide and to have the same 2nd nucleotide A. So from the assumption that the middle nucleotide is the one which strongly determines the physical-chemical properties, comparable mutation rates should be expected. On the contrary in our model, from eq.(8), we expect different rates, except for a numerical coincidence for at most two of the considered mutations. The experimental rates are different (resp.: 150 | 478 | 233). So we derive the following inequality:

$$|60\gamma - 4\beta - 10\alpha| > |12\gamma - 2\alpha| > |36\gamma - 4\beta - 2\alpha| \qquad (17)$$

Let us note that our analysis puts into evidence:
a) a dissimilarity between the transversions $C \Leftrightarrow A$ and $U \Leftrightarrow G$, which apparently has not before either remarked;
b) a "penalty", in the form of an increase of the distance, appears for mutations between codons with $|J_{3,H}|$ or $|J_{3,V}| > 1/2$.

Let us recall once more that in the determination of the mutation rate the mutability, the frequency of occurrence and the codon distribution frequency of the considered amino acid play a role.

5. Conclusions

It is believed that the mutations are essentially random effects, especially in the non coding sequences. For the coding sequences it is known the presence of evolutionary bias. Our analysis concerns only the coding sequences and provides indication of the presence of general pattern and symmetry, not before observed. By trial and errors, following the leading idea to incorporate in a suitable **metric** in a n-dim. space the effects of the near neighbours and the influence of the physical-chemical properties of the different amino acids in the rate mutation, we have build a simple model which is able to reproduce in a semi-quantitative way the hierarchy of the most frequently observed mutation between amino acids. The predictions well agree with the experimental data of PET91. One should check that no inconsistency appears in the computed inequalities. This is true for the reported set, but it has to be carefully checked for all the mutations rates. It should also be noticed that the model is able to explain some puzzling features, for example:

a the almost equality of the rates $R(Gly \Leftrightarrow Asp)$ and $R(Gly \Leftrightarrow Arg)$ (Exp.: 70), the first mutation resulting from the transition of the 1st nucleotide, $GGR \Leftrightarrow AGR$, and the second from the transitions of the 2rd nucleotide, $GGR \Leftrightarrow GAR$;

b the fact that $R(Gln \Leftrightarrow His)$ ($CAR \Leftrightarrow CAY$, transversion of the 3rd nucleotide) is lower than $R(Gln \Leftrightarrow Glu)$ ($CAR \Leftrightarrow GAR$, transversion of the 1st nucleotide) c the fact that $R(Ser \Leftrightarrow Thr)$ is lower than $R(Ser \Leftrightarrow Ala)$ although any codon of the sextet Ser can go into the multiplet encoding Thr by single nucleotide change while only the codons of the quartet UCN can go into the multiplet encoding Ala, by single nucleotide change.

A more quantitative analysis requires to take into account the normalisation of the transition matrix

$$\sum_j Q_{ji} = 1 \qquad \forall i \tag{18}$$

and to evaluate the function F of eq.(8). Moreover one should know the codon usage frequency. The parametrization in terms of only 4 parameters (which indeed can be reduced to 3 as one can be absorbed in the function F) and the identification of a codon with a real number may be a too simple choice. Going on with the analysis, likely, one will face some inconsistencies between the theoretical relations. Hopefully these pathologies can be cured with slight modifications of eqs.(8) and (10).

It is appropriate to underline that this approach can be easily generalized to describe more complex phenomena, neglected in this paper, as the multiple nucleotide changes, the observed presence of hotspots for the mutations, the variation of the mutations with the type of proteins, the probable occurrence of spurts in the evolution, the scaling behavior of the mean parameter substitution in function of the total length of genome [14], etc. A criticism can be raised against this model: it is essentially based on the properties of the genetic code while the accepted mutations are the replacement of an amino acid by a similar one. Some of the chemical properties which mostly influence the chances of mutations, like the hydrophobicity, charge, size, are related to the genetic code, [10], but many of the physical chemical properties of the amino acids are believed to have been more imposed by natural selection than by genetic code constraints. If the plausibility of the model is confirmed, this arises a puzzling question. The comparison for the mutation rates between the predicted values of the theoretical time evolution operator **P(t)** and the experimental values of the evolution distance matrix **M**, which can be criticized from many points of view, has been done as the amino acid mutation matrix is, at my knowledge, the only source of mutation data with a large statistics, obtained by analysing many thousands of proteins.

References

[1] Gamow G. Possible Relation between Deoxyribonucleic Acid and Protein Structures. *Nature*, 173:318 1954.

[2] Singer M. and Berg P. *Genes and Genomes*. Editions Vigot, Paris, 1990.

[3] Rodriguez F., Oliver J.L., Marin A. and Medina J.R. The General Stochastic Model for Nucleotide Substitution. *J.Theor.Biol.*, 142:485, 1990.

[4] Wen-Hsiung Li. *Molecular Evolution.* Sinauer Associates Incorporates, Sunderland, 1997.

[5] Blake R. D., Hess S. T. and J. Nicholson-Tuell, The Influence of Nearest Neighbors on the Rate and Pattern of Spontaneous Point Mutations. *J.Mol.Evol.*, 34:189, 1992

[6] Arndt Peter F., Burge Christopher B. and Hwa Terence. DNA Sequence Evolution with Neighbor-Dependent Mutation. *J.Comput.Biol.*, 10:313, 2003.

[7] Konopel'chenko, B. G. and Yu.B. Rumer. Classification of Codons in the Genetic Code . *Translated from Doklady Akademi Nauk SSSR* , 223(2):471, 1975.

[8] Frappat, L., Sciarrino, A. and P. Sorba. A crystal base for the genetic code . *Phys. Lett. A*, 250:214, 1998.

[9] Kashiwara, M. Crystalizing the q-analogue of universal enveloping algebras. *Commun.Math.Phys.*, 133:249, 1990.

[10] Frappat, L., Sciarrino, A. and P. Sorba. Predictions of Physical-Chemical Properties of Amino Acids from Genetic Code. *J.Biol.Phys.*, 28:17, 2002.

[11] Dayhoff, M. O., Schwartz R. M. and B. C. Orcutt. in *Atlas of Protein Sequence and Structure*. National Biomedical Research Foundation, Washigton D.C, Vol. 5, Suppl. 3, 345 (1978).

[12] Jones, D. T., Taylor, W. R. and J. M. Thornton. The rapid generation of mutation data matrix from protein sequences. *CABIOS*, 8:275, 1992.

[13] Henikoff S. and J. G. Henikoff. Amino acid substitution matrices from protein blocks *Proc.Natl.Acad.Sci USA*, 89:10915, 1992.

[14] Nilsson M. and Snoad N. Optimal Mutation Rates in Dynamic Environments. *physics/0004042*, 2000.

NONCOMPACT QUANTUM ALGEBRA $u_q(2,1)$: POSITIVE DISCRETE SERIES OF IRREDUCIBLE REPRESENTATIONS

Yu. F. Smirnov
Skobeltsyn Institute of Nuclear Physics,
Moscow State University,
Moscow, 119992, Russia

Yu. I. Kharitonov*

Abstract The structure positive of unitary irreducible representations of the noncompact $u_q(2,1)$ quantum algebra that are related to a positive discrete series is examined. With the aid of projection operators for the $su_q(2)$ subalgebra, a q-analog of the Gel'fand–Graev formulas is derived in the basis corresponding to the reduction $u_q(2,1) \to su_q(2) \times u(1)$. Projection operators for the $su_q(1,1)$ subalgebra are employed to study the same representations for the reduction $u_q(2,1) \to u(1) \times su_q(1,1)$. The matrix elements of the generators of the $u_q(2,1)$ algebra are computed in this new basis. A general analytic expression for an element of the transformation bracket $\langle U|T\rangle_q$ between the bases associated with above two reductions (the elements of this matrix are referred to as q-Weyl coefficients) is obtained for a general case where the deformation parameter q is not equal to a root of unity. It is shown explicitly that, apart from a phase, q-Weyl coefficients coincide with the q-Racah coefficients for the $su_q(2)$ quantum algebra.

1. Introduction

It is well known that the group theory methods are widely used in the theory of nucleus. They form the basis of nuclear spectroscopy and of various nuclear models, including the shell model, models dealing with collective degrees of freedom, and the interacting boson model. Since group theory or algebraic models usually admit an analytic solution, they are employed to study various

*Deceased.

B. Gruber (ed.), Symmetries in Science XI, 505–526.

properties of nuclear systems in particular, some of their asymptotic properties. For example, the popular Elliott model, based on $SU(3)$ symmetry, was successfully employed by Belyaev and his colleagues [1] to analyze the asymptotic properties of the generalized density matrix. The discovery of quantum algebras and groups that was made more than 20 years by mathematical physicists of Leningrad school [2]–[4] gave new impetus to the development of algebraic methods in theoretical physics, in particular, to searches for applications of the representations of quantum groups and algebras in physics. For example, the construction of q analogs of various nuclear models became a new field of research in theoretical nuclear physics. The point is that quantum algebras involve an additional variable parameter, the deformation parameter q. This renders models based on quantum algebras more adaptable and extends their potential in describing physical systems (see, for example, the study of Raychev et al [5] and the review article of Bonatsos and Daskaloyannis [6]) which is devoted to applications of quantum algebras in the theoretical nuclear physics). However, the searches for physical applications of quantum algebras must be preceded by a detailed investigations of their irreducible representations. In this connection, the structure of unitary irreducible representations of the compact $u_q(3)$ algebra was examined in details in [7]–[16]. In our opinion it is important to extend these results on the noncompact $u_q(2,1)$ quantum algebra. The classical algebra $u(2,1)$ describes the dynamical symmetry of a two-dimensional harmonic oscillator and of some other physical systems. In view of this, a comprehensive analysis of unitary irreducible representations of its quantum analogs may be helpful in constructing respective physical models. In the present study, we restrict our consideration to the unitary irreducible representations associated with a positive discrete series.

Unitary irreducible representations of conventional (nondeformed) $u(n,m)$ algebras were studied by Gel'fand and Graev [17] (see also [18]), who showed, among other things, that the unitary irreducible representations of the $u(2,1)$ algebra can be divided into three discrete series. The series of the highest weight unitary irreducible representations or a negative discrete series consists of representations such that each includes the highest weight vector $|H\rangle$ that is, a vector annihilated by any raising generator of the algebra.

A positive discrete series is the series of representations such that each includes the lowest weight vector $|L\rangle$ that is, a vector annihilated by any lowering generator. There is yet another series, that is referred to as an intermediate one and which is formed by unitary irreducible representations having neither the highest weight vector $|H\rangle$ nor the lowest weight vector $|L\rangle$. For this reason, this series deserves a dedicated consideration.

Gel'fand and Graev [17] presented explicit expressions for the matrix elements of generators associated with the above representations that is, the matrix elements of the generators A_{ik} of the $u(n,m)$ algebra in the basis corresponding to the following reduction of this algebra to the chain of subalgebras:

$$u(n,m) \to u(n,m-1) \to \dots \to u(n) \to \dots \to u(2) \to u(1), n \geq m. \quad (1)$$

However these authors did not give a regular procedure for deriving the expressions that they quoted in [17]. For the $u(n,1)$, these formulas were derived in [19]–[21], but there is no derivation of such formulas for the general case of the $u(n,m)$ algebras. In this study, we extend, the approach proposed by Vilenkin in [22] for the case of the $u(2,1)$ classical algebra and examine the structure of its unitary irreducible representations associated with the positive series. These results obtained in this way are readily generalized to the case of negative discrete series. The intermediate discrete series of unitary irreducible representations will be considered in a separate paper. As in [7]–[16], we assume that the deformation parameter q is specified by an arbitrary positive number and define q-numbers and q-factorials as follows:

$$[n] = \frac{q^n - q^{-n}}{q - q^{-1}}, \quad (2)$$

$$[n]! = [n][n-1]\dots[2][1], \quad [0]! = 1. \quad (3)$$

Below, we employ brackets to denote q-numbers, enclose the signatures of unitary irreducible representations in Dirac brackets, and reserve parentheses for the weight of a vector, for example, the symbol $|\langle f_1 f_2 f_3\rangle(m_1 m_2 m_3)\rangle$ stands for a basis vector of a weight $(m_1 m_2 m_3)$ in the unitary irreducible representation $D^{\langle f_1 f_2 f_3\rangle} = D^{\langle f\rangle}$.

2. Positive discrete series of unitary irreducible representations

The $u(2,1)$ algebra is known to involve nine generators A_{ik} $(i,k = 1,2,3)$ satisfying the same commutation relations that the corresponding generators of the compact $u(3)$ classical Lie algebra. However, properties of the $u(2,1)$ generators under Hermitian conjugations differ from those of the $u(3)$ generators. The "compact" generators A_{11}, A_{22}, A_{33}, A_{12} and A_{21} of the $u(2,1)$ algebra have the same Hermitian properties, as the $u(3)$ generators,

$$A_{ik}^{+} = A_{ki}, \quad (4)$$

whereas the "noncompact" generators A_{13}, A_{23}, A_{31} and A_{32} satisfy the relations

$$A_{13}^{+} = -A_{31}, \quad (5)$$

$$A_{23}^{+} = -A_{32}. \quad (6)$$

The minus sign in formulas (5) and (6) generates a fundamental distinction between the structure of any unitary irreducible representation of the $u(2,1)$

algebra and the structure of the corresponding unitary irreducible representation of the $u(3)$ algebra: all unitary irreducible representations of the compact $u(3)$ algebra are finite-dimensional, whereas all unitary irreducible representations of the noncompact $u(2,1)$ algebra(with the exception of the trivial identity representation) are infinite-dimensional. The noncompact $u_q(2,1)$ quantum algebra is also specified by nine generators A_{ik} $(i,k=1,2,3)$ satisfying the same commutation relations as the generators of the $u_q(3)$ compact quantum algebra. The explicit expressions for these commutators can be found in [7].

As to their properties with respect to Hermitian conjugation, those in (4) and (6) remain valid, whereas, in view of the relations

$$A_{13}^{+} = \tilde{A}_{31} = A_{32}A_{21} - qA_{21}A_{32} \neq A_{31}\,, \tag{7}$$

$$A_{31}^{+} = \tilde{A}_{13} = A_{12}A_{23} - q^{-1}A_{23}A_{12} \neq A_{13}\,, \tag{8}$$

for the $u_q(3)$ algebra, that in (5) must be replaced by

$$A_{13}^{+} = -\tilde{A}_{31}\,. \tag{9}$$

With the aid of (7) and (8), this relation can be recast into either of the following two equivalent forms:

$$A_{13}^{+} = -A_{31} + (q-q^{-1})A_{21}A_{12}\,, \tag{10}$$

or

$$A_{13}^{+} = -q^2A_{31} + (1-q^2)A_{32}A_{21}\,. \tag{11}$$

For the $u_q(2,1)$ algebra, we will consider the unitary irreducible representation $D^{\langle f\rangle}$ of the lowest weight $(f)=(f_1f_2f_3)$:that is, we assume that, in the space of this representation, there is the lowest weight vector $|L\rangle$ that satisfies the relations

$$A_{ii}|L\rangle = f_i|L\rangle\,, \qquad (i=1,2,3) \tag{12}$$

annihilated by a pair of lowering generators

$$A_{31}|L\rangle = 0 \quad \text{and} \quad A_{32}|L\rangle = 0\,. \tag{13}$$

Also it is annihilated by one raising generator

$$A_{12}|L\rangle = 0. \tag{14}$$

It is assumed that this vector is normalized by the relation

$$\langle L|L\rangle = 1\,. \tag{15}$$

All the other basis vectors $|X\rangle$ of this unitary irreducible representation can be derived by applying the generators A_{13}, A_{23} and A_{21} to $|L\rangle$

$$|X\rangle = A_{21}^{g}A_{23}^{k}A_{13}^{\ell}|H\rangle. \tag{16}$$

In order to construct a basis of any unitary irreducible representation of the $u_q(2,1)$ algebra, it is necessary to specify a chain of subalgebras, and this can be done, as it is well known, in three ways. The first way is to use the U-spin subalgebra involving the generators A_{11}, A_{12}, A_{21}, and A_{22}, in which case the respective reduction is

$$u_q(2,1) \to u_q(2) \to u_q(1)\,. \tag{17}$$

One can also use the generators A_{22}, A_{23}, A_{32}, and A_{33} forming the basis of the T-spin subalgebra or the generators A_{11}, A_{13}, A_{31}, and A_{33} generating the V-spin subalgebra. Either of these two subalgebras correspond to the reduction

$$u_q(2,1) \to u_q(1,1) \to u_q(1)\,. \tag{18}$$

In this study, we restrict our consideration to the case of U and T-spin bases.

3. Basis vectors and matrix elements of the generators in the basis associated with U-spin reduction

First, we consider that the generators A_{11}, A_{12}, A_{21}, and A_{22} form a basis of the U-spin algebra, which is a compact subalgebra of the noncompact quantum $u_q(2,1)$ algebra, the generators

$$U_+ = A_{12}\,, \quad U_- = A_{21}\,, \quad U_0 = \frac{1}{2}(A_{11} - A_{22}) \tag{19}$$

generating the compact $su_q(2)$ subalgebra.

In the case of U-spin reduction, the basis of an unitary irreducible representation of the $u_2(2,1)$ algebra can be derived in the same way as the basis of $u_q(3)$ algebra [11]:

$$|\langle f_1 f_2 f_3\rangle m_3 U M_U\rangle_q = \frac{1}{N(k\ell)N(UM_U)} A_{21}^{U-M_U} P^U A_{23}^k A_{13}^\ell |H\rangle \tag{20}$$

where

$$m_3 = f_3 - k - \ell\,, \tag{21}$$

$$U = \frac{1}{2}(f_1 - f_2 - k + \ell)\,, \tag{22}$$

$$M_U = \frac{1}{2}(m_1 - m_2)\,, \qquad -U \le M_U \le U\,, \tag{23}$$

$$P^U = \sum_{r=0}^{\infty} (-1)^r \frac{[2U+1]!}{[r]![2U+r+1]!} A_{21}^r A_{12}^r \tag{24}$$

is the projection operator for the $su_q(2)$ algebra [23],

$$N(UM_U) = \sqrt{\frac{[2U]![U - M_U]!}{[U + M_U]!}}, \tag{25}$$

$N(k\ell)$ are normalization factors, and

$$|L\rangle = |\langle f\rangle f_3 U_L U_L\rangle, \quad U_L = \frac{1}{2}(f_1 - f_2) \tag{26}$$

is the lowest weight vector. The main distinction between the $u_q(2,1)$ and $u_q(3)$ algebras lies in the normalization factor $N(k\ell)$. In the Appendix, it is shown that, in the latter case, the square of the normalization factor has the form:

$$N^2(k\ell) = \frac{[k]![\ell]![f_1 - f_2 - k + \ell + 1]![f_1 - f_2]!}{[f_1 - f_2 - k]![f_1 - f_2 + \ell + 1]!} \times \frac{[f_2 - f_3 + k - 2]![f_1 - f_3 + \ell - 1]!}{[f_1 - f_3 - 1]![f_2 - f_3 - 2]!}. \tag{27}$$

Here, we impose the conventional requirement that the arguments of all q-factorials be nonnegative integers. This requirement ensures that the square of the norm of basis vectors is positive. It also follows that a nonzero vector exists only under the conditions from

$$f_1 \geq f_2,$$

$$f_1 - f_3 \geq 1, \tag{28}$$

$$f_2 - f_3 \geq 2, \tag{29}$$

$$0 \leq k \leq f_1 - f_2.$$

At the same time, no condition is imposed on the exponent ℓ ($\ell = 0, 1, 2, ...$), with the result that, in the case of a U-basis, an unitary irreducible representation of the $u_q(2,1)$ algebra is infinite-dimensional.

In [17], each basis vector of the lowest weight unitary irreducible representation was characterized by the scheme

$$\left| \begin{matrix} m_{13} & m_{23} & m_{33} \\ m_{12} & m_{22} & \\ m_{11} & & \end{matrix} \right\rangle, \tag{30}$$

where the integers m_{ij} satisfy the conditions:

$$m_{13} \geq m_{23} \geq m_{33} \geq 0, \tag{31}$$

$$m_{12} \geq m_{13} + 1 \geq m_{22} \geq m_{23} + 1, \tag{32}$$

$$m_{12} \geq m_{11} \geq m_{22}. \tag{33}$$

The numbers in the first row in (30) represent the signature of a unitary irreducible representation of the $u_q(2,1)$ algebra. They are related to the components of the lowest weight by the equations

$$f_1 = m_{13} + 1, \tag{34}$$

$$f_2 = m_{23} + 1, \tag{35}$$

$$f_3 = m_{33} - 2. \tag{36}$$

The numbers in the second row in (30) represent the signature of a unitary irreducible representation of the $u_q(2)$ subalgebra. In our notation,

$$m_{12} = f_1 + \ell, \tag{37}$$

$$m_{22} = f_2 + k. \tag{38}$$

The number in the third row is

$$m_{11} = U + M_U + m_{22}. \tag{39}$$

From the condition $f_1 \geq f_2$, it follows

$$m_{13} \geq m_{23}. \tag{40}$$

The condition $f_2 - f_3 - 2 \geq 0$ means that

$$m_{23} \geq m_{33} - 1. \tag{41}$$

Combining these conditions, we obtain

$$m_{13} \geq m_{23} \geq m_{33} - 1. \tag{42}$$

At the same time, the condition $0 \leq k \leq f_1 - f_2$ is equivalent to the constraints

$$m_{13} + 1 \geq m_{22} \geq m_{23} + 1. \tag{43}$$

With regard for the allowed values of the exponent ℓ, $\ell = 0, 1, 2, \ldots$, we derive

$$m_{12} \leq m_{13} + 1. \tag{44}$$

The condition $-U \leq M_U \leq U$ leads to the constraints

$$m_{12} \geq m_{11} \geq m_{22}. \tag{45}$$

A comparison of formulas (42)–(45) with (31)–(33) demonstrates that our constraints on the structure of basis vectors are identical to the constraints on the values of m_{ij} in the Gel'fand–Graev schemes, with the only exception that, in our case, there exists a unitary irreducible representation for which $m_{23} = m_{33} - 1$. This means that there are unitary irreducible representations corresponding to the Gel'fand–Graev signature,

$$\{m_{13}m_{23}m_{33}\} = \{m_{13}, m_{33} - 1, m_{33}\}, \tag{46}$$

which are beyond the standard constraints (31). The existence of such nonstandard discrete series of unitary representations of the $u(2,1)$ algebra was indicated in [20], and [21]. The $u_q(2,1)$ algebra has analogous special series of unitary irreducible representations.

Further, it should be noted that at $f_1 = f_2$, in which case $k = 0$, the condition that the norm $N^2(0\ell)$ is positive requires fulfillment of inequality

$$f_3 - f_2 - 1 + \ell > 0 \tag{47}$$

for all values of ℓ, including $\ell = 1$. Therefore, the lowest weights corresponding to $f_1 - f_3 > 0$ are allowed at $f_1 = f_2$ (that is, at $m_{13} = m_{33}$). Therefore, there is an additional nonstandard series of the lowest weight unitary irreducible representations such that condition (31) is violated for them. Those are characterized by Gel'fand–Graev signatures $\{m_{23} - 2, m_{23} - 2, m_{23}\}$.

Let us now consider the matrix elements of generators in the U basis. In the basis specified by (20), the weight generators A_{ii} (($i = 1, 2, 3$)) naturally have a diagonal form are diagonal form, their matrix elements being given by

$$m_1 = f_1 + \ell - (U - M_U), \tag{48}$$

$$m_2 = f_2 + k + (U - M_U), \tag{49}$$

$$m_3 = f_3 - k - \ell, \tag{50}$$

where

$$m_1 + m_2 + m_3 = f_1 + f_2 + f_3. \tag{51}$$

The action of the generators $A_{12} = U_+$ and $A_{21} = U_-$ are well known from the theory of angular momenta:

$$U_\pm|\langle f\rangle m_3 U M_U\rangle_q = \sqrt{[U \mp M_U][U \pm M_U + 1]}\ |\langle f\rangle m_3 U M_U \pm 1\rangle_q. \tag{52}$$

The matrix elements of the generators $A_{13}, A_{23}, A_{31}, A_{32}$ are given by the q-analogs of Gel'fand–Graev formulas

$$A_{ij}|\langle f\rangle m_3 U M_U\rangle = \sum_{U'} a_{ij}(m_3' U' M_U')\ |\langle f\rangle m_3' U' M_U'\rangle, \tag{53}$$

where

$$U' = U \pm 1/2, \quad M'_U = M_U \pm 1/2 \tag{54}$$

and

$$a_{ij}(m'_3 U' M'_U) = {}_q\langle m'_3 U' M'_U | A_{ij} | m_3 U M_U \rangle_q . \tag{55}$$

The list of these matrix elements is given in the Table 1; their derivation is given in [24].

4. Basis vectors and matrix elements of the generators in the basis associated with T-spin reduction.

Let us consider the structure of the unitary irreducible representations $D^{\langle f \rangle}$ of the lowest weight $(f_1 f_2 f_3)$ in the case of the reduction

$$u_q(2,1) \rightarrow u_q(1,1) \tag{56}$$

of the $u_q(2,1)$ algebra to the $u_q(1,1)$ subalgebra specified by generators A_{22}, A_{23}, A_{32}, and A_{33}, or to the $su_q(2)$ subalgebra of a noncompact T-spin, the generators in latter case being

$$T_+ = A_{23}, \quad T_- = A_{32}, \quad T_0 = \frac{1}{2}(A_{22} - A_{33}). \tag{57}$$

We note that the condition (1) imposed in [17] on a chain of subalgebras is not satisfied in formulas (56). For this reason, the results obtained in [17] are not valid in the case of the T-spin basis even for classical $u(2,1)$ algebra, not to mention its deformation $u_q(2,1)$.

Before proceedings to discuss the $u_q(2,1)$ algebra as a whole, it is reasonable to recall general information about the $su_q(1,1)$ subalgebra and its unitary irreducible representations. The generators of the $su_q(1,1)$ subalgebra satisfy the well-known commutation relations

$$[T_0, T_+] = T_+, \tag{58}$$

$$[T_0, T_-] = -T_-, \tag{59}$$

$$[T_+, T_-] = [2T_0]. \tag{60}$$

Under Hermitian conjugation, they transform as follows:

$$T_0^+ = T_0, \tag{61}$$

$$T_+^+ = -T_-. \tag{62}$$

The unitary irreducible representations D^T of the positive discrete series are infinite-dimensional; the respective T-spin is given by

$$T = -\frac{1}{2},\ 0,\ \frac{1}{2},\ 1,\ \frac{3}{2},\ 2,\ \ldots \tag{63}$$

Table 1. Matrix elements of the generators of the noncompact $u_q(2,1)$ algebra for the unitary irreducible representation $D^{\{\langle f\rangle +\}}$ associated with the positive discrete series (U-basis used here was derived from the lowest weight vector $|L\rangle$).

$a_{13}\left(m_3-1, U+\frac{1}{2}, M_U+\frac{1}{2}\right) = q^{-U+M_U}\left[\frac{[\ell+1][f_1-f_3+\ell][2U+k+2][U+M_U+1]}{[2U+1][2U+2]}\right]^{1/2}$
$a_{23}\left(m_3-1, U+\frac{1}{2}, M_U-\frac{1}{2}\right) = \left[\frac{[\ell+1][f_1-f_3+\ell][2U+k+2][U-M_U+1]}{[2U+1][2U+2]}\right]^{1/2}$
$a_{13}\left(m_3-1, U-\frac{1}{2}, M_U+\frac{1}{2}\right) = -q^{U+M_U+1}\left[\frac{[k+1][f_2-f_3+k-1][2U-\ell][U-M_U]}{[2U][2U+1]}\right]^{1/2}$
$a_{23}\left(m_3-1, U-\frac{1}{2}, M_U-\frac{1}{2}\right) = \left[\frac{[k+1][f_2-f_3+k-1][2U-\ell][U+M_U]}{[2U][2U+1]}\right]^{1/2}$
$a_{31}\left(m_3+1, U-\frac{1}{2}, M_U-\frac{1}{2}\right) = -q^{U-M_U}\left[\frac{[\ell][f_1-f_3+\ell-1][2U+k+1][U+M_U]}{[2U][2U+1]}\right]^{1/2}$
$a_{32}\left(m_3+1, U-\frac{1}{2}, M_U+\frac{1}{2}\right) = -\left[\frac{[\ell][f_1-f_3+\ell-1][2U+k+1][U-M_U]}{[2U][2U+1]}\right]^{1/2}$
$a_{31}\left(m_3+1, U+\frac{1}{2}, M_U-\frac{1}{2}\right) = q^{-U-M_U-1}\left[\frac{[k][f_2-f_3+k-2][2U-\ell+1][U-M_U+1]}{[2U+1][2U+2]}\right]^{1/2}$
$a_{32}\left(m_3+1, U+\frac{1}{2}, M_U+\frac{1}{2}\right) = -\left[\frac{[k][f_2-f_3+k-2][2U-\ell+1][U+M_U+1]}{[2U+1][2U+2]}\right]^{1/2}$

The T-spin projection M (or the weight of a vector) is an eigenvalue of the operator of the T-spin projectionT_0, takes the positive values:

$$M = T+1,\ T+2,\ \ldots \tag{64}$$

the lowest weight being $T+1$. We assume that the lowest weight vector $|H\rangle = |T, T+1\rangle$ is known and that it satisfies the requirements

$$T_{-}|L\rangle = 0, \tag{65}$$

$$T_0|L\rangle = (T+1)|L\rangle, \tag{66}$$

and the normalization condition

$$\langle L|L\rangle = 1. \tag{67}$$

The basis vectors of a higher weight can be obtained from the lowest weight vector by the formula

$$|TM\rangle = \frac{1}{N(TM)}\, T_{+}^{M-T-1}\, |TT+1\rangle. \tag{68}$$

The square norm of a vector is derived in this way has a form $(x = M-T-1)$

$$\begin{aligned} N^2(TM) &= (-1)^x \langle L|A_{32}^x A_{23}^x|L\rangle = [x][2T+x+1]N^2(T, M+1) \\ &= \frac{[-T+M-1]![T+M]!}{[2T+1]!}. \end{aligned} \tag{69}$$

It can be seen that the condition $N^2(TM) > 0$ imposes no constraints on $x = 0, 1, 2, \ldots$; therefore, the unitary irreducible representation is infinite-dimensional. Nonzero matrix elements of the generators in the basis specified (68) are given by

$$\langle TM|T_0|TM\rangle = M, \tag{70}$$

$$a_{23} = \langle TM+1|A_{23}|TM\rangle = \{[-T+M][T+M+1]\}^{1/2}, \tag{71}$$

$$a_{32} = \langle TM-1|A_{32}|TM\rangle = -\{[T+M][-T+M-1]\}^{1/2}. \tag{72}$$

The Casimir operator for the $su_q(1,1)$ algebra has the same form as for the $su_q(2)$ algebra:

$$C_2(su_q(1,1)) = T_{-}T_{+} + [T_0 + 1/2]^2. \tag{73}$$

All vectors in (64) are the eigenvectors of this operator and correspond to the same eigenvalue:

$$C_2(su_q(1,1))|TM\rangle = [T+1/2]^2|TM\rangle. \tag{74}$$

We also need the extremal projection operator $P^T = P^T_{T+1.T+1}$ for the discrete series of the lowest weight unitary irreducible representation. As in the case of the $su_q(2)$ algebra, we seek the expression for the extremal projection operator in the form of series:

$$P^T = \sum_{r=0} C_r T_+^r T_-^r. \tag{75}$$

In what follows, we apply this projection operator only to those vectors $|T+1\rangle$ that have a specific weight $M = T+1$, but which, in general, do not have a specific value of T-spin that is, to vectors that are represented by linear combinations of

$$|T+1\rangle = \sum_{T'} |T', T+1\rangle. \tag{76}$$

In contrast to the case of $su_q(2)$ algebra, however, the sum over T' is finite in the case under study, because the inequality $T' \leq M' - 1$ must hold for the basis vectors $|T'M'\rangle$ of the positive discrete series. In the case (76) it means that $T' \leq T$. Hence the variable T' in the sum (76) runs through the values from $T_{min} = -1/2$ or 0 up to T, depending on whether T is an integer or a half-integer. By applying the operator in (75) to the vectors in (76), can show that only a finite number of terms in (75) make a non-vanishing contribution, namely, those that satisfy $T + 1 - r \leq 1$ or $1/2$ (that is, $r \leq T$ or $T + 1/2$). Hence, the terms in (75) that involve higher powers r can be disregarded.

The projection operator P^T satisfies the equations

$$T_- P^T = 0, \tag{77}$$

$$P^T |T, T+1\rangle = |T, T+1\rangle. \tag{78}$$

From (77) it follows that the coefficients C_r satisfy the recursion relation

$$C_{r-1} + [r][-2T + r - 1] C_r = 0. \tag{79}$$

From this relation, we obtain

$$C_r = C_0 \frac{[2T - r]!}{[r]![2T]!}, \quad r \leq 2T. \tag{80}$$

From the condition (78), it follows that

$$C_0 = 1. \tag{81}$$

At $r = 2T + 1$, relation (80) is meaningless, but we have shown above that we do need the coefficients C_r for $r > T$ or $T + 1/2$. Thus, it is sufficient, for our purposes, to use the simple projection operator

$$P^T = \sum_{r=0}^{r=2T} \frac{[2T - r]!}{[r]![2T]!} T_+^r T_-^r. \tag{82}$$

A projection operator of a more general form can be represented as

$$P^T_{MM'} = \frac{(-1)^{-T-M'-1}}{N(TM)N(TM')} T_-^{-T+M-1} P^T T_+^{-T+M'-1}. \tag{83}$$

As a matter of fact, Vilenkin [22] used similar projection operators (of course, only for $q = 1$) long ago to derive the harmonic projections of polynomials depending on n Cartesian variables.

Let us present yet another relation helpful for subsequent computations

$$P^T_{T+1,M} P^T_{M,T+1} = \frac{(-1)^{T+M+1}}{N^2(TM)} P^T T_-^{M-T-1} T_+^{M-T-1} P^T = P^T. \tag{84}$$

From this equation, it follows

$$P^T T_-^{M-T-1} T_+^{M-T-1} P^T = (-1)^{M-T-1} N^2(TM) P^T. \tag{85}$$

We now return to a consideration of the lowest weight unitary irreducible representations of the $u_q(2,1)$ algebra. As in the case of the $u_q(3)$ algebra, the basis vectors of the unitary irreducible representation $D^{\langle f \rangle}$ of the $u_q(2,1)$ algebra that correspond to the lowest weight $(f) = (f_1 f_2 f_3)$ will be represented in a form

$$|\langle f \rangle m_1 T M_T\rangle_q = \frac{1}{N(sp)N(TM_T)} A_{23}^{M-T-1} P^T A_{13}^s A_{21}^p |H\rangle, \tag{86}$$

where

$$T = \frac{1}{2}(f_2 - f_3 + p + s - 2), \tag{87}$$

$$M = T+1, T+2, \ldots. \tag{88}$$

The normalization factor $N(TM_T)$ is determined by formula (69), while the projection operator P^T is given by (82). The normalization factor $N(sp)$ for the vectors of T-spin basis is calculated by a method similar to that used for the norm of the vectors of the U-spin basis described in the Appendix (see also [24]). The square of this norm is

$$N^2(sp) = \frac{[s]![p]![f_1 - f_2]![f_1 - f_3 + s - 1]!}{[f_1 - f_2 - p]![f_1 - f_3 - 1]!} \times \frac{[f_2 - f_3 + s - 2]![f_2 - f_3 + p - 2]!}{[f_2 - f_3 - 2]![f_2 - f_3 + p + s - 2]!}. \tag{89}$$

From the analysis of the norm of the basis vectors, we derive the conditions

$$f_1 \geq f_2, \tag{90}$$

$$f_2 - f_3 - 2 \geq 0, \tag{91}$$

$$0 \leq p \leq f_1 - f_2. \tag{92}$$

There is no constraints on the exponent s; that is, $s = 0, 1, 2, ...$ Since the number of values of the projections M is infinite, this means that the representations under study are infinite-dimensional. The constraints in (91) and (92) on the the signature of unitary irreducible representations are identical to those obtained for the U-spin basis. For this reason, the classification of the standard and nonstandard discrete series for the T-basis remains unchanged, as might have been expected.

Let us now proceed to discuss the matrix elements of generators. For the weight generators A_{ii} in the T-spin basis, only diagonal matrix elements do not vanish. They are given by

$$a_{11}(m_1TM) = a_{11}(m_1m_2m_3) = m_1 = f_1 - p + s, \tag{93}$$

$$a_{22}(m_1TM) = a_{22}(m_1m_2m_3) = m_2 = f_2 + p - T + M - 1, \tag{94}$$

$$a_{33}(m_1TM) = a_{33}(m_1m_2m_3) = m_3 = f_3 - s + T - M + 1. \tag{95}$$

The matrix elements of the generators $A_{23} = T_+$ and $A_{32} = T_-$ can be determined by formulas (71) and (72).

The remaining four non-diagonal generators act on the T-basis vectors as follows

$$A_{ij}|\langle f\rangle m_1TM >_q = \sum_{T'} a_{ij}(m_1'T'M')\ |\langle f\rangle m_1'T'M'\rangle_q \tag{96}$$

where

$$T' = T \pm 1/2, \quad M' = M \pm 1/2, \tag{97}$$

and

$$a_{ij}(m_1'T'M') = {}_q\langle m_1'T'M'|A_{ij}|m_1TM\rangle_q. \tag{98}$$

These matrix elements are presented in Table 2(the derivation of these expressions is given in [24].

5. Weyl coefficients $\langle U|T\rangle_q$ for the positive discrete series of unitary irreducible representations of the $u_q(2,1)$ quantum algebra

By definition, the Weyl coefficient $\langle U|T\rangle_q$ for an irreducible representation $\langle f\rangle$ of the $u_q(2,1)$ quantum algebra has a form

$$\langle U|T\rangle_q = {}_q\langle\{f\}m_3UM_U|\{f\}m_1TM_T\rangle_q$$

$$= \frac{(-1)^{k+\ell}}{N(k\ell)N(UM_U)N(sp)N(TM)}\ {}_q\langle L|A_{31}^{\ell}A_{32}^{k}P^UA_{12}^{a}A_{23}^{b}P^TA_{21}^{p}A_{13}^{s}|L\rangle_q, \tag{99}$$

Table 2. Matrix elements of the generators of the noncompact $u_q(2,1)$ quantum algebra for the unitary irreducible representation $D^{\{\langle f\rangle +\}}$ of the positive discrete series (T-spin basis used here was constructed with the aid of the lowest weight vector $|L\rangle$).

$a_{12}\left(m_1+1, T+\frac{1}{2}, M-\frac{1}{2}\right) = \left[\frac{[s+1][f_1-f_3+s][2T-p+1][-T+M-1]}{[2T+1][2T+2]}\right]^{1/2}$
$a_{13}\left(m_1+1, T+\frac{1}{2}, M+\frac{1}{2}\right) = q^{T-M+1}\left[\frac{[s+1][f_1-f_3+s][2T-p+1][T+M+1]}{[2T+1][2T+2]}\right]^{1/2}$
$a_{12}\left(m_1+1, T-\frac{1}{2}, M-\frac{1}{2}\right) = \left[\frac{[p][f_1-f_2-p+1][2T-s][T+M]}{[2T][2T+1]}\right]^{1/2}$
$a_{13}\left(m_1+1, T-\frac{1}{2}, M+\frac{1}{2}\right) = q^{-T-M}\left[\frac{[p][f_1-f_2-p+1][2T-s][-T+M]}{[2T][2T+1]}\right]^{1/2}$
$a_{21}\left(m_1-1, T-\frac{1}{2}, M+\frac{1}{2}\right) = \left[\frac{[s][f_1-f_3+s-1][2T-p][-T+M]}{[2T][2T+1]}\right]^{1/2}$
$a_{31}\left(m_1-1, T-\frac{1}{2}, M-\frac{1}{2}\right) = -q^{-T+M-1}\left[\frac{[s][f_1-f_3+s-1][2T-p][T+M]}{[2T][2T+1]}\right]^{1/2}$
$a_{21}\left(m_1-1, T+\frac{1}{2}, M+\frac{1}{2}\right) = \left[\frac{[p+1][f_1-f_2-p][2T-s+1][T+M+1]}{[2T+1][2T+2]}\right]^{1/2}$
$a_{31}\left(m_1-1, T+\frac{1}{2}, M-\frac{1}{2}\right) = -q^{T+M}\left[\frac{[p+1][f_1-f_2-p][2T-s+1][-T+M-1]}{[2T+1][2T+2]}\right]^{1/2}$

where $|L\rangle$ is the lowest weight vector of the irreducible representation $\langle f\rangle$:

$$a = U - M_U, \qquad b = -T + M - 1, \tag{100}$$

and the normalization factors $N(k\ell)$, $N(UM_U)$, $N(sp)$, and $N(TM)$ and the projection operators P^U and P^T were defined in the foregoing. Since the weight in the left hand side of Eq. (99) for the matrix element is equal to the weight of the right hand side, we conclude that the parameters k and ℓ are related to s and p by the equations

$$U - M_U = p - s + \ell, \tag{101}$$

$$T - M + 1 = s - \ell - k. \tag{102}$$

The computation of the above matrix element is performed by making use of the commutation relations between the generators raised to a power. The scheme of the computations is identical to that in the case of the $u_q(3)$ algebra [25]. Taking into account the explicit form of projection operator P^U, we arrive at

$$\begin{aligned} B &= {}_q\langle L|A_{13}^{\ell}A_{32}^{k}P^U A_{21}^{r}A_{12}^{a}A_{23}^{b}P^T A_{21}^{p}A_{13}^{s}|L\rangle_q \\ &= \sum_r (-1)^r \frac{[2U+1]!}{[r]![2U+r+1]!}\, {}_q\langle L|A_{31}^{\ell}A_{32}^{k}A_{21}^{r}A_{12}^{a+r}A_{23}^{b}P^T A_{21}^{p}A_{13}^{s}|L\rangle_q. \end{aligned} \tag{103}$$

With the aid of commutation relations, we transfer the operator A_{21}^r in the matrix element to the left until it appears immediately after vector $\langle L|$ and consider that $\langle L|A_{21} = 0$. The expression for B then takes the form

$$B = \sum_r \frac{[2U+1]![k]!}{[r]![k-r]![2U+r+1]!} B_1, \tag{104}$$

where

$$B_1 = {}_q\langle L|A_{31}^{\ell+r}A_{12}^{r+a}A_{32}^{k-r}A_{23}^{b}P^T A_{21}^{p}A_{13}^{s}|L\rangle_q. \tag{105}$$

To compute the matrix element B_1, the generators A_{32} must be transferred to the right until they appear immediately before the projection operator P^T, whereupon the equation $A_{32}P^T = 0$ is taken into account. As a result the matrix element B_1 reduces to the expression

$$B_1 = \frac{[k]![b]!}{[k-r]![b-k+r]!} \prod_t [f_3 - f_2 - a + b - k - \ell - r - t]\, B_2, \tag{106}$$

where

$$B_2 = {}_q\langle L|A_{31}^{\ell+r}A_{12}^{a+r}A_{23}^{b-k+r}P^T A_{21}^{p}A_{13}^{s}|L\rangle_q. \tag{107}$$

Considering that, in the case of T-spin basis,

$$2T = f_2 - f_3 + p + s - 2 \tag{108}$$

and that all factors in the product $\prod_t$ are negative, we recast the product into the form

$$\prod_t [f_3 - f_2 - a + b - k - \ell - r - t] = (-1)^{k-r} \frac{[f_2 - f_3 + p + \ell + k - 1]!}{[f_2 - f_3 + p + l + r - 1]!}. \tag{109}$$

Further, we transfer the generators A_{23}^{b-k+r} in the expression for the matrix element B_2 to the left until they appear immediately after the vector $\langle L|$, which annihilates them, and consider that $\langle L|A_{21}^y = \delta_{y,0}\langle L|$. As a result we have

$$B_2 = \frac{[a+r]!}{[a-b+k]!}\, {}_q\langle L|A_{31}^{\ell+r} A_{13}^{b-k+r} A_{12}^{a-b+k} P^T A_{21}^p A_{13}^s |L\rangle_q. \tag{110}$$

The commutation of generators $A_{31}^{\ell+r}$ and A_{13}^{b-k+r} whereupon the condition $\langle L|A_{13}^z = \delta_{z,0}\langle L|$ is taken into account, gives the ultimate expression for the matrix element B_2:

$$\begin{aligned} B_2 &= (-1)^{\ell+r+s} \frac{[a+r]![\ell+r]![f_1 - f_3 + \ell + r - 1]!}{[p]![s]![f_1 - f_3 + s - 1]!} \\ &\quad \times {}_q\langle L|A_{31}^s A_{12}^p P^T A_{21}^p A_{13}^s |L\rangle_q \\ &= (-1)^{\ell+r} \frac{[a+r]![\ell+r]![f_1 - f_3 + \ell + r - 1]!}{[p]![s]![f_1 - f_3 + s - 1]!} N^2(sp). \end{aligned} \tag{111}$$

Combining the above results, we reduce the expression for the Weyl coefficients $\langle U|T\rangle_q$ of the form

$$\begin{aligned} \langle U|T\rangle_q = &\left\{ \frac{[2U+1][2T+1][k]![-T-1+M][U+M_U]![T+M]!}{[p]![s]![\ell]![U-M_U]![f_1 - f_3 + s - 1]!} \right. \\ &\times \left. \frac{[f_1 - f_2 - k]![f_1 - f_2 + \ell + 1]![f_2 - f_3 + s - 2]![f_2 - f_3 + p - 2]!}{[f_1 - f_2 - p]![f_2 - f_3 + k - 2]![f_1 - f_3 + \ell - 1]!} \right\}^{1/2} \\ &\times \sum_r (-1)^r \frac{[U - M_U + r]![\ell + r]![f_1 - f_3 + \ell + r - 1]!}{[r]![2U + r + 1]![k - r]![\ell - s + r]![f_2 - f_3 + p + \ell + r - 1]!}. \end{aligned} \tag{112}$$

The substitution $r = k - n$ allows to rewrite the last formula as follows

$$\langle U|T\rangle_q = \left\{ \frac{[2U+1][2T+1][k]![-T+M-1]![U+M_U]![T+M]!}{[s]![p]![\ell]![U-M_U]![f_1-f_3+s-1]!} \right.$$

$$\times \left. \frac{[f_1-f_2-k]![f_1-f_2+\ell+1]![f_2-f_3+s-2]![f_2-f_3+p-2]!}{[f_1-f_2-p]![f_2-f_3+k-2]![f_1-f_3+\ell-1]!} \right\}^{1/2}$$

$$\times \sum_n \frac{(-1)^{k+n}[U-M_U+k-n]![\ell+k-n]!}{[n]![k-n]![2U+1+k-n]![\ell-s+k-n]!}$$

$$\times \frac{[f_1-f_3+\ell+k-n-1]!}{[f_2-f_3+p+\ell+k-n-1]!}. \qquad (113)$$

6. Relation between the q-Weyl coefficients for the $u_q(2,1)$ quantum algebra and q-Racah coefficients for the $su_q(2)$ quantum algebra.

The explicit expression for the q-Weyl coefficient for the $u_q(3)$ algebra was obtained in [25], and its relation to the Racah coefficient for the $su_q(2)$ quantum algebra was established there. Here, we show that the expression (113) for the q-Weyl coefficient for the $u_q(2,1)$ quantum algebra can also be related to the q-Racah coefficients for the $su_q(2)$ quantum algebra. Our consideration is based on one of five general formulas in [25] for the q-Racah coefficient for the $su_q(2)$ quantum algebra [namely, formula (5.31) in [25]]:

$$U_q(abed;cf) = (-1)^{a+d-c-f} \left\{ \frac{[2c+1][2f+1][a+b+c+1]!}{[a+e+f+1]![c+d+e+1]!} \right.$$

$$\times \frac{[b+d+f+1]![a-b+c]![-a+b+c]![a+e-f]![b-d+f]!}{[a+b-c]![a-e+f]![b+d-f]![c+d-e]![c-d+e]!}$$

$$\times \left. \frac{[-b+d+f]![-c+d+e]!}{[-a+e+f]!} \right\}^{1/2}$$

$$\times \sum_n \frac{(-1)^n[2b-n]![b+c-e+f-n]!}{[n]![-a+b+c-n]![b-d+f-n][a+b+c+1-n]!}$$

$$\times \frac{1}{[b+d+f+1-n]!}. \qquad (114)$$

A comparison of the expression (114) with formula (113) for the Weyl coefficient for the positive discrete series of the representations of $u_q(2,1)$ reveals the summands in the two formulas coincide, provided

$$a = T = \frac{1}{2}(f_2 - f_3 + p + s - 2), \quad b = j_3 = \frac{1}{2}(\ell + k),$$

$$c = j_2 = \frac{1}{2}(f_2 - f_3 + p - s + \ell + k - 2)), \tag{115}$$

$$d = U = \frac{1}{2}(f_1 - f_2 + \ell - k), \qquad e = j_1 = \frac{1}{2}(f_1 - f_3 - p + s - 2),$$

$$f = j = \frac{1}{2}(f_1 - f_2). \tag{116}$$

Further, the substitution of parameters a, b, c, d, e and f from the formulas (115) and (116) gives the relation between the q-Weyl coefficient (113) for the $u_q(2,1)$ quantum algebra and q-Racah coefficient for the $su_q(2)$ quantum algebra,

$$\langle U \mid T \rangle_q = (-1)^s \sqrt{\frac{[2U+1][2T+1]}{[2j_2+1][2j+1]}}\, U(Tj_3j_1U; j_2j)_q$$
$$= (-1)^k U(j_1j_2jj_3; UT)_q. \tag{117}$$

7. Conclusion

In this study, the projection operators for the $su_q(2)$ subalgebra have used to explore the positive discrete series of unitary irreducible representations of the noncompact $u_q(2,1)$ quantum algebra. The q-analog of the Gel'fand–Graev formulas has been derived in the basis associated with the reduction $u_q(2,1) \to su_q(2) \times u(1)$. It seems that the reduction $u_q(2,1) \to u(1) \times su_q(1,1)$ for the discrete series of the lowest weight representations has been considered for the first time in the present study. With the aid of the projection operator for the $su_q(1,1)$ subalgebra, we constructed the basis of the representation for this reduction and calculated the matrix elements of the generators. We have obtained analytic expressions for the elements of the transformation brackets $\langle U|T\rangle_q$ relating the U-spin and T-spin bases of the lowest weight irreducible representations. By the analogy with q-Weyl coefficients for the $u_q(3)$ algebra [25], they can be called the q-Weyl coefficients for the noncompact $u_q(2,1)$ algebra. It has been explicitly shown that these q-Weyl coefficients are equivalent (apart from phase factor) to specific q-Racah coefficient for the $u_q(2)$ algebra or are proportional to the q–6j symbol for the $su_q(2)$ algebra. The negative discrete series was discussed by us in [26]. The intermediate discrete series requires a dedicated investigation, and this will be done in our further publication.

Acknowledgments

This work was supported by the Russian Foundation of Basic Research (project No 02-01-00668).

Appendix: Normalization of the U-spin basis vectors of the $u_q(2,1)$ algebra (positive discrete series).

The structure of the U-basis vectors is described by formulas (19)–(26).

Here, we use the transformation properties of the "noncompact" generators under Hermitian conjugation and the properties of projection operator P^U:

$$(P^U)^+ = P^U, \tag{A.1}$$

$$(P^U)^2 = P^U, \tag{A.2}$$

$$A_{12}^{U-M_U} A_{21}^{U-M_U} P^U = N^2(UM_U)P^U. \tag{A.3}$$

With allowance for these formulas, the square of the norm $N^2(k\ell)$ takes a form:

$$N^2(k\ell) = (-1)^{k+\ell} \langle L|A_{31}^\ell \tilde{A}_{32}^k P^U A_{23}^k A_{13}^\ell|L\rangle, \tag{A.4}$$

where

$$\tilde{A}_{31} = A_{32}A_{23} - qA_{23}A_{32}. \tag{A.5}$$

Since, by definition, the relation

$$A_{31} = A_{32}A_{21} - q^{-1}A_{21}A_{32} \tag{A.6}$$

holds we can represent the generator $\tilde{A}_{31}$ in the form

$$\tilde{A}_{31} = A_{31} - (q - q^{-1})A_{21}A_{32}. \tag{A.7}$$

From the relations

$$A_{12}P^U = P^U A_{21} = 0 \tag{A.8}$$

it follows that

$$N^2(k\ell) = (-1)^{k+\ell} \langle L|A_{23}^\ell A_{13}^k P^U A_{31}^k A_{32}^\ell|L\rangle. \tag{A.9}$$

A straightforward computation of $N^2(k\ell)$ by transferring of lowering generators to the lowest vector $|L\rangle$ is rather cumbersome. In view of this, we will try to construct a recursion relation between the expressions for $N^2(k\ell)$ for various values of k and ℓ, bearing in mind that

$$\langle L|P^U|L\rangle = \langle L|L\rangle = 1. \tag{A.10}$$

We begin by establishing a relation between $N^2(k\ell)$ and $N^2(k-1,\ell)$. In the expression for $N^2(k\ell)$, we replace, for this purpose, A_{32}^k by $A_{32}^{k-1}P^{U+1/2}A_{32}$. This is legitimate because, in (A.9), the projection operator $P^{U+1/2}$ taken in this combination is equivalent to the identity operator. Indeed, we have

$$P^{U+1/2}A_{32}P^U = \sum_r (-1)^r \frac{[2U+2]!}{[r]![2U+r+2]!} A_{21}^r A_{12}^r A_{32}P^U = A_{32}P^U, \tag{A.11}$$

since the generators A_{12}^r and A_{32} commute and since $A_{12}^r P^U = \delta_{r,0} P^U$. We now consider the application of the generator A_{32} to the projection operator P^U:

$$A_{32} P^U = \sum_r (-1)^r \frac{[2U+1]!}{[r]![2U+r+1]!} A_{32} A_{21}^r A_{12}^r =$$

$$= \sum_r (-1)^r \frac{[2U+1]!}{[r]![2U+r+1]!} \left(q^{-r} A_{21}^r A_{32} + [r] A_{21}^{r-1} A_{23}\right) A_{12}^r. \quad (A.12)$$

From here on, we use the commutation relations from [7, 25] for generators raised to a power. In view of the relation $P^{U+1/2} A_{21} = (A_{12} P^{U+1/2})^+ = 0$, the application of this operator on the projection operator $P^{U+1/2}$ from the left yields

$$P^{U+1/2} A_{32} P^U = P^{U+1/2} \left(A_{32} - \frac{A_{31} A_{12}}{[2U+2]} \right), \quad (A.13)$$

where

$$[2U+2] = [f_1 - f_2 - k + \ell + 2]. \quad (A.14)$$

As a result, the square of the norm becomes

$$N^2(k\ell) = (-1)^{k+\ell} \left\langle L \left| A_{31}^\ell A_{32}^{k-1} P^{U+1/2} \left(A_{32} - \frac{A_{32} A_{21}}{[f_1 - f_2 - k + \ell + 2]} \right) A_{23}^k A_{13}^\ell \right| L \right\rangle. \quad (A.15)$$

Commuting the generators A_{32} and A_{23}^k, we arrive at

$$A_{32} A_{23}^k A_{13}^\ell \, |L >= [k][f_3 - f_2 - k - \ell + 1] A_{23}^{k-1} A_{13}^\ell \, |L > . \quad (A.16)$$

and

$$A_{31} A_{12} A_{23}^k A_{13}^\ell \, |L\rangle = [k] A_{31} A_{23}^{k-1} A_{13}^{\ell+1} \, |L > . \quad (A.17)$$

The commutation of the generators A_{32} and A_{31}^ℓ makes it possible to derive the relation

$$A_{32} A_{13}^\ell = \left(A_{13}^\ell - [\ell] q^{\ell-1} A_{13}^{\ell-1} A_{12} q^{-(A_{22} - A_{33} - 1)} \right). \quad (A.18)$$

Transferring the generator A_{31} to the right until it appears immediately in front of the lowest weight $|L\rangle$, which annihilates it, we obtain

$$P^{U+1/2} A_{32} A_{12} A_{23}^k A_{13}^\ell |L\rangle = [k][\ell+1][f_3 - f_1 - \ell] P^{U+1/2} A_{23}^{k-1} A_{13}^\ell |L\rangle; \quad (A.18)$$

therefore, we have

$$P^{U+1/2} A_{32} P^U A_{23}^k A_{13}^\ell \, |L\rangle$$

$$= [k]([f_3 - f_2 - k - \ell + 1] - \frac{[\ell+1][f_3 - f_1 - \ell]}{[f_1 - f_2 - k + \ell + 2]} P^{U+1/2} A_{23}^{k-1} A_{13}^\ell \, |L\rangle$$

$$= -\frac{[k][f_1 - f_2 - k + 1][f_2 - f_3 + k - 2]}{[f_1 - f_2 - k + \ell + 2]} P^{U+1/2} A_{23}^{k-1} A_{13}^\ell \, |L\rangle. \quad (A.19)$$

Thus, the square of the norm, $N^2(k\ell)$, takes the form

$$N^2(k\ell) = (-1)^{k+\ell-1} \frac{[k][f_1 - f_2 - k + 1][f_2 - f_3 + k - 2]}{[f_1 - f_2 - k + \ell + 2]}$$

$$\times \langle L | A_{31}^\ell A_{32}^{k-1} P^{U+1/2} A_{23}^{k-1} A_{23}^\ell | L\rangle. \quad (A.20)$$

In other words we have derived a recursion relation between $N^2(k\ell)$ and $N^2(k-1,\ell)$,

$$N^2(k\ell) = \frac{[k][f_1 - f_2 - k + 1][f_2 - f_3 + k - 2]}{[f_1 - f_2 - k + \ell + 2]} N^2(k-1,\ell). \tag{A.21}$$

The recursion relation

$$N^2(0\ell) = (-1)^\ell \langle H|A_{31}^\ell P^U A_{13}^\ell|H\rangle = [\ell][f_1 - f_3 + \ell - 1]\, N^2(0,\ell-1) \tag{A.22}$$

can be obtained in a similar way.

Using these recursion relations, we arrive at an ultimate expression for for the square of the norm in (A.9):

$$N^2(k\ell) = \frac{[k]![\ell]![f_1-f_2]![f_1-f_2-k+\ell+1]![f_2-f_3+k-2]![f_1-f_3+\ell-1]!}{[f_1-f_2-k]![f_1-f_2+\ell+1]![f_1-f_3-1]![f_2-f_3-2]!}. \tag{A.23}$$

References

[1] S.T. Belyaev, I.M. Pavlichenkov, Yu. F. Smirnov, Nucl. Phys. **444**, 36 (1985).
[2] P.P. Kulish, N. Yu. Reshetikhin, Zapiski Nauchn. Seminarov LOMI **101**, 101, (1981).
[3] E.K. Sklyanin, Funkzionalny Analiz i Pril. **16**(4), 27 (1982).
[4] L.D. Faddeev, L.A. Takhtajan, Lect. Notes. Phys. **246**, 183 (1986).
[5] P.P. Raychev, R.P. Roussev, Yu. F. Smirnov, J. Phys. G **16**, L137 (1990).
[6] D. Bonatsos, C. Daskaloyannis, Progr. Part. Nucl. Phys. **43**, 537 (1990).
[7] Yu. F. Smirnov, V.N. Tolstoy, Yu. I. Kharitonov, Yadern. Fis. **54**, 721 (1991).
[8] Yu. F. Smirnov, Yu. I. Kharitonov, Yadern. Fis **56**, 223 (1991).
[9] A.A. Malashin, Yu. F. Smirnov, Yu. I. Kharitonov, Yadern. Fis **58**, 665 (1995).
[10] Yu. F. Smirnov, Yu. I. Kharitonov, Yadern. Fis. **58**, 749 (1995).
[11] A. A. Malashin, Yu. F. Smirnov, Yu. I. Kharitonov, Yadern. Fis. **58**, 1105 (1995).
[12] Yu. F. Smirnov, Yu. I. Kharitonov, Yadern. Fis. **59**, 379 (1996).
[13] Yu. F. Smirnov, Yu. I. Kharitonov, Preprint PIYaF-2140 (St. Petersbourg, 1996).
[14] Yu. F. Smirnov, Yu. I. Kharitonov, Preprint PIYaF-2251, (St. Petersbourg, 1998).
[15] Yu. F,Smirnov, Yu. I. Kharitonov, Preprint PIYaF-2301, (St. Petersbourg, 1999).
[16] Yu. F.Smirnov, Yu. I. Kharitonov, Preprint PIYaF-2345, (St. Petersbourg, 2000).
[17] I.M. Gel'and, M.I. Graev, Izv. AN SSSR, Ser. Mat. **29**, 1329 (1965).
[18] A. Barut, R. Raczka, *Theory of Group Representations and Applications.* Polish Scientific Publishers, Warszava 1977.
[19] U. Ottoson, Comm. Math. Phys. **10**, 114 (1968).
[20] Yu. F. Smirnov, V.N. Tolstoy, V.A. Knyr, L. Ya. Stotland, *Group Theory Methods in Physics,* (Nauka, Moscow,1986) vol. 2, p. 77.
[21] I.T. Todorov, Preprint IC/66/71, ICTP (Triest, 1966).
[22] N. Ya. Vilenkin, *Special functions and the theory of group representations.* Translation of Mathematical Monographs, vol. 22, American Mathematical Society, 1968.
[23] Yu. F. Smirnov, V.N. Tolstoy, Yu. I. Kharitonov, Yadern. Fis. **53**, 959 (1991).
[24] Yu. F. Smirnov, Yu. I. Kharitonov, Preprint PIYaF-2446, (St. Petersbourg, 2000).
[25] R.M. Asherova, Yu. F. Smirnov, V.N. Tolstoy, Yadern. Fis. **59**, 1859 (1996).
[26] Yu. F. Smirnov, Yu. I. Kharitonov, R.M. Asherova, Yadern. Fis. **66**, 1969 (2003).

COMBINATORIAL PHYSICS, NORMAL ORDER AND MODEL FEYNMAN GRAPHS

A.I. Solomon[†⋆], P. Blasiak[†‡], G. Duchamp[⋆⋆], A. Horzela[‡] and K.A. Penson[†]

[†] *Université Pierre et Marie Curie*
Laboratoire de Physique Théorique des Liquides, CNRS UMR 7600
Tour 16, 5ième étage, 4, place Jussieu, F 75252 Paris, Cedex 05, France
penson@lptl.jussieu.fr

[‡] *H. Niewodniczański Institute of Nuclear Physics,*
Polish Academy of Sciences
Department of Theoretical Physics
ul. Radzikowskiego 152, PL 31-342 Kraków, Poland
pawel.blasiak@ifj.edu.pl, andrzej.horzela@ifj.edu.pl

[⋆] *The Open University*
Physics and Astronomy Department
Milton Keynes MK7 6AA
a.i.solomon@open.ac.uk

[⋆⋆] *LIFAR, Université de Rouen*
76821 Mont-Saint Aignan Cedex, France
gduchamp2@free.fr

Abstract The general normal ordering problem for boson strings is a combinatorial problem. In this talk we restrict ourselves to single-mode boson monomials. This problem leads to elegant generalisations of well-known combinatorial numbers, such as Bell and Stirling numbers. We explicitly give the generating functions for some classes of these numbers. Finally we show that a graphical representation of these combinatorial numbers leads to sets of model field theories, for which the graphs may be interpreted as Feynman diagrams corresponding to the bosons of the theory. The generating functions are the generators of the classes of Feynman diagrams.

Keywords: Combinatorics, normal order, Feynman diagrams.

B. Gruber (ed.), Symmetries in Science XI, 527–536.

1. Boson Normal Ordering

In this note we give a brief review of the combinatorial properties associated with the normal ordering of bosons, and the model Feynman graphs which result.

Combinatorial sequences appear naturally in the solution of the boson normal ordering problem [1], [2].

The normal ordering problem for canonical bosons $[a, a^\dagger] = 1$ is related to certain combinatorial numbers $S(n, k)$ called Stirling numbers of the second kind through [3]

$$(a^\dagger a)^n = \sum_{k=1}^{n} S(n, k)(a^\dagger)^k a^k, \tag{1}$$

with corresponding numbers $B(n) = \sum_{k=1}^{n} S(n, k)$ called Bell numbers. In fact, for physicists, these equations may be taken as the *definitions* of the Stirling and Bell numbers. For quons (q-bosons) satisfying $[a, a^\dagger]_q \equiv aa^\dagger - qa^\dagger a = 1$ a natural q-generalisation [4] of these numbers is

$$(a^\dagger a)^n = \sum_{k=1}^{n} S_q(n, k)(a^\dagger)^k a^k. \tag{2}$$

In the canonical boson case, for integers $n, r, s > 0$ we define generalized Stirling numbers of the second kind $S_{r,s}(n, k)$ through ($r \geq s$):

$$[(a^\dagger)^r a^s]^n = (a^\dagger)^{n(r-s)} \sum_{k=s}^{ns} S_{r,s}(n, k)(a^\dagger)^k a^k, \tag{3}$$

as well as generalized Bell numbers $B_{r,s}(n)$

$$B_{r,s}(n) = \sum_{k=s}^{ns} S_{r,s}(n, k). \tag{4}$$

For both $S_{r,s}(n, k)$ and $B_{r,s}(n)$ exact and explicit formulas have been found [1, 2]. We refer the interested reader to these sources for further information on those extensions. However, in this note we shall only deal with the classical Bell and Stirling numbers, corresponding to $B_{1,1}(n)$ and $S_{1,1}(n)$ in our notation.

2. Generating Functions

In general, for combinatorial numbers $g(n)$ we may define an *exponential generating function* $G(x)$ through [6]

$$G(x) = \sum_{n=0}^{\infty} g(n) \frac{x^n}{n!}. \tag{5}$$

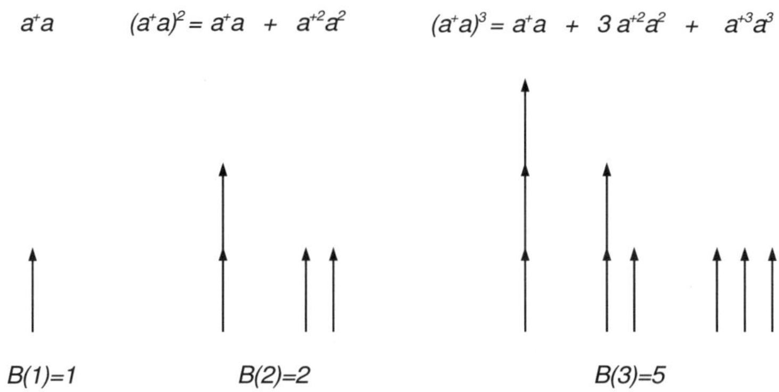

Figure 1. Arrow graphs for $(a^\dagger a)^n \quad n = 1, 2, 3.$

For the Bell numbers, the generating function takes the particularly nice form [5]

$$G(x) = \sum_{n=0}^{\infty} B(n)\frac{x^n}{n!} = \exp(\exp(x) - 1). \tag{6}$$

Some initial terms of the sequence $\{B(n)\}$ are $\{1, 2, 5, 15, 52, 203, 877, \ldots\}$.

3. Graphs

A convenient way of representing combinatorial numbers is by means of *graphs*. To illustrate this, we now consider a graphical method for illustrating the combinatorial numbers associated with the normal order expansion of $(a^\dagger a)^n$.

A single arrow represents a "time-segment" of a line corresponding to the "propagator" $(a^\dagger a)$. Thus we may concatenate one, two or more arrows to form a single line, or propagator $(a^\dagger a)$. However, two lines correspond to two distinct propagators ${a^\dagger}^2 a^2$, and so on. Further, the constitutent arrows are labelled, for example by time, and so they may only be concatenated respecting the time ordering. These rules are illustrated by the diagrams of Figure 1, in which we consider the cases of 1, 2, and 3 arrows respectively. We have pre-emptively labelled these numbers as Bell numbers - which fact we demonstrate below.

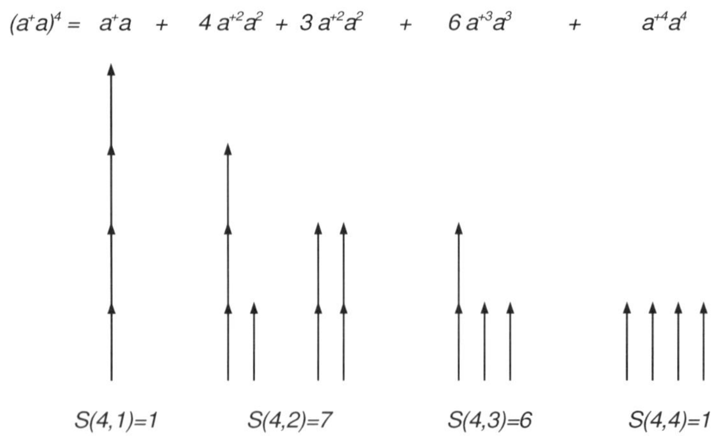

Figure 2. Arrow graphs for $(a^\dagger a)^4$.

For the case of 4 arrows (Figure 2) we have additionally given the individual associated symmetry factors (in fact Stirling numbers) which add to $B(4) = 15$. It should be clear from this illustration how the time-ordering rules are applied to give the symmetry coefficients. Thus there is only one way in which we can concatenate 4 arrows respecting time-ordering (first grouping), 4 ways in which we can divide the arrows into a set of 3 and 1, and so on.

From these first few examples it would seem that these graphs are essentially like the Feynman Diagrams of a zero-dimensional (no integration) Model Field Theory associated with $H = a^\dagger a$. In other words, at order n the total number of graphs would appear to be $B(n)$ while the individual coefficients of ${a^\dagger}^k a^k$ are given by $S(n,k)$. In order to show that this is indeed the case, we must be able to count the number of graphs associated with a given number n of arrows. To do this we can use the First of Three Great Results.

4. First Great Result

This First Great Result is sometimes known as the *connected graph theorem* [7]. It states that if $C(x) = \sum_{n=1}^{\infty} c(n)x^n/n!$ is the generating function of *labelled connected* graphs, *viz.* $c(n)$ counts the number of connected graphs of order n, then

$$A(x) = \exp(C(x)) \tag{7}$$

is the generating function for *all* graphs.

We may apply this very simply to the case of the arrow graphs above. For each order n, the connected graphs consist of the single graph obtained by concatenating all the arrows into one propagator. Therefore for each n we have $c(n) = 1$; whence, $C(x) = exp(x) - 1$. It follows that the generating function for all the arrow graphs $A(x)$ is given by

$$A(x) = \exp(\exp(x) - 1) \tag{8}$$

which is the generating function for the Bell numbers.

Such graphs may be generalised to give graphical representations for the extensions $B_{r,s}(n)$ [8].

However, just as an abstract group is capable of more than one presentation, there are many graphical representations for a given combinatorial sequence; and we now give an alternative one for the numbers $B(n)$, $S(n,k)$ due to Bender and collaborators [9], [10].

5. Second Great Result

As before, we shall be counting lines. A line starts from a white dot, the *origin*, and ends at a black dot, the *vertex*. What we refer to as *origin* and *vertex* is, of course, arbitrary. At this point there are no other rules, although we are at liberty to impose further restrictions; a white dot may be the origin of 1,2,3,... lines, and a black dot the vertex for 1,2,3,... lines. We may further associate *strengths* V_s with each vertex receiving s lines, and multipliers L_m with a white dot which is the origin of m lines. Again $\{V_s\}$ and $\{L_m\}$ play symmetric roles; in this note we shall only consider cases where the L_m are either 0 or 1.

We illustrate these rules for four different graphs corresponding to $n = 4$.

There is an generating function $G(x, V, L)$ which counts the number $g(n)$ of graphs with n lines arising from the above rules [11]:

$$\begin{aligned} G(x,V,L) &= \exp(\sum_{m=1}^{\infty} L_m \frac{x^m}{m!}\frac{d^m}{dy^m}) \exp(\sum_{s=1}^{\infty} V_s \frac{y^s}{s!})\bigg|_{y=0} \\ &\equiv \sum_{n=0}^{\infty} g(n)\frac{x^n}{n!} \end{aligned} \tag{9}$$

Consider the following example: $L_m = 0$ for all $m \neq 1$, that is, we allow only one line from each origin (with multiplier 1); there is no restriction on the number of lines to a vertex, and $V_s = 1$ for all s. We give an example of the cases $n = 1, 2, 3, 4$ in Figure 4. Note that for correct counting the lines should be labelled, as they were in the case of the arrows above.

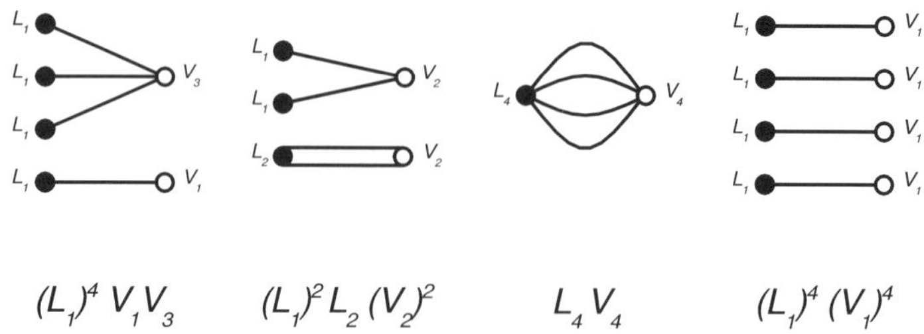

Figure 3. Some examples of 4-line graphs.

The generating function $G(x)$ which counts the lines corresponding to the above rules follows immediately from Eq.(9)

$$\begin{aligned} G(x) &= \exp(\sum_{m=1}^{\infty} L_m \frac{x^m}{m!}\frac{d^m}{dy^m}) \exp(\sum_{s=1}^{\infty} V_s \frac{y^s}{s!})\bigg|_{y=0} \\ &= \exp(\frac{x}{1!}\frac{d}{dy}) \exp(\sum_{s=1}^{\infty} \frac{y^s}{s!})\bigg|_{y=0} \\ &= \exp(xd/dy)\exp(e^y - 1)|_{y=0} \\ &= \exp(e^x - 1) \equiv \sum_{n=0}^{\infty} B(n)\frac{x^n}{n!}. \end{aligned} \tag{10}$$

The penultimate step is a consequence of the Taylor expansion.

We thus have yet another representation of the integer sequence $\{B(n)\}$. Note that when $L \equiv \{L_m\}$ and $V \equiv \{V_s\}$ are integer sequences we obtain an integer sequence from Eq.(9). A convenient method of obtaining the resulting integer sequence is afforded by the next useful result.

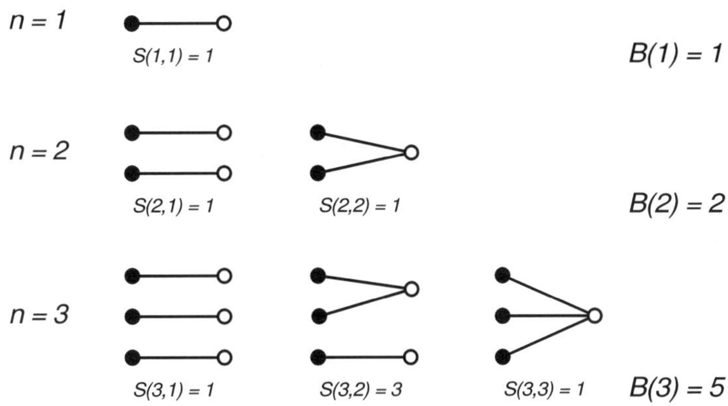

Figure 4. Graphs of second type for $B(n), \quad n = 1, 2, 3.$

6. Third Great Result

Straightforward manipulation of series shows the following: Define

$$A_i(x) = \sum_{n=0}^{\infty} a_i(n) \frac{x^n}{n!} \qquad i = 1, 2, 3. \tag{11}$$

Then if

$$A_1(xd/dy) A_2(y)|_{y=0} = A_3(x) \tag{12}$$

we have

$$a_3(n) = a_1(n) a_2(n). \tag{13}$$

This is a useful and rather surprising equality.

Using the results of the previous section, Eq.(13) enables us to create graphical representations of *products* of integral sequences. For example: if in the case above we chose $L_m = 1$ for all m, enabling any number of lines from each origin (with multiplier 1), the resulting sequence of graphs would have given us a representation of the integer sequence $\{B(n)^2\}$ [9].

We exemplify the possibilities offered by application of the Third Great Result by our two final examples.

Example 1: Generating function for the sequence $\{B(n)B(n+1)\}$.

From Eqs.(11) and (13), the required generating function is given by

$$G(x, V, L) = A_1(xd/dy)A_2(y)\big|_{y=0} \tag{14}$$

where $A_1(x)$ is the generating function for $\{B(n)\}$ and $A_2(x)$ is that of $\{B(n+1)\}$. Note that $A_1(x) = \exp(\exp(x) - 1)$ from Eq.(8), while $A_2(x) = (d/dx)A_1(x) = \exp(\exp(x) - 1 + x)$.

This shows that the graphs for the sequence $\{B(n)B(n+1)\}$ may be obtained by putting $L_m = 1$ for all m, so that there are any number of lines emanating from an origin, and with multiplicity 1. For the sequence $\{B(n+1)\}$ we have $V_s = 1$ for all $s \neq 2$ and $V_2 = 2$, so that any number of lines may end at a vertex, and all have strength 1 except for the case where *two* lines meet at a vertex, when the strength is 2. The generating function is, according to Eq.(13),

$$G_1(x, V, L) = \sum_{n=0}^{\infty} B(n)B(n+1)\frac{x^n}{n!}. \tag{15}$$

It may be explicitly obtained after some formal algebraic manipulation based on the Dobiński formula [12]

$$B(n) = \frac{1}{e}\sum_{k=0}^{\infty}\frac{k^n}{k!} \tag{16}$$

as

$$G_1(x, V, L) = \sum_{k=0}^{\infty}\frac{1}{k!}\exp(\exp((k+1)x) - 2). \tag{17}$$

The formal series (17) diverges for all $x > 0$, although having finite Taylor coefficients for $x = 0$. Such formal series are nevertheless useful in representing integer sequences.

Example 2: In our last example we retain all the derivative terms in Eq.(11) but choose $V_1 = V_2 = 1; V_s = 0, s > 2$, thus allowing vertices where at most *two* lines meet. The corresponding generating function is defined by

$$G_2(x, V, L) = \sum_{n=0}^{\infty} B(n)I(n)\frac{x^n}{n!} \tag{18}$$

where the *Involution numbers* $I(n) = 1, 2, 4, 10, 26, 76, \ldots$ are defined through their generating function

$$G_I(x) = \exp(x + \frac{x^2}{2}) = \sum_{n=0}^{\infty} I(n)\frac{x^n}{n!} \tag{19}$$

and are special values of the Hermite polynomials $H_n(x)$

$$I(n) = H_n(\frac{1}{\sqrt{2}i})/(-\sqrt{2}i)^n.$$

Consequently

$$G_2(x, V, L) = \sum_{n=0}^{\infty} B(n) H_n(\frac{1}{\sqrt{2}i})(\frac{-x}{\sqrt{2}i})^n/n! \tag{20}$$

$$= \sum_{k=0}^{\infty} \frac{1}{k!} \exp(kx(1+\frac{kx}{2}) - 1). \tag{21}$$

In obtaining Eq.(21) we have used the standard form of the generating function of the Hermite polynomials. Again, we must consider the series Eqs.(20) and (21) as formal power series, since for example they diverge for $x > 0$.

In conclusion, we emphasize that the expressions of Eqs.(17) and (21) constitute exact solutions of Model Field Theories defined by the appropriate sets $\{V_s\}$ and $\{L_m\}$. Many other applications and extensions of the ideas sketched in this note will be found in [8].

Acknowledgments

We thank Carl Bender and Itzhak Bars for interesting discussions.

References

[1] Blasiak, P., Penson, K.A. and Solomon, A.I.: The general boson normal ordering problem, *Phys. Lett. A* **309** (2003), 198.

[2] Blasiak, P., Penson, K.A. and Solomon, A.I.: The boson normal ordering problem and generalized Bell numbers, *Ann. Comb.* **7** (2003), 127.

[3] Katriel, J.: Combinatorial aspects of boson algebra, *Lett. Nuovo Cimento* **10** (1974), 565.

[4] Katriel, J.: Bell numbers and coherent states, *Phys. Lett. A.* **273** (2000), 159.

[5] Comtet, L.: *Advanced Combinatorics*, Reidel, Dordrecht, 1974.

[6] Wilf, H.S.: *Generatingfunctionology*, Academic Press, New York, 1994.

[7] Stanley, R.P: *Enumerative Combinatorics* vol.2, Cambridge University Press, 1999; Riddel, R.J. and Uhlenbeck, G.E.: On the theory of the virial development of the equation of state of monoatomic gases, *J. Chem. Phys.* **21** (1953) 2056; Aldrovandi, R.: *Special Matrices of Mathematical Physics*, World Scientific, Singapore, 2001.

[8] Blasiak, P., Duchamp, G., Horzela, A., Penson, K.A. and Solomon, A.I.: Model combinatorial field theories, to be published.

[9] Bender, C.M, Brody, D.C. and Meister, B.K.: Quantum field theory of partitions, *J.Math. Phys.* **40** (1999) 3239; Bender, C.M., Brody, D.C., and Meister, B.K.: Combinatorics and field theory, *Twistor Newsletter* **45** (2000) 36.

[10] Bender, C.M. and Caswell, W.E.: Asymptotic graph counting techniques in ψ^{2N} field theory, *J. Math. Phys* **19** (1978) 2579; Bender, C.M., Cooper, F., Guralnik, G.S., Sharp, D.H.,

Roskies, R. and Silverstein, M.L.: Multilegged propagators in strong-coupling expansions, *Phys. Rev. D* **20** (1979) 1374.

[11] Vasiliev, N.A.: *Functional Methods in Quantum Field Theory and Statistical Physics*, Gordon and Breach Publishers, Amsterdam, 1998.

[12] Blasiak, P., Penson, K.A. and Solomon, A.I.: Dobiński-type relations and the log-normal distribution, *J. Phys. A: Math. Gen.* **36** (2003), L273.

BOHR'S SYMMETRY AND THE QUANTUM NUMBERS FOR THE TRIAXIALLY SUPERDEFORMED BANDS IN ODD MASS NUCLEI

K. Tanabe
Department of Physics,
Saitama University,
Saitama 338-8570, Japan

K. Sugawara-Tanabe
Otsuma Women's University, Tama,
Tokyo 206-8540, Japan

Abstract A problem of quantum number assignment is solved for the particle-rotor model of an odd mass nucleus with one high-j valence nucleon coupled to the triaxially superdeformed core. An algebraic method is developed by applying the Holstein-Primakoff transformation both to the total angular momentum and to the single-particle angular momentum. The allowed nuclear states are restricted by Bohr's symmetry and D_2 symmetry, which are characteristic to the nuclear Hamiltonian. A set of quantum numbers is assigned to each physical level by comparing the algebraic expression with the exact solution from the diagonalization of the rotor Hamiltonian.

Keywords: Bohr's symmetry, D_2 symmetry, triaxial rotor, odd nucleus

1. Introduction

Recently, triaxial superdeformed bands in ^{163}Lu nucleus have been observed [1], which are interpreted in terms of the wobbling motion [2] as proposed by Bohr and Mottelson [34]. Our present investigation is motivated by this experimental result on the one hand, and by the theoretical interest in the quantum numbers specifying the physical states for the triaxially superdeformed (TSD) bands in connection with our previous algebraic approach [4] on the other hand. We have already proposed the level scheme for the triaxially deformed rotor

B. Gruber (ed.), Symmetries in Science XI, 537–552.

nearly thirty years ago. This was the first application of the Holstein-Primakoff (HP) transformation to nuclear physics. The difference between our treatment and the wobbling motion in Bohr-Mottelson's textbook is in the order of approximation. We took account the whole effect coming from the next to leading order in $1/I$, in addition to the leading one, where I is the total angular momentum. This treatment gives an algebraic expression for the energy eigenvalue, which reproduces the exact values in the axially symmetric limit for the prolate shape nucleus only if we choose the z-axis as quantization axis [5, 6].

In this paper we extend our scheme to the odd mass nuclei by introducing two kinds of HP bosons for the total angular momentum $\mathbf{I}$ and for the single-particle angular momentum $\mathbf{j}$. Special attention is paid to Bohr's symmetry as well as the D_2 symmetry. Our interest is focused rather on the theoretical aspects, but it is meaningful to mention that some reasonable agreement with the experimental data has been attained for the energy difference between two superdeformed bands (i.e. TSD1 and TSD2) as functions of angular momentum, and also for the ratio of E2 transitions among these bands [8]. Our model gives two quantum numbers corresponding to the rotor angular momentum $\mathbf{R} = \mathbf{I} - \mathbf{j}$ and its component along the quantization axis. Sufficient accuracy of the algebraic expression allows us to assign the quantum numbers to the levels obtained from the exact diagonalization of the rotor Hamiltonian. We discuss a further extension of our treatment to the case of the rotor-plus-single particle Hamiltonian.

2. Holstein-Primakoff boson expansion

2.1 z-axis as a quantization axis

We consider the case where the z-axis is chosen as the quantization axis as in our old paper [4]. Then, the total angular momentum $\mathbf{I}$ and the single-particle angular momentum $\mathbf{j}$ operators are expressed in terms of bosons through the Holstein-Primakoff (HP) transformations.

$$\begin{aligned}
I_+ &= I_x + iI_y = -a^\dagger(2I - \hat{n})^{1/2}, \quad I_- = I_+^\dagger, \\
I_z &= I - a^\dagger a = I - \hat{n}, \\
j_+ &= j_x + ij_y = (2j - \hat{k})^{1/2} b, \qquad j_- = j_+^\dagger, \\
j_z &= j - b^\dagger b = j - \hat{k},
\end{aligned} \tag{1}$$

where a and b are two kinds of boson operators which commute with each other. It can easily be confirmed that the components along the principal axes of the rotor, I_i ($i = 1, 2, 3$; or x, y, z), satisfy the commutation relations $[I_i, I_j] = -iI_{i\times j}$ with $-$ sign, while j_i satisfy $[j_i, j_j] = ij_{i\times j}$ with $+$ sign. The operator I_i commutes with j_j, i.e. $[I_i, j_j] = 0$.

We express the rotor Hamiltonian

$$H_{\rm rot} = \sum_i \frac{(I_i - j_i)^2}{2\mathcal{J}_i} \equiv \sum_i A_i (I_i - j_i)^2 \tag{2}$$

in terms of these bosons. Then we diagonalize $H_{\rm rot}$ using Eq. (1) in two steps. We expand $H_{\rm rot}$ up to the order $\hat{n}/I$ and $\hat{k}/j$. In the first step, the bosons a and b are transformed into other bosons α and β in order to diagonalize the terms of 0th order in $\hat{n}/I$ and $\hat{k}/j$, and in the second step α and β are transformed into the final bosons ρ and σ to include also the terms in first order $\hat{n}/I$ and $\hat{k}/j$ of $H_{\rm rot}$. The transformation in the first step is given by

$$\begin{aligned} \alpha &= \eta_+ \frac{\sqrt{I}a + \sqrt{j}b^\dagger}{\sqrt{I-j}} - \eta_- \frac{\sqrt{I}a^\dagger + \sqrt{j}b}{\sqrt{I-j}}, \\ \beta &= \frac{\sqrt{j}a^\dagger + \sqrt{I}b}{\sqrt{I-j}}, \end{aligned} \tag{3}$$

with

$$\eta_\pm^2 = \frac{1}{2}\left(\frac{\xi}{\zeta} \pm 1\right), \tag{4}$$

where

$$\xi = A_z - \frac{1}{2}(A_x + A_y), \quad \eta = \frac{1}{2}(A_y - A_x), \quad \zeta = \sqrt{\xi^2 - \eta^2}. \tag{5}$$

In the symmetric limit of $A_x = A_y$ (prolate shape), η_+^2 becomes 1 and η_-^2 vanishes. However, in the limit of $A_z = A_y$ (oblate shape), both of $\eta_\pm^2$ diverge because of $\xi = \eta$. Thus, the transformation (1) is applicable near the prolate shape nucleus, but not near the oblate shape.

In the second step, α and β are transformed into the final boson operators ρ and σ through the following boson Bogoliubov transformations with coefficients x, y, z and t regarded as infinitesimals,

$$\begin{aligned} \rho &= \alpha + x\alpha^\dagger - z\beta - t\beta^\dagger, \\ \sigma &= \beta + y\beta^\dagger + z\alpha - t\alpha^\dagger. \end{aligned} \tag{6}$$

We finally arrive at the energy eigenvalues of $H_{\rm rot}$ as

$$\begin{aligned} E_{In_\sigma\kappa} &= \xi\left(I - j + n_\sigma + \frac{1}{2} - \kappa\right)^2 - \frac{1}{8}(A_x + A_y) \\ &\quad + A_z\left(I - j + n_\sigma + \frac{1}{2}\right)^2 \\ &\quad - 2\zeta\left(I - j + n_\sigma + \frac{1}{2}\right)\left(I - j + n_\sigma + \frac{1}{2} - \kappa\right). \end{aligned} \tag{7}$$

In Eq. (7), κ is defined by the relation,

$$\kappa = I - j + n_\sigma - n_\rho, \tag{8}$$

where n_ρ and n_σ are the eigenvalues of the boson numbers $\hat{n}_\rho = \rho^\dagger\rho$ and $\hat{n}_\sigma = \sigma^\dagger\sigma$, respectively. In the axially symmetric limit for the prolate shape ($\eta = 0, A_\perp \equiv A_x = A_y$), Eq. (7) continuously goes to

$$E_{In_\sigma\kappa} = (A_z - A_\perp)\kappa^2 + A_\perp(I - j + n_\sigma)(I - j + n_\sigma + 1). \tag{9}$$

This expression is equivalent to the energy eigenvalue of $H_{\rm rot}$ in the limit of prolate shape, i.e. $(A_z - A_\perp)R_z^2 + A_\perp R(R+1)$. Thus, we find that R corresponds to $I - j + n_\sigma$ and κ to R_z.

2.2 D_2 symmetry and Bohr's symmetry

In this subsection, we consider the basic symmetry in the nuclear Hamiltonian and in the nuclear states. The rotor itself is composed of many nucleons and the last odd nucleon is coupled to this rotor. There is the D_2-symmetry ([34]), i.e. the rotor Hamiltonian is invariant with respect to rotations through the angle π about each of three principal axes, i.e. $\hat{R}_i = \exp\{-i\pi(I_i - j_i)\}$ $(i = 1, 2, 3)$. This symmetry group is a point group composed of three rotations together with an identity. When we define body-fixed principal axes of a nuclear system with quadrupole deformation described in terms of deformation parameters, β and γ, the nuclear state is described in the 5-dimensional space of collective coordinates $(\theta_i, \beta, \gamma)$, where the set of Euler angles θ_i $(i = 1, 2, 3)$ corresponds to the direction of the deformed body. Then, the number of different ways in choosing the direction of principal axes for a nuclear state is twenty-four, and the set of finite rotations among these directions composes the octahedral group O_8. The effect of such a transformation on the nuclear wavefunction can be compensated by a suitable change of the deformation parameter γ to leave the nuclear state invariant. These transformations represented in the 5-dimensional space compose Bohr's symmetry group [7]. The invariance of the nuclear state under these symmetry transformations restricts κ to even integer values (i.e. $\kappa = 0, \pm 2, \pm 4, \cdots$). Thus, among four kinds of D_2-representation classified by the set of eigenvalues r_i of the rotation operators $\hat{R}_i$, only the eigenstate characterized by $(r_1, r_2, r_3) = (+, +, +)$ (i.e. A-type) is allowed as a physical state of the nucleus. Consequently, the nuclear state with the quantum number $K = 0$ does not exist for the angular momentum I =odd, and the state of $I = 1$ is excluded from rotational bands. Such a strong condition imposed on the nuclear state is in sharp contrast to the rotational motion of a general rigid body [9].

The Hamiltonian in Eq. (2) with three moments of inertia, which are given by the formulae in Eq. (19) or (20) below, is invariant under D_2 symmetry transformation. However, the approximate Hamiltonian up to the order of $1/I$ and $1/j$ violates the D_2 symmetry in general. If we choose $\hat{R}_2 = \exp\{-i\pi(I_y - j_y)\}$, then there appear the following relations:

$$\hat{R}_2 I_\pm \hat{R}_2^\dagger = -I_\mp, \;\; \hat{R}_2 I_z \hat{R}_2^\dagger = -I_z,$$
$$\hat{R}_2 j_\pm \hat{R}_2^\dagger = -j_\mp, \;\; \hat{R}_2 j_z \hat{R}_2^\dagger = -j_z. \tag{10}$$

As I_k commutes with $j_{k'}$ for any k and k', the operators a and b in Eq. (1) satisfy the following relations:

$$\begin{aligned} \hat{R}_2 \hat{n} \hat{R}_2^\dagger &= 2I - \hat{n}, \;\; \hat{R}_2 \hat{k} \hat{R}_2^\dagger = 2j - \hat{k}, \\ \hat{R}_2 a^\dagger \hat{R}_2^\dagger &= -\sqrt{\frac{2I - \hat{n}}{\hat{n}+1}}\, a, \;\; \hat{R}_2 b^\dagger \hat{R}_2^\dagger = -\sqrt{\frac{2j - \hat{k}}{\hat{k}+1}}\, b. \end{aligned} \tag{11}$$

The combinations $I_+I_- + I_-I_+$, I_z^2, $j_+j_- + j_-j_+$ and j_z^2 are of diagonal form when written in terms of $\hat{n}$ and $\hat{k}$, and have nothing to do with the expansion in $1/I$ and $1/j$. As for the combinations $I_+^2 + I_-^2$, $j_+^2 + j_-^2$ and $I_+j_- + I_-j_+$, we expand the square roots up to the first order of the small quantities $(I - \hat{n})/I$ and $(j - \hat{k})/j$, i.e.

$$\begin{aligned} I_+^2 + I_-^2 &\cong 2(I - \frac{1}{2})(a^\dagger a^\dagger - \frac{a^\dagger a^\dagger a a}{2I}) + h.c., \\ j_+^2 + j_-^2 &\cong 2(j - \frac{1}{2})(b^\dagger b^\dagger - \frac{b^\dagger b^\dagger b b}{2j}) + h.c., \\ I_+j_- + I_-j_+ &\cong -a^\dagger b^\dagger \sqrt{Ij}(2 - \frac{\hat{n}}{2I} - \frac{\hat{k}}{2j}) + h.c.\,. \end{aligned} \tag{12}$$

With the help of Eq. (11), these terms are proven to be invariant under the transformation $\hat{R}_2$. Thus, the approximated Hamiltonian up to the order of $1/I$ and $1/j$ is proved to be D_2-invariant. However, the D_2-invariance is not guarateed if we stop the expansion at the lowest order in Eq. (1). To see this concisely, we consider the following lowest order expansions for the even-mass nucleus as given in the textbook of Bohr-Mottelson ([34]).

$$\begin{aligned} &I_+ \cong -\sqrt{2I} a^\dagger, \;\; I_- \cong -\sqrt{2I} a, \;\; I_z = I - a^\dagger a \\ &I_+I_- + I_-I_+ \cong 2I(1 + 2\hat{n}) \end{aligned} \tag{13}$$

We can easily find that the term $\hat{R}_2(I_+^2 + I_-^2)\hat{R}_2^\dagger$ equals to $I_+^2 + I_-^2$, but the transformation $\hat{R}_2(I_+I_- + I_-I_+)\hat{R}_2^\dagger$ yields $2I(1 + 4I - 2\hat{n})$, which is not equal to $(I_+I_- + I_-I_+)$ in Eq. (13). If we introduce $H' = (H + \hat{R}_2 H \hat{R}_2^\dagger)/2$ to restore D_2 invariance of the Hamiltonian given by Eq. (13), H' becomes

$(A_x - A_y)I(a^\dagger a^\dagger + h.c.)/2 - A_z I^2 + (A_x + A_y)(I + 2I^2)/2$. As the linear order in $\hat{n}$ disappears, H' can not be diagonalized by a linear transformation of the boson a.

Next, we investigate the difference between the expansion in $\hat{n}/I$, which is used in deriving Eq. (7), and the expansion in $(I - \hat{n})/I$, which is used in Eq. (12). We apply the latter to the Hamiltonian in Eq. (2) and get

$$\begin{aligned} E_{In_\sigma\kappa} &= (\xi - C)(I - j + n_\sigma + \frac{1}{2} - \kappa)^2 - \frac{1}{8}(A_x + A_y) - \frac{C}{4} \\ &+ A_z(I - j + n_\sigma + \frac{1}{2})^2 \\ &- 2\zeta'(1 - D)(I - j + n_\sigma + \frac{1}{2})(I - j + n_\sigma + \frac{1}{2} - \kappa) \end{aligned} \quad (14)$$

with

$$\xi' = 2A_z - \frac{1}{2}(A_x + A_y), \;\; \zeta' = \sqrt{\xi'^2 - \eta^2}, \;\; C = \frac{3}{2}\frac{A_z}{\zeta'^2}, \;\; D = \frac{A_z\xi'}{\zeta^2}. \quad (15)$$

The expression given by Eq. (14) coincides with the exact one in the limit of a prolate nucleus, and provides a better approximation than Eq. (7). However, the difference between Eq. (7) and Eq. (14) is small. For example, for $\gamma = 40°$ and I =39/2, the difference is at most 3.6%. For the larger I and for the smaller γ this difference becomes negligible, so that we employ the expression given by Eq. (7) in what follows.

As $\mathbf{R} = \mathbf{I} + (-\mathbf{j})$, the allowed value of $R = I - j + n_\sigma$ runs over $I + j, I + j - 1, \cdots, |I - j|$. Correspondingly, the quantum number n_σ takes the values $0, 1, \cdots, 2j$. However, as explicitly shown by Eq. (25) in subsection 2.5., independent nuclear states are labeled by $\Omega(= j_z) > 0$ in the axially symmetric limit. This causes a further restriction of n_σ to the values $n_\sigma = 0, 1, \cdots, j - 1/2$.

From D_2 symmetry, $R_z = \kappa = 0, \pm 2, \pm 4 \cdots$ for even values of R, and $\pm 2, \pm 4 \cdots$ for odd values of R. Since R_z must be less than R, the maximum value of κ is R for even R, and $R - 1$ for odd R. As a result, the D_2 symmetry selects special levels as the physical ones from all the states of the triaxially deformed rotor [9], i.e. the level of higher (lower) energy among the two with $\kappa = |\kappa|$ and $\kappa = -|\kappa|$ for the case of even (odd) R.

2.3 x-axis as a quantization axis

In the case where the x-axis is chosen as the quantization axis, we follow the same argument as in subsection 2.1. by exchanging x, y and z into z, x and y, respectively, in Eq. (1). If we use the same definition of $\eta_\pm$ as in Eq. (4), the quantities ξ and η in Eq. (5) are replaced by

$$\xi = -A_x + \frac{1}{2}(A_z + A_y), \;\; \eta = \frac{1}{2}(A_z - A_y), \;\; \zeta = \sqrt{\xi^2 - \eta^2}. \quad (16)$$

Then, $\eta_\pm$ diverges in the symmetric limit of $A_x = A_y$ (prolate shape), because of $\xi^2 - \eta^2 = 0$. However, η_+^2 continuously goes to 1 and η_-^2 goes to 0 in the symmetric limit of $A_y = A_z$ (oblate shape). Thus, we can apply the boson Bogoliubov transformation near the oblate shape nucleus. For the triaxial nucleus between the prolate and the oblate limit, the energy eigenvalue corresponding to Eq. (2) in the previous case is replaced, in the present case, by

$$\begin{aligned} E_{In_\sigma n_\rho} &= -\xi\left(n_\rho + \frac{1}{2}\right)^2 - \frac{1}{8}(A_z + A_y) \\ &+ A_x\left(I - j + n_\sigma + \frac{1}{2}\right)^2 + 2\zeta\left(I - j + n_\sigma + \frac{1}{2}\right)\left(n_\rho + \frac{1}{2}\right). \end{aligned} \quad (17)$$

In this expression, we have not employed κ but n_ρ. In the axially symmetric limit of oblate shape, the energy $E_{In_\sigma n_\rho}$ in Eq. (17) becomes $A_x(I - j + n_\sigma - n_\rho)^2 + A_z[(I - j + n_\sigma)(I - j + n_\sigma + 1) - (I - j + n_\sigma - n_\rho)^2]$. On the other hand, the eigenvalue of H_{rot} in the oblate limit is $A_x R_x^2 + A_z[R(R+1) - R_x^2]$. Thus, we find that $R = I - j + n_\sigma$ and $R_x = I - j + n_\sigma - n_\rho$. Therefore, the quantum number n_ρ is interpreted as the wobbling quantum number describing the precession of the rotor around x-axis. However, we must notice that n_ρ does not take the values $0, 1, \cdots$ as proposed in the textbook ([34]). As we have seen in the previous subsection, the rotor states are realized only for $R_x \equiv \kappa = R - n_\rho =$ even due to Bohr's symmetry. Hence, $n_\rho = 0, 2, \cdots, R$ for $R \equiv I - j + n_\sigma =$ even, corresponding to the rotor levels with negative R_x, which are lower than those with positive R_x. On the other hand, $n_\rho = 1, 3, \cdots, R-2$ for $R \equiv I - j + n_\sigma =$ odd, corresponding to the rotor levels with positive R_x, which are higher than those with negative R_x. The smaller n_ρ (larger κ) is favourable for the yrast states.

Thus, the successive increase of the wobbling quantum number n_ρ by one unit is not allowed, and moreover the lowest order approximation of the boson expansion in n_ρ violates the D_2-invariance. Therefore, any concept of harmonic approximation cannot arise from the wobbling motion of a rotor with quadrupole deformation.

2.4 Moments of inertia

The components of the quadrupole moment are assumed to be proportional to the moments of inertia $\mathcal{J}_i$ for $i = x, y$ and z, which are determined by the mass distribution, i.e.

$$\begin{aligned} Q_0 &\propto \mathcal{J}_x + \mathcal{J}_y - 2\mathcal{J}_z, \\ Q_2 &\propto \sqrt{\frac{3}{2}}(\mathcal{J}_y - \mathcal{J}_x). \end{aligned} \quad (18)$$

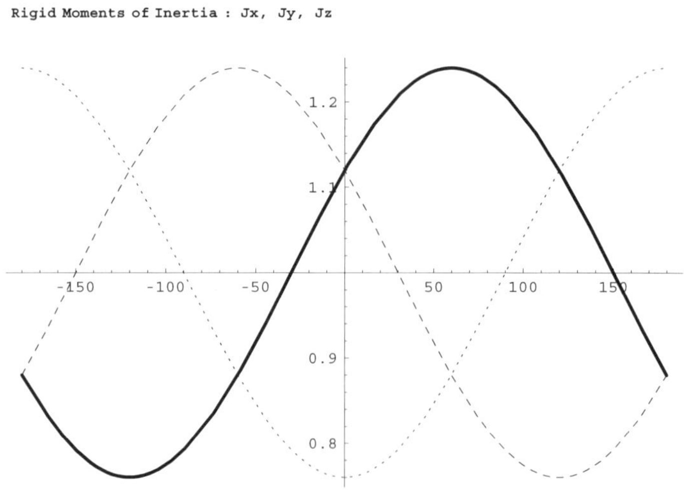

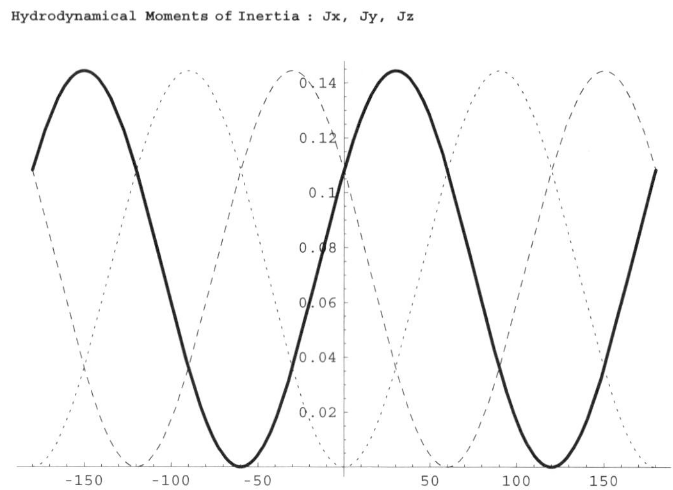

Figure 1. The comparison between the rigid-body moments of inertia (top) and the hydrodynamical moments of inertia (bottom). The solid lines correspond to $\mathcal{J}_x$, the dashed lines to $\mathcal{J}_y$ and the dotted lines to $\mathcal{J}_z$. The ordinate is in arbitrary units and the abscissa is γ.

We consider two models to define the functional dependence of the moments of inertia on the deformation parameters β and γ.

(a) Irrotational flow model:

$$\mathcal{J}_i \propto \beta^2 \sin^2(\gamma - \frac{2\pi}{3}i), \quad \frac{Q_2}{Q_0} = -\frac{\tan(2\gamma)}{\sqrt{2}} \tag{19}$$

(b) Rigid-body model:

$$\mathcal{J}_i \propto 1 - \sqrt{\frac{5}{4\pi}}\beta\cos(\gamma + \frac{2\pi}{3}i), \quad \frac{Q_2}{Q_0} = -\frac{\tan(\gamma)}{\sqrt{2}}. \tag{20}$$

In both Eqs. (19) and (20), i on the r.h.s. runs over 1, 2 and 3, corresponding to $i = x, y$ and z on the l.h.s. of these equations. The behavior of the moments of inertia for the irrotational flow model is more sensitive to γ than for the rigid-body model. The phase of γ is chosen so that $\mathcal{J}_x$ takes its maximum within the region of $0 \leq \gamma \leq 60°$ in both models. We compare Eqs (19) and (20) in Fig. 1.

2.5 The single-particle Hamiltonian

We consider the particle-plus-rotor model for an odd mass nucleus with one valence nucleon in a high-j orbital coupled to the triaxially deformed core. We assume that the single-particle Hamiltonian $H_{\rm sp}$ is given by

$$\begin{aligned} H_{\rm sp} &\propto \cos\gamma Y_{20} - \frac{\sin\gamma}{\sqrt{2}}(Y_{22} + Y_{2-2}) \\ &= \frac{V}{j(j+1)}\{\cos\gamma(3j_z^2 - \mathbf{j}^2) - \sqrt{3}\sin\gamma(j_x^2 - j_y^2)\}. \end{aligned} \tag{21}$$

Our interest is in the total Hamiltonian describing the microscopic behavior of the odd nucleon moving in the deformed mean-field produced by the core nucleons, i.e.

$$H_{\rm tot} = H_{\rm rot} + H_{\rm sp}. \tag{22}$$

It is obvious that the total Hamiltonian is invariant under the D_2 transformations as well as one of Bohr's transformations defined in the 5-dimensional space, i.e.

$$\mathcal{R}_2(x, y, z, \beta, \gamma) = (y, -x, z, \beta, -\gamma). \tag{23}$$

Note that two successive operations of $\mathcal{R}_2$, i.e. $\mathcal{R}_2^2$, is realized by one of the D_2 operations, $\hat{R}_3 = \exp\{-i\pi(I_z - j_z)\}$.

For the case where the odd nucleon is in a certain single-j shell, the diagonalization of $H_{\rm tot}$ can be attained in principle with the physical space spanned by $\{|IMK\rangle|j\Omega\rangle; K = -I, -I+1, \cdots, I; \Omega = -j, -j+1, \cdots, j\}$, where the quantum numbers $K = I_z$ and $\Omega = j_z$ are the projections of the total spin $\mathbf{I}$ and the single-particle spin $\mathbf{j}$ onto the intrinsic z-axis, respectively. However, almost 3/4 of the resultant eigenstates turn out to be of zero-norm due to the symmetry of the Hamiltonian, and those unphysical solutions should be eliminated.

In order to avoid such a redundancy, we project out relevant states from the beginning, i.e.

$$\begin{aligned} & (1+e^{-i\pi(I_y-j_y)})(1+e^{-i\pi(I_z-j_z)})|IMK\rangle|j\Omega\rangle \\ = & \{1+(-1)^{K-\Omega}\}\{|IMK\rangle|j\Omega\rangle+(-1)^{I-j}|IM-K\rangle|j-\Omega\rangle\}. \end{aligned} \tag{24}$$

In this way we provide a complete set of relevant wave functions as

$$\sqrt{\frac{2I+1}{16\pi^2}}\{\mathcal{D}^I_{MK}(\theta_i)\chi^j_\Omega+(-1)^{I-j}\mathcal{D}^I_{M-K}(\theta_i)\chi^j_{-\Omega}\}; \\ |K-\Omega|=\text{even},\quad \Omega>0. \tag{25}$$

Employing this complete set, we diagonalize H_{tot} to obtain the λ-th eigenstates in the form [10]

$$\Psi(jIM\lambda)=\sum_{\Omega>0}\sum_{|K-\Omega|}C^{jI\lambda}_{\Omega K}\{\mathcal{D}^I_{MK}(\theta_i)\chi^j_\Omega+(-1)^{I-j}\mathcal{D}^I_{M-K}(\theta_i)\chi^j_{-\Omega}\}. \tag{26}$$

The problem of the assignment of quantum numbers cannot be solved only by the exact diagonalization of H_{tot}. For this purpose we need a further step, which will be discussed in the next section.

3. Application of boson expansion method to H_{tot}

Here we discuss a possible extension of the HP boson expansion method to H_{tot}. As for the model in which one valence nucleon is in the single high-j orbital, we can apply the HP transformation defined in Eq. (1) to H_{tot}. Then, expanding $\sqrt{2I-\hat{n}}$ and $\sqrt{2j-\hat{k}}$ in Eq. (22) and retaining up to the first order in $1/I$ and $1/j$, we rewrite H_{tot} in terms of bilinear and quadratic forms of boson operators a, $a^\dagger$, b and $b^\dagger$. By the boson Bogoliubov transformation from the set of original boson operators $(a, a^\dagger, b, b^\dagger)$ to the new set of boson operators $(\alpha, \alpha^\dagger, \beta, \beta^\dagger)$, we transform H_{tot} into the new form as

$$H_{\text{tot}}=H_0+H_2+H_4, \tag{27}$$

where H_0 is the part independent of the new boson operators $(\alpha, \alpha^\dagger, \beta, \beta^\dagger)$, and H_2 is composed of bilinear forms like

$$\begin{aligned} & \alpha^\dagger\alpha+\alpha\alpha^\dagger,\quad \beta^\dagger\beta+\beta\beta^\dagger,\quad \alpha^\dagger\beta^\dagger+\beta^\dagger\alpha^\dagger+\alpha\beta+\beta\alpha \\ \text{and}\quad & \alpha^\dagger\beta+\beta\alpha^\dagger+\alpha\beta^\dagger+\beta^\dagger\alpha. \end{aligned} \tag{28}$$

H_4 is composed of all possible quadratic forms of the new boson operators.

Here we consider the most general boson Bogoliubov transformation in the space of two kinds of bosons as

$$\begin{pmatrix} a \\ a^\dagger \\ b \\ b^\dagger \end{pmatrix} = \begin{pmatrix} \cos\chi & 0 & \sin\chi & 0 \\ 0 & \cos\chi & 0 & \sin\chi \\ -\sin\chi & 0 & \cos\chi & 0 \\ 0 & -\sin\chi & 0 & \cos\chi \end{pmatrix} \begin{pmatrix} \cosh\vartheta & \sinh\vartheta & 0 & 0 \\ \sinh\vartheta & \cosh\vartheta & 0 & 0 \\ 0 & 0 & \cosh\varphi & \sinh\varphi \\ 0 & 0 & \sinh\varphi & \cosh\varphi \end{pmatrix}$$
$$\times \begin{pmatrix} \cos\omega & 0 & \sin\omega & 0 \\ 0 & \cos\omega & 0 & \sin\omega \\ -\sin\omega & 0 & \cos\omega & 0 \\ 0 & -\sin\omega & 0 & \cos\omega \end{pmatrix} \begin{pmatrix} \alpha \\ \alpha^\dagger \\ \beta \\ \beta^\dagger \end{pmatrix}. \quad (29)$$

This transformation introduces 4 parameters, i.e. $\vartheta, \varphi, \omega$ and χ. These are just enough to eliminate the 4 non-diagonal terms given by Eq. (28). Since in practice such a simultaneous elimination in an algebraic way is difficult, we propose an alternative method as follows.

In the first step, we reduce artificially the number of parameters by requiring $\vartheta = -\varphi$, for example, and impose the vanishing of two terms, $\alpha^\dagger\beta^\dagger + \beta^\dagger\alpha^\dagger + \alpha\beta + \beta\alpha$ and $\alpha^\dagger\beta + \beta\alpha^\dagger + \alpha\beta^\dagger + \beta^\dagger\alpha$, in order to separate two kinds of bosons, α and β. Hence, two parameters, ϑ and ω, can be solved as functions of χ. In the second step, we diagonalize H_2 in the variable-separated form as follows.

$$\begin{aligned} H_2 &= A(\alpha^\dagger\alpha + \alpha\alpha^\dagger) + C(\alpha^\dagger\alpha^\dagger + \alpha\alpha) \\ &\quad + B(\beta^\dagger\beta + \beta\beta^\dagger) + D(\beta^\dagger\beta^\dagger + \beta\beta) \\ &= \frac{A}{|A|}\sqrt{A^2 - C^2}(\rho^\dagger\rho + \rho\rho^\dagger) + \frac{B}{|B|}\sqrt{B^2 - D^2}(\sigma^\dagger\sigma + \sigma\sigma^\dagger), \end{aligned} \quad (30)$$

where new bosons, ρ and σ, are introduced through the new boson Bogoliubov transformation given by

$$\alpha = \xi_+\rho + \xi_-\rho^\dagger, \quad \beta = \eta_+\sigma + \eta_-\sigma^\dagger \quad (31)$$

with

$$\xi_\pm = \{\frac{1}{2}(\frac{|A|}{\sqrt{A^2 - C^2}} \pm 1)\}^{1/2}, \quad \eta_\pm = \{\frac{1}{2}(\frac{|B|}{\sqrt{B^2 - D^2}} \pm 1)\}^{1/2}. \quad (32)$$

In the variable-separated form of H_2 in the first line of Eq. (30), the 4 coefficients, A, B, C and D are already determined as functions of χ. Notice that these two steps are meaningful only when the following inequalities are satisfied.

$$-1 < \tanh 2\vartheta < 1, \quad |A| > |C|, \quad \text{and} \quad |B| > |D|. \tag{33}$$

In practice, shifting the value of χ, we look for a certain region of χ, where all of these inequalities are satisfied, and determine the boson transformations. In the final step, we apply the transformation in Eq. (31) to the quadratic terms in H_4, and retain only the diagonal contributions written in terms of $\hat{n}_\rho = \rho^\dagger \rho$ and $\hat{n}_\sigma = \sigma^\dagger \sigma$. Practice of such a calculation is in progress. Preliminary calculations have been successful in finding the region of χ where we get desirable solutions.

We expect that such a treatment provides formulas which are good approximations to the energy levels given by the exact diagonalization. Comparison between both results enables us to assign a set of quantum numbers (n_σ, n_ρ) or

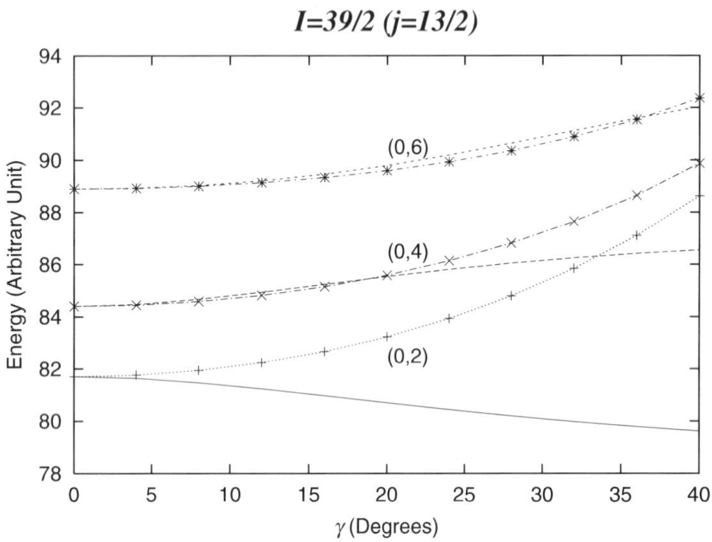

Figure 2. The comparison of the energy levels derived from the case of z-axis as a quantization axis with the exact results for I = 39/2 and j =13/2 as functions of γ from 0° to 40°. The ordinate is in an arbitrary unit. The numerals inside the parenthesis denote (n_σ, κ). The lines without marks are exact results, and the lines with the signs of pluses, crosses and asterisks are on Eq. (7).

(n_σ, κ) $(\kappa = I - j + n_\sigma - n_\rho)$ to each physical state determined by the total Hamiltonian H_{tot}.

4. Numerical Results

In our single-j model, one valence nucleon is assumed to be in the unique-parity orbital of $i_{13/2}$ with $j = 13/2$. We will compare the exact results attained

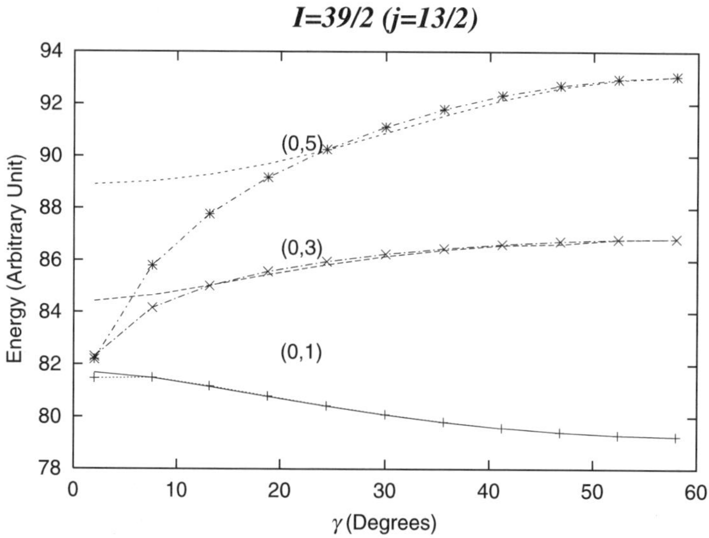

Figure 3. The comparison of the energy levels derived from the case of x-axis as a quantization axis with the exact results for I = 39/2 and j =13/2 as functions of γ from 0° to 60°. The numerals inside the parenthesis denote (n_σ, n_ρ). The lines without marks are exact results, and the lines with the signs of pluses, crosses and asterisks are on Eq. (17).

from diagonalization of H_{rot} with two approximate solutions, i.e. Eqs. (7) and (17). In Fig. 3, both the approximate energy levels calculated from Eq. (7) and the exact ones for $I = 39/2$ are plotted as functions of γ ($0° \leq \gamma \leq 40°$). Similarly, in Fig. 2, both the approximate energy levels calculated from Eq. (17) and the exact ones for the same I value are plotted as functions of γ ($0° < \gamma < 60°$). The prolate limit corresponds to $\gamma = 0°$ and the oblate limit to $\gamma = 60°$. In both figures, the rigid-body moments of inertia, i.e. Eq. (20), are adopted with a common proportional constant (=1). The ordinates are in arbitrary units, but the scales of all figures are common. The numerals inside the parenthesis in Fig. 2 denote the assigned pair of quantum numbers (n_σ, κ). In this case, $I-j$ is odd as $39/2-13/2 = 13$, so that κ runs over $2, 4, 6, \cdots, 12$ for $n_\sigma = 0$. For the case of $n_\sigma = 1$, κ starts from 0 and increases by 2 units. In Fig. 2, $\kappa = 2$ corresponds to $n_\rho = 11$, $\kappa = 4$ to $n_\rho = 9$, $\kappa = 6$ to $n_\rho = 7$, since $\kappa = I - j + n_\sigma - n_\rho$. The reason why the approximate energies become better for larger κ (smaller n_ρ) is that the approximate energies given by Eq. (7) becomes better for smaller n_ρ.

In Fig. 3 the assigned pair of quantum numbers inside the parenthesis stands for (n_σ, n_ρ). Since $I - j = 13$ is odd, n_ρ runs over 1, 3, 5, $\cdots$, 12 for the case of $n_\sigma = 0$. As is seen in the figure, Eq. (17) becomes a better approximation for larger γ (near oblate shape) and for smaller n_ρ. Comparing Figs. 2 and 3, we

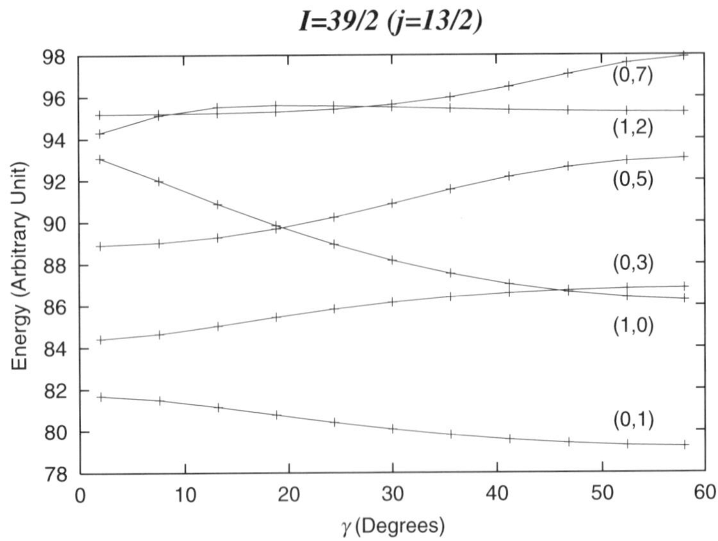

Figure 4. The energy levels derived from the exact diagonalization of the rotor Hamiltonian for I = 39/2 and j =13/2 as functions of γ. The numerals inside the parenthesis denote (n_σ, n_ρ).

recognize that the approximation given by Eq. (17) is preferable, i.e. choosing the x-axis for the quantization, as far as the levels near yrast in the region of triaxial deformation, typically $\gamma \sim 20°$, are concerned.

In Fig. 4, the energy eigenvalues obtained from the exact diagonalization of the rotor Hamiltonian are shown as a function of γ for I =39/2 together with the quantum numbers (n_σ, n_ρ), which are assigned from Eq. (17). For the case n_σ=1, n_ρ takes the values 0, 2 $\cdots$. Similarly, in Fig. 5 the energy eigenvalues from the exact diagonalization of $H_{\rm rot}$ are shown as a function of γ for I =37/2 together with the assigned quantum numbers (n_σ, n_ρ). As $I - j = 12$ is an even number in Fig. 5, n_ρ starts from 0, 2, $\cdots$ for the n_σ=0 case, and from 1, 3, $\cdots$ for the n_σ=1 case. As the scales of the ordinates are the same in Figs. 3 and 4, we see that the yrast level with (0,1) for I =39/2 lies near to the yrare level with (1,1) for I =37/2. Thus, we adopt (0,1) for the odd R band (the triaxially deformed superdeformed band 2, i.e. TSD2) and (1,1) for the even R band (the triaxially deformed superdeformed band 1, i.e. TSD1) in the analysis of ^{163}Lu [8].

5. Summary

We have derived an algebraic expression for the triaxially deformed rotor in an odd mass nucleus by applying the Holstein-Primakoff (HP) transformation to the angular momenta **I** and **j**. We have considered two alternative cases. One is

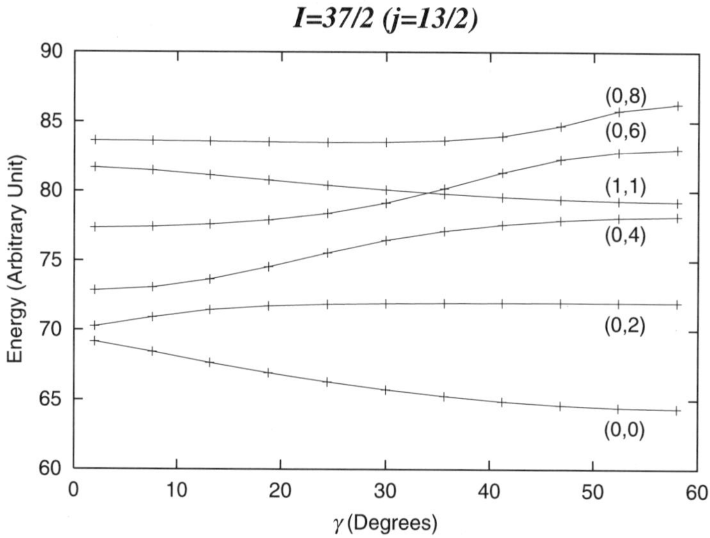

Figure 5. The energy levels derived from the exact diagonalization of the rotor Hamiltonian for I = 37/2 and j =13/2 as functions of γ. The numerals inside the parenthesis denote (n_σ, n_ρ).

the case where the z-axis is chosen as the quantization axis so that the algebraic expression coincides with the exact formula in the prolate limit. Another is the case where the x-axis is chosen as the quantization axis so that the algebraic expression coincides with the exact formula in the oblate limit.

We have clarified that the selection of physical quantum numbers from the HP boson numbers is uniquely determined by the D_2-symmetry together with Bohr's symmetry. The quantum number assignment is established by the comparison of the energy levels derived from the exact diagonalization of the rotor Hamiltonian with two algebraic formulas expressed in terms of HP boson numbers.

We have proposed a possible extension of our method to the general case which includes a single-particle potential. Such an approach is now in progress.

References

[1] Ödegard, S.W. et al., *Nucl. Phys.*, A682, 427C, 2001; *Phys. Rev. Lett.*, 86, 5866, 2001.

[2] Hamamoto, I., *Phys. Rev.*, C65, 044305, 2002.

[3] Bohr, A. and B.R. Mottelson, *Nuclear Structure* vol. II (Benjamin, 1975).

[4] Tanabe, K. and K. Sugawara-Tanabe, *Physics Letters*, B34, 575, 1971; *Nucl. Phys.*, A208, 317, 1973.

[5] Tanabe, K., *J. Math. Phys.*, 14, 618, 1973.

[6] Tanabe, K. and K. Sugawara-Tanabe, *Phys. Rev.*, C14, 1963, 1976.

[7] Bohr, A., *Mat. Fys. Medd. Dan. Vid. Selsk.*, 26, no.14, 1952; *Mat. Fys. Medd. Dan. Vid. Selsk.*, 27, no.16, 1953.

[8] Sugawara-Tanabe, K. and K. Tanabe, *Proc. of Int. Conf. on "Electromagnetic Interactions in Nuclear and Hadron Physics"*, (World Scientific, 2001) 468.

[9] King G.W., R.M. Hainer and P.C. Cross, J. Chem. Phys., 11, 27, 1943; 17, 826, 1949.

[10] Mayer ter Vehn, J., S.S. Stephens and R.M. Diamond, *Phys. Rev. Letters*, 32, 1383, 1974.

UNDERSTANDING BRAIN AND CONSCIOUSNESS?

G. Vitiello
Dipartimento di Fisica "E.R. Caianiello"
Università di Salerno, 84100 Salerno, Italia
INFN, Gruppo collegato, Salerno and INFM, Sezione di Salerno
vitiello@sa.infn.it

Abstract This is a review of the dissipative quantum model of brain in the form of an extended abstract of recent works addressing to the question of the scientific understanding of brain and consciousness in the frame of quantum field theory. The intrinsic dissipative character of the brain dynamics appears to be a possible root of consciousness mechanisms.

Keywords: Brain, consciousness, quantum dissipation, entanglement, quantum field theory

1. Introduction

The study of the brain is a real challenge to physiologists, to biologists, to physicians, to psychologists and today it is very difficult to sustain the point of view that such a challenge does not call into game also the physicists. Actually, the same is true for the study of the biological systems in general. However, it has been not always an accepted fact the one that Physics may bring its own specific contribution to the understanding of living systems. By myself I had to work not a little to accept the view that "understanding" living matter is a problem to be solved not only by the biologists, but *also* by the physicists. The thought that everything is *encoded* in the ordered molecular patterns of the DNA was quite consoling to me: understanding living matter is a problem for biologists, I thought. However, in a natural way, being a physicist, I asked the question: "What is the dynamical mechanism generating ordered patterns?" Then, I was no more able to shift that problem to my friends in biology. Their job is to analyze and to list in all possible details the molecular components of living matter and see how they fit together. Molecular engineers can do almost everything today. But even biological engineers do not deal with the *generation* of ordering. They are only engineers. One cannot ask them questions about the *dynamical* generation of the ordering. Such kinds of questions have to be asked

B. Gruber (ed.), Symmetries in Science XI, 553–574.

to physicists. Thus I realized how important is to mark the distinction between naturalism and science: naturalism is *necessary* to the progress of science, which means that without the efforts in collecting data and making detailed observations you cannot even think of making any progress in knowledge. But naturalism is *not sufficient*. It is only phenomenology. Soon or later you will ask questions about the "dynamics", questions which naturalism cannot answer. Of course, also asking questions about the dynamics is a *necessary* but *not sufficient* condition for making science. So you need both, phenomenology *and* dynamics. The Italian writer Italo Calvino writes in his book "Le città invisibili" (The invisible towns) that Marco Polo describes a bridge stone by stone. Then the Kublai Kan asks: "Which one is the stone that sustains the bridge?". Polo: "The bridge is not sustained by one specific stone, but by the line of the arch formed by the stones". Kublai Kan remains silent, reflecting upon those words. Then he says: "Why are you telling me about the stones? I only care about the arch", and Polo: "Without stones there is no arch". Thus we need to know both, the stones and the *line of the arch*. Today we know the "stones" of brain and living matter in great details. However, we know practically nothing about the "line of the arch" of life.

In one of the previous edition of this series of Conferences (Symmetries in Science II, in 1986) I had the occasion to report about an interesting and intriguing aspect of the mechanism of spontaneous breakdown of symmetry in quantum field theory (QFT) consisting in the change of scale, from microscopic to macroscopic scale. Such a change of scale is possible in quantum theory since the theory internal consistency requires the existence of collective modes, whose coherent behavior manifests as a property of the system as a whole, as a *macroscopic quantum system*. Could this scenario possibly apply also to living matter and to brain?

Herbert Fröhlich in the middle of the 1960s proposed the model of living matter as a collective system of coherent electric dipole waves. Since 1982, with Emilio Del Giudice, Marziale Milani and Sivia Doglia, and later on with the late Giuliano Preparata, we have been working at a quantum field theoretical approach to living matter, inspired by the Fröhlich proposal. Some time later, in 1995 I have been attracted by the Ricciardi and Umezawa paper [23] on the brain treated as a many-body system in condensed matter physics. I have extended that quantum model of brain to the dissipative dynamics and the inclusion of such a distinctive treat of the brain, namely that it is an open system continuously interacting with the environment (this is the meaning of dissipation), has revealed useful also in approaching some discussion on Consciousness.

The present paper is in a good part an extended abstract of the qualitative description of the dissipative quantum model of brain presented in ref. [33]. I also shortly comment here on some new developments concerning quantum noise, entanglement and chaos in the dissipative model. In my exposition I will closely follow few sections of my book *My Double unveiled* [33]. I will completely avoid to present the mathematical formalism. It can be found in [30, 2, 18, 19].

2. Statistical order and dynamical order

In many respects living matter appears to be a real mystery. It presents several levels of spatial organization (cells, tissues and other ordered domains), time ordering (sequentially ordered chains of chemical reactions), functional organization (functional differentiation among different parts and compartments, hierarchical and temporal sequences of functions). Thus, from one side, there is the high level of space and time ordering, and the high and stable functional efficiency; on the other side, there is the randomness of kinematics which rules any chemical reaction.

Macroscopic laws exhibiting ordering and regularities in the behavior of ensembles of large number of entities, say atoms or molecules, are predicted by statistical mechanics.

In his book *What is life?* [24] Schrödinger however points out that such an order, or, in his words, such "regularities only in the average" (ibidem p.78) emerging from the "statistical mechanisms" is not enough to explain the "enigmatic biological stability" (ibidem p.47). Pretending to explain the biological functional stability in terms of the regularities of statistical origin would be the "classical physicist's expectation" that "far from being trivial, is wrong"(ibidem p.19).

Schrödinger calls it the "naïve physicist" answer and he argues that it is wrong since there is biological evidence (he refers to hereditary phenomena) which shows that very small groups of atoms, "much too small to display exact statistical laws"(ibidem p.20), have control of observable large scale features, very sharply and strictly determined, of the organism. According to him, this is the point where the "Quantum Mechanics evidence" enters into play: namely, by explaining the stability of configurations of a small number of atoms, which has no explanation in classical physics, Quantum Mechanics (QM) explains the stability of certain biological features.

Although the data available to Schrödinger have drastically changed due to the enormous progress of molecular biology, this progress in fact supports his arguments on the "smallness" of the number of the atoms controlling the system macroscopic features in a highly stable way; the most striking example is the one of the DNA: its strict and stable atomic ordering has a determinant rôle in the biological macroscopic organization.

I want here to stress Schrödinger's distinction (ibidem p.80) between ordering generated by the "statistical mechanisms" and ordering generated by "dynamical" quantum (necessarily quantum!) interactions among the atoms and the molecules.

Molecular biology has collected so many successes; we know so much about so many components of biological systems. The question is now how to put together all these data so to derive the complex behavior of the whole system.

Chemical efficiency and functional stability to the degree observed in living matter seem to be out of reach of any probabilistic approach *solely* based on microscopic random kinematics. It is a fact that there is no available computation or even abstract proof which shows how to obtain the characteristic chemical efficiency and stability of living matter by resorting uniquely to statistical concepts.

Classical statistical mechanics and short range forces of molecular biology, although necessary, do not seem to be completely adequate tools. It appears to be necessary to supplement them with a further step so to include underlying quantum *dynamical* features. In Schrödinger words: "it needs no poetical imagination but only clear and sober scientific reflection to recognize that we are here obviously faced with events whose regular and lawful unfolding is guided by a "mechanism" entirely different from the "probability mechanism" of physics" (ibidem p.79).

3. The quantum model of brain

The model of brain as a many-body system was conceived between 1966 and 1967, in the exciting scientific atmosphere of the Istituto di Fisica Teorica in Naples. In one of his last papers, dedicated to Eduardo Caianiello, Umezawa writes: "His Institute was not just an institute of theoretical physics, but included mathematical and experimental section for information and brain science. This gave me a very enjoyable environment. Practically everyday I met theorists and experimentalists on brain science". He then adds, "Since I was deeply involved in the subject of order and long range correlation in many-body systems, I naturally asked myself the question "is there any long range correlation associated to brain? If there is long range correlation, each constituent of the system should be trapped by this correlation and its individual behavior should not be freely exhibited and should instead be controlled by the correlation. In that case we do not observe individual cells, but the quasi-cells (in analogy to the term quasi-particle)... " [29].

It was clear to Umezawa that the mechanism of the *dynamical generation* of long range correlation in spontaneous breakdown of symmetry was of such a general validity and so relevant that it could not be "confined" to the domains of particle physics and solid state physics. For the first time there was the possibility to give a *quantitative* description of *collective* modes for a physical system not on a purely statistical, kinematic basis, but on a dynamical ground.

These collective modes do not describe in fact the collective behavior of an ensemble of elements of the kind described by Statistical Mechanics. They are dynamically generated as long range correlation among the system components. For example, the phonon in the crystal is not the cooperative mode of a large number of atoms in the statistical sense. It is a truly long range interaction mode among the atoms. The ordered patterns observed in crystals, superconductors, superfluids, ferromagnets and other solid state systems are not collective phenomena of statistical origin. They are macroscopic manifestations of the quantum dynamics.

Lashley's experimental work was suggesting that "masses of excitations... within general fields of activity, without regard to particular nerve cells" [16, 21] were involved in the determination of behavior. In the middle of the 1960's Karl Pribram, motivated by experimental observations, started to formulate his holographic hypothesis. Information appears indeed in such observations to be spatially uniform "in much the way that the information density is uniform in a hologram" [7, 9]. While the activity of the single neuron is experimentally observed in form of discrete and stochastic pulse trains and point processes, the "macroscopic" activity of large assembly of neurons appears to be spatially coherent and highly structured in phase and amplitude [8, 9].

The formulation of the quantum brain model was motivated by the experimental findings confirming the existence of almost simultaneous responses in several regions of the brain to some external stimuli and that these responses could not be explained in terms of single neuron activity [20, 21].

Brain functioning cannot depend too strictly on the functioning of each single neuron since specific activities of the brain manifestly persist in spite of destructive action on local parts of the brain or after treatments with electric shock or with drugs and in spite of the continuous changes in the number of living neurons. In the brain metabolic activity constituent macromolecules undergo chemical changes or disassembly within a couple of weeks and are then replaced by new ones. Despite such a continuous molecular "turn over", the brain functions appear to be highly stable over long period of time. Still in Lashley words, in "all behavior [...] it is the pattern and not the element that counts" [16].

Ricciardi and Umezawa write in the Introduction of their paper: ... "in the case of natural brain, it might be pure optimism to hope to determine the numerical values for the coupling coefficients and the thresholds of all neurons by means of anatomical or physiological methods" ... "many questions immediately arise ... is it essential to know the behavior in time of any single neuron in order to understand the behavior of natural brain? Probably the answer is negative. The behavior of any single neuron should not be significant for functioning of the whole brain, otherwise a higher and higher degree of malfunctioning should be observed ...".

The observed non-local, diffuse activity of the brain thus suggest that the brain states are characterized by dynamical long range correlation among the constituents. The generation of such long range correlation occurs through the dynamical mechanism of spontaneous breakdown of symmetry: the resulting mathematical model is a QFT model where brain is described as a macroscopic quantum system.

The quantum model of brain was formulated in order to describe short-term and long-term memory and the brain's capability to recall stored information.

Although it is estimated that in the brain there are about 10^{10} neurons, interconnected by a myriad of dendritic branches and synaptic connections in an intricate series of neuron nets, and ten times as many glia cells, it appears that memory is not "wired" into individual neuron nets: incoming information seems to involve large regions of brain cells aggregates [20, 21, 9, 10].

Another feature of memory activity is the lack of "conscious simultaneous recall" of several recorded information. Rather, it is often our common experience that once some information has been recalled, another sometimes completely different information is subsequently recalled, too, in a mechanism of "association of ideas" through a path or sequence of memories. This suggests that stored information can be recalled according to serial, rather than parallel, processes [23].

The above features of memory activity are taken to be experimental evidences on which the quantum brain model is to be based. The formal apparatus of the model is the QFT of many-body physics and the starting point is that *the brain is a system in interaction with the external world* from which it receives stimuli carrying information.

Stimuli coming to the brain from the external world should be *coded* and their effects on the brain should persist also after they have ceased; this means that stimuli should be able to change the state of the brain pre-existing the stimulation into another state where the information has been "printed" in a stable fashion. This means that the state where information is recorded under the action of the stimuli must be a ground state in order to realize the *stability*, of the recorded information; and that symmetry is broken in that state in order to allow the coding of the information.

4. Spontaneous breakdown of symmetry and collective modes

In QFT the dynamics (i.e. the Lagrangian or the Hamiltonian, or simply the field equations) is in general invariant under some group, say G, of continuous transformations. Spontaneous breakdown of symmetry occurs when the minimum energy state (the ground state or vacuum) of the system is not invariant under the full group G, but under one of its subgroups. Then it can be shown [14, 28] that collective modes, the so-called Nambu-Goldstone boson modes, are dynamically generated. Propagating over the whole system, these modes are the carrier of the ordering information (*long range correlation*): order manifests itself as a global property dynamically generated. The long range correlation modes are responsible for keeping the ordered pattern: they are coherently *condensed* in the ground state. In the crystal case they keep the atoms trapped in their lattice sites. The long range correlation thus forms a sort of net, extending over all the system volume, which traps the system components in the ordered pattern. This explain the macroscopic collective behavior of the components as a "whole".

It is very important to remark that the spontaneous breakdown of symmetry is possible since in QFT there exist infinitely many ground states or vacua which are physically distinct (technically speaking, they unitarily inequivalent). In QM, on the contrary, all the vacua are physically equivalent and thus there cannot be symmetry breakdown.

The brain is modeled by Ricciardi and Umezawa by following the above scheme. Stimuli coming to the brain from the external world should be *coded* and their effects on the brain should persist also after they have ceased; this means that stimuli should be able to change the state of the brain pre-existing the stimulation into another state where the information has been "printed" in a stable fashion. This means that the state where information is recorded under the action of the stimuli must be a ground state in order to realize the *stability* of the recorded information; and that symmetry is broken in that state in order to allow the coding of the information. Recording of information is represented by coherent condensation of collective modes in the ground state.

Since collective modes are massless bosons, their condensation in the vacuum does not add energy to it: the stability of the ordering, and therefore of the registered information, is thus insured. Long-term memory is modelled in this way.

The observable specifying the ordered state is called the order parameter. It is a measure of the condensation of the Nambu-Goldstone modes in the ground state and acts as a macroscopic variable.

The order parameter is specific to the kind of symmetry of the dynamics and its value is considered to be the *code* specifying the information printed in that ordered vacuum. Non-local properties, related to a code specifying the system state, are dynamical features of quantum origin: it is in this way that the stable and diffuse, non-local character of memory is represented in the quantum model; it is derived as a dynamical feature rather than as a property of specific neural nets (which would be critically damaged by local destructive actions).

It may also happen that under the action of external stimuli the brain may be put into an excited state, i.e. a quasi-stationary state of greater energy than the one of the ground state. Such an excited state also carries collective modes in their non-minimum energy state. Thus this state also can support recording some information. However, due to its higher energy such a state and the collective modes are not stable and it will sooner or later decay: short-term memory is then modelled by the condensation of long range correlation modes in the excited states. Different types of short-term memory are represented by different excitation levels in the brain state.

Another possibility is the excitation of collective modes out of the ground state. This brings us to the mechanism of recall of the stored information. In Umezawa's words: "I noticed that this could provide a remarkable mechanism for memory recollection. Suppose that an ordered pattern was printed on the brain by condensation mechanism in the vacuum which was induced by certain external stimuli. Though an order is stored, brain is not conscious of this because it is in the ground state. However, when a similar external stimulation comes in, it easily excites the massless boson associated with the long range correlation. Since the boson is massless, any small amount of energy can cause its excitation. During the time of excitation, brain becomes conscious of the stored order (memory). This explains recollection mechanism." [29]. The excited modes have finite life-time and thus the recall mechanism is a temporary activity of the brain, according indeed to our common experience. This also suggests that the capability to be "alert" or "aware" or to keep our "attention" focused on certain subjects (information) for a short or a long time may have to do with the capability of the brain to be put into an excited state with short or long life-time.

The short-term memory mechanism has been further analyzed in terms of non-equilibrium phase transitions in the context of the quantum model [25].

5. Brain as a mixed system and the overprinting problem

The brain model should explain how memory remains stable and well protected within a highly excited system, as indeed the brain is. Such a "stability" must be realized in spite of the permanent electrochemical activity and the continual response to external stimulation. The electrochemical activity must also, of course, be coupled to the correlation modes which are triggered by external stimuli. It is indeed the electrochemical activity observed by neurophysiology that provides [26, 27] a first response to external stimuli.

This has suggested to model the memory mechanism as a separate mechanism from the electrochemical processes of neuro-synapic dynamics: the brain is then a "mixed" system involving two separate but interacting levels. The memory level is a quantum dynamical level, the electrochemical activity is at a classical level. The interaction between the two dynamical levels is possible because of the specificity of the quantum dynamics: the memory state is a *macroscopic quantum state* due to the *coherence* of the correlation modes.

The problem of the coupling between the quantum dynamical level and the classical electrochemical level is then reduced to the problem of the coupling of two macroscopic entities. Such a coupling is analogous to the coupling between classical acoustic waves and phonons in crystals. Acoustic waves are classical waves; phonons are quantum long range modes. Nevertheless, their coupling is possible since the macroscopic behavior of the crystal "resides" in the phonon modes, so that the coupling acoustic waves-phonon is equivalently expressed as the coupling acoustic wave-crystal (which is a perfectly acceptable coupling from a classical point of view).

The quantum model of brain fits the neurophysiological observations of memory nonlocality and stability. However, several problems are left open. One is that of memory capacity, the *overprinting problem*: Suppose a specific code corresponding to a specific information has been printed in the vacuum. The brain then sets in that state and successive recording of a new, distinct (i.e. of different code) information, under the action of a subsequent external stimulus, is possible only through a new condensation process, corresponding to the new code. This last condensation will superimpose itself on the former one (*overprinting*), thus destroying the first registered information.

In the following section I will discuss how the dissipative character of brain dynamics may solve the problem of memory capacity.

Let me finally stress that the quantum variables in the quantum model of brain are basic field variables (the electrical dipole field) and the brain is described as a macroscopic quantum system. Stuart, Takahashi and Umezawa [26] have indeed remarked that "it is difficult to consider neurons as quantum objects". In other models of brain the relevant variables are binary variables describing the neuron's on/off activity. However, in the quantum model "we do not intend", Ricciardi and Umezawa say "to consider necessarily the neurons as the fundamental units of the brain".

6. Dissipation and brain

In the quantum brain model spontaneous breakdown of dipole rotational symmetry is triggered by the coupling of the brain with external stimuli. Once the dipole rotational symmetry has been broken (and information has thus been recorded), *then, as a consequence*, time-reversal symmetry is also broken: *Before* the information recording process, the brain can in principle be in anyone of the infinitely many (unitarily inequivalent) vacua. *After* information has been recorded, the brain state is completely determined and the brain cannot be brought to the state configuration in which it was *before* the information printing occurred. What I am saying is nothing but the content of the well known warning *...NOW you know it!...*, which tells you that since *now* you know, you are *another* man, not the same one as *before*...Once you have known, you cannot go back in time.

Thus, the same fact of getting information introduces *the arrow of time* into brain dynamics. Due to the memory printing process, time evolution of the brain states is intrinsically irreversible: getting information introduces a partition in the time evolution, it introduces the *distinction* between the past and the future, a distinction which did not exist *before* the information recording. There is thus an irreversible exchange of energy between the brain and the environment.

The available quantum formalism is not well suited to treat problems where the energy is dependent on time. Thus one only can study systems which are isolated from other systems, so that their energy cannot be exchanged and thus it remains constant in time.

When the system under study is not an isolated system one has to incorporate in the treatment also the other systems (which constitute *the environment*) to which the original system is coupled. The full set of systems then behaves as a single isolated (closed) one. At the end of the required computations, one extracts the information regarding the evolution of the original system by neglecting the changes in the remaining systems.

In many cases, the specific details of the coupling of our system with the environment may be very intricate and changeable so that they are difficult to be measured and known. One possible strategy is to average the effects of the coupling and represent them, at some degree of accuracy, by means of some "effective" interaction. Another possibility is to take into account the environmental influence on the system by a suitable *choice* of the vacuum state (the minimum energy state or ground state). The chosen vacuum thus carries the *signature* of the reciprocal system-environment influence at a given time under given boundary conditions. A change in the system-environment reciprocal influence then would correspond to a change in the choice of the system vacuum : the system ground state evolution or "story" is thus the story of the trade of the system with its environment. The theory should then provide the equations describing the system evolution "through the vacua", each vacuum corresponding to the system ground state at each time of its history.

In conclusion, in order to describe open quantum systems first of all one needs to use QFT (Quantum Mechanics does not have the many "inequivalent" vacua!). Then one also needs to use the time variable as a label for the set of ground states of the system [3]: as the time (the label value) changes, the system moves to a "new" (physically inequivalent) ground state (assuming continuous changes in the boundary conditions determining the system-environment coupling). Here, "physically inequivalent" means that the system observables, such as the system energy, assume different values in different inequivalent vacua, as is expected to happen in the case of open systems.

One thus gets a description for the open systems which is similar to a collection of photograms: each photogram represents the "picture" of our system at a given instant of time (a specific time label value). Putting together these photograms in "temporal order" one gets a movie, i.e. the story (the evolution) of our open system, which includes the system-environment interaction effects.

The evolution of the $\mathcal{N}$-coded memory can be represented as a trajectory of given initial condition running over time-dependent states $|0(t) >_{\mathcal{N}}$, each one minimizing the free energy functional. Recent results [19] show that such trajectories have chaotic character. This is a feature which fits experimental observations by Freeman [7, 8, 9] who indeed finds characteristic chaotic behavior in neural aggregates of the olfactory system of laboratory pets.

The mathematical representation of the environment must explicitly satisfy the requirement that the energy lost by the system must match the energy gained by the environment, and vice-versa. All other details of the system-environment interaction may be taken into account by the vacuum structure of the system, in the sense above explained. Then the environment may be represented in the simplest way one likes, provided the energy flux balance is preserved. One possible choice is to represent the environment as the "time-reversed copy" of the system: time must be reversed since the energy "dissipated" by the system is "gained" by environment.

Summarizing, the system has thus been *doubled*. The environment is mathematically represented as the *time-reversed image* of the system, i.e. as the system "double". What the system loses, the environment gains. Let me denote the system degrees of freedom by A_k, and the "doubled" degrees of freedom by $\tilde{A}_k$. The suffix k here generically denotes kinematical variables (*e.g.* spatial momentum) or intrinsic variables of the fields fully specifying the field degree of freedom. The structure of the vacuum turns out to be a condensate of couples of A_k and $\tilde{A}_k$.

I stress that the "tilde" or doubled mode is not just a mathematical fiction. It corresponds to a real excitation mode (quasiparticle) living in the system as an effect of its interaction with the environment: the couples $A_k\tilde{A}_k$ represent the correlation modes dynamically created in the system as a response to the system-environment reciprocal influence. It is the interaction between tilde and non-tilde modes that controls the time evolution of the system: the collective modes $A_k\tilde{A}_k$ are confined to live *in* the system. They vanish as soon as the links between the system and the environment are cut. Technically speaking, the modes A_k and $\tilde{A}_k$ are *entangled* modes, which means that the memory states cannot be factorized in terms of states of A_k modes alone and of $\tilde{A}_k$ modes alone. In other words this entanglement mathematically represents the impossibility to cut the links between the brain and the external world (a closed, i.e. fully isolated, brain is a dead brain according to physiology).

I observe that corresponding to different subjects (systems) we will have "different" representations of the environment, each of them being indeed a (time-reversed) "copy" of the corresponding subject. Therefore we have that the environment is "subjectively represented" by each subject. I will discuss more on the subjective representation of the world in the following Sections.

7. Dissipative quantum brain dynamics

Let us now see how quantum dissipation solves the overprinting problem in the quantum model of brain [30].

Let A_κ denotes the dipole wave quantum (dwq) mode, namely the Nambu-Goldstone mode associated to the spontaneous breakdown of rotational electrical dipole symmetry. $\tilde{A}_\kappa$ will denote its "doubled mode". The $\tilde{A}$ mode is the "time-reversed mirror image" of the A mode and represents the environment mode. Let $\mathcal{N}_{A_\kappa}$ and $\mathcal{N}_{\tilde{A}_\kappa}$ denote the number of A_κ modes and $\tilde{A}_\kappa$ modes, respectively.

Taking into account dissipativity requires [30] that the memory state, identified with the vacuum $|0>_\mathcal{N}$, is a condensate of *equal number* of A_κ and $\tilde{A}_\kappa$ modes, for any κ: such a requirement ensures that the flow of the energy exchanged between the system and the environment is balanced. Thus, the difference between the number of tilde and non-tilde modes must be zero: $\mathcal{N}_{A_\kappa} - \mathcal{N}_{\tilde{A}_\kappa} = 0$, for any κ. Note that the label $\mathcal{N}$ in the vacuum symbol $|0>_\mathcal{N}$ specifies the set of integers $\{\mathcal{N}_{A_\kappa},\ for\ any\ \kappa\}$ which indeed defines the "initial value" of the condensate, namely the *code* number associated to the information recorded at time $t_0 = 0$. Note now that the requirement $\mathcal{N}_{A_\kappa} - \mathcal{N}_{\tilde{A}_\kappa} = 0$, for any κ, does not uniquely fix the set $\{\mathcal{N}_{A_\kappa},\ \text{for any}\ \kappa\}$. Also $|0>_{\mathcal{N}'}$ with $\mathcal{N}' \equiv \{\mathcal{N}'_{A_\kappa}; \mathcal{N}'_{A_\kappa} - \mathcal{N}'_{\tilde{A}_\kappa} = 0,\ for\ any\ \kappa\}$ ensures the energy flow balance and therefore also $|0>_{\mathcal{N}'}$ is an available memory state: it will correspond, however, to a different code number (i.e. $\mathcal{N}'$) and therefore to a different information than the one of code $\mathcal{N}$.

The conclusion is that fixing to zero the difference $\mathcal{N}_{A_\kappa} - \mathcal{N}_{\tilde{A}_\kappa} = 0$, for any κ, leaves completely open the choice for the value of the code $\mathcal{N}$.

Thus, infinitely many memory (vacuum) states, each one of them corresponding to a different code $\mathcal{N}$, may exist: A huge number of sequentially recorded information data may *coexist* without destructive interference since infinitely many vacua $|0>_\mathcal{N}$, for all $\mathcal{N}$, are *independently* accessible in the sequential recording process. Recording information of code $\mathcal{N}'$ does not necessarily produce destruction of previously printed information of code $\mathcal{N} \neq \mathcal{N}'$, contrary to the non-dissipative case. In the dissipative case the "brain (ground) state" may be represented as the collection (or the superposition) of the full set of memory states $|0>_\mathcal{N}$, for all $\mathcal{N}$. In the non-dissipative case the "$\mathcal{N}$-freedom" is missing and consecutive information printing produces overprinting.

Let me remind that there does not exist in the infinite volume limit any unitary transformation which may transform one vacuum of code $\mathcal{N}$ into another one of code $\mathcal{N}'$: this fact, which is a typical feature of QFT, guarantees that the corresponding printed information data are indeed *different* or *distinguishable* ones ($\mathcal{N}$ is a *good* code) and that each information printing is also *protected* against interference from other information printing (absence of *confusion* among information data).

The effect of finite (realistic) size of the system may however spoil unitary inequivalence. In the case of open systems, in fact, transitions among (would be) unitary inequivalent vacua may occur (phase transitions) for large but finite volume, due to coupling with the external environment. The inclusion of dissipation leads thus to a picture of the system "living over many ground states" (continuously undergoing phase transitions). Note that even very weak (although above a certain threshold) perturbations may drive the system through its macroscopic configurations. In this way, occasional (random) weak perturbations are recognized to play an important rôle in the complex behavior of the brain activity.

The possibility of transitions among different vacua is a feature of the model which is not completely negative: smoothing out the exact unitary inequivalence among memory states has the advantage of allowing the familiar phenomenon of the "association" of memories: once transitions among different memory states are "slightly" allowed the possibility of associations ("following a path of memories") becomes possible. Of course, these "transitions" should only be allowed up to a certain degree in order to avoid memory "confusion" and difficulties in the process of storing "distinct" informational inputs [30, 2]. It is interesting to observe that Freeman, on the basis of experimental observations, shows that noisy fluctuations at a microscopic level may have a stabilizing effect on brain activity, noise preventing to fall into some unwanted state (attractor) and being an essential ingredient for the neural chaotic perceptual apparatus [8, 9].

I also observe that the dwq may acquire an effective non-zero mass due to the effects of the system finite size [2, 30, 32]. Such an effective mass will then act as a threshold for the excitation energy of dwq so that, in order to trigger the recall process, an energy supply equal or greater than such a threshold is required. When the energy supply is lower than the required threshold a "difficulty in recalling" may be experienced. At the same time, however, the threshold may positively act as a "protection" against unwanted perturbations (including thermalization) and contributes to the stability of the memory state. In the case of zero threshold any replication signal could excite the recalling and the brain would fall into a state of "continuous flow of memories" [30].

Summarizing, the brain system may be viewed as a complex system with (infinitely) many macroscopic configurations (the memory states). Dissipation, which is intrinsic to the brain dynamics, is recognized to be the root of such a complexity, namely of the huge memory capacity.

Of course, the brain has several structural and dynamical levels (the basic level of coherent condensation of dwq, the cellular cytoskeleton level, the neuronal dendritic level, and so on) which coexist, interact among themselves and influence each other's functioning. Dissipation introduces the further richness of the replicas or degenerate vacua at the basic quantum level. The crucial point is that the different levels of organization are not simply structural features of the brain, their reciprocal interaction and their evolution is intrinsically related to the basic quantum dissipative dynamics.

The brain's functional stability is ensured by the system's "coherent response" to the multiplicity of external stimuli. Thus dissipation also seems to suggest a solution to the so called *binding problem*, namely the understanding of the unitary response and behavior of apparently separated units and physiological structures of the brain.

I finally note that, when considering dwq with time-dependent frequency, modes with longer life-time are found to be the ones with higher momentum. Since the momentum is proportional to the reciprocal of the distance over which the mode can propagate, this means that modes with shorter range of propagation will survive longer. On the contrary, modes with longer range of propagation will decay sooner. The scenario becomes then particularly interesting since this mechanism may produce the formation of ordered domains of finite different sizes with different degree of stability: smaller domains would be the more stable ones. Remember now that the regions over which the dwq propagate are the domains where ordering (i.e. symmetry breakdown) is produced. Thus we arrive at the dynamic formation of a hierarchy (according to their life-time or equivalently to their sizes) of ordered domains [2].

8. Understanding Consciousness?

Is it possible to apply "scientific methods" to the study of consciousness?

Together with Schrödinger, a physicist or a neuroscientist would ask the question: What kind of material process is directly associated with consciousness [24]? Thus, for a physicist or a neuroscientist the starting point to investigate consciousness is the physical brain. But "where" in the brain? In which one of its regions should one look in order to find some special tissue or neuronal circuit or anything special out of which consciousness comes out?

In recent years proposals have been advanced [15, 13] that consciousness finds its root in the activity of the cytoskeleton. A possible hint in such a direction is provided by the fact that anesthetic substances produce the loss of consciousness by interfering with the normal cytoskeleton activity. Even if much is known, a fully understanding of the activity of anesthetic molecules has not yet been reached.

According to Hameroff and Penrose [13], consciousness might find its origin in the non-computational nature of the brain quantum state, which is supposed to be characterized by the coherent collective state of the microtubules. Here I will not consider the Hameroff-Penrose model, neither other models. Rather, I would like to report about some physiological aspects of the brain which, according to Susan Greenfield [10], might be relevant in understanding consciousness. On the other hand, the dissipative quantum model of brain seems to be in quite well agreement with the physiologically based Greenfield's conclusions.

The problem of the "location" of functions in the brain is a recurring problem in neuroscience. However, although the search of the location of functions such as memory, vision and other functions relating to the external world, has been carried on for a long time, such functions have not been found to be related to respective single brain regions. Neuroscientists now know from laboratory observations that several brain regions act cooperatively and simultaneously, as a connected whole [16, 21, 9, 10, 11] in performing functions of the brain which relate it to the outside world. On the basis of such experimental observations, it has been suggested [4, 10] that the same would also be the case for consciousness, so that there would be no specific neuronal circuits or neuronal populations committed to consciousness generation. Thus, a first property of consciousness, consistent with these observations is the one of being "spatially multiple" and "temporally unitary" [10]

But consciousness, according to Greenfield, also appears to be a continuum and it derives from a specific stimulus. The property of being a continuum means that consciousness grows as the brain develops: consciousness is not "all or none", it is more "like a dimmer switch that grows as the brain does" [10, 12]. The manifestation of consciousness in different degrees (as values in a continuous scale) is not only referred to different beings (animals versus humans) or to different stages of the growth (children versus adults), but also to different moments of one's mental experience, perhaps consequent to some external action.

On the other hand, it is a common experience that one is always conscious of something, never of nothing and not of everything at once. This is the content of the third property of consciousness. One is always conscious of "some kind of focus, epicentre or trigger" [10, 12].

Which ones are the physiological features which may possibly support these last two properties of consciousness? In order to answer to such a question, one should consider that the brain presents an extremely dense network of connections between the neurons (ranging from 10 to 100,000 connections between each of the 100 billion brain cells) and that a distinction has to be made between structural or anatomical connectivity, which is quite stable (quasi-stationary), and functional or effective connectivity, which, on the contrary, may be highly dynamic with modulation times of the order of hundreds of milliseconds. The dynamic cooperativity of neurons is sustained by such an intricate net of connections: neural cooperativity is thus an emergent property of neurons which could not be inferred by single neuron observation [1, 9]. Observations also show that the connectivity of non-specialized neurons grows as the brain develops and relates to the external world. The growth of the connectivity is observed to be related to brain age, with training in performing certain operations and in general with the brain's experience in relating to its environment. For example, the production of a large quantity of tubulin is observed in the visual cortex of baby rats when they first open their eyes (starting of the critical learning activity). Such a production decreases when the critical learning activity is over [5]. Moreover, the plasticity of the functional neural connections, the fact that they are not "hard wired" [34], but dynamical, implies that the brain can *learn*, it is an "adaptive" system, able to perform a large and rich spectrum of activities within a wide range of changeable boundary conditions.

The connected domains of neurons thus change in time by assembling and disassembling, and recruiting time by time a different number of neurons. The Libet experiments [17] show that under the external stimulus a (relatively) restricted number of neurons are first recruited in the brain response. Such a response was clearly recorded by electroencephalogram, but the subject did not report any conscious feeling of the stimulus at this stage. Only after 500 milliseconds or so, when the instrumentation recorded the neural activity spreading over a much larger region of the brain, did the subject report "the feeling" of the external stimulus. The set of neurons firstly responding to the external stimulus, which act as a quite robust "seed" for the connection spreading (as a "stone thrown in a puddle" generating spreading rings of waves on the water surface, in the Greenfield picture), may physically represent the epicentre. This is a pre-conscious stage. Consciousness sets in with the observed time-delay, when the larger assembly of correlated neurons is formed. The emergence of consciousness is thus described by this growing population of neurons, gradually recruited in about half a second. The continuous, experience-related, formation of neural connectivity supports the view of consciousness as a continuum.

The dynamics of the connection formation is such that the same number of neurons are never correlated in exactly the same extent in exactly the same way. Thus one never has the same consciousness. The physical property which can sustain such a scenario is the modulation of the neuronal activity. And this may be obtained by means of some chemical messengers able to influence the degree of sensibility of the neurons to a certain input signal. The chemical action does not produce the cell excitation, but makes it more receptive to the recruiting signal from the epicentre. These chemicals thus "cooperate" to the enlargement of the connected neural domain starting from the epicentre. Apparently, these chemicals are indeed the target of those drugs (such as prozac, amphetamine, LSD) which are known to modify our consciousness states. Consciousness is triggered when the assembly of correlated neurons is sufficiently extended.

I will consider in the next section a few more results of the dissipative quantum model which appear to fit particularly well with the above physiological observations.

9. Life-time and localizability of correlated domains

The dissipative quantum model of brain provides a first understanding of how opposite features, such as "non-locality" from one side and "localization" from the other side, may be not mutually incompatible. Rather they correspond to different dynamical regimes, continuously merging one into the other, in dependence on the behavior of well specified variables and parameters. In the quantum model, domains of correlated dwq may be dynamically generated. The size and the life-time of these domains appear to depend on the number of links that the brain sets with its environment and on internal parameters, in agreement with the observed plasticity of the brain.

It is also interesting that physiological observations show that the recruitment of neurons in correlated domains occurs not in a process of "one cell to another one at a time", as in the traditional view of neuronal communication through electrical signals, but "all at once over an ever larger group" and is mediated by "a chemical, than can bias large number of neurons to be activated simultaneously" [10]. *The emergence of such a simultaneous coherent cooperativity among neurons is exactly what the dissipative quantum model predicts.* The quantum model provides the dynamical ground which manifests in simultaneous neuronal activation under the "recruiting signal". The further, necessary step forward to be made is to disclose how the underlying dynamics controls the details of the chemical scenario.

Let me comment on what it can happen in the case of very small or very large neuronal assembly.

Very small neuronal assembly can occur, for example, when the epicentre is so weak (the external driving stimulus is not so strong) that a large number of neurons cannot be excited in a collective mode. In the quantum model this corresponds to having very few "links" interlaced with the external world, and in this model we expect sudden shifts from one vacuum (one memory) to another one (another memory) of the too small correlated domain, a typical situation of "confusion". When one is dreaming the external inputs are not so strong, the number of links with the external world is in fact small. Since the EEG of a dreaming subject turns out to be similar to when the subject is awake, some (low) degree of consciousness may be attached to the dreaming activity. It is the case of "fragile, little bits of consciousness" [10], I would say of very short and unstable permanence in one single vacuum of the correlated domain, with sudden transitions to other vacua; in fact in dreams we experience sudden change of scenarios, of facts, we feel flooded by a rapid succession of emotions. It would be interesting to study other cases of an extreme low number of links with the external world such as in autism and coma states.

Another cause of too small neuronal assembly may be in the weak neuronal recruitment; in other words, in the low level of connectivity. Physiology tells us that this is the case of the brain with a low level of relation and experience with the external world. These are also the cases where the dissipative model predicts small correlation lengths (correlated domains of small size) due to lack of links to the outside world, or else a low coherence in the correlation. Typically, in such cases the subject appears to be easily "distracted" from a certain object by another upcoming sensory input. He can be emotionally taken by a new scenario, apparently dominated by any epicentre triggered from the outside.

What can be the phenomenology in the opposite case of abnormally large neuronal assembly? The dissipative model would predict a more strict inequivalence among the vacua, namely a stronger resistance in switching from one to another one of them, an inclination to "fixations", to be trapped in one of the vacua and to remain fixed on the information, image or idea, there coded. This is also expected on physiological grounds, where, in the absence of competing epicenters, the subject is expected to show a strong continuity, a perseverance in what he has come to think. Such a sort of control on the incoming changeable stimuli will make the world appear to him as remote, grey, as in clinical depression cases where the patient appears unable to get emotionally involved, "the opposite of the glowing bright colors of the child's perspective" [10].

10. A trade with my Double

The mathematical and physical meaning of the tilde-system is to describe the environment to which the brain is permanently coupled (linked). Since the brain is intrinsically an open system, the tilde-system can *never* be neglected. We have seen that this is the meaning of the entanglement. The tilde-modes thus might play a rôle in the conscious as well in the unconscious brain activity. The tilde-modes might tell us also something about that fuzzy region between fragile consciousness and the obscure unconscious core of the dream activity.

The nonlinearity of the coupling of A with $\tilde{A}$ describes the self-inter - action or back-reaction process for the A system [30]. $\tilde{A}$ thus plays a rôle in such self-coupling or "self-recognition" processes. The $\tilde{A}$ system is the "mirror in time" image, or the "time-reversed copy" of the A system. It actually duplicates the A system, it is the A system's *Double*.

The rôle of the $\tilde{A}$ modes in the self-interaction processes has led me to conjecture that the tilde-system is actually involved in consciousness mechanisms [30]. Thus the specific dissipative character of the dynamics strongly point to consciousness as a "time mirror", as a "reflection in time" which manifests as nonlinear coupling or *dialogue* [6] with the inseparable own *Double* [30]. In some sense, the unavoidable coupling with the external world is "internalized" in the dialectic, permanent relation with the Double.

The "doubling" of the self is actually a very old literary metaphor. Sosia in the Plautus comedy Amphitruo, or the falling in love of Narcissus with himself mediated by his "reflection" in the water, are famous examples of such a metaphoric use of the "doubling ". The ancient Vedic tradition consciousness also flows between two poles: an identity of self and an identity with the processes of the Universe.

Consciousness seems thus to emerge as a manifestation of the dissipative dynamics of the brain. In this way, consciousness appears to be not solely characterized by a subjective dynamics; its roots, on the contrary, seem to be grounded in the permanent "trade" of the brain (the subject) with the external world, on the dynamical relation between the system A and its Sosia or Double $\tilde{A}$, permanently joined (conjugate) to it.

I am absolutely not saying that A acts as a mirror of the outside world. On the contrary, consciousness is reached "through" the opening to the external world. The crucial rôle of dissipation is that self-mirroring is not anymore a "self-trap" (as for Narcissus), the conscious subject *cannot* be a closed system. Consciousness is only possible if dissipation, openness onto the outside world is allowed. Without the "objective" external world there would be no possibility for the brain to be an open system, and no $\tilde{A}$ system would at all exist. The very same existence of the external world is the *prerequisite* for the brain to build up its own "subjective simulation" or representation of the external world.

The informational inputs from the external world are the "images" of the world. Once they are recorded by A they become the "image" of A: $\tilde{A}$ is the "address" of A, it is identified with (is a copy of) A. We have seen that such a process implies a "breakdown", a "lack" of symmetry: memory as "negation" of the symmetry which makes things indistinguishable among themselves [31]; memory as "non-oblivion", literally the $\alpha\lambda\eta\theta\epsilon\iota\alpha$, i.e. the word used by the ancient Greeks to denote the "truth".

As already mentioned, the finiteness of the correlated domains implies that recording memories requires some expense of energy. Thus, unavoidably, we are led to make a "choice", an "active" selection among the many inputs we receive: we record only those that we judge worthwhile to expend some energy for, the ones to which we attribute a "value", which involve our "emotion". It is the specific information received through those selected inputs which then becomes "our memory", it becomes "our truth" ($\alpha\lambda\eta\theta\epsilon\iota\alpha$, indeed). It is here, in such a map of values, that our memory depicts our "identity" and, since our choice is unavoidable, the emergence of the identity is also a "necessary" event. In fact, mathematically speaking, in the model the brain state is "identified" by the collection of the memory codes.

We are not simply spectators or victims of "passive perceptions". Plasticity allows agency, volition and intentionality which have a non-negligible rôle in the raising of our consciousness. Pribram remarks [22] that there is always an "attention" content in the input, an "intention" content in the output, and a "thought" content in the memory processes and all of it participates in a "vast unconscious processing". Freeman stresses that brain actually processes *meanings* rather than information, meanings are "intended actions", namely the meaning *is in* the subject and arises from the "active" perception of that subject. In the light of Freeman's suggestion, the tilde-modes express meanings or "meaningful representations" rather than just representations.

The conclusion is that one reaches "an active point of view" of the world [6, 31], which naturally carries in it the "unfaithfulness" of subjectivity. But such unfaithfulness is precious. It is exactly in such an unfaithfulness that the map of the values (Freeman meanings) which *identify* the subject has to be searched.

The openness to the external world, dissipation, thus implies the capability of the brain to respond to the external stimuli *at each specific instant of time*, to be "present", namely to singling out at each specific instant of time one specific vacuum among those entering the superposition of the brain state. From one side, openness thus guaranties against the risk of remaining "trapped" in one single vacuum, without updating the vacuum "choice" to the present time, it guaranties "presence", "conscious feeling" of that specific, "actual" vacuum. On the other hand, it also avoids blind travelling, running without "looking inside", over the superposition of memory states which makes the brain state. In the absence of the healthy or normal state of openness, too small or too large neuronal correlated assembly may turn into "low level of consciousness". The unconscious brain activity may be then related to lack of openness or to too high level of openness (too many inputs in a too rapid succession), which paradoxically may correspond to "closure", producing too slow, or even absent, adaptive capability to the present or too high emotional arousal. In the dissipative quantum model unconscious brain activity and consciousness appear to be merging dynamical regimes, different modulations in the dialogue with the Double.

On the basis of the above discussion it appears that the conscious identity emerges *at any instant of time*, in the "present", as the minimum energy brain state which separates the past from the future, that "point" on the "mirror of time" where the conjugate images A and $\tilde{A}$ join together. In the absence of such a mirroring there is neither consciousness of the past nor its projection in the future. The subject identity may only emerge in the dissipative dynamics, in its interplay with the objective external world. Eventually, the intrinsic dissipative nature of the brain excludes any model of consciousness centered exclusively on "first person" inner activity. Dissipation manifests itself as a "second person", the Double or Sosia, to dialogue with [31, 33].

Acknowledgments

I am very thankful to Professor Bruno Gruber for inviting me to present this paper at the XIII International Conference Symmetry in Science. I also thank Murst and INFN for partial financial support.

References

[1] Aertsen, A. and G.L. Gerstein. Dynamic aspects of neuronal cooperativity. In J. Kruger, editor, *Neuronal cooperativity*. Springer Verlag, Berlin, 1991

[2] Alfinito, E. and G. Vitiello. Formation and life-time of memory domains in the dissipative quantum model of brain. *Int. J. Mod. Phys.* **B 14**, 853–868, 2000.

[3] Celeghini, E., M. Rasetti and G. Vitiello. Quantum dissipation. *Ann.Phys. (N.Y.)* **215**, 156–170, 1992.

[4] Crick, F. and C. Koch. Towards a neurobiological theory of consciousness. *Seminars in the Neurosciences* **2**, 263–275, 1990.

[5] Cronley–Dillon, J., D. Carden and C. Birks. The possible involvement of brain microtubules in memory fixation. *J. Exp. Biol.* **61**, 443–454, 1974.

[6] Desideri, F. *L'ascolto della coscienza*. Feltrinelli, Milano, 1998.

[7] Freeman, W.J. On the the problem of anomalous dispersion in chaotic phase transitions of neural masses, and its significance for the management of perceptual information in brains. In H. Haken and M. Stadler, editors, *Synergetics of cognition* **45**, 126–143. Springer Verlag, Berlin, 1990.

[8] Freeman, W.J. Random activity at the microscopic neural level in cortex ("noise") sustains and is regulated by low dimensional dynamics of macroscopic cortical activity ("chaos"). *Intern. J. of Neural Systems* **7**, 473–480, 1996.

[9] Freeman, W.J. *Neurodynamics: An exploration of mesoscopic brain dynamics*. Springer, Berlin, 2000.

[10] Greenfield, S.A. How might the brain generate consciousness? *Communication and Cognition* **30**, 285–300, 1997.

[11] Greenfield, S.A. *The brain: a guided tour*. Freeman, New York, 1997.

[12] Greenfield, S.A. A Rosetta Stone for mind and brain. In Hameroff, S.R., A.W. Kaszniak and A.C. Scott, editors, *Toward a science of consciousness II. The second Tucson Discussions and debates* p. 231–236. MIT Press, Cambridge, 1998.

[13] Hameroff S.R. and R. Penrose. Conscious events as orchestrated space-time selections. *J. of Consciousness Studies* **3**, 36–53, 1996.

[14] Itzykson, C. and J. Zuber. *Quantum field theory*. McGraw-Hill, New York, 1980.

[15] Jibu, M. and K. Yasue. *Quantum brain dynamics and consciousness*. John Benjamins, Amsterdam, 1995.

[16] Lashley, K.S. The problem of cerebral organization in vision. In *Biological Symposia, VII, Visual mechanisms* p.301–322. Jaques Cattell Press, Lancaster, 1942.

[17] Libet, B., E.W. Wright, B. Feinstein and D.K. Pearl. Subjective referral of the timing for a conscious experience. *Brain* **102**, 193–224, 1979.

[18] Pessa, E. and G. Vitiello. Quantum dissipation and neural net dynamics. *Biolectrochemistry and bioenergetics* **48**, 339–342, 1999.

[19] Pessa, E. and G. Vitiello. In preparation, 2003.

[20] Pribram, K.H. *Languages of the brain*. Prentice-Hall, Englewood Cliffs, N.J., 1971.

[21] Pribram, K.H. *Brain and perception*. Lawrence Erlbaum, Hillsdale, N. J., 1991.

[22] Pribram, K.H. Brain and quantun holography: Recent ruminations. In M. Jibu, T. Della Senta and K.Yasue, editors, *Toward a Science of Consciousness - Fundamental Approaches*. John Benjamins, Amsterdam, 2000.

[23] Ricciardi, L.M. and H. Umezawa. Brain physics and many-body problems. *Kibernetik* **4**, 44–48, 1967.

[24] Schrödinger, E. *What is life?*, 1944. Cambridge University Press, Cambridge, 1967 [reprint].

[25] Sivakami S. and V. Srinivasan. A model for memory. *J. Theor. Biol.* **102**, 287–294, 1983.

[26] Stuart, C.I.J., Y. Takahashi and H. Umezawa. On the stability and non-local properties of memory. *J. Theor. Biol.* **71**, 605–618, 1978.

[27] Stuart, C.I.J., Y. Takahashi and H. Umezawa. Mixed system brain dynamics: neural memory as a macroscopic ordered state. *Found. Phys.* **9**, 301–327, 1979.

[28] Umezawa, H. *Advanced field theory: micro, macro and thermal concepts*. American Institute of Physics, New York, 1993.

[29] Umezawa, H. Development in concepts in quantum field theory in half century. *Math. Japonica* **41**, 109–124, 1995.

[30] Vitiello, G. Dissipation and memory capacity in the quantum brain model. *Int. J. Mod. Phys.* **9**, 973–989, 1995.

[31] Vitiello, G. Dissipazione e coscienza. *Atque* **16**, 171–198, 1997 [in Italian].

[32] Vitiello, G. The arrow of time and consciousness. *Cognitive Processing* **1**, 35–43, 2000.

[33] Vitiello, G. *My Double unveiled*, John Benjamins, Amsterdam, 2001.

[34] Zheng, J.Q., M. Felder, J.A. Connor and M.M. Poo. Tuning of nerve growth cones induced by neurotransmitters. *Nature* **368**, 140–144, 1994.

BETA-DECAY IN ODD-*A* CS TO XE IN THE INTERACTING BOSON-FERMION MODEL

N. Yoshida
Faculty of Informatics, Kansai University,
2-1-1 Ryozenji-cho, Takatsuki-shi, 569-1095 Japan

L. Zuffi
Dipartimento di Fisica dell'Università di Milano
and Istituto Nazionale di Fisica Nucleare, Sezione di Milano
Via Celoria 16, Milano 20133, Italy

S. Brant
Department of Physics, Faculty of Science,
University of Zagreb,
10000 Zagreb, Croatia

Abstract Beta-decay is studied in the odd-mass isotopes of Cs and Xe in the proton-neutron interacting boson-fermion model (IBFM2). The model provides a consistent description of the energy levels and the electromagnetic properties as well as the beta-decay rates.

Keywords: Beta-decay, interacting boson-fermion model

1. Introduction

The interacting boson model (IBM) [2, 10] and the interacting boson-fermion model (IBFM) [11, 12] have been successful in describing a variety of nuclear structure phenomena in even-even and odd-even nuclei. One remarkable character of these models is their symmetry limits. The IBM is known to have its symmetry limits like SU(5), SU(3) and O(6) limits corresponding to the vibrational, the rotational and the γ-unstable nuclei. The symmetries in IBFM are known as the nuclear supersymmetries. Many of the even-even isotopes of Xe have been known as showing the O(6) limit of the IBM, i.e., the γ-unstable limit. They have been extensively studied by using the IBM, together with their

B. Gruber (ed.), Symmetries in Science XI, 575–588.

neighboring odd-even nuclei by the IBFM [1, 3, 5, 24]. As one of the possible applications of the IBFM, we consider here the description of beta-decay rates in odd mass nuclei. Beta-decay rates are very sensitive to details of the wave functions and therefore can provide a test of the nuclear model.

The application of the proton-neutron interacting boson-fermion model (IBFM2) to the beta decay was proposed for nuclei in the region $52 \leq Z \leq 58$ [6]. Later the study was extended to Ru and Tc nuclei [16, 17] and for beta transitions from even-even to odd-odd nuclei [18]. For beta transitions from spherical Rh to Pd odd mass nuclei, this approach was very successful [25, 26, 27]. In that case (near the SU(5) limit of the IBM) the wave functions were dominated by few big components in the SU(5), i.e., the vibrational basis, and the calculation of beta-decay rates tested those components.

In the present report, we analyze the beta-decay from Cs to Xe isotopes of mass number $A = 125, 127, 129$. These isotopes can be considered to be around the O(6) limit of the IBM[28]. The wave functions are very complex, therefore one can have a more sensitive test of the model than in the spherical case. We calculate the energy levels and electromagnetic matrix elements. We then calculate the beta-decay rates. This final step is parameter free and provides a unique test of the model.

2. The IBFM2 model

In the IBFM2, an odd-A nucleus is described by coupling an odd nucleon to an even-even core of proton- and neutron-bosons. The Hamiltonian is written as

$$H = H^{\mathrm{B}} + H^{\mathrm{F}} + V^{\mathrm{BF}}, \tag{1}$$

where H^{B} is an IBM2 Hamiltonian consisting of the single d-boson energy, the quadrupole interaction, the Majorana interaction and so on:

$$\begin{aligned} H^{\mathrm{B}} &= \epsilon_d n_d + \kappa\,(Q_\nu^{\mathrm{B}} \cdot Q_\pi^{\mathrm{B}}) \\ &+ \frac{1}{2}\xi_2\left((d_\nu^\dagger s_\pi^\dagger - d_\pi^\dagger s_\nu^\dagger)\cdot(\tilde{d}_\nu s_\pi - \tilde{d}_\pi s_\nu)\right) \\ &+ \sum_{K=1,3} \xi_K\left([d_\nu^\dagger d_\pi^\dagger]^{(K)} \cdot [\tilde{d}_\pi \tilde{d}_\nu]^{(K)}\right) \\ &+ \frac{1}{2}\sum_{L=0,2,4} c_L^\nu\left([d_\nu^\dagger d_\nu^\dagger]^{(k)} \cdot [\tilde{d}_\nu \tilde{d}_\nu]^{(k)}\right) \end{aligned} \tag{2}$$

where n_d is the total d-boson number,

$$Q_\nu^{\mathrm{B}} = d_\nu^\dagger s_\nu + s_\nu^\dagger \tilde{d}_\nu, + \chi_\nu\,[d_\nu^\dagger \tilde{d}_\nu]^{(2)}, \tag{3}$$

$$Q_\pi^{\mathrm{B}} = d_\pi^\dagger s_\pi + s_\pi^\dagger \tilde{d}_\pi + \chi_\pi\,[d_\pi^\dagger \tilde{d}_\pi]^{(2)} \tag{4}$$

are the boson quadrupole operators, where $s_\rho^\dagger$ and $d_\rho^\dagger$ are the s-boson and the d-boson creation operators, s_ρ is the s-boson annihilation operator. The symbol ρ represents either ν (neutron) or π (proton). The modified d-boson annihilation operator $\tilde{d}_\rho$ is related to the d-boson annihilation operator by $\tilde{d}_{\rho,m} = (-1)^m d_{\rho,-m}$. The Hamiltonian of the odd fermion H^{F} is

$$H^{\mathrm{F}} = \sum_i \epsilon_i n_i \tag{5}$$

where ϵ_i is the quasi-particle energy of the ith orbital while n_i is its number operator. The interaction between the bosons and the odd particle V^{BF} includes the quadrupole interaction, the monopole interaction and the exchange interaction:

$$\begin{aligned} V^{\mathrm{BF}} &= \sum_{i,j} \Gamma_{ij} \left([a_i^\dagger \tilde{a}_j]^{(2)} \cdot Q_{\rho'}^{\mathrm{B}} \right) \\ &\quad + A \sum n_i n_{d_{\rho'}} \\ &\quad + \sum_{i,j} \Lambda_{ki}^{j} \left\{ : \left[\left[d_\rho^\dagger \tilde{a}_j \right]^{(k)} a_i^\dagger s_\rho \right]^{(2)} : \cdot \left[s_{\rho'}^\dagger \tilde{d}_{\rho'} \right]^{(2)} \right. \\ &\qquad\qquad \left. + \text{Hermitian conjugate} \right\}, \end{aligned} \tag{6}$$

where the symbols ρ and ρ' denote π (ν) and ν (π) respectively, if the odd fermion is a proton (a neutron). The creation operator of the odd particle is written as $a_{jm}^\dagger$, while the modified annihilation operator is defined as $\tilde{a}_{jm} = (-1)^{j-m} a_{j-m}$. We adopt the orbital dependence based on a microscopic derivation[22]:

$$\Gamma_{i,j} = (u_i u_j - v_i v_j)\, Q_{i,j}\, \Gamma \tag{7}$$

$$\Lambda_{k,i}^{j} = -\beta_{k,i} \beta_{j,k} \left(\frac{10}{N_\rho (2j_k + 1)} \right)^{1/2} \Lambda \tag{8}$$

where u_i and v_i are the BCS unoccupation and occupation amplitudes, and

$$\beta_{i,j} = (u_i v_j + v_i u_j)\, Q_{i,j} \tag{9}$$

$$Q_{i,j} = \left\langle l_i, \frac{1}{2}, j_i || Y^{(2)} || l_j, \frac{1}{2}, j_j \right\rangle. \tag{10}$$

The parameters Γ, A and Λ determine the strengths of the quadrupole, monopole and the exchange interactions.

Table 1. IBM2 parameters. The unit is MeV except for dimensionless χ_ν. The parameters $\chi_\pi = -0.80$ and $\xi_1 = \xi_2 = 0.24$ MeV, $\xi_3 = -0.18$ MeV are fixed.

odd nuclei	core nucleus	ϵ_d	κ	χ_ν	c_0^ν	c_2^ν
^{125}Cs	^{124}Xe	0.70	-0.145	0.00	0.05	-0.10
^{125}Xe, ^{127}Cs	^{126}Xe	0.70	-0.155	0.20	0.10	-0.10
^{127}Xe, ^{129}Cs	^{128}Xe	0.70	-0.170	0.33	0.30	0.00
^{129}Xe	^{130}Xe	0.76	-0.190	0.50	0.30	0.10

Table 2. Single-particle energies (MeV) of the proton orbitals in Cs.

$d_{5/2}$	$g_{7/2}$	$s_{1/2}$	$d_{3/2}$	$h_{11/2}$	$h_{9/2}$	$f_{7/2}$
0.05	0.00	3.35	3.00	1.50	7.00	8.00

Table 3. Parameters in the boson-fermion interaction (MeV) for Cs.

isotope	Γ	A	Λ
^{125}Cs	0.90	-0.60	1.65
^{127}Cs	0.76	-0.66	2.30
^{129}Cs	0.74	-0.80	2.90

3. Calculations

3.1 Hamiltonian and energy levels

The IBM2 parameters for the even-even Xe isotopes are taken from Ref. [20]. Table 1 shows the cores for the considered Cs and Xe isotopes and the IBM2 parameters.

The odd-mass Cs isotopes are described by coupling an odd proton to the even-even Xe cores. The proton single-particle energies are taken from Ref. [8] (only the energy of $d_{5/2}$ has been reduced from the original value of 0.20 MeV to 0.05 MeV). The BCS equations are solved with the orbitals $g_{7/2}$, $d_{5/2}$, $s_{1/2}$, $d_{3/2}$, $h_{11/2}$, $h_{9/2}$ and $f_{7/2}$ with the gap energy of $\Delta = 12/\sqrt{A}$ MeV. The adopted single-particle energies are shown in Table 2. The quasi-particle energies and the u, v-factors calculated with them have been used in $H^{\rm F}$ and $V^{\rm BF}$. In the IBFM2 calculation for positive-parity states we include the fermion orbitals with positive parity. In the boson-fermion interaction, the quadrupole and the monopole interactions are included between the odd proton and the neutron bosons, in addition to the exchange interaction of the quadrupole type. We allow the interaction strengths Γ, A, Λ vary gradually depending on the mass number. In Table 3 we present the parameter values in $V^{\rm BF}$ interaction potential.

The results of the calculation are shown in Fig. 1. The experimental data are taken from Refs. [13, 14, 23]. A generally reasonable agreement is seen. To see more detail, the calculated $3/2_1^+$ systematically comes lower than the

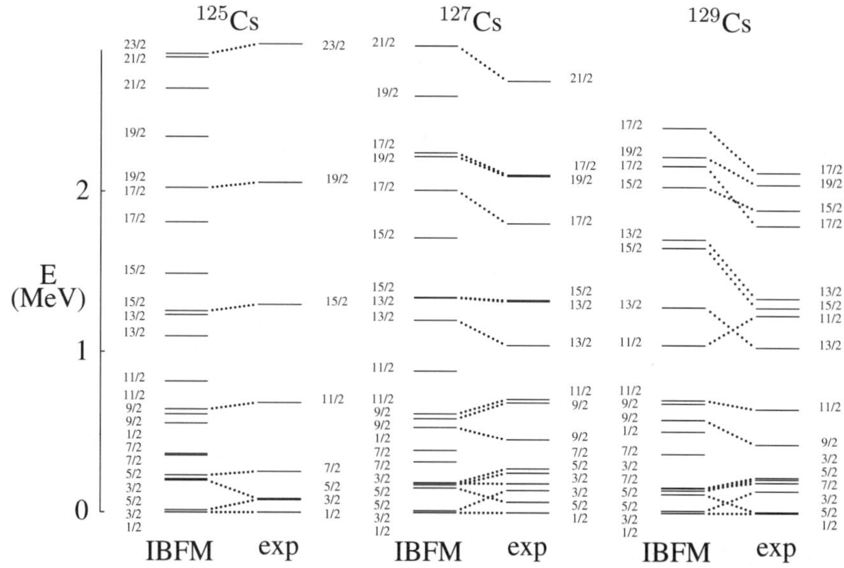

Figure 1. Comparison between the calculated (IBFM) and the exp. energy levels of positive parity in ^{125,127,129}Cs.

Table 4. Single-particle energies (MeV) of the neutron orbitals in Xe. The energy of $g_{7/2}$ has been changed.

	$d_{5/2}$	$g_{7/2}$	$s_{1/2}$	$d_{3/2}$	$h_{11/2}$
^{125}Xe	0.00	0.30	1.55	2.00	1.30
^{127}Xe	0.00	0.35	1.55	2.00	1.30
^{129}Xe	0.00	0.40	1.60	2.00	1.30

experimental counterpart, while the calculated $5/2_1^+$ lies higher. This difference may be explained by the Coriolis effect. The locations of the yrast states with $I \geq 7/2$ are reasonably well reproduced. Nevertheless, we notice that the $\Delta I = 1$ structure consisting of $9/2^+$, $13/2^+$, $17/2^+$, $21/2^+$ yrast and $11/2^+$, $15/2^+$, $19/2^+$ yrare states, was proposed as intruder proton $g_{9/2}$ configuration in ^{125}Cs [9], as well as some levels in ^{127}Cs [15]. These levels are not presented in Fig. 1, but we notice that the excitation energies of these levels, calculated in the valence shell space, are very close to the $g_{9/2}$ experimental ones. This could be the reason why they have not been observed.

The odd-mass Xe isotopes are described by coupling an odd neutron hole to the neighboring even-even Xe cores. The single-particle energies are taken from Ref.[5], except small modifications for $g_{7/2}$ and $h_{11/2}$.

The adopted single-particle energies are shown in Table 4.

Table 5.]

Parameters in the boson-fermion interaction (MeV) for Xe.

isotope	Γ	A	Λ
^{125}Xe	0.39	-0.42	0.40
^{127}Xe	0.44	-0.42	0.40
^{129}Xe	0.50	-0.42	0.40

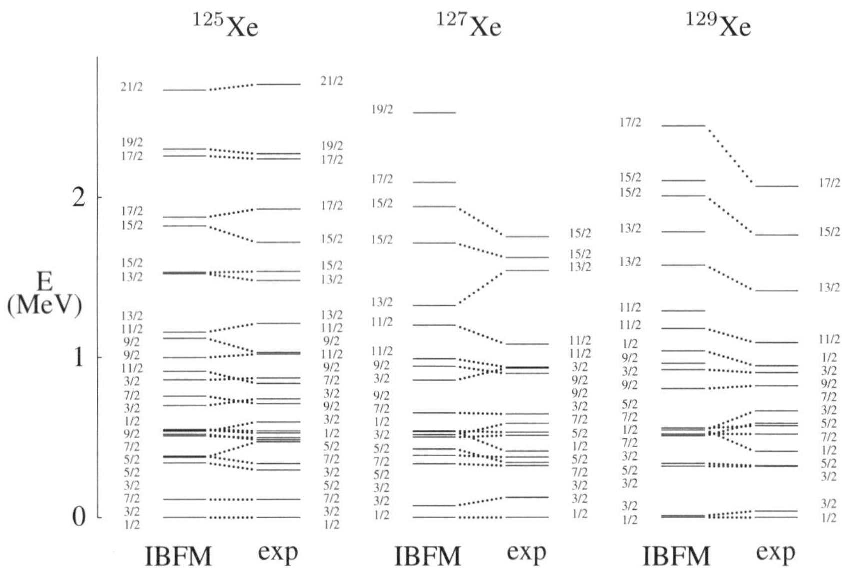

Figure 2. Comparison between the calculated (IBFM) and the exp. energy levels of positive parity in ^{125,127,129}Xe.

The parameter values used in V^{BF} are shown in Table 5. These values are almost identical to the approximate projection from the IBFM1 values in Ref. [5].

The results of the calculation are shown in Fig. 2. The experimental data are taken from Refs. [19, 13, 14, 23]. A reasonable agreement is seen. In ^{127}Xe, there are two different interpretations about the spin of the 510 keV level. Although Ref.[14] adopts $I = 3/2$, Refs.[5, 19] insist on $I = 5/2$ because of very weak beta-decay from $I = 1/2$ in ^{127}Cs and the level systematics in neighboring nuclei. We have chosen the latter on the basis of level systematics. However, the spin of 510 keV is still an open problem. The wave functions and organization of the levels into bands in the present IBFM2 calculation are in very good agreement with the recent analysis in IBFM1 [5].

3.2 Electromagnetic properties

The electromagnetic transition operators are

$$T^{(\mathrm{E2})} = e_\pi^{\mathrm{B}}\, Q_\pi^{\mathrm{B}} + e_\nu^{\mathrm{B}}\, Q_\nu^{\mathrm{B}} + \sum_{i,j} e'_{i,j}\, [a_i^\dagger \tilde{a}_j]^{(2)} \tag{11}$$

where

$$e'_{i,j} = -\frac{1}{\sqrt{5}}(u_i u_j - v_i v_j) \left\langle i||r^2 Y^{(2)}||j \right\rangle . \tag{12}$$

For M1,

$$T^{(\mathrm{M1})} = \sqrt{\frac{3}{4\pi}} \left(g_\pi^{\mathrm{B}}\, L_\pi^{\mathrm{B}} + g_\nu^{\mathrm{B}}\, L_\nu^{\mathrm{B}} + \sum_{i,j} e_{i,j}^{(1)}\, [a_i^\dagger \tilde{a}_j]^{(1)} \right) \tag{13}$$

where

$$e_{i,j}^{(1)} = -\frac{1}{\sqrt{3}}(u_i u_j + v_i v_j) \left\langle i||g_l \mathbf{l} + g_s \mathbf{s}||j \right\rangle . \tag{14}$$

The boson effective charges $e_\pi^{\mathrm{B}} = e_\nu^{\mathrm{B}} = 0.150$ eb that we take in order to explain the experimental data in these odd-mass nuclei are somewhat larger than the ones determined from the corresponding even-even cores (≈0.108 eb) [21]. This difference may be due to a polarization effect caused by the odd fermion. In fact the polarization of the boson core by the odd fermion was also observed in previous calculations in this region, in the signature dependence of the energy levels[24]. For the odd proton in Cs $e_\pi^{\mathrm{F}} = 1.5$ e, while for the odd neutron in Xe $e_\nu^{\mathrm{F}} = 0.5$ e. For the magnetic dipole operator, the boson g-factors for all the isotopes are: $g_\nu^{\mathrm{B}} = 0$, $g_\pi^{\mathrm{B}} = 0.8\ \mu_{\mathrm{N}}$. For the odd proton in Cs, the spin g-factor is reduced by the factor of 0.85, while for the odd neutron in Xe, the spin g-factor is reduced by the factor of 0.5. Figure 3 shows the results of the calculations. On the basis of calculated excitation energies and electromagnetic properties we may conclude that the IBFM2 description of the analysed Cs and Xe isotopes is realistic. The calculated electromagnetic transitions and static moments are in reasonable agreement with experimental data. However, the beta-decay rates, where both the parent and the daughter wave functions are involved, can give a severer test.

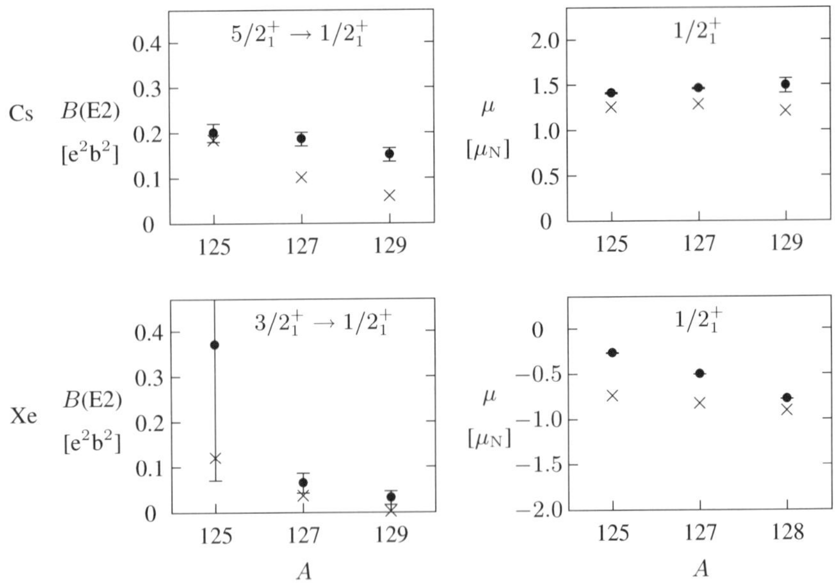

Figure 3. B(E2) values and magnetic moments. The symbols • with the error bars denote the experimental data, while the symbols × show the calculated values.

3.3 Beta-decay

The Fermi $\sum_k t^{\pm}(k)$ and the Gamow-Teller $\sum_k t^{\pm}(k)\sigma(k)$ transition operators of the IBFM2 can be constructed by the transfer operators [22, 4, 6, 12]:

$$A_m^{\dagger(j)} = \zeta_j a_{jm}^{\dagger} + \sum_{j'} \zeta_{jj'} s^{\dagger} [\tilde{d} a_{j'}^{\dagger}]_m^{(j)}, \quad (\Delta n_j = 1,\ \Delta N = 0) \tag{15}$$

$$B_m^{\dagger(j)} = \theta_j s^{\dagger} \tilde{a}_{jm} + \sum_{j'} \theta_{jj'} [d^{\dagger} \tilde{a}_{j'}]_m^{(j)} \quad (\Delta n_j = -1,\ \Delta N = 1). \tag{16}$$

The former creates a fermion, while the latter annihilates a fermion simultaneously creating a boson. Either operator increases the sum $n_j + 2N$ by one unit.

The conjugate operators are:

$$\begin{aligned}\tilde{A}^{(j)}_{m} &= (-1)^{j-m}\left\{A^{\dagger(j)}_{-m}\right\}^{\dagger} \\ &= \zeta^{*}_{j}\tilde{a}_{jm} + \sum_{j'} \zeta^{*}_{jj'} s[d^{\dagger}\tilde{a}_{j'}]^{(j)}_{m} \\ &\qquad (\Delta n_j = -1,\ \Delta N = 0) \qquad (17)\\ \tilde{B}^{(j)}_{m} &= (-1)^{j-m}\left\{B^{\dagger(j)}_{-m}\right\}^{\dagger} \\ &= -\theta^{*}_{j} s a^{\dagger}_{jm} - \sum_{j'} \theta^{*}_{jj'}[\tilde{d}a^{\dagger}_{j'}]^{(j)}_{m} \\ &\qquad (\Delta n_j = 1,\ \Delta N = -1) \qquad (18)\end{aligned}$$

where the asterisks mean complex conjugate. These decrease $n_j + 2N$ by one unit. The IBFM image of the Fermi $\sum_k t^{\pm}(k)$ and the Gamow-Teller transition operator $\sum_k t^{\pm}(k)\sigma(k)$ are written as

$$O^{\mathrm{F}} = \sum_{j} -\sqrt{2j+1}\left[P^{(j)}_{\nu}P^{(j)}_{\pi}\right]^{(0)}, \qquad (19)$$

$$O^{\mathrm{GT}} = \sum_{j'j} \eta_{j'j}\left[P^{(j')}_{\nu}P^{(j)}_{\pi}\right]^{(1)} \qquad (20)$$

respectively, where

$$\begin{aligned}\eta_{j'j} &= -\frac{1}{\sqrt{3}} < l'\frac{1}{2}; j'||\sigma||l\frac{1}{2}; j > \\ &= -\delta_{l'l}\sqrt{2(2j'+1)(2j+1)}\, W\left(lj'\frac{1}{2}1; \frac{1}{2}j\right). \qquad (21)\end{aligned}$$

The transfer operators $P^{(j)}_{\rho}$ are chosen from Eqs. (15)-(18) depending on the nuclei. In the present case,

$$P^{(j)}_{\pi} = \tilde{A}^{(j)}_{\pi}, \qquad (22)$$

$$P^{(j)}_{\nu} = \tilde{B}^{(j)}_{\nu}. \qquad (23)$$

The square of the beta-decay matrix elements are

$$\langle M_{\mathrm{F}}\rangle^{2} = \frac{1}{2I_i+1}|\langle I_f||O^{\mathrm{F}}||I_i\rangle|^{2} \qquad (24)$$

$$\langle M_{\mathrm{GT}}\rangle^{2} = \frac{1}{2I_i+1}|\langle I_f||O^{\mathrm{GT}}||I_i\rangle|^{2} \qquad (25)$$

from which the ft value is calculated by

$$ft = \frac{6163}{\langle M_{\rm F}\rangle^2 + (G_{\rm A}/G_{\rm V})^2 \langle M_{\rm GT}\rangle^2} \tag{26}$$

in units of second where $(G_{\rm A}/G_{\rm V})^2 = 1.59$. The coefficients η_j, $\eta_{jj'}$, θ_j, $\theta_{jj'}$ appearing in Eqs. (15)- (18) are calculated by the formulation of Refs. [22, 12]:

$$\zeta_j = u_j \frac{1}{K'_j}, \tag{27}$$

$$\zeta_{jj'} = -v_j \beta_{j'j} \left(\frac{10}{N(2j+1)}\right)^{1/2} \frac{1}{KK'_j}, \tag{28}$$

$$\theta_j = \frac{v_j}{\sqrt{N}} \frac{1}{K''_j}, \tag{29}$$

$$\theta_{jj'} = u_j \beta_{j'j} \left(\frac{10}{2j+1}\right)^{1/2} \frac{1}{KK''_j}. \tag{30}$$

where N is N_π or N_ν, depending on the transfer operator, and K, K'_j, K''_j are determined by

$$K = \left(\sum_{jj'} \beta^2_{jj'}\right)^{1/2}, \tag{31}$$

and the conditions

$$\sum_{\alpha J} \left\langle \text{odd;}\alpha J || A^{\dagger j} || \text{even;}0^+_1 \right\rangle^2 = (2j+1)u_j^2, \tag{32}$$

$$\sum_{\alpha J} \left\langle \text{even;}0^+_1 || B^{\dagger j} || \text{odd;}\alpha J \right\rangle^2 = (2j+1)v_j^2. \tag{33}$$

The formulas (27)-(33) are valid when the odd nucleon is a particle. If the latter half of a major shell is partly occupied (e.g., $N = 73$), we consider the fully closed shell as the vacuum (e.g., $N = 82$ core to deal with $66 < N < 82$), according to the convention of IBM and IBFM. In this case, the microscopic derivations are done for the holes in the shell. For example, Eq. (15) is an IBFM image of a hole creation operator. Then the formulas corresponding to (27)-(33) can be obtained by interchanging u_j and v_j. The scheme is essentially the same as in Ref.[6]. Although some previous works introduced overall normalization factors to account for the absolute values of the beta-transition rates, we did not introduce any adjustable parameters in the beta-decay operators. For Xe, because the odd neutron is a hole in respect to the boson core, u_j and v_j are interchanged in Eqs. (5)-(33).

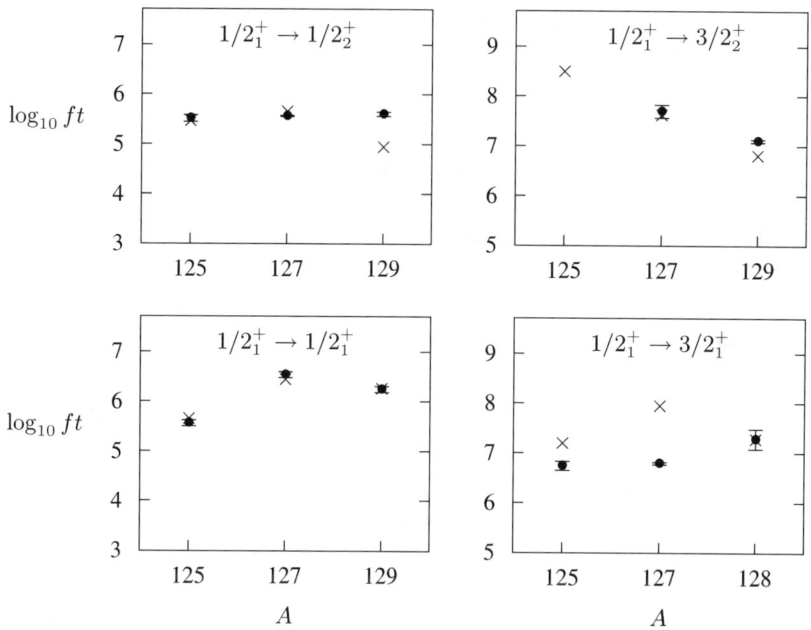

Figure 4. The beta-decay rates from ^{A}Cs to ^{A}Xe shown in terms of $\log_{10} ft$ values. The symbols • with the error bars denote the experimental data, while the symbols × show the calculated values.

Figure 4 shows the $\log_{10} ft$ values of the beta-decay rates from the ground states $1/2_1^+$ in Cs to the yrast and the yrare $1/2^+$ and $3/2^+$ levels in Xe. The experimental values have been derived by electron capture and β^+ experiments in Refs.[13, 14, 23]. By taking the wave functions from our calculations, the decay rates from the ground states $(1/2_1^+)$ to the ground states $(1/2_1^+)$ are reproduced very well. In addition, we obtain reasonable agreement in decays to $1/2_2^+$, $3/2_1^+$ and $3/2_2^+$, except for the decay to $3/2_1^+$ in ^{127}Xe. The decay rates to higher excited levels are very sensitive to details in wave functions. In that sense, we notice a reasonable agreement in $\log_{10} ft$ values for decays to the $3/2_3^+$ levels in A = 125, 127 and 129: theoretical values 6.402, 6.887 and 7.147, compared to the experimental data 6.360(70), 6.308(12) and 6.400(50), respectively.

4. Discussion

It is remarkable that once the wave functions are determined from the energy levels and the electromagnetic moments, the beta-decay rates values are obtained in a parameter-free calculation. In fact, in contrast to shell model calculations of the beta-decay rates, we do not use any additional normalization. Concerning the structures of wave functions, the ground states of ^{125,127,129}Cs are dominated by two orbitals $g_{7/2}$ and $d_{3/2}$ (30$\sim$40% each). The orbital $d_{5/2}$ has comparable amount of mixture, too. The mixture of the component $s_{1/2}$ (10$\sim$15%) is small. In the daughter nuclei ^{125,127,129}Xe, the dominant component of the ground state is $s_{1/2}$ (80$\sim$90%). The main contribution to the Gamow-Teller matrix elements comes from the term: $[[\tilde{d}_\nu\ \nu s_{1/2}]^{(3/2)}\pi d_{3/2}]^{(1)}$ in Eq. (20). Concerning the decay to the first excited states, the states $3/2_1^+$ in Xe have the main component of $d_{3/2}$. The main contributions to the Gamow-Teller matrix elements come from the terms: $[[\tilde{d}_\nu\ \nu d_{3/2}]^{(3/2)}\pi d_{3/2}]^{(1)}$ and $[[\tilde{d}_\nu\ \nu d_{3/2}]^{(5/2)}\pi d_{3/2}]^{(1)}$, but cancellation occurs in these two contributions. That is one reason why $\log ft$ values to the $3/2_1^+$ states are larger (i.e., the matrix elements are smaller) than $\log ft$ values to the ground states $(1/2_1^+)$. It would be interesting if simple explanation of the above difference is found based on the O(6) symmetry basis.

On the basis of calculated beta-decay rates, the present calculation strongly supports the IBFM description of positive parity states in odd Xe isotopes, reported in Ref.[5]. On the other side, it does not confirm the structure of low-lying positive parity states in odd Cs isotopes, as described in the first calculation for these nuclei [3]. In fact, the main difference between that calculation and the present work is that in Ref.[3], due to a weak, almost negligible exchange interaction, the wave functions are not very much mixed, while in the present calculation characterized by a strong exchange interaction, they show a strong configuration mixing. In the case of a weak exchange interaction for Cs isotopes, the wave functions of the parent nuclei give beta-decay rates that are one to two orders of magnitude different, even for the decays to the daughter ground states. Recent calculations for odd mass [8] and odd-odd [7] Cs isotopes, with strong exchange interaction seem to be far more realistic, in the light of the present calculation.

5. Conclusions

We have performed an IBFM2 analysis of A = 125, 127 and 129 isotopes of Cs and Xe. The agreement of the calculated energy levels and electromagnetic transition properties with the experimental data suggest that the choice of interaction parameters is realistic. In order to test how realistic the wave functions are, we have calculated the beta-decay rates from Cs to Xe nuclei. In our approach this type of calculation is a very sensitive test of wave functions, because it is parameter free, without any normalization of theoretical results. In addition to transition rates to ground states, it also gives transition rates to excited levels of daughter nuclei. The results of beta-decay calculations are in very good agreement with observed data. The present analysis of beta-decay in O(6) like nuclei, together with recent calculations for spherical nuclei, shows that the IBFM2 is appropriate for calculation of beta-decay properties in odd nuclei. The next subjects will be to extend the calculations to odd nuclei of other mass regions including the Gamow-Teller transition strength distributions to higher-excited states, and also to the beta decay between even-even and odd-odd nuclei.

References

[1] Alonso, C.E., J.M. Arias, R. Bijker, and F. Iachello, Phys. Lett. **B144**, 141 (1984).

[2] Arima, A. and F. Iachello, Phys. Rev. Lett. **35**, 1069 (1975).

[3] Arias, J.M., C.E. Alonso, and R. Bijker, Nucl. Phys. **A445**, 333 (1985).

[4] Bijker, R., Ph.D. thesis, University of Groningen, 1984 (unpublished).

[5] Cata-Danil, Gh., D. Bucurescu, A. Gizon, and J. Gizon, J. Phys. G **20**, 1051 (1994).

[6] Dellagiacoma, F., Ph.D. thesis, Yale University, 1988; Dellagiacoma, F. and F. Iachello, Phys. Lett. **B218**, 299 (1989).

[7] Gizon, A., J. Timár, J. Gizon, B. Weiss, D. Barnéoud, C. Foin, J. Genevey, F. Hannachi, C.F. Liang, A. Lopez-Martens, P. Paris, B.M. Nyakó, L. Zolnai, J.C. Merdinger, S. Brant, and V. Paar, Nucl. Phys. **A694**, 63 (2001).

[8] Gizon, A., B. Weiss, P. Paris, C.F. Liang, J. Genevey, J. Gizon, V. Barch, Gh. Cata-Danil, J.S. Dionisio, J.M. Lagrange, M. Pautrat, J. Vanhorenbeeck, Ch. Vieu, L. Zolnai, J. M. Arias, J. Barea, and Ch. Droste, Eur. Phys. J. A **8**, 41 (2000).

[9] Hughes, J.R., D.B. Fossan, D.R. LaFosse, Y. Liang, P. Vaska, and M.P. Waring, Phys. Rev. C **44**, 2390 (1991).

[10] Iachello F. and A. Arima, *The interacting boson model*, (Cambridge Univ. Press, Cambridge, 1987).

[11] Iachello, F. and O. Scholten Phys. Rev. Lett. **43**, 679 (1979).

[12] Iachello, F. and P. Van Isacker, *The interacting boson-fermion model*, (Cambridge Univ. Press, Cambridge, 1991).

[13] Katakura, J., Nucl. Data Sheets **86**, 955 (1999).

[14] Kitao, K. and M. Oshima, Nucl. Data Sheets **77**, 1 (1996).

[15] Liang, Y., R. Ma, E.S. Paul, N. Xu, D.B. Fossan, and R.A. Wyss, Phys. Rev. C **42**, 890 (1990).

[16] Maino, G. and L. Zuffi, in *Proceedings of the 7th International Conference on Nuclear Reaction Mechanisms, Varenna, 1994*, edited by E. Gadioli (University of Milan, Milan, 1994), p. 765.

[17] Maino, G., in *Proceeding of the International Symposium on Perspectives for the Interacting Boson Model, Padova, 1994*, edited by R.F. Casten, A. Vitturi, A.B. Balantekin, B.R. Barrett, Ginocchio, J.N., G. Maino and T. Otsuka (World Scientific, Singapore, 1995), p. 617.

[18] Maino, G. and L. Zuffi, in *Proceedings of the 5th International Spring Seminar on Nuclear Physics, Ravello, 1996*, edited by A. Covello (World Scientific, Singapore, 1996), p. 611.

[19] Mantica, P.F., Jr., B.E. Zimmerman, W.B. Walters, H.K. Carter, D. Rupnik, E.F. Zganjar, W.L. Croft, and Y.-S. Xu, Phys. Rev. C **42**, 902 (1990).

[20] Puddu, G., O. Scholten, and T. Otsuka, Nucl. Phys. **A348**, 109 (1980),

[21] Raman, S., C.H. Malarkey, W.T. Milner, C.W. Nestor, Jr., and P.H. Stelson, At. Data Nucl. Data Tables **36**, 1 (1987).

[22] Scholten, O., Ph.D. thesis, University of Groningen, 1980 (unpublished).

[23] Tendow Y., Nucl. Data Sheets **77**, 631 (1996).

[24] Yoshida, N., A. Gelberg, T. Otsuka, I. Wiedenhöver, H. Sagawa, and P. von Brentano, Nucl. Phys. **A619**, 65 (1997).

[25] Yoshida, N. and L. Zuffi, in *Proceedings of the of the Conference: Bologna 2000–Structure of the Nucleus at the Dawn of the Century*, edited by G.C. Bonsignori, M. Bruno, A. Ventura and, D. Vretenar (World Scientific, Singapore, 2001), p. 233.

[26] Yoshida, N., L. Zuffi, and S. Brant, Phys. Rev. C **66**, 014306 (2002).

[27] Yoshida, N., L. Zuffi, and A. Arima, Czech. J. Phys. **52** Suppl. C615 (2002).

[28] Zuffi, L., S. Brant and N. Yoshida, Phys. Rev. C, in press.

TRIAXIALITY AND CHIRALITY IN NUCLEI AROUND MASS 130

N. Yoshinaga and K. Higashiyama
Department of Physics, Saitama University,
Saitama City
338-8570, Japan

Abstract Nuclei around mass 130 have many interesting features such as high-spin isomers, backbending phenomena, even-odd energy staggering of quasi-γ band caused by a soft triaxial deformation, and features recently referred to as "chiral bands". Moreover, the beta-decays and electron-captures in this region provide us with necessary information for predicting the abundance of nuclei in the environment of super-nova explosions. The nuclei in this region are neither vibrational nor rotational. Thus it is very difficult to treat them in terms of conventional mean field theories. To overcome this difficulty, we construct many-body states in terms of collective nucleon pairs which have angular momenta zero (S) and two (D). The purpose of this paper is to understand many seemingly different features in this region in a self-consistent and unified way. In order to understand systematics, we introduce effective interactions which depend smoothly on the neutron and proton numbers. It is found that energy spectra of the yrast and quasi-γ-bands of Xe, Ba and Ce isotopes are nicely reproduced along with inter- and intra-band $E2$ transitions, which simulate the typical features of the O(6) limit of the interacting boson model.

The description of even-even nuclei in terms of S and D pairs is, however, not enough for the backbending phenomena. For the description of these phenomena we introduce another pair made of neutrons only in the $h_{11/2}$ orbital. The extended model quite well reproduces the backbending phenomena of yrast bands and also the nature of 10^+ isomers in this region.

For the description of the phenomena in nuclei with odd mass number, we need to extend our model space to include an unpaired particle in addition to the even-even core, and for odd-odd nuclei we need both an unpaired proton and an unpaired neutron. Energy spectra of odd-A Xe, Ba and Ce isotopes are found to be quite well reproduced by this simple extension of the model.

In some of odd-odd nuclei, chiral bands have been reported, where the three angular momenta of the neutron, the proton and the even-even core can form both right-handed and left-handed systems, which cannot be transformed into each other by rotations. Our model can reproduce spectra of these experimental doublet bands quite naturally, and predicts the $E2$ and $M1$ features of these bands, most of which are not observed experimentally.

B. Gruber (ed.), Symmetries in Science XI, 589–609.

Keywords: Triaxiality, collective states, chiral bands, pair-truncated shell model

1. Introduction

Xe, Ba and Ce isotopes around mass $A = 130$ belong to a typical transitional region between spherical and deformed shapes. The even-even nuclei in this region seem to be soft with regard to the γ-deformation with an almost maximum effective triaxiality of $\gamma \sim 30°$ [1, 2]. The low-lying states, showing a rich collective structure in this region, were investigated extensively in terms of various models, such as the interacting boson model (IBM) [3, 4, 5, 6], the fermion dynamical symmetry model (FDSM) [7], the pair-truncated shell model (PTSM) [8, 9, 10, 11, 12, 13] and the nucleon-pair shell model [14, 15, 16]. As shown in Ref. [4], the excitation spectra of the even-even nuclei in the Xe-Ba mass region can be well approximated by the O(6) dynamical symmetry limit of the IBM.

In many nuclei of this region the backbending phenomena have been observed. The basic mechanism of these phenomena has been regarded as the crossing of the ground state band with the s band originating from the alignment of two $\nu h_{11/2}$ quasi-particles. A few theoretical studies were made using an extended IBM, where one of the IBM sd-bosons is replaced by a pair of nucleons with high spin [17, 18, 19, 20]. The description was quite successful, but the Pauli effect had to be incorporated rather implicitly into the interactions between the bosons. In order to construct an effective shell-model space by retaining the successful aspects of the IBM, but eliminating its boson image, we introduce an SD+H version of the PTSM, which consists of a high-spin pair made of nucleons in $0h_{11/2}$ orbitals, in addition to the collective S and D nucleon pairs [13]. The PTSM exactly conserves rotational invariance and nucleon number.

Recently, experimental investigations of odd-odd nuclei in the $A \sim 130$ mass region [21, 22, 23, 24, 25, 26, 27] have resulted in the observation of systematic doublet bands built on unique-parity $0h_{11/2}$ valence proton and neutron orbitals. These structures are interpreted as a manifestation of chirality in the angular momentum coupling, which was predicted by Frauendorf and Meng [28] in the context of the tilted axis cranking model. In odd-odd nuclei with mass around 130, we can think of two physically different configurations (right and left), where the angular momenta of the triaxial core, the proton and the neutron are perpendicular to each other. Originating from these two configurations in the intrinsic frame, it has been predicted that two energetically degenerate bands called "chiral bands" appear in experimental odd-odd spectra. Up to now theoretical investigations of chiral bands have been made through semi-classical approaches using mean field approximations. However, there was no microscopic study which did not break rotational symmetry and particle number conservation of the interactions. In this paper we clarify the difference of internal structures of the doublet bands by applying the PTSM.

The paper is organized as follows: In section 2, the framework of the PTSM and its effective interaction in the model space are presented, and the PTSM calculations are carried out for Xe, Ba and Ce isotopes with mass $A \sim 130$, where the effective interactions are smoothly changed as functions of valence particles. In section 3 we extend the PTSM to incorporate a high-spin pair in $0h_{11/2}$ orbitals for the study of backbending phenomena. In section 4 we extend the PTSM for the application of odd-A nuclei, and carry out calculations for Xe, Ba and Ce odd-A isotopes. In section 5 we apply the PTSM to odd-odd nuclei, and reproduce energy levels of doublet bands and electromagnetic transitions, and analyze the internal structure of the doublet bands. Principal results are summarized in section 6.

2. Framework of the PTSM and its SD-pair truncation

In the shell model calculation, the number of configurations increases exponentially with the number of particles, and the treatment becomes soon infeasible for the present computers. Therefore, if the number of valence nucleons is large, we need to abandon the treatment by the full-fledged shell model, and have to think about a truncation of the space.

In the first stage of the PTSM, we truncate the collective subspace to the space which is constructed only in terms of collective S and D pairs. The S and D pairs, as building blocks of the model, are defined in terms of pair-creation operators as,

$$S^\dagger = \sum_j \alpha_j A_0^{\dagger(0)}(jj), \tag{1}$$

$$D_M^\dagger = \sum_{j_1 j_2} \beta_{j_1 j_2} A_M^{\dagger(2)}(j_1 j_2), \tag{2}$$

where the structure coefficients α and β are determined by variation in the present approach for each nucleus. Here the creation operator of a pair of nucleons with total spin J and projection M is defined as follows,

$$A_M^{\dagger(J)}(j_1 j_2) = [c_{j_1}^\dagger c_{j_2}^\dagger]_M^{(J)}. \tag{3}$$

Here $c_j^\dagger$ represents a single particle creation operator in orbital j. The many-body states are thus constructed in terms of collective S and D pairs,

$$|S^{n_s} D^{n_d} J\eta\rangle = \left(S^\dagger\right)^{n_s} \left(D^\dagger\right)^{n_d} |-\rangle, \tag{4}$$

where J indicates the total spin of a many-body state, η is a quantum number which is necessary to uniquely specify the state, and $n_s + n_d$ represents

Table 1. Adopted single-particle energies for neutron-holes and proton-particles, which are extracted from experiment [29, 30, 31] (in MeV).

j	$2s_{1/2}$	$0h_{11/2}$	$1d_{3/2}$	$1d_{5/2}$	$0g_{7/2}$
ε_ν	0.332	0.242	0.000	1.655	2.434
ε_π	2.990	2.793	2.708	0.962	0.000

the number of valence pairs for a specific nucleus. Here angular momentum coupling is carried out exactly, but for simplicity it is not denoted explicitly.

In order to describe open-shell nuclei, we use the above states in both neutron and proton spaces to couple them to a total angular momentum state. The total state is expressed as follows,

$$|\Phi(J\eta)\rangle = \left[\left|S_\nu^{\bar{n}_s} D_\nu^{\bar{n}_d} J_\nu \eta_\nu\right\rangle \otimes \left|S_\pi^{n_s} D_\pi^{n_d} J_\pi \eta_\pi\right\rangle\right]^{(J)}, \quad (5)$$

where $\bar{n}_\nu = \bar{n}_s + \bar{n}_d$ and $n_\nu = n_s + n_d$ are numbers of neutron-hole pairs and proton-particle pairs, respectively. In this atomic mass region we treat neutrons as holes and protons as particles so that N=82 and Z=50 become the nearest closed shells.

We adopt pairing plus quadrupole interactions as effective interactions ($\tau = \nu$ or π),

$$\begin{aligned} H = &\sum_{jm\tau} \varepsilon_{j\tau} c^\dagger_{jm\tau} c_{jm\tau} \\ &+ \sum_\tau \left[-G_{0\tau} P_\tau^{\dagger(0)} P_\tau^{(0)} - G_{2\tau} P_\tau^{\dagger(2)} \cdot \tilde{P}_\tau^{(2)} - \kappa_\tau : Q_\tau \cdot Q_\tau :\right] \\ &+ \kappa_{\nu\pi} Q_\nu \cdot Q_\pi, \end{aligned} \quad (6)$$

where $\varepsilon_{j\tau}$, $G_{0\tau}$, $G_{2\tau}$, κ_τ and $\kappa_{\nu\pi}$ represent the single particle energies and the strengths of monopole-pairing, quadrupole-pairing and quadrupole-quadrupole interactions of like particles, and that of the quadrupole-quadrupole interaction between neutrons and protons, respectively. The detailed definitions of the interactions are given in Refs. [8, 9, 10, 11, 12, 13].

As for single-particle levels, all the five orbitals $0g_{7/2}$, $1d_{5/2}$, $1d_{3/2}$, $0h_{11/2}$ and $2s_{1/2}$ are considered in the major shell of $50 \leq N(Z) \leq 82$, where valence neutrons (protons) are treated as holes (particles). The adopted single-particle energies, listed in Table 1, have been extracted from experimental excitation energies in Refs. [29, 30, 31].

In order to study the systematics of Xe, Ba and Ce isotopes, the strengths of the effective interactions are smoothly changed as functions of the numbers of

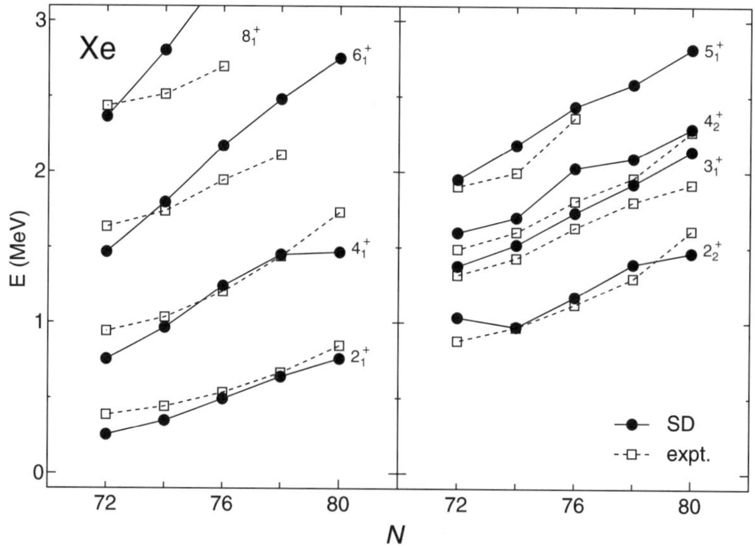

Figure 1. Energy spectra of the yrast and quasi-γ bands for Xe isotopes as a function of neutron number N. Experimental data are taken from Ref. [32].

pairs as follows (in MeV),

$$
\begin{aligned}
G_{0\nu} &= 0.140 - 0.010\,\bar{n}_\nu, &(7)\\
G_{2\nu} &= 0.000 + 0.005\,\bar{n}_\nu + 0.003\,n_\pi, &(8)\\
\kappa_\nu &= 0.050 + 0.005\,\bar{n}_\nu, &(9)\\
G_{0\pi} &= 0.150 - 0.010\,n_\pi, &(10)\\
G_{2\pi} &= 0.018 - 0.002\,n_\pi + 0.002\,\bar{n}_\nu, &(11)\\
\kappa_\pi &= 0.030 + 0.005\,n_\pi, &(12)\\
\kappa_{\nu\pi} &= 0.070. &(13)
\end{aligned}
$$

Figure 1 shows the energy spectra of yrast and quasi-γ bands for Xe isotopes. We have obtained a good agreement with experiment up to spin 6 except N=80. For N=80 a higher spin pair seems necessary because the quadrupole collectivity is not so dominant near the closed shell. The first 8^+ states are not well reproduced, which implies the effect of higher spin pairs originating from the $h_{11/2}$ orbital. Energy spectra for Ba isotopes are shown in figure 2. We see a better agreement with experiment, compared to Xe isotopes, because quadrupole collectivity becomes now dominant. The experimental energy staggering for the even-spin and odd-spin members of the quasi-γ-band, indicating γ-instability, is well reproduced theoretically. The spectra for Ce isotopes are shown in figure 3. All three figures 1 to 3 show that theoretical quasi-γ-band

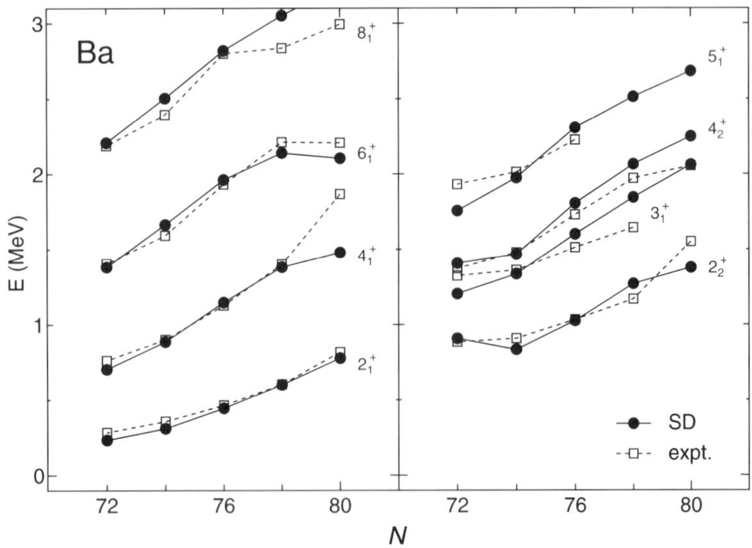

Figure 2. Energy spectra of the yrast and quasi-γ bands for Ba isotopes. Experimental data are taken from Ref. [32].

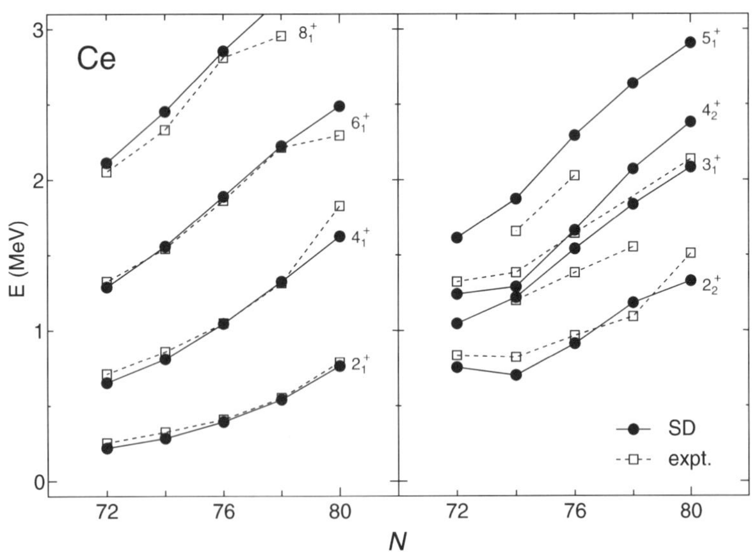

Figure 3. Energy spectra of the yrast and quasi-γ bands for Ce isotopes. Experimental data are taken from Ref. [32].

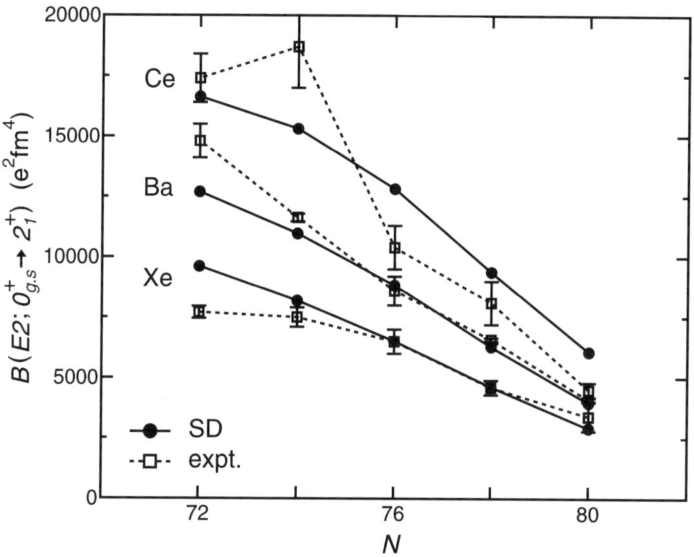

Figure 4. $B(E2)$ values from the ground state to the first 2^+ state for Xe, Ba, Ce isotopes. Experimental data are taken from Ref. [33].

energies decrease as a function of the number of neutron-hole increases, but they cease to decrease at $N = 74$, which well reproduces the experimental trend.

Next let us turn to the $E2$ transition rates. The $E2$ transition operator is defined as

$$T(E2;\mu) = e_\nu Q_{\nu\mu} + e_\pi Q_{\pi\mu}, \tag{14}$$

where e_τ represents the effective charge of the nucleon, and the operator Q_τ is the quadrupole operator with the oscillator parameter $b = 1.005A^{1/6}$ fm. The effective charges are assumed to follow the conventional relation e_ν=$-\delta e$ and e_π=$(1+\delta)e$ [34], and the adopted values are δ=$0.70 + 0.10(\bar{n}_\nu + n_\pi)$.

In figure 4 the calculated $B(E2)$ values from the ground state to the first 2^+ state for Xe, Ba and Ce isotopes are compared with experiment. The overall trend is well reproduced, but we have no good agreement with experiment for ${}^{132}_{58}\mathrm{Ce}_{74}$. We infer that the calculated deformation is small compared to the experimental one. After N=74 the nuclei rapidly develop deformation, and we may need higher spin pairs such as G-pairs to get large deformation.

Table 2 shows relative $B(E2)$ values between low-lying states for ^{134}Ba, ^{132}Ba and ^{130}Ba. It is seen that the theoretical results reproduce very well the experimental data, which simulate the O(6) limit prediction of the IBM. Especially, transitions from the 5_1^+ state to others are quite important for the appearance of the O(6) symmetry.

Table 2. Comparison of relative $B(E2)$ values between low-lying states for ^{134}Ba, ^{132}Ba and ^{130}Ba. Experimental data are taken from Refs. [35, 36, 37].

	^{134}Ba		^{132}Ba		^{130}Ba		O(6)
$J_i^\pi \to J_f^\pi$	SD	expt.	SD	expt.	SD	expt.	
$2_2^+ \to 2_1^+$	100	100	100	100	100	100	100
$\to 0_1^+$	0.17	0.9(2)	0.063	2.7(4)	4.9	6.2(7)	0
$3_1^+ \to 2_2^+$	100	100	100	100	100	100	100
$\to 4_1^+$	15	≥2.6	30	38(6)	36	22(3)	40
$\to 2_1^+$	0.29	1.1	2.7	2.6(4)	14	4.5(6)	0
$4_2^+ \to 2_2^+$	100	100	100	100	100	100	100
$\to 3_1^+$	21		1.3	≤50(11)	9.9		0
$\to 4_1^+$	24	73	65	73(10)	79	54(10)	91
$\to 2_1^+$	23	2.4	1.6	1.8(3)	0.25	2.3(4)	0
$5_1^+ \to 3_1^+$	100	100	100	100	100	100	100
$\to 4_2^+$	67		42	≤45(7)	43		46
$\to 6_1^+$	9.1		17		33		45
$\to 4_1^+$	1.8		0.49	≤2.2(3)	6.8		0

3. High-spin states

Although the SD version of the PTSM well describes the low-lying states of nuclei as shown in section 2, it cannot describe an irregular yrast sequence in high-spin states, i.e., the backbending phenomenon. This phenomenon, in mass around 130, is caused by the neutron $0h_{11/2}$ orbital. Therefore it is absolutely necessary to incorporate this orbital and to extend the SD version of the PTSM to the SD+H version of the PTSM by including a pair in the $0h_{11/2}$ orbital. In the second stage of the PTSM, we extend the PTSM to incorporate this high-spin pair (H pair).

The H pair creation operator is defined in terms of the creation operator for the $h_{11/2}$ orbital,

$$H_M^{\dagger(K)} = [c_{11/2}^\dagger \, c_{11/2}^\dagger]_M^{(K)}. \tag{15}$$

Here K takes on the angular momenta 0, 2, 4, 6, 8 and 10. Using this H pair, the many-body states are now expressed as follows,

$$|S^{n_s} D^{n_d} H^{n_h} J\eta\rangle = \left(S^\dagger\right)^{n_s} \left(D^\dagger\right)^{n_d} \left(H^\dagger\right)^{n_h} |-\rangle\,, \tag{16}$$

where J and η are the total spin and another quantum number as before. Here we assume $n_h = 0$ or $n_h = 1$ for computational simplicity. In this paper we investigate N=76 isotones as examples. In Table 3 we show the strengths of two-body interactions, which are slightly different from those used for the SD version of the PTSM because the introduction of the H pair suggests a different effective interaction. Here for ^{134}Ce we assume only the K=10 component for computational reasons, so that the effective interactions for this nucleus are

Table 3. Force strengths used for N=76 isotones (in MeV).

	$G_{0\nu}$	$G_{2\nu}$	κ_ν	$G_{0\pi}$	$G_{2\pi}$	κ_π	$\kappa_{\nu\pi}$
^{130}Xe	0.150	0.026	0.100	0.150	0.030	0.025	0.070
^{132}Ba	0.150	0.026	0.100	0.170	0.040	0.030	0.080
^{134}Ce	0.120	0.032	0.140	0.120	0.042	0.050	0.060

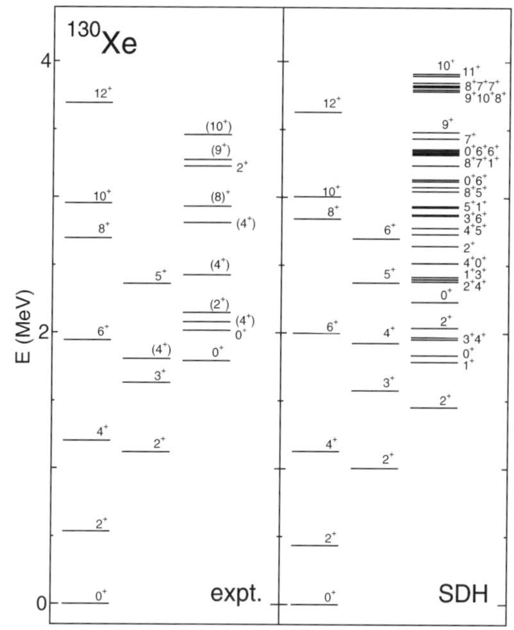

Figure 5. Comparison of experimental energy spectrum (expt.) with the SD+H version of the PTSM results (SDH) for the nucleus ^{130}Xe. The experimental data are taken from Ref. [32].

very different from those used for ^{130}Xe and ^{132}Ba. Single-particle energies are the same as used before.

Figure 5 shows a comparison of calculated results and experimental data of energy levels of ^{130}Xe. Our calculation reproduces quite well the energy levels for the yrast and quasi-γ bands. Especially the staggering seen in 2_2^+, 3_1^+, 4_2^+, 5_1^+ in the quasi-γ band is well reproduced. Figure 6 shows a comparison of calculated results and experimental data of energy levels of ^{132}Ba. Our PTSM calculation reproduces the energy levels of yrast and quasi-γ bands [13]. The backbending phenomena in ^{132}Ba were studied in terms of the extended IBM [20]. In that work only yrast states were calculated and the backbending at 10^+ was reproduced. However, their agreement with experiment is not so good compared with ours. Figure 7 shows a comparison of calculated results and experimental data of energy levels for ^{134}Ce. We can fairly reproduce the yrast

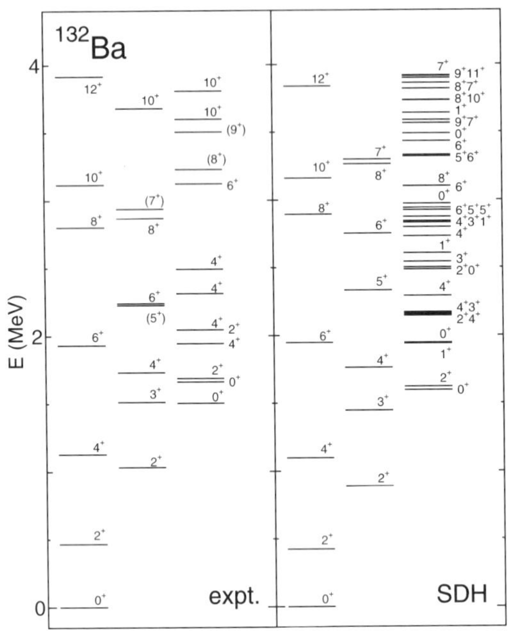

Figure 6. Comparison of experimental energy spectrum (expt.) with the $SD+H$ version of the PTSM results (SDH) for the nucleus ^{132}Ba. The experimental data are taken from Refs. [36, 38, 39].

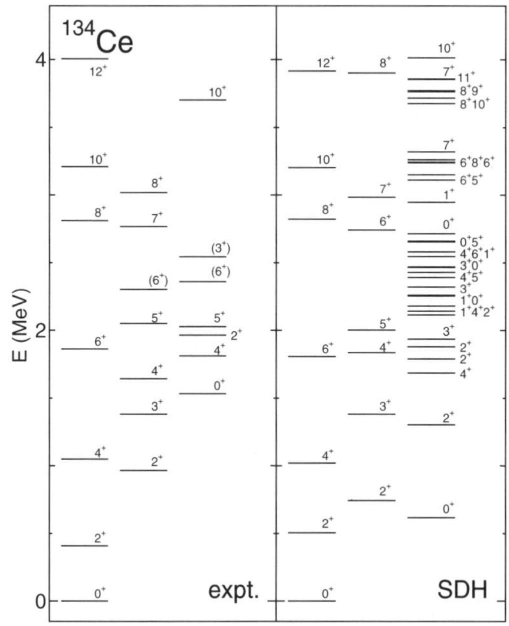

Figure 7. Comparison of experimental energy spectrum (expt.) with the SD+H version of the PTSM results (SDH) for the nucleus ^{134}Ce. The experimental data are taken from Refs. [40, 41].

and quasi-γ bands, but the agreement is not so good compared to ^{132}Ba and ^{130}Xe. This is due to the fact that we only consider K=10 angular momentum for the H pair in this nucleus. Because of this truncation, the dimensions of 2^+ states for ^{132}Ba and ^{130}Xe are 5138 and 928 respectively, whereas it is 484 for ^{134}Ce, and the nuclear model spaces are completely different. Nevertheless, we can reproduce the yrast states, and the effect of the H pair is very important to describe the backbending phenomena. The yrast states of Ce isotopes were also studied by the extended IBM [18], and good results were obtained like ours.

Backbending plots for $N = 74$ isotones are shown in figure 8. As seen from the figure, the experimental level spacings between 8^+ and 10^+ are very small, and they are well reproduced. In figure 9 the $B(E2)$ values between yrast states are shown as a function of spin J. Here the effective charges are taken as e_ν=$-1.00e$ and e_π=$2.00e$. The broken line for ^{134}Ce uses the effective charges e_ν=$-1.90e$ and e_π=$2.90e$, which are determined to fit experimental $B(E2; 2_1^+ \rightarrow 0_1^+)$ value. The rapid drop of $B(E2)$ values occurs at the point of backbending, and our calculation reproduces the decrease of $B(E2; 10_1^+ \rightarrow 8_1^+)$ values for ^{132}Ba and ^{134}Ce. For ^{130}Xe, the $B(E2; 8_1^+ \rightarrow 6_1^+)$ value is reduced instead of $B(E2; 10_1^+ \rightarrow 8_1^+)$. We infer that the number of proton pairs is only two and the hexadecapole degree of freedom is effective. Unfortunately we

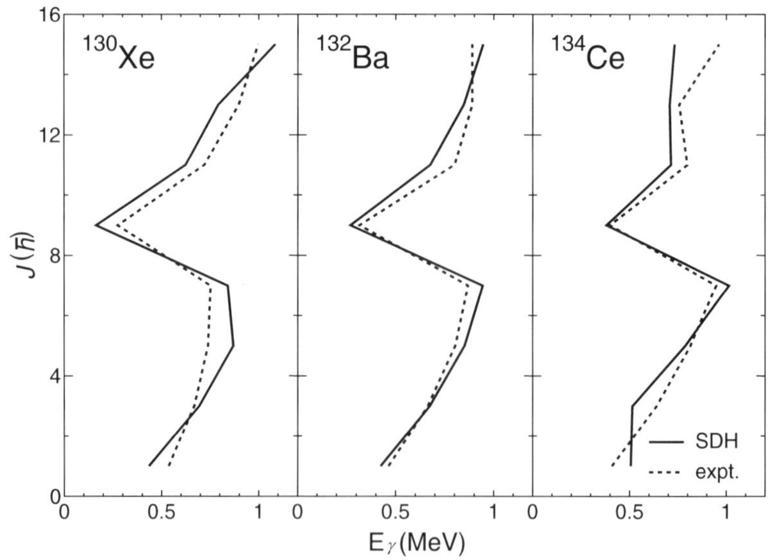

Figure 8. Comparison of γ-ray energies E_γ $[E_\gamma = E(J+1) - E(J-1)]$ versus spin J in experiment (expt.) with the SD+H version of the PTSM results (SDH) for ^{130}Xe, ^{132}Ba and ^{134}Ce. The experimental data are taken from Refs. [32, 38, 40].

cannot reproduce the experimental $B(E2)$ values for ^{134}Ce, but we believe that experimental data might be erroneous because theoretically it is quite hard to produce a value of $B(E2; 4_1^+ \to 2_1^+)$ which is smaller than $B(E2; 2_1^+ \to 0_1^+)$ in any existing collective models.

4. Odd-A

For a description of odd-A nuclei, we add an unpaired particle to the even-even SD pair core and consider a SD pair+1 particle state. The state is now written as

$$|jS^{n_s}D^{n_d}J\eta\rangle = \left[c_j^\dagger \left|S^{n_s}D^{n_d}J'\eta'\right\rangle\right]^{(J)}, \tag{17}$$

where J is the total spin and η is an additional quantum number. The number $2n_s$+$2n_d$+1 represents the total number of valence particles. Here we use Eqs. (7) $\sim$ (13) for the strengths of the interactions by counting the number of the last particle as one half of the pair. Figure 10 shows energy levels of the odd Xe isotopes. The orderings of $11/2^-$ and $9/2^-$ and of $13/2^-$ and $15/2^-$ are reversely predicted for ^{129}Xe, but pretty well reproduced for ^{131}Xe and ^{133}Xe. For the ordering and position of these negative-parity states we may need an octupole interaction, which is missing in the present calculation. Our calculation successfully reproduces the mild change of the ordering of $3/2^+$

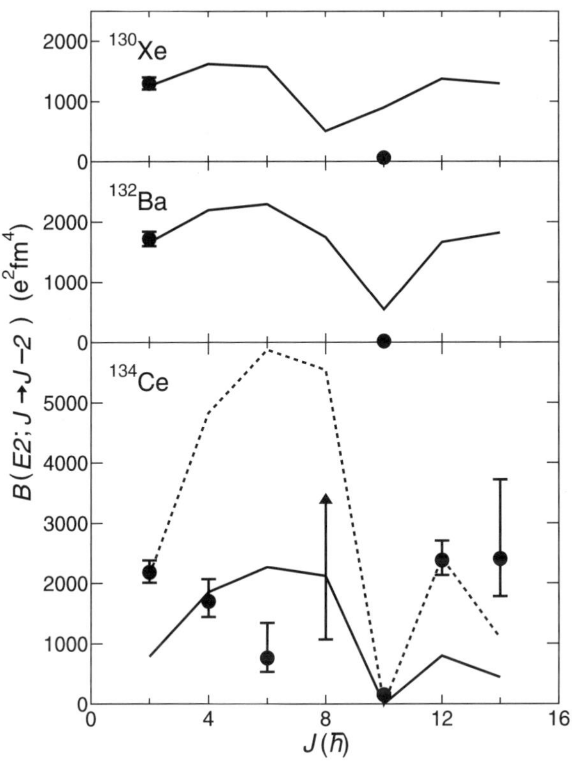

Figure 9. Comparison of the yrast $B(E2)$ values in the SD+H version of the PTSM with the measured values (expt.) for ^{130}Xe, ^{132}Ba and ^{134}Ce. The experimental data are taken from Refs. [33, 42, 43, 44, 45, 46].

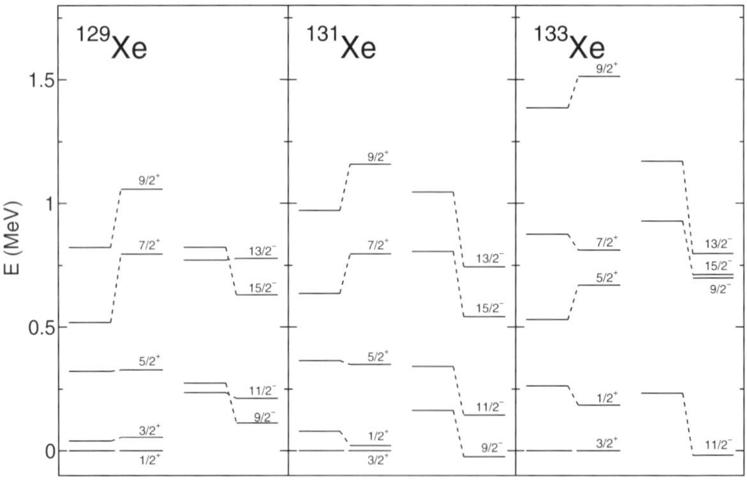

Figure 10. Spectra of odd-A Xe isotopes. Left hand shows experiment and Right hand shows theory. The experimental data are taken from Ref. [32].

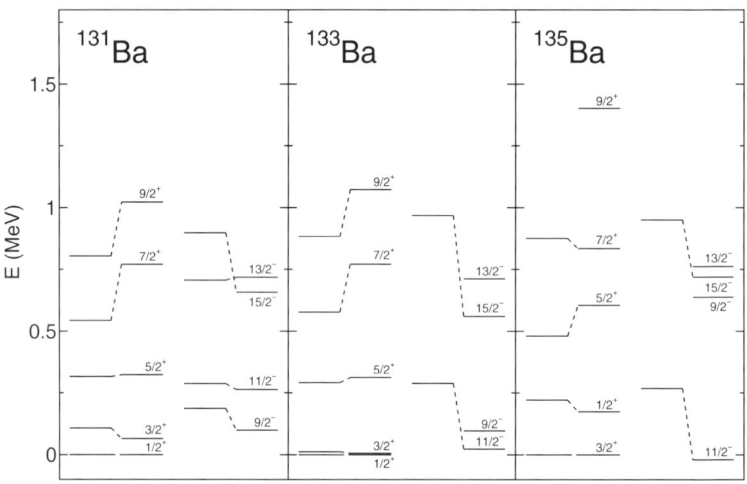

Figure 11. Spectra of odd-A Ba isotopes. Left hand shows experiment and Right hand shows theory. The experimental data are taken from Ref. [32].

and $1/2^+$ states between the three Xe nuclei. Figure 11 shows energy levels of the odd Ba isotopes. For ^{131}Ba and ^{133}Ba the energy levels of $7/2^+$ and $9/2^+$ states are predicted high in energy, but the other states are reproduced well. Like Xe isotopes, the smooth change of ordering is seen for the $3/2^+$ and $1/2^+$ states. Some calculations were also done for Ba and Xe isotopes by the FDSM [7]. In these calculations, the energies of positive-parity states were well

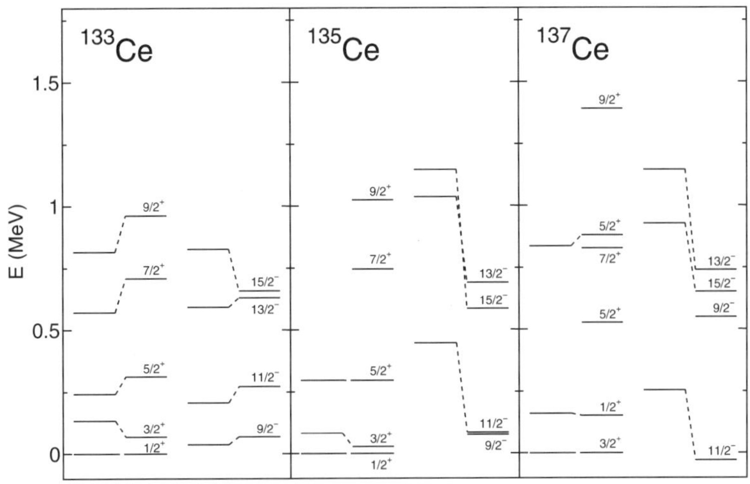

Figure 12. Spectra of odd-A Ce isotopes. Left hand shows experiment and Right hand shows theory. The experimental data are taken from Ref. [32].

reproduced, but no energy levels of negative-parity states were given. Figure 12 shows energy levels of odd Ce isotopes. Concerning the first $5/2^+$ state of ^{137}Ce, our calculation seems to fail in reproducing the experimental energy, but theoretically we predict another $5/2^+$ state at 880 keV, and the experimental observation might correspond to this theoretically predicted level. Although reproducing energy levels of odd-A nuclei is much more difficult compared to even-even nuclei, the agreement is rather well, considering the fact that the effective interactions are solely determined for even-even nuclei and no further adjustment is made for odd-A nuclei.

5. Chiral bands

In the final stage of the extension of the PTSM, we add a neutron and a proton to the even-even core made of collective S and D nucleon pairs. Experimentally one observes pairs of almost degenerate bands, members of which differ by $\Delta J = 1$. They are interpreted to arise from the angular momentum coupling of a neutron and a proton in the $0h_{11/2}$ orbital to the triaxial core, and called chiral bands [21, 22, 23, 24, 25, 26, 27]. Thus we only take into account the $\nu h_{11/2} \otimes \pi h_{11/2}$ state, which corresponds to Eq. (17) with j=11/2,

$$|\Phi(\eta J)\rangle = \left[\left|j_\nu S_\nu^{\bar{n}_{s\nu}} D_\nu^{\bar{n}_{d\nu}} \eta_\nu J_\nu\right\rangle \otimes \left|j_\pi S_\pi^{n_{s\pi}} D_\pi^{n_{d\pi}} \eta_\pi J_\pi\right\rangle\right]^{(J)}, \quad (18)$$

where the interactions are assumed exactly the same as before (Eqs. (7) $\sim$ (13)). The comparison between experimental energy spectrum corresponding to the $\nu h_{11/2} \otimes \pi h_{11/2}$ configuration and that in the PTSM is given in figure 13

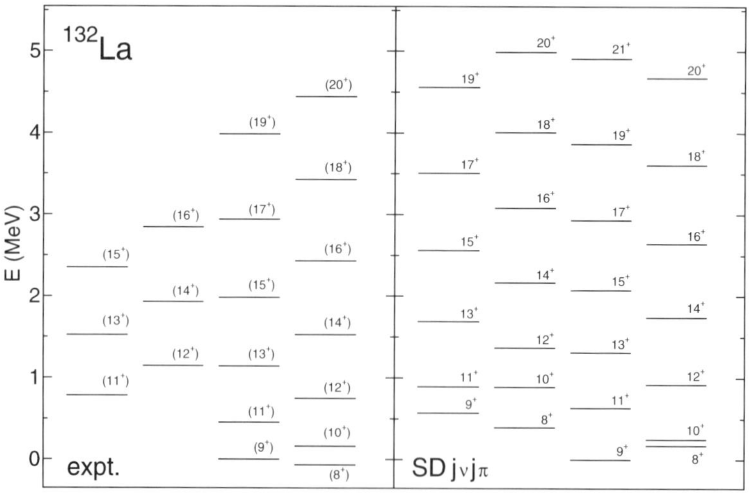

Figure 13. Comparison of energy spectrum in experiment (expt.) with the PTSM ($\mathrm{SDj}_\nu \mathrm{j}_\pi$) results for the odd-odd nucleus ^{132}La. The experimental data are taken from Ref. [26].

for ^{132}La. It is noticed that good agreement between theoretical spectrum of the PTSM and experimental data is achieved. The PTSM result predicts the existence of the 8_2^+, 9_2^+, and 10_2^+ levels in the low-lying states.

The calculated $B(E2)$ values of the yrast states and partner states for ^{132}La and those of the yrast states of ^{132}Ba are shown as a function of spin J in figure 14. It is seen that in the yrast states of ^{132}La the theoretical values of $B(E2)$ are very similar to those of ^{132}Ba. Meanwhile, in the partner states of ^{132}La, theoretical $B(E2)$ values are smaller than those of ^{132}Ba. This suggests that the yrast band has a pure structure made of a neutron and a proton in $h_{11/2}$ orbital which are coupled to the ^{132}Ba core, while the partner band has a rather different structure compared to the yrast band.

The $M1$ transition operator is defined as follows,

$$T(M1;\mu) = \mu_N \sqrt{\frac{3}{4\pi}} \sum_{\tau=\nu,\pi} [g_{\ell\tau} \boldsymbol{j}_\tau + (g_{s\tau} - g_{\ell\tau}) \boldsymbol{s}_\tau]_\mu, \qquad (19)$$

where μ_N is the nuclear magneton, and $g_{\ell\tau}$ and $g_{s\tau}$ are g factors for orbital and spin angular momentum, respectively. The $\boldsymbol{j}_\tau$ and $\boldsymbol{s}_\tau$ are the total and spin angular momenta, respectively. In this study we use the free value for $g_{\ell\tau}$ and 0.70 times the free value for $g_{s\tau}$. The adopted values are $g_{\ell\nu} = 0$, $g_{\ell\pi} = 1.00$, $g_{s\nu} = -2.68$ and $g_{s\pi} = 3.91$.

The calculated $B(M1)$ values of the yrast states and partner states for ^{132}La are shown as a function of spin J in figure 15. The $B(M1)$ between yrast states are large from odd spin to even spin, and small from even spin to odd

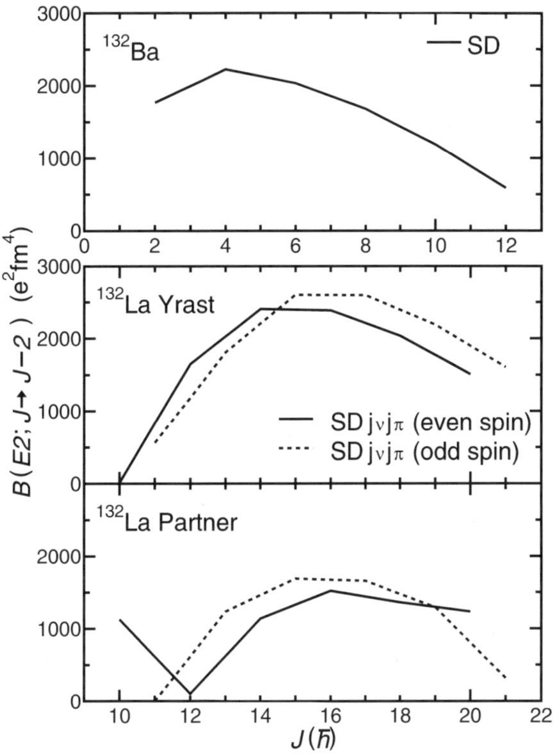

Figure 14. Comparison of $B(E2)$ values between ^{132}La and ^{132}Ba. The solid and dashed lines are the even-spin states and the odd-spin states, respectively.

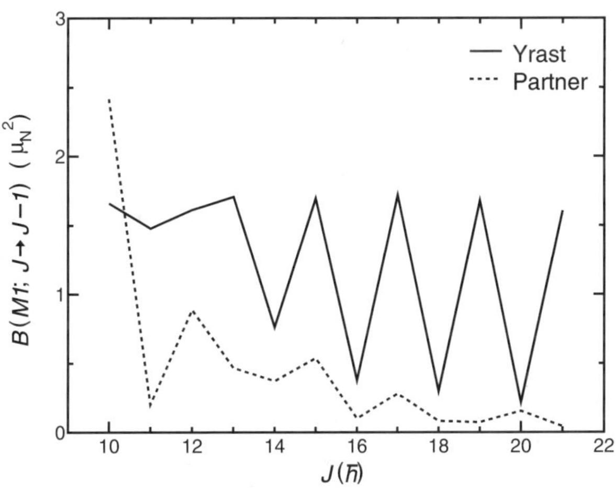

Figure 15. Theoretical prediction of $B(M1)$ values of ^{132}La. The solid and dashed lines are the yrast band and the partner band, respectively.

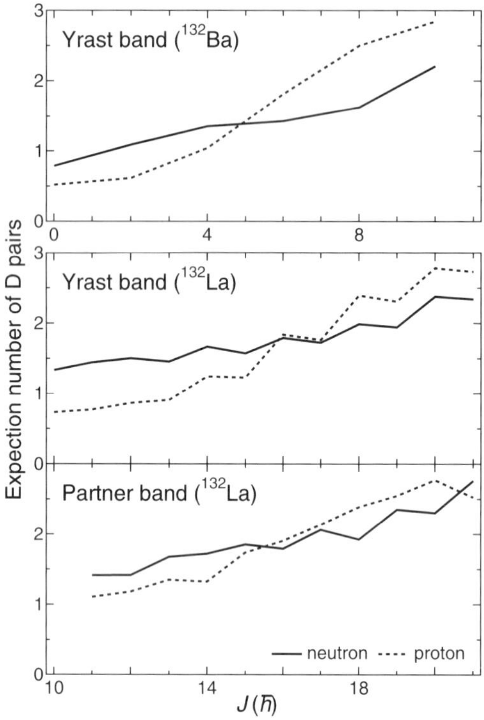

Figure 16. The expectation numbers of D pairs calculated in the PTSM. The solid line is for neutrons, and the dashed line is for protons.

spin. On the contrary, for the partner bands, both $B(M1)$ values are found to be very small. This result tells us that yrast states and partner states, which form $\Delta J = 1$ bands, have very different structures. Furthermore, staggering of $B(M1)/B(E2)$ for the yrast states are observed for nuclei in this region [25], and the calculations predict that a similar situation holds also in ^{132}La. The experimental confirmation is urgently necessary.

The calculated expectation numbers of D pairs of the yrast states and partner states for ^{132}La and those of the yrast states of ^{132}Ba are shown as a function of spin J in figure 16. The yrast band in ^{132}La is interpreted as having the structure of a neutron-hole and a proton-particle in addition to the ^{132}Ba core. Comparing the number of D pairs in ^{132}La and that in ^{132}Ba, we can understand the structure of the core. The results show that the yrast states in ^{132}La have a structure which is very similar to ^{132}Ba for angular momentum states larger than 12^+, whereas the partner band has a slight different structure compared to ^{132}Ba, which suggests that the partner band consists of a mixture of many excited bands.

6. Summary and Conclusions

Many theoretical investigations have been carried out on nuclei with mass numbers around 130, which exhibit many interesting features coming from the soft triaxial deformation. However, there are only very few researches on odd-A or odd-odd nuclei due to the difficulty of a theoretical treatment. Especially there is no work which systematically treats odd-A and odd-odd nuclei in a framework consistent with even-even nuclei. Here we have proposed the PTSM which can systematically treat even-even, odd-A and odd-odd nuclei on the equal footing. As another aspect, the model has the feature that it drastically and efficiently truncates the gigantic shell model space.

For even-even nuclei, the subspace of the shell model space is built by the angular momentum zero (S) and two (D) pairs. For a description of odd-A or odd-odd nuclei, an unpaired neutron or proton is added to the even-even core. As realistic applications of the PTSM to Xe, Ba and Ce isotopes, we have used an effective interaction which varies smoothly as a function of neutron and proton numbers. Spectra of both yrast and quasi-γ bands are reproduced very well. We have also applied the same interaction to odd-A and odd-odd isotopes. For the odd-A nuclei the reproduction is very good except for negative-parity states which implies the necessity of an octupole interaction. For odd-odd nuclei, our model has revealed that in fact those almost degenerate bands have quite different structures. Our model can reproduce spectra of these experimental doublet bands quite naturally and predict $E2$ and $M1$ features of those bands, most of which are not observed experimentally. Further experimental investigation is urgently needed.

Finally we would like to assert that researches by the IBM, the IBFM and the IBFFM are still quite important and necessary for the study of nuclei especially in the deformed region, since at present we have a limitation on the number of pairs for computational reasons. We hope that our present framework gives the microscopic foundation of these phenomenological models in future.

Acknowledgments

We would like to thank Professors A. Arima, K. Tanabe, N. Yoshida and Y. M. Zhao for their valuable discussions, and also thank Prof. W. Bentz for careful reading of the manuscript.

References

[1] J. Yan, O. Vogel, P. von Brentano, A. Gelberg, Phys. Rev. C **48**, (1993) 1046.

[2] O. Vogel, P. van Isacker, A. Gelberg, P. von Brentano, A. Dewald, Phys. Rev. C **53**, (1996) 1660.

[3] G. Puddu, O. Scholten, and T. Otsuka, Nucl. Phys. **A348**, 109 (1980).

[4] R.F. Casten and P. von Brentano, Phys. Lett. **152B**, 22 (1985).

[5] F. Iachello and A. Arima, *The Interacting Boson Model* (Cambridge University, Cambridge, 1987).

[6] T. Mizusaki and T. Otsuka, Prog. Theor. Phys. Suppl. **125**, 97 (1996).

[7] X.W. Pan, J.L. Ping, D.H. Feng, J.Q. Chen, C.L. Wu, and M.W. Guidry, Phys. Rev. C **53**, 715 (1996).

[8] N. Yoshinaga, Nucl. Phys. **A503**, 65 (1989).

[9] N. Yoshinaga and D.M. Brink, Nucl. Phys. **A515**, 1 (1990).

[10] N. Yoshinaga, Nucl. Phys. **A570**, 421 (1994).

[11] N. Yoshinaga, T. Mizusaki, A. Arima, and Y.D. Devi, Prog. Theor. Phys. Suppl. **125**, 65 (1996).

[12] N. Yoshinaga, Y.D. Devi, and A. Arima, Phys. Rev. C **62**, 024309 (2000).

[13] K. Higashiyama, N. Yoshinaga, and K. Tanabe, Phys. Rev. C **67**, 044305 (2003).

[14] Y.M. Zhao, S. Yamaji, N. Yoshinaga, and A. Arima, Phys. Rev. C **62**, 014315 (2000).

[15] Y.A. Luo and J.Q. Chen, Phys. Rev. C **58**, 589 (1998).

[16] Y.A. Luo, J.Q. Chen, and J.P. Draayer, Nucl. Phys. **A669**, 101 (2000).

[17] A. Gelberg and A. Zemel, Phys. Rev. C **22**, 937 (1980).

[18] N. Yoshida, A. Arima, and T. Otsuka, Phys. Lett. **114B**, 86 (1982).

[19] H. Kusakari and M. Sugawara, Z. Phys. **A317**, 287 (1984).

[20] A. Gelberg, N. Yoshida, T. Otsuka, A. Arima, A. Gade, A. Dewald, and P. von Brentano, in *Quasiparticle and Phonon Excitations in Nuclei (Soloviev 99)* edited by N. Dang and A. Arima (World Scientific, Singapore, 1999), p.100.

[21] K. Starosta et al., Phys. Rev. Lett. **86**, 971 (2001).

[22] A.A. Hecht et al., Phys. Rev. C **63**, 051302 (2001).

[23] T. Koike, K. Starosta, C.J. Chiara, D.B. Fossan, and D.R. LaFosse, Phys. Rev. C **63**, 061304 (2001).

[24] D. J. Hartley et al., Phys. Rev. C **64**, 031304 (2001).

[25] R.A. Bark, A.M. Baxter, A.P. Byrne, G.D. Dracoulis, T. Kibedi, T.R. McGoram, and S.M. Mullins, Nucl. Phys. **A691**, 577 (2001).

[26] K. Starosta, C.J. Chiara, D.B. Fossan, T. Koike, T.T.S. Kuo, D.R. LaFosse, S.G. Rohozinski, Ch. Droste, T. Morek, and J. Srebrny, Phys. Rev. C **65**, 044328 (2002).

[27] T. Koike, K. Starosta, C.J. Chiara, D.B. Fossan, and D.R. LaFosse, Phys. Rev. C **67**, 044319 (2003)

[28] S. Frauendorf and Jie Meng, Nucl. Phys. **A617**, 131 (1997).

[29] B. Fogelberg and J. Blomqvist, Nucl. Phys. **A429**, 205 (1984).

[30] M. Sanchez-Vega, B. Fogelberg, H. Mach, R.B.E. Taylor, A. Lindroth, J. Blomqvist, A. Covello, and A. Gargano, Phys. Rev. C **60**, 024303 (1999).

[31] W.J. Baldridge, Phys. Rev. C **18**, 530 (1978).

[32] NUDAT database, National Nuclear Data Center, http://www.nndc.bnl.gov/nndc/nudat/

[33] S. Raman, C.W. Nestor, JR., and P. Tikkanen, At. Data Nucl. Data Tables **78**, 1 (2001).

[34] A. Bohr and B. Mottelson, *Nuclear Structure* (Benjamin, New York, 1975) Vol.1.

[35] A.M. Kleinfeld, A. Bockisch, and K.P. Lieb Nucl. Phys. **A283**, 526 (1977).

[36] A. Gade, I. Wiedenhover, H. Meise, A. Gelberg, and P. von Brentano, Nucl. Phys. **A697**, 75 (2002).

[37] K. Kirch, G. Siems, M. Eschenauer, A. Gelberg, R. Luhn, A. Mertens, U. Neuneyer, O. Vogel, I. Wiedenhover, P. von Brentano, and T. Otsuka, Nucl. Phys. **A587**, 211 (1995).

[38] S. Juutinen, S. Tormanen, P. Ahonen, M. Carpenter, C. Fahl-ander, J. Gascon, R. Julin, A. Lampinen, T. Lonnroth, J. Ny-berg, A. Pakkanen, M. Piiparinen, K. Schiffer, P. Simecek, G. Sletten, and A. Virtanen, Phys. Rev. C **52**, 2946 (1995).

[39] R. Kuhn, K. Kirch, I. Wiedenhover, O. Vogel, M. Wilhelm, U. Neuneyer, M. Luig, A. Gelberg, and P. von Brentano, Nucl. Phys. **A597**, 85 (1996).

[40] M. Muller-Veggian, H. Beuscher, D.R. Haenni, R.M. Lieder, and A. Neskakis, Nucl. Phys. **A417**, 189 (1984).

[41] A. Gade, I. Wiedenhover, M. Luig, A. Gelberg, H. Meise, N. Pietralla, V. Werner, and P. von Brentano, Nucl. Phys. **A673**, 45 (2000).

[42] H. Kusakari, K. Kitao, K. Sato, M. Sugawara, and H. Katsuragawa, Nucl. Phys. A401, 445 (1983).

[43] E.S. Paul, D.B. Fossan, Y. Liang, R. Ma, and N. Xu, Phys. Rev. C **40**, 1255 (1989).

[44] A. Dewald, S. Harissopulos, G. Bohm, A. Gelberg, K.P. Schmittgen, R. Wirowski, K.O. Zell, and P. von Brentano, Phys. Rev. C **37**, 289 (1988).

[45] S.M. Burnett, A.M. Baxter, S. Hinds, F. Pribac, R.H. Spear, and W.J. Vermeer, Nucl. Phys. **A432**, 514 (1985).

[46] D. Husar, S.J. Mills, H. Graf, U. Neumann, D. Pelte, and G. Seiler–Clark, Nucl. Phys. **A292**, 267 (1977).